Nonaqueous Electrolytes Handbook

VOLUME II

Nonaqueous Electrolytes Handbook

VOLUME II

**G. J. JANZ AND
R. P. T. TOMKINS**

*Rensselaer Polytechnic Institute
Troy, New York*

With contributions by

D. A. AIKENS
J. N. BUTLER
K. DOBLHOFER
R. J. GILLESPIE
R. J. JASINSKI
P. V. JOHNSON
A. A. PILLA
H. V. VENKATASETTY

ACADEMIC PRESS, INC.   1973
New York and London
A Subsidiary of Harcourt Brace Jovanovich, Publishers

COPYRIGHT © 1973, BY ACADEMIC PRESS, INC.
ALL RIGHTS RESERVED.
NO PART OF THIS PUBLICATION MAY BE REPRODUCED OR
TRANSMITTED IN ANY FORM OR BY ANY MEANS, ELECTRONIC
OR MECHANICAL, INCLUDING PHOTOCOPY, RECORDING, OR ANY
INFORMATION STORAGE AND RETRIEVAL SYSTEM, WITHOUT
PERMISSION IN WRITING FROM THE PUBLISHER.

ACADEMIC PRESS, INC.
111 Fifth Avenue, New York, New York 10003

United Kingdom Edition published by
ACADEMIC PRESS, INC. (LONDON) LTD.
24/28 Oval Road, London NW1

Library of Congress Cataloging in Publication Data

Janz, George J
    Nonaqueous electrolytes handbook.

    Includes bibliographies.
    1. Electrolyte solutions—Tables. 2. Nonaqueous solvents—Tables. I. Tomkins, Reginald P. T., joint author. II. Title.
QD560.J36   541'.372   72-77364
ISBN 0—12—380402—7 (v. 2)

PRINTED IN THE UNITED STATES OF AMERICA

Errata

Nonaqueous Electrolytes Handbook
Volume II

G. J. Janz and R. P. T. Tomkins

On page 770 add the following references:

205. V. A. Gushchin, Optics and Spectroscopy, **17**. 205 (1964).
206. R. V. Orye, F. E. Weimer, and J. M. Prausnitz, Science, **148**, 74 (1965).
207. W. J. Considine, J. Chem. Phys., **42**, 1130 (1965).
208. R. G. Gordon, J. Chem. Phys., **43**, 1307 (1965).
209. G. Haugen and R. Hardwick, J. Phys. Chem., **69**, 2988 (1965).
210. R. F. Pasternack and R. A. Plane, Inorg. Chem., **4**, 1171 (1965).
211. S. C. Mohr, W. D. Wilk, and G. M. Barrow, J. Amer. Chem. Soc., **87**, 3048 (1965).
212. A. DaSilveira, M. A. Marques, and N. M. Marques, Mol. Phys., **9**, 271 (1965).
213. D. Feakins and D. J. Turner, Chem. and Ind., 2056 (1964).
214. V. P. Petro Diss. Abs., **25**, 6218 (1965).
215. E. Glueckauf, Trans. Faraday Soc., **61**, 914 (1965).
216. S. D. Christian, J. R. Johnson, H. E. Affsprung, and P. J. Kilpatrick, J. Phys. Chem., **70**, 3376 (1966).
217. M. Chabanel, J. Chim. Phys., **63**, 1143 (1966).
218. D. F. Burow, Diss. Abs., **27**, 395 (1966).
219. B. W. Clare, D. Cook, E. C. F. Ko, Y. C. Mac, and A. J. Parker, J. Amer. Chem. Soc., **88**, 1911 (1966).
220. Yu. M. Kessler and L. E. Fadeeva, Elektrokhimiya, **2**, 413 (1966).
221. G. A. Krestov, and V. I. Klopov, Izvest. K. U. Z. M. V. O. S. S. R., Khim. i Khim. Tekhnol., **9**, 34 (1966).
222. R. Gopal and S. A. Rizvi, J. Indian Chem. Soc., **43**, 104 (1966).
223. J. T. Nelson, R. E. Cuthrell, and J. J. Lagowski, J. Phys. Chem., **70**, 1492 (1966).
224. H. Schneider and H. Strehlow, Z. Phys. Chem. (Frankfurt), **49**, 44 (1966).
225. R. J. Gillespie and J. B. Milne, Inorg. Chem., **5**, 1233 (1966).
226. G. Fraenkel and J. P. Kim, J. Amer. Chem. Soc., **88**, 4203 (1966).
227. T. M. Jenkins, Diss. Abs., **27**, 129 (1966).
228. W. H. Baur, Acta Cryst., **19**, 909 (1965).
229. H. Schneider and H. Strehlow, Ber. Bunsen Gesellschaft Phys. Chem., **69**, 674 (1965).
230. D. J. Glover, J. Amer. Chem. Soc., **87**, 5275 (1965).
231. J. R. Johnson, S. D. Christian, and H. E. Affsprung, J. Chem. Soc. **(A)**, 77 (1966).
232. K. F. Purcell and R. S. Drago, J. Amer. Chem. Soc., **88**, 919 (1966).
233. B. E. Conway, R. E. Verral and J. E. Desnoyers, Z. Phys. Chem. (Leipzig), **230**, 157 (1965).
234. J. C. Justice and R. M. Fuoss, J. Chim. Phys., **62**, 1366 (1965).
235. K. Bowden, A. Buckley, and R. Stewart, J. Amer. Chem. Soc., **88**, 947 (1966).
236. A. Mohammad and D. P. N. Satchell, Chem. and Ind., 2069 (1965).
237. D. C. Luehrs, R. T. Iwamoto, and J. Kleinberg, Inorg. Chem., **5**, 201 (1966).
238. V. Plichon, Bull. Soc. Chim. France, 2382 (1967).
239. R. Alexander, E. C. F. Ko, Y. C. Mac, and A. J. Parker, J. Amer. Chem. Soc., **89**, 3703 (1967).
240. R. Alexander and A. J. Parker, J. Amer. Chem. Soc., **89**, 5549 (1967).
241. R. I. Mostkova and Yu. M. Kessler, Zhur. strukt. Khim., **8**, 692 (1967).
242. L. L. Chan and J. Smid, J. Amer. Chem. Soc., **89**, 4547 (1967).
243. T. J. Swift and H. H. Lo, J. Amer. Chem. Soc., **89**, 3988 (1967).
244. J. F. Coetzee, Progr. Phys. Org. Chem., **4**, 45 (1967).

245. J. T. Nelson and J. J. Lagowski, Inorg. Chem., **6**, 1292 (1967).
246. H. D. Hardt and M. Eckle, Z. anorg. Chem., **350**, 300 (1967).
247. H. G. Heal and J. Kane, J. Inorg. Nuclear Chem., **29**, 1539 (1967).
248. J. T. Nelson and J. J. Lagowski, Inorg. Chem., **6**, 862 (1967).
249. M. Herlem, Bull. Soc. Chim. France, 1687 (1967).
250. E. C. Steiner and J. D. Starkey, J. Amer. Chem. Soc., **89**, 2751 (1967).
251. C. D. Ritchie and R. E. Uschold, J. Amer. Chem. Soc., **89**, 2752 (1967).
252. R. Gaboriaud, Compt. rend., **265**, 425 (1967).
253. V. Gutmann and U. Mayer, Monatsh., **98**, 294 (1967).
254. M. K. Chantooni, Jr., and I. M. Kolthoff, J. Amer. Chem. Soc., **89**, 1582 (1967).
255. R. P. Eswein, E. S. Howald, R. A. Howald, and D. P. Keeton; J. Inorg. Nuclear Chem., **29**, 437 (1967).
256. W. R. Gilkerson and J. B. Ezell, J. Amer. Soc., **89**, 808 (1967).
257. C. L. De Ligny and M. Alfenaar, Rec. Trav. Chim., **86**, 1182 (1967).
258. M. Alfenaar and C. L. De Ligny, Rec. Trav. Chim., **86**, 1185 (1867).
259. J. J. Campion, Diss. Abs., **28**, 517 (1967).
260. R. V. Slates and M. Szwarc, J. Amer. Chem. Soc., **89**, 6043 (1967).
261. G. H. Megerle, Diss. Abs. **28**, 521 (1967).
262. G. P. Roshchina, A. S. Kaurova, and I. D. Kosheleva, Zhur. strukt. Khim., **9**, 3 (1968).
263. Ya. D. Fridman and R. A. Veresova, Zhur. neorg. Khim., **13**, 762 (1968).
264. J. Vedel, Ann. Chim. (France), **2**, 335 (1967).
265. O. W. Kolling, Analyt. Chem., **40**, 956 (1968).
266. R. M. Detters, Diss. Abs., **28**, 2319 (1967).
267. S. Guiot, and B. Tremillon, J. Electroanalyt. Chem. Interfacial Electrochem., **18**, 261 (1968).
268. H. F. Henneike and R. S. Drago, J. Amer. Chem. Soc., **90**, 5112 (1968).
269. C. Barraque, J. Vedel, and B. Tremillon, Bull. Soc. Chim. France, 3421 (1968).
270. I. Abrahamer and Y. Marcus, J. Inorg. Nuclear Chem., **30**, 1563 (1968).
271. L. L. Chan and J. Smid, J. Amer. Chem. Soc., **90**, 4654 (1968).
272. N. Gorski and H. Koch, Z. Naturforsch., **23a**, 629 (1968).
273. A. P. Brunetti, M. C. Lim, and G. H. Nancollas, J. Amer. Chem. Soc., **90**, 5120 (1968).
274. J. Fajer and H. Linschitz, J. Inorg. Nuclear Chem., **30**, 2259 (1968).
275. G. Djega-Mariadassou, R. Giovanoli, and G. Pannetier, Compt. rend., **267**, 677 (1968).
276. C. Courty, Compt. rend., **267**, 701 (1968).
277. L. V. Lanshina and L. G. Serveeva, Vestnik. Moskov. Univ., No. 4, 12 (1968).
278. A. M. Golub and V. I. Golovorushkin, Zhur. fiz. Khim., **42**, 1902 (1968).
279. D. J. LaFolette, Diss. Abs., **28**, 4069 (1968).
280. R. Scott, D. De Palma, and S. Vinogradov, J. Phys. Chem., **72**, 3192 (1968).
281. J. N. Butler, J. Phys. Chem., **72**, 3288 (1968).
282. D. Bauer, Bull. Soc. Chim. France, 4313 (1968).
283. A. Collumeau, Bull. Soc. Chim. France, 4317 (1968).
284. R. L. Benoit, Inorg. Nuclear Chem. Letters, **4**, 723 (1968).
285. O. Redlich, R. W. Duerst, and A. Merbach, J. Chem. Phys., **49**, 2986 (1968).
286. R. Waack, L. D. McKeever, and M. A. Doran, Chem. Comm., 117 (1969).
287. W. Gerrard, Chem. and Ind., 295 (1969).
288. Ya. D. Fridman and N. V. Dolgashova, Zhur. neorg. Khim., **14**, 2094 (1969).
289. H. Ohtaki, Bull. Chem. Soc. Japan, **42**, 1573 (1969).
290. Y. Hermodsson, Arkiv Kemi, **31**, 119 (1969).
291. C. Barraque, J. Vedel, and B. Tremillon, Analyt. Chim. Acta, **46**, 263 (1969).
292. B. W. Maxey, Diss. Abs., **29**, 3704 (1969).
293. D. C. Luehrs, J. Inorg. Nuclear Chem., **31**, 3517 (1969).
294. P. V. Johnson and E. C. Baughan, J. Chem. Soc. (A), 2686 (1969).
295. A. P. Kreshkov, L. N. Bykova, and V. D. Ardashnikova, Zhur. Analit. Khim., **24**, 1453 (1969).
296. V. Plichon; Bull. Soc. Chim. France, 3369 (1969).
297. S. I. Drakin, M. Kh. Karapet'yants, and R. Kh-Kurmalieva, Zhur. neorg. Khim., **14**, 2697 (1969).
298. V. Gutmann and K. H. Wegleitner, Z. Phys. Chem. Frankfurt, **77**, 77 (1972).
299. I. M. Kolthoff, Pure and Appl. Chem., **25**, 305 (1971).
300. H. Strehlow and H. Schneider, Pure and Appl. Chem., **25**, 327 (1971).
301. H. Koffer, Ber. Bunsen-Gesell. Phys. Chem., **75**, 1245 (1971).
302. H. Grasdalen, and I. Svare, Acta Chem. Scand., **25**, 1089 (1971).
303. J. L. Dye, M. T. Kok, F. J. Tehan, R. B. Coolen, N. Papadakis, J. M. Ceraso and M. G. DeBacker, Ber. Bunsengesell. Phys. Chem., **75**, 659 (1971).
304. J. P. Badiali, H. Cachet and J. C. Lestrade, Electrochim, Acta, **16**, 731 (1971).

CONTENTS

CONTRIBUTORS	ix
PREFACE	xi
ACKNOWLEDGMENTS	xiii

I. SOLUBILITIES OF ELECTROLYTES ... 1

(a)	Single Solvents	3
(b)	Additional Data	92
(c)	pK_{sp} Data	110
	(1) Single Solvents	110
	(2) Nonaqueous-Aqueous Solvents	123

II. EMF DATA ... 127

(a)	Equilibrium Measurements in Single Solvents	129
(b)	Equilibrium Measurements in Nonaqueous-Aqueous Mixed Solvents	233
(c)	Equilibrium Measurements in Nonaqueous-Nonaqueous Mixed Solvents	265
(d)	Potentiometric Titrations in Protic Solvents	275
(e)	Potentiometric Titrations in Aprotic Solvents	293
(f)	Potentiometric Titrations in Mixed Solvents	319

III. VAPOR PRESSURE ... 333

(a)	Single Solvents	335
(b)	Mixed Solvents	385

IV. CRYOSCOPY ... 401

CONTENTS

V. HEATS OF SOLUTION CALORIMETRY 413

    (a) Single Solvents 415
    (b) Nonaqueous-Aqueous Mixed Solvents 445
    (c) Nonaqueous-Nonaqueous Mixed Solvents 458

VI. POLAROGRAPHY 463

    (a) Inorganic Electrolytes 466
    (b) Organic Electrolytes 523
    (c) Organometallic Compounds 611

VII. LIGAND EXCHANGE RATES AND ELECTRODE REACTIONS 669

    (a) Ligand Exchange Rates 671
    (b) Electrode Reactions in Single Solvents 687
    (c) Electrode Reactions in Mixed Solvents 699

VIII. ELECTRICAL DOUBLE LAYER 725

IX. NONAQUEOUS SPECTROSCOPY AND STRUCTURE OF ELECTROLYTES 743

X. ORGANIC ELECTROLYTE BATTERY SYSTEMS 771

XI. ADDITIONAL REFERENCES AND DATA SOURCES .. 781

    (a) Antimony Trichloride 784
    (b) Electrical Conductance 800
    (c) Viscosity 814
    (d) Transference Numbers 815
    (e) Solubility 816
    (f) EMF 830
    (g) Potentiometric Titrations 845
    (h) Vapor Pressure 849
    (i) Cryoscopy 850

CONTENTS

	(j)	Heats of Solution Calorimetry	851
	(k)	Polarography	854
		(1) Inorganic Compounds	854
		(2) Organic Compounds	858
		(3) Organometallic Compounds	885

XII. COMPOUND INDEX ... **895**

- (a) Solvent ... 897
- (b) Solute ... 905

ERRATA FOR VOLUME I ... **931**

CONTRIBUTORS

D. A. AIKENS, Rensselaer Polytechnic Institute, Troy, New York

J. N. BUTLER, Harvard University, Cambridge, Massachusetts

K. DOBLHOFER,* Electrochemistry Laboratory, ESB Inc., Yardley, Pennsylvania

R. J. GILLESPIE, McMaster University, Hamilton, Ontario, Canada

G. J. JANZ, Rensselaer Polytechnic Institute, Troy, New York

R. J. JASINSKI, Texas Instruments, Inc., Dallas, Texas

P. V. JOHNSON, Royal Military College of Science, Shrivenham, Swindon, England

A. A. PILLA, Bioelectrochemistry Laboratory, Orthopedic Research Laboratories, Columbia Medical Center, New York, New York and Electrochemistry Laboratory, ESB Inc., Yardley, Pennsylvania

R. P. T. TOMKINS, Rensselaer Polytechnic Institute, Troy, New York

H. V. VENKATASETTY, Research Center, Honeywell Inc., Bloomington, Minnesota

---

* Present address: Max-Planck-Institut, Berlin, West Germany

PREFACE

This volume includes data for some 310 solvent systems and covers the literature to 1973. As in Volume I, we have drawn extensively on the earlier studies as well as the more recent contributions in preparing the material for this volume. For nonaqueous polarography and potentiometric titrations, the focus has been only on the more recent literature owing to the relatively vast number of publications since 1940.

This volume has been organized to include eleven well-defined areas: Solubilities of Electrolytes; EMF and Potentiometric Titrations; Vapor Pressures; Cryoscopy; Heats of Solution Calorimetry; Polarography; Ligand Exchange Rates and Electrode Reactions; Electrical Double Layer; Spectroscopy and Structure of Electrolytes; Organic Electrolyte Battery Systems; and Additional References and Data Sources.

The section on polarography is divided further according to inorganic electrolytes, organic electrolytes, and organometallic compounds, in order to present the wealth of data in a concise and orderly manner.

As in Volume I, the last section covers additional data sources, reviews, and data and references that were received too late to include in the earlier sections.

The method of presentation of material is briefly described in the introduction to each section to facilitate the use of the tabulated information and bibliographies are given at the end of each section. A Compound Index is included.

We are indebted to our research co-workers for assistance in all phases of the preparation of this volume.

G. J. Janz
R. P. T. Tomkins

ACKNOWLEDGMENTS

We are pleased to acknowledge the following for their authoritative contributions to this volume: Dr. H. V. Venkatasetty for preparation of the tables relating to nonaqueous polarography; Professor J. N. Butler and Dr. R. J. Jasinski for use of their extensive records, particularly their Government Technical Reports file; Dr. A. A. Pilla and Dr. K. Doblhofer for preparation of the section on electrical double layer; Professor D. A. Aikens for contributions to the section on ligand exchange rates and electrode reactions; Professor R. J. Gillespie for the section on cryoscopy; and Dr. P. V. Johnson for the special section on antimony trichloride.

Informative suggestions, discussions, and materials were received during the preparation of this volume from the following: Professor R. A. Robinson (University of Florida, Gainesville); Professor R. H. Wood (University of Delaware); Dr. J. T. Nelson (Harry Diamond Laboratories, Washington, D.C.); Professor J. A. Friend (University of the West Indies, Trinidad); Professor C. M. Criss (University of Miami, Florida); Professor R. G. Bates (University of Florida, Gainesville); and Dr. J. F. Campbell (Rensselaer Polytechnic Institute). The continued interest of Dr. J. Ambrose (Carleton University, Ottawa) is also acknowledged.

At Rensselaer the source material for this handbook is set up as a cumulative file to which predoctoral and postdoctoral researchers interested in nonaqueous electrolytes have contributed on a continuing basis over a number of years. We wish to take this opportunity to thank the following members of this "team" for their material assistance: E. J. Andalaft, Sarada Balasubrahmanyam, B. D. Briggs, F. W. Dampier, S. S. Danyluk, F. J. Kelly, Myrna P. Klotzkin, G. R. Lakshminarayanan, A. E. Marcinkowsky, G. E. Mayer, J. Meier, Sarah Singer, M. J. Tait, W. A. Tracinski, and P. J. Turner.

Support for basic research on nonaqueous electrolytes from the following sources is also acknowledged: U.S. Air Force, Office of Scientific Research; U.S. Navy, Office of Naval Research; U.S. Atomic Energy Commission; U.S. Department of Interior, Office of Saline Water and the National Science Foundation to one of us (G.J.J.); and the Themis Program, U.S. Department of Defense, to both of us.

This handbook is an outgrowth of the studies thus supported, and its completion was greatly assisted by the impetus received in the final phases of this work from the Themis program.

I. SOLUBILITIES OF ELECTROLYTES

(a) Single Solvents

Solubility data are presented for a selected range of common electrolyte and solvent systems. The classification adopted lists the electrolytes alphabetically by cation and for any given cation the anions are arranged alphabetically. Within each electrolyte table the solvents in which solubility measurements have been reported are also arranged alphabetically. The solubilities are expressed in wt. %; vol. % or g/l. The temperature or temperature range of each study is reported. References follow immediately after each solvent and a complete bibliography is presented at the end of this section.

For a more complete survey of solubilities, particularly with reference to ternary systems the reader is referred to the authoritative compilation by Stephen and Stephen.[a]

[a] Solubilities of Inorganic and Organic Compounds, Vols. I and II, edited by H. Stephen and T. Stephen, Pergamon Press Ltd; (1963).

AMMONIUM ACETATE, $NH_4C_2H_3O_2$

Solvent	Solubility wt. %	Temp.	Ref.
Acetone	2.7*	19	124
Ammonia	71.68	25	64
Methanol	7.31	15	33
	56.75	94.2	
Sulfur dioxide	1.08	0	36

* Solubility given in g/l.

AMMONIUM BENZOATE, $NH_4C_7H_5O_2$

Solvent	Solubility wt. %	Temp.	Ref.
Glycerol	9.09	20	31
Methanol	6.00	15	33
	11.56	66	

AMMONIUM BROMIDE, NH_4Br

Solvent	Solubility g/l	Temp.	Ref.
Ammonia	471	−50	119
	482	−44	
	489	−41.2	
	490	−42.8	
	497	−40.5	
	506	−38.2	
	505	−33.9	
	520	−31.8	
	57.96*	0	
	70.40*	25	
Ethanol	78	15	120
	2.97*	15	44
	3.12*	19	
	9.50*	78	
Ethyl ether	0.122*	15	44
Hydrazine	1100	20	8
Methanol	11.1*	19	44
Sulfur dioxide	0.059*	0	115
	0.052*	25	17

* Solubility given in wt. %.

AMMONIUM CARBONATE, $(NH_4)_2CO_3$

Solvent	Solubility wt. %	Temp.	Ref.
Ethanol	21.2*	25	31
Glycerol	16.7	15	50

* Solubility given in g/l.

I. SOLUBILITIES OF ELECTROLYTES

AMMONIUM CHLORIDE, NH$_4$Cl

Solvent	Solubility wt. %	Temp.	Ref.
Acetic acid	0.7*	16.6	118
	0.053†	16.5	39
	0.053†	21	
	0.065†	25	
	0.084†	32	
	0.095†	38	
	0.110†	43	
	0.134†	53	
	0.150†	58	
	0.178†	65	
	0.209†	72	
	0.224†	77	
	0.259†	84	
	0.282†	87	
	0.312†	92	
	0.348†	98	
Ammonia	52*	−49.6	
	59*	−46.8	
	65*	−44.2	
	70*	−42.2	
	76*	−40.1	
	86*	−37.2	
	96*	−34.6	
	96.3*	−33.9	
	39.9	0	
	50.6	25	
	5.23	−52.9	78
	10.00	−39.3	
	22.60	−20.5	
	29.50	−11.8	
	40.65	0.9	
	44.20	4.2	
	49.20	8.2	
	50.50	9.1	
	51.50	9.8	
	52.50	9.3	
	54.00	8.3	
	55.40	36.9	
Hydrazine	750*	20	8
Methanol	3.24	19	44, 48
Sulfur dioxide	0.009	0	36
	0.0031	25	17

* Solubility given in g/l.
† Solubility given in mol. %.

AMMONIUM FORMATE, NH$_4$CHO$_2$

Solvent	Solubility wt. %	Temp.	Ref.
Formic acid	35.3	−3	123
	40.6	8.5	
	50.0	21.5	
	57.8	39	
	73.1	78	
	100.0	116	

AMMONIUM IODIDE, NH$_4$I

Solvent	Solubility g/l	Temp.	Ref.
Ammonia	621	−50.0	119
	633	−47.5	
	646	−45.2	
	667	−42.0	
	692	−38.6	
	713	−35.3	
	76.8*	0.0	
	78.7*	25.0	
Ethyl N-ethylcarbamate	90.0	60	46
Sulfur dioxide	7.74*	0	36

* Solubility given in wt. %.

AMMONIUM NITRATE, NH$_4$NO$_3$

Solvent	Solubility wt. %	Temp.	Ref.
Acetic acid	0.0741	16.57	65
	0.1287	16.47	
	0.2832	17.7	
	0.3249	21.4	
	0.3916	27.0	
	0.5143	33.6	

I. SOLUBILITIES OF ELECTROLYTES

AMMONIUM NITRATE, NH₄NO₃ (Continued)

Solvent	Solubility wt. %	Temp.	Ref.
Acetic acid	0.8745	45.8	
	1.634	61.2	
	1.887	63.5	
	2.468	67.6	
	2.636	69.0	
	3.239	71.4	
	3.447	72.8	
	4.710	78.3	
	5.508	80.9	
	7.255	85.7	
	8.620	89.0	
	13.68	97.1	
	17.15	101.0	
	19.36	102.6	
	23.30	106.3	
	27.64	108.9	
	31.25	110.6	
	37.98	113.1	
	43.31	115.8	
	47.56	117.0	
	51.67	118.3	
	55.5	120.0	
	60.1	121.4	
	63.1	122.9	
	66.8	124.8	
	71.6	128.9	
	75.0	131.4	
	78.6	136.9	
	82.3	143.1	
	86.3	149.7	
	89.6	157.8	
	100.0	167.5	
Ammonia	701*	−50.6	119
	726*	−46.5	
	734*	−45.0	
	735*	−44.0	
	751*	−40.8	
	770*	−36.6	
	779*	−24.0	
	2355*	25	
	0.0**	−80	122
	6.25**	−60	
	13.9**	−44.5	
	32.3**	−30	

AMMONIUM NITRATE, NH$_4$NO$_3$ (Continued)

Solvent	Solubility wt. %	Temp.	Ref.
Ammonia (cont.)	36.9**	−10.5	
	38.3**	0	
	45.9**	33.3	
	47.0**	35.9	
	53.8**	68.8	
	67.3**	94.0	
	74.2**	190.8	
Hydrazine	43.82†	18	8
Nitric acid	21.1†	8.0	90
	28.7†	23.0	
	38.8†	29.5	
	44.6†	27.5	
	49.4†	23.5	
	54.0†	17.5	
	54.3†	16.5	
	51.7†	11.0	
	57.6†	11.5	
	54.7†	17.0	
	56.2†	27.0	
	60.4†	49.0	
	68.1†	79.0	
Pyridine	3.4*	20	26

* Solubility given in g/l.
** Solubility given in mol. %.
† Solubility given in wt. %.

AMMONIUM OXALATE, (NH$_4$)$_2$C$_2$O$_4$

Solvent	Solubility wt. %	Temp.	Ref.
Formic acid	5.83	21	67
Hydrazine	440*	20	8

* Solubility given in g/l.

I. SOLUBILITIES OF ELECTROLYTES

AMMONIUM PERCHLORATE, NH_4ClO_4

Solvent	Solubility wt. %	Temp.	Ref.
Acetone	2.21	25	24
Ammonia	57.97	25	64
1-Butanol	0.017	25	24
Ethanol	1.872	25	24
Ethyl acetate	0.032	25	24
Methanol	6.41	25	24
2-Methyl-1-propanol	0.127	25	24
1-Propanol	0.385	25	24
Sulfur dioxide	0.025	0	36

AMMONIUM PYROSULFITE, $(NH_4)_2S_2O_5$

Solvent	Solubility wt. %	Temp.	Ref.
Sulfur dioxide	0.048	0	36

AMMONIUM SUCCINATE, $(NH_4)_2C_4H_4O_4$

Solvent	Solubility wt. %	Temp.	Ref.
Acetone	0.47	15	33
Methanol	1.59	15	33
	5.52	65.6	

AMMONIUM SULFATE, $(NH_4)_2SO_4$

Solvent	Solubility wt. %	Temp.	Ref.
Sulfur dioxide	0.067	0	36

AMMONIUM SULFIDE, $(NH_4)_2S$

Solvent	Solubility wt. %	Temp.	Ref.
Ammonia	54.5	25	64

AMMONIUM THIOCYANATE, NH_4CNS

Solvent	Solubility wt. %	Temp.	Ref.
Ammonia	75.73	25	37
Ethanol	19.07	18.45	121
	21.54	33.25	
	22.16	36.93	
	23.46	43.36	
	26.72	57.62	
	28.63	64.20	
Methanol	37.11	24.58	121
	40.05	32.94	
	44.70	44.80	
	49.30	54.76	
	54.55	64.55	
Sulfur dioxide	31.88	0	36

BERYLLIUM ACETATE, $Be_4O(C_2H_3O_2)_6$

Solvent	Solubility wt. %	Temp.	Ref.
Chloroform	25.0	18	127

BERYLLIUM BROMIDE, $BeBr_2$

Solvent	Solubility g/l	Temp.	Ref.
Pyridine	185.6	25	26

I. SOLUBILITIES OF ELECTROLYTES

BERYLLIUM CHLORIDE, BeCl$_2$

Solvent	Solubility g/l	Temp.	Ref.
Ethanol	151.1	20	126
Methanol	256.7	20	126
1-Pentanol	153.6	20	126
Pyridine	133.3	20	126
Sulfur dioxide	0.046*	0	36

* Solubility given in wt. %.

BERYLLIUM NITRATE, Be(NO$_3$)$_2 \cdot$ H$_2$O

Solvent	Solubility g/l	Temp.	Ref.
Ethyl ether	2.21	20	31

CALCIUM BROMIDE, CaBr$_2$

Solvent	Solubility wt. %	Temp.	Ref.
Acetone	2.82	0	1
	2.74	10	
	2.67	20	
	2.65	25	
	2.65	30	
	2.71	35	
	2.84	40	
Ammonia	0.009	0	5, 128
Benzyl alcohol	11.4	10	10
	13.0	20	
	14.5	30	
	15.0	40	
	14.8	50	
	14.7	60	
	14.5	70	

CALCIUM BROMIDE, CaBr$_2$ (*Continued*)

Solvent	Solubility wt. %	Temp.	Ref.
1-Butanol	15.5	0	10
	20.2	10	
	25.3	20	
	30.4	30	
	35.2	40	
	39.3	50	
	42.9	60	
Ethanol	31.8	0	53
	32.4	10	
	33.5	15	
	34.8	20	
	35.1	25	
	35.7	30	
	37.5	40	
	39.6	50	
	43.2	60	
	48.4	70	
	50.6	75	
	50.8	80	
	51.5	85	
Methanol	33.5	0	10
	34.2	10	
	35.9	20	
	38.6	30	
	41.8	40	
	45.4	50	
	49.4	60	
	51.3	65	
3-Methyl-1-butanol	17.8	10	10
	20.4	20	
	23.0	30	
	26.9	40	
	30.3	50	
	33.3	60	
	36.0	70	
1-Pentanol	14.8	10	10
	20.2	20	
	25.4	30	
	30.0	40	
	34.0	50	
	38.4	60	
	42.6	70	

I. SOLUBILITIES OF ELECTROLYTES

CALCIUM BROMIDE, CaBr$_2$ (Continued)

Solvent	Solubility wt. %	Temp.	Ref.
1-Propanol	6.2	0	10
	11.5	10	
	18.4	20	
	25.9	30	
	32.9	40	
	38.8	50	
	43.4	60	

CALCIUM CHLORIDE, CaCl$_2$

Solvent	Solubility wt. %	Temp.	Ref.
Acetamide	0.0	82	51
	3.1	78	
	5.9	74	
	10.4	66	
	15.1	54	
	17.3	46	
	18.7	58	
	21.0	62	
	23.9	64	
	25.3	100	
	27.1	150	
	38.8	165	
	31.0	175	
	32.9	180	
	34.8	184	
	36.4	186	
	37.5	200	
	38.5	210	
Acetic acid	0.0	16.2	134
	5.7	15	
	8.5	14	
	10.7	13	
	13.3	11.1	
	15.0	30	
	15.8	35	
	17.3	40	
	19.9	45	
	21.9	50	

CALCIUM CHLORIDE, CaCl₂ (Continued)

Solvent	Solubility wt. %	Temp.	Ref.
Acetic acid	25.1	60	
	26.7	65	
	28.8	70	
	31.6	73	
Acetone	0.0062	0	1
	0.0073	10	
	0.0086	15	
	0.0101	20	
	0.0118	25	
	0.0131	30	
	0.0154	35	
	0.0173	40	
	0.0190	45	
	0.0213	50	
Benzil alcohol	2.06	10	10
	1.79	20	
	1.61	30	
	1.43	40	
	1.33	50	
	1.24	60	
1-Butanol	13.93	0	10
	17.62	10	
	20.43	20	
	22.55	30	
	24.00	40	
	25.10	50	
	25.80	60	
Ethanol	15.5	0	134
	20.5	20	
	26.1	40	
	32.5	60	
	36.0	70	
	38.7	80	
	39.7	85	
	40.8	90	
	42.8	95	
	44.5	97	
Formic acid (95%)	30.1	19	67
Hydrazine	160*	20	8

I. SOLUBILITIES OF ELECTROLYTES

CALCIUM CHLORIDE, CaCl$_2$ (Continued)

Solvent	Solubility wt. %	Temp.	Ref.
Methanol	17.85	0	134
	20.5	10	
	22.6	20	
	25.2	30	
	27.8	40	
	30.7	50	
	32.1	55	
	32.8	56	
	32.4	55	
	33.8	75	
	35.5	95	
	37.6	115	
	40.3	135	
	43.8	155	
	46.2	165	
	47.9	170	
	50.1	174	
	53.6	177	
	55.7	190	
	57.7	215	
3-Methyl-1-butanol	2.95	10	10
	6.53	20	
	10.79	30	
	14.53	40	
	18.55	50	
	21.20	60	
	24.00	70	
1-Pentanol	6.54	10	10
	10.32	20	
	14.53	30	
	18.43	40	
	22.17	50	
	25.58	60	
	29.25	70	
1-Propanol	7.66	0	10
	10.62	10	
	13.63	20	
	16.67	30	
	19.69	40	
	22.55	50	
	25.85	60	
Pyridine	1.63	25	26
Selenium oxychloride	6.11	25	16

* Solubility given in g/l.

CALCIUM IODIDE, CaI_2

Solvent	Solubility wt. %	Temp.	Ref.
Acetone	42.1	0	1
	44.6	10	
	47.1	20	
	49.3	30	
	51.4	40	
	53.0	50	
	54.5	60	
Ammonia	3.85	0	5, 128
Methanol	53.9	0	10
	54.3	10	
	55.4	15	
	55.8	20	
	56.3	30	
	57.7	40	
	58.7	50	
	59.8	60	
Sulfur dioxide	0.75	25	17

CALCIUM NITRATE, $Ca(NO_3)_2$

Solvent	Solubility wt. %	Temp.	Ref.
Acetic acid	8.90*	33.9	25
	8.71*	33.5	
	8.27*	32.0	
	7.87*	30.3	
	2.70*	15.86	
	3.65*	15.72	
	4.59*	15.60	
	6.57*	15.18	
	7.87*	14.83	
	8.90*	14.37	
Acetone	17.3	0	1
	14.5	10	
	14.4	20	
	14.6	30	
	14.7	40	
	15.5	50	
	58.5	25	136

I. SOLUBILITIES OF ELECTROLYTES

CALCIUM NITRATE, Ca(NO₃)₂ (Continued)

Solvent	Solubility wt. %	Temp.	Ref.
Ammonia	44.5	25	64
	45.1	0	5
Ethanol	46.1	25	137
	31.6	10	10
	33.9	20	
	36.0	30	
	38.6	40	
	42.3	50	
	45.1	60	
	47.4	70	
	47.8	80	
Methanol	57.3	10	10
	59.0	40	
	61.3	60	
	62.8	70	
	63.1	72	
	63.2	73	
	62.9	80	
	65.5	25	136
Methyl acetate	41.0	18	95
2-Methyl-1-propanol	25.0	25	136
1-Pentanol	56.0†	25	138
	13.3	25	136
1-Propanol	36.5	25	136
2-Propanol	2.60	25	137
Urea	47.2	142.2	139
	45.6	144.6	
	44.2	147.2	
	40.5	151.5	
	40.13	151.3	
	34.55	143.1	
	26.88	17.7	
	23.18	96.3	
	21.00	88.0	
	20.04	90.7	
	18.04	100.5	
	12.62	111.7	
	7.45	121.6	
	3.90	127.0	
	0.00	132.2	

* Solubility given in mol. %.
† Solubility given in g/l.

CALCIUM NITRITE, $Ca(NO_2)_2 \cdot H_2O$

Solvent	Solubility g/l	Temp.	Ref.
Ethanol	11.0	20	135

CALCIUM OXIDE, CaO

Solvent	Solubility wt. %	Temp.	Ref.
Calcium chloride	13.9	910	133

CALCIUM PERCHLORATE, $Ca(ClO_4)_2$

Solvent	Solubility wt. %	Temp.	Ref.
Acetone	38.18	25	24
1-Butanol	53.17	25	24
Diethyl ether	0.26	25	24
Ethanol	62.44	25	24
Ethyl acetate	43.06	25	24
Methanol	70.36	25	24
2-Methyl-1-propanol	36.29	25	24
1-Propanol	59.17	25	24

CESIUM BROMIDE, CsBr

Solvent	Solubility wt. %	Temp.	Ref.
Acetone	0.00403	18	18
	0.00406	37	
Acetonitrile	0.10	18	117
	0.14	25	
Formic acid	41.04	18	117
	41.76	25	
Methanol	2.12	18	117
	2.20	25	

I. SOLUBILITIES OF ELECTROLYTES

CESIUM CARBONATE, Cs_2CO_3

Solvent	Solubility wt. %	Temp.	Ref.
Ethanol	10	19	102
	16.7	b.p.	

CESIUM CHLORIDE, CsCl

Solvent	Solubility wt. %	Temp.	Ref.
Acetone	0.040	18	18
	0.044	37	
	0.032	25	116
Acetonitrile	0.0083	18	117
	0.0084	25	
Ammonia	0.0381	0	5
1-Butanol	0.621	25	104
Ethanol	7.697	25	104
Formic acid	51.85		117
	56.62		
Methanol	26.590		104
	3.28		117
	2.92		
3-Methyl-1-butanol	0.263		104
Selenium oxychloride	3.83		16
Sulfur dioxide	0.294		17

CESIUM DIBROMOIODIDE, $CsIBr_2$

Solvent	Solubility g/l	Temp.	Ref.
Carbon tetrachloride	0.059	25	84

CESIUM FLUORIDE, CsF

Solvent	Solubility wt. %	Temp.	Ref.
Acetone	0.00077	18	18
	0.00087	37	

CESIUM IODIDE, CsI

Solvent	Solubility wt. %	Temp.	Ref.
Acetone	0.159	18	18
	0.139	37	
	0.36	−78.5	54
	0.38	−60	
	0.37	−40	
	0.35	−20	
	0.29	0	
	0.22	20	
	0.20	25	
	0.18	30	
	0.13	40	
Acetonitrile	0.75	18	117
	0.98	25	117
Formic acid	21.93	18	117
	22.78	25	117
Methanol	2.96	18	117
	3.65	25	117

CESIUM m-NITROPHENOXIDE, $CsC_6H_4NO_3$

Solvent	Solubility wt. %	Temp.	Ref.
Acetone	0.176	25	104
1-Butanol	0.117	25	104
Ethanol	0.623	25	104
Methanol	3.806	25	104
3-Methyl-1-butanol	0.063	25	104

I. SOLUBILITIES OF ELECTROLYTES

CESIUM o-NITROPHENOXIDE, $CsC_6H_4NO_3$

Solvent	Solubility wt. %	Temp.	Ref.
Acetone	0.148	25	104
1-Butanol	0.066	25	104
Ethanol	0.881	25	104
Methanol	3.780	25	104
3-Methyl-1-butanol	0.055	25	104

CESIUM p-NITROPHENOXIDE, $CsC_6H_4NO_3$

Solvent	Solubility wt. %	Temp.	Ref.
Acetone	0.059	25	104
1-Butanol	0.232	25	104
Ethanol	0.889	25	104
Methanol	3.956	25	104
3-Methyl-1-butanol	0.126	25	104

CESIUM PERCHLORATE, $CsClO_4$

Solvent	Solubility wt. %	Temp.	Ref.
Acetone	0.15	25	24
	1.183*	25	88
1-Butanol	0.006	25	24
	0.046*	25	88
Ethanol	0.011	25	24
	0.093*	25	88
Methanol	0.093	25	24
	0.742*	25	88
3-Methyl-1-butanol	0.046*	25	88
3-Methyl-1-propane	0.007	25	24
1-Propanol	0.006	25	24
	0.046*	25	88

* Solubility given in g/l.

CESIUM PICRATE, $CsC_6H_2N_3O_7$

Solvent	Solubility wt. %	Temp.	Ref.
1-Butanol	0.044	25	104
Ethanol	0.370	25	104
Methanol	1.121	25	104
3-Methyl-1-butanol	0.017	25	104

CESIUM PYROSULFITE, $Cs_2S_2O_5$

Solvent	Solubility wt. %	Temp.	Ref.
Sulfur dioxide	0.047	0	115

CESIUM TETRAPHENYLBORATE, $Cs(C_6H_5)_4B$

Solvent	Solubility g/l	Temp.	Ref.
Acetone	14.22	20	93
Ethanol	0.36	20	93
Ethyl ether	0.50	20	93

COPPER(II) ACETATE, $CuC_4H_6O_4$

Solvent	Solubility wt. %	Temp.	Ref.
Acetic acid	0.0000*	16.65	100
	0.0307*	16.63	
	0.0471*	16.50	
	0.0831*	16.43	
	0.0994*	16.57	
	0.169*	25.3	
	0.236*	30.1	

I. SOLUBILITIES OF ELECTROLYTES

COPPER(II) ACETATE, CuC$_4$H$_6$O$_4$ (Continued)

Solvent	Solubility wt. %	Temp.	Ref.
Acetic acid	0.266*	32.1	
	0.305*	35.0	
	0.408*	41.5	
	0.487*	45.8	
	0.584*	48.7	
	0.768*	56.0	
	0.773*	55.7	
	0.976*	61.0	
	0.844*	91.3	
	0.912*	95.5	
	0.976*	99.0	
Acetone	0.28	15	33
Glycerol (96%)	9.1	15	79
Methanol	0.48	15	33
	0.48	16	
Pyridine	0.37	−11.6	101
	0.60	2.0	
	1.03	13.0	
	1.61	26.45	
	2.83	37.4	
	3.12	41.9	
	3.39	43.2	
	4.17	45.2	
	4.13	55.7	
	4.48	64.3	
	4.83	76.2	
	5.40	83.3	
	6.31	95.4	

* Solubility given in mol. %.

COPPER BENZOATE, CuC$_{14}$H$_{10}$O$_4$

Solvent	Solubility wt. %	Temp.	Ref.
Acetone	1.99	15	33
Methanol	0.49	15	33
	1.93	66	

COPPER(I) BROMIDE, CuBr

Solvent	Solubility wt. %	Temp.	Ref.
Acetonitrile	3.72	18	86

COPPER(II) BROMIDE, $CuBr_2$

Solvent	Solubility wt. %	Temp.	Ref.
Acetonitrile	19.63	18	86
Formic acid (95%)	0.16	21	67

COPPER(I) CHLORIDE, CuCl

Solvent	Solubility wt. %	Temp.	Ref.
Acetonitrile	11.8	18	86

COPPER(II) CHLORIDE, $CuCl_2$

Solvent	Solubility wt. %	Temp.	Ref.
Acetic acid	0.0070*	30	39
	0.0081*	35	
	0.0150*	50	
	0.0163*	62	
	0.0218*	75	
Acetone	2.88	18	31
	1.40	56	
Acetonitrile	1.57	18	86

I. SOLUBILITIES OF ELECTROLYTES

COPPER(II) CHLORIDE, CuCl$_2$ (Continued)

Solvent	Solubility wt. %	Temp.	Ref.
Benzyl alcohol	1.38	10	10
	1.62	20	
	2.10	30	
	2.93	40	
	4.21	50	
	5.76	60	
1-Butanol	15.0	15	31
	15.3	20	
	16.0	40	
	16.5	80	
Ethanol	32.0	0	31
	35.0	15	
	35.7	20	
	39.0	40	
	29.7	0	10
	31.5	10	
	33.3	20	
	35.0	30	
	36.8	40	
	39.0	50	
	41.5	60	
Ethyl acetate	0.4	18	94
	3.0	20	31
	2.5	40	
	1.3	72	
Ethyl ether	0.043	11	31
	0.11	20	
Ethyl formate	10	0	31
	9	20	
	8	40	
Methanol	36.1	0	10
	36.4	10	
	36.9	20	
	37.5	30	
	38.2	40	
	39.2	50	
	39.9	60	
	36.0	0	31
	40.5	15	
	36.5	20	
	37.0	40	

COPPER(II) CHLORIDE, CuCl$_2$ (*Continued*)

Solvent	Solubility wt. %	Temp.	Ref.
Methyl acetate	0.55	18	95
3-Methyl-1-butanol	9.7	10	10
	10.8	20	
	12.4	30	
	13.9	40	
	17.4	50	
	20.1	60	
1-Propanol	16.7	10	10
	19.8	20	
	22.7	30	
	25.4	40	
	27.4	50	
	29.1	60	
	29	0	31
	30.5	20	
	30.5	40	
2-Propanol	16.0	40	31
	30.0	80	
2-Propen-1-ol	23	0	31
	23	20	
Pyridine	0.140	−17.3	96
	0.195	−12.1	
	0.295	−10.0	
	0.270	−8.9	
	0.275	2	
	0.293	10	
	0.348	25	
	0.382	35	
	0.422	45	
	0.493	53	
	0.565	60	
	0.616	62	
	0.543	63	
	0.631	75	
	0.917	95	

* Solubility given in mol. %.

I. SOLUBILITIES OF ELECTROLYTES

COPPER(II) FORMATE, $C_2H_2O_4Cu$

Solvent	Solubility mol. %	Temp.	Ref.
Formic acid	0.004	25	98
	0.0045	35	
	0.0048	42	
	0.0065	49	
	0.0075	52	

COPPER(I) IODIDE, CuI

Solvent	Solubility wt. %	Temp.	Ref.
Acetonitrile	3.4	18	86
Pyridine	17.4*	25	26

* Solubility given in g/l.

COPPER(II) NITRATE, $Cu(NO_3)_2$

Solvent	Solubility g/l	Temp.	Ref.
Hydrazine	10	20	8

COPPER(II) PERCHLORATE, $Cu(ClO_4)_2$

Solvent	Solubility wt. %	Temp.	Ref.
2-Ethoxyethanol	1000†	20	97
Furfural	200*	20	97
	600†	20	

* $Cu(ClO_4)_2 \cdot 2H_2O$.
† $Cu(ClO_4)_2 \cdot 6H_2O$.

COPPER(II) SULFATE, CuSO$_4$

Solvent	Solubility wt. %	Temp.	Ref.
Methanol	0.76	15	99
	1.40	25	
	2.15	35	
	2.90	45	

ETHYLAMMONIUM CHLORIDE, NH$_4$C$_2$H$_4$Cl

Solvent	Solubility wt. %	Temp.	Ref.
Chloroform	21.19	20	125

LITHIUM ACETATE, C$_2$H$_3$O$_2$Li

Solvent	Solubility mol. %	Temp.	Ref.
Acetic acid	3.56	14.85	32
	5.01	14.16	
	6.86	13.15	
	8.45	12.10	
	8.61	16.7	
	9.15	22.1	
	9.53	25.0	
	12.28	51.0	
	16.24	70.0	
	20.25	83.0	
	27.64	98.0	
	35.34	106.5	
	39.83	109.0	
	41.84	110.0	
	45.74	112.0	
	48.76	112.5	
	50.06	112.5	
	43.95	136.0	
	46.04	147.5	
	47.57	156.0	
	48.22	161.0	
	52.41	178.5	

I. SOLUBILITIES OF ELECTROLYTES

LITHIUM ACETATE, C₂H₃O₂Li (Continued)

Solvent	Solubility mol. %	Temp.	Ref.
Acetic acid	55.89	188.0	
	56.90	192.0	
	70.87	221.0	
	100.0	272.0	
Methanol	23.3*	15	33
	24.5*	67.2	

* Solubility given in wt. %.

LITHIUM BENZOATE, LiC₇H₅O₂

Solvent	Solubility wt. %	Temp.	Ref.
Methanol	17.0	15	33
	16.46	67	

LITHIUM BROMIDE

Solvent	Solubility wt. %	Temp.	Ref.
Acetone	11.8	10	1
	15.4	20	
	17.6	30	
	18.1	32	
	18.3	35	
	18.4	35.5	
	19.1	37	
	20.8	40	
	25.7	50	
	28.4	60	
Acetonitrile	7.65	18	117
	8.10	25	
Benzaldehyde	11.527	25	20

LITHIUM BROMIDE (Continued)

Solvent	Solubility wt. %	Temp.	Ref.
Ethanol	24.59	0	21
	26.48	10	
	41.17	13.2	
	41.24	15	
	41.31	16	
	41.45	20	
	41.59	23	
	41.62	23.8	
	41.89	25	
	42.03	30	
	42.21	40	
	43.67	50	
	45.31	60	
	47.13	70	
	47.68	75	
	49.77	80	
Formic acid	45.32	18	117
	44.72	25	
Glycol	37.5	14.7	22
4-Hydroxy-4-methyl-2-pentanone	10.12	20	9
	12.68	35	
	16.93	50	
	22.58	65	
Methanol	53.92	18	117
	58.33	25	
Sulfur dioxide	0.067	25	17

LITHIUM CHLORIDE

Solvent	Solubility wt. %	Temp.	Ref.
Acetone	1.70	0	1
	1.46	10	
	0.94	18	
	1.17	20	
	0.86	30	
	0.61	37	
	0.68	40	
	0.61	50	

I. SOLUBILITIES OF ELECTROLYTES

LITHIUM CHLORIDE (Continued)

Solvent	Solubility wt. %	Temp.	Ref.
Acetonitrile	0.15	18	117
	0.14	25	
Ammonia	0.535	−33.9	4
	1.39	0	5
1-Butanol	11.49	25	2
	9.56	25	3
Ethanol	12.61	0	6
	13.08	5	
	14.36	10	
	15.80	15	
	16.91	17	
	19.57	20	
	20.06	30	
	20.22	40	
	19.61	50	
	18.99	60	
	2.48	25	3
Formic acid	21.01	18	117
	21.56	25	
Glycerol	10.0	15	7
	4.14	25	3
Hydrazine	160	20	8
4-Hydroxy-4-methyl-2-pentanone	6.14	20	9
	6.82	35	
	8.08	50	
	9.93	65	
Methanol	29.75	25	6
	31.1	0	10
	30.7	10	
	30.5	15	
	30.4	20	
	30.4	30	
	30.5	40	
	30.7	50	
	30.8	60	
	30.79	18	117
	29.08	25	
1-Pentanol	8.26	25	3
	8.28	25	6
	6.71	20	11
Phenol	1.89	53	3

LITHIUM CHLORIDE (Continued)

Solvent	Solubility wt. %	Temp.	Ref.
1-Propanol	3.72	25	3
	13.95	25	6
	13.68	20	12
2-Propen-1-ol	4.19	25	3
Pyridine	7.22	15	13
	12.48	100	
	11.31*	8	14
	11.87*	28	
	11.60*	40	
	11.38*	60	
	11.71*	80	
	13.01*	100	
	12.50†	22	
	13.79†	32	
	15.58†	45	
	16.72†	58	
	17.12†	72	
	18.35†	97	
Quinoline	0.1515	0	15
	0.3537	25	
	0.6138	40	
	1.0228	45	
	1.0986	50	
	1.1597	56.4	
	1.2218	67	
	0.8116	75	
	0.4563	96	
Selenium oxychloride	3.21	25	16
Sulfur dioxide	0.00062	25	17

* Solubility in anhydrous pyridine.
† Solubility in pyridine containing 3 vol. % of water.

LITHIUM FLUORIDE

Solvent	Solubility wt. %	Temp.	Ref.
Acetone	3.3×10^{-7}	18	18
	4×10^{-7}	37	
Bromine trifluoride	0.125	25	19
	0.081	70	

I. SOLUBILITIES OF ELECTROLYTES

LITHIUM FORMATE, CHO_2Li

Solvent	Solubility mol. %	Temp.	Ref.
Formic acid	1.58	7.0	30
	3.47	5.2	
	5.23	3.2	
	7.09	1.1	
	8.93	−1.3	
	10.75	−3.5	
	12.23	−5.6	
	13.99	−8.2	
	18.19	−14.6	
	19.56	−17.1	
	21.25	−19.8	
	22.24	−21.7	
	23.49	−23.5	
	24.33	−25.0	
	23.49	18.0	
	23.98	34.0	
	25.31	80.0	
	25.91	90.5	
	26.38	97.9	
	27.71	113.1	
	29.87	131.2	
	31.98	145.1	
	33.04	150.4	
	35.01	159.1	
	36.13	163.5	
	0.00	8.4	31
	1.58	7.0	
	3.47	5.2	
	5.23	3.2	
	7.09	1.1	
	8.93	−1.3	
	10.75	−3.5	
	12.23	−5.6	
	13.99	−8.2	
	18.19	−14.6	
	19.56	−17.1	
	21.25	−19.8	
	22.24	−21.7	
	23.49	−23.5	
	24.33	−25.0	
	23.49	18.0	
	23.98	34.0	
	25.31	80.0	
	25.91	90.5	
	26.38	97.9	
	27.71	113.1	
	29.87	131.2	

LITHIUM FORMATE, CHO_2Li (Continued)

Solvent	Solubility mol. %	Temp.	Ref.
Formic acid	31.98	145.1	
	33.04	150.4	
	35.01	159.1	
	36.13	163.5	

LITHIUM HEXAFLUOTITANATE, Li_2TiF_6

Solvent	Solubility g/l	Temp.	Ref.
Ethanol	0.30	20	29

LITHIUM IODIDE

Solvent	Solubility wt. %	Temp.	Ref.
Acetone	29.85	18	18
	42.94	37	
Acetonitrile	59.51	18	117
	60.63	25	
Ethanol	71.50	25	6
Formic acid	58.16	18	117
	59.35	25	
Furfural	459*	25	23
Glycol	28.00	15.3	7
Methanol	77.44	25	6
	58.85	18	
	63.10	25	117
Nitromethane	12.2*	0	23
	25.2*	25	
1-Pentanol	52.94	25	6
1-Propanol	32.21	25	6

* Solubility in g/l.

I. SOLUBILITIES OF ELECTROLYTES

LITHIUM NITRATE

Solvent	Solubility mol. %	Temp.	Ref.
Acetic acid	0.0	16.6	25
	0.156	16.3	
	0.983	16.0	
	2.013	15.7	
	4.115	15.0	
	5.305	14.5	
	9.38	39.0	
	10.31	46.3	
	11.26	54.2	
	12.64	66.7	
	14.66	82.6	
	15.66	91.6	
	17.08	101.0	
	19.84	118.3	
	22.34	132.2	
Acetone	23.67*	20	27
Acetonitrile	2.90†	25	26
3-Methyl-1-butanol	5.13†	0	26
	8.68†	25	
Pyridine	21.51†	0	26
	27.10†	25	

* Solubility in g/l.
† Solubility in wt. %.

LITHIUM PERCHLORATE

Solvent	Solubility wt. %	Temp.	Ref.
Acetone	57.72	25	24
1-Butanol	44.23	25	24
Diethyl ether	53.21	25	24
Ethanol	60.28	25	24
Ethyl acetate	48.75	25	24
Methanol	64.57	25	24
2-Methyl-1-propanol	36.73	25	24
Propanol	51.22	25	24

LITHIUM PERCHLORATE TRIHYDRATE, $LiClO_4 \cdot 3H_2O$

Solvent	Solubility wt. %	Temp.	Ref.
Acetone	49.04	25	24
1-Butanol	21.40	25	24
Diethyl ether	0.196	25	24
Ethanol	42.16	25	24
Ethyl acetate	26.35	25	24
Methanol	60.95	25	24
2-Methyl-1-propanol	18.85	25	24
1-Propanol	26.82	25	24

LITHIUM SALICYLATE, $LiC_7H_5O_3$

Solvent	Solubility wt. %	Temp.	Ref.
1-Propanol	15.75	20	12

LITHIUM SULFATE

Solvent	Solubility g/l	Temp.	Ref.
Sulfuric acid	271.9	12	28

MAGNESIUM ACETATE, $MgC_4H_6O_4$

Solvent	Solubility wt. %	Temp.	Ref.
Methanol	5.00	15	33
	6.98	68.2	

I. SOLUBILITIES OF ELECTROLYTES

MAGNESIUM BENZOATE, $MgC_{14}H_{10}O_4$

Solvent	Solubility wt. %	Temp.	Ref.
Acetone	2.32	15	33
Methanol	1.21	15	33
	41.51	71.9	

MAGNESIUM BROMIDE, $MgBr_2$

Solvent	Solubility wt. (g)	Temp.	Ref.
Acetonitrile	133.5*	25	26
Ammonia	0.004	0	128
Ethyl ether	0.22	−20	129
	0.40	−10	
	0.70	0	
	1.17	10	
	1.55	14	
	1.81	16	
	2.10	18	
	2.44	20	
	2.83	22	
Pyridine	4.9*	18	26
	5.4*	25	
	25.0*	60	

* Solubility given in g/l.

MAGNESIUM CHLORIDE, $MgCl_2$

Solvent	Solubility wt. %	Temp.	Ref.
Ethanol	3.48	0	10
	4.16	10	
	5.30	20	
	6.89	30	
	9.09	40	
	11.35	50	
	13.71	60	
	14.67	65	
	15.70	70	

MAGNESIUM CHLORIDE, MgCl$_2$ (Continued)

Solvent	Solubility wt. %	Temp.	Ref.
Methanol	13.4	0	10
	13.6	10	
	13.8	20	
	14.3	30	
	15.1	40	
	15.9	50	
	16.9	60	

MAGNESIUM METHOXIDE, C$_2$H$_6$O$_2$Mg

Solvent	Solubility wt. %	Temp.	Ref.
Methanol	6.39	−20	131
	10.18	20	
	6.82	66	

MAGNESIUM PERCHLORATE, Mg(ClO$_4$)$_2$

Solvent	Solubility wt. %	Temp.	Ref.
Acetone	30.01	25	24
1-Butanol	39.16	25	24
Diethyl ether	0.20	25	24
Ethanol	19.33	25	24
Ethyl acetate	41.49	25	24
Methanol	34.14	25	24
2-Methyl-1-propanol	31.27	25	24
1-Propanol	42.33	25	24

I. SOLUBILITIES OF ELECTROLYTES

MAGNESIUM SALICYLATE, $MgC_{14}H_{10}O_6$

Solvent	Solubility wt. %	Temp.	Ref.
Ethanol (90%)	0.6	20	132

MAGNESIUM SULFATE, $MgSO_4$

Solvent	Solubility wt. %	Temp.	Ref.
Ethanol	1.3*	3	44, 48
	0.025	15	99
	0.020	35	
	0.016	55	
Formic acid (95%)	0.34	19	67
Glycerol	20.8	25	130
Methanol	22.5*	3	44, 48
	29.1*	17	
	0.276	15	99
	0.224	25	
	0.180	35	
	0.153	45	
	0.123	55	

* Solubility of the 7-hydrate, $MgSO_4 \cdot 7H_2O$.

POTASSIUM ACETATE

Solvent	Solubility mol. %	Temp.	Ref.
Acetic acid	0.00	16.50	32
	1.22	15.82	
	6.70	10.95	
	9.10	7.45	
	10.03	5.95	
	10.97	14.76	
	12.71	28.03	
	15.75	49.90	
	18.71	64.1	
	20.48	73.5	

POTASSIUM ACETATE (Continued)

Solvent	Solubility mol. %	Temp.	Ref.
Acetic acid	23.85	83.6	
	28.44	99.0	
	30.67	110.0	
	32.47	118.0	
	34.41	124.5	
	35.27	126.0	
	39.50	137.0	
	44.88	145.0	
	48.30	147.5	
	50.22	148.0	
	52.32	147.5	
	58.45	170.0	
	64.16	206.0	
	76.50	245.0	
	100.0	292.0	
	0.00	16	73
	5.23	9	
	10.33	3	
	16.22	14	
	23.43	64	
	29.43	101	
	37.27	112	
	42.88	134	
	50.06	138.6	
	58.05	147	
	60.20	175	
	71.83	234	
	81.73	258	
	98.24	292	
	100.00	297	
Ammonia	1.015*	−33.9	4
Methanol	19.51*	15	33
	34.87	73.4	
Sulfur dioxide	0.006*	0	36

* Solubility given in wt. %.

POTASSIUM AMIDE

Solvent	Solubility wt. %	Temp.	Ref.
Ammonia	3.47	25	64

I. SOLUBILITIES OF ELECTROLYTES

POTASSIUM AZIDE

Solvent	Solubility wt. %	Temp.	Ref.
Benzene	0.15	80.2	60
Ethanol	0.16	0	60
	0.54	78.4	
	1.8*	0	
	5.6*	78.4	

* 80% ethanol.

POTASSIUM BENZOATE

Solvent	Solubility wt. %	Temp.	Ref.
Methanol	6.60	15	33
	7.69	66	

POTASSIUM BROMIDE

Solvent	Solubility wt. %	Temp.	Ref.
Acetone	0.0359*	18	18
	0.0328*	37	
	0.023	25	80
Acetonitrile	0.023	18	117
	0.024	25	
1-Butanol	0.0112	20	38
	0.0130	30	
	0.0137	40	
	0.0148	50	
	0.0148	55	
	0.0132	25	43
2-Butanol	0.0044	25	43
Ethanol	0.142	25	6
	0.135	25	43
	0.453	20	38

POTASSIUM BROMIDE (Continued)

Solvent	Solubility wt. %	Temp.	Ref.
Ethanol	0.501	30	
	0.563	40	
	0.568	50	
	0.537	55	
Ethyl-N-ethylcarbamate	3.87*	60	46
Formic acid	18.8	18.5	67
	18.30	18	117
	18.50	25	
Furfural	0.12	25	23
Hydrazine	37.5	20	8
Hydroxylamine	30.9	18	44
Methanol	2.12	25	6
	2.07	25	43
	1.79	0	10
	1.89	10	
	1.96	15	
	2.04	20	
	2.18	30	
	2.32	40	
	2.48	50	
	2.66	60	
	2.542	20	38
	2.496	30	
	2.440	40	
	2.316	50	
	2.241	55	
	1.95	18	117
	2.05	25	
3-Methyl-1-butanol	0.00175	25	81
2-Methyl-1-propanol	0.0076	25	43
1-Pentanol	0.003	25	6
	0.0048	25	43
1-Propanol	0.035	25	6
	0.0314	25	43
iso-Propanol	0.0110	25	43
Sulfur dioxide	2.73	0	36
	0.50	25	17

* Solubility given in g/l.

I. SOLUBILITIES OF ELECTROLYTES

POTASSIUM BUTANOATE

Solvent	Solubility wt. %	Temp.	Ref.
Methanol	33.79	15	33
	54.71	70.9	

POTASSIUM CARBONATE

Solvent	Solubility g/l	Temp.	Ref.
Hydrazine	10	20	8

POTASSIUM CHLORATE

Solvent	Solubility wt. %	Temp.	Ref.
Ammonia	2.46	0	64

POTASSIUM CHLORIDE

Solvent	Solubility wt. %	Temp.	Ref.
Acetic acid	0.168*	16.45	39
	0.157*	24	
	0.170*	30	
	0.188*	39	
	0.207*	47	
	0.229*	55	
	0.239*	59	
	0.257*	63	
	0.272*	70	
	0.295*	76	
	0.336*	87	
	0.367*	93	
Acetone	0.000087	18	18
	0.000097	37	

POTASSIUM CHLORIDE (Continued)

Solvent	Solubility wt. %	Temp.	Ref.
Acetonitrile	0.0024	18	117
	0.0024	25	
Ammonia	0.078	−76.6	78
	0.115	−76.6	
	0.167	−76.7	
	0.193	−76.8	
	0.209	−76.9	
	0.219	−77.0	
	0.252	−77.2	
	0.219	−57.9	
	0.209	−45.0	
	0.193	−35.2	
	0.167	−19.4	
	0.141	−0.6	
	0.132	0.0	
	0.115	15.0	
	0.102	18.9	
	0.089	31.2	
	0.078	44.2	
1-Butanol	0.00822*	20	42
	0.00852*	30	
	0.00904*	40	
	0.00925*	50	
	0.0030	25	43
2-Butanol	0.00084	25	43
Ethanol	0.034	18.5	44, 48
	0.1270*	20	42
	0.1378*	30	
	0.1443*	35	
	0.1454*	40	
	0.1277*	45	
	0.0845*	50	
	0.022	25	6
	0.0294	25	43
Ethyl carbamate	1.05†	60	46
Formic acid	16.04	18	117
	16.10	25	
Formic acid (95%)	16.25	19.7	67
Furfural	0.85†	25	23
Glycerol	3.58	15	79

I. SOLUBILITIES OF ELECTROLYTES

POTASSIUM CHLORIDE (Continued)

Solvent	Solubility wt. %	Temp.	Ref.
Hydrazine	90†	20	8
Hydroxylamine	10.95	18	44
Methanol	0.50	18.5	44, 48
	0.53	25	6
	0.5363	25	43
	0.826*	20	42
	0.724*	30	
	0.696*	35	
	0.638*	40	
	0.526*	45	
	0.412*	50	
	0.48	18	117
	0.52	25	
2-Methyl-1-propanol	0.0020	25	43
	0.00326*	20	42
	0.00356*	30	
	0.00400*	40	
	0.00407*	50	
1-Pentanol	0.0022	25	43
	0.0008	25	6
1-Propanol	0.004	25	43
	0.0061	25	42
	0.00700*	20	42
	0.00796*	30	
	0.00793*	35	
	0.00773*	40	
	0.00683*	45	
	0.00473*	50	
2-Propanol	0.1235*	20	42
	0.1300*	30	
	0.1340*	35	
	0.1390*	40	
	0.1295*	45	
	0.1060*	50	
iso-Propanol	0.0023	25	43
Selenium oxychloride	2.89	25	16
Sulfur dioxide	0.041	0	36
	0.0126	25	17

* Solubility given in mol. %.
† Solubility given in g/l.

POTASSIUM CYANATE

Solvent	Solubility wt. %	Temp.	Ref.
Ammonia	1.67	25	37
Benzene	0.18	b.p.	60
Ethanol	0.16	0	60
	0.53	b.p.	

POTASSIUM CYANIDE

Solvent	Solubility wt. %	Temp.	Ref.
Ammonia	4.36	−33.9	31
Ethanol	0.86	19.5	31
Glycerol	24.2	15.5	31
Hydroxylamine	29.1	17.5	31
Methanol	4.67	19.5	31
Sulfur dioxide	0.017	0	31

POTASSIUM DIBROMOIODIDE

Solvent	Solubility g/l	Temp.	Ref.
Carbon tetrachloride	15.42	25	84

POTASSIUM DIHYDROGENHYPOPHOSPHITE, KH_2PO_2

Solvent	Solubility wt. %	Temp.	Ref.
Ethanol	12.5	25	91

I. SOLUBILITIES OF ELECTROLYTES

POTASSIUM ETHYL SULFATE, $KSO_4C_2H_5$

Solvent	Solubility wt. %	Temp.	Ref.
Methanol	1.69	15	33
	12.18	65.8	

POTASSIUM FLUORIDE

Solvent	Solubility mol. %	Temp.	Ref.
Acetone	0.000022*	18	18
	0.000025	37	
Acetonitrile	0.0036	18	117
	0.0024	25	
Ethanol	0.106*	20	38
	0.096*	30	
	0.068*	40	
	0.023*	50	
	0.016*	55	
Hydrogen fluoride	380†	0	76
	0.00	−83.7	77
	1.25	−85.2	
	2.68	−86.9	
	4.20	−89.5	
	5.34	−92.8	
	6.89	−97.0	
	8.57	−45.0	
	11.16	8.0	
	14.28	48.0	
	16.45	63.2	
	17.59	67.7	
	19.14	71.8	
	20.07	72.0	
	20.99	71.0	
	22.17	67.8	
	23.24	64.4	
	24.17	65.4	
	25.10	65.8	
	25.62	65.5	
	26.58	64.5	
	27.22	62.6	
	27.50	62.7	

POTASSIUM FLUORIDE (Continued)

Solvent	Solubility mol. %	Temp.	Ref.
Hydrogen fluoride (continued)	28.27	64.0	
	28.85	64.3	
	29.60	63.4	
	30.31	61.8	
	30.68	62.4	
	32.23	70.0	
	33.30	71.7	
	33.94	71.1	
	35.65	84.0	
	38.03	128	
	39.86	148	
	43.05	175	
	45.12	189	
	46.18	195	
	47.82	217	
	48.97	231	
	49.25	234	
	49.68	236.8	
	50.04	238.8	
	51.53	236	
	51.40	229.5	
	52.25	292	
	53.54	346	
Methanol	0.192*	20	38
	0.168*	30	
	0.150*	40	
	0.125*	50	
	0.092*	55	
1-Propanol	0.34*	20	49
	9.09	18	117
	9.26	25	
Sulfur dioxide	0.018*	0	36

* Solubility given in wt. %.
† Solubility given in g/l.

POTASSIUM HYDROGEN SULFATE

Solvent	Solubility wt. %	Temp.	Ref.
Formic acid (95%)	12.7	19.3	67

I. SOLUBILITIES OF ELECTROLYTES

POTASSIUM HYDROXIDE

Solvent	Solubility wt. %	Temp.	Ref.
Ethanol	27.9	28	34
Methanol	35.5	28	34

POTASSIUM IODIDE

Solvent	Solubility wt. %	Temp.	Ref.
Acetamide	0.0	82	51
	6.5	78	
	12.8	74	
	17.8	70	
	21.5	66	
	26.2	58	
	28.4	53	
	28.75	70	
	29.1	85	
	29.45	100	
	30.15	130	
	30.5	145	
	30.8	160	
	31.1	175	
Acetone	2.98	−2.5	13
	2.32	22	
	2.84	25	
	1.19	26	
	1.106	18	18
	0.795	37	
	2.105	0	23
	1.302	25	
Acetonitrile	2.259	0	23
	2.003	24	
	2.06	18	117
	2.06	25	
Ammonia	64.8	0	37
	64.5	25	
Benzaldehyde	0.328	23	23

POTASSIUM IODIDE (Continued)

Solvent	Solubility wt. %	Temp.	Ref.
Benzonitrile	0.0506	25	23
1-Butanol	0.200	25	43
2-Butanol	0.0582	25	43
Ethanol	1.60	13.6	6, 82
	2.11	25.0	
	1.479	0	23
	1.922	25	
	1.84	25	43
	1.72	20.5	44
Ethyl acetate	0.00012	25	23
Ethyl cyanoacetate	0.930	25	23
Ethyl N-ethyl carbamate	58.1*	60	46
Ethylene glycol	31.03	0	23
	33.01	25	
Formic acid	25.92	18	117
	26.09	25	
Formic acid (95%)	27.6	18.5	67
Furfural	12.58	0	23
	4.94	25	
Glycerol	28.6	15.5	50
Hydrazine	63.5	20	8
Hydroxylamine	52.4	17.5	44
Methanol	11.89	11.4	6, 82
	12.74	12.2	
	13.79	13.5	
	15.28	25.0	
	12.95	0	23
	14.97	25	
	14.56	25	43
	14.16	20.5	44
	12.6	15	83
	13.9	30	
	15.9	50	
	18.4	80	
	20.0	100	

I. SOLUBILITIES OF ELECTROLYTES

POTASSIUM IODIDE (*Continued*)

Solvent	Solubility wt. %	Temp.	Ref.
Methanol	21.4	120	
	22.6	140	
	23.4	160	
	23.5	180	
	22.5	200	
	21.6	220	
	19.9	240	
	18.5	245	
	17.3	247	
	12.1	250	
	7.1	252.5	
	14.09	18	117
	14.53	25	
Methoxybenzaldehyde	1.355	0	23
	0.644	25	
Methyl cyanacetate	2.827	0	23
	2.165	25	
2-Methyl-1-propanol	0.0954	25	43
Nitrobenzene	0.00016	25	23
Nitromethane	0.315	0	23
	0.307	25	
1-Pentanol	0.098	25.0	6, 82
	0.0893	25	43
Propanenitrile	0.0429	0	23
	0.0404	25	
1-Propanol	0.73	12.2	6, 82
	0.43	25.0	
	0.442	25	43
	0.46	15	45
iso-Propanol	0.177	25	43
Pyridine	0.26	10	13
	0.11	119	
Salicylaldehyde	0.483	25	23
	1.093	0	
Sulfur dioxide	29.2	0	36

* Solubility given in g/l.

POTASSIUM NITRATE

Solvent	Solubility wt. %	Temp.	Ref.
Ammonia	9.52	0	5
	9.7	0.1	89
	9.59	0	40
	10.57	25	
	9.42	25	37
Hydrazine	12.3	20	8
Nitric acid	24.4	−6.0	90
	32.6	14.0	
	34.8	17.0	
	37.2	19.5	
	44.5	22.0	
	47.2	22.5	
	47.8	23.5	
	48.6	25.5	
	49.4	27.0	
	50.1	29.0	
	50.9	39.0	
	51.7	50.0	
Trichloroethylene	0.01	15	62

POTASSIUM PERCHLORATE

Solvent	Solubility wt. %	Temp.	Ref.
Acetone	0.155	25	24
	1.173*	25	88
1-Butanol	0.0045	25	24
	0.041*	25	88
Ethanol	0.012	25	24
	0.966*	25	88
Ethyl acetate	0.0015	25	24
	13.0*	25	87
	0.014*	25	88
Methanol	0.105	25	24
	0.828*	25	88
2-Methyl-1-propanol	0.005	25	24
	0.055	25	88
1-Propanol	0.010	25	24
	0.083*	25	88

* Solubility given in g/l.

I. SOLUBILITIES OF ELECTROLYTES

POTASSIUM PHENYL SULFATE

Solvent	Solubility wt. %	Temp.	Ref.
Ethanol	0.70	17	75

POTASSIUM PHTHALATE, $C_8H_4O_4K_2$

Solvent	Solubility g/l	Temp.	Ref.
Ethanol	0.161	20	92
Ethanol (95%)	0.252	20	92
Methanol	7.92	20	92

POTASSIUM PROPANOATE

Solvent	Solubility wt. %	Temp.	Ref.
Methanol	28.06	15	33
	35.62	71.3	

POTASSIUM SUCCINATE

Solvent	Solubility wt. %	Temp.	Ref.
Methanol	3.06	15	33
	3.61	66.6	

POTASSIUM SULFATE

Solvent	Solubility wt. %	Temp.	Ref.
Formic acid (95%)	26.7	21	67
Hydrazine	50*	20	8
Hydroxylamine	3.38	18	44

* Solubility given in g/l.

POTASSIUM TETRAPHENYLBORATE

Solvent	Solubility g/l	Temp.	Ref.
Acetone	45.16	20	93
Ethanol	0.40	20	93
Ethyl ether	0.55	20	93

POTASSIUM THIOCYANATE

Solvent	Solubility wt. %	Temp.	Ref.
Acetone	17.20	22	13
	16.97	58	
Acetonitrile	10.23	18	86
Ethyl acetate	0.44	0	13
	0.40	14	
	0.20	79	
1-Pentanol	0.18	13	13
	1.32	65	
	2.10	100	
	3.06	133.5	
Pyridine	6.32	0	13
	5.79	20	
	4.73	58	
	3.73	97	
	3.11	115	
	0.0	−42	85
	0.5	−42.1	
	1.33	−42.4	
	2.4	−42.8	
	3.1	−43.3	
	2.2	10	
	1.23	70–71	
	0.89	116–117	
Sulfur dioxide	4.64	0	36

I. SOLUBILITIES OF ELECTROLYTES

RUBIDIUM BROMIDE, RbBr

Solvent	Solubility wt. %	Temp.	Ref.
Acetone	0.00505	18	18
	0.00470	37	
Acetonitrile	0.061	18	117
	0.047	25	
Ammonia	18.23	0	5
Ethyl carbamate	5.66*	60	46
Formic acid	33.24	18	117
	33.60	25	
Methanol	2.51	18	117
	2.46	25	

* Solubility given in g/l.

RUBIDIUM CARBONATE, Rb_2CO_3

Solvent	Solubility wt. %	Temp.	Ref.
Ethanol	0.73	20	102

RUBIDIUM CHLORIDE, RbCl

Solvent	Solubility wt. %	Temp.	Ref.
Acetone	0.00021	18	18
	0.00024	37	
Acetonitrile	0.0034	18	117
	0.0036	25	
Ammonia	0.289	0	5
Ethanol	0.078	25	6
Formic acid	37.58	18	117
	36.26	25	
Hydrazine	50*	20	8
Methanol	1.39	25	6
	1.28	18	117
	1.32	25	
1-Pentanol	0.0025	25	6
1-Propanol	0.015	25	6
Selenium oxychloride	3.56	25	16
Sulfur dioxide	0.329	0	36
	0.402	25	17

* Solubility given in g/l.

RUBIDIUM FLUORIDE, RbF

Solvent	Solubility wt. %	Temp.	Ref.
Acetone	0.00036	18	18
	0.00039	37	

RUBIDIUM IODIDE, RbI

Solvent	Solubility wt. %	Temp.	Ref.
Acetone	3.77	−78.5	54
	2.98	−60	
	2.25	−40	
	1.64	−20	
	1.15	0	
	0.79	20	
	0.72	25	
	0.65	30	
	0.648	18	18
	0.492	25	
	0.960*	0	23
	0.674*	25	
Acetonitrile	1.478*	0	23
	1.350*	25	
	1.80	18	117
	1.65	25	
Ammonia	65.15	0	5
Ethyl carbamate	40.79*	60	46
Formic acid	31.55	18	117
	32.16	25	
Furfural	4.930*	25	23
Methanol	10.00	18	117
	10.00	25	
Nitromethane	0.567*	0	23
	0.518*	25	
Propanenitrile	0.274*	0	23
	0.305*	25	

* Solubility given in g/l.

I. SOLUBILITIES OF ELECTROLYTES

RUBIDIUM PERCHLORATE, RbClO$_4$

Solvent	Solubility wt. %	Temp.	Ref.
Acetone	0.095	25	24
	0.740*	25	88
1-Butanol	0.002	25	24
	0.018*	25	88
Ethanol	0.009	25	24
	0.074*	25	88
Ethyl acetate	0.0016	25	24
	0.018*	25	88
Methanol	0.060	25	24
	0.462*	25	88
2-Methyl-1-propanol	0.004	25	24
	0.037*	25	88
1-Propanol	0.006	25	24
	0.056*	25	88

RUBIDIUM SULFATE, Rb$_2$SO$_4$

Solvent	Solubility g/l	Temp.	Ref.
Sulfuric acid	588.1	20	28

RUBIDIUM TETRAPHENYLBORATE, Rb(C$_6$H$_5$)$_4$B

Solvent	Solubility g/l	Temp.	Ref.
Acetone	14.86	20	93
Ethanol	0.44	20	93
Ethyl ether	0.58	20	93

SILICON DIOXIDE, SiO_2

Solvent	Solubility wt. %	Temp.	Ref.
Calcium chloride	2.5	800	133
	3.8	850	
	5.4	900	
	7.6	950	

SILVER ACETATE, $AgC_2H_3O_2$

Solvent	Solubility wt. %	Temp.	Ref.
Acetone	0.00048	25	104
1-Butanol	0.00187	25	104
2-Butanone	0.00036	25	104
Ethanol	0.00755	25	104
1-Heptanol	0.00088	25	104
1-Hexanol	0.00235	25	104
Methanol	0.0167	25	104
3-Methyl-1-butanol	0.00200	25	104
4-Methyl-2-pentanone	0.00018	25	104
Sulfur dioxide	0.017	0	115

SILVER BENZOATE, $AgC_7H_5O_2$

Solvent	Solubility wt. %	Temp.	Ref.
Acetone	0.0024	25	104
1-Butanol	0.0081	25	104
2-Butanone	0.0015	25	104
Ethanol	0.0221	25	104
4-Heptanone	0.0006	25	104
1-Hexanol	0.0069	25	104
Methanol	0.0334	25	104
3-Methyl-1-butanol	0.0053	25	104

I. SOLUBILITIES OF ELECTROLYTES

SILVER BROMIDE, AgBr

Solvent	Solubility wt. %	Temp.	Ref.
Ammonia	2.31	0	5
	5.59	25	64
Ethanol	9.013×10^{-5}*	25	105
Methanol	0.56×10^{-5}*	25	105

* Solubility given in g/l.

SILVER BUTANOATE, AgC$_4$H$_7$O$_2$

Solvent	Solubility wt. %	Temp.	Ref.
Acetone	0.00132	25	104
1-Butanol	0.0399	25	104
2-Butanone	0.00125	25	104
Ethanol	0.0892	25	104
Methanol	0.1343	25	104
3-Methyl-1-butanol	0.0244	25	104
4-Methyl-2-pentanone	0.00081	25	104

SILVER CHLORIDE, AgCl

Solvent	Solubility wt. %	Temp.	Ref.
Acetone	0.0000013	25	104
Ammonia	0.215	−33.9	4
	0.28	0	5
	0.82	25	64
1-Butanol	0.0000014	25	104
2-Butanone	0.0000011	25	104
Ethanol	0.0000063	25	104
	0.0000138*	25	105

SILVER CHLORIDE, AgCl (Continued)

Solvent	Solubility wt. %	Temp.	Ref.
Methanol	0.0000116	25	104
	0.0000559*	25	105
3-Methyl-1-butanol	0.0000012	25	104
2-Pentanone	0.0000003	25	104
Pyridine	0.764	−49	106
	0.980	−35	
	1.341	−30	
	1.768	−25	
	2.152	−22	
	3.614	−20	
	3.707	−18	
	4.169	−10	
	4.807	−5	
	5.303	−1	
	4.169	0	
	3.072	10	
	1.874	20	
	1.185	30	
	0.793	40	
	0.525	50	
	0.399	60	
	0.317	70	
	0.249	80	
	0.219	90	
	0.180	100	
	0.120	110	
Sulfur dioxide	0.001	0	36

* Solubility given in g/l.

SILVER p-CHLOROBENZOATE, AgC$_7$H$_4$O$_2$Cl

Solvent	Solubility wt. %	Temp.	Ref.
1-Butanol	0.0007	25	104
Ethanol	0.0019	25	104
Methanol	0.0036	25	104
3-Methyl-1-butanol	0.0004	25	104

I. SOLUBILITIES OF ELECTROLYTES

SILVER 2,5-DINITROPHENOXIDE, $AgC_6H_3N_2O_5$

Solvent	Solubility wt. %	Temp.	Ref.
1-Butanol	0.0660	25	104
Ethanol	0.2790	25	104
1-Hexanol	0.0275	25	104
Methanol	0.2965	25	104
1-Pentanol	0.0302	25	104

SILVER FLUORIDE, AgF

Solvent	Solubility g/l	Temp.	Ref.
Hydrogen fluoride	330.0	−15	103

SILVER p-HYDROXYBENZOATE, $AgC_7H_5O_3$

Solvent	Solubility wt. %	Temp.	Ref.
1-Butanol	0.0260	25	104
Ethanol	0.0371	25	104
Methanol	0.1511	25	104

SILVER IODIDE, AgI

Solvent	Solubility wt. %	Temp.	Ref.
Ammonia	84.1	0	5
	67.4	25	64
Ethanol	0.049×10^{-6}*	25	105
Methanol	1.519×10^{-6}*	25	105
Sulfur dioxide	0.017	0	36

* Solubility given in g/l.

SILVER 3-METHYLBUTANOATE, AgC$_5$H$_9$O$_2$

Solvent	Solubility wt. %	Temp.	Ref.
1-Butanol	0.0087	25	104
Ethanol	0.0182	25	104
1-Heptanol	0.0034	25	104
Methanol	0.0207	25	104
3-Methyl-1-butanol	0.0075	25	104

SILVER NITRATE, AgNO$_3$

Solvent	Solubility wt. %	Temp.	Ref.
Acetic acid	0.0274*	30	65
	0.0398*	36	
	0.0439*	40	
	0.0708*	54	
	0.1035*	64	
	0.1332*	78	
	0.1361*	81	
	0.1940*	93	
Acetone	5.0†	20	26
	0.35	14	13
	0.44	18	
Acetonitrile	52.9	25	31
	1382†	25	26
	18.7	20	109
	28.6	81.6	
	74.4	18	86
Acetophenone	0.007	25	31
Aniline	18.0	18	31
	185†	20	26
Ammonia	46.25	25	64
Benzene	0.022	35	31
	0.044	40.5	110

I. SOLUBILITIES OF ELECTROLYTES

SILVER NITRATE, AgNO$_3$ (*Continued*)

Solvent	Solubility wt. %	Temp.	Ref.
Benzonitrile	51.2	18	111
	154†	20	26
2-Butanone	0.22	25	31
Ethanol	16.7†	20	26
	3.00	19	44, 48
Ethyl acetate	24.2†	20	26
Ethyl cyanoacetate	25.9	18	31
Ethylene glycol	33.2	25	31
Hydrazine	10†	20	8
Methanol	3.58	19	44, 48
	28.4†	20	26
3-Methyl-1-butanol	5.4†	25	26
Phenol	11.9	35.5	112
	22.1	27.2	
	31.8	11.4	
	37.5	0.0	
	39.8	1.4	
	41.5	16.0	
	43.2	30.2	
	47.4	7.6	
Phenylacetonitrile	42.8	18	31
Piperidine	4.46	25	31
Propanenitrile	64.3	18	31
iso-Propanol	13.8†	20	26
Pyridine	300.2†	10	26
	452.4†	25	
	19.58	−30	113
	19.57	0	
	21.20	3	
	22.86	5	
	26.03	11	
	26.60	30	

SILVER NITRATE, AgNO₃ (*Continued*)

Solvent	Solubility wt. %	Temp.	Ref.
Pyridine	38.03	35	
	49.80	40	
	57.70	44	
	57.74	100	
	0.0	−48.5	114
	2.9	−50.5	
	5.7	−53	
	8.3	−59	
	10.0		
	10.5	−44	
	10.9	−40	
	11.2	−35	
	12.2	−30	
	15.0	−25	
	15.8	−22	
	16.7	−10	
	18.2	0	
	21.4	10	
	25.2	20	
	29.0	30	
	34.8	40	
	38.3	45	
	38.7	46	
	39.8	47	
	41.5	48	
	41.9	50	
	44.0	60	
	47.1	70	
	54.8	80	
	68.3	87	
	69.6	80	
	69.7	74	
	69.2	74	
	69.7	80	
	70.4	87	
	70.8	90	
	71.7	100	
	73.1	110	
Quinoline	55.8†	25	26
Toluidine	0.012	25	31

* Solubility given in mol. %.
† Solubility given in g/l.

I. SOLUBILITIES OF ELECTROLYTES

SILVER p-NITROBENZOATE, AgC₇H₄NO₄

Solvent	Solubility wt. %	Temp.	Ref.
1-Butanol	0.0049	25	104
Ethanol	0.0164	25	104
Methanol	0.0296	25	104
3-Methyl-1-butanol	0.0023	25	104

SILVER PERCHLORATE, AgClO₄

Solvent	Solubility wt. %	Temp.	Ref.
Benzene	3.44	5.12	107
	5.00	25.0	
	10.07	50.0	
	32.2	80.3	
	40.1	92.0	
	50.5	115.5	
	60.0	138.5	
	62.6	140.0	
	63.0	145.0	
	64.6	159.0	
	65.6	160.0	
2-Ethoxyethanol	55.5	20	97
Furfural	28.5	20	97
Pyridine	0.0	−40.3	108
	4.0	−41.5	
	7.2	−43.0	
	8.3	−35.0	
	12.23	−11.5	
	14.53	−1.3	
	20.90	25.0	
	24.52	36.1	
	29.4	49.2	
	34.5	60.4	
	41.1	66.8	
	41.7	68.0	
	42.4	71.0	
	43.0	75.0	
	45.8	86.3	
	47.8	95.6	
	50.0	110	

SILVER PHENYLACETATE, $AgC_8H_7O_2$

Solvent	Solubility wt. %	Temp.	Ref.
1-Butanol	0.0040	25	104
Ethanol	0.0158	25	104
Methanol	0.0204	25	104

SILVER 2-PHENYLBUTANOATE, $AgC_{10}H_{11}O_2$

Solvent	Solubility wt. %	Temp.	Ref.
1-Butanol	0.0051	25	104
Ethanol	0.0109	25	104
Methanol	0.0221	25	104
3-Methyl-1-butanol	0.0032	25	104

SILVER SALICYLATE, $AgC_7H_5O_3$

Solvent	Solubility wt. %	Temp.	Ref.
Acetone	0.0033	25	104
1-Butanol	0.0027	25	104
2-Butanone	0.0024	25	104
Ethanol	0.0157	25	104
1-Heptanol	0.0040	25	104
4-Heptanone	0.0011	25	104
1-Hexanol	0.0047	25	104
Methanol	0.0160	25	104
3-Methyl-1-butanol	0.0027	25	104

I. SOLUBILITIES OF ELECTROLYTES

SILVER THIOCYANATE, AgCNS

Solvent	Solubility wt. %	Temp.	Ref.
Sulfur dioxide	0.014	0	36

SODAMIDE, NaNH$_2$

Solvent	Solubility wt. %	Temp.	Ref.
Ammonia	0.004	0	64

SODIUM ACETATE

Solvent	Solubility mol. %	Temp.	Ref.
Acetic acid	0.83	16.1	31
	3.59	14.3	
	5.40	13.1	
	7.11	25.3	
	8.92	36.7	
	12.17	54.3	
	15.27	66.9	
	16.58	71.9	
	21.55	85.7	
	26.86	93.2	
	30.72	96.1	
	33.03	96.25	
	33.16	96.3	
	34.03	112.0	
	36.87	132.0	
	39.06	145.2	
	42.54	157.0	
	44.25	160.6	
	46.28	162.3	
	48.76	174.0	
	49.49	195.5	
Acetone	0.05*	15	33
Glycol	22.50*	25	71

SODIUM ACETATE (Continued)

Solvent	Solubility mol. %	Temp.	Ref.
Hydrazine	60†	20	8
Methanol	13.79*	15	33
	14.20*	67.7	
1-Propanol	0.96*	20	12
Sulfur dioxide	0.073*	0	36

* Solubility given in wt. %.
† Solubility given in g/l.

SODIUM AZIDE

Solvent	Solubility wt. %	Temp.	Ref.
Benzene	0.10	80.1	60
Ethanol	0.22	0	60
	0.46	78.4	

SODIUM BENZENESULFONATE, $NaC_6H_5SO_3$

Solvent	Solubility wt. %	Temp.	Ref.
Methanol	6.11	15	33
	7.41	66.6	

SODIUM BENZOATE

Solvent	Solubility wt. %	Temp.	Ref.
Methanol	7.69	15	33
	7.02	66.2	

I. SOLUBILITIES OF ELECTROLYTES

SODIUM BORATE

Solvent	Solubility wt. %	Temp.	Ref.
Glycerol	37.6	15.5	63
	50.0	80	
Ethanol	2.42	15.5	63
Trichloroethylene	0.011	15	62

SODIUM BROMIDE

Solvent	Solubility wt. %	Temp.	Ref.
Acetamide	2.8	80	51
	5.4	78	
	7.6	76	
	9.4	74	
	10.7	72	
	11.6	70	
	12.6	80	
	13.7	90	
	15.0	100	
	16.4	110	
	18.0	120	
	19.8	130	
	21.1	135	
	21.6	155	
	22.1	175	
Acetone	0.095*	18	31
	0.079*	25	
	0.122*	25	
	0.075*	37	
	0.04	18	117
Acetonitrile	0.04	25	
Ammonia	9.1	−45	31
	12.0	−40	
	16.0	−34	
	19.0	−30	
	26.5	−20	
	34.0	−10	
	41.5	0	

SODIUM BROMIDE (*Continued*)

Solvent	Solubility wt. %	Temp.	Ref.
Ammonia	49.0	10	
	58.0	25	
	57.5	35	
	57	50	
	54.5	75	
	52.5	100	
	52	120	
	50	140	
	41	160	
1-Butanol	0.245	25	43
2-Butanol	0.0341	25	43
Ethanol	2.349	25	43
	5.93†	15	52
	2.385	0	53
	2.323	10	
	2.289	15	
	2.269	20	
	2.261	25	
	2.240	30	
	2.225	40	
	2.209	50	
	2.259	60	
	2.291	70	
Ethyl ether	0.08†	15	52
Ethyl *N*-ethylcarbamate	50.81*	60	46
Formic acid	16.18	18	117
	16.25	25	
Hydrazine	370*	20	8
Methanol	14.74	0	10
	14.53	10	
	14.45	15	
	14.38	20	
	14.16	30	
	13.86	40	
	13.64	50	
	13.27	60	
	14.79	25	43
	14.78	19.5	44, 48
	14.02	18	117
	14.82	25	

I. SOLUBILITIES OF ELECTROLYTES

NaBr (Continued)

Solvent	Solubility wt. %	Temp.	Ref.
2-Methyl-1-propanol	0.0950	25	43
1-Pentanol	0.1101	25	43
1-Propanol	2.00	20	12
	0.454	25	43
iso-Propanol	0.1310	25	43
Sulfur dioxide	0.0038	25	17
	0.016	0	36

* Solubility given in g/l.
† NaBr·2H$_2$O.

SODIUM BUTANOATE

Solvent	Solubility wt. %	Temp.	Ref.
Acetone	0.14	15	33
Butanoic acid	7.11*	−6	73
	14.56*	−13	
	22.12*	−22	
	29.76*	20	
	39.51*	83.1	
	47.22	127	
	53.14*	155	
	67.80*	196	
	72.63*	206	
	84.12*	225	
	92.90*	238	
	100.00*	286	
Methanol	13.27	15	33
	17.15	68.2	

* Solubility given in mol. %.

SODIUM CARBONATE

Solvent	Solubility wt. %	Temp.	Ref.
Glycerol	49.57	15	50
	44–50	20	31
Glycol	3.34	20	7

SODIUM CHLORATE

Solvent	Solubility wt. %	Temp.	Ref.
Ethanol	0.99	25	50
	2.44	78.4	
Glycerol	16.7	15.5	50
Hydrazine	660*	20	8

* Solubility given in g/l.

SODIUM CHLORIDE

Solvent	Solubility wt. %	Temp.	Ref.
Acetic acid	0.076*	30	39
	0.081*	41	
	0.085*	45	
	0.097*	54	
	0.105*	60	
	0.115*	68	
	0.130*	76	
	0.145*	84	
	0.166*	92	
Acetone	0.00032†	18	18
	0.00035†	37	
Acetonitrile	0.00015	18	117
	0.00025	25	
Ammonia	11.39	0	40
	0.28	−76.6	41
	0.40	−70	
	0.55	−60	
	1.15	−50	
	2.10	−40	
	4.0	−30	
	7.5	−20	
	10.6	−15	
	12.6	−12.5	

I. SOLUBILITIES OF ELECTROLYTES 73

SODIUM CHLORIDE (Continued)

Solvent	Solubility wt. %	Temp.	Ref.
Ammonia	14.5	−10	
	15.4	−9.5	
	13.4	−5.0	
	11.5	0	
	7.85	10	
	5.3	20	
	3.2	30	
	2.0	40	
	1.6	45	
1-Butanol	0.0050	25	43
	0.014	25	24
	0.00695*	20	42
	0.00710*	30	
	0.00762*	40	
	0.00774	50	
2-Butanol	0.00047	25	43
Ethanol	0.065	25	6
	0.0648	25	43
	0.065	18.5	44
	0.176	20	45
	0.1146*	20	42
	0.1196*	30	
	0.1241*	35	
	0.1257*	40	
	0.1292*	45	
	0.1141*	50	
Ethyl N-ethylcarbamate	0.132†	60	46
Formic acid	4.95	18	117
	4.95	25	
Glycol	31.7	14.8	7
Hydrazine	80†	20	8
Hydrogen peroxide	1.08	−2.37	47
	3.14	−3.62	
	5.42	−5.07	
	8.61	−7.67	
	11.81	−10.7	
	15.63	−10.37	
	17.0	0.0	
Hydroxylamine	12.81	17.5	48

SODIUM CHLORIDE (*Continued*)

Solvent	Solubility wt. %	Temp.	Ref.
Methanol	1.29	25	6
	1.381	25	43
	0.772*	20	42
	0.752*	30	
	0.738*	35	
	0.728*	40	
	0.715*	45	
	0.706*	50	
	1.39	18.5	44
	1.31	20	45
	1.42	18	117
	1.38	25	
2-Methyl-1-propanol	0.00280*	20	42
	0.00292*	30	
	0.00341*	40	
	0.00356*	50	
	0.0020	25	43
1-Pentanol	0.002	25	6
	0.00177	25	43
1-Propanol	0.012	25	6
	0.0124	25	43
	0.00446*	20	42
	0.00434*	30	
	0.00421*	35	
	0.00417*	40	
	0.00384*	45	
	0.00370*	50	
	0.809	20	45
	0.04	25	49
2-Propanol	0.096*	20	42
	0.100*	30	
	0.108*	35	
	0.102*	40	
	0.102*	45	
	0.093*	50	
iso-Propanol	0.0027	25	43
Selenium oxychloride	0.57	25	16
Sulfur dioxide	0.0004	25	17
	0.016	0	36

* Solubility given in mol. %.
† Solubility given in g/l.

I. SOLUBILITIES OF ELECTROLYTES

SODIUM CHROMATE

Solvent	Solubility wt. %	Temp.	Ref.
Methanol	0.344	25	44, 48

SODIUM CITRATE, $NaC_6H_7O_7$

Solvent	Solubility g/l	Temp.	Ref.
Ethanol	0.0047[a]	25	31
	0.0207[b]	25	
	0.0527[c]	25	

[a] $Na_3C_6H_5O_7$.
[b] $Na_2C_6H_6O_7$.
[c] $NaC_6H_7O_7$.

SODIUM CYANATE

Solvent	Solubility wt. %	Temp.	Ref.
Benzene	0.13	80.1	60
Ethanol	0.22	0	60
	0.52	78.4	

SODIUM CYANIDE

Solvent	Solubility wt. %	Temp.	Ref.
Methanol	6.05	15	33
	3.94	67.4	
Sulfur dioxide	0.018	0	36

SODIUM DECANOATE

Solvent	Solubility wt. %	Temp.	Ref.
Glycol	13.77	25	71
1,2-Propanediol	10.76	25	72

SODIUM DICHROMATE, $Na_2Cr_2O_7 \cdot 2H_2O$

Solvent	Solubility wt. %	Temp.	Ref.
Ethanol	51.32	19.4	69

SODIUM DODECANOATE

Solvent	Solubility wt. %	Temp.	Ref.
Glycol	6.34	25	71
1,2-Propanediol	10.86	25	72

SODIUM FLUORIDE

Solvent	Solubility wt. %	Temp.	Ref.
Acetone	0.0010*	18	18
	0.0011*	37	
Acetonitrile	0.0014	18	117
	0.003	25	
Ammonia	0.35	25	37
Bromine trifluoride	2.08	25	19
	2.55	70	

I. SOLUBILITIES OF ELECTROLYTES

SODIUM FLUORIDE (Continued)

Solvent	Solubility wt. %	Temp.	Ref.
1-Butanol	0.0030	20	38
	0.0041	30	
	0.0043	40	
	0.0049	50	
	0.0054	55	
Ethanol	0.095	20	38
	0.108	30	
	0.119	40	
	0.158	50	
	0.179	55	
Hydrogen peroxide	3.8†	−5.2	35
	7.4†	−9.1	
	8.0†	−10.6	
	9.8†	−12.1	
	11.1†	−13.9	
	12.3†	−15.2	
Methanol	0.413	20	38
	0.440	30	
	0.458	40	
	0.476	50	
	0.484	55	
	0.041	18	117
Sulfur dioxide	0.03	25	
Sulfur dioxide	0.029	0	36

* Solubility given in g/l.
† Solubility given in mol. %.

SODIUM FORMATE

Solvent	Solubility wt. %	Temp.	Ref.
Methanol	3.40	15	33
	3.56	66.6	

SODIUM HEXANOATE

Solvent	Solubility wt. %	Temp.	Ref.
Glycol	26.71	25	71
1,2-Propanediol	21.53	25	72

SODIUM HYDROGEN CARBONATE

Solvent	Solubility wt. %	Temp.	Ref.
Ethanol	1.18	15.5	50
Glycerol	7.4	15.5	50
	7.28	20	31

SODIUM HYDROGEN SULFATE

Solvent	Solubility wt. %	Temp.	Ref.
Ethanol (95%)	1.4	25	63
Formic acid (95%)	23.08	19.5	67

SODIUM HYDROXIDE

Solvent	Solubility wt. %	Temp.	Ref.
Ethanol	14.7	28	34
Methanol	23.6	28	34

I. SOLUBILITIES OF ELECTROLYTES

SODIUM IODIDE

Solvent	Solubility wt. %	Temp.	Ref.
Acetamide	5.32	78	51
	10.08	74	
	14.00	70	
	17.86	66	
	20.90	62	
	23.44	58	
	25.80	54	
	28.0	50	
	30.1	46	
	32.3	41.5	
	33.0	50	
	33.9	60	
	34.8	70	
	35.9	80	
	37.2	90	
	38.7	100	
	40.6	110	
	44.0	120	
	47.4	125	
	47.7	150	
	47.9	175	
Acetone	10.3	0	1
	15.4	10	
	18.8	15	
	19.8	18	
	23.0	20	
	28.6	25	
	28.0	30	
	26.8	35	
	24.9	37	
	25.8	40	
	24.8	45	
	23.6	50	
	22.6	55	
	1.55	−76.5	54
	2.20	−67.3	
	1.77	−62.6	
	2.51	−46.3	
	5.38	−21.5	
	10.83	0.5	
	11.60	0.9	
	23.75	20.7	
	3.2	−34.0	55
	7.4	−12.3	
	11.6	0.0	

SODIUM IODIDE (Continued)

Solvent	Solubility wt. %	Temp.	Ref.
Acetone	21.0	15.9	
	28.5	25.0	
	29.2	25.7	
	29.4	25.0	
	32.2	32.2	
	26.8	40.2	
	25.0	50.0	
	23.6	57.2	
	23.0	59.8	
	5.2	−20.0	56
	7.2	−10.0	
	10.5	0.0	
	15.4	10.0	
	23.1	20.0	
	28.9	25.7	
	28.2	30.0	
	26.3	40.0	
	24.2	50.0	
	22.2	60.0	
	20.0	70.0	
	17.9	80.0	
Acetonitrile	220.9*	0	23
	184.3*	25	
	20.94	18	117
	19.93	25	
Ammonia	448*	−42.2	31
	448*	−37.8	
	513*	−35.2	
	546*	−31.5	
	56.88	0	
	59.40	25	
Benzyl alcohol	12.59	25	59
1-Butanol	17.76	25	43
2-Butanol	13.05	25	43
2-Butanone	6.8	−70	56
	9.1	−60	
	12.3	−50	
	21.9	−30	
	20.0	−20	
	18.6	−10	
	16.7	0	

I. SOLUBILITIES OF ELECTROLYTES

SODIUM IODIDE (Continued)

Solvent	Solubility wt. %	Temp.	Ref.
2-Butanone	14.8	10	
	13.0	20	
	11.3	30	
	9.6	40	
	8.2	50	
	6.9	60	
	5.8	70	
Ethanol	31.51	25	6
	30.23	25	43
	29.86	25	57
	30.44	10	58
	30.67	30	
	30.79	50	
	31.03	80	
	31.08	100	
	31.13	120	
	31.03	160	
	30.70	180	
	29.73	200	
	27.80	220	
	26.58	230	
	24.64	240	
	20.76	250	
	17.35	255	
	9.74	260	
	7.92	261.5	
Formic acid	38.54	18	117
	38.20	25	
Furfural	251.0*	25	23
Hydrazine	640*	20	8
Methanol	47.46	25	6
	39.39	10	10
	42.16	20	
	43.82	25	
	44.84	27	
	44.91	28	
	44.87	30	
	44.66	40	
	44.47	50	
	44.26	60	
	44.61	25	43
	41.86	18	117
	45.35	25	

SODIUM IODIDE (Continued)

Solvent	Solubility wt. %	Temp.	Ref.
2-Methyl-1-propanol	15.02	25	43
Nitromethane	3.4*	0	23
	4.8*	25	
1-Pentanol	14.01	25	6
	14.02	25	43
1-Propanol	22.01	25	6
	22.66	25	43
	22.17	25	59
iso-Propanol	20.84	25	43
Propanenitrile, C_3H_5N	90.9*	0	23
	62.3*	25	
Sulfur dioxide	13.0	0	36
	1.59	25	17

* Stability given in g/l.

SODIUM NITRATE

Solvent	Solubility wt. %	Temp.	Ref.
Acetic acid	0.031	16.53	65
	0.150	16.45	
	0.122	27.0	
	0.150	31.5	
	0.166	36.0	
	0.173	40.0	
	0.202	46.5	
	0.246	53.7	
	0.448	78.5	
	0.573	88.3	
	0.641	93.4	
	0.677	95.7	
	0.854	103.0	

I. SOLUBILITIES OF ELECTROLYTES

SODIUM NITRATE (Continued)

Solvent	Solubility wt. %	Temp.	Ref.
Ethanol	0.036	25	44
Hydrazine	1000*	20	8
Hydrogen peroxide	1.62	−2.47	47
	3.23	−2.82	
	5.11	−3.72	
	7.58	−4.72	
	10.11	−5.62	
	13.19	−6.97	
	16.66	−8.12	
	20.19	−9.52	
	22.25	−10.82	
	25.22	11.8	
	28.25	32.3	
	31.53	49.3	
Hydroxylamine	11.6	17	48
Methanol	0.41	25	44
Urea	6.9	121.4	66
	16.3	106.8	
	23.2	95.8	
	27.0	88.8	
	29.5	83.9	
	30.5	88.4	
	31.1	92.5	
	33.2	109.0	
	36.7	131.7	
	41.3	156.0	

* Solubility given in g/l.

SODIUM NITRITE

Solvent	Solubility wt. %	Temp.	Ref.
Ethanol	0.31	19.5	44
Methanol	4.24	19.5	44

SODIUM OCTANOATE

Solvent	Solubility wt. %	Temp.	Ref.
Glycol	22.47	25	71
1,2-Propanediol	18.23	25	72

SODIUM OXALATE, $Na_2C_2O_4$

Solvent	Solubility wt. %	Temp.	Ref.
Formic acid (95%)	8.1	19.3	67

SODIUM PENTANOATE

Solvent	Solubility wt. %	Temp.	Ref.
Glycol	23.74	25	71
1,2-Propanediol	17.76	25	72

SODIUM PERCHLORATE

Solvent	Solubility wt. %	Temp.	Ref.
Acetone	8.80	25	24
1-Butanol	1.83	25	24
Ethanol	12.82	25	24
Ethyl acetate	8.80	25	24
Methanol	33.93	25	24
2-Methyl-1-propanol	0.78	25	24
1-Propanol	4.66	25	24

I. SOLUBILITIES OF ELECTROLYTES

SODIUM PHENYL SULFATE, NaC$_6$H$_5$O$_4$S·3H$_2$O

Solvent	Solubility wt. %	Temp.	Ref.
Ethanol	7.0	17	75

SODIUM PROPANOATE

Solvent	Solubility wt. %	Temp.	Ref.
Glycol	23.62	25	71
Methanol	11.72	15	33
	12.10	68	
1,2-Propanediol	16.59	25	72

SODIUM SALICYLATE, NaC$_7$H$_5$O$_2$

Solvent	Solubility wt. %	Temp.	Ref.
Methanol	20.81	15	33
	25.78	67.2	
1-Propanol	1.15	20	12

SODIUM SUCCINATE, Na$_2$C$_4$H$_4$O$_4$

Solvent	Solubility wt. %	Temp.	Ref.
Methanol	0.72	15	33
	1.77	66.6	

SODIUM SULFATE

Solvent	Solubility wt. %	Temp.	Ref.
Ethanol	0.00143*	20	42
	0.00154*	30	
	0.00156*	40	
	0.00114*	50	
Hydrogen peroxide	1.19	−2.17	47
	2.56	−2.87	
	4.47	−3.12	
	6.92	−4.27	
	10.23	−5.52	
	13.51	−6.77	
	16.77	−9.02	
	19.47	−10.57	
	22.64	19.6	
	23.59	29.1	
	25.62	39.1	
Methanol	0.00555*	20	42
	0.00544*	30	
	0.00523*	40	
	0.00514*	50	
2-Propanol	0.000886*	20	42
	0.000939*	30	
	0.000928*	40	
	0.000739*	50	
Sulfuric acid	299.9†	20	28

* Solubility given in mol. %.
† Solubility given in g/l.

SODIUM SULFITE

Solvent	Solubility wt. %	Temp.	Ref.
Sulfur dioxide	0.023	0	36

I. SOLUBILITIES OF ELECTROLYTES

SODIUM TETRAETHOXYBORATE, $C_8H_{20}O_4BNa$

Solvent	Solubility g/l	Temp.	Ref.
Ethanol	96	0	74
	113	25	

SODIUM TETRAFLUOROBORATE

Solvent	Solubility wt. %	Temp.	Ref.
Ethanol	0.28	0	70
	0.47	25	
	0.64	40	
	0.82	50	
	1.00	60	
Methanol	2.85	0	70
	4.17	25	
	5.15	40	

SODIUM TETRAMETHOXYBORATE, $C_4H_{12}O_4BNa$

Solvent	Solubility g/l	Temp.	Ref.
Methanol	272	0	74
	334	25	

SODIUM THIOCARBONATE

Solvent	Solubility wt. %	Temp.	Ref.
Ethanol	5.25	15.5	50

SODIUM THIOCYANATE

Solvent	Solubility wt. %	Temp.	Ref.
Acetone	6.41	18.8	61
	8.67	29.2	
	12.34	41.9	
	15.69	51.0	
	17.62	56.0	
Ethanol	15.52	18.8	61
	16.00	35.8	
	17.36	52.8	
	18.44	61.8	
	19.63	70.9	
Methanol	25.92	15.8	61
	28.59	24.7	
	31.10	34.6	
	33.76	48.0	
	34.87	52.3	

SODIUM THIOSULFATE

Solvent	Solubility wt. %	Temp.	Ref.
Ethanol	0.0025	20	68

References

1. W. R. G. Bell, C. B. Rowlands, J. J. Bamford, W. G. Thomas and W. J. Jones, J. Chem. Soc., 1927 (1930).
2. H. H. Willard and G. F. Smith, J. Am. Chem. Soc., **44**, 2816 (1922).
3. H. E. Patten and W. R. Mott, Z. physik. Chem., **8**, 153 (1904).
4. W. C. Johnson and O. F. Krumboltz, Z. physik. Chem., **167**, 249 (1933).
5. M. Linhard and M. Stephan, Z. physik. Chem., **167**, 87 (1934).
6. W. E. S. Turner and C. C. Bissett, J. Chem. Soc., **103**, 1904 (1913).
7. O. De Connick, Ann. Chim. Phys.; (7) **28**, 7 (1903).
8. T. W. B. Welsh and H. J. Broderson, J. Am. Chem. Soc., **37**, 816 (1915).
9. E. T. J. Fuge, S. T. Bowden and W. G. Jones, J. Phys. Chem., **56**, 1013 (1952).
10. E. Lloyd, C. B. Brown, D. Glynwyn, R. Bonnell and W. J. Jones, J. Chem. Soc., 658 (1928).
11. L. W. Andrews and C. Ende, Z. physik. Chem., **17**, 136 (1895).
12. A. Schlamp, Z. physik. Chem., **14**, 272 (1894).

I. SOLUBILITIES OF ELECTROLYTES

13. S. Laszczyncki, Ber., **27,** 2285 (1894).
14. L. Rahlenberg and F. C. Kranskopf, J. Am. Chem. Soc., **30,** 1104 (1908).
15. J. H. Walton and C. R. Wise, J. Am. Chem. Soc., **44,** 103 (1922).
16. C. R. Wise, J. Am. Chem. Soc., **45,** 1233 (1923).
17. A. I. Shatenshtein and M. M. Viktorov, Zhur. Fiz. Khim., **11,** 18 (1938).
18. A. Lannung, Z. physik. Chem., (A) **161,** 255 (1932).
19. I. Sheft, H. H. Hyman and J. J. Katz, J. Am. Chem. Soc., **75,** 5221 (1953).
20. R. Müller, V. Raschka and M. Wittemann, Monatsh. Chem., **48,** 660 (1927).
21. J. P. Simmons, H. Freimuth and H. Russell, J. Am. Chem. Soc., **58,** 1692 (1936).
22. O. De Connick, Chem. Zbl., **76,** 883 (1905).
23. P. T. Walden, Z. physik. Chem., **55,** 712 (1906).
24. H. H. Willard and G. F. Smith, J. Am. Chem. Soc., **45,** 286 (1923).
25. A. W. Davidson and H. A. Geer, J. Am. Chem. Soc., **60,** 1211 (1938).
26. R. Müller, Z. anorg. u. allgem. Chem., **142,** 130 (1924–1925).
27. A. Roshdestwensky and W. C. Lewis, J. Chem. Soc., **99,** 2144 (1911).
28. F. Bergius, Z. physik. Chem., **72,** 338 (1910).
29. H. Ginsberg, Z. anorg. Chem., **204,** 225 (1932).
30. J. Kendall and A. W. Davidson, J. Am. Chem. Soc., **43,** 979 (1921).
31. A. Seidell, Solubility of Inorganic, Metalorganic and Organic Compounds, 3rd Ed., New York (1940).
32. A. W. Davidson and W. H. MacAllister, J. Am. Chem. Soc., **52,** 507 (1930).
33. H. Henstock, J. Chem. Soc., 1340 (1934).
34. A. G. Murray, J. Ass. Offic. Agr. Chem., **12,** 309 (1929).
35. G. L. Matheson, O. Maas, J. Am. Chem. Soc., **51,** 674 (1929).
36. G. Jander and W. Ruppolt, Z. physik. Chem., (A) **179,** 43 (1937).
37. H. Hunt, J. Am. Chem. Soc., **54,** 3509 (1932).
38. F. G. Germuth, J. Franklin Inst., **212,** 346 (1931).
39. A. W. Davidson, W. Chappell, J. Am. Chem. Soc., **60,** 2043 (1938).
40. M. M. Viktorov and A. I. Shatenshtein, Zhur Fiz. Khim., **8,** 260 (1936).
41. G. Patscheke, Z. physik. Chem., **163A,** 340 (1933).
42. E. R. Kirn and H. E. Dunlop, J. Am. Chem. Soc., **53,** 391 (1931).
43. R. G. Larson and H. Hunt, J. Phys. Chem., **43,** 417 (1939).
44. C. A. L. De Bruyn, Z. physik. Chem., **10,** 781 (1892).
45. P. Rohland, Z. anorg. Chem., **18,** 327 (1898).
46. M. Stuckgold, J. Chim. Phys., **15,** 502 (1917).
47. O. Maass and W. A. Hatcher, J. Am. Chem. Soc., **44,** 2473 (1922).
48. C. A. L. De Bruyn, Rec. trav. chim., **11,** 29 112 (1892).
49. G. B. Frankforter and F. C. Frary, J. Phys. Chem., **17,** 402 (1913).
50. A. M. Ossendowski, Pharm. J., **79,** 575 (1907).
51. B. N. Menshutkin, Izv. SPb. Politekh. Inst., **9,** 200 (1908).
52. Eder, Dinglers Polytech. J. **221,** 89, 189 (1876).
53. D. K. G. Bonnell and W. J. Jones, J. Chem. Soc., **129,** 318 (1926).
54. C. R. Evertz and R. Livingston, J. Phys. et Colloid Chem., **53,** 1330 (1949).
55. R. Macy and E. W. Thomas, J. Am. Chem. Soc., **48,** 1547 (1926).
56. A. E. Wadsworth and H. M. Dawson, J. Chem. Soc., **129,** 2784 (1926).
57. F. E. King and J. R. Partington, J. Chem. Soc., **129,** 20 (1926).
58. D. Tyrer, J. Chem. Soc., **71,** 850 (1897).
59. I. R. Partington and R. I. Winterton, Trans. Faraday. Soc., **30,** 619 (1934).
60. J. A. Cranston and A. J. Livingstone, J. Chem. Soc., **129,** 501 (1926).
61. O. L. Hughes and T. H. Mead, J. Chem. Soc., 2282 (1929).
62. D. J. Wester, Pharm. Weekblad, **51,** 1443 (1914).

63. U. S. Pharmacopia, 8. Ed. (1907).
64. H. Hunt and L. Boncyk, J. Am. Chem. Soc., **55**, 3528 (1933).
65. A. W. Davidson and H. A. Geer, J. Am. Chem. Soc., **55**, 642 (1933).
66. W. J. Howells, J. Chem. Soc., 2010 (1930).
67. O. Aschan, Chem. Zig, **37**, 1–17 (1913).
68. E. Bödtker, Z. physik. Chem., **22**, 510, 570 (1897).
69. D. Reinitzer, Z. angew. Chem., **26**, (1913).
70. I. G. Ryss and V. A. Plit, Zhur. Obshchei Khim., **25**, 19 (1955).
71. S. R. Palit and J. W. McBain, Ind. Eng. Chem., **38**, 741 (1946).
72. S. R. Palit, J. Am. Chem. Soc., **69**, 3120 (1947).
73. M. Bakunin and E. Vitale, Gazz. chim. ital., **65**, 593 (1935).
74. H. C. Brown and E. J. Mead, J. Am. Chem. Soc., **78**, 3614 (1956).
75. G. N. Burkhardt and A. Lapworth, J. Chem. Soc., **129**, 684 (1926).
76. K. Fredenhagen, Z. physik. Chem., **165A**, 179 (1933).
77. G. H. Cady, J. Am. Chem. Soc., **56**, 1431 (1934).
78. G. Patscheke and C. Tanne, Z. physik. Chem., **174**, 135 (1935).
79. A. M. Ossendowski, J. pharm. chim., (6), **26**, 162 (1907).
80. W. H. Krug and K. P. MacElroy, J. Anal. Chem., **6**, 184 (1892).
81. J. H. Walton and C. R. Wise, J. Am. Chem. Soc., **44**, 103 (1922).
82. V. Timofeev, Dissertatsiya, Khar'kov (1894).
83. A. E. H. Tutton, J. Chem. Soc., **71**, 850 (1897).
84. H. W. Cremer and D. R. Duncan, J. Chem. Soc., 2243 (1931).
85. K. L. Wagner and E. Zerner, Monatsh. Chem., **31**, 833 (1911).
86. A. Naumann and A. Schier, Ber., **47**, 249 (1914).
87. G. F. Smith, J. Am. Chem. Soc., **47**, 762 (1925).
88. J. E. Ricci and T. W. Davis, J. Am. Chem. Soc., **62**, 407 (1940).
89. A. I. Shatenshtein and A. Monoszon, Z. anorg. Chem., **207**, 204 (1932).
90. E. Groschuff, Ber., **37**, 1488 (1904).
91. Salzer, Ann. Chem., Liebig's, **211**, 1 (1882).
92. J. A. Handy, L. F. Hout, J. Am. Pharm. Assoc., **16**, 7 (1927).
93. G. P. Mwzharaup and A. F. Ievin'sh, Uchebnye zapiski Latv. Univ., **9**, 49 (1956).
94. A. Naumann, Ber., **37**, 3600, 4328 (1904).
95. A. Naumann, Ber., **42**, 3789 (1909).
96. J. H. Matthews and S. Spero, J. Phys. Chem., **21**, 402 (1917).
97. A. L. Channey and C. A. Mann, J. Phys. Chem., **35**, 2289 (1931).
98. A. W. Davidson and V. Holm, J. Am. Chem. Soc., **53**, 1350 (1931).
99. G. C. Gibson, J. O. Driscoll and W. J. Jones, J. Chem. Soc., 1440 (1929).
100. A. W. Davidson, E. Griswold, J. Am. Chem. Soc., **53**, 1341 (1941).
101. J. H. Mathews and E. B. Benger, J. Phys. Chem., **18**, 264 (1914).
102. R. Bunsen, Gasometrische Methoden. 2. Ed. (1877).
103. K. Fredenhagen and G. Cadenbach, Z. Physik. Chem., **146A**, 245 (1930).
104. N. A. Izmailov and V. S. Chernyi, Trudy Komissii Anal. Khim., T. I. (XII), page 44, AN-SSSR, 1958.
105. F. K. Koch, J. Chem. Soc., 1551 (1930).
106. L. Kahlenberg and W. J. Wittich, J. Phys. Chem., **13**, 421 (1909).
107. A. E. Hill, J. Am. Chem. Soc., **50**, 2678 (1928).
108. R. Macy, J. Am. Chem. Soc., **47**, 1031 (1925).
109. R. Scholl and W. Steinkopf, Ber., **39**, 4393 (1906).
110. C. E. Linebarger, Am. J. Sci., **49**, 48 (1895).
111. A. Naumann, Ber., **47**, 1370 (1914).
112. C. R. Bailey, J. Chem. Soc., 1534 (1930).

I. SOLUBILITIES OF ELECTROLYTES

113. R. Müller, Z. Elektrochem., **38,** 450 (1932).
114. L. Kahlenberg and R. K. Brewer, J. Phys. Chem., **12,** 283 (1908).
115. G. Jander and K. Wickert, Z. physik. Chem., **178,** 63 (1936).
116. H. W. Foote, J. Am. Chem. Soc., **34,** 880 (1912).
117. T. Pavlopoulos and H. Strehlow, Z. physik. Chem., **202,** 474 (1954).
118. W. C. Eichelberger, J. Am. Chem. Soc., **56,** 801 (1934).
119. P. C. Scherer, J. Am. Chem. Soc., **53,** 3694 (1931).
120. H. G. Greenish, Pharm. J., **65,** 190 (1900).
121. L. Shinidman, J. Phys. Chem., **38,** 901 (1934).
122. B. Kuriloff, Z. physik. Chem., **25,** 109 (1898).
123. E. Groschuff, Ber., **36,** 1791, 4351 (1903).
124. A. Rosdestwensky and W. C. Lewis, J. Chem. Soc., **101,** 2098 (1912).
125. W. E. Henry, Compt. rend., **99,** 1157 (1884).
126. J. M. Schmidt, Ann. chim., (10), **11,** 351 (1929).
127. F. Wirth, Z. anorg. Chem., **87,** 1 (1914).
128. M. Linhard and M. Stephan, Z. physik. Chem., **163,** 185 (1933).
129. H. H. Rowley, J. Am. Chem. Soc., **58,** 1337 (1936).
130. W. Schnellbach and J. Rosin, J. Am. Pharm. Ass., **20,** 227 (1931).
131. L. Quinet, Bull Soc. chim. (France), (5), **2,** 1201 (1935).
132. P. W. Squire and C. M. Caines, Pharm. J., **74,** 720, 784 (1905).
133. K. Arndt, Z. Elektrochem., **15,** 784 (1909).
134. B. N. Menshutkin, Izvest. SPb. Politekh. Inst., **5,** 355 (1906).
135. F. Vogel, Z. anorg. Chem., **35,** 389 (1903).
136. J. d'Ans and R. Siegler, Z. physik. Chem., **82,** 35 (1913).
137. S. W. Ferner and M. G. Mellon, Ind. Eng. Chem. (Anal. Ed.) **6,** 345 (1934).
138. R. Muller, E. Printer and K. Prett, Monatsh. Chem., **48,** 660 (1925).
139. W. J. Howells, J. Chem. Soc., 3208 (1931).

(b) Additional Data

Some additional solubility studies are reported in this section. Numerical values are not given. The electrolytes are listed in alphabetical order according to cation and the solvents in which solubility measurements have been made for each electrolyte are included. References are provided in each case.

SOLUBILITIES—ADDITIONAL STUDIES

Solute	Solvent	Ref.
Aluminium (III)		
chloride	carbon tetrachloride	50
	chloroform	
	m-chloronitrobenzene	94
	o-chloronitrobenzene	
	p-chloronitrobenzene	
	m-bromonitrobenzene	
	o-bromonitrobenzene	
	p-bromonitrobenzene	
	nitrobenzene	
	benzene–cyclohexane	95, 96
	methylcyclopentane	96
	hexane	
	benzoyl chloride	94
	m-nitrotoluene	
	o-nitrotoluene	
	p-nitrotoluene	
	toluene	95, 96
	1,3-dimethylbenzene	96
	mesitylene	
	benzophenone	94
bromide	carbon disulfide	97
	bromoethane	94, 98
	pyridine	99
	m-chloronitrobenzene	94, 98
	o-chloronitrobenzene	
	p-chloronitrobenzene	
	m-bromonitrobenzene	94, 98
	o-bromonitrobenzene	

I. SOLUBILITIES OF ELECTROLYTES 93

SOLUBILITIES—ADDITIONAL STUDIES (*Continued*)

Solute	Solvent	Ref.	
bromide	p-bromonitrobenzene		
	nitrobenzene		
	benzene		
	benzoyl chloride		
	benzonitrile	99	
	m-nitrotoluene	94,	98
	o-nitrotoluene		
	p-nitrotoluene		
	toluene		
	1,4-dimethylbenzene		
	benzophenone		
iodide	sulfur dioxide	11	
	pyridine	19	
sulfate	glycol	101	
octadecanoate	methanol–acetone benzene	3	
Antimony			
trifluoride	hydrogen fluoride	9	
	methanol-acetone	131	
	1-propanol		
	2-butanone		
	dioxane		
	chlorobenzene	131	
	nitrobenzene		
	benzene		
	heptane		
	butoxybutane		
trichloride	tetrachloroethane	132	
	acetic acid	133	
	acetone–ethyl acetate	14	
	p-dichlorobenzene	98,	133
	p-dibromobenzene		
	m-dinitrobenzene		
	fluorobenzene		
	chlorobenzene		
	bromobenzene		
	iodobenzene		
	nitrobenzene		
	benzene		
	phenol		
	benzenesulfonic acid	133	
	aniline		
	tetrahydrobenzene		

SOLUBILITIES—ADDITIONAL STUDIES (Continued)

Solute	Solvent	Ref.
trichloride	cyclohexane	98, 133
	benzyl chloride	133
	benzonitrile	98, 133
	benzaldehyde	133
	benzoic acid	
	m-chlorotoluene	98, 134
	o-chlorotoluene	
	p-chlorotoluene	
	m-nitrotoluene	
	o-nitrotoluene	
	p-nitrotoluene	
	toluene	
	methoxybenzene	133
	acetophenone	
	1,3-dimethylbenzene	98
	1,2-dimethylbenzene	
	1,4-dimethylbenzene	
	ethylbenzene	98, 133
	ethoxybenzene	98, 134
	mesitylene	
	pseudocumene	98, 133
	propylbenzene	
	1-chloronaphthalene	
	2-chloronaphthalene	133
	1-bromonaphthalene	98, 133
	2-nitronaphthalene	
	naphthalene	
	p-cymene	
	isoamylbenzene	
	biphenyl	
	benzophenone	133
	diphenylmethane	98, 133
	triphenylmethane	
pentachloride	selenium oxychloride	12
tribromide	acetic acid	133
	p-dichlorobenzene	98
	p-dibromobenzene	
	m-dinitrobenzene	
	fluorobenzene	
	chlorobenzene	
	bromobenzene	
	iodobenzene	

I. SOLUBILITIES OF ELECTROLYTES

SOLUBILITIES—ADDITIONAL STUDIES (*Continued*)

Solute	Solvent	Ref.
tribromide	nitrobenzene	
	m-nitrophenol	135
	benzene	98
	phenol	133
	benzenesulfonic acid	
	tetrahydrobenzene	98, 134
	cyclohexane	133
	benzoyl chloride	133
	benzonitrile	
	benzaldehyde	
	benzoic acid	
	m-chlorotoluene	98, 134
	o-chlorotoluene	
	p-chlorotoluene	
	m-nitrotoluene	
	o-nitrotoluene	
	p-nitrotoluene	
	toluene	
	methoxybenzene	133
	acetophenone	
	1,3-dimethylbenzene	98, 133
	1,2-dimethylbenzene	
	1,4-dimethylbenzene	
	ethylbenzene	98
	ethoxybenzene	133
	propylbenzene	98
	mesitylene	133
	1-chloronaphthalene	98, 134
	2-chloronaphthalene	
	1-bromonaphthalene	98, 134
	nitronaphthalene	133
	naphthalene	
	p-cymene	98
	amylbenzene	133
	biphenyl	
	benzophenone	
	diphenylmethane	
	triphenylmethane	98, 133
tri-iodide	di-iodomethane	130
	nitrobenzene	135
	nitrotoluene	
	naphthalene	

SOLUBILITIES—ADDITIONAL STUDIES (*Continued*)

Solute	Solvent	Ref.	
Arsenic			
trioxide	carbon disulfide	126	
	formic acid	22	
	ethanol	126	
	glycerol	127	
	ethyl ether	126	
	ethyl malonate	128	
pentoxide	formic acid (95%)	22	
tri-iodide	carbon disulfide	129	
	di-iodomethane	130	
Barium			
fluoride	hydrogen fluoride	9	
chloride	hydrazine	10	
	selenium oxychloride	12	
	formic acid	22	
	methanol	23,	43
	acetic acid	29	
	glycerol	2	
	nitrobenzene	50	
bromide	methanol	43,	31
	ethanol	43,	51
	3-methyl-1-butanol	33	
iodate	ethanol (95%)	53	
iodide	sulfur dioxide	11	
	formic acid (95%)	22	
	ethanol (97%)	52	
	ethanol	51	
	pyridine	52	
perchlorate	various solvents	54	
nitrate	hydroxylamine	23,	43
	hydrazine	10	
	methanol	55,	31
	acetic acid	56	
	ethanol	38	
	acetone	55	
	2-propanol	38	

I. SOLUBILITIES OF ELECTROLYTES

SOLUBILITIES—ADDITIONAL STUDIES (*Continued*)

Solute	Solvent	Ref.
sulfate	sulfuric acid	57
	formic acid (95%)	22
succinate	ethanol	5
maleate	ethanol	5
tartrate	ethanol (95%)	5
acetate	methanol	3
	acetic acid	6
	ethanol (97%)	58
propionate	methanol (95%)	58
lactate	methanol	3
benzene sulfonate	methanol	3
phenyl sulfate	ethanol	59
citrate	ethanol (95%)	5
p-bromobenzoate	acetone (94%)	42
benzoate	methanol	3
Bismuth		
fluoride	hydrogen fluoride	9
trichloride	hydrazine	10
	antimony trichloride	137
	acetone	138
	ethyl acetate	67(a)
tribromide	nitrobenzene	135
oxychloride	formic acid (95%)	22
nitrate	acetone	66
diethylthiolthionocarbamate	various solvents	88
Boron		
trifluoride	sulfuric acid	36
	benzene–toluene	92
trichloride	hydrogen chloride	93, 100

SOLUBILITIES—ADDITIONAL STUDIES (Continued)

Solute	Solvent	Ref.
Cadmium		
chloride	selenium oxychloride	12
	methanol–ethanol	23, 43
	methanol–ethanol	31
	pyridine	16
	benzonitrile	44
bromide	ethanol–methanol	31
	various solvents	45, 14, 10
iodide	sulfur dioxide	11
	methanol	36, 31
	methyl formate–ethanol	36
	ethanol	31
	acetone	46
	ethyl formate	36
	1-propanol	
	2-propanol	
	ethyl acetate	
	propyl formate	
	ethyl ether	47, 48
	pyridine	36, 19
	aniline–quinoline	36
	various solvents	
perchlorate	2-ethoxyethanol furfural	49
sulfate	formic acid (95%)	22
	methanol–ethanol	24
acetate	ethanol	3
Calcium		
sulfate	glycerol (97%)	1, 2
formate	methanol	3
glycero–phosphate	glycerol (98.5%)	4
malate	ethanol (95%)	5
tartrate	ethanol (95%)	5
acetate	methanol	3
	acetic acid	6
cacodylate	alcohols	7
lactate	methanol	3

I. SOLUBILITIES OF ELECTROLYTES

SOLUBILITIES—ADDITIONAL STUDIES (*Continued*)

Solute	Solvent	Ref.
propionate	methanol	3
citrate	ethanol (95%)	8
benzoate	methanol	3
Cerium		
chloride	hydrazine	10
	pyridine	19
bromide	pyridine	99
nitrate	ethyl ether	36
acetate	methanol	3
tartrate	various solvents	36
Erbium		
nitrate	ethyl ether	105
Lanthanum		
nitrate	ethyl ether	36
cobalt nitrate	nitric acid	163
nickel nitrate	nitric acid	103
acetate	methanol	3
Lead		
oxide	hydrazine	10
fluoride	hydrazine	10
	sulfur dioxide	
chloride	pyridine	116
bromide	sulfur dioxide	11
	pyridine	116
iodide	hydrazine	10
	sulfur dioxide	11
	formic acid (95%)	22
	pyridine	116
	various solvents	66
cyanide	sulfur dioxide	11

SOLUBILITIES—ADDITIONAL STUDIES (*Continued*)

Solute	Solvent	Ref.
thiocyanate	sulfur dioxide	11
perchlorate	2-ethoxyethanol furfural	21
borate	hydrazine	10
nitrate	hydrazine	10
	methanol–ethanol	43
	pyridine	117
alkyl fluorides	methanol–ethanol benzene	118
salts	blood serum	119
formate	formic acid	36
succinate	ethanol (95%)	5
malate	ethanol (95%)	5
tartrate	ethanol (95%)	5
acetate	formic acid (95%)	22
	methanol	3
	acetic acid	6, 40, 120
	glycerol	2
	lanoline	87
styphnate	glycol diacetate	36
diethylthiolthiono-carbamate	various solvents	88
citrate	ethanol (95%)	5
Dicyclohexyllead		
dichloride	various solvents	121
dibromide	various solvents	121
Lead		
benzoate	methanol	3
benzoate	acetone	3
tricyclohexyl-lead	chloroform–ethanol-benzene	122
tetraphenyl-lead	various solvents	121
diphenyldicyclohexyl-lead	various solvents	121

I. SOLUBILITIES OF ELECTROLYTES

SOLUBILITIES—ADDITIONAL STUDIES (Continued)

Solute	Solvent	Ref.
tetracyclohexyl-lead	various solvents	121
lead hexadecanoate	turpentine	104
	various solvents	123
octadecenoate	ethyl ether	36
	various solvents	123
Mercury(II) chloride	sulfur dioxide	11
	selenium oxychloride	12
	carbon disulfide	60
Mercury(I) chloride	tribromomethane	61
Mercury(II) chloride	methanol	23, 43
	dichloroethane	62
	acetic acid	29, 63
	methanol	64
	ethanol	23, 43, 64
	2-propen-1-ol	64
	acetone	14, 65
		66, 46
	methyl acetate	67
	1-propanol	64
	ethyl acetate	65, 66,
		62, 67(a)
	1-butanol	64
	2-methyl-1-propanol	64
	ethyl ether	36, 68
	pyridine	69
	benzene	70, 62, 36
	various solvents	61
	chlorinated paraffins	71
bromide	sulfur dioxide	11
	methanol	72, 31, 73
	acetic acid	29
	ethanol	72, 31, 73
	acetone	74, 46
	1-propanol	36
	2-methyl-1-propanol	
	pyridine	75
	benzene	76
	aniline–quinoline	75
iodide	hydrazine	10
	carbon disulfide	109, 77, 78

SOLUBILITIES—ADDITIONAL STUDIES (*Continued*)

Solute	Solvent	Ref.		
iodide	carbon tetrachloride	78		
	chloroform	79		
	di-iodomethane	130		
	acetic acid	29		
	acetone	66,	79,	74
	dimethoxymethane	36		
	ethyl acetate	66,	67(a)	
	ethyl ether	66		
	pyridine	80,	75	
	benzene	66,	79,	76
	aniline	81		
	quinoline	75		
	alcohols	4		
	various solvents	82,	83,	61
cyanide	sulfur dioxide	11		
	methanol	84		
	acetonitrile	85		
	ethanol	86		
	acetone	3		
	glycerol	76		
	ethyl ether	76		
	pyridine–aniline	75		
	benzonitrile	44		
	quinoline	75		
	various solvents	61		
thiocyanate	sulfur dioxide	11		
Mercury(I)	hydrazine			
nitrate	hydrazine	10		
	lanoline	87		
Mercury(II)				
sulfate	sulfur dioxide	11		
acetate	methanol	3		
	sulfur dioxide	11		
	methanol–acetone	3		
diethylthiolthiono-carbamate	various solvents	3,	88	

I. SOLUBILITIES OF ELECTROLYTES

SOLUBILITIES—ADDITIONAL STUDIES (*Continued*)

Solute	Solvent	Ref.
Neodymium		
chloride	ethanol-pyridine	102
camphor carbonate	various solvents	26
Phosphorus		
trisulfide	various solvents	124
heptasulphide	carbon disulfide	124
pentasulfide	carbon disulfide	124
selenide	carbon disulfide	125
Praseodymium		
chloride	pyridine	102
nitrate	ethyl ether	36
Samarium		
chloride	pyridine	102
sulfate	hydrazine	10
metal nitrates	nitric acid	103
Strontium		
fluoride	hydrogen fluoride	9
chloride	deuterium oxide (98.1%)	28
	hydrazine	10
	formic acid (95%)	22
	methanol	23
	acetic acid	29
	ethanol	23
bromide	ammonia	30
	methanol–ethanol	31
	ethanol	32
	acetone	18
	3-methyl-1-butanol	33
iodide	sulfur dioxide	34
	ethanol	35

SOLUBILITIES—ADDITIONAL STUDIES (*Continued*)

Solute	Solvent	Ref.
perchlorate	various solvents	36
nitrate	ethanol	37
	hydrazine	10
	ethanol-2-propanol	38
	pyridine	19
	2-butoxyethanol	39
sulphate	formic acid	22
acetate	methanol	3
	acetic acid	40
cacodylate	methanol–ethanol	41
halogen benzoates	acetone	42
Sulfur Dioxide		
	sulfuric acid	139
	carbon tetrachloride	140
Tantalum		
chloride	carbon disulfide	136
	carbon tetrachloride	
	chloroform	
	bromoethane	
	ethylenediamine	
	nitrobenzene	
	carbon tetrachloride	
	bromoethane	
Thallium		
chloride	methanol	106
bromide	sulfur dioxide	11
iodide	sulfur dioxide	11
cyanide	sulfur dioxide	107
thiocyanate	sulfur dioxide	11
perchlorate	sulfur dioxide	11
carbonate	sulfur dioxide	11
sulfite	sulfur dioxide	11
sulfate	sulfur dioxide	11

I. SOLUBILITIES OF ELECTROLYTES

SOLUBILITIES—ADDITIONAL STUDIES (Continued)

Solute	Solvent	Ref.
methoxide	methanol–benzene	108
acetate	sulfur dioxide	11
ethoxide	ethanol	108
picrate	methanol	109
dodecanoate	ethyl ether–acetone	110
tetradecanoate	ethanol (96%)	110
	acetone	
	ethyl ether	
hexadecanoate	ethanol (96%)	111
	acetone	110
Tin		
tetrabromide	sulfur dioxide	112
tetraiodide	carbon disulfide	109
	hexane	113
	various solvents	114
alkyl fluorides	methanol–ethanol benzene	115
oxalate	formic acid (95%)	22
Zinc		
fluoride	hydrogen fluoride	9
chloride	hydrazine	10
	sulfur dioxide	11
	selenium oxychloride	12
	acetic acid	13
	acetone	14
	glycerol	15
	pyridine	16
tetrachloride	sulfur dioxide	17
bromide	acetone	18
	pyridine	19
iodide	ammonia	20
	sulfur dioxide	11
	glycerol	15
	pyridine	19

SOLUBILITIES—ADDITIONAL STUDIES (*Continued*)

Solute	Solvent	Ref.
thiocyanate	sulfur dioxide	11
perchlorate	2-ethoxyethanol	21
	furfural	21
arsenite	formic acid (95%)	22
arsenate	formic acid	22
sulfate	methanol	23
	methanol–ethanol	24
	glycerol	15
acetate	hydrazine	10
	methanol	3
	acetic acid	25
benzoate	various solvents	3
camphorcarbonate	various solvents	26
dodecanoate	toluene	27
octadecanoate	toluene–dodecanoic acid–octadecanol	27

References

1. E. Asselin, Compt. rend., **76**, 884 (1873).
2. A. M. Ossendowski, J. pharm. chim., (6), **26**, 162 (1908).
3. H. Henstock, J. Chem. Soc., 1340 (1934).
4. L. B. Lane, Ind. Eng. Chem., **17**, 924 (1925).
5. Partheil, Hübner, Arch. Pharm., **241**, 413 (1903).
6. A. W. Davidson and W. H. MacAllister, J. Am. Chem. Soc., **52**, 507 (1930).
7. A. Travers, Malaprade, Bull. Soc. chim., (4), **39**, 1543 (1926).
8. Partheil, Hübner, Arch. Pharm., **241**, 554 (1903).
9. A. W. Jache and G. H. Cady, J. Phys. Chem., **56**, 1106 (1952).
10. T. W. B. Welsh and H. J. Broderson, J. Am. Chem. Soc., **37**, 816 (1915).
11. G. Jander and W. Ruppolt, Z. physik. Chem., (A), **179**, 43 (1937).
12. C. R. Wise, J. Am. Chem. Soc., **45**, 1233 (1923).
13. A. W. Davidson and W. Chappell, J. Am. Chem. Soc., **61**, 2164 (1939).
14. A. Naumann, Ber., **37**, 3600, 4328 (1904).
15. A. M. Ossendowski, Pharm. J., **79**, 575 (1907).
16. R. B. Mason and J. H. Mathews, J. Phys. Chem., **29**, 1179 (1925).
17. P. A. Bond, W. R. Stephens, J. Am. Chem. Soc., **51**, 2910 (1929).
18. W. R. G. Bell, C. B. Rowlands, J. J. Bamford, W. G. Thomas and W. J. Jones, J. Chem. Soc., 1927 (1930).

I. SOLUBILITIES OF ELECTROLYTES

19. R. Muller, Z. anorg. u. allgem. Chem., **142**, 130 (1924–1925).
20. H. Hunt and L. Boncyk, J. Am. Chem. Soc., **55**, 3528 (1933).
21. A. L. Cheney and C. A. Mann, J. Phys. Chem., **35**, 2289 (1931).
22. O. Aschan, Chem. Zig, **37**, 1117 (1913).
23. C. A. De Bruyn, Rec. trav. chim., **11**, 29, 112 (1892).
24. G. C. Gibson, J. O. Driscoll and W. J. Jones, J. Chem. Soc., 1440 (1929).
25. A. W. Davidson and W. H. MacAllister, J. Am. Chem. Soc., **52**, 507 (1930).
26. M. Picon, Bull. Soc. chim. (France), (4), **49**, 399 (1931).
27. E. P. Martin and C. R. Pink, J. Chem. Soc., **11**, 1750 (1948).
28. F. T. Miles and A. W. Menzies, J. Am. Chem. Soc., **56**, 2392 (1937).
29. A. Davidson and W. Chappell, J. Am. Chem. Soc., **60**, 2043 (1938).
30. M. Linhard and M. Stephan, Z. physik. Chem., **167**, 87 (1934).
31. E. Lloyd, C. B. Brown, D. Glynwyn, R. Bonnell and W. J. Jones, J. Chem. Soc., 658 (1928).
32. Diacon-Fonzes, J. pharm. chim., (6), **1**, 59 (1895).
33. H. Yagoda, J. Am. Chem. Soc., **52**, 3068 (1930).
34. A. I. Shatenshtein and M. M. Viktorov, Zhur. Fiz. Khim., **11**, 18 (1938).
35. Etard, A. Compt. rend., **84**, 1090 (1877).
36. A. Seidell, Solubility of Inorganic, Metalorganic and Organic Compounds, 3 Ed., New York (1940).
37. F. Vogel, Z. anorg. Chem., **35**, 389 (1903).
38. S. W. Ferner and M. G. Mellon, Ind. Eng. Chem. (Anal. Ed.) **6**, 345 (1934).
39. K. A. Kobe and W. L. Motsch, J. Phys. Chem., **56**, 185 (1952).
40. A. W. Davidson and W. Chappell, J. Am. Chem. Soc., **55**, 3531, 4524 (1933).
41. R. Tiollais, Bull. Soc. chim. (France), (5), **3**, 70 (1936).
42. J. C. Bailar, Ind. Eng. Chem., Anal. Ed., **3**, 362 (1931).
43. C. A. L. De Bruyn, Z. physik. Chem., **10**, 781 (1892).
44. A. Naumann, Ber., **47**, 1370 (1914).
45. J. Eder, Prakt. Chem., (2), **17**, 45 (1878).
46. C. Zapata, Zapata, Anal. Soc. esp. fis. quim., **28**, 603 (1930).
47. C. E. Linebarger, Am. J. Sci., **49**, 48 (1895).
48. O. Guemple, Bull. Soc. chim. (Belg.), **38**, 443 (1929).
49. R. Salvadori, Gass. chim. ital., **42**, 1, 458 (1912).
50. S. J. Lloyd, J. Phys. Chem., **22**, 300 (1918).
51. D. K. G. Bonnell and W. J. Jones, J. Chem. Soc., **129**, 318 (1926).
52. P. Rohland, Z. anorg. Chem., **15**, 412 (1897).
53. A. E. Hill and W. A. H. Zink, J. Am. Chem. Soc., **31**, 44 (1909).
54. H. H. Willard and G. F. Smith, J. Am. Chem. Soc., **45**, 286 (1923).
55. J. d'Ans and R. Siegler, Z. physik. Chem., **82**, 35 (1913).
56. A. W. Davidson and H. A. Geer, J. Am. Chem. Soc., **55**, 642 (1933).
57. F. Bergius, Z. physik. Chem., **72**, 338 (1910).
58. R. D. Crowell, J. Am. Chem. Soc., **40**, 455 (1918).
59. G. N. Burkhardt and A. Lapworth, J. Chem. Soc., **129**, 684 (1926).
60. H. Arctowski, Z. anorg. Chem., **6**, 267, 404 (1894).
61. Sulc, Z. anorg. Chem., **25**, 401 (1900).
62. M. P. Dukel'skii, Z. anorg. Chem., **53**, 327; **54**, 45 (1907).
63. Etard, A. Ann. Chim. et (Phys.), (7), **2**, 526; **3**, 275 (1894).
64. Etard, A. Ann. Chim. et (Phys.), (7), **2**, 563 (1894).
65. A. H. W. Aten, Z. physik. Chem., **54**, 86, 124 (1905–1906).
66. S. Laszczyncki, Ber., **27**, 2285 (1894).

67. A. Naumann, Ber., **42,** 3789 (1909).
67a. A. Naumann, Ber., **43,** 313 (1910).
68. F. Richard, J. Pharm. Chim., (8), **4,** 306 (1926).
69. R. S. MacBride, J. Phys. Chem., **14,** 189 (1910).
70. A. Lannung, Z. physik. Chem., (A), **161,** 255 (1932).
71. K. A. Hoffmann, K. Kirmireuther and A. Thal, Ber., **43,** 183 (1910).
72. V. Timofeev, Kissertatsia, Khar'kov (1894).
73. K. L. Malhotra, J. Indian Chem. Soc., **5,** 545 (1928).
74. W. Reinders, Z. physik. Chem., **32,** 514 (1900).
75. W. Staronka, Anzeiger Akad. Wiss. Krakau, Ser. A, 372 (1910).
76. R. Abegg and M. S. Sherrill, Elektrochem., **9,** 550 (1903).
77. C. E. Linebarger, J. Am. Chem. Soc., **16,** 214 (1894).
78. H. M. Dawson, J. Chem. Soc., **95,** 874 (1909).
79. E. Beckman and A. Stock, Z. physik. Chem., **17,** 130 (1895).
80. J. H. Mathews and P. A. Ritter, J. Phys. Chem., **21,** 269 (1917).
81. H. J. Cole, Philippine J. Sci., **47,** 351 (1932).
82. Mehu, J. pharm. chim., (5), **12,** 249 (1885).
83. Anon, Pharm. J., **72,** 77 (1904).
84. M. P. Dukel'skii, Zhur. Russ. Fiz. Khim. Obshchestva, **39,** 975 (1907).
85. A. Naumann and A. Schier, Ber., **47,** 249 (1914).
86. W. Herz and F. Kuhn, Z. anorg. Chem., **58,** 159; **60,** 152 (1908).
87. G. Klose, Arch. intern, Pharmacodynamie, **17,** 459 (1907).
88. J. Sedlitzki, Monatsh. Chem., **8,** 563–573 (1887).
89. D. H. Wester, Pharm. Weekblad, **51,** 1443 (1914).
90. R. Hooper, Pharm. J. Trans., (3), **13,** 258 (1892).
91. M. T. Robinson, J. Phys. Chem., **61,** 120 (1957).
92. C. M. Wheeler and H. P. Keating, J. Phys. Chem., **58,** 1171 (1954).
93. W. Graff, Compt. rend., **196,** 1930 (1933).
94. B. N. Menshutkin, Izv. SPb. Politekh. Inst., **11,** 261, 567, **12,** (1909).
95. B. N. Menshutkin, Zhur. Russ. Fiz. Khim., Obshchestva, **41,** 1089 (1909).
96. F. Fairbrother, N. Scott and H. Prophet, J. Chem. Soc., 1164 (1956).
97. H. H. Kaveler and C. J. Monroe, J. Am. Chem. Soc., **50,** 2421 (1928).
98. B. N. Menshutkin, Izv. SPb. Politekh. Inst., **13,** 1, 263, 411, 565, **14,** 251 (1910).
99. R. Müller, Z. Elektrochem., **38,** 227 (1932).
100. L. Hackspill, Helv. chim. acta, **16,** 1096 (1933).
101. O. De Connick, Bull acad. roy. (Belg.), **257,** 359 (1905).
102. C. Matignon, Ann. chim. et phys., (8), **8,** 249, 388, 407 (1906).
103. G. Jantsch. Z. anorg. Chem., **76,** 321 (1912).
104. R. S. Morrell, J. Chem. Soc., **113,** 111 (1918).
105. F. Wirth, Z. anorg. Chem., **76,** 174 (1912).
106. P. Buckley and H. Hartley, Phil. Mag., **8,** 320 (1929).
107. G. Jander and K. Wickert, Z. physik. Chem., **178,** 63 (1936).
108. N. V. Sidgwick and L. E. Sutton, J. Chem. Soc., 146 (1930).
109. H. Arctowski, Z. anorg. Chem., **11,** 273 (1895).
110. G. Canneri and D. Bigalli, Ann. chim. appl., **26,** 430 (1936).
111. D. Holde and M. Selim, Ber., **58,** 523 (1925).
112. P. A. Bond and H. T. Beach, J. Am. Chem. Soc., **48,** 348 (1926).
113. M. E. Dice and J. H. Hildebrand, J. Am. Chem. Soc., **50,** 3023 (1928).
114. M. E. Dorfman and J. H. Hildebrand, J. Am. Chem. Soc., **49,** 729 (1927).
115. E. Krause, Ber., **51,** 1447 (1918).

I. SOLUBILITIES OF ELECTROLYTES

116. G. W. Heise, J. Phys. Chem., **16**, 373 (1912).
117. J. H. Walton and R. C. Judd, J. Am. Chem. Soc., **33**, 1036 (1911).
118. E. Krause and E. Pohland, Ber., **55B**, 1282 (1922).
119. L. T. Fairhall, J. Biol. Chem., **60**, 481 (1924).
120. W. Crookes, Chem. News, **9**, 37, 205 (1864).
121. G. Grüttner, Ber., **27**, 3259 (1914).
122. B. N. Menshutkin, Efiraty i drugie molekulyarnye soedineniya bromistogo i iodistoga magniya. SPb. (1907).
123. C. A. Jacobson and A. Holmes, J. Biol. Chem., **25**, 29 (1916).
124. A. Stock, Ber., **43**, 156, 1227 (1910).
125. J. Mai, Ber., **61**, 1808 (1928).
126. L. Winkler, J. Pract. Chem., (2), **31**, 347 (1885).
127. W. Schnellbach and J. Rosin, J. Am. Pharm. Ass., **20**, 227 (1931).
128. E. V. Zappi and A. Manini, Ann. Assoc. quim. Argent., **17**, 90 (1929).
129. P. W. Squire and C. M. Caines, Pharm. J., **74**, 720, 784 (1905).
130. J. W. Retgers, Z. anorg. Chem., **3**, 253, 344 (1893).
131. D. W. Breck, J. L. Harwey and H. M. Haendler, J. Phys. & Colloid Chem., **53**, 906 (1949).
132. P. F. M. Pauw, Tabell. Annuel., **5**, 929 (1926).
133. B. N. Menshutkin, Izv SPb. Politekh. Inst., **15**, 65, 397, 613, 647, 757 (1911).
134. B. N. Menshutkin, Izv. SPb. Politekh. Inst., **16**, 33, 397 (1912).
135. N. A. Pushin, Zhur. Obshchei Khim. **18**, 1599 (1948).
136. Z. G. Namoradze and O. E. Zvyagintsev, Zhur. Priklad. Khim., **12**, 603 (1939).
137. E. R. Natsvlishvilli and A. G. Bergman, Zhur. Obshchei Khim., **9**, 642 (1939).
138. A. Naumann, Ber., **38**, 2293 (1905).
139. Dunn, Chem. News, **45**, 272 (1882).
140. J. Horiuti, Sci. Papers Inst. Phys. Chem. Research (Tokyo), **17**, 125 (1931).

(c) pK$_{sp}$ Data

(1) SINGLE SOLVENTS

The data in this section consists of K_{sp} and pK_{sp} values for electrolytes in single solvents. Electrolytes are arranged alphabetically by chemical symbol. Additional information relative to ionic strengths, supporting electrolytes and temperatures is provided in footnotes.

ACETONE

Electrolyte	pK_{sp}	Comments	Ref.
AgBr	18.7	$I = 0.100F$ in $LiClO_4$; $23 \pm 1°C$	1, 4
AgCl	16.4	$I = 0.100F$ in $LiClO_4$; $23 \pm 1°C$	1, 4
	14.1	$I = 0.100F$ in $LiClO_4$; $23 \pm 1°C$	2
AgI	20.9	$I = 0.100F$ in $LiClO_4$; $23 \pm 1°C$	1, 4
	22.0	$I = 0.100F$ in $LiClO_4$; $23 \pm 1°C$	3

ACETONITRILE

Electrolyte	K_{sp}	pK_{sp}	Comments	Ref.
$AgB(C_6H_5)_4$		7.2	$I = 0.10-0.05M$	8, 13
		7.2		9
		7.2	$I \sim 0.01M$ in $(C_2H_5)_4NNO_3$	23
AgBr		13.2	$I = 0.100$ F in $(C_2H_5)_4NClO_4$; $23 \pm 1°C$	1, 4
		12.9		8
		12.9	$I = 0.01-0.05M$	13
		12.9	$I \sim 0.01M$ in $(C_2H_5)_4 NNO_3$	23
			K_{sp}, 3×10^{-14}	25

I. SOLUBILITIES OF ELECTROLYTES

ACETONITRILE (Continued)

Electrolyte	K_{sp}	pK_{sp}	Comments	Ref.
AgCl		12.4	$I = 0.100$ F in $(C_2H_5)_4NClO_4$; $23 \pm 1°C$	1, 4
		12.9		8
		12.9	$I = 0.01 = 0.05M$	13
		12.9	$I \sim 0.01M$ in $(C_2H_5)_4NNO_3$	23
		13		25
AgI		14.2	$I = 0.100$ F with $(C_2H_5)_4NClO_4$; $23 \pm 1°C$	1, 4
		14.2		8, 9
		14.2	$I = 0.01–0.05M$	13
		15		25
AgN_3		9.6		8
		9.6	$I = 0.01–0.05M$	13
		9.6	$I \sim 0.01M$ $(C_2H_5)NNO_3$	23
AgO_2CCH_3		7.4	$I = 0.10–0.05M$	8
		7.4	$I = 0.01–0.05M$	13
		7.4	$I \sim 0.01M$ in $(C_2H_5)_4NNO_3$	23
AgSCN		10.0	$I = 0.01–0.05M$	8, 9
		10.0	$I = 0.01–0.05M$	13
		10.0	$I \sim 0.01M$ in $(C_2H_5)_4NNO_3$	23
$(C_6H_5)_4AsB(C_6H_5)_4$		5.2		9
$(C_6H_5)_4AsI$		2.0		9
$(C_6H_5)_4As[OC_6H_2(NO_2)_3]$		2.1		9
$(C_6H_5)_4C$		3.2		9
$(C_4H_9)_4NB(C_6H_5)_4$		1.9		9
$CsB(C_6H_5)_4$		3.1		9
CsBr		4.4		8
CsCl		6.8		8, 9
CsI		3.0		8, 9
$Cs[OC_6H_2(NO_2)_3]$		4.4		9
	1.9×10^{-5}			24
Ferrocene		0.9		9

ACETONITRILE (Continued)

Electrolyte	K_{sp}	pK_{sp}	Comments	Ref.
Ferrocene-B$(C_6H_5)_4$		3.8		9
KB$(C_6H_5)_4$		2.7		9
KBr		5.60		8
KCH$_3$SO$_3$	8.0×10^{-8}			24
KCl		7.20		8, 9
	8.6×10^{-9}			24
KClO$_4$	6.1×10^{-5}			24
K(3,5-dinitro-benzoate)	6.0×10^{-8}			24
K(3,5-dinitro-phenolate)	6.7×10^{-8}			24
KI		2.00		8, 9
KIO$_4$	7.5×10^{-7}			24
KNO$_3$	1.4×10^{-6}			24
K[OC$_6$H$_2$(NO$_2$)$_3$]		4.55		8
	2.1×10^{-5}			24
		3.92		25
K Salicylate	9.2×10^{-6}			24
LiCH$_3$SO$_3$	1.0×10^{-8}			24
NaCH$_3$SO$_3$	1.2×10^{-8}			24
NaCl		9.0		8
	0.9×10^{-8}			24
NaIO$_4$	1.8×10^{-5}			24
NaNO$_3$	1.1×10^{-6}			24
NaOC$_6$H$_4$NO$_2$-p		5.42		8
TlBr		12.74		25
TlCl		12.99		25
TlClO$_4$		3.34		25
TlI		12.19		25
TlNO$_3$		6.66		25
TlSCN		8.11		25

I. SOLUBILITIES OF ELECTROLYTES

DIETHYL ETHER

Electrolyte	pK_{sp}	Ref.
AgBr	17.75 ±0.14	5*
	17.8	5*
AgCl	14.30 ±0.11	5*
	14.3	
AgI	22.60 ±0.11	5*
	23.5	6

* Supporting electrolyte $LiClO_4$; temp. 23.5°C.

DIMETHYLACETAMIDE

Electrolyte	pK_{sp}	Ref.
$AgB(C_6H_5)_4$	5.9	8, 13*
	5.9	9
AgBr	14.5	8
	14.5	13*
AgCl	14.3	8
	14.3	13*
AgI	14.7	8, 9
	14.7	13*
AgN_3	10.8	8
	10.8	13*
AgO_2CCH_3	9.7	8
	9.7	13*
AgSCN	10.5	8, 9
	10.5	13*
$(C_6H_5)_4AsB(C_6H_5)_4$	3.4	9
$(C_6H_5)_4C$	2.3	9
KBr	−0.74	8
KCl	1.84	8
$KClO_4$	−0.76	8
KI	+0.26	8
NaCl	−0.94	8

* $I = 0.10$–0.05M.

DIMETHYLFORMAMIDE

Electrolyte	pK_{sp}	Ref.
$AgB(C_6H_5)_4$	6.7	8*
	6.7	9, 13
AgBr	15.0	8,* 13
AgCl	14.49 ±0.01	17†
	14.55 ±0.05	17†
	14.7 ±0.1	17†
AgI	15.8	8,* 9, 13
	16.4	18
AgN_3	11.0	8,* 13
$AgO\ CCH_3$	10.2	8,* 13
AgOTs	1.3	8*
AgSCN	11.5	8,* 9, 13
$(C_6H_5)_4AsB(C_6H_5)_4$	3.7	9
$(C_6H_5)_4AsI$	1.1	9
$(C_6H_5)_4C$	2.1	9
$(C_2H_5)_4N[OC_6H_2(NO_2)_3]$	0.0	8*
$(C_6H_5)_4Sn$	2.0	9
CsBr	3.3	8*
CsCl	4.9	8,* 9
CsI	1.7	8,* 9
$Cs[OC_6H_2(NO_2)_3]$	0.5	8,* 9
KBr	2.4	8*
KCl	5.4	8,* 9
$KClO_4$	0.1	8,* 9
KI	−0.5	8,* 9
$K[OC_6H_4NO_2\text{-}p]$	0.4	8*
$K[OC_6H_2(NO_2)_3]$	−0.2	8,* 9
NaN_3	1.9	8*
$Na[OC_6H_4NO_2\text{-}p]$	−0.58	8*

* $I = 0.10$–0.05M.
† $T = 29.8°C$.

I. SOLUBILITIES OF ELECTROLYTES

DIMETHYL SULFOXIDE

Electrolyte	pK_{sp}	Comments	Ref.
$AgB(C_6H_5)_4$	4.6		8, 9
	4.6	$I = 0.1$–$0.05M$	13
AgBr	10.6	$I = 0.100$ F with $(C_2H_5)NClO_4$; $23 \pm 1°C$	1, 4
	10.6		8
	10.6	$I = 0.01$–$0.05M$	19
AgCl	10.4	$I = 0.100$ F with $(C_2H_5)NClO_4$; $23 \pm 1°C$	1
	10.5	$23 \pm 1°C$	4
	10.4	$I = 0.01$–$0.05M$	19
	10.4		8, 22
	10.279 ±0.003		20
$AgClO_4$	−0.8		9
AgCN	15.0	$23 \pm 1°C$	4
AgI	12.0	$I = 0.100$ F with $(C_2H_5)NClO_4$; $23 \pm 1°C$	1, 4
	11.4		8, 9
	11.4	$I = 0.01$–$0.04M$	19
AgN_3	6.5	$I = 0.10$–$0.05M$	8
	6.5	$I = 0.01$–$0.05M$	19
AgO_2CCH_3	4.4	$I = 0.10$–$0.05M$	8, 13
AgSCN	7.4	$23 \pm 1°C$	4
	7.1	$I = 0.10$–$0.05M$	8, 19
	7.1		9
KBr	0.60		8
$KClO_4$	−0.80		8
NaN_3	0.64		8

ETHANOL

Electrolyte	pK_{sp}	Ref.
AgBr	16.1	10
	16.0	11
AgCl	12.7	2
	14.0	10, 11
AgI	19.4	10
	18.9	11
AgSCN	14.3	11

ETHYLENE GLYCOL

Electrolyte	K_{sp}	Comments	Ref.
TlCl	0.188×10^{-4}	Supporting electrolyte KCl	21

FORMAMIDE

Electrolyte	pK_{sp}	Ref.
AgB(C$_6$H$_5$)$_4$	10.3	8, 9
	10.3	13*
AgBr	11.4	8
	11.4	13*
AgCl	9.4	8
	9.4	13*
AgI	14.5	8, 9
	14.5	13*
AgN$_3$	7.7	8
	7.7	13*

FORMAMIDE (Continued)

Electrolyte	pK_{sp}	Ref.
AgSCN	9.9	8, 9
	9.9	13*
$(C_6H_5)_4AsB(C_6H_5)_4$	8.3	9
$(C_6H_5)_4AsI$	1.1	9
$CsB(C_6H_5)_4$	3.6	9
CsBr	0.29	8
CsCl	0.53	8
	−0.5	9
CsI	0.23	8
	−0.2	9

*$I = 0.1$–$0.05M$.

HEXAMETHYLPHOSPHOROTRIAMIDE

Electrolyte	pK_{sp}	Ref.
$AgB(C_6H_5)_4$	4.7	8,* 9
	4.7	16*
AgBr	12.3	8*
	12.3	19*
AgCl	11.9	8*
	11.9	19*
AgN_3	8.5	8,* 19*
AgSCN	7.4	8,* 19*
	7.4	9
$(C_6H_5)_4AsB(C_6H_5)_4$	3.1	9
KBr	4.0	8*
NaCl	4.0	8*

*$I = 0.01$–$0.05M$.

HYDROGEN FLUORIDE

Electrolyte	K_{sp}	Comments	Ref.
$AgBF_4$	2.96×10^{-3}	0°C	7
KPF_6	1.32×10^{-3}	0°C	7
$KSbF_6$	5.84×10^{-2}	0°C	7
$TlSbF_6$	2.76×10^{-4}	0°C	7

METHANOL

Electrolyte	pK_{sp}	Ref.
$AgB(C_6H_5)_4$	13.2	8, 13*
	13.9	9
AgBr	15.2	1,† 4†
	15.2	8, 12
	15.0	10
	15.3	11
	15.2	13*
AgCl	13.0	1,† 4†
	13.1	8, 11, 12, 22
	12.2	2
	12.8	10
	13.1	13*
AgI	18.2	1,† 4†
	18.3	8, 9
	16.4	10
	18.2	11, 12
	18.3	13*
AgN_3	11.2	8
	11.2	13*
AgO_2CCH_3	6.1	8, 13*
$Ag[OC_6H_2(NO_2)_3]$	2.9	9
AgOTs	3.2	8
AgSCN	13.9	8, 9
	13.8	11, 12
	13.9	13*

I. SOLUBILITIES OF ELECTROLYTES

METHANOL (Continued)

Electrolyte	pK_{sp}	Ref.
$(C_6H_5)_4AsB(C_6H_5)_4$	8.5	9
$(C_6H_5)_4AsClO_4$	4.7	9
$(C_6H_5)_4AsI$	1.4	9
$(C_6H_5)_4As[OC_6H_2(NO_2)_3]$	3.6	9
$(C_6H_5)_4AsSCN$	1.2	9
$(C_6H_5)_4C$	3.7	9
$(i-C_5H_{11})_3C_4H_9NB(C_6H_5)_4$	4.9	9
$(C_6H_5)_4Sn$	3.6	9
$CsB(C_6H_5)_4$	6.1	9
CsBr	2.2	8
CsCl	1.7	8, 9
CsI	1.9	8, 9
$Cs[OC_6H_2(NO_2)_3]$	4.2	8, 9
Ferrocene	1.2	9
Ferrocene $B(C_6H_5)_4$	5.4	9
$KB(C_6H_5)_4$	5.0	8, 9
KBr	1.7	8
KCl	2.5	8, 9
$KClO_4$	4.5	8, 9
KI	0.2	8, 9
$KOC_6H_2NO_2$-p	0.2	8
$K[OC_6H_2(NO_2)_3]$	4.2	8, 9
$N(C_4H_9)_4B(C_6H_5)_4$	5.2	9
$N(C_2H_5)_4[OC_6H_2(NO_2)_3]$	2.0	8
NaCl	1.5	8
NaN_3	0.9	8
$NaOC_6H_4NO_2$-p	−0.2	8

* $I = 0.01$–0.05M.
† $I = 1.00$ F in $LiClO_4$, $T = 23 \pm 1°C$.

N-METHYL FORMAMIDE

Electrolyte	pK_{sp}	Ref.
AgCl	10.4	14

METHYL NITRITE

Electrolyte	pK_{sp}	Ref.*
AgCl	19.8	13
AgI	22.6	13
AgN_3	17.5	13
AgSCN	18.0	13

*$I = 0.01-0.05$M.

N-METHYL PROPIONAMIDE

Electrolyte	pK_{sp}	Ref.*
AgCl	14.4	13
AgI	14.5	13
AgSCN	10.0	13

*$I = 0.01-0.05$M.

NITROETHANE

Electrolyte	pK_{sp}	Ref.*
AgBr	21.8	1, 4
AgCl	21.1	1, 4
AgI	22.6	1, 4

*$I = 0.100$ F in $LiClO_4$; $T = 23 \pm 1°C$.

I. SOLUBILITIES OF ELECTROLYTES

PROPYLENE CARBONATE

Electrolyte	pK_{sp}	Ref.
$AgB(C_6H_5)_4$	12.4	13
AgCl	19.87 ±0.02	16
AgI	20.3	13
AgSCN	16.3	13

SULFOLANE

Electrolyte	pK_{sp}	Ref.*
$AgB(C_6H_5)_4$	9.5	13
AgCl	16.9	13
AgI	18.7	13
AgN_3	14.7	13
AgSCN	14.5	13

* $T = 30°C$.

TETRAHYDROFURAN

Electrolyte	pK_{sp}	Ref.
AgBr	19	15
AgCl	16.5	15

References

1. R. T. Iwamoto, J. Kleinberg and D. C. Luehrs, Inorg. Chem., **5,** 201 (1966).
2. N. A. Izmailov and V. S. Chernyi, Zhur. fiz. Khim., **34,** 127 (1960).
3. E. L. Mackov, Rec. trav. chim., **76,** 457 (1951).
4. D. C. Luehrs, Ph.D. Thesis, University of Kansas, (1965).
5. B. Althin, E. Wahlin and L. G. Sillen, Acta Chem. Scand., **3,** 321 (1949).
6. B. Althin, E. Wahlin and L. G. Sillen, Acta Chem. Scand., **6,** 759 (1952).

7. A. F. Clifford and S. Kongpricha, J. Inorg. Nuclear Chem., **20,** 147 (1961).
8. R. Alexander, E. C. F. KO, Y. C. Mac and A. J. Parker, J. Amer. Chem. Soc., **89,** 3703 (1967).
9. R. Alexander and A. J. Parker, J. Amer. Chem. Soc., **89,** 5549 (1967).
10. F. K. V. Koch, J. Chem. Soc., 1551 (1930).
11. A. MacFarlane and H. Hartley, Phil. Mag., **13,** 425 (1932).
12. P. S. Buckley and H. Hartley, Phil. Mag., **8,** 320 (1929).
13. R. Alexander, Ph.D. Thesis, The University of Western Australia, 1968.
14. Yu. M. Povarov, V. E. Kazarinov, Yu. M. Kessler and A. I. Gorbanev, Russ. J. Inorg. Chem., **9,** 550 (1964).
15. J. Badoz-Lambling and M. Sato, Compt. Rend., **254,** 3354 (1962).
16. J. N. Butler, Analyt. Chem., **39,** 1799 (1967).
17. J. N. Butler, J. Phys. Chem., **72,** 3288 (1968).
18. H. Chateau and M. C. Moncet, J. Chim. Phys., **60,** 1060 (1963).
19. E. C. J. Ko, Ph.D. Thesis, The University of Western Australia, 1968.
20. J. N. Butler and J. C. Synnott, J. Phys. Chem., **73,** 1470 (1969).
21. O. D. Black and A. B. Garrett, J. Amer. Chem. Soc., **65,** 862 (1943).
22. J. P. Morel, Bull. Soc. Chim., 896 (1968).
23. J. N. Butler, Adv. Electrochem., **7,** 106 (1970).
24. I. M. Kolthoff and M. K. Chantoonie, J. Amer. Chem. Soc., **89,** 1582 (1967).
25. J. J. Campion, Ph.D. Thesis, University of Pittsburgh, 1966.

(2) NONAQUEOUS–AQUEOUS SOLVENTS

The following tables include pK_{sp} values for electrolytes in binary solvent systems. The solvents are arranged alphabetically by the nonaqueous component. The salts included in each table are listed alphabetically according to chemical symbol. The temperature corresponding to each set of values is 25°C unless otherwise stated.

ACETONE–WATER
AgCl

Ref. 1

Wt. % Acetone	0	10	20	30	40
pK_{sp}	9.750	10.01	10.26	10.57	10.94

Wt. % Acetone	50	60	70	80	100
pK_{sp}	11.35	11.90	12.67	13.61	16.4

AgO_2CCH_3

Ref. 2

Wt. % Acetone	0	10	20	30	40
pK_{sp}	2.71	2.99	3.35	3.79	4.3 (extrapolated)

DIMETHYL SULFOXIDE–WATER
AgCl

Ref. 1

Wt. % DMSO	0	10	20	30	40	50
pK_{sp}	9.750	9.78	9.81	9.84	9.86	9.84

Wt. % DMSO	60	70	80	100
pK_{sp}	9.82	9.82	9.82	10.4

Ref. 3 supporting electrolyte $LiClO_4$

Concn. H_2O (moles l^{-1})	<0.01	0.03	0.07	0.55
pK_{sp}	10.279 ±0.003	10.272 ±0.002	10.234 ±0.001	10.179 ±0.003

Concn. H_2O (moles l^{-1})	5.55	13.87	27.75	41.63
pK_{sp}	9.711 ±0.001	9.558 ±0.002	9.535 ±0.003	9.530 ±0.008

Ref. 3 supporting electrolyte Et_4NClO_4

Concn. H_2O (moles l^{-1})	0.03	0.03	0.75
pK_{sp}	11.0	10.26 ±0.01	10.205 ±0.003

DIMETHYL SULFOXIDE–METHANOL

AgCl

Ref. 5

Wt. % DMSO	80
pK_{sp}	9.9

AgI

Ref. 4

Wt. % DMSO	80
pK_{sp}	12.4

AgSCN

Ref. 5

Wt. % DMSO	80
pK_{sp}	8.1

DIOXANE–WATER

AgCl

Ref. 6

Wt. % Dioxane	20
pK_{sp}	10.216

ETHANOL–WATER

AgCl

Ref. 7

Wt. % EtOH	10	20	30	40	50
pK_{sp}	10.04	10.22	10.49	10.77	11.11

AgO_2CCH_3

Ref. 2

Wt. % EtOH	0	10	20	30
pK_{sp}	2.71	2.99	3.29	3.62

ETHYLENE GLYCOL–WATER

TlCl

Ref. 8

Wt. % Glycol	0	20	40	60
K_{sp}	1.92×10^{-4}	1.25×10^{-4}	0.831×10^{-4}	0.558×10^{-4}

Wt. % Glycol	80	100
K_{sp}	0.352×10^{-4}	0.188×10^{-4}

I. SOLUBILITIES OF ELECTROLYTES

METHANOL–WATER

AgCl

Ref. 1

Wt. % MeOH	0	10	20	43.12
pK_{sp}	9.755	10.05; 9.981	10.27	10.848

Wt. % MeOH	50	75	100
pK_{sp}	11.13	12.08	13.0

Ref. 6

Wt. % MeOH	10	43.12
K_{sp} (mole²-kg⁻²)	1.79×10^{-10}	1.66×10^{-11}
pK_{sp}	9.981	

Ref. 9

Wt. % MeOH	20	30	40	50
pK_{sp}	10.27	10.51	10.79	11.13

Wt. % MeOH	60	70	80
pK_{sp}	11.45	11.85	12.30

PROPYLENE CARBONATE–WATER

AgCl

Ref. 10

Concn. H₂O (moles l⁻¹)	0.004	0.0094	0.0147	0.0308
pK_{sp}	19.90 ±0.05	19.72 ±0.03	19.46 ±0.02	19.64 ±0.02

Concn. H₂O (moles l⁻¹)	0.25	0.25	0.845	2.09	3.57
pK_{sp}	18.17 ±0.04	18.15 ±0.01	16.91 ±0.01	15.36 ±0.02	13.79 ±0.02

Additional information: The above set of data is described by the empirical equation $-\log K_{sp} = 20.0 - 3.25\, (C_{H_2O})^{1/2}$ where C_{H_2O} is the concentration of water in moles l⁻¹.

TETRAHYDROFURAN–WATER

Co(dithenoyl-2)methane

Ref. 11

pK_{sp}, 15.6 at unspecified concentration of THF

Ni(dithenoyl-2)methane

Ref. 11

pK_{sp}, 15.9 at unspecified concentration of THF

References

1. J. P. Morel, Bull. Soc. Chim., 896 (1968).
2. F. M. MacDougall and L. E. Topol, J. Phys. Chem., **56,** 1090 (1952).
3. J. N. Butler and J. C. Synnott, J. Phys. Chem., **73,** 1470 (1969).
4. R. Alexander, Ph.D. Thesis, The University of Western Australia, 1968.
5. T. J. Broxton, Ph.D. Thesis, The University of Western Australia, 1967.
6. D. Feakins, K. G. Lawrence and R. P. T. Tomkins, J. Chem. Soc., A, 1909 (1967).
7. K. P. Anderson, E. A. Butler, D. R. Anderson and E. M. Wolley, J. Phys. Chem., **71,** 3567 (1967).
8. O. D. Black and A. B. Garrett, J. Amer. Chem. Soc., **65,** 862 (1943).
9. H. N. Parton and D. D. Perrin, Trans. Faraday Soc., **41,** 579 (1945).
10. J. N. Butler, D. R. Cogley and W. Zurosky, J. Electrochem. Soc., **115,** 445 (1968).
11. C. Caullet, Bull. Soc. Chim. France, 3459 (1965).

II. EMF DATA

(a) Equilibrium Measurements in Single Solvents

The following section indicates the various reference electrodes and cell configurations that have been used for equilibrium-type measurements in nonaqueous solvents. The information is arranged alphabetically by solvent and for each solvent table the electrodes are listed alphabetically by name. Information relative to the concentration range employed and the values of E^0 are given when quoted. Additional comments indicating the purpose of the study, extrapolation procedures used, stability of electrodes and other pertinent information are provided. An indication relative to the availability of activity coefficient data is included. References are provided for each cell studied and a complete bibliography is given at the end of this section.

For authoritative discussions on the use and behavior of reference electrodes in nonaqueous solvent systems the reader is referred to excellent reviews by Hills,[a] Butler,[b] and Covington.[c]

[a] G. J. Hills; in Reference Electrodes, Theory and Practice, D. J. G. Ives and G. J. Janz, Eds., Academic Press, New York, (1961), pp. 433–463.

[b] J. N. Butler; in Advances in Electrochemistry and Electrochemical Engineering Vol. 7; P. Delahay and C. W. Tobias, Eds; Wiley-Interscience (1970), pp. 77–175.

[c] A. K. Covington; in Electrochemistry Vol. 1, Specialist Periodicals Reports, G. J. Hills, ed; The Chemical Society (1970), pp. 56–72.

ACETAMIDE

Electrode	Cell
Molybdenum	Mo | HCl || pure solvent | Mo K or Na acetate

ACETIC ACID

Electrode	Cell
Antimony	Sb | NaOAc, HClO$_4$ | sat. aq. Calomel
Calomel	Pt | Chloranil / Hydrochloranil | H$_2$SO$_4$(m) or HClO$_4$ | sat. aq. Calomel
	Pt | Chloranil / Hydrochloranil | NaOAc HClO$_4$ | sat. aq. Calomel
	Pt | Chloranil / Hydrochloranil | HClO$_4$ or H$_2$SO$_4$ or HCl | sat. aq. Calomel
	H$_2$, Pt | NaOAc || KCl / $N/10$ | sat. aq. Calomel
	Glass | NaOAc / HClO$_4$ | sat. aq. Calomel
	Pt | Quinhydrone | NaOAc / HClO$_4$ | sat. aq. Calomel
	Ag | AgCl (base H)$^+$ Cl$^-$ | sat. aq. Calomel
	Pt | Chloranil / Hydrochloranil H$^+$ | LiCl | Hg$_2$Cl$_2$ | Hg sat.
	Pt | Ar$_3$C· / Ar$_3$COH | LiCl | Hg$_2$Cl$_2$ | Hg sat. (buffered)

II. EMF DATA

H_2NCOCH_3

Conc. range	E^0	Comments	Ref.
0.075–0.13M	0	94°C; titrations; $K_s = 3.2 \times 10^{-11}$	34

CH_3CO_2H

Conc. range	E^0	Comments	Ref.
$\simeq N/10$		18°±2; titrations; pH scale in HOAc	1
0.064–1M	0.83	E^0 relates to 1M $HClO_4$; titration	2, 3
$\simeq N/10$		18°–20; titrations	4
0.0004–0.1	0.9095	K_A of a series of acids, bases and salts	5
0.094–1.7	0.5068	Room temp.	6
		18°; titrations	4
		18°; titrations	4
0.0004–0.1	−0.3928		5
$\mu = 0.2$		Titrations for redox potls.	7
$\mu = 0.2$		Redox potls. and ΔG of carbinols	8

ACETIC ACID

Electrode	Cell
Calomel	Pt | Chloranil / Hydrochloranil / Titration mixture | KCl sat. in H$_2$O | Hg$_2$Cl$_2$ | Hg
Carbinol	Pt | Ar$_3$C· / Ar$_3$COH / (buffered cell) | LiCl sat. | Hg$_2$Cl$_2$ | Hg
Chloranil	Pt | Chloranil / Hydrochloranil sat. | H$_2$SO$_4$(m) or HClO$_4$ | sat. aq. | Calomel
	Pt | Chloranil / Hydrochloranil sat. | H$^+$ | H$^+$ | Pt, H$_2$
	Pt | Chloranil / Hydrochloranil sat. | HCl | AgCl | Ag
	Pt | Chloranil / Hydrochloranil sat. | NaOAc / HClO$_4$ | sat. aq. | Calomel
	Pt | Chloranil / Hydrochloranil sat. | HClO$_4$ or H$_2$SO$_4$ or HCl | sat. aq. | Calomel
	Pt | Chloranil / Hydrochloranil / H$_2$SO$_4$ (1M) | LiCl sat. | Gelatin / Amino acids / HClO$_4$ | H$_2$(Pd)
	Pt | Chloranil / Hydrochloranil / H$^+$ | LiCl sat. | Hg$_2$Cl$_2$ | Hg

II. EMF DATA

CH$_3$CO$_2$H *(Continued)*

Conc. range	E^0	Comments	Ref.
~0–0.2M		23°C; titrations; strength of weak bases	9
$\mu = 0.2$		Redox potls. and ΔG of carbinols	8
0.064–1M	0.83	E^0 related to 1M HClO$_4$; titration	2, 3
	0.680	E^0 vs H$^+$ in HOAc	10
0.0005–0.13	1.045	E^0 vs AgCl; γ_{HCl}	11
$\simeq N/10$		18°–20°; titrations	4
0.0004–0.1	0.9095	K_A of a series of acids, bases and salts	5
~0.01M		Room temp; titrations; solvent contains ~2% H$_2$O (to dissolve gelatin)	12
$\mu = 0.2$		Titrations for redox potls.	7

ACETIC ACID

Electrode	Cell
Chloranil	Pt | Chloranil / Hydrochloranil (Titration mixture) ‖ KCl sat. in H$_2$O | Hg$_2$Cl$_2$ | Hg
Hydrogen	H$_2$, Pt | NaOAc ‖ KCl $N/10$ | sat. aq. Calomel
	H$_2$, Pd | Gelatin / Amino acids / HClO$_4$ | LiCl sat. | Chloranil / Hydrochloranil / H$^+$ (1M) | Pt
Glass	Glass | NaOAc / HClO$_4$ | sat. aq. Calomel
Lead amalgam	(Hg)Pb | Pb(OAc)$_2$, ½HOAc sat. | Pb(OAc)$_2$ | Hg$_2$(OAc)$_2$ | Hg ½HOAc
Quinhydrone	Pt | Quinhydrone | NaOAc / HClO$_4$ | sat. aq. Calomel
Mercury–mercurous acetate	Hg | Hg$_2$(OAc)$_2$ | Pb(OAc)$_2$ | Pb(OAc)$_2$ | Pb(Hg)
Mercury–mercurous sulphate	Hg | Hg$_2$SO$_4$ | H$_2$SO$_4$ | H$_2$, Pt
Silver–silver chloride	Ag | AgCl | HCl | Chloranil / Hydrochloranil | Pt
	Ag | AgCl | (Base H)$^+$ Cl$^-$ | sat. aq. Calomel
	H$_2$, Pt | HCl | AgCl | Ag
Tellurium	Te | NaOAc, HClO$_4$ | sat. aq. Calomel

II. EMF DATA

CH_3CO_2H (*Continued*)

Conc. range	E^0	Comments	Ref.
~0–0.2M		23°C; titrations; strength of weak bases	9
0.094–1.7	0.5086	Room temp.	6
≃0.01M		Room temp; titrations	12
		18°, titrations	4
sat.	0.7359	16°–50°, ΔH	13
		18°, titrations	4
	0.7359	16°–50°, ΔH	13
0.002–0.87	0.181	γ values	14
0.0005–0.13	1.045	γ_{HCl}	11
0.0004–0.1	−0.3928		5
0.00035–0.01	−0.61800	35°	15
≃N/10		18°±2; titrations; pH scale in HOAc	1

ACETONE

Electrode	Cell
Cadmium	Cd | CdI$_2$ | AgNO$_3$ | Ag (0.009M) (0.007M)
Calomel	Cd | Cd^{++} | Calomel (0.1M) Glass $\begin{vmatrix} \text{HCl} \\ \text{NaCl,} \end{vmatrix}$ base | Hg$_2$Cl$_2$ | Hg Cd | Cd^{++} | Calomel (0.1M) Ag | Ag$^+$ | Calomel (0.1M) Also see Glass
Cesium	Cs(Hg) | CsI | AgI | Ag
Copper	Cu | CuCl$_2$ | AgNO$_3$ | Ag (0.05M) (0.007M)
Glass	Glass $\begin{vmatrix} \text{HCl} \\ \text{NaCl,} \end{vmatrix}$ base | Hg$_2$Cl$_2$ | Hg Glass | RϕNR$_2$, HClO$_4$ || KCl | Hg$_2$Cl$_2$ | Hg Glass | RϕNR$_2$, HClO$_4$ || KCl | AgCl | Ag Glass | HCl | AgCl | Ag
Hydrogen	H$_2$, Pt | HCl, AgCl | Ag
Iron-phenanthroline	Pt | Fe(phen)$_3^{\text{II}\|\text{III}}$ || AgClO$_4$ | Ag
Lead	Pb | Pb(NO$_3$)$_2$ | AgNO$_3$ | Ag (0.003M) (0.007M)
Silver	Ag | AgNO$_3$ | AgNO$_3$ | Ag (0.007M) (0.0035M) Ag | AgNO$_3$ | Pb(NO$_3$)$_2$ | Pb (0.007M) (0.003M)

II. EMF DATA

$(CH_3)_2CO$

Conc. range	E^0	Comments	Ref.
	−0.916	No temp. stated; no attempt to correlate C and E; 0.007 $AgNO_3$ | Ag used as a ref. electrode	35
	0.35		35
$\simeq$1M	−0.82	Titration; K_a of ϕNH_2 and $\phi NHMe$	36
	0.35	Also see Cadmium	35
	0.574	Also see Cadmium	35
3×10^{-4}–3×10^{-3}m	1.1430	Also MeOH and mixtures	16
	0.31	Also see Cadmium	35
$\simeq$1M	−0.82	Titration; K_a of ϕNH_2 and $\phi NHMe$	36
$I = 10^{-3}$–10^{-1} m		Base strengths as a function of solvent	17
5×10^{-3}–0.15 m	−0.53	γ_{HCl}	37
5×10^{-3} M	−0.465	30°	18
	0.974	Also see Cadmium	35
	0.016	Also see Cadmium	35
	0.974	Also see Cadmium	35

ACETONE

Electrode	Cell
Silver	Ag | AgNO$_3$ | CuCl$_2$ | Cu (0.007M) (0.05M)
	Ag | AgNO$_3$ | CdI$_2$ | Cd (0.007M) (0.009M)
	Ag | Ag$^+$ | Calomel (0.1M)
	Ag | AgNO$_3$ | ZnCl$_2$ | Zn (0.007M) (0.2M)
	Ag | AgNO$_3$ | AgNO$_3$ | Ag (aq.) (Me$_2$CO)
Silver–silver chloride	H$_2$, Pt | HCl, AgCl | Ag Also see Glass
Silver–silver iodide	Also see Cesium
Silver–silver perchlorate	Also see Phenanthroline
Sodium	Na(Hg) | NaI | Na(Hg) (c_1) (0.1N) (c_2)
Zinc	Zn | ZnCl$_2$ | AgNO$_3$ | Ag (0.27) (0.007M)

ACETONITRILE

Electrode	Cell

Anthracene	Pt | ArH / ArH$^+$ | AgClO$_4$ | Ag
	   Et$_4$NClO$_4$ (0.001M)
	   (.1M) Et$_4$NClO$_4$
	              (.1M)

II. EMF DATA

$(CH_3)_2CO$ (*Continued*)

Conc. range	E^0	Comments	Ref.
	0.31	Also see Cadmium	35
	−0.916	Also see Cadmium	35
	0.574	Also see Cadmium	35
	−1.219	Also see Cadmium	35
$0.09N-10^{-3}N$		λ solutions measured; $E = A \log (\lambda_w/\lambda_s) + B$; $B = 0.227$	38
$5 \times 10^{-3}-0.15m$	−0.53	γ_{HCl}	37
$8 \times 10^{-3}-0.63\%$ Na		20°; γ_{Na} in Hg	19
	−1.219	Also see Cadmium	35

CH_3CN

Conc. range	E^0	Comments	Ref.
	0.99	Calcd. from polarographic measurements (pyrene, chrysene, phenanthrene and triphenylene electrodes were also used)	39

NONAQUEOUS ELECTROLYTES

ACETONITRILE

Electrode	Cell
Calomel	H_2, Pt | H_2SO_4 || KCl | Calomel $\quad$ Et_4NHSO_4 $\quad$ sat. aq.
	Pt $\begin{vmatrix} Fe(phen)_3^{3+} \\ Fe(phen)_3^{2+} \end{vmatrix}$ KCl $\begin{vmatrix} \text{Calomel} \\ \text{sat. aq.} \end{vmatrix}$
	Pt $\begin{vmatrix} Fc \\ Fc^+Pi^- \end{vmatrix}$ KCl $\begin{vmatrix} \text{Calomel} \\ \text{sat. aq.} \end{vmatrix}$
	Ag | $AgNO_3$ | KCl $\begin{vmatrix} \text{Calomel} \\ \text{sat. aq.} \end{vmatrix}$ $\quad$ (0.01M)
	Hg | Hg_2Cl_2, MCl, AgCl | Ag $\quad$ sat. $\quad$ sat.
Calomel aq.	Hg | Hg_2Cl_2 | KCl $\begin{vmatrix} Et_4NClO_4 \\ \text{(MeCN)} \end{vmatrix}$ KCl, Hg_2Cl_2 | Hg $\quad$ aq.
Calomel	Hg | Hg_2Br_2, MBr, AgBr | Ag
	Hg | Hg_2I_2, MI, AgI | Ag
Copper [I and II]	Ag | $AgNO_3$ | $LiClO_4$ | $CuClO_4$, $Cu(ClO_4)_2$ | Pt $\quad$ (0.01M)
$Fe(bipy)_3$	Pt | $Fe(bipy)_3^{II,III}$ || $AgNO_3$ | Ag
$Fe(phen)_3$	Pt $\begin{vmatrix} Fe(phen)_3^{3+} \\ Fe(phen)_3^{2+} \\ (ClO_4^-) \end{vmatrix}$ $AgNO_3$ | Ag $\quad$ (0.01M)
	Pt $\begin{vmatrix} Fe(phen)_3^{3+} \\ Fe(phen)_3^{2+} \end{vmatrix}$ KCl $\begin{vmatrix} \text{Calomel} \\ \text{sat. aq.} \end{vmatrix}$
	Ag | $AgNO_3$ | Et_4NClO_4 | $Fe(phen)_3^{II-III}$ | Pt $\quad$ T_1 $\quad\quad$ T_2
	Pt | $Fe(phen)_3^{II,III}$ || $AgNO_3$ | Ag

II. EMF DATA

CH₃CN (*Continued*)

Conc. range	E^0	Comments	Ref.
	$\simeq 0.3$	Value not quoted as E^0	20
	$\simeq 1.1$		20
	$\simeq 0.3$		20
	$\simeq 0.27$		20
10^{-3}–10^{-1}m	0.0486	35°; instability attributed to complex formation M = Li, Et₂NH₂, Na or Et₄N	21
10^{-3}–10^{-1}m	0.0622		21
10^{-3}–10^{-1}m	0.1330	Instability due to complex formation	21
Ionic strength 10^{-3}–10^{-2}m	0.7956	Considers effect of water on solvent	40
10^{-4}–10^{-3}M	−0.8069	30°; electrochemical series of complexes	18
10^{-4}–10^{-1}m	0.8459	Debye–Hückel extrap.	20
10^{-4}–10^{-1}m	$\simeq 1.1$		20
			41
10^{-4}–10^{-3}M	−0.8301	30°; electrochemical series of complexes	18

ACETONITRILE

Electrode	Cell
Ferrocene	Pt $\begin{vmatrix} \text{Ferrocene} \\ \text{Fecp}_2^+\text{Pi}^- \end{vmatrix}$ AgNO$_3$ | Ag (0.01M)
	Pt | FeC$_{10}$H$_{10}$$^{\text{II, III}}$ || AgNO$_3$ | Ag
Glass	Ag | AgCl, KCl | Glass | HA | AgClO$_4$ | Ag (0.01m)
	Glass | H$^+$ $\begin{vmatrix} \text{acids} \\ \text{bases} \end{vmatrix}$ NEt$_4$Br | NEt$_4$Br, AgBr | Ag (0.1M) sat. sat.
	Ag, Ag$^+$ | Glass | H$^+$B$^-$ || Ag$^+$ | Ag
	Glass | H$^+$A$^-$ || AgNO$_3$ | Ag (0.1M) Et$_4$NClO$_4$
	Glass | R$_3$N, R$_3$NH$^+$ClO$_4^-$ || AgNO$_3$ | Ag
	Glass | Et$_4$N$^+$Pi$^-$, HPi || AgNO$_3$ | Ag
	Glass | ArOH, ArO$^-$Na$^+$ || AgNO$_3$ | Ag
Hydrogen	Ag | AgNO$_3$ | H$^+$ | Pt, H$_2$ (0.01M) (a = 1)
	H$_2$, Pt | H$^+$B$^-$ || Ag$^+$ | Ag (10^{-2}M)
	H$_2$, Pt $\begin{vmatrix} \text{H}_2\text{SO}_4 \\ \text{Et}_4\text{NHSO}_4 \end{vmatrix}$ AgNO$_3$ | Ag (0.01M)
	H$_2$, Pt $\begin{vmatrix} \text{H}_2\text{SO}_4 \\ \text{Et}_4\text{NHSO}_4 \end{vmatrix}$ | KCl | Calomel sat. aq.
	H$_2$, Pt | H$^+$ $\begin{vmatrix} \text{acids} \\ \text{bases} \end{vmatrix}$ NEt$_4$Br | NEt$_4$Br, AgBr | Ag (0.1M)
	H$_2$, Pt | H$^+$ $\begin{vmatrix} \text{acids} \\ \text{bases} \end{vmatrix}$ NEt$_4$Br | NEt$_4$Br, H$^+$ | Pt (0.1M) sat. PiH (0.1M)

II. EMF DATA

CH$_3$CN (*Continued*)

Conc. range	E^0	Comments	Ref.
10^{-4}–10^{-1}m	0.0740	Polarography	20
10^{-4}–10^{-3}M	-0.0716	30°; electrochemical series of complexes	18
$\simeq$0.1M ionic strength		K_a organic acids	44
0.1M		Titration curves; K_a of various acids and bases	45
10^{-4}–2×10^{-2}M	0.82 (graphical extrapolation)	Glass electrode calibration	42 43
10^{-4}–10^{-3}M		K_b of bases	22
10^{-3}–10^{-2}M		K_s of MeCN	22
10^{-4}–10^{-3}M		K_a of phenols	22
	0.637V	Indirect measurement via Pleskov method	41
		Acid strength; titrations	42
	0.03	Polarography	20
	$\simeq$0.3		20
		Titration curves, K_a of various acids and bases	45
			45

ACETONITRILE

Electrode	Cell
Hydrogen	H_2, Pt | H^- $\begin{vmatrix} \text{acids} \\ \text{bases} \end{vmatrix}$ | NEt_4Br | NEt_4Br, HPi | Pt, H_2 (0.1M) sat. sat. (0.1M)
	H_2, Pt | H^+, | Ag^+ | Ag
Iodine	I_2, Pt | KI | NEt_4Pi | I_2, KI | Pt (c_1) (c_2) sat. (c_3)(c_4)
	Pt | I_2, I^- || I^-, I_2 | Pt
Mercury	Also see Silver
Nickelocene	Pt | $NiC_{10}H_{10}^{II,III}$ || $AgNO_3$ | Ag
Osmium–bipyridyl	Pt | $Os(bipy)_3^{II,III}$ || $AgNO_3$ | Ag
Osmium–phenanthroline	Pt $Os(phen)_3^{II,III}$ || $AgNO_3$ | Ag
Osmium–terpyridyl	Pt | $Os(trpy)_2^{II,III}$ || $AgNO_3$ | Ag
Platinum	Pt | $Cu,^{I,II}(ClO_4)_{1,2}$, $M^{II}(ClO_4)_2$ | $Cu^{I,II}(ClO_4)_2$ | Pt M = Cd or Ni
Silver	Ag | $AgNO_3$ | $AgNO_3$ | Ag
	Ag | $AgNO_3$ || $AgNO_3$ | Ag
	Ag | $AgNO_3$ || Et_4NClO_4, $TlClO_4$ | (Hg)Tl
	Ag | $AgNO_3$ || $Hg(ClO_4)_2$ | Hg
	Ag | $AgNO_3$ || $HgClO_4$ | Hg
	Ag | $AgNO_3$ || $Hg(ClO_4)_2$, $HgClO_4$ | Pt
	Ag | $AgNO_3$ || Calomel (0.01M)
	Ag | $AgNO_3$, MX | Et_4NPi | $AgNO_3$ | Ag

II. EMF DATA

CH₃CN (*Continued*)

Conc. range	E^0	Comments	Ref.
		Titration curves	45
	≃0.25	H₂ solubility, H⁺ diffusion, exchange currents	46
$c_1 = 0.01$M $c_2 = 0.005$M $c_3 = 0.01$M $c_4 = 0.005$M		Stability consts.	47
			22
			23
10^{-4}–10^{-3}M	+0.3660	30°; electrochemical series of complexes	18
10^{-4}–10^{-3}M	−0.5833	30°; electrochemical series of complexes	18
10^{-4}–10^{-3}M	−0.5680	30°; electrochemical series of complexes	18
10^{-4}–10^{-3}M	−0.6832	30°; electrochemical series of complexes	18
0.02–0.15m		$\gamma_\pm$ in mixed electrolyte solutions	24
2×10^{-3}–0.1m		t_+	25
10^{-4}–10^{-3}M	0.128		23
10^{-4}–10^{-3}M	−0.639		23
			23
			23
		Disproportionation constant	23
			26
∼10^{-2}M		Titrations; K_{sp}	22

146 NONAQUEOUS ELECTROLYTES

ACETONITRILE

Electrode	Cell
Silver	Ag, Ag$^+$ | Glass | H$^+$B$^-$ || Ag$^+$ | Ag $\qquad\qquad\qquad\qquad\qquad$ (10^{-2}M)
	Ag | AgNO$_3$ | KCl | Calomel $\quad$ (0.01M) $\qquad$ sat. aq.
	Ag | Ag$^+$ | AgNO$_3$ | Ag $\qquad\qquad$ (0.01M)
	Glass | H$^+$A$^-$ || AgNO$_3$ | Ag $\qquad\qquad\qquad$ (0.01M) $\qquad$ (0.1M)Et$_4$NClO$_4$
	$\quad$ | ArH $\quad$ | Pt $\quad$ | $\qquad$ | AgClO$_4$ | $\quad$ | ArH$^+$ | (0.001M) Ag Et$_4$NClO$_4$ $\quad$ Et$_4$NClO$_4$ $\quad$ (0.1M) $\qquad$ (0.1M)
	Ag | MX $\quad$ | NEt$_4$Pi | AgNO$_3$ | Ag $\qquad$ (c$_1$) $\quad$ sat. $\qquad$ (c$_2$)
	Ag | AgCl, KCl | Glass | HA AgClO$_4$ | Ag $\qquad\qquad\qquad\qquad\qquad\qquad$ (0.01m)
	Ag | AgClO$_4$ | AgClO$_4$ | Ag $\qquad$ (x) $\qquad$ (10^{-2})
	H$_2$, Pt | H$^+$ | Ag$^+$ | Ag
	Ag | AgNO$_3$ | Et$_4$NClO$_4$ | AgNO$_3$ | Ag $\quad$ (0.01M) $\quad$ (0.1M) $\qquad$ (0.01M) $\quad$ T$_1$ $\qquad\qquad\qquad\qquad$ T$_2$
	Ag | AgNO$_3$ | Et$_4$NClO$_4$ | AgClO$_4$ | Ag $\qquad$ M $\qquad$ 0.1m $\qquad$ M
	Ag | AgNO$_3$ | H$^+$ | Pt, H$_2$ $\quad$ (0.01M) $\quad$ (a = 1)
	Ag | AgNO$_3$ | LiClO$_4$ | CuClO$_4$, Cu(ClO$_4$)$_2$ | Pt $\quad$ (0.01M) LiNO$_3$ | LiClO$_4$ (supporting electrolytes)
	Ag | AgNO$_3$ | AgNO$_3$ | AgNO$_3$ | Ag $\quad$ (0.01M) (0.01M) (0.01M) $\quad$ T$_1$ $\qquad\qquad$ T$_2$
	Ag | AgNO$_3$ | Et$_4$NClO$_4$ | Fe(phen)$_3$$^{\text{II–III}}$ | Pt $\quad$ T$_1$ $\qquad\qquad\qquad$ T$_2$
	Ag | AgNO$_3$ | AgNO$_3$ | Ag $\qquad$ (C$_1$) $\qquad$ (C$_2$)

II. EMF DATA

CH_3CN *(Continued)*

Conc. range	E^0	Comments	Ref.
		Acid strengths, titrations	42
10^{-4}–10^{-1}M	$\simeq 0.27$		20
	0.1301		20
10^{-4}–2×10^{-3}M	0.82 (graphical interpolation)	Glass electrode calibration	43
	0.99	Calcd. from polarographic measurements	39
$c_1 = 0.01$M $c_2 = 0.01$M		Stability consts.	47
$\simeq 0.1$M ionic strength		K_a organic acids	44
10^{-4}–10^{-2}M			48
	$\simeq 0.25$	H_2 solubility, H^+ diffusion, exchange current	46
$\partial E^0/\partial T = 0.9$ mV/°C		$\partial E/\partial T = +0.43$	41
2×10^{-3}–10^{-2}M		$K_A' = AgNO_3/(Ag)(NO_3) = 74 \pm 5$	41
	0.637	Indirect measurement via Pleskov method	41
10^{-3}–10^{-2} (ionic strength)	0.7956	Potentiometric titrations	40
		$\partial E/\partial T = 0.35$ mV/°C	41
			41
10^{-1}–2×10^{-3}m		$t_+ = 0.458$	49

ACETONITRILE

Electrode	Cell
Silver	Ag | AgNO$_3$ | AgNO$_3$ | Ag aq. (MeCN) Also see Glass, Fe-phenanthroline, Fe-bipyridyl, Os-bipyridyl,

Silver–silver bromide	H$_2$, Pt | H$^+$ | acids/bases | NEt$_4$Br | NEt$_4$Br, AgBr | Ag
	           (0.1M) sat. sat.
	
	Glass | H$^+$ | acids/bases | NEt$_4$Br | NEt$_4$Br, AgBr | Ag
	         (0.1m) sat. sat.
	
	Hg | Hg$_2$Br$_2$, MBr, AgBr | Ag

Silver–silver chloride	Hg | Hg$_2$Cl$_2$, MCl, AgCl | Ag
Silver–silver iodide	Hg | Hg$_2$I$_2$, MI, AgI | Ag
Sodium	Na(Hg) | NaI | Na(Hg) (C$_1$) (0.1N) (C$_2$)
Thallium	Also see Silver

AMMONIA

Electrode	Cell

Cadmium	(Hg)Cd | CdCl$_2$·6NH$_3$, NH$_2$Cl, ZnCl$_2$, 6NH$_3$ | Zn(Hg)
	               sat. sat.
	
	H$_2$, Pt | NH$_4$Cl, CdCl$_2$·6NH$_3$ | Cd
	
	Hg | HgI$_2$ || CdCl$_2$ | Cd
	           sat.

Calcium	Ca | Ca(SCN)$_2$, KSCN, Pb(SCN)$_2$ | Pb (20 mole%)

II. EMF DATA

CH₃CN (*Continued*)

Conc. range	E^0	Comments	Ref.
0.1N–0.001N		$E = A \log \lambda_{aq}/\lambda_{solv} + B$ $B = 0.091$	38
Os-terpyridyl, Os-phenanthroline, Ferrocene, and Nickelocene			
0.1m		Titration curves, K_a of various acids and bases	45
			45
10^{-3}–10^{-1}m	0.0622	35°; electrode unstable due to complex formation	21
10^{-3}–10^{-1}m	0.0486	35°; instability of electrodes due to complex formation	21
10^{-3}–10^{-1}m	0.1330	35°;	21
8×10^{-3}–0.63% Na		20°; γ_{Na} in Hg	19
			23

NH₃

Conc. range	E^0	Comments	Ref.
6×10^{-3}–10^{-2}m 8×10^{-4}–5×10^{-2}(Hg)	0.3605	$\Delta G; \Delta H; \Delta S$	27
	0.3688		27
	−0.93	−36°;	28
0.05–0.3M	not less than 2.185	−34°	29

AMMONIA

Electrode	Cell
Calomel (aq)	Hg | Hg_2Cl_2 | KCl || NH_4NO_3 | Hg
Cesium	Also see Lead
Hydrogen	H_2, Pt | NH_4Cl | NH_4Cl | Pt, H_2 C_1 C_2
	H_2, Pt | NH_4NO_3 | NH_4Cl | Pt, H_2 (0.1N) sat.
	H_2, Pt | NH_4Cl, TlCl | Tl(Hg) sat.
	H_2, Pt | NH_4Cl, TlCl | Tl
	H_2, Pt | NH_4Cl, $ZnCl_2 \cdot 6NH_3$ | Zn
	H_2, Pt | NH_4Cl, $CdCl_2 \cdot 6NH_3$ | Cd
	H_2 | H^+ || NH_4NO_3 | Hg aq. (NH_3)
	H_2 | NH_4Cl | N_2
	Pt | H_2 | NH_4I | NH_4I | H_2 | Pt
Lead	Pb | $Pb(NO_3)_2$, NH_4NO_3 | Glass | NH_4NO_3, $Pb(NO_3)_2$ | Pb (0.1M) (C) (0.1M) (0.1M)
	Ca | $Ca(SCN)_2$, KSCN, $Pb(SCN)_2$ | Pb (C) (0.1N) 20 mole%
	Pb | $Pb(NO_3)_2$ | KNO_3 | $CsNO_3$ | Cs(Hg) (0.1N) sat.
	Pb | $Pb(NO_3)_2$ || NH_4NO_3 | Hg
	Pb | Pb^{++} || H^+ | H_2 (NH_3) aq.
	Pb | $Pb(NO_3)_2$ | KNO_3 | $NaNO_3$ | Na(Hg)
	Pb | $Pb(NO_3)_2$ | KNO_3 | KNO_3 | K(Hg) sat.
Mercury	Hg | $HgCl_2$, KCl || KI, HgI_2 | Hg

II. EMF DATA

NH₃ (*Continued*)

Conc. range	E^0	Comments	Ref.
	−0.30	20°; dropping Hg electrode used as ref. electrode	32
			95
		−50°; no values	30
	0.218	−70°;	30
2.65×10^{-2}N amal.	≃0.42	Very unstable	31
	−0.1000	formal potentials	27
	0.7293	by calculation	
	0.3688		
		−50°; dropping Hg electrode used as ref. electrode	32
	0.082	−50°; theoretical from ΔG_f^0	33
10^{-3}–10^{-1}m	(0)	−40°; $\gamma_\pm$; t_+ See also under Lead	93
		−35°; resistance and stability	94
0.05–0.3M	not less than +2.185	−34°	29
0.01N	−1.0785	−35°; E^0 refers to potl. at stated concentrations	95
0.1N	−0.351	−35°; dropping Hg electrode used as ref.	32
	−0.42	−50°	32
0.1N, 0.09524% in Hg	−1.3375	−35°; standard potentials calculated	96
0.1N, 0.2216% in Hg	−1.2873	−35°; standard potentials calculated	96
	−0.068	−36°; titrations	28

AMMONIA

Electrode	Cell
Mercury	Hg | HgI_2 | HgI_2 | Hg (C) C_{Ref} Hg | HgI_2, KI || KI, HgI_2 | Hg Hg | HgI_2 || $CdCl_2$ | Cd sat. Also see Calomel, Hydrogen, and Lead
Nitrogen	
Platinum	Na(Hg) | NaI || Na | Pt
Potassium	(Hg)K | KCl, $ZnCl_2 \cdot 6NH_3$ | Zn(Hg) (M) sat. sat. (0.275%) Also see Lead
Quinhydrone	Pt | QH, NH_4Cl : NH_4Cl, QH | Pt (C_1) (C_1) (C_2) (C_2)
Silver	Ag | $AgNO_3$, NH_4NO_3 | Membrane | NH_4NO_3, $AgNO_3$ | Ag (10^{-3}M) (C) (10^{-3}M) (10^{-3}M)
Sodium	Na | Na | Na | Na (C_1) (C_2) (Hg)Na | NaCl, $ZnCl_2 \cdot 6NH_3$ | Zn(Hg) (0.2%) sat. Na | NaI | Na(Hg) (solid) ($EtNH_2$) Na(Hg) | NaI || Na | Pt Also see Lead
Thallium	(Hg)Tl | TlCl, NH_4Cl, $ZnCl_2 \cdot 10NH_3$ | Zn(Hg) sat. sat. Tl | TlCl, NH_4Cl, $ZnCl_2 \cdot 10NH_3$ | Zn (solid) (solid) H_2 | NH_4Cl, TlCl | Tl

II. EMF DATA

NH₃ (*Continued*)

Conc. range	E^0	Comments	Ref.
		−40°;	28
		−40°; titrations	28
	−0.93	−36°	28
			33
10^{-4}–10^{-2}M	−0.933	−50°; assumes NaHg₄ in amal-	97
	−0.947	−70° gams, γ is const.	
10^{-3}–10^{-2}M	0.1052	−36°; γ data	98
			96
10^{-4}–10^{-3}M		−50°; $E \simeq 30$mV	30
10^{-3}–10^{-2}M		−50°; cation exchange membranes; γ data	99
3×10^{-3}–1.0M		−33°; transport numbers	100
2×10^{-3}–0.5M	0.2208	−36°; γ data	98
in Hg 0.3–0.6	0.828	−50°; Liquid junction potentials estimated	97
in NH₃ 10^{-4}–10^{-2}	0.835		
	−0.933	−50°; assumes NaHg₄ in amal-	97
	−0.947	−70°; gams, γ is const.	
			96
	0.9016		31
	0.8293	Born Haber cycle; ΔG; ΔH; ΔS	31
	−0.1000		

AMMONIA

Electrode	Cell
Tungsten	Na(Hg) | NaI ‖ Na | W
Zinc	(Hg)Tl | TlCl, NH$_4$Cl, ZnCl$_2$·10NH$_3$ | Zn(Hg) sat. sat.
	Tl | TlCl, NH$_4$Cl, ZnCl$_2$·10NH$_3$ | Zn (solid) (solid)
	(Hg)Zn | ZnCl$_2$·6NH$_3$, NH$_4$Cl, CdCl$_2$·6NH$_3$ | Cd(Hg) sat. sat.
	H$_2$ | NH$_4$Cl, ZnCl$_2$·6NH$_3$ | Zn
	(Hg)K | KCl, ZnCl$_2$·6NH$_3$ | Zn(Hg) (0.275%) (M) sat. sat.
	(Hg)Na | NaCl, ZnCl$_2$·6NH$_3$ | Zn(Hg) (0.2%) sat.

i-AMYL ALCOHOL

Electrode	Cell
Silver–silver iodide	(Hg)Na | NaI | AgI | Ag
Sodium (Hg)	Also see Silver–silver iodide

BENZOL

Electrode	Cell
Copper	Cu | — | Cu^{II} (oleate)$_2$ | Cu

II. EMF DATA

NH₃ (*Continued*)

Conc. range	E^0	Comments	Ref.
	−0.933 −0.947	−50°; assumes NaHg₄ in amalgams −70°;	97
	0.9016	activity in amalgams	31
	0.8293	Born Haber cycle; ΔG	31
	0.3605	Refers to solid electrodes	
	0.7293		
10^{-3}–10^{-2} M	0.1052	−36°; γ data	98
2×10^{-3}–0.5 M	0.2208	−36°; γ data	98

i-C₅H₁₁OH

Conc. range	E^0	Comments	Ref.
3×10^{-3}–0.4 m	1.2460	$\gamma_{\pm}$; solvent effects	101

C₆H₆

Conc. range	E^0	Comments	Ref.
0.01 M, solvent side not defined	−0.05	Intersolvent potentials; no temp. given	50

BENZONITRILE

Electrode	Cell
Silver	Ag $\mid$ AgNO$_3$: AgNO$_3$ $\mid$ Ag (C_1) (C_2)
	Ag $\mid$ AgNO$_3$: AgNO$_3$ $\mid$ Ag aq. (ϕCN)
	Ag $\mid$ AgNO$_3$ $\mid$ AgNO$_3$ $\mid$ Ag

BROMINE TRIFLUORIDE

Electrode	Cell
Aluminum	Pt $\mid$ BrF$_3$ $\mid$ Al
Cadmium	Pt $\mid$ BrF$_3$ $\mid$ Cd
Copper	Pt $\mid$ BrF$_3$ $\mid$ Cu
Gold	Pt $\mid$ BrF$_3$ $\mid$ Au
Lead	Pt $\mid$ BrF$_3$ $\mid$ Pb
Magnesium	Pt $\mid$ BrF$_3$ $\mid$ Mg
Monel	Pt $\mid$ BrF$_3$ $\mid$ Monel
Nickel	Pt $\mid$ BrF$_3$ $\mid$ Ni
Platinum	Also see Al, Cd, Cu, Au, Pb, Mg, Monel, Ni, Ag, and Zn
Silver	Pt $\mid$ BrF$_3$ $\mid$ Ag
Zinc	Pt $\mid$ BrF$_3$ $\mid$ Zn

II. EMF DATA

C_6H_5CN

Conc. range	E^0	Comments	Ref.
10^{-1}–10^{-3}m		0°C; $t_+ = 0.461$ 25°C; $t_+ = 0.466$	49
0.1N–10^{-3}N		λ solutions measured $E = A \log (\lambda_w/\lambda_s) + B$ $B = 0.087$	38
2×10^{-3}–0.1m		t_+	25

BrF_3

Conc. range	E^0	Comments	Ref.
	$E = -0.27$	Open circuit potentials, battery design	102
	$E = -2.24$	Open circuit potentials, battery design	102
	$E = -0.47$	Open circuit potentials, battery design	102
	$E = -0.15$	Open circuit potentials, battery design	102
	$E = -2.36$	Open circuit potentials, battery design	102
	$E = -1.30$	Open circuit potentials, battery design	102
	$E = -0.09$	Open circuit potentials, battery design	102
	$E = -0.13$	Open circuit potentials, battery design	102
	$E = -1.26$	Open circuit potentials, battery design	102
	$E = -0.53$	Open circuit potentials, battery design	102

n-BUTANOL

Electrode	Cell
Cadmium	(Hg)Cd | CdCl$_2$, AgCl | Ag (0.1%)
Hydrogen	H$_2$, Pt | HCl, AgCl | Ag
Silver–silver chloride	H$_2$, Pt | HCl, AgCl | Ag
	(Hg)Cd | CdCl$_2$, AgCl | Ag (0.1%)

CHLORINE TRIFLUORIDE

Electrode	Cell
Cadmium	Pt | ClF$_3$ | Cd
	Pt | ClF$_3$, BF$_3$ | Cd
Platinum	Also see Cadmium

CYCLOHEXANOL

Electrode	Cell
Calomel	Hg | Hg$_2$Cl$_2$MCl, AgCl | Ag
	Hg | Hg$_2$Br$_2$, MBr, AgBr | Ag
Silver–silver chloride	Hg | Hg$_2$Cl$_2$, MCl, AgCl | Ag
	Hg | Hg$_2$Br$_2$, MBr, AgBr | Ag
	[M = Li, Et$_2$NH$_2$, Na, K, Me$_2$EtϕN]

II. EMF DATA

n–BuOH

Conc. range	E^0	Comments	Ref.
4×10^{-4}–2×10^{-2} m	0.3330	γ; Pleskov theory	103
10^{-3}–10^{-1} m	-0.1430	γ; diss const.	104
10^{-3}–10^{-1} m	-0.1430	γ; diss const.	104
4×10^{-4}–2×10^{-2} m	0.3330	γ; Pleskov theory	103

ClF_3

Conc. range	E^0	Comments	Ref.
	$E = 2.88$	0°C; $E_{Calc} = 3.14$	102
$X_{BF_3} = 0.05$	$E = 2.73$	0°C	102
$= 0.10$	$E = 2.69$	0°C	102

ϕOH

Conc. range	E^0	Comments	Ref.
10^{-3}–10^{-1} m	0.0622	75°; Conc. independent emf. both electrodes reversible to same ion	105
10^{-3}–10^{-1} m	0.0866	75°	105
10^{-3}–10^{-1} m	0.0622	75°; also see calomel	105
10^{-3}–10^{-1} m	0.0866	75°; also see calomel	105

DECANOL

Electrode	Cell
Silver–silver iodide	(Hg)Na | NaI | AgI | Ag
Sodium (Hg)	Also see Silver–silver iodide

N,N-DIBUTYLACETAMIDE

Electrode	Cell
Calomel	Hg | Hg_2Cl_2 | Cu(I), Cu(II) | Pt
	Hg | Hg_2Cl_2 | Cu(I), Cu(II) | Hg
	Hg | Hg_2Cl_2 | Cu(I) | Pt
	Hg | Hg_2Cl_2 | Cu(I) | Hg
Copper	Also see Calomel

N,N-DIETHYLACETAMIDE

Electrode	Cell
Calomel	Hg | Hg_2Cl_2 | Cu(I), Cu(II) | Pt
	Hg | Hg_2Cl_2 | Cu(I), Cu(II) | Hg
	Hg | Hg_2Cl_2 | Cu(I) | Pt
	Hg | Hg_2Cl_2 | Cu(I) | Hg
Copper	Also see Calomel

II. EMF DATA

$C_{10}H_{21}OH$

Conc. range	E^0	Comments	Ref.
6×10^{-4}–5×10^{-3}m	0.9480	γ_+; solvent effects	101

Bu_2NCOMe

Conc. range	E^0	Comments	Ref.
0.1M $LiClO_4$ 10^{-4}–10^{-3}M Cu(I)	0.31	Polarography; minimal exper. details	51
0.1M $LiClO_4$ 10^{-4}–10^{-3}M Cu(I)	0.16	Polarography; minimal exper. details	51
0.1M $LiClO_4$ 10^{-4}–10^{-3}M Cu(I)	0.01	Polarography; minimal exper. details	51
0.1M $LiClO_4$ 10^{-4}–10^{-3}M Cu(I)	0.07	Polarography; minimal exper. details	51

Et_2NCOMe

Conc. range	E^0	Comments	Ref.
0.1M $LiClO_4$ 10^{-4}–10^{-3}M Cu(I)	0.17	Polarography; minimal exper. detail	51
0.1M $LiClO_4$ 10^{-4}–10^{-3}M Cu(I)	0.12	Polarography; minimal exper. detail	51
0.1M $LiClO_4$ 10^{-4}–10^{-3}M Cu(I)	0.07	Polarography; minimal exper. detail	51
0.1M $LiClO_4$ 10^{-4}–10^{-3}M Cu(I)	0.02	Polarography; minimal exper. detail	51

DIETHYLAMINE

Electrode	Cell
Calcium	Ca(Hg) | CaI$_2$ | Ca

DIETHYL ETHER

Electrode	Cell
Magnesium	Mg | MgBr$_2$(Et$_2$O)$_2$, Hg$_2$Br$_2$ | Hg sat.
Mercury–mercurous bromide	Also see Magnesium
Potassium	(Hg)K | KCϕ_3 | K
Silver–silver bromide	Ag | AgBr, LiBr | LiBr, AgBr | Ag (C$_1$) (C$_2$) + LiClO$_4$ (conc.) added

N,N-DIISOPROPYLACETAMIDE

Electrode	Cell
Calomel	Hg | Hg$_2$Cl$_2$ $\left|\begin{array}{c}\text{Cu(I)}\\ \text{Cu(II)}\end{array}\right|$ Pt
	Hg | Hg$_2$Cl$_2$ $\left|\begin{array}{c}\text{Cu(I)}\\ \text{Cu(II)}\end{array}\right|$ Hg
	Hg | Hg$_2$Cl$_2$ | Cu(I) | Pt
	Hg | Hg$_2$Cl$_2$ | Cu(I) | Hg
Copper	Also see Calomel

II. EMF DATA

Et_2NH

Conc. range	E^0	Comments	Ref.
0.1–1N	0.89±0.01	17.5°C	106

$(C_2H_5)_2O$

Conc. range	E^0	Comments	Ref.
	1.561	V.P. of solns; ΔG formation of $MgBr_2$	52
			52
	Not measurable	K (but not K $\mid$ Hg) attacks $KC\phi_3$ in Et_2O	53
	0	Titrations for Grignard studies; Mg(Hg) fails to work well	54

iPr_2NCOMe

Conc. range	E^0	Comments	Ref.
0.1M $LiClO_4$ 10^{-4}–10^{-3}M Cu(I)	0.24	Polarography; minimal exper. detail	51
0.1M $LiClO_4$ 10^{-4}–10^{-3}M Cu(I)	0.12	Polarography; minimal exper. detail	51
0.1M $LiClO_4$ 10^{-4}–10^{-3}M Cu(I)	0.02	Polarography; minimal exper. detail	51
0.1M $LiClO_4$ 10^{-4}–10^{-3}M Cu(I)	0.02	Polarography; minimal exper. detail	51

N,N-DIISOPROPYLPROPIONAMIDE

Electrode	Cell
Calomel	Hg | Hg$_2$Cl$_2$ | Cu(I), Cu(II) | Pt
	Hg | Hg$_2$Cl$_2$ | Cu(I), Cu(II) | Hg
	Hg | Hg$_2$Cl$_2$ | Cu(I) | Pt
	Hg | Hg$_2$Cl$_2$ | Cu(I) | Hg
Copper	Also see Calomel

N,N-DIMETHYLACETAMIDE

Electrode	Cell
Calomel	Hg | Hg$_2$Cl$_2$ | Cu(I), Cu(II) | Pt
	Hg | Hg$_2$Cl$_2$ | Cu(I), Cu(II) | Hg
	Hg | Hg$_2$Cl$_2$ | Cu(I) | Pt
	Hg | Hg$_2$Cl$_2$ | Cu(I) | Hg
Copper	Also see Calomel
Iodine	Pt | I$_2$, I$^-$ | Et$_4$N Picrate | I$^-$, I$_2$ | Pt
Silver	Ag | AgNO$_3$, MX || AgNO$_3$ | Ag

II. EMF DATA

iPr$_2$NCOEt

Conc. range	E^0	Comments	Ref.
0.1M LiClO$_4$ 10^{-4}–10^{-3}M Cu(I)	0.33	Polarography; minimal exper. detail	51
0.1M LiClO$_4$ 10^{-4}–10^{-3}M Cu(I)	0.13	Polarography; minimal exper. detail	51
0.1M LiClO$_4$ 10^{-4}–10^{-3}M Cu(I)	0.06	Polarography; minimal exper. detail	51
0.1M LiClO$_4$ 10^{-4}–10^{-3}M Cu(I)	−0.03	Polarography; minimal exper. detail	51

Me$_2$NCOMe

Conc. range	E^0	Comments	Ref.
0.1M LiClO$_4$ 10^{-4}–10^{-3}M Cu(I)	0.16	Polarography; minimal exper. details	51
0.1M LiClO$_4$ 10^{-4}–10^{-3}M Cu(I)	0.14	Polarography; minimal exper. details	51
0.1M LiClO$_4$ 10^{-4}–10^{-3}M Cu(I)	0.07	Polarography; minimal exper. details	51
0.1M LiClO$_4$ 10^{-4}–10^{-3}M Cu(I)	0.05	Polarography; minimal exper. details	51
∼10^{-2}M		K_{sp}; complexes	22
∼10^{-2}M		Solvent effects; titrations	22

DIMETHYLAMINE

Electrode	Cell
Bismuth	H_2, Pt | Me_2NH_2Cl || Me_2NH_2Cl, MCl_n | M $M = Bi$; $n = 3$
Cadmium	Also see above; $M = Cd$; $n = 2$
Cobalt	Also see above; $M = Co$; $n = 2$
Copper	See above; $M = Cu(I)$; $n = 1$, $M = Cu(II)$; $n = 2$
Hydrogen	H_2, Pt | Me_2NH_2Cl || Me_2NH_2Cl | Pt, H_2 $HCl(x)$ Also see all other entries
Lead	Also see Bismuth; $M = Pb$; $n = 2$
Mercury	Also see Bismuth; $M = Hg$; $n = 2$
Nickel	Also see Bismuth; $M = Ni$; $n = 2$
Platinum	Also see Bismuth; $M = Pt$; $n = 2$
Silver–silver chloride	Also see Bismuth; $M = Ag$; $n = 1$
Sodium	Na | NaI | Na(Hg)
Tin	Also see Bismuth; $M = Sn$; $n = 2$

II. EMF DATA

Me$_2$NH

Conc. range	E^0	Comments	Ref.
10^{-3}–10^{-1}m	-0.1952	180°C; strictly fused salt; Ag | AgCl Ref.	107
10^{-3}–10^{-1}m	-0.789	180°C; strictly fused salt; Ag | AgCl Ref.	107
10^{-3}–10^{-1}m	-0.3443	180°C; strictly fused salt; Ag | AgCl Ref.	107
10^{-3}–10^{-1}m	-0.2875	180°C; strictly fused salt; Ag | AgCl Ref.	107
	$+0.4278$	180°C; strictly fused salt; Ag | AgCl Ref.	107
10^{-3}–10^{-1}m	0.1165	180°C; strictly fused salt; Ag | AgCl Ref.	107
10^{-3}–10^{-1}m	-0.5113	180°C; strictly fused salt; Ag | AgCl Ref.	107
10^{-3}–10^{-1}m	-0.1952	180°C; strictly fused salt; Ag | AgCl Ref.	107
10^{-3}–10^{-1}m	-0.2557	180°C; strictly fused salt; Ag | AgCl Ref.	107
10^{-3}–10^{-1}m	0.3397	180°C; strictly fused salt; Ag | AgCl Ref.	107
10^{-3}–10^{-1}m	0.0000	180°C; strictly fused salt; Ag | AgCl Ref.	107
0.06514% wt.	0.88442 0.88467 0.88500 0.88537 0.88577 0.88626 0.88667 0.88720	5°; E_{Na}^0 vs aq. AgCl 10°; 15°; 20°; 25°; 30°; 35°; 40°;	108
10^{-3}–10^{-1}m	-0.5405	180°C; strictly fused salt; Ag | AgCl Ref.	107

N,N-DIMETHYLBUTYRAMIDE

Electrode	Cell
Calomel	Hg | Hg$_2$Cl$_2$ | Cu(I), Cu(II) | Pt
	Hg | Hg$_2$Cl$_2$ | Cu(I), Cu(II) | Hg
	Hg | Hg$_2$Cl$_2$ | Cu(I) | Pt
	Hg | Hg$_2$Cl$_2$ | Cu(I) | Hg
Copper	Also see Calomel

N,N-DIMETHYLFORMAMIDE

Electrode	Cell
Bromine	Pt | Br$_2$, KBr || Br$_2$, KBr | Pt
	Pt | Br$_2$, KBr (DMF) || Br$_2$, KBr (MeOH) | Pt
Cadmium	Cd | Cd(Hg) | CdCl$_2$ || NaCl || CdCl$_2$·H$_2$O | Cd(Hg) | Cd sat.
	Cd | Cd(Hg) | CdCl$_2$, NaCl, NaClO$_4$ | Cd(Hg) | Cd sat.
	Cd(Hg) | CdCl$_2$, Et$_4$NCl, CdCl$_2$ | Cd(Hg) sat. (1M) sat.
	(Hg)Li | LiCl, CdCl$_2$ | Cd(Hg) (1.22 mole%) sat. (4%)
Calomel	Calomel sat. | Cu(I), Cu(II) | Pt
	Calomel sat. | Cu(I) | Pt

II. EMF DATA

$Me_2NCOC_3H_7$

Conc. range	E^0	Comments	Ref.
0.1M $LiClO_4$ 10^{-4}–10^{-3}M Cu(I)	0.29	Polarography; minimal exper. detail	51
0.1M $LiClO_4$ 10^{-4}–10^{-3}M Cu(I)	0.19	Polarography; minimal exper. detail	51
0.1M $LiClO_4$ 10^{-4}–10^{-3}M Cu(I)	0.11	Polarography; minimal exper. detail	51
0.1M $LiClO_4$ 10^{-4}–10^{-3}M Cu(I)	0.06	Polarography; minimal exper. detail	51

Me_2NCHO

Conc. range	E^0	Comments	Ref.
10^{-3}–0.1M		DMF used as ref. solvent for γ_x^- in MeOH	55
Sat.	$0 \pm 0.002V$	27°C; bias pots; polarization. ref. electrode	56
≃5% sat.		Polarization; solub. of MCl_n; bias potentials; No E^0	109
0.1M LiCl	E = 1.652	See above	109
10^{-4}–10^{-3}M Cu(I)	0.04	Polarography; few exper. details	51
	0.0		

N,N-DIMETHYLFORMAMIDE

Electrode	Cell
Calomel	Calomel $\begin{vmatrix} Cu(II) \\ Cu(I) \end{vmatrix}$ Hg sat.
	H_2, Pt | HCl or ROLi || KCl | Hg_2Cl_2 | Hg sat.
	Nonaq. | glass | KCl Hg_2Cl_2 | Hg ref. sat.
	Calomel || KI, $AgNO_3$ | Ag aq. (DMF)
	Hg | Hg_2Cl_2 | KCl || H^+, RCH_2py | Glass
	Hg | Hg_2Cl_2 | KCl || HA, BA | Glass
	Hg | Hg_2Cl_2 | KCl || HA, BA, B'A' | Glass
	Hg | Hg_2Cl_2 | KCl || MA, HA | Glass
Copper	Also see Calomel
Glass	Also see Calomel and Silver
Hydrogen	H_2, Pt | HCl or ROLi || KCl | Hg_2Cl_2 | Hg sat.
Iodine	Pt | I_2, KI || I_2, KI | Pt
	Pt | I_2, KI || I_2, KI | Pt DMF MeOH
	Pt | I_2, KI | NEt_4Pi | I_2, KI | Pt $(C_1)(C_2)$ sat. $(C_3)(C_4)$
	Pt | I_2, I^- | Et_4NPi | I^-, I_2 | Pt
Lead	(Hg)Li | LiCl, $PbCl_2$ | Pb(Hg) (1.22 mole%) sat. (1%)

II. EMF DATA

Me₂NCHO (*Continued*)

Conc. range	E^0	Comments	Ref.
	0.05		
5×10^{-3}–0.1M	0.090	20°C; γ_{HCl}; acid and base strengths	57
5×10^{-3}–0.1M	$\simeq$0.080	20°C; drifts badly	57
10^{-3}–1.69M (KI)		K_{sp} of AgI; complexes	110
10^{-4}–10^{-2}M		ΔG, ΔH, ΔS of ionization	111
$I = 10^{-3}$–10^{-1}		EtNClO₄ bridge; aq. and nonaq. glass electrode fillings	112
$I = 10^{-3}$–10^{-1}		pH and K_a	
$\sim 10^{-2}$M		K_a of a range of acids and phenols	113
5×10^{-3}–0.1M	0.090	20°C; γ_{HCl}; acid and base strengths	57
10^{-3}–10^{-1}M	0	DMF used as ref. solvent for γ_{x^-} in MeOH	55
	(E $\sim$ 0.05)		55
$C_1 = 10^{-2}$m $C_2 = 5 \times 10^{-3}$m $C_3 = 10^{-2}$m $C_4 = 5 \times 10^{-3}$m		K_{sp} and K(stability) for a range of MX	47
$\sim 10^{-2}$M		K_{sp}; complexes; solvent effects	22
	$E = 1.845$	No E^0; polarization, solub. of MCl$_n$; bias potentials; unstable	109

N,N-DIMETHYLFORMAMIDE

Electrode	Cell
Lead	Pb(Hg) | PbCl$_2$, Et$_4$NCl, PbCl$_2$ | Pb(Hg) sat. (1M) sat.

Lithium	(Hg)Li | LiCl, AgCl | Ag
	        sat.
	
	(Hg)Li | LiCl, TlCl | Tl(Hg)
	        sat. (12.9%)
	
	(Hg)Li | LiCl, PbCl$_2$ | Pb(Hg)
	        sat. (1%)
	
	(Hg)Li | LiCl, CdCl$_2$ | Cd(Hg)
	(1.22 mole%) sat. (4%)
	
	(Hg)Tl | TlCl, LiCl | Li(Hg)
	(1.01 mole%) sat. (1.06 mole%)

Silver	Ag | Et$_4$NCl | Et$_4$NClO$_4$ | Et$_4$NClO$_4$ | Ag
	    (C$_1$) (C$_2$) (C$_3$)
	    AgCl ↑
	    sat. Titrate with
	                   AgClO$_4$(10^{-1}m)
	                   Et$_4$NCl(10^{-1}m)
	
	Ag | MX | NEt$_4$Pi | AgNO$_3$ | Ag
	   (C$_1$) sat. (C$_2$)
	
	Ag | AgNO$_3$ || MA, HA | Glass
	
	Ag | AgNO$_3$, MX || AgNO$_3$ | Ag
	
	Ag | AgNO$_3$ || HA, BA, B'A' | Glass
	HA = R$_2\phi$OH, RϕCO$_2$H,
	B$^+$ = Et$_4$N$^+$, Li$^+$, Na$^+$

Silver–silver chloride	Also see Silver (Hg)Li | LiCl, AgC | Ag (1.22 mole%) sat.
Thallium	Tl(Hg) | TlCl, Et$_4$NCl, TlCl | Tl(Hg) sat. (1M) sat. Also see Lithium

II. EMF DATA

Me$_2$NCHO (*Continued*)

Conc. range	E^0	Comments	Ref.
1%		Polarization; solubility of MCl$_n$	109
0.1M LiCl	$E = 1.982$	Polarization; solubility of MCl$_n$	109
0.1M LiCl	$E = 1.377$	Polarization; solubility of MCl$_n$	109
0.1M LiCl	$E = 1.845$	Polarization; solubility of MCl$_n$	109
0.1M LiCl	$E = 1.652$	Polarization; solubility of MCl$_n$	109
10^{-3}–2m	−1.3310	$\Delta G°$; $\Delta H°$; $\Delta S°$; γ_{LiCl};	58
$C_1 = 10^{-1}$m $C_2 = 10^{-1}$m $C_3 = 10^{-1}$m	0.6585 0.6265 0.685 0.637	Equil. constants of complexes; K_{sp}; E^0 depends on residual AgCl	59
$C_1 = 10^{-2}$m $C_2 = 10^{-2}$m		K_{sp} and K(stability) for a range of MX	47
∼10^{-2}M		K_a of a range of acids and phenols	113
∼10^{-2}M		K_{sp}; complexes; solvent effects	22
		Et$_4$NClO$_4$ bridge; aq. and nonaq. glass electrode fillings	112
0.1M LiCl	$E = 1.982$	No E^0; polarization; solub. of MCl$_n$; bias potentials	109
12.9% amalgam		Polarization; solub. of MCl$_n$	109

N,N-DIMETHYLPROPIONAMIDE

Electrode	Cell
Calomel	Hg $\mid$ Hg$_2$Cl$_2$ $\mid$ Cu(I), Cu(II) $\mid$ Pt
	Hg $\mid$ Hg$_2$Cl$_2$ $\mid$ Cu(I), Cu(II) $\mid$ Hg
	Hg $\mid$ Hg$_2$Cl$_2$ $\mid$ Cu(I) $\mid$ Pt
	Hg $\mid$ Hg$_2$Cl$_2$ $\mid$ Cu(I) $\mid$ Hg
Copper	Also see Calomel

DIMETHYL SULFOXIDE

Electrode	Cell
Cadmium	(Hg)Cd $\mid$ CdCl$_2$, Et$_4$NCl, CdCl$_2$ $\mid$ Cd(Hg) sat. (1M) sat.
	(Hg)Li $\mid$ LiCl, CdCl$_2$ $\mid$ Cd(Hg) (1.22 mole%) (0.1M) sat. (4%)
Calcium	Ca $\mid$ CaCl$_2$, TlCl $\mid$ Tl(Hg) $\mid$ Pt
Calomel	Ag $\mid$ NH$_4$NO$_3$, AgNO$_3$, AgCl$_x$, LiCl $\parallel$ LiCl, Hg$_2$Cl$_2$ $\mid$ Hg
Glass	glass $\mid$ HCl $\parallel$ NaCl, AgCl $\mid$ Ag sat. sat.
	Ag $\mid$ AgClO$_4$, TosOH $\mid$ Glass $\mid$ HA, $\parallel$ Et$_4$NClO$_4$, CsCH$_2$MeSO $\parallel$ AgClO$_4$ $\mid$ Ag (2 $\times$ 10^{-2}M) (10^{-1}M) (5 $\times$ 10^{-2}M)
Hydrogen	H$_2$, Pt $\mid$ H$^+$, Et$_4$NClO$_4$ $\parallel$ Ref. electrode (C) (0.1M)

II. EMF DATA

Me₂NCOEt

Conc. range	E^0	Comments	Ref.
0.1M LiClO₄ 10⁻⁴–10⁻³M Cu(I)	0.31	Polarography; minimal exper. detail	51
0.1M LiClO₄ 10⁻⁴–10⁻³M Cu(I)	0.11	Polarography; minimal exper. detail	51
0.1M LiClO₄ 10⁻⁴–10⁻³M Cu(I)	0.11	Polarography; minimal exper. detail	51
0.1M LiClO₄ 10⁻⁴–10⁻³M Cu(I)	−0.08	Polarography; minimal exper. detail	51

(CH₃)₂SO

Conc. range	E^0	Comments	Ref.
		Polarization, solubility of CdCl$_n$	109
	$E = 1.550$	Polarization and solubility given; bias potentials	109
4 × 10⁻⁴–0.8m	2.17	Estimated	60
≃10⁻¹M		Solubilities of AgX; K of AgX₂⁻.	61
4.4 × 10⁻²–10⁻⁴M	−0.392	pH studies	62
<10⁻³M acid		pK for a series of organic acids	63
Dilute	0.070	Temp. not given; K_a organic acids	64

DIMETHYL SULFOXIDE

Electrode	Cell
Hydrogen	H_2, Pt | H^+B^-, Et_4NClO_4 | Ref. electrode (C) (0.1M)
Iodine	I_2, Pt | KI | NEt_4Pi | I_2, KI | Pt (C_1) (C_2) (C_3) (C_4)
	Pt | I_2, I^- | Et_4NPi | I^-, I_2 | Pt
Lead	(Hg)Pb | $PbCl_2$, Et_4NCl, $PbCl_2$ | Pb(Hg) sat. (1M) sat.
	(Hg)Li | LiCl, $PbCl_2$ | Pb(Hg) (1.22 mole%) (0.1M) sat. (1%)
Lithium	(Hg)Li | LiCl | TlCl | Tl(Hg) sat.
	Li | LiX | Li(Hg) (solid) X = Cl, ClO_4
	(Pt)Li | LiCl, TlCl | Tl(Hg) | Pt sat. liq.
	Li | LiCl, TlCl | Tl(Hg) | Pt sat. liq.
	(Hg)Li | LiCl, TlCl | Tl(Hg) (0.818 mole%) sat. (6.5 mole%)
	Li | LiCl | Li(Hg)
	Li | LiCl | Ni(Hg)(Li) sat.
	(Hg)Li | LiCl, AgCl | Ag (1.22 mole%) (0.1M) sat.
	(Hg)Li | LiCl, TlCl | Tl(Hg) (0.825 mole%) (0.1M) sat. (6.5%)
	(Hg)Li | LiCl, $PbCl_2$ | Pb(Hg) (1.22 mole%) (0.1M) sat. (1%)
	(Hg)Li | LiCl, $CdCl_2$ | Cd(Hg) (1.22 mole%) (0.1M) sat. (4%)

II. EMF DATA

$(CH_3)_2SO$ *(Continued)*

Conc. range	E^0	Comments	Ref.
Dilute	-1.900	Temp. not given, K_a organic acids; K(solvent)	64
$C_1 = 0.01M$ $C_2 = 0.005M$ $C_3 = 0.01M$ $C_4 = 0.005M$		K_{sp}	47
$\sim 10^{-2}M$		K_{sp}; complexes	22
1%Pb		Solubility of $PbCl_n$	109
	$E = 1.777$	No E^0 values; Polarization and solubility given; bias potentials	109
10^{-2}–1.0M		Complex formation constants; solb. prod. $E \simeq 1.5V$	65
1.33 mole % Li in (Hg); up to saturation	0.8438	γ_{Li} in (Hg)	66
4×10^{-4}–0.8m	2.27234 2.26665 2.26083	25°C; γ, $\Delta G°$, $\Delta S°$, 30°C; $\Delta H°$ reaction 35°C;	67
4×10^{-4}–0.8m	2.4212 2.4176 2.4137	25°C; $\gamma = f(conc)$ 30°C; 35°C;	60
5×10^{-3}–2.0m	1.3963 (± 0.001)	22–26°C; check with cryoscopic method; γ values	68
	0.8438	Calcd. from E_{Li}^0 in H_2O	69
	0.950	Kinetic studies; appears to be a sat. double amalgam	70
	$E = 2.265$	No E^0 values; polarization and solubility; bias potentials	109
	$E = 1.543$	No E^0 values; polarization and solubility; bias potentials	109
	$E = 1.777$	No E^0 values; polarization and solubility; bias potentials	109
	$E = 1.550$	No E^0 values; polarization and solubility; bias potentials	109

DIMETHYL SULFOXIDE

Electrode	Cell
Lithium	Li | LiBr, TlBr | Tl(Hg) (m) sat.
	(Hg)Li | LiCl ┊ LiCl | Li(Hg) (Hg)Li | LiNO₃ ┊ LiNO₃ | Li(Hg)
	Li | LiI | TlI | Tl(Hg)
Mercury glass	Pt | Hg | Glass $\begin{vmatrix} \text{HA,} \\ \text{CsCH}_2\text{MeSO} \end{vmatrix}$ Et₄NClO₄ | AgClO₄ | Ag
Potassium	(Hg)K | KPF₆ : KPF₆ | K(Hg) (0.1M)
Silver	Ag | NH₄NO₃, AgNO₃, AgCl$_x$ $\begin{Vmatrix} \text{LiCl} \\ \text{Hg}_2\text{Cl}_2 \end{Vmatrix}$ Hg
	Glass | HCl $\begin{Vmatrix} \text{NaCl} \\ \text{AgCl} \end{Vmatrix}$ Ag
	Ag | MX | NEt₄Pi | AgNO₃ | Ag (0.01M) sat. (0.01M)
	Ag | AgClO₄, TosOH | Glass $\begin{vmatrix} \text{HA,} \\ \text{CsCH}_2\text{MeSO} \end{vmatrix}$ Et₄NClO₄ | AgClO₄ | Ag (C₁) (C₂) (C₃) (C₄)
	Pt | Hg | Glass $\begin{vmatrix} \text{HA,} \\ \text{CsCH}_2\text{MeSO} \end{vmatrix}$ Et₄NClO₄ | AgClO₄ | Ag
	Ag | AgNO₃, MX ‖ AgNO₃ | Ag Ag | AgNO₃ ‖ AgNO₃, LiCl, NH₄NO₃ | Ag (0.01M) (0.02M) (0.03M) (I)
	(Hg)Li | LiCl, AgCl | Ag (1.22 mole%) (0.1M) sat.

II. EMF DATA

$(CH_3)_2SO$ *(Continued)*

Conc. range	E^0	Comments	Ref.
2×10^{-2}–0.8m	2.28381	15°C; $\gamma_\pm$; $\Delta G_t°$;	71
	2.27301	25°C; $\Delta H°$; $\Delta S°$	
	2.26214	35°C;	
	2.25120	45°C;	
0.01–1M		$\gamma_\pm$; $t_\pm$	114
0.01–1M		$\gamma_\pm$; $t_\pm$	
0.05–0.1m	2.3152	25°; $\gamma_\pm$; $\Delta G_t°$; $\Delta H_t°$;	115
	2.3065	35°; $\Delta S_t°$	
	2.2964	45°	
$<10^{-3}$M acid		pK of series of organic acids	63
0.01–1M		$\gamma_\pm$; $t_\pm$	114
$\sim 10^{-1}$M		Solubilities of AgX; K of AgX_2^-; temp. range (25–85°C)	61
4.4×10^{-2}–10^{-4}M	-0.392	pH	62
		K_{sp}, K(stability)	47
$C_1 = 2 \times 10^{-2}$M		pK of a series of organic acids	63
$C_2 = 6 \times 10^{-3}$M			
$C_3 = 10^{-1}$M			
$C_4 = 5 \times 10^{-2}$M			
$<10^{-3}$M			63
$\sim 10^{-2}$M $I = 0$–0.01M		K_{sp}; complexes; solvent effects	22
	$E = 2.265$	No E^0 values; polarization and solubility; bias potentials	109

DIMETHYL SULFOXIDE

Electrode	Cell
Silver–silver chloride	Glass | HCl ‖ NaCl / AgCl sat. | Ag
Thallium	(Hg)Li | LiCl | TlCl | Tl(Hg) sat.
	(Hg)Tl | TlCl | Tl(Hg) sat.
	(Pt)Li | LiCl, TlCl | Tl(Hg) | Pt sat. liq.
	Li | LiCl, TlCl | Tl(Hg) | Pt sat. liq.
	Ca | CaCl$_2$, TlCl | Tl(Hg) | Pt sat.
	(Hg)Li | LiCl, TlCl | Tl(Hg) (0.818 mole%) sat. (6.5 mole%)
	(Hg)Tl | TlCl, Et$_4$NCl, TlCl | Tl(Hg) sat. (1M) sat.
	(Hg)Li | LiCl , PbCl$_2$ | Pb(Hg) (1.22 mole%) (0.1M) sat. (1%)
	Li | LiBr, TlBr | Tl(Hg) (m) sat.
	Also see Lithium

DIOXANE

Electrode	Cell
Glass	Glass | Ha ¦ HA | Glass (C$_1$) (C$_2$)

II. EMF DATA

$(CH_3)_2SO$ *(Continued)*

Conc. range	E^0	Comments	Ref.
4.4×10^{-2}–10^{-4}M	-0.392	pH	62
10^{-2}–1.0M	$E \simeq 1.5$V	Complex formation constants; solb. prod.	65
10^{-2} mole fraction amalgam, 10^{-2}M solutions		Bias potentials	72
4×10^{-4}–0.8m	2.27234 2.26665 2.26083	25°C; γ; $\Delta G°$; $\Delta S°$ 30°C; $\Delta H°$ for reaction 35°C;	67
4×10^{-4}–0.8m	2.4212 2.4176 2.4137	25°C; $\gamma = f$(conc) 30°C; 35°C;	60
	2.17	Estimated	60
5×10^{-3}–2m	1.3963	Temp. range (22–26°C); checked by cryoscopy	68
1%–40% Tl		Polarization; solubility of MCl_n	109
	$E = 1.777$	No E^0 values; polarization and solubility; bias potentials	109
2×10^{-2}–0.8m	2.28381 2.27301 2.26214 2.25120	15°C; $\gamma_\pm$; $\Delta G_t°$; $\Delta H°$; 25°C; $\Delta S°$ 35°C; 45°C;	71

$C_4H_8O_2$

Conc. range	E^0	Comments	Ref.
		No temp. given; K_a; t_H^+; Λ	73

ETHANOL

Electrode	Cell
Cadmium	(Hg)Cd | $CdCl_2$, AgCl | Ag (11%)
	Cd | $CdCl_2$ ⋮ $AgNO_3$ | Ag (0.066M) (0.1M)
	(Hg)Cd | $CdCl_2$, AgCl | Ag
Calcium	(Hg)Ca | $CaCl_2$ | Ca(Hg) (0.11%) sat.
Calomel	H_2, Pt | HCl, Hg_2Cl_2 | Hg
	Ag | Ag^+ | Calomel (0.1M)
	H_2, Pt | HBr, Hg_2Br_2 | Hg
Copper	Cu | $CuCl_2$ ⋮ $AgNO_3$ | Ag (0.158M)
Hydrogen	H_2, Pt | HCl, AgCl | Ag
	H_2, Pt | HCl, Hg_2Cl_2 | Hg
	H_2, Pt | HBr, Hg_2Br_2 | Hg
	H_2, Pt | HBr, AgBr | Ag
	H_2, Pt | HI, AgI | Ag
	H_2, Pt | HCl, AgCl | Ag
	H_2, Pt | HCl, AgCl | Ag
	H_2, Pt | HCl, AgCl | Ag
Lead	Pb | $Pb(ClO_3)_2$ ⋮ $AgNO_3$ | Ag (0.042M) (0.1M)
Lithium	Also see Silver–silver chloride
Mercury	Hg | $Hg_2(NO_3)_2$ ⋮ $AgNO_3$ | Ag (0.027M) (0.1M)
Silver	Ag | $AgNO_3$ ⋮ $AgNO_3$ | Ag (0.1M) (0.05M)

II. EMF DATA

C_2H_5OH

Conc. range	E^0	Comments	Ref.
5×10^{-4}–10^{-2}M	0.33711	$\Delta G°$; $\Delta H°$; $\Delta S°$	116
	0.33183	30°; γ, extrap. methods	
	0.32430	35° discussed	
	−0.124	No temp. given Ag | $AgNO_3$ used as ref. electrode	117
5×10^{-4}–10^{-1}m	0.26506	K_{diss}; $\gamma_\pm$	118
	0.25897	30°;	
	0.25233	35°;	
		20°	118
4×10^{-4}–3×10^{-2}	−0.0638	$\gamma_\pm$	119
	0.508		
10^{-3}–2×10^{-2}M	−0.02320	$\gamma_\pm$	120
	−0.545	No temp. given	117
5×10^{-3}–5×10^{-2}N	−0.0886	γ_{H^+, Cl^-, OH^-}	121
4×10^{-4}–3×10^{-2}m	−0.0638	$\gamma_\pm$	119
10^{-3}–2×10^{-2}M	−0.02320	$\gamma_\pm$	120
10^{-4}–10^{-2}M	−0.06895	35°; ΔG_{trans}	122
4×10^{-4}–2×10^{-2}M	−0.24047	35°; cf ion size	122
10^{-4}–10^{-1}M	0.02190	35°; ΔG_{trans}	123
5×10^{-3}–10^{-1}M	−0.08138	ion pairs	124
10^{-3}–3.5M	not given	$\gamma_\pm$; t_{H^+}; t_{Cl^-}	125
	−0.878	No temp. given	126
	0.010	No temp. given	126
	0.011	No temp. given	126

ETHANOL

Electrode	Cell
	Ag | Ag$^+$ | Calomel (0.1M)
Silver–silver bromide	H$_2$, Pt | HBr, AgBr | Ag
	(Hg)Cd | CdCl$_2$, AgCl | Ag (11%)
	(Hg)Zn | ZnCl$_2$, AgCl | Ag (1%)
	(Hg)Cd | CdCl$_2$, AgCl | Ag
	Ag | AgCl | LiCl | Li(Hg) | LiCl | AgCl | Ag
	Also see Hydrogen
Silver–silver iodide	H$_2$, Pt | HI, AgI | Ag
	Ag | AgI, NaI | NaI, AgI | Ag (KI) (KI) sat. C$_1$ C$_2$ sat.
Silver–silver nitrate	Ag | AgNO$_3$ ┆ AgNO$_3$ | Ag (0.1M) (0.05M)
	Pb | Pb(ClO$_3$)$_2$: AgNO$_3$ | Ag (0.042M) (0.1M)
	Cu | CuCl$_2$: AgNO$_3$ | Ag (0.158M) (0.1M)
	Hg | Hg$_2$(NO$_3$)$_2$: AgNO$_3$ | Ag (0.027M) (0.1M)
	Cd | CdCl$_2$: AgNO$_3$ | Ag (0.066M) (0.1M)
	Zn | ZnCl$_2$: AgNO$_3$ | Ag (0.064M) (0.1M)
Zinc	Zn | ZnCl$_2$: AgNO$_3$ | Ag (0.064M) (0.1M)
	(Hg)Zn | ZnCl$_2$, AgCl | Ag (1%)
	(Hg)Zn | ZnCl$_2$ | Zn(Hg) (1.138%) (0.0977%)

II. EMF DATA

C_2H_5OH *(Continued)*

Conc. range	E^0	Comments	Ref.
	0.508	No temp. given	126
10^{-4}–10^{-2}M	−0.06895	35°; ΔG_{trans}	122
5×10^{-4}–10^{-2}M	0.33711	$\Delta G°$; $\Delta H°$; $\Delta S°$;	116
	0.33183	30°; $\gamma_\pm$	
	0.32430	35°;	
10^{-2}–1M	Not given	15°; ΔG	127
5×10^{-4}–10^{-1}m	0.26506	K_{diss}; $\gamma_\pm$	118
	0.25897	30°;	
	0.25233	35°;	
0.006–0.6M		25, 30, 35°; ΔG of dilution; ΔG of transfer; solvent effects	128
4×10^{-4}–2×10^{-2}	−0.24047	35°; ΔG_{trans}	122
10^{-3}–0.5M	0	Conc. cells; transport nos.	129
	0.011	No temp. given	117
	−0.878	No temp. given	117
	−0.545	No temp. given	117
	0.010	No temp. given	117
	−0.124	No temp. given	117
	−1.44	No temp. given	117
	−1.44	No temp. given	117
10^{-2}–1m	Not given	15°; ΔG	127
10–120g/l	Not calcd.	17°	118

ETHYLAMINE

Electrode	Cell
Potassium	K(Hg) | KI | K 0.2216% solid
Sodium	Na(Hg) | NaI | Na 0.206% solid
	Na(Hg) | NaI | Na (C)

ETHYL CYANOACETATE

Electrode	Cell
Silver	Ag | AgNO$_3$ | AgNO$_3$ | Ag aq. solv.

ETHYLENE GLYCOL

Electrode	Cell
Hydrogen	H$_2$, Pt | HCl, AgCl | Ag
	H$_2$, Pt | HBr, AgBr | Ag
	H$_2$, Pt $\begin{vmatrix} \text{NaHa,} \\ \text{LiBr} \\ \text{Na}_2\text{A,} \end{vmatrix}$ AgBr | Ag
Lead	Pb(Hg) | PbCl$_2$, AgCl | Ag KCl, m
	Pb(Hg) | PbCl$_2$, AgCl | Ag
Silver–silver bromide	Also see Hydrogen
Silver–silver chloride	Also see hydrogen
	Also see lead
	Ag | AgCl | KCl, TlCl | Tl(Hg) sat. (m) sat.
Thallium	Also see Silver–silver chloride

II. EMF DATA

EtNH$_2$

Conc. range	E^0	Comments	Ref.
	1.0481	For H$_2$O–E^0 of K	130
	0.8456	For H$_2$O–E^0 of Na	131
	0.828	$-50°$; for NH$_3$–E^0 of Na	97
	0.835	$-70°$; for NH$_3$–E^0 of Na	97

EtCOCH$_2$CN

Conc. range	E^0	Comments	Ref.
0.1N–10^{-3}N		Unstable solutions, λ solutions measured; $E = A \log(\lambda_w/\lambda_s) + B$; $B = 0.095$	38

HOCH$_2$CH$_2$OH

Conc. range	E^0	Comments	Ref.
4 × 10^{-3}–10^{-1}m	0.022	30°; assumes complete dissociation	132
4 × 10^{-3}–10^{-1}m	-0.104	30°;	132
$\mu = 0.02$–0.1m	-0.104	30°; K_a phthalic acid	133
4 × 10^{-4}m–sat.	Not given	$\gamma_\pm$; solubility of PbCl$_2$	134
Sat.	0.4963	ΔG formation of complexes	135
			132
			133
			132
			134, 135
0–0.5m in KCl	0.39461	Gives as E/0.1183 ± log m solubility data for TlCl	136
			136

FORMAMIDE

Electrode	Cell
Cadmium	H$_2$, Pt	HCl, CdCl$_2$	Cd sat.
	(Hg)Rb	RbCl, CdCl$_2$	Cd (0.231%) sat.
	(Hg)K	KCl, CdCl$_2$	Cd (0.212%) sat.
	Zn	ZnCl$_2$, CdCl$_2$	Cd sat.
	Cd	Cd(NO$_3$)$_2$, CdCl$_2$	Cd sat.
	Tl	TlNO$_3$, CdCl$_2$	Cd sat.
	Pb	Pb(NO$_3$)$_2$, CdCl$_2$	Cd
	Cu	CuBr$_2$, CdCl$_2$	Cd
	H$_2$, Pt	HCl, KCl, CdCl$_2$	Cd sat. sat.

II. EMF DATA

H_2NCHO

Conc. range	E^0	Comments	Ref.
10^{-2}–1M	−0.617 −0.619	25°C; $K_{sp}CdCl_2$; 18°C; electrochem series in formamide: refer all values to standard Hydrogen electrode	74
10^{-2}–1M	−1.163 −1.159	25°C; $K_{sp}CdCl_2$; 18°C; electrochem series in formamide: refer all values to standard Hydrogen electrode	74
10^{-2}–1M	−1.205 −1.201	25°C; $K_{sp}CdCl_2$; 18°C; electrochem series in formamide: refer all values to standard Hydrogen electrode	74
10^{-2}–1M	−0.149 −0.150	25°C; $K_{sp}CdCl_2$; 18°C; electrochem series in formamide: refer all values to standard Hydrogen electrode	74
10^{-2}–1M	+0.200 +0.200	25°C; $K_{sp}CdCl_2$; 18°C; electrochem series in formamide: refer all values to standard Hydrogen electrode	74
10^{-2}–1M	+0.264 +0.271	25°C; $K_{sp}CdCl_2$; 18°C; electrochem series in formamide: refer all values to standard Hydrogen electrode	74
10^{-2}–1M	+0.415 +0.416	25°C; $K_{sp}CdCl_2$; 18°C; electrochem series in formamide: refer all values to standard Hydrogen electrode	74
10^{-2}–1M	+0.887 +0.889	25°C; $K_{sp}CdCl_2$; 18°C; electrochem series in formamide: refer all values to standard Hydrogen electrode	74
10^{-2}–1M	−0.608 −0.609	25°C; $K_{sp}CdCl_2$; 18°C; electrochem series in formamide: refer all values to standard Hydrogen electrode	74

FORMAMIDE

Electrode	Cell
Cadmium	Cd | CdCl$_2$ | HCl or KCl | AgCl | Ag
	(Hg)Cd | CdCl$_2$ | HCl or KCl | AgCl | Ag
	(Hg)Cd | CdCl$_2$ | AgCl | Ag (m)
	Cd | Cd^{++} || H$^+$ | H$_2$, Pt
Calomel	Also see Hydrogen
Copper	Also see Cadmium
Hydrogen	H$_2$, Pt | HCl, AgCl | Ag
	H$_2$, Pt | HOAc, NaOAc, NaCl, AgCl | Ag
	H$_2$, Pt | HA, NaA, NaCl, AgCl | Ag
	H$_2$, Pt | HCl, AgCl | Ag
	H$_2$, Pt | HBr, AgBr | Ag
	H$_2$, Pt $\left\| \begin{array}{l} \text{KCl, K Salicylate} \\ \text{H Salicylate} \\ \text{(0.05M)} \end{array} \right\|$ KCl $\left\| \begin{array}{l} \text{HA, KA} \\ \text{KCl} \\ \text{(C)} \end{array} \right\|$ Pt, H$_2$

II. EMF DATA

H_2NCHO (Continued)

Conc. range	E^0	Comments	Ref.
0.05–0.5m(KCl)	0.8222	5°; thermodynamics of solvate formation	137
0.016–0.072m(HCl)	0.8195	10°	
	0.8164	15°	
	0.8133	20°	
	0.8105	25°	
	0.8075	30°	
	0.7603	25°	137
$m = 0.000114$–0.01706	0.612	Correction to E^0 of Cd Electrode	138
	−0.412	$CdCl_2$ association and solubility	138
			74
$\times 10^{-3}$–5×10^{-2}M	0.204	Least squares extrap.	75
2×10^{-3}–5×10^{-2}M		No E values K_A of HOAc	
10^{-3}–10^{-1}m	0.208	15°C; series of carboxylic	
	0.207	20°C; acids; $K_a = f(T)$;	
	0.204	25°C; Hitchcock extrap.	76
	0.198	30°C	
	0.191	35°C	
	0.181	40°C	
	0.172	45°C	
2×10^{-2}–3×10^{-1}M	0.1986	25°C; $\Delta H°$, $\Delta G°$, $\Delta S°$,	77
	0.1937	30°C γ_{HCl}	
	0.1888	35°C	
	0.1853	40°C	
	0.1801	45°C	
	0.1753	50°C	
	0.1715	55°C	
10^{-2}–10^{-1}m	0.0967	Debye–Hückel extrap.	78
0.02–0.1M		20°C; K_a of a series of organic acids	79

FORMAMIDE

Electrode	Cell
Hydrogen	H_2, Pt | HCl, $CdCl_2$ | Cd sat.
	H_2, Pt | HCl, KCl, $CdCl_2$ | Cd sat. sat.
	Pt, H_2 | HCl | AgCl | Ag
	Pt, H_2 | HCl | Hg_2Cl_2 | Hg
	Also see Cadmium
Iodine	Pt | I_2, KI | Et_4NPi | I_2, KI | Pt $(c_1)(c_2)$ sat. $(c_3)(c_4)$
	Pt, I_2 | I^- Et_4NPi | I^-, I_2, Pt
Lead	Also see Cadmium
Potassium	Also see Cadmium
Quinhydrone	Pt | QH, KCl, K Salicylate, H Salicylate | (0.05M) | KCl sat. | HA, KA KCl (C) | QH | Pt
Rubidium	Also see Cadmium
Silver	Ag | MX | Et_4NPi | $AgNO_3$ | Ag $(10^{-2}M)$ sat. $(10^{-2}M)$
	Ag | $AgNO_3$, MX || $AgNO_3$ | Ag
Silver–Silver bromide	Also see Hydrogen
Silver–silver chloride	Also see Hydrogen and Cadmium
Thallium	Also see Cadmium
Zinc	Also see Cadmium

II. EMF DATA

H₂NCHO (*Continued*)

Conc. range	E^0	Comments	Ref.
10^{-2}–1M	−0.617 −0.619	25°C; K_{sp} CdCl₂; 18°C; Electrochem. series in formamide	74
	−0.608	25°C; Electrochem. series in formamide 18°C	74
0.01–0.1m	0.2214 0.2100 0.2002	5°; also solubility of AgCl 15° 25°	139
	0.2453 (±0.0002)	$\gamma_\pm$	140
$c_1 = 0.01$M $c_2 = 0.005$M $c_3 = 0.01$M $c_4 = 0.005$M		K_{sp} and K(stability) of complexes for range of MX	47
∼10^{-2}M		K_{sp}; complexes; solvent effects	22
			74
			74
2×10^{-2}–10^{-1}M		20°C; attacks solvent, unsatisfactory	79
			74
		K_{sp} and K(stability) of complexes for range of MX	47
∼10^{-2}M		K_{sp}; complexes; solvent effects	22
			78
			75, 76, 77
			74
			74

FORMIC ACID

Electrode	Cell
Cadmium	Cd | CdCl$_2$ | AgPi | Ag
Calcium	Ca(Hg) | Ca(O$_2$CH)$_2$ | AgPi | Ag
Calomel	Pt | QH HX sat. | KCl sat. | KCl sat. H$_2$O | KCl sat. Hg$_2$Cl$_2$ H$_2$O | Hg
	Pt | QH (0.05M) HCO$_2$Na 0.25M | KCl sat. | Hg$_2$Cl$_2$ sat. | Hg
Cesium	Cs(Hg) | CsO$_2$CH | AgPi | Ag
Copper	Cu | Cu(picrate)$_2$ | AgPi | Ag
Gold	Au | QH HX sat. | Na$^+\phi$SO$_3^-$ sat. | QH sat. | Na$^+\phi$SO$_3^-$ sat. 0.098M HCO$_2$Na | Au
Hydrogen	H$_2$, Pt | HCl, AgCl | Ag
	Pt, H$_2$ | HCO$_2$H | AgPi | Ag
Lead	Pb | Pb(O$_2$CH)$_2$ | AgPi | Ag
Lithium	Li(Hg) | LiO$_2$CH | AgPi | Ag
Mercury	Hg | Hg$_2$(O$_2$CH)$_2$ | AgPi | Ag
Potassium	K(Hg) | KO$_2$CH | AgPi | Ag
Quinhydrone	Au | QH HX sat. | Na$^+\phi$SO$_3^-$ sat. | QH sat. | Na$^+\phi$SO$_3^-$ sat. 0.098M HCO$_2$Na | Au
	Pt | QH HX sat. | KCl sat. | KCl sat. H$_2$C | KCl sat. Hg$_2$Cl$_2$ H$_2$O | Hg
	HX = HClO$_4$, HCl, H$_2$SO$_4$, TosOH AcOH, Cl$_3$CO$_2$H, Cl$_2$CHCO$_2$H, PiOH, HOϕCO$_2$H
	Pt | QH (0.05M) HCO$_2$Na (0.25M) | KCl, sat. | Hg$_2$Cl$_2$ sat. | Hg

II. EMF DATA

HCO_2H

Conc. range	E^0	Comments	Ref.
0.01N	−0.924(3)	All potls. referred to Ag^+ 0.01N and M^+ 0.01N	144
0.01N; 0.0253%	−2.527	Electrochem. series; reduct. potls. temp. coeffs. (15–25°)	144
$\simeq 10^{-1}$M	$E \simeq 0.8$V	Titrations; K_a values	141
0.25M	$E = 0.5384$ (of Pt | QH)	Polarography	142
0.01N Cs(Hg) = 0.2827 atom %	−2.492		144
0.01N	−0.314(7)		144
$\simeq 10^{-2}$M		Titrations; no E^0	143
10^{-3}–10^{-2}M	−0.11995	35°C; molar and molal scales discussed	15
0.001–0.01N	−0.170(3)		144
0.01N	−0.889(5)		144
0.01N; 0.035% (Li in Hg)	−2.684		144
0.01N	−0.0012		144
0.01N 0.2216% K in Hg	−2.486		144
$\simeq 10^{-2}$M		Titrations; no E^0	143
$\simeq 10^{-1}$M	$E \simeq 0.8$ V	Titrations; K_a values	141
0.25M	$E = 0.5384$ (of Pt | QH)	Polarography	142

FORMIC ACID

Electrode	Cell
Rubidium	Rb(Hg) | RbO$_2$CH | AgPi | Ag
Silver	Ag | AgPi | AgPi | Ag
Silver–silver chloride	H$_2$, Pt | HCl, AgCl | Ag
Silver–silver picrate	Also see other entries
Sodium	Na(Hg) | NaO$_2$CH | AgPi | Ag
Zinc	Zn | ZnCl$_2$ | AgPi | Ag

GASOLINE

Electrode	Cell
Copper	Cu | — | Cu^{II}(oleate)$_2$ | Cu

HEPTANOL

Electrode	Cell
Silver–silver iodide	Also see Sodium
Sodium	(Hg)Na | NaI | AgI | Ag

HEXAMETHYLPHOSPHOROTRIAMIDE

Electrode	Cell
Iodine	Pt | I$_2$, KI | NEt$_4$Pi | I$_2$, KI | Pt (0.01M) (0.005M) sat. (0.01M) (0.005M) (+titrant)
Silver	Ag | MX | NEt$_4$Pi | AgNO$_3$ | Ag (0.01M) sat. (0.01M) (+titrant)

II. EMF DATA

HCO_2H (*Continued*)

Conc. range	E^0	Comments	Ref.
0.01N;	−2.547		144
0.0005–0.1N	0		144
10^{-3}–10^{-2}M	−0.11995	35°; molar and molal scales discussed	15
0.01N; 0.205% of Na in Hg	−2.747		144
0.01N	−1.326(4)		144

Conc. range	E^0	Comments	Ref.
0.01M. solvent side not defined	0–0.03	No temp. given; intersolvent potentials	50

$C_7H_{15}OH$

Conc. range	E^0	Comments	Ref.
0.0015–0.2m	1.0902	$\gamma_\pm$; solvent effects	101

HMPT

Conc. range	E^0	Comments	Ref.
		K_{sp} and K(stability) for a range of MX	47
		K_{sp} and K(stability) for a range of MX	47

HEXANOL

Electrode	Cell
Silver–silver iodide	Also see Sodium
Sodium	(Hg)Na | NaI | AgI | Ag

HYDRAZINE

Electrode	Cell
Cadmium	H_2, Pt | $H_2SO_4 \cdot N_2H_4$ | $CdSO_4$ | Cd(Hg) sat. sat.
	Zn(Hg) | $ZnSO_4$, H_2SO_4, $CdSO_4$ | Cd(Hg) sat. sat.
Hydrogen	H_2, Pt | $H_2SO_4 \cdot N_2H_4$ | $ZnSO_4$ | Zn(Hg)
	H_2, Pt | $H_2SO_4 \cdot N_2H_4$ | $CdSO_4$ | Cd(Hg) sat. sat.
Zinc	Also see Cadmium and Hydrogen

HYDROGEN FLUORIDE

Electrode	Cell
Cadmium fluoride	(Hg)Cd | CdF_2, HF NaF, HgF | Hg sat. sat.
	Pb | PbF_2, NaF, CdF_2 | Cd sat. sat.
Copper fluoride	Cu | CuF_2, HF, NaF, HgF | Hg sat. sat.
	Cu | CuF_2, TlF, TlF_3 | Pt sat. sat.
	Cu | CuF_2, $KTlF_4$, TlF | Pt

II. EMF DATA

$C_6H_{13}OH$

Conc. range	E^0	Comments	Ref.
0.0024–0.3m	1.1501	$\gamma_{\pm}$; solvent effects	101

N_2H_4

Conc. range	E^0	Comments	Ref.
3×10^{-4}–0.06m	∼0.3225	0°C; $\gamma_{H_2SO_4}$	145
	0.3875	20°C	145
3×10^{-4}–0.06m	0.7100	0°C; $\gamma_{H_2SO_4}$	145
3×10^{-4}–0.06m	∼0.3225	0°C; $\gamma_{H_2SO_4}$	145
			145

HF

Conc. range	E^0	Comments	Ref.
10^{-2}–4 $\times 10^{-2}$M	1.0337	0°; includes vapor pressure studies. Addition compounds. Thermodynamic functions.	146
10^{-2}–4 $\times 10^{-2}$M	0.0300	0°	146
10^{-2}–4 $\times 10^{-2}$M	0.274	0°	146
	0.283	10°	
3×10^{-2}–1 $\times$ 2M	0.9282	0°; acidity of TlF_3, solubility	147
	≃0.7	0°; erratic behavior	147

HYDROGEN FLUORIDE

Electrode	Cell
Copper fluoride	H$_2$, Pt | HF, CuF$_2$ | Cu sat.
	Cu | CuF$_2$, TlF, TlF$_3$ | Pt
	Cu | CuF$_2$ | TlF | TlF$_3$ | Pt solid solid
Hydrogen	H$_2$, Pt | HF ¦ NiF$_2$ | Ni
	H$_2$, Pt | HF, CuF$_2$ | Cu sat.
Lead	Pb | PbF$_2$, HF, NaF, HgF | Hg sat. sat.
	Pb | PbF$_2$, NaF, CdF$_2$ | Cd sat. sat.
Mercury–mercuric fluoride	(Hg)Cd | CdF$_2$, HF, NaF, HgF | Hg
	(Hg)Cu | CuF$_2$, HF, NaF, HgF | Hg sat. sat.
	(Hg)Pb | PbF$_2$, HF, NaF, HgF | Hg sat. sat.
	Hg | HgF, NaF || H$_3$OF, NaF, AgF | Ag sat.
Nickel	H$_2$, Pt | HF ¦ NiF$_2$ | Ni
Silver–silver fluoride	Hg | HF, NaF || H$_3$OF, NaF, AgF | Ag sat.
	Ag | AgF, TlF$_3$ | Pt sat.
	Ag | AgF, AgF$_2$ | Pt sat.

II. EMF DATA

HF (*Continued*)

Conc. range	E^0	Comments	Ref.
	0.276		148
3×10^{-3}–1×2m	0.9269	0°; ionization constants	149
	0.9269	0°	
		No temp. given. Reversibility tests and Polarization curves.	150
	0.276	25°	148
10^{-2}–4×10^{-2}M	1.0637	0°	146
10^{-2}–4×10^{-2}M	0.0300	0°	146
10^{-2}–4×10^{-2}M		0°; Vapor pr. studies	146
	1.0337	10²; Addition compounds	
10^{-2}–4×10^{-2}M	0.274	0°	146
	0.283	10°	
10^{-2}–4×10^{-2}M	1.0637	0°	
10^{-2}–5×10^{-2}M	0.3510	0°	
		No temp. given. Reversibility tests and Polarization curves	150
10^{-2}–5×10^{-2}	0.3510	0°	146
6×10^{-2}–1×7m	1.3816	0°; Ionization constants; $\Delta G°$; $\gamma_\pm$	149
1.7×10^{-2}–0×78m	0.5654	0°; $\Delta G°$; $\gamma_\pm$	149

HYDROGEN FLUORIDE

Electrode	Cell
Silver–silver fluoride	Ag | AgF | AgF$_2$ | Pt
	Ag | AgF, KF | AgF$_2$ | Pt
	Ag | AgF, TlF | TlF$_3$ | Pt
Thallium	Cu | CuF$_2$, TlF, TlF$_3$ | Pt sat. sat.
	Cu | CuF$_2$, KTlF$_4$, TlF | Pt
	Cu | CuF$_2$, TlF, TlF$_3$ | Pt sat. sat.
	Ag | AgF, TlF$_3$ | Pt sat.
	Ag | AgF, TlF | TlF$_3$ | Pt
	Cu | CuF$_2$ | TlF | TlF$_3$ | Pt solid solid

KEROSENE

Electrode	Cell
Copper	Cu | — | Cu$^{(II)}$(oleate)$_2$ | Cu

METHANOL

Electrode	Cell
Barium amalgam	(Hg)Ba | BaCl$_2$, AgCl | Ag
Bromine	Br$_2$, Pt | KBr || KBr, Pt | Br$_2$ in DMF in MeOH

II. EMF DATA

HF (*Continued*)

Conc. range	E^0	Comments	Ref.
	0.7714	0°; ΔG; K_b; γ	151
	0.7725	0°; ΔG; K_b; γ	151
	0.5654	0°; ΔG; K_b; γ	151
3×10^{-2}–1.2m	0.9282	0°; acidity of TlF$_3$ solubility	147
3×10^{-2}–1.2m	$\simeq$0.7	0°; erratic	147
3×10^{-3}–1.2m	0.9269	0°; ionization constants; $\Delta G°$	149
6×10^{-2}–1.7m	1.3816	0°	149
	0.7725	0°; ΔG; K_b; γ	151
	0.9269	0°; ΔG; K_b; γ	151

Conc. range	E^0	Comments	Ref.
0.01M. solvent side not defined	−0.035	No temp. given; intersolvent potentials	50

CH$_3$OH

Conc. range	E^0	Comments	Ref.
10^{-3}–4×10^{-2}M	1.6262	25°C	
	1.6100	35°C	152
	1.5938	45°C	
	1.5775	55°C	
		γ_{BaCl_2}; dissociation constants	
10^{-2}–10^{-1}M		γ in DMF	153

METHANOL

Electrode	Cell
Cadmium	Cd | CdCl$_2$ ┆ AgNO$_3$ | Ag (0.052M) (0.1M)
Cadmium amalgam	(Hg)Cd | CdCl$_2$, AgCl | Ag (11%)

Calomel	Hg | Hg$_2$Cl$_2$, LiCl or CsCl | NaCl, Hg$_2$Cl$_2$ | Hg
	
	Hg | Hg$_2$Cl$_2$, KCl ┆ KBr, HgBr | Hg
	    sat. (C) (C) sat.
	
	Hg | Hg$_2$Cl$_2$ / KCl sat. / H$_2$O | AgNO$_3$ (0.1M) | Ag
	
	H$_2$, Pt | HBr, Hg$_2$Br$_2$ | Hg
	                sat.
	(99.5% MeOH)
	
	Also see glass

Cesium amalgam	(Hg)Cs | CsI, AgI | Ag (0.1%) (Hg)Cs | CsI | AgI | Ag
Copper	Cu | Cu(NO$_3$)$_2$ ┆ AgNO$_3$ | Ag (0.01M) (0.1M)
Glass	Glass | HA, MA || KCl | Hg$_2$Cl$_2$ | Hg Also see Silver–silver nitrate

Hydrogen	Ag | AgCl, NaCl, NaHSucc, H$_2$Succ | ......
	   sat. (m) (m) (m)
	
	H$_2$, Pt | NaI, NaHSucc, H$_2$Succ, AgI | Ag
	      (m) (m) (m) sat.
	
	H$_2$, Pt | HCl, AgCl | Ag
	
	H$_2$, Pt | HCl, AgCl | Ag
	
	H$_2$, Pt | HBr, Hg$_2$Br$_2$ | Hg
	(99.5% MeOH)

II. EMF DATA

CH$_3$OH (Continued)

Conc. range	E^0	Comments	Ref.
	-1.194	Imprecise measurements	117
5×10^{-4}–10^{-1}M	0.41033 0.40598 0.40050	25°C, γ(CdCl$_2$, ΔG), 30°C, 35°C. ΔH, ΔS.	154
10^{-3}–10^{-2}M		γ_{LiCl}, γ_{CsCl} with NaCl as standard	155
C = 0.025M	0.1053	8–30°C; no extrap. to zero ionic strength, extrap. to 0°C	156
	-0.525	Imprecise measurements	117
2×10^{-3}–12×10^{-2}M	0.01071 0.02072 0.03131	15°C, pH, γ, 25°C, ΔS, ΔH, 35°C, ΔCp	157
10^{-2}–10^{-1}M	1.6640	24.5°C ± 0.5 γ, medium effects	103
0.003–0.01m	1.4820	Also Me$_2$CO and mixtures	16
	0.418 (±0.005)	Imprecise measurements	117
~10^{-2}M		K_a of a range of acids and phenols	113
0.025–4 × 10^{-4}	-0.3075		158
	-0.0101	By extrapolation of MeOH	H$_2$O	159
10^{-3}–10^{-1}M	-0.0099	γ_{HCl}, ΔG transfer, ion size	160
2×10^{-3}–12×10^{-3}M	0.01071 0.02072 0.03131	15°C, pH, 25°C, ΔS, ΔH 35°C, ΔCp	157

METHANOL

Electrode	Cell

Hydrogen	Ag | AgCl, NaCl, NaHSucc, H₂Succ | ......
	      sat. (m) (m) (m)
	
	H₂, Pt | NaBr, NaHSucc, H₂Succ, AgBr | Ag
	      (m) (m) (m) sat.
	
	H₂, Pt | HBr, AgBr | Ag
	         sat.
	
	H₂, Pt | AgI | Ag
	     sat.

Iodine	I₂, Pt | KI || KI | Pt, I₂ in DMF in MeOH Pt, I₂ | I⁻ || I⁻, | I₂, Pt
Lead	Pb | Pb(NO₃)₂ ⁞ AgNO₃ | Ag (0.01M) (0.1M)
Lithium	Also see Silver–silver chloride
Magnesium amalgam	(Hg)Mg | MgCl₂, AgCl | Ag
Mercury	Hg | Hg₂(NO₃)₂ AgNO₃ | Ag (0.069M) (0.1M)
Mercury–mercurous bromide and mercury–mercurous iodide	Hg | HgBr, KBr KI, HgI | Hg sat. (C) (C) sat. Also see Calomel
Potassium amalgam	Ag | AgCl, KCl | K(Hg)
Rubidium amalgam	Ag | AgCl, RbCl | Rb(Hg)
Silver	Ag | AgNO₃ AgNO₃ | Ag (0.1M) (0.05M) Also see Cadmium, Copper, Lead, Mercury, and Zinc
Silver–silver bromide	H₂, Pt | HBr, AgBr | Ag sat.

II. EMF DATA

CH$_3$OH (*Continued*)

Conc. range	E^0	Comments	Ref.
0.025–4 × 10^{-4}m	−0.1286		158
	−0.1387	Calc. from above cell	158
	−0.3176	Calc. from above cell	158
10^{-2}–10^{-1}M		γ in MeOH relative to γ in DMF	153
~10^{-2}M		K_{sp}; complexes	22
	−0.961 (±0.005)	Imprecise measurements	117
3 × 10^{-3}–4 × 10^{-2}M	1.8350	25°C, γ_{MgCl_2}	152
	1.8220	35°C, dissociation constants	
	1.8052	45°C,	
	1.7828	55°C,	
	−0.021	Imprecise measurements	117
C = 0.05M	0.1662	20.1°C; no extrap. to zero μ;	156
0.02–0.3%K Ref. state 0.2216% K	−1.850	γ_{KCl}	161
0.2% Rb	−1.813	γ_{RbCl}	161
	0.015	Imprecise measurements	117
	−0.1387	Calc. from buffer cell	158

METHANOL

Electrode	Cell
Silver–silver chloride	Ag | AgCl, LiCl or CsCl | NaCl, AgCl | Ag
	H$_2$, Pt | HCl, AgCl | Ag
	H$_2$, Pt | HCl, AgCl | Ag
	Ag | AgCl, NaCl | Na(Hg) | NaCl
	Ag | AgCl | LiCl | Li(Hg) | LiCl | AgCl | Ag
	Ag | AgCl | NaCl | Na(Hg)
	Also see Barium, Cadmium, Magnesium, Potassium, Rubidium, and Strontium amalgams
Silver–silver iodide	(Hg)Cs | CsI, AgI | Ag (0.1%)
	H$_2$, Pt | HI, AgI | Ag sat.
	Also see Cesium
Silver–silver nitrate	Ag | AgNO$_3$, MX | Et$_4$NPi | AgNO$_3$ | Ag
	Ag | AgNO$_3$ || MA, HA | Glass
Sodium amalgam	(Hg)Na | NaI | Na(Hg) C$_1$ C$_2$ Also see Silver–silver chloride
Strontium amalgam	(Hg)Sr | SrCl$_2$, AgCl | Ag

Zinc	Zn | ZnCl$_2$: AgNO$_3$ | Ag
	   (1.175M) (0.1M)
	   (0.107M) (0.1M)

II. EMF DATA

CH$_3$OH (Continued)

Conc. range	E^0	Comments	Ref.
10^{-3}–10^{-2}M		γ_{LiCl} and γ_{CsCl} with NaCl as standard	155
	-0.0101	By extrapolation of MeOH | H$_2$O	159
10^{-3}–10^{-1}M	-0.0099	γ_{HCl}, ion size ΔG transfer	160
10^{-3}–10^{-2}M		20°C; γ_{NaCl}	162
0.006–0.6M		ΔG of dilution; ΔG of transfer; solvent effects 25, 30, 35°	128
0.0005–0.15M	-2.697	t_+; $\gamma_\pm$	163
10^{-2}–10^{-1}M	-1.6640	γ_{CsI},	103
	-0.3176	Medium effects Calc. from Buffer cell	158
~10^{-2}M		K_{sp}; complexes; solvent effects	22
~10^{-2}M		K_a of a range of acids and phenols	113
10^{-2}–0.6%		20°C, γ_{Na} in Hg	162
10^{-3}–4 × 10^{-2}M	1.9305 1.9105 1.8905 1.8705	25°C, γ_{SrCl_2}, 35°C, dissociation constants 45°C, 55°C,	152
	-1.490	Imprecise measurements	126
	-1.520		

N-METHYL ACETAMIDE

Electrode	Cell
Hydrogen	H$_2$, Pt | HCl, AgCl | Ag sat.
	H$_2$, Pt | HCl, AgCl | Ag
Silver–silver chloride	Also see Hydrogen

N-METHYLFORMAMIDE

Electrode	Cell
Cesium	Ag | AgCl | CsCl | Cs(Hg) | CsCl | AgCl | Ag (m) (0.5%)
Silver–silver chloride	Also see Cesium

N-METHYL-2-PYRROLIDONE

Electrode	Cell
Iodine	Pt, I$_2$ | I$^-$ | Et$_4$NPi | I$^-$ | I$_2$, Pt
Silver–silver nitrate	Ag | AgNO$_3$, MX || AgNO$_3$ | Ag

II. EMF DATA

MeNHCOMe

Conc. range	E^0	Comments	Ref.
2×10^{-3}–0.1m	0.20573	40°C; γ_{HCl}	80
5×10^{-3}–0.15m	0.21187	35°C; γ_{HCl}; $\overline{C}_{pHCl}$	81
	0.20573	40°C; $\overline{H}^°_{HCl}$ relative	81
	0.20091	45°C	
	0.19456	50°C; $\Delta G°$; $\Delta H°$; $\Delta S°$	81
	0.18972	55°C	81
	0.18357	60°C	81
	0.19194	70°C	81

MeNHCHO

Conc. range	E^0	Comments	Ref.
4×10^{-4}–0.2m		γ_{CsCl}	82

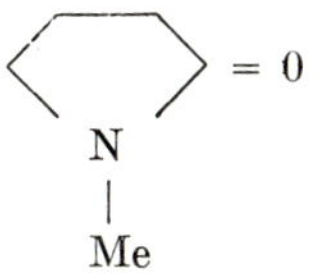

$$\left\langle \begin{array}{c} \\ N \\ | \\ Me \end{array} \right\rangle = 0$$

Conc. range	E^0	Comments	Ref.
$\sim 10^{-2}$M		K_{sp}; complexes	22
$\sim 10^{-2}$M		K_{sp}; complexes; solvent effects	22

NITROMETHANE

Electrode	Cell
Iodine	Pt, I_2	I^-	Et_4NPi	I^-	I_2, Pt
Silver–silver nitrate	Ag	$AgNO_3$, MX		$AgNO_3$	Ag

OCTANOL

Electrode	Cell
Silver–silver iodide	Ag	AgI	NaI	Na(Hg)
Sodium	Also see Silver–silver iodide

PHENYLACETONITRILE

Electrode	Cell
Silver	Ag	$AgNO_3$ aq.	$AgNO_3$ (ϕCH_2CN)	Ag

n-PROPANOL

Electrode	Cell
Hydrogen	H_2, Pt	HCl, AgCl	Ag
	H_2, Pt	H_2SO_4, Hg_2SO_4	Hg
	Pt, H_2	HCl	AgCl	Ag

II. EMF DATA

Conc. range	E^0	Comments	Ref.
$\sim 10^{-2}$M		K_{sp}; complexes; solvent effects	997
$\sim 10^{-2}$M		K_{sp}; complexes	997

$C_8H_{17}OH$

Conc. range	E^0	Comments	Ref.
0.0015–0.2m	1.0400	$\gamma_\pm$; solvent effects	783

$C_6H_5CH_2CN$

Conc. range	E^0	Comments	Ref.
0.1N–10^{-3}N		λ solutions measured $E = A \log(\lambda_w/\lambda_s) + B$ $B = 0.088$	38

$n = $ PrOH

Conc. range	E^0	Comments	Ref.
10^{-3}–10^{-1}m	0.102	25°; Debye–Hückel extrap.	164
	0.092	20°	
0.022–4m	0.2687	32°; $\gamma_{H_2SO_4}$	165
0.006–0.1m	−0.0854	5°; ΔH, ΔS and ΔG of transfer from water; $\lambda_\pm$; $\gamma_\pm^\circ$	166
	−0.0943	10°	
	−0.1043	15°	
	−0.1115	20°	
	−0.1200	25°	
	−0.1278	30°	
	−0.1333	35°	
	−0.1420	40°	
	−0.1500	45°	

n-PROPANOL

Electrode	Cell
Mercury-mercuric sulphate	Also see Hydrogen
Silver-silver chloride	Also see Hydrogen

PROPIONITRILE

Electrode	Cell
Silver	Ag | AgNO$_3$ ⋮ AgNO$_3$ | Ag aq. (EtCN)

n-PROPYLAMINE

Electrode	Cell
Lithium	$\quad\quad+\quad\quad\quad-$ Li(Hg) | LiI | Li (0.035%) solid

PROPYLENE CARBONATE

Electrode	Cell
Cadmium	(Hg)Cd | CdCl$_2$, Et$_4$NCl, CdCl$_2$ | Cd(Hg) $\quad\quad\quad$ sat. (1M) sat. Also see Thallium halide and Lead halide
Iodine	Pt, I$_2$ | I$^-$ | Et$_4$NPi | I$^-$ | I$_2$, Pt

II. EMF DATA

$n = \text{PrOH}$ (*Continued*)

Conc. range	E^0	Comments	Ref.

C_2H_5CN

Conc. range	E^0	Comments	Ref.
$0.1N{-}10^{-3}N$		λ solutions measured $E = A \log(\lambda_w/\lambda_s) + B$, $B = -0.012$	38

$n\text{PrNH}_2$

Conc. range	E^0	Comments	Ref.
	0.9502	25°C	167
	0.9532	35°C	

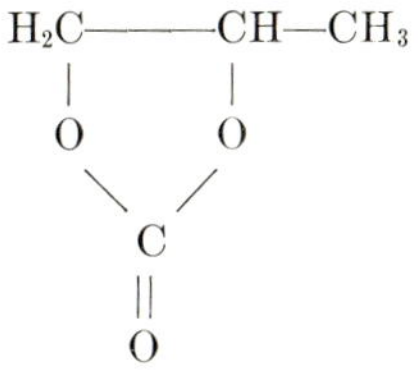

$$\begin{array}{c} H_2C\text{——}CH\text{—}CH_3 \\ || \\ OO \\ \diagdown\diagup \\ C \\ \| \\ O \end{array}$$

Conc. range	E^0	Comments	Ref.
4% Cd		Polarization; solubility of MCl_n	109
$\sim 10^{-2}M$		K_{sp} complexes; solvent effects	22

PROPYLENE CARBONATE

Electrode	Cell
Lead	(Hg)Pb | PbCl$_2$, Et$_4$NCl, PbCl$_2$ | Pb(Hg) sat. (1M) sat.
Lead bromide	(Hg)Pb | PbBr$_2$ | LiBr | CdBr$_2$ | Cd(Hg) Also see Thallium halide
Lead chloride	Pb | PbCl$_2$ | LiCl | CdCl$_2$ | Cd(Hg) Also see Thallium halide
Lead iodide	Pb(Hg) | PbI$_2$ | LiI | CdI$_2$ | Cd(Hg) Also see Thallium halide
Lithium	Li | LiClO$_4$ | Pt
	Li | LiClO$_4$ | NiS$_n$(C)
	Li | LiCl | TlCl | Tl(Hg)
	Li | LiBr | TlBr | Tl(Hg)
	Li | LiAlCl$_4$ | LiAlCl$_4$ | Li
	Li | LiAlCl$_4$ | AgCl | Ag
Nickel–sulfide	Also see Lithium
Silver	Ag | AgClO$_4$, NEt$_4$ClO$_4$ $\begin{vmatrix} Ag^+Cl^- \\ NEt_4ClO_4 \end{vmatrix}$ Ag (0.004M) (0.1M) (0.1M)
Silver–silver nitrate	Ag | AgNO$_3$, MX || AgNO$_3$ | Ag
Thallium	(Hg)Tl | TlCl, LiCl | Tl(Hg) sat. 10^{-3}m

II. EMF DATA

(*Continued*)

Conc. range	E^0	Comments	Ref.
1% Pb		Polarization; solubility of MCl_n	109
0.125m	−0.085	Solvate formation	171
Sat.	−0.260	Electrode reversibility studies	171
	−0.1377	25°; solvate formation	171
	−0.1413	35°	
1M	$E = -3.0$	28°C; used Li powder, foil, ribbon, wire; bias potentials	83
1M	$E = 1-3v$	E obtained from charging curves	84
0.0015–0.032m LiCl	1.84641	15°; $\gamma_{\pm}$; $\Delta G°$; $\Delta H°$	168
Tl ~ 3–7%	1.83480	25°; $\Delta S°$	
	1.82316	35°	
	1.81143	45°	
0.0025–0.65m LiBr	1.83958	15°	168
Tl ~ 3–7%	1.82874	25°	
	1.81807	35°	
	1.80734	45°	
0.1–1M		Also electrode kinetics	169
	~2.8	−55° to 125°; battery tests	170
	0.0687	Titration in RH compartment with Ag^+(0.1M) and Cl^-(0.1M); complex equilibrium	85
~10^{-2}M		K_{sp} complexes; solvent effects	22
10^{-3}–0.13m Amalgam 0.8–25%	0	30°C; bias potentials	86

PROPYLENE CARBONATE

Electrode	Cell
Thallium	(Hg)Tl | TlBr, LiBr | Tl(Hg) sat. (0.13m)
	(Hg)Tl | TlI, KI | Tl(Hg) sat. (0.1m)
	(Hg)Tl | TlCl, Et$_4$NCl, TlCl | Tl(Hg) sat. (1M) sat.
	(Hg)Tl | TlCl | LiCl | Hg$_2$Cl$_2$ | Hg
	(Hg)Tl | TlBr | LiBr | PbBr$_2$ | Pb(Hg)
	(Hg)Tl | TlBr | LiBr | CdBr$_2$ | Cd(Hg)
	(Hg)Tl | TlI | LiI | PbI$_2$ | Pb(Hg)
	(Hg)Tl | TlI$_2$ | LiI | CdI$_2$ | Cd(Hg)
	(Hg)Tl | TlBr | LiBr | Glass
	(Hg)Tl | TlBr | LiBr | LiBr | TlBr | Tl(Hg)
	(Hg)Tl | TlBr | LiBr | Glass | LiBr | Tl(Hg)
	(Hg)Tl | TlCl | LiCl | PbCl$_2$ | Pb(Hg)
	(Hg)Tl | TlCl | LiCl | CdCl$_2$ | Cd(Hg)

PROPYLENE GLYCOL

Electrode	Cell
Hydrogen	H$_2$, Pt | HCl, AgCl | Ag
	H$_2$, Pt | HBr, AgBr | Ag
	H$_2$, Pt $\begin{vmatrix} \text{NaHa} \\ \text{Na}_2\text{A} \end{vmatrix}$ LiBr | AgBr | Ag (H$_2$A = phthalic acid)
Silver–silver bromide	Also see Hydrogen
Silver–silver chloride	Also see Hydrogen

II. EMF DATA

(*Continued*)

Conc. range	E^0	Comments	Ref.
10^{-3}–0.13m Amalgam 0.8–25%	0	30°C; bias potentials	86
10^{-3}–0.13m Amalgam 0.8–25%			86
Amalgam 2 and 20%		Polarization, solubility of MCl_n	109
Sat.	0.905		171
0.038–0.073m	0.450	Solvate formation	171
0.038m	0.540	Solvate formation	171
0.207m	0.4906 0.4861	25°; ΔS of cell reaction 35°	171
0.207m	0.6284	25°	171
0.003–0.02m			
0.005–0.15m		t_+	171
0.003–0.15m		γ	171
Sat.	0.310	Electrode reversibility studies;	171
Sat.	0.560		171

HOCHCH$_2$OH
   CH$_3$

Conc. range	E^0	Comments	Ref.
4×10^{-3}–10^{-1}m	−0.029	30°; assumes complete dissociation	
4×10^{-3}–10^{-1}m	−0.152	30°	
$\mu = 0.02$–0.1m	−0.152	30°C; K of phthalic acid	133
			133
			132

PYRIDINE

Electrode	Cell
Aluminum	Ag $\left\vert \begin{array}{c} AgNO_3 \\ AlBr_3 \end{array} \right\vert$ AlBr$_3$ | Al(Hg) (C)
Barium	Ba(Hg) | BaI$_2$ | Ba(Hg)
Cadmium	Cd | CdI$_2$ ┊ AgNO$_3$ | Ag (10^{-3}M) (0.25M)
Calcium	Ca(Hg) | CaI$_2$ | Ca Ca(Hg) | CaI$_2$ | Ca(Hg) (0.11%) (0.0546%)
Copper	Cu | CuCl ‖ AgNO$_3$ | Ag Pt $\left\vert \begin{array}{c} CuCl \\ CuCl_2 \end{array} \right\vert\!\!\vert$ AgNO$_3$ | Ag Cu | CuCl ┊ AgNO$_3$ | Ag (0.14M) (0.1M)
Hydrogen	H$_2$ | HNO$_3$, AgNO$_3$ | Ag (739mm) (0.048M)

Lead	Pb | Pb(NO$_3$)$_2$ ┊ AgNO$_3$ | Ag
	    (0.1M) (0.1M)
	Pb | Pb(NO$_3$)$_2$ ┊ Pb(NO$_3$)$_2$ | Pb
	  (0.1M)py (0.1M)H$_2$O

Lithium	Li | LiNO$_3$ ┊ AgNO$_3$ | Ag (0.1M) (0.1M) Li(Hg) | LiCl | Li(Hg)
Mercury	Hg | HgI$_2$ ┊ AgNO$_3$ | Ag (0.25M) (0.5M) (0.25M) (0.25M) (0.1M) (0.25M)

II. EMF DATA

C_5H_5N

Conc. range	E^0	Comments	Ref.
0.100–3.6% amalgams		100°C; $E \sim 0.375$–0.400	87
0.05–0.1% amalgam		19°C; No E given; py. not highly purified	88
	$E = 0.72$	Temp. not given; unstable; no correlation of E and C	35
0.1–1N	0.89	17.5°C	89
Sat.		20°C; No $E°$ given; py. not highly purified	88
0.0064–0.05m	−0.538	18°C; comparison of py. and water systems	90
	−0.477	95°C; comparison of py. and water systems	
	−0.025	18°C; comparison of py. and water systems	90
	0.026	95°C; comparison of py. and water systems	
	$E = -0.592$	Temp. not given; no attempt to correlate E and C	35
	0.534	18°C	90
	$E = 0.38$	Temp. not given; no attempt to correlate E and C	35
	$E = 0.087$	Temp. not given; no attempt to correlate E and C	35
	$E \sim 3.185$	Temp. not given; no attempt to correlate E and C	35
0.4–2% amalgam		20°C; py. not highly purified	88
	$E = 0.004$ $=0.003$ $=0.023$	Temp. not given; no attempt to correlate E and C	35

PYRIDINE

Electrode	Cell
Potassium	K(Hg) | KI | K(Hg)
Silver	H_2 | HNO_3, $AgNO_3$ | Ag (739mm) (0.048M)
	Cu | CuCl || $AgNO_3$ | Ag (0.048m)
	Ag $\begin{vmatrix} AgNO_3 \\ AlBr_3 \end{vmatrix}$ $AlBr_3$ | Al(Hg) (C)
	Ag | $AgNO_3$ ⋮ $AgNO_3$ | Ag (0.25M) (0.025M)
	(0.025M) (0.0025M) (0.0025M) (0.00125M) (0.00125M) (0.000625M)
	Pb | $Pb(NO_3)_2$ ⋮ $AgNO_3$ | Ag (0.1M) (0.1M)
	Cu | CuCl ⋮ $AgNO_3$ | Ag (0.14M) (0.1M)
	Hg | HgI_2 ⋮ $AgNO_3$ | Ag (0.25M) (0.5M)
	Cd | CdI_2 ⋮ $AgNO_3$ | Ag (10^{-3}M) (0.25M)
	Zn | $ZnCl_2$ ⋮ $AgNO_3$ | Ag (0.125M) (0.1M)
	Li | $LiNO_3$ ⋮ $AgNO_3$ | Ag (0.1M) (0.1M)
	Ag | $AgNO_3$ | $AgNO_3$ | Ag aq. (py)
Sodium	Na(Hg) | NaI | Na(Hg)
	Na(Hg) | NaI | Na(Hg) (C_1) (0.1N) (C_2)
Zinc	Zn | $ZnCl_2$ ⋮ $AgNO_3$ | Ag (0.125M) (0.1M)
	Zn(Hg) | $ZnCl_2$ | Zn(Hg) (1.138%) (0.0977%)

II. EMF DATA

C_5H_5N (Continued)

Conc. range	E^0	Comments	Ref.
0.04–0.2% amalgam		20°C; py. not highly purified	88
	0.534	18°C	90
0.0064–0.05	−0.538	18°C; compares py. and water systems	90
	−0.477	95°C	
0.1–3.6% amalgam	$E \sim 0.375$	100°C	87
	$E = 0.039$	Temp. not given; no attempt to correlate E and C	35
	$= 0.039$		
	$= 0.014$		
	$= 0.017$		
	$E = 0.38$	Temp. not given; no attempt to correlate E and C	35
	$E = -0.592$	Temp. not given; no attempt to correlate E and C	35
	$E = 0.004$	Temp. not given; no attempt to correlate E and C	35
	$E = 0.72$	Temp. not given; no attempt to correlate E and C	35
	$E = \sim 0.83$	Temp. not given; no attempt to correlate E and C	35
	$E \sim 3.185$	Temp. not given; no attempt to correlate E and C	35
10^{-2}–2×10^{-3} N		λ solutions measured $E = A \log(\lambda_w/\lambda_s) + B$ $B = -0.352$	38
		4–22°C; py. not highly purified	88
8×10^{-3}–0.63% Na		20°C; λ_{Na} in Hg	19
	$E = 0.83$	Temp. not given; no attempt to correlate E and C	35
50 g/l		18°C; py. not highly purified	88

SULFOLANE

Electrode	Cell
Iodine	Pt, I_2	I^- Et_4NPi	I^-	I_2, Pt
Silver–silver nitrate	Ag	$AgNO_3$, MX ‖ $AgNO_3$	Ag

SULFUR

Electrode	Cell
Cadmium	Cd	CdI_2 ⁝ HgCl	Hg
Calomel	Pb	$PbCl_2$	Solns	HgCl	Hg
	Pb	$PbCl_2$	Et_2NH_2Cl	HgCl	Hg
	Ag	AgCl	Et_2NH_2Cl	HgCl	Hg
	H_2, Pt	Et_2NH_2Cl	HgCl	Hg
	H_2, Pt	KCl	HgCl	Hg
	Pb	$PbCl_2$	HgCl	Hg sat. sat.
	Zn	$ZnBr_2$	HgCl	Hg sat. sat.
	Cd	CdI_2	HgCl	Hg sat. sat.
Hydrogen	H_2, Pt	Et_2NH_2Cl	HgCl	Hg
	H_2, Pt	KCl	AgCl	Ag
	H_2, Pt	KCl	HgCl	Hg
	H_2, Pt	KBr	AgBr	Ag
Iron	Na	$KI \cdot IBr_3$	Fe

II. EMF DATA

Conc. range	E^0	Comments	Ref.
$\sim 10^{-2}$M		K_{sp} complexes; solvent effects	22
$\sim 10^{-2}$M		K_{sp} complexes; solvent effects	22

DIOXIDE SO₂

Conc. range	E^0	Comments	Ref.
	0.425	$-35°$C; incomplete attempt at E^0 of metal electrodes	91
	$E = 0.433$	$-35°$C; ΔH	172
0.1m	$E = 0.361$	$-38°$C; ΔG	172
0.1m	$E = 0.016$	$-35°$C; cf theory $E = 0.06$	172
0.1–1m		$-25°$C; H⁺ provided by HCl gas	172
0.00475m		$-25°$C	
	0.435	$-35°$C; incomplete attempt at E^0 of metal electrodes	91
	0.39	$-35°$C; incomplete attempt at E^0 of metal electrodes	91
	0.425	$-35°$C; incomplete attempt at E^0 of metal electrodes	91
0.1–1m		$-25°$C; H⁺ provided by HCl gas	172
0.00475m		$-25°$C;	172
0.00475m		$-25°$C	172
0.03m		$-25°$C	172
5gmKI/100 cc 10 cc IBr₃	$E = 3.0+$	15°C; very rough values, battery techniques	92
	$= 3.0+$	$-21°$C;	
	$= 3.0$	$-38°$C;	
	$= 2.75$	$-60°$C	

SULFUR

Electrode	Cell
Iron	Mg | KI·IBr$_3$ | Fe
Lead–lead chloride	Pb | PbCl$_2$ | (Solns) | HgCl | Hg
	Pb | PbCl$_2$ | Et$_2$NH$_2$Cl | HgCl | Hg
	Pb | PbCl$_2$ | HgCl | Hg
Magnesium	Mg | KI·IBr$_3$ | Fe
Mercury–mercurous bromide	O$_2$, Pt | O$_2^{2-}$, HgBr | Hg sat.
	Ag | AgBr | KBr | HgBr | Hg
Oxygen	O$_2$, Pt | O$_2^{2-}$, HgBr | Hg sat.
	Ag | AgBr | KBr | Pt, O$_2$
Silver–silver bromide	Ag | AgBr | KBr | HgBr | Hg
	H$_2$, Pt | KBr | AgBr | Ag
	Ag | AgBr | KBr | Pt, O$_2$
Silver–silver chloride	Ag | AgCl | Et$_2$NH$_2$Cl | HgCl | Hg
	H$_2$, Pt | KCl | AgCl | Ag
Sodium	Na | KI·IBr$_3$ | Fe
Zinc–zinc bromide	Zn | ZnBr$_2$ | HgCl | Hg sat. sat.

II. EMF DATA

DIOXIDE SO$_2$ (*Continued*)

Conc. range	E^0	Comments	Ref.
	$E = 1.10$	21°C; very rough values, battery techniques	92
	$= 1.00$	−31°C;	
	$= 0.98$	−43°C;	
	$= 0.92$	−60°C	
	$E = 0.433$	−35°C; ΔH	172
0.1m	$E = 0.361$	−38°C; ΔG	172
	0.435	−35°C; incomplete attempt at metal electrodes	91
2gKI 10 cc IBr$_3$	$E = 1.10$	21°C; very rough values, battery techniques	92
	$= 1.00$	−31°C;	
	$= 0.98$	−43°C;	
	$= 0.92$	−60°C	
	$\simeq 0.2$	80°C; [O^{-2}] not apparently measured; no E^0 in the true sense	173
0.5m	$E = 0.060$	−32°C; cf theory 0.05	172
	$\simeq 0.2$	80°C; see entry for Mercury–mercurous bromide	173
0.135M		−39°C	172
0.5m	$E = 0.060$	−32°C	172
0.03m		−25°C	172
0.135M		−39°C	172
0.1m	$E = 0.016$	−35°C	172
0.00475m		−25°C; H$^+$ provided by HCl gas	172
5 g KI/100 cc 10 cc IBr$_3$	$E = 3.0+$	15°C; very rough values, battery techniques	92
	$= 3.0+$	−21°C;	
	$= 3.0$	−38°C	
	$= 2.75$	−60°C	
	0.39	−35°C; incomplete attempt at E^0 of metal electrodes	91

XYLENE

Electrode	Cell
Copper	Cu | — | Cu$^{(II)}$(oleate)$_2$ | Cu
	Cu | — | Cu$^{(II)}$(oleate)$_2$ +2.5% ϕOH | Cu
	Cu | Cu$^{(II)}$(oleate)$_2$ | Cu$^{(II)}$(oleate)$_2$ | Cu (0.01M) (0.001M)
Glass	Glass | HA ⦙ HA | Glass (C$_1$) (C$_2$)

References

1. A. Heyrovsky and O. Tomicek, Coll. Czech. Chem. Comm., **15**, 984 (1950).
2. J. B. Conant and N. F. Hall, J. Amer. Chem. Soc., **49**, 3047 (1927).
3. N. F. Hall and T. H. Werner, J. Amer. Chem. Soc., **50**, 2367 (1928).
4. O. Tomicek, Coll. Czech. Chem. Comm., **13**, 116 (1948).
5. S. Bruckenstein and I. M. Kolthoff, J. Amer. Chem. Soc., **78**, 2974 (1956).
6. N. Isgarischew and S. A. Pletenew; Z. Electrochem., **36**, 457 (1950).
7. B. F. Chow and J. B. Conant, J. Amer. Chem. Soc., **55**, 3745 (1933).
8. B. F. Chow and J. B. Conant, J. Amer. Chem. Soc., **55**, 3752 (1933).
9. J. B. Conant and T. H. Werner, J. Amer. Chem. Soc., **52**, 4436 (1930).
10. N. F. Hall and B. O. Heston, J. Amer. Chem. Soc., **55**, 4729 (1933).
11. N. F. Hall and B. O. Heston, J. Amer. Chem. Soc., **56**, 1462 (1934).
12. A. E. Cameron and J. Russell, J. Amer. Chem. Soc., **58**, 774 (1936).
13. G. Tarbutton and W. C. Vosburgh, J. Amer. Chem. Soc., **55**, 618 (1933).
14. G. C. Chandlee and A. W. Hutchison, J. Amer. Chem. Soc., **53**, 2881 (1931).
15. L. M. Mukherjee, J. Amer. Chem. Soc., **79**, 4040 (1957).
16. L. Ya. Shapovalova and A. M. Shkodin, Electrokhimiya., **6**, 1757 (1970).
17. F. Aufauvre, D.Sc. Thesis, Univ. Clermont—Ferrand; (1969).
18. W. Ward, Ph.D. Thesis, Univ. of Iowa (1958).
19. H. Ulich and G. Spiegel, Z. Phys. Chem., **177**, 103 (1936).
20. I. M. Kolthoff and F. G. Thomas, J. Phys. Chem., **69**, 3049 (1965).
21. K. Cruse, E. P. Goertz and H. Petermoller, Z. Electrochem., **55**, 405 (1951).
22. R. Alexander, Ph.D. Thesis, Univ. of Western Australia (1968).
23. J. J. Campion, Ph.D. Thesis; Univ. of Pittsburgh (1966).
24. H. Strzelecki and W. Libus, Electrochim. Acta., **17**, 577 (1972).
25. F. K. V. Koch, J. Chem. Soc., 524 (1928).
26. G. R. Padmanabhan, Ph.D. Thesis; Univ. of Pittsburgh, (1964).
27. C. S. Gardner, E. W. Green and D. M. Yost, J. Amer. Chem., Soc., **57**, 2055 (1935).
28. D. M. Sowards and G. W. Watt, J. Electrochem. Soc., **102**, 545 (1955).
29. F. E. Rosztoczy and C. W. Tobias, Electrochim. Acta., **11**, 857 (1966).
30. S. Neumayr and E. Zintl, Bull. Soc. Chim. France **63**, 237 (1930).
31. N. Eeliot and D. M. Yost, J. Amer. Chem. Soc., **56**, 1057 (1934).
32. A. M. Monosson and V. A. Pleskov, Acta. Physicochim. U.S.S.R., **2**, 621 (1935).

II. EMF DATA

Conc. range	E^0	Comments	Ref.
0.01M conc. in solvent side undefined	$E = -0.03$	No temp. given; intersolvent potentials	50
	$E = -0.09$	No temp. given; intersolvent potentials	50
	$E = 0.099$	No temp. given; intersolvent potentials	50
		No temp. given; K_a; t_H^+; Λ	73

33. V. Pleskov, Acta Physicochim. U.S.S.R., **20**, 578 (1945).
34. G. Jander and G. Winkler, J. Inorg. Nucl. Chem., **9**, 32 (1959).
35. J. Neustadt and R. Abegg, Z. Phys. Chem., **69**, 486 (1909).
36. F. Aufauvre and M. L. Dondon, Memories Presentes a la Sov. Chim., 716 (1963).
37. D. H. Everett and S. E. Rasmussen, J. Chem. Soc. 2812 (1954).
38. F. K. V. Koch, J. Chem. Soc., 269 (1928).
39. B. Case, N. S. Hush, R. Parsons and M. E. Peover, J. Electroanal. Chem., **10**, 360 (1965).
40. D. A. Zatko, Ph.D. Thesis, University of Wisconsin, 1967.
41. B. Kratochvil, E. Lorah and C. Garber, Anal. Chem., **41**, 1793 (1969).
42. J. Desbarres, Bull. Soc. Chim. Fr., 2103 (1962).
43. I. M. Kolthoff and M. K. Chantooni, J. Amer. Chem. Soc., **87**, 4428 (1965).
44. J. Badoz-Lambling, J. Desbarres and J. Tacussel, Memoires Presentes a la Soc. Chim., **53** (1962).
45. E. Romberg and K. Cruse, Z. Elektrochem., **63**, 404 (1959).
46. C. Papon and J. Jacq. Bull. Soc. Chim. Fr., **13**, (1965).
47. R. Alexander, E. C. F. Ko, Y. C. Mac and A. J. Parker, J. Amer. Soc., **89**, 3703 (1967).
48. J. P. Billon, J. Electroanal. Chem., **1**, 486 (1960).
49. F. K. V. Koch, J. Chem. Soc., 524 (1928).
50. A. Gemant, Trans. Electrochim. Soc., **78**, 49 (1940).
51. K. W. Boyer and R. T. Iwamoto, J. Electroanal. Chem., **7**, 458 (1964).
52. G. A. Scherer and R. F. Newton, J. Amer. Chem. Soc., **56**, 18 (1934).
53. H. E. Bent and E. S. Gilfillan, J. Amer. Chem. Soc., **55**, 247 (1933).
54. U. Berglund and L. G. Sillen, Acta Chem. Scand., **2**, 116 (1948).
55. A. J. Parker, J. Chem. Soc. (A), 220 (1966).
56. L. W. Marple, Anal. Chem., **39**, 844 (1967).
57. M. Teze and R. Schaal, Bull. Soc. Chim. Fr., 1372 (1962).
58. J. N. Butler and J. C. Synnott, J. Amer. Chem. Soc.; in press, (July 1969).
59. J. N. Butler, J. Phys. Chem., **72**, 3288 (1968).
60. W. H. Smyrl, Ph.D. Thesis, University of California, Berkeley, 1966.
61. N. A. Rumbaut and H. L. Peeters, Bull. Soc. Chim; Belges., **76**, 33 (1967).
62. I. M. Kolthoff and T. B. Reddy, Inorg. Chem., **1**, 189 (1962).

63. C. D. Ritchie and R. E. Uschold, J. Amer. Chem. Soc., **89,** 1721 (1967).
64. J. Courtot-Coupez and M. Le Demezet, Compt. Rend., **266C,** 1438 (1968).
65. D. R. Cogley and J. N. Butler, J. Electrochem. Soc., **113,** 1074 (1966).
66. D. R. Cogley and J. N. Butler, J. Phys. Chem., **72,** 1017 (1968).
67. W. H. Smyrl and C. W. Tobias, J. Electrochem. Soc., **115,** 33 (1968).
68. G. Holleck, D. R. Cogley and J. N. Butler, J. Electrochem. Soc., **116,** 952 (1969).
69. R. Huston and J. N. Butler, J. Phys. Chem., **72,** 4263 (1968).
70. D. R. Cogley and J. N. Butler, J. Phys. Chem., **72,** 4568 (1968).
71. M. Salomon, J. Electrochem. Soc., **116,** 1392 (1969).
72. W. H. Smyrl and C. W. Tobias, J. Electrochem. Soc., **113,** 754 (1966).
73. A. Gemant, J. Chem. Phys., **12,** 79 (1944).
74. T. Pavlopoulos and H. Strehlow, Z. Phys. Chem; N.F., **2,** 89 (1954).
75. M. Mandel and P. Decroly, Nature., **182,** 794 (1958).
76. M. Mandel and P. Decroly, Trans. Faraday Soc., 29 (1960).
77. R. K. Agarwal and B. Nayak, J. Phys. Chem., **71,** 2062 (1967).
78. K. W. Morcom and B. L. Muju, Nature, **217,** 1046 (1968).
79. F. H. Verhoek, J. Amer. Chem. Soc., **58,** 2577 (1936).
80. L. R. Dawson, R. C. Sheridan and H. C. Eckstrom., J. Phys. Chem., **65,** 1829 (1961).
81. L. R. Dawson, W. H. Zuber and H. C. Eckstrom, J. Phys. Chem., **69,** 1335 (1965).
82. Yu. M. Povarov, A. N. Gorbaiev, Yu. M. Kessier and I. V. Safonova, Dokl. Akad. Nauk. S.S.S.R., **142,** 1128 (1962).
83. B. Burrows and R. Jasinski, J. Electrochem. Soc., **115,** 365 (1968).
84. R. Jasinski and B. Burrows, J. Electrochem. Soc., **116,** 422 (1969).
85. J. N. Butler, Anal. Chem., **39,** 1799 (1967).
86. F. G. K. Baucke and C. W. Tobias, J. Electrochem. Soc., **116,** 34 (1969).
87. R. Muller, Z. Elektrochem., **35,** 240 (1929).
88. H. P. Cady, J. Phys. Chem., **2,** 551 (1898).
89. M. Tamele, J. Phys. Chem., **28,** 502 (1924).
90. A. K. Gupta, J. Chem. Soc., 3473 (1952).
91. L; S. Bagster and B. D. Steele, Chem. News, 171 (1912).
92. E. Schaschl and H. J. McDonald, J. Electrochem. Soc., **94,** 299 (1948).
93. J. Baldwin, J. B. Gill and J. Evans, J. Chem. Soc., **A,** 3389 (1971).
94. M. J. Bergin and A. H. A. Heyn, J. Amer. Chem. Soc., **75,** 5120 (1953).
95. V. Pleskov; Acta Physiochim. U.S.S.R., **21,** 235 (1946).
96. A. M. Monosson and V. A. Pleskov, Acta Physicochim. U.S.S.R., **2,** 615 (1935).
97. J. B. Russell and M. J. Sienko, J. Amer. Chem. Soc., **79,** 4051 (1957).
98. I. Devries and J. Sedlet, J. Amer. Chem. Soc., **73,** 5808 (1951).
99. M. J. Bergin and A. H. A. Heyn, J. Amer. Chem. Soc., **76,** 4765 (1954).
100. C. A. Kraus, J. Amer. Chem. Soc., **36,** 864 (1914).
101. V. A. Podolyanko and A. M. Shkodin, Izv. Vyssh. Ucheb. Zaved. Khim. Tekhnol., **12,** 244 (1969).
102. M. S. Toy and W. A. Cannon, Electrochem. Technol., **4,** 520 (1966).
103. V. V. Aleksandrov and E. F. Ivanova, Russ. J. Phys. Chem; (Zhur. Fiz. Khim.) **38,** 476 (1964).
104. E. M. Ryzhkov and A. M. Sukhotin, Russ. J. Phys. Chem; (Zhur. Fiz. Khim.), **34,** 672 (1960).
105. K. Cruse and E. Romberg, Z. Elektrochem., **63,** 404 (1959).
106. M. Tamele, J. Phys. Chem., **28,** 502 (1924).
107. A. Kisza, Z. Phys. Chem., **237,** 97 (1966).
108. E. R. Smith and J. K. Taylor, J. Res. Nat. Bur. Stand; Sect. A, **25,** 731 (1940).
109. J. C. Synnott and J. N. Butler, Anal. Chem., **41,** 1890 (1969).

II. EMF DATA

110. K. P. Mischenko and V. V. Sokolov, J. Struct. Chem., **5,** 760 (1964).
111. G. H. Mergele and C. D. Ritchie, J. Amer. Chem. Soc., **89,** 1447 (1967).
112. J. Julliard, Bull. Soc. chim. France., **63,** 1190 (1966).
113. J. Julliard, D.S.C. Thesis; Univ. of Clermont-Ferrnad (1968).
114. R. M. Reeves, Ph.D. Thesis, Univ. of Southampton (1969).
115. M. Salomon, J. Electrochem. Soc., **117,** 325 (1970).
116. E. S. Amis and J. D. Hefley, J. Phys. Chem., **69,** 2802 (1965).
117. R. Abegg and J. Neustadt, Z. Phys. Chem., **69,** 486 (1909).
118. H. P. Cady, J. Phys. Chem., **2,** 551 (1898).
119. M. Kunz and K. Schwabe, Z. Elektrochem., **64,** 1188 (1960).
120. A. Ferse, K. Schwabe and R. Urlass, Ber. Bunsengesellschaft Phys. Chem., **68,** 46 (1964).
121. E. Grunwald and E. F. Sieckmann, J. Amer. Chem. Soc., **76,** 3855 (1954).
122. L. M. Mukherjee, J. Phys. Chem., **60,** 974 (1956).
123. L. M. Mukherjee, J. Phys. Chem., **58,** 1042 (1954).
124. G. J. Janz and H. Tanigucki, J. Phys. Chem., **61,** 688 (1957).
125. M. H. Fleysher and H. S. Harned, J. Amer. Chem. Soc., **47,** 82 (1925).
126. R. Abegg and J. Neustadt, Z. Phys. Chem., **69,** 486 (1909).
127. J. A. V. Butler and R. T. Hamilton, Proc. Roy. Soc., **138A,** 450 (1932).
128. H. B. Hart and J. N. Pearce, J. Amer. Chem. Soc., **42,** 2411 (1922).
129. G. F. Isaacs and J. R. Partington, Trans. Faraday Soc., **25,** 53 (1929).
130. F. G. Keys and G. N. Lewis, J. Amer. Chem. Soc., **34,** 119 (1912).
131. C. A. Kraus and G. N. Lewis, J. Amer. Chem. Soc., **32,** 1459 (1910).
132. M. N. Das and K. K. Kundu, J. Chem. and Eng. Data, **9,** 87 (1964).
133. Chemistry and Industry., **30,** 1336 (1965).
134. R. Bryant, A. B. Garrett and G. F. Kiefer, J. Amer. Chem. Soc., **65,** 1905 (1943).
135. R. Bryant, A. B. Garrett, G. F. Kiefer and M. V. Noble, J. Amer. Chem. Soc., **65,** 293 (1943).
136. O. D. Black and A. B. Garrett, J. Amer. Chem. Soc., **65,** 862 (1943).
137. B. L. Muju, K. W. Morcom and R. W. C. Broadbank, Trans. Faraday Soc., **64,** 3318 (1968).
138. L. W. Bahe, G. A. Strack and S. K. Swanda, J. Chem. and Eng. Data, **9,** 416 (1964).
139. S. Dhabanandana, R. W. C. Broadbank and K. W. Morcom, Trans. Faraday Soc., **64,** 3311 (1968).
140. D. K. Sahu, Electrochim. Acta, **16,** 757 (1971).
141. N. P. Dzyuva, N. A. Izmailov and A. M. Shkodin, Zhur. Obschei Khim; (J. Gen. Chem. U.S.S.R.) **23,** 27 (1953).
142. T. A. Pinfold and F. Sebba, J. Amer. Chem. Soc., **78,** 2095 (1956).
143. N. Dietz and L. P. Hammett, J. Amer. Chem. Soc., **52,** 4792 (1930).
144. V. Pleskov, Acta Physicochim U.S.S.R., **21,** 41 (1946).
145. K. Biastoch and H. Ulich, Z. Phys. Chem. **A178,** 306 (1937).
146. J. Devries and G. G. Koerber, J. Amer. Chem. Soc., **74,** 5008 (1952).
147. A. F. Clifford and E. Zamona, Trans. Faraday Soc., **57,** 1963 (1961).
148. B. Burrows and R. J. Jasinski, J. Electrochem. Soc., **115,** 348 (1968).
149. A. F. Clifford, W. D. Pardieck and M. W. Wadley, J. Phys. Chem., **70,** 3241 (1966).
150. TS. S. Adzhemyan, L. M. Grubina and G. I. Kaurova, Elektrokhimiya, **3,** 1222 (1967).
151. M. W. Wadley, Ph.D. Thesis, Purdue Univ., (1963).
152. V. V. Aleksandrov and N. Van-Mao, Elektrokhimiya, **1,** 297 (1965).
153. A. J. Parker, J. Chem. Soc., **A,** 220 (1966).
154. E. S. Amis and J. D. Hefley, J. Electrochem. Soc., **112,** 337 (1965).
155. J. Jastrzebska and S. Minc, Roczniki. Chem., **42,** 719 (1968).

156. A. E. Brodsky, Z. Phys. Chem., **121,** 1 (1926).
157. E. Ferse and K. Schwabe, Ber. Bunsengesellscbalt. Phys. Chem., **70,** 849 (1966).
158. M. Alfenaar, C. L. DeLigny and A. G. Remijnse, Rec. Trav. Chim., **86,** 555 (1967).
159. H. S. Harned and H. C. Thomas, J. Amer. Chem. Soc., **58,** 761 (1936).
160. I. T. Diwa; J. Phys. Chem.; **60,** 754 (1956).
161. K. Braver and H. Strehlow, Z. Phys. Chem., **17,** 346 (1958).
162. G. Spiegel and H. Ulich, Z. Phys. Chem., **177,** 103 (1936).
163. C. A. Kraus, J. Amer. Chem. Soc., **36,** 864 (1914).
164. R. Schaal and A. Teze, Compt. Rend., 252 4020 (1961).
165. S. S. Haque and F. Hussain, Pakistan J. Sci. Ind. Res., **6,** 251 (1963).
166. R. N. Roy, A. L. M. Bothwell, J. J. Gibbons and W. Vernon, J. Chem. Thermodynamics, **3,** 883 (1971).
167. F. G. Keys and G. N. Lewis, J. Amer. Chem. Soc., **35,** 340 (1913).
168. M. Salomon, J. Phys. Chem., **73,** 3219 (1969).
169. S. G. Meibuhr, J. Electrochem. Soc., **118,** 1320 (1971).
170. G. M. Cook, J. Electrochem. Soc., **117,** 559 (1970).
171. R. C. Murray, Ph.D. Thesis, Rensselaer Polytechnic Institute, Troy, N. Y. (1971).
172. K. Cruse, Z. Electrochem., **46,** 571 (1940).
173. K. Wickert, Z. Elektrochem., **44,** 410 (1938).

(b) Equilibrium Measurements in Nonaqueous-Aqueous Mixed Solvents

The information in this section surveys the various reference electrodes and cell configurations that have been used for equilibrium-type measurements in nonaqueous-aqueous mixed solvent systems. The tables are arranged alphabetically according to the nonaqueous component. For each solvent table the reference electrodes are listed alphabetically. Information relative to the composition range of the solvent mixture, the concentration range of the electrolyte and the values of E^0 are given when quoted. Additional comments indicating the purpose of the study, extrapolation procedures used, stability of electrodes and other pertinent information are provided. The availability of actively coefficient data is indicated. References are provided for each cell studied and a complete bibliography is given at the end of this section.

ACETIC ACID–

Electrode	Cell
Hydrogen and Silver–silver halide	Pt, H_2 | HCl | AgCl | Ag
	Pt, H_2 | HCl | AgCl | Ag
	Pt, H_2 | HBr | AgBr | Ag
	Pt, H_2 | HBr | AgBr | Ag
	Pt, H_2 | HI | AgI | Ag
	Pt, H_2 | HI | AgI | Ag

II. EMF DATA

WATER MIXTURES

Conc. range	% HOAc	E^0	Comments	Ref.
0.005–0.16m	10	0.2105	ΔG transfer	1
	20	0.1968		
	40	0.1621		
	60	0.1115		
0.017M	10	−0.43	Acidity H_0; graphical presentation	2
	30	−0.41		
	50	−0.39		
	70	−0.32		
	80	−0.28		
	90	−0.2		
	95	−0.08		
0.005–0.15m	10	0.0606	ΔG transfer	1
	20	0.0474		
	40	0.0140		
	60	0.0385		
0.01M	10	−0.30	Acidity H_0; graphical presentation	2
	30	−0.27		
	50	−0.24		
	70	−0.17		
	80	−0.12		
	95	−0.02		
0.005–0.16m	10	−0.1609	ΔG transfers	1
	20	−0.1719		
	40	−0.2019		
	60	−0.2464		
0.01M	10	−0.08	Acidity H_0; graphical presentation	2
	30	−0.05		
	50	−0.03		
	70	+0.04		
	80	+0.11		
	90	+0.22		
	95	+0.26		

ACETONE–WATER

Electrode	Cell
Calomel	Also see Glass and Hydrogen
Glass	Glass | HCl | AgCl | Ag
	Glass | HOAc, KOAc, KCl | AgCl | Ag
	Glass $\begin{vmatrix} \phi CO_2H \\ \phi CO_2K \\ KCl \end{vmatrix} \begin{vmatrix} HOAc \\ KOAc \\ KCl \end{vmatrix}$ Glass
	Glass | HCl | AgCl | Ag
	Glass | RϕNR$_2$, HClO$_4$ || KCl | Hg$_2$Cl$_2$ | Hg
	Glass | RϕNR$_2$, HClO$_4$ || KCl | AgCl | Ag
	Hg | Hg$_2$Cl$_2$, HCl | glass | HCl, Hg$_2$Cl$_2$ | Hg (m) (m$_1$)
Hydrogen	Pt, H$_2$ | HCl | Hg$_2$Cl$_2$ | Hg
	Pt, H$_2$ | HCl | AgCl$_2$ | Ag
Potassium	Also see Silver–silver chloride
Silver–silver chloride	Ag | AgCl | KCl | K(Hg) | KCl | AgCl | Ag (m) (0.04%) (m)
	Ag | AgCl | KCl || AgNO$_3$ | AgCl | Ag

II. EMF DATA

MIXTURES

Conc. range	% Me$_2$CO	E^0	Comments	Ref.
0.5–20 × 10^{-3}M	10, 20, 30, 40, 50, 60, 70, 80		ΔG transfer; K_a; K_{sp}	3
0.5–20 × 10^{-3}M	10, 20, 30, 40, 50, 60, 70, 80		ΔG transfer; K_a; K_{sp}	3
0.5–20 × 10^{-3}M	10, 20, 30, 40, 50, 60, 70, 80		ΔG transfer; K_a; K_{sp}	3
10^{-3}–10^{-1}m	0, 15.6, 20, 30, 60, 80, 90, 100		Base strength as f(solvent)	4
10^{-3}–10^{-1}m	0, 15.6, 20, 30, 60, 80, 90, 100		Base strength as f(solvent)	4
10^{-3}–10^{-1}m	0, 15.6, 20, 30, 60, 80, 90, 100		Base strength as f(solvent)	4
0.002–0.2m	0	0.2754	$\gamma_{\pm}$; medium effects	5
	40	0.31135		
	60	0.3491		
	80	0.4290		
0.002–0.2m	0	−0.26821	$\gamma_{\pm}$; pH medium effects	5
	20	−0.25340		
	40	−0.23170		
	60	−0.1939		
	80	−0.1164		
	95	−0.0125		
0.004–0.01m	5	0.2190	$\gamma_{\pm}$, pH, medium effects	6
	10	0.2156		
	20	0.2079		
	40	0.1859		
0.05–0.5m	20	−0.0386		7
	40	−0.0834		
	60	−0.1401		
	80	−0.2233		
0.5–20 × 10^{-3}M	10, 20, 30, 40, 50, 60, 70, 80		ΔG transfer; K_a; K_{sp}	3

t-BUTANOL–

Electrode	Cell
Glass and Silver–silver chloride	Glass | HCl | AgCl | Ag

DIMETHYLSULFOXIDE–

Electrode	Cell
Glass	Glass | HCl | AgCl | Ag
	Glass | HOAc, KOAc, KCl | AgCl | Ag
	Glass | ϕCO_2H, ϕCO_2K, KCl | HOAc, KOAc, KCl | Glass
Hydrogen	Pt, H_2 | HCl | AgCl | Ag
	Pt, H_2 | HBr | AgBr | Ag
	Pt, H_2 | HI | AgI | Ag
	Pt, H_2 | HI | AgI | Ag
	Pt, H_2 | HCl | AgCl | Ag
	Pt, H_2 | HBr | AgBr | Ag

II. EMF DATA

WATER MIXTURES

% t-BUOH	E⁰	Comments	Ref.
0	0.0159	ΔG transfer	8
10	0.0110		
20	0.0080		
30	0.0040		
40	0.0015		
50	0.0120		
60	0.0275		

WATER MIXTURES

Conc. range	% Me₂SO	E⁰	Comments	Ref.
0.5–20 × 10⁻³M	10, 20, 30, 40, 50, 60, 70, 80		ΔG transfer; K_a, K_{sp}	3
0.5–20 × 10⁻³M	10, 20, 30, 40, 50, 60, 70, 80		ΔG transfer; K_a, K_{sp}	3
0.5–20 × 10⁻³M	10, 20, 30, 40, 50, 60, 70, 80		ΔG transfer; K_a, K_{sp}	3
0.003–0.05m	20	0.21992	$\gamma_{\pm}$; pK_a	9
0.003–0.05m	40	0.21768	$\gamma_{\pm}$; pK_a	
0.003–0.05m	20	0.0765	$\gamma_{\pm}$; pK_a	
	40	0.0896	$\gamma_{\pm}$; pK_a	
0.003–0.05m	20	−0.1343	$\gamma_{\pm}$; pK_a	9
	40	−0.1059	$\gamma_{\pm}$; pK_a	
0.003–0.03m	10	0.2210	ΔG transfer of HX	10
	20	0.2199	ΔG transfer of HX	
	40	0.2177	ΔG transfer of HX	
	60	0.2117	ΔG transfer of HX	
	80	0.1771	ΔG transfer of HX	
0.003–0.03m	10	0.0736	ΔG transfer of HX	10
	20	0.0765	ΔG transfer of HX	
	40	0.0896	ΔG transfer of HX	
	60	0.1040	ΔG transfer of HX	
	70	0.1014	ΔG transfer of HX	
	80	0.0950	ΔG transfer of HX	

DIMETHYLSULFOXIDE-

Electrode	Cell
Hydrogen	Pt, H_2 | HI | AgI | Ag
Silver–silver bromide	Also see Hydrogen
Silver–silver chloride	Ag | AgCl | KCl || $AgNO_3$ | AgCl | Ag
	Also see Glass and Hydrogen
Silver–silver iodide	Also see Hydrogen

DIOXANE–WATER

Electrode	Cell
Amalgams	
Cesium	Also see Silver–silver chloride
Lithium	Also see Silver–silver chloride
Potassium	Also see Silver–silver chloride
Sodium	Also see Silver–silver chloride
Rubidium	Also see Silver–silver chloride
Calomel	Also see Glass and Hydrogen
Glass	Glass M^{++}, R_1NHCOR_2 | Calomel
	$M^{++} = 2H^+, Zn^{++}, Cu^{++}, Ni^{++}$
	$R_1 = 2H^+, Zn^{++}, Cu^{++}, Ni^{++}$
	$R_1 = \phi$ or Meϕ
	$R_2 = \phi$ or Pr
Hydrogen	Pt, H_2 | HCl | AgCl | Ag

II. EMF DATA

WATER MIXTURES (*Continued*)

Conc. range	% Me$_2$CO	E^0	Comments	Ref.
0.003–0.03m	10	−0.1436	ΔG transfer of HX	10
	20	−0.1343		
	40	−0.1059		
	60	−0.0653		
0.5–20 × 10^{-3}M	10, 20, 30, 40, 50, 60, 70, 80		ΔG transfer; K_a; K_{sp}	3

MIXTURES

Conc. range	% Dioxane	E^0	Comments	Ref.
0.001M in M^{++} (50 vol)			Chelate stability constants	19
0.001–0.01M	82	0.0025	5°; ΔH transfer	11
		−0.0078	10°	
		−0.0185	15°	
		−0.0295	20°	
		−0.0415	25°	
		−0.0536	30°	
		−0.0658	35°	
		−0.0787	40°	
		−0.0925	45°	

DIOXANE–WATER

Electrode	Cell
Hydrogen	Pt, H_2 | HCl | AgCl | Ag
	Pt, H_2 | HCl | AgCl | Ag
	Pt, H_2 | HCl | AgCl | Ag
	Pt, H_2 | HCl | AgCl | Ag

II. EMF DATA

MIXTURES (*Continued*)

Conc. range	% Dioxane	E^0	Comments	Ref.
0.001–0.01M	70	0.10584	0°; ΔC_p; ΔH	12
		0.09784	5°	
		0.08970	10°	
		0.08123	15°	
		0.07267	20°	
		0.06395	25°	
		0.05500	30°	
		0.04587	35°	
		0.03661	40°	
		0.02705	45°	
		0.01746	50°	
0.001–0.01	45	0.18938	0°; ΔC_p; ΔH	13
		0.18468	5°	
		0.17972	10°	
		0.17454	15°	
0.001–0.01	45	0.16916	20°; ΔC_p; ΔH	13
		0.16358	25°	
		0.15778	30°	
		0.15182	35°	
		0.14560	40°	
		0.13925	45°	
		0.13282	50°	
	20	0.21975	0°	13
		0.20674	20°	
		0.20303	25°	
		0.19080	40°	
0.003–0.2m	20	0.21982	0°; other temperatures studied	14
		0.20682	20°	
		0.20315	25°	
		0.19092	40°	
	45	0.1894	0°; other temperatures studied	14
		0.16910		
		0.16344		
		0.14612		
	70	0.10783	0°	14
		0.07475	20°	
		0.06620	25°	
		0.03930	40°	

DIOXANE–WATER

Electrode	Cell
Hydrogen	Pt, H_2 | HCl, Hg_2Cl_2 | Hg
	Pt, H_2 | HBr | AgBr | Ag
	Pt, H_2 | HBr | AgBr | Ag
	Pt, H_2 | HCl | AgCl | Ag
	Pt, H_2 | HI | AgI | Ag
Silver–silver chloride	Ag | AgCl | LiCl | Li(Hg) | LiCl | AgCl | Ag
	Ag | AgCl | NaCl | Na(Hg) | NaCl | AgCl | Ag
	Ag | AgCl | KCl | K(Hg) | KCl | AgCl | Ag
	Ag | AgCl | RbCl | Rb(Hg) | RbCl | AgCl | Ag
	Ag | AgCl | CsCl | Cs(Hg) | CsCl | AgCl | Ag mixed aq. solvent
	Ag | AgCl | KCl | K(Hg) | KCl | AgCl | Ag (0.04%)
	Also see Hydrogen

II. EMF DATA

MIXTURES (*Continued*)

Conc. range	% Dioxane	$E°$	Comments	Ref.
0.001–0.05m	20	−0.2501	$\gamma_\pm$; solvent effects	15
	45	−0.2104		
	70	−0.1150		
	82	−0.0180		
0.001–0.01m	5	0.06785	Temp. range studied for each comp; K_c; $\gamma_\pm$; $\gamma_\pm^0$; ΔG transfer	16
	10	0.06542		
	15	0.06271		
	20	0.06015		
	45	0.03447		
	70	−0.05898		
	82	−0.1785		
~4 × 10⁻³m	5	0.06785		17
	10	0.06542		
	15	0.06271		
	20	0.06015		
~4 × 10⁻³m	5	0.21704		17
	10	0.21396		
	15	0.20812		
	20	0.20308		
~4 × 10⁻³m	5	−0.15274		17
	10	−0.15423		
	15	−0.15426		
	20	−0.15452		
0.03–0.2m	20	0.02701	$\Delta G°$ transfer; $\Delta H_t°$; $\Delta S_t°$	18
0.03–0.2m	20	0.03427	$\Delta G°$ transfer; $\Delta H_t°$; $\Delta S_t°$	18
0.03–0.2m	20	0.03439	$\Delta G°$ transfer; $\Delta H_t°$; $\Delta S_t°$	18
0.03–0.2m	20	0.03541	$\Delta G°$ transfer; $\Delta H_t°$; $\Delta S_t°$	18
0.03–0.2m	20	0.03593	ΔG transfer; $\Delta H_t°$; $\Delta S_t°$	18
0.05–0.5m	20	−0.0338		7
	45	−0.0996		
	70	−0.2277		

ETHANOL–WATER

Electrode	Cell
Cadmium	Also see Hydrogen and Glass
Calomel	Hg | HgCl, KCl | KBr, HgBr | Hg sat. (C) (C) sat.
	Hg | HgBr, KBr | KI, HgI | Hg
	Hg | HgCl, KCl | KI, HgI | Hg (C) (C)
	Glass | HA | KCl | KCl, Hg_2Cl_2 | Hg buffers sat.
Chloranil	Also see Hydrogen and Glass
Glass	Glass | HCl | KCl | Hg_2Cl_2 | Hg sat.
	Glass | allyl isothiocyanate, py or quinoline | Calomel
	Glass | Allyl NCS, R_2NH | Hg_2Cl_2 | Hg
	Glass | ϕCO_2H, ϕCO_2K, KCl || HOAc, KOAc, KCl | Glass
	Glass | MX | Calomel
	M = H, NH_4, Li, Na, K, Rb, Cs
	Also see Silver–silver chloride
Hydrogen	Pt, H_2 | H^+ | Chloranil, Hydrochloranil | Pt
	Pt, H_2 | HCl | AgCl | Ag

II. EMF DATA

MIXTURES

Conc. range	% EtOH	E^0	Comments	Ref.
C = 0.1M	25.07	0.1241	8 to 40°	36
	50.01	0.1164		
	75.03	0.1095		
	97.3	0.1043		
C = 0.1M	50.01	0.1552	20.1°	36
C = 0.1M	75.03	0.2706	19.1°	36
10^{-4}–10^{-1}M			18°; acidity potentials	32
10^{-2}–10^{-1}m	50, 75, 95, 98		pK_a	32
0.2–0.4m	90			33
0.2–0.4m	90			34
0.5–20 × 10^{-3}m	10, 20, 30, 40, 50, 60, 70, 80			3
0–0.1M	0, 10, 30, 50, 70		30°; selectivity ratios	35
	50	0.664		
3 × 10^{-3}–10^{-1}m	10	0.22726	0°; other temperatures studied	20
		0.21467	25°	
		0.20783	40°	

ETHANOL–WATER

Electrode	Cell
Hydrogen	
	Pt, H_2 | HCl | AgCl | Ag
	Pt, H_2 | HCl | AgCl | Ag
	Pt, H_2 | HCl | Hg_2Cl_2 | Hg
	Pt, H_2 | HBr | Hg_2Cl_2 | Hg
	Pt, H_2 | HCl | AgCl | Ag
	Pt, H_2 | HCl | AgCl | Ag
	Pt, H_2 | HCl | AgCl | Ag
Potassium	Also see Hydrogen and Glass
Silver–silver chloride	(Hg)Cd | $CdCl_2$ | AgCl | Ag (11%)
	Ag | AgCl, KCl | KBr, AgBr | Ag (m)　　(m) solvent I　solvent II

II. EMF DATA

MIXTURES (*Continued*)

Conc. range	% EtOH	E^0	Comments	Ref.
10^{-2}–10^{-1}m	20	0.21606 0°		20
		0.20757 25°		
		0.19962 40°		
0.005–0.05m	30	0.20033		21
	40	0.19454		
	50	0.18588		
10^{-2}–10^{-3}m	0	0.2215		22
	20	0.2080		
	35	0.1958		
	50	0.1800		
	65	0.1579		
	80	0.1192		
	100	−0.0886		
0.0004–0.03m	20	−0.25390	$\gamma_\pm$	23
	40	−0.24060		
	65	−0.21110		
	90	−0.13865		
	98.09	−0.06480		
0.001–0.05m	20	−0.13023	$\gamma_\pm$	24
	40	−0.12435		
	65	−0.10560		
	90	−0.06450		
	95	−0.02210		
			Effect of traces of water on EtOH	25
0.002–0.265m	(50 wt)		γ_{HCl}; t_{HCl}	26
0.005–2m	20	−0.20736		27
0.006–2m	10	0.21442		27
0.0005–0.01M	0	0.57162	$\gamma_\pm$; $\Delta H°$, $\Delta G°$; other temperatures studied	28
	30	0.52742		
	60	0.46930		
	90	0.37498		
	100	0.33711		
10^{-3}–10^{-1}N	24.67	∼0.1480	15°±3°	29
	50.6	∼0.142		
	75.06	∼0.141		

250 NONAQUEOUS ELECTROLYTES

ETHANOL–WATER

Electrode	Cell
Silver–silver chloride	Zn(Hg) | ZnCl$_2$ | AgCl | Ag (1%)
	(Hg)Cd | CdCl$_2$ | AgCl | Ag
	Ag | AgCl | KCl | K(Hg) | KCl | AgCl | Ag (aq.m) (m)
	Ag | AgCl | KCl || AgNO$_3$ | AgCl | Ag
	Glass | HCl | AgCl | Ag
	Glass | HOAc, KOAc, KCl | AgCl | Ag
	Also see Hydrogen
Zinc	Also see Hydrogen and Glass

ETHYLENE GLYCOL–

Electrode	Cell
Calomel	Also see Hydrogen
Hydrogen	Pt, H$_2$ | HCl | AgCl | Ag
	Pt, H$_2$ | HCl | Hg$_2$Cl$_2$ | Hg
	Pt, H$_2$ | HCl | AgCl | Ag
Silver–silver chloride	Also see Hydrogen

II. EMF DATA

MIXTURES (*Continued*)

Conc. range	% EtOH	E^0	Comments	Ref.
0.01–1m	25 mole % 50 mole % 90 mole %		15°; ΔG	30
0.0005–0.1m	30 60 90 100	0.52586 0.46962 0.31848 0.26506	K_{diss}; $\gamma_\pm$; temp. range studied	31
0.05–0.5m	30	−0.0608		7
0.5–20 × 10⁻³m	10, 20, 30, 40, 50, 60, 70, 80		ΔG transfer; K_a; K_{sp}	3
0.5–20 × 10⁻³m	10, 20, 30, 40, 50, 60, 70, 80		ΔG transfer; K_a; K_{sp}	3
0.5–20 × 10⁻³m	10, 20, 30, 40, 50, 60, 70, 80		ΔG transfer; K_a; K_{sp}	3

WATER MIXTURES

Conc. range	% Glycol	E^0	Comments	Ref.
0.002–0.01m 0.002–0.09m	10 20	0.216 0.2101		37
0.002–0.05m	19.25 50.0 77.91	−0.2570 −0.2364 −0.20125	$\gamma_\pm$; solvent effects	38
0.002–1m	10 20 40 60	0.21484 0.20936 0.19720 0.18070	$\gamma_\pm$	39

ETHYL METHYL KETONE–

Electrode	Cell
Hydrogen	Pt, H_2 | HCl | AgCl | Ag
Silver–silver chloride	Also see Hydrogen

GLYCEROL–WATER

Electrode	Cell
Hydrogen	Pt, H_2 | HCl | AgCl | Ag
	Pt, H_2 | HCl | AgCl | Ag
Silver–silver chloride	Also see Hydrogen

p-HYDROXYBENZOIC ACID–

Electrode	Cell
Hydrogen	Pt, H_2 | HCl | AgCl | Ag
Silver–silver chloride	Also see Hydrogen

II. EMF DATA

WATER MIXTURES

Conc. range	% EtCOMe	E^0	Comments	Ref.
0.003–0.1m	10	0.2153	$\gamma_\pm$	40
	20	0.2078		

MIXTURES

Conc. range	% Glycerol	E^0	Comments	Ref.
0.003–0.1m	50	0.20065	0°; other temps. studied	41
		0.18398	25°	
		0.16341	50°	
		0.12280	90°	
0.005–0.2m	10	0.2166	$\Delta G°$ transfer; $\gamma_\pm$; other	42
	20	0.2090	temps. studied	
	40	0.1900		
	70	0.1490		
	90	0.0947		

WATER MIXTURES

Conc. range	% HOϕCO$_2$H	E^0	Comments	Ref.
0.02–0.05m	$m = 0.006$	0.2219	$\gamma_\pm$	43
	$m = 0.012$	0.2217		
	$m = 0.018$	0.2215		

ISOPROPANOL–

Electrode	Cell
Calomel	Also see Hydrogen
Hydrogen	Pt, H_2 | HCl | AgCl | Ag
	Pt, H_2 | HCl | Hg_2Cl_2 | Hg
	Pt, H_2 | HCl | AgCl | Ag
	Pt, H_2 | HCl | AgCl | Ag
	Pt, H_2 | HCl | AgCl | Ag
	Glass | HCl | AgCl | Ag
Silver–silver halide	Also see Hydrogen and Glass

MANNITOL–

Electrode	Cell
Hydrogen	Pt, H_2 | HCl, KCl | AgCl | Ag
	Pt, H_2 | HCl | AgCl | Ag
Silver–silver chloride	Also see Hydrogen

II. EMF DATA

WATER MIXTURES

Conc. range	% i PrOH	E^0	Comments	Ref.
0.005–0.02	20	0.2060	$\gamma_{\pm}$	21
8×10^{-4}–2×10^{-2}m	20	−0.25226	$\gamma_{\pm}$	63
	40	−0.23690		
	60	−0.20810		
	90	−0.10260		
	97.5	−0.02300		
0.003–0.1m	5	0.21807	Temp. range studied	64
0.005–0.2m	95	−0.02269	Temp. range studied; $\gamma_{\pm}$; ΔG transfer	65
0.002–1m	10	0.21363		27
	0	0.0159	ΔG transfer	8
	10	0.0115		
	20	0.0080		
	30	0.0055		
	40	0.0010		
	50	−0.0065		
	60	−0.0180		

WATER MIXTURES

Conc. range	% Mannitol	E^0	Comments	Ref.
0–0.09m	5	0.21948	Medium effects	44
	10	0.21639		
	15	0.21310		
0.005–0.05m	5	0.21950	Temp. range studied (0–60°)	45

METHANOL–WATER

Electrode	Cell
Cadmium	Also see Silver–silver halide
Calomel	Glass | HA, MA || KCl | Hg_2Cl_2 | Hg
	Hg | Hg_2Cl_2, KCl | KBr, Hg_2Br_2 | Hg
	Hg | Hg_2Cl_2, KCl | KI, Hg_2I_2 | Hg
	Also see hydrogen and Silver–silver halide
Glass	Also see Calomel, Silver–silver halide and Silver–silver nitrate
Hydrogen	Pt, H_2 | HCl | AgCl | Ag
	Pt, H_2 | HCl | AgCl | Ag
	Pt, H_2 | HCl | AgCl | Ag
	Pt, H_2 | HCl | AgCl | Ag
	Pt, H_2 | HCl | AgCl | Ag
	Pt, H_2 | HBr | Hg_2Br_2 | Hg
	Pd, H_2 | HBr | AgBr | Ag

II. EMF DATA

MIXTURES

Conc. range	% MeOH	E^0	Comments	Ref.
$\sim 10^{-2}$M	20, 40, 60, 80		K_a of a range of acids and phenols	46
0.1M	25.95	0.1269	Temp. range (10–35°)	47
	45.13	0.1204		
	74.88	0.1102		
0.1M	45.13	0.2887	17.3°	47
0.005–0.1M	10	0.21535	Temp. range studied; $\gamma_\pm$; $\Delta G°$ transfer	48, 49
	20	0.20881		
10^{-3}–10^{-1}M	0	0.22216	$\gamma_\pm$; $\Delta G°$ transfer	
	20	0.2094		
	40	0.1968		
	60	0.1818		
	80	0.1492		
	90	0.1135		
	100	−0.0099		
10^{-3}–10^{-1}M	20.22	−0.2545	$\gamma_\pm$; medium effects	50
	43.12	−0.2415		
	68.33	−0.2173		
	97.29	−0.1027		
10^{-2}–1M	33.4	−0.20167	$\gamma_\pm$	51
10^{-2}–10^{-1}M	50	−0.20167	10°	52
		−0.19826	15°	
0.002–0.12M	0	−0.13940	Temp. range studied; $\gamma_\pm$; ΔG, ΔH	53
	20	−0.13161		
	40	−0.12662		
	65	−0.11170		
	90	−0.05825		
	99.5	+0.02072		
10^{-3}–10^{-1}M	20.22	0.0634	$\gamma_\pm$; ΔG transfer	54
	33.40	0.0585		
	50.00	0.0538		
	68.33	0.0385		
	90.00	0.0171		

METHANOL–WATER

Electrode	Cell
Hydrogen	Pd, H_2 | HI | AgI | Ag
	Pt, H_2 | HBr | AgBr | Ag (buffer)
	Pt, H_2 | HI | AgI | Ag
	Pt, H_2 | HBr | AgBr | Ag
	Ag | AgCl | NaCl, NaH Succ, H_2 Succ | Pt, H_2 | ...... NaBr, NaH Succ, H_2 Succ | AgBr | Ag
	Similar cell with I^- and Ag | AgI
	Pt, H_2 | HI | AgI | Ag
	Pt, H_2 | HCl, NaCl | AgCl | Ag
	Pt, H_2 | HCl | AgCl | Ag
Lithium	Also see Silver–silver halide
Potassium	Also see Silver–silver halide
Silver–silver halide	(Hg)Cd | $CdCl_2$ | AgCl | Ag
	Ag | AgCl $\left\| \begin{array}{c} KCl \\ KNO_3 \end{array} \right\|$ KNO_3 $\left\| \begin{array}{c} AgNO_3 \\ KNO_3 \end{array} \right\|$ AgCl | Ag
	(Hg)Cd | $CdCl_2$, AgCl | Ag (m) sat.
	Glass | $H_2N(CH_2)NH_2$ || NaCl | AgCl | Ag NaOA, HCl, NaCl
	Ag | AgCl | KCl || $AgNO_3$ | AgCl | Ag
	Glass | HCl | AgCl | Ag

II. EMF DATA

MIXTURES (*Continued*)

Conc. range	% MeOH	E^0	Comments	Ref.
10^{-3}–10^{-1}M	20.22	−0.1566	$\gamma_\pm$; ΔG transfer	54
	33.40	−0.1518		
	50.00	−0.1496		
	68.33	−0.1555		
	90.00	−0.2053		
0.004–0.1m	10	0.06655	ΔG transfer	55
	43.12	0.0560		
0.004–0.01m	43.12	−0.15385	ΔG transfer	55
		−0.1489		
0.025–0.0004m	87.68	−0.0077		56
	87.68	−0.1302		56
	87.68	−0.3110		56
	87.68	−0.1885		56
0.1–1m	10, 20, 30, 40, 50, 60		Temp. range studied; $\gamma_\pm$	57
0.02–0.5m	0, 20, 30, 40, 50, 60, 70, 80, 90		$\gamma_\pm$	58
0.005–0.1M	30	0.53611	Temp. range studied; ΔG°; ΔH°	58
	60	0.49510		
	90	0.44253		
	100	0.41033		
0.01–0.05m	10	0.58939	Owen cell	55
	43.12	0.03763		
0.0005–0.1M	30	0.53241	$\gamma_\pm$; temp. range studied	31
	60	0.48634		
	0, 10, 20, 30, 40, 50, 60, 70, 80		K_a	59
0.5–20 × 10^{-3}M	10, 20, 30, 40, 50, 60, 70, 80		ΔG transfer	3
0.5–20 × 10^{-3}M	10, 20, 30, 40, 50, 60, 70, 80		ΔG transfer	3

METHANOL–WATER

Electrode	Cell
Silver–silver halide	Glass | HOAc, KOAc, KCl | AgCl | Ag
	Glass $\begin{vmatrix} \phi CO_2H \\ \phi CO_2K \\ KCl \end{vmatrix} \begin{Vmatrix} HOAc \\ KOAc \\ KCl \end{Vmatrix}$ Glass
	Ag | AgCl | LiCl | Li(Hg) | LiCl | AgCl | Ag
	Ag | AgCl | NaCl | Na(Hg) | NaCl | AgCl | Ag
	Ag | AgCl | KCl | K(Hg) | KCl | AgCl | Ag
	Also see under Hydrogen
Silver–silver nitrate	Ag | AgNO$_3$ | KNO$_3$ | AgNO$_3$, X | Ag X = MeNH$_2$, Me$_2$NH, S$_2$O$_3^{-2}$
	Glass | HA, MA || AgNO$_3$ | Ag
Sodium	Also see Silver–silver halide

PROPANOL–

Electrode	Cell
Hydrogen	Pt, H$_2$ | HCl | AgCl | Ag
Silver–silver halide	Also see Hydrogen

PROPYLENE CARBONATE–

Electrode	Cell
Hydrogen	Pt, H$_2$ | HCl | AgCl | Ag
Silver–silver chloride	Also see Hydrogen

II. EMF DATA

MIXTURES (*Continued*)

Conc. range	% MeOH	E^0	Comments	Ref.
0.5–20×10^{-3}M	10, 20, 30, 40, 50, 60, 70, 80		ΔG transfer	3
0.5–20×10^{-3}M	10, 20, 30, 40, 50, 60, 70, 80		ΔG transfer	3
0.02–0.5m	0, 20, 30, 40, 50, 60, 70, 80, 90		$\gamma_\pm$	60
				60
				60
$(0.2$–$2) \times 10^3$	0, 50		ΔG, ΔH, ΔS of complex formation	61
10^{-2}M	20, 40, 60, 80			

WATER MIXTURES

Conc. range	% Propanol	E^0	Comments	Ref.
0.005–0.15m	95	0.04542 0.03363 0.02105 −0.00319	5°; $\gamma_\pm$; ΔG_t° 15° 25° 35°	62

WATER MIXTURES

Conc. range	% PC	E^0	Comments	Ref.
0.005–0.2m	5 10 20	0.2209 0.2188 0.2132	Temp. range studied; $\gamma_\pm$; ΔG transfer	66

SORBITOL–

Electrode	Cell
Hydrogen	Pt, H_2	HCl	AgCl	Ag
Silver–silver chloride	Also see Hydrogen

SUCCINIC ACID–

Electrode	Cell
Hydrogen	Pt, H_2	HCl	AgCl	Ag
Silver–silver chloride	Also see Hydrogen

TRIETHYLENE GLYCOL–

Electrode	Cell
Hydrogen	Pt, H_2	HCl	AgCl	Ag
Silver–silver chloride	Also see Hydrogen

References

1. H. P. Bennetto, D. Feakins and D. J. Turner, J. Chem. Soc., **A,** 1211 (1966).
2. G. Schwarzenbach and P. Stensby, Helv. Chim. Acta., **42,** 2342 (1959).
3. J. P. Morel, D.Sc. Thesis, Univ. of Clermont-Ferrand (1969).
4. F. Aufauvre, D.Sc. Thesis; Univ. of Clermont-Ferrand (1969).
5. K. Schwabe and K. Wankmuller, Ber. Bunsengesellschaft Phys. Chem., **69,** 528 (1965).
6. D. Feakins and C. M. French, J. Chem. Soc., 3168 (1956).
7. D. Bax, C. L. De Ligny and M. Alfenaar, Rec. Trav. Chim., **90,** 1002 (1971).
8. J. Morin and J. P. Morel, Bull. Soc. Chim. France., 2019 (170).
9. K. H. Khoo, J. Chem. Soc., **A,** 1177 (1971).
10. N. A. Izmailov, Russ. J. Phys. Chem., **34,** 1142 (1960).

II. EMF DATA

WATER MIXTURES

Conc. range	% Sorbilol	E^0	Comments	Ref.
0.01–0.02M	5	0.21968	Temp. range 0–60°	45

WATER MIXTURES

Conc. range	% H_2Succ. (m)	E^0	Comments	Ref.
0.02–0.05m	0.04	0.2216	$\gamma_\pm$	43
	0.08	0.2211		
	0.12	0.2208		
	0.16	0.2201		
	0.20	0.2197		

WATER MIXTURES

Conc. range	% Glycol	E^0	Comments	Ref.
0.003–0.1m	10	0.2149		40
	20	0.2086		

11. C. Calmon, J. G. Donelson, H. S. Harned, J. O. Morrison and F. Walker, J. Amer. Chem. Soc., **61,** 44 (1939).
12. C. Calmon, J. G. Donelson and H. S. Harned, J. Amer. Chem. Soc., **60,** 2128 (1938).
13. C. Calmon and H. S. Harned, J. Amer. Chem. Soc., **60,** 334 (1938).
14. H. S. Harned and J. O. Morrison, J. Amer. Chem. Soc., **58,** 1908 (1936).
15. K. Schwabe and W. Schwenke, Z. Elektrochem., **63,** 441 (1959).
16. P. Andrigo, C. Massarani-Formaro and T. Mussini, J. Electroanalyt. Chem. Interfacial Electrochem., **33,** 177 (1971).
17. C. Massarani-Formaro, T. Mussini and P. Andrigo, J. Electroanalyt. Chem. Interfacial Electrochem., **33,** 189 (1971).
18. K. G. Lawrence, D. Feakins and H. P. Bennetto, J. Chem. Soc., **A,** 1493 (1968).
19. J. P. Shukla and S. G. Tandon, J. Electroanalyt. Chem. Interfacial Electrochem., **33,** 195 (1971).

20. W. A. Felsing and A. Patterson, J. Amer. Chem. Soc., **64,** 1478 (1942).
21. D. S. Allen and H. S. Harned, J. Phys. Chem., **58,** 191 (1954).
22. E. Grunwald and E. F. Sieckmann, J. Amer. Chem. Soc., **76,** 3855 (1954).
23. M. Kunz and K. Schwabe, Z. Elektrochem., **64,** 1188 (1960).
24. A. Ferse, K. Schwabe and R. Urlass, Ber. Bunsengesellschaft Phys. Chem., **68,** 46 (1964).
25. M. C. Day and C. L. LeBas, J. Phys. Chem., **64,** 465 (1960).
26. M. H. Fleysher and H. S. Harned, J. Amer. Chem. Soc., **47,** 82 (1925).
27. C. Calmon and H. S. Harned, J. Amer. Chem. Soc., **61,** 1491 (1939).
28. E. S. Amis and J. D. Hefley, J. Phys. Chem., **69,** 2082 (1965).
29. J. N. Butler, D. R. Cogley and G. Holleck, J. Electrochem. Soc., **116,** 952 (1969).
30. J. A. V. Butler and R. T. Hamilton, Proc. Roy. Soc., **138A,** 450 (1932).
31. J. D. Hefley, Ph.D. Thesis, Univ. of Arkansas, (1963).
32. P. Decroly and M. Mandel, Trans. Faraday Soc., **29,** (1960).
33. F. N. Kozlenko and S. P. Miskidzh'yan, Russ. J. Phys. Chem., **39,** 506 (1965).
34. F. N. Kozlenko and S. P. Miskidzh'yan, Russ. J. Phys. Chem., **37,** 1181 (1963).
35. G. Kugler and G. A. Rechnitz, Z. Analyt. Chem., **214,** 405 (1965).
36. A. E. Brodsky, Z. Phys. Chem., **121,** 1 (1926).
37. B. H. Claussen and C. M. French, Trans. Faraday Soc., **51,** 1124 (1955).
38. R. Hertzsch and K. Schwabe, Z. Elektrochem., **63,** 445 (1945).
39. S. B. Knight, J. F. Masi and D. Roesel, J. Amer. Chem. Soc., **68,** 661 (1946).
40. D. Feakins and C. M. French, J. Chem. Soc., 2284 (1957).
41. H. S. Harned and F. H. M. Nestler, J. Amer. Chem. Soc., **68,** 665 (1946).
42. A. L. M. Bothwell, W. Vernon and R. N. Roy, J. Electrochem. Soc., **118,** 1302 (1971).
43. T. F. Tadros and H. Sadek, Trans. Faraday Soc., **58,** 2192 (1962).
44. M. Paabo and R. A. Robinson, J. Phys. Chem., **67,** 2861 (1963).
45. P. A. Anderson, J. Amer. Chem. Soc., **48,** 2285 (1926).
46. J. Julliard, D.Sc. Thesis; Univ. of Clermont-Ferrand (1969).
47. A. E. Brodsky, Z. Phys. Chem., **121,** 1 (1926).
48. H. S. Harned and H. C. Thomas, J. Amer. Chem. Soc., **58,** 761 (1936).
49. H. S. Harned and H. C. Thomas, J. Amer. Chem. Soc., **57,** 1666 (1935).
50. K. Schwabe and S. Ziegenbalg, Z. Elektrochem., **42,** 172 (1958).
51. R. G. Bates and D. Rosenthal, J. Phys. Chem., **67,** 1088 (1963).
52. R. G. Bates, M. Paabo and R. A. Robinson, J. Chem. and Eng. Data, **9,** 374 (1964).
53. E. Ferse and K. Schwabe, Ber. Bunsengesellschaft Phys. Chem., **70,** 849 (1966).
54. D. Feakins and R. P. T. Tomkins, J. Chem. Soc., **A,** 1458 (1967).
55. D. Feakins and P. Watson, J. Chem. Soc., **A,** 4686 (1963).
56. M. Alfenaar, C. L. De Ligny and A. G. Remijnse, Rec. Trav. Chim., **86,** 555 (1967).
57. G. Akerlof, J. W. Teare and H. Turck, J. Amer. Chem. Soc., **59,** 1916 (1937).
58. E. S. Amis and J. D. Hefley, J. Electrochem. Soc., **112,** 337 (1965).
59. N. Tonaka and H. Otaki, J. Phys. Chem., **75,** 90 (1971).
60. G. Akerlof, J. Amer. Chem. Soc., **52,** 2353 (1930).
61. C. Luca, V. Mageu and G. R. Popa, Anal. Univ. C. I. Parron, Bucuresti, **14,** 107 (1965).
62. R. N. Roy, W. Vernon and A. L. M. Bothwell, Electrochim. Acta, **17,** 5 (1972).
63. N. F. Nhuan and K. Schwabe, Z. Elektrochem., **65,** 891 (1961).
64. W. A. Felsing and R. L. Moore, J. Amer. Chem. Soc., **69,** 1076 (1947).
65. R. N. Roy, A. L. M. Bothwell and W. Vernon, J. Chem. Thermodynamics, **3,** 769 (1971).
66. R. N. Roy, J. J. Gibbons and W. Vernon, J. Chem. Soc., **A,** 3589 (1971).

(c) Equilibrium Measurements in Nonaqueous-Nonaqueous Mixed Solvents

The information in this section surveys the various reference electrodes and cell configurations that have been used for equilibrium-type measurements in nonaqueous-nonaqueous mixed solvent systems. The tables are arranged alphabetically according to the particular component concerned. For each solvent table the reference electrodes are listed alphabetically. Information relative to the composition range of the solvent mixture, the concentration range of the electrolyte and the values of E^0 are given when reported. Additional comments indicating the purpose of the study, extrapolation procedures used, stability of electrodes and other pertinent information are provided. The availability of activity coefficient data is indicated. References are provided for each cell studied and a complete bibliography is given at the end of this section.

ACETIC ACID–ACETIC ANHYDRIDE MIXTURES

Electrode	Cell	Conc. range	E^0	Comments	Ref.
Calomel	Pt $\|$ Chloranil, Hydrochloranil $\|$ H$^+$ $\|$ LiCl sat. $\|$ Hg$_2$Cl$_2$ $\|$ KCl sat. H$_2$O $\|$ Hg		0.566 vs aq. calomel		1
Chloranil	H$_2$, Pd $\|$ LiCl sat. $\|$ H$^+$ $\|$ Chloranil, Hydrochloranil $\|$ Pt	10^{-3} – 1M	0.680	Electrode used as pH meter; gives only pH as f (acid) and as f (Ac$_2$O): systems studied were m-AcSO$_3$H, HClO$_4$, H$_2$SO$_4$, 2,5-Cl$_2$ ϕSO$_3$H, 2ClϕCH$_3$-5-SO$_2$H, 2NO$_2\phi$Br-4-SO$_3$H, 4-NO$_2$ Clϕ-2-SO$_3$H, m-NO$_2\phi$SO$_3$H, all in 30% Ac$_2$O	1
	Pt $\|$ Chloranil, Hydrochloranil $\|$ H$^+$ $\|$ LiCl sat. $\|$ Hg$_2$Cl$_2$ $\|$ KCl sat. H$_2$O $\|$ Hg		0.566 vs aq. calomel		
Hydrogen	H$_2$, Pd $\|$ LiCl sat. $\|$ H$^+$ $\|$ Chloranil, Hydrochloranil $\|$ Pt	10^{-3} – 1M	0.680	Also see Clhoranil	1

II. EMF DATA

ACETIC ACID–ANILINE MIXTURES

Electrode	Cell	Conc. range	Comments	Ref.
Calomel	Glass $\|\|$ Hg_2Cl_2 $\|$ Hg sat. aq.	5–98 mole % HOAc	Phase diag; E values	2
Glass and Hydrogen	Pt, H_2 $\|\|$ Hg_2Cl_2 $\|$ Hg sat. aq.			

ACETIC ACID–ETHYLANILINE MIXTURES

Electrode	Cell	Conc. range	Comments	Ref.
Calomel	Glass $\|\|$ Hg_2Cl_2 $\|$ Hg sat. aq.	5–98 mole % HOAc	Phase diag; E values	2
Glass and Hydrogen	Pt, H_2 $\|\|$ Hg_2Cl_2 $\|$ Hg sat. aq.			

ACETIC ACID–PYRIDINE MIXTURES

Electrode	Cell	Conc. range	Comments	Ref.
Calomel	Glass $\|\|$ Hg_2Cl_2 $\|$ Hg sat. aq.	5–98 mole % HOAc	Phase diag; E values	2
Glass and Hydrogen	Pt, H_2 $\|\|$ Hg_2Cl_2 $\|$ Hg sat. aq.			

ACETIC ACID–QUINOLINE MIXTURES

Electrode	Cell	Conc. range	Comments	Ref.
Calomel	Glass ∥ Hg$_2$Cl$_2$ ∣ Hg sat. aq.	5–98 mole % HOAc	Phase diag; E values	2
Glass and Hydrogen	Pt, H$_2$ ∥ Hg$_2$Cl$_2$ ∣ Hg sat. aq.			

ACETONE–METHANOL MIXTURES

Electrode	Cell	Conc. range	% Me$_2$CO	E^0	Comments	Ref.
Cesium and silver-silver iodide	(Hg)Cs ∣ CsI ∣ AgI ∣ Ag	0.03–0.1m	0	1.4820	$\gamma_\pm$; ΔG solvation	3
			10	1.4640		
			20	1.4510		
			30	1.4440		
			48	1.4370		
			50	1.4330		
			60	1.4230		
			64.4	1.4050		
			70	1.3830		
			80	1.3240		
			90	1.2720		
			100	1.1430		

II. EMF DATA

ACETONITRILE–METHANOL MIXTURES

Electrode	Cell	Conc. range	% MeOH	E^0	Comments	Ref.
Silver–silver chloride and Hydrogen	Pt, H_2	HCl	AgCl	Ag	0.005–0.02m		−0.04456	ΔG_{tr}^0 from MeOH	4

AMMONIA–ETHYLAMINE MIXTURES

Electrode	Cell	% $EtNH_2$	E^0	Comments	Ref.
Rubidium	(Hg)Rb	RbI	Rb 2.698×10^{-5} solid eq/gm	92.1	1.0745		5

BROMINE TRIFLUORIDE–CHLORINE TRIFLUORIDE MIXTURES

Electrode	Cell	X_{ClF_3}	E^0	Comments	Ref.
Cadmium	Pt	BrF_3, ClF_3	Cd	0.6	−2.45	23°	6
		0.6	−2.45	20°	
		0.6	−2.37	−17°	
Platinum	Pt	BrF_3, ClF_3, BF_3	Cd	0.998	−2.57	−8°	6

CARBON TETRACHLORIDE–ETHANOL MIXTURES

Electrode	Cell	Comp. range	% CCl	$E°$	Comments	Ref.
Silver–silver iodide and Sodium	(Hg)Na | NaI | AgI | Ag 0.15% (m)	10^{-3}–10^{-1} m	90.4	−0.062	$\gamma_\pm$; å	7

DIOXANE–ETHANOL MIXTURES

Electrode	Cell	Conc. range (m)	% EtOH	E^0	Comments	Ref.
Hydrogen and Silver–silver chloride	Pt, H$_2$ | HCl | AgCl | Ag	0.002–0.1	90	−0.0820	$\gamma_\pm$; solvent effects	8
		0.003–0.2	85	−0.0980		
		0.002–0.25	80	−0.1060		
		0.0014–0.17	78.2	−0.1180		
		0.0014–0.36	72.5	−0.1260		
		0.0015–2.6	67.6	−0.1400		
		0.003–0.35	65	−0.1480		
		0.0017–0.2	61	−0.1540		
		0.002–0.06	55	−0.1840		
		0.001–1.5	51	−0.2000		
		0.0005–0.25	47	−0.2060		
		0.0006–1	43	−0.2130		

II. EMF DATA

DIOXANE–METHANOL MIXTURES

Electrode	Cell	Conc. range m	% Dioxane	E^0	Comments	Ref.
Hydrogen and Silver–silver chloride	Pt, H_2 | HCl | AgCl | Ag	0.003–0.2m	30	−0.072		9
			35	−0.088		
			40	−0.012		
			45.6	−0.112		
			50	−0.148		
			55	−0.162		
			60	−0.198		
Lithium and Silver–silver chloride	Li(Hg) | LiCl | AgCl | Ag	0.0004–5m	10	2.0230	γ_{LiCl}; ΔG transfer of LiCl; other solvent compositions also studied	10
			20	2.0070		
			30	1.9810		
			40	1.9500		
			50	1.8980		
			60	1.8300		
			70	1.7560		
			80	1.6850		
			90	1.6080		

METHANOL–PROPYLENE CARBONATE MIXTURES

Electrode	Cell	Conc. range	% MeOH	E^0	Comments	Ref.
Lithium	(Hg)Li | LiCl | AgCl | Ag	0.03–0.1	100	1.9908	ΔG of transfer	13
			70	1.9780		
			50	1.9720		
			30	1.9490		
			10	1.9304		
			0	1.9496		
Potassium	(Hg)K | KCl | AgCl | Ag	0.03–0.1	100	1.7740		13
			70	1.7784		
			50	1.7626		
			30	1.7940		
			10	1.8164		
			0	1.8080		
Silver–silver bromide	Pt, H_2 | HoAc, NaOAc, NaBr | AgBr | Ag	0.006–0.05	100	−0.1340	Medium effects	11
		0.01–0.05	90	−0.1310		
		0.008–0.04	70	−0.1320		
		0.005–0.04	30	−0.1445		
		0.005–0.04	10	−0.1633		
			0	−0.1633		

II. EMF DATA

Silver–silver Chloride	Pt, H₂	HCl	AgCl	Ag	0.005–0.08	100	−0.0090	Medium effects; pK_a of HOAc	11
		0.004–0.06	90	−0.0045	
		0.005–0.08	70	−0.0070	
		0.005–0.08	50	−0.01	
		0.004–0.08	30	−0.0170	
		0.005–0.06	10	−0.0275	
		0.004–0.06	0	−0.0320	
	Pt	H₂	NaS, NaCl	AgCl	Ag	0.015–0.1		pK_s	12
			100	16.69	
			90	16.61	
			70	16.68	
			50	16.75	
			30	16.88	
			10	17.05	
			0	17.15	
Sodium	(Hg)Na	NaCl	AgCl	Ag	0.03–0.1	100	1.6784		13
			70	1.7804	
			50	1.7116	
			30	1.7900	
			0	1.7620	

References

1. A. E. Cameron and J. Russell, J. Amer. Chem. Soc., **60**, 1345 (1938).
2. F. N. Kozlenko and S. P. Miskidzn'yan, Russ. J. Phys. Chem., **34**, 163 (1960).
3. L. Ya Shapovalova and A. M. Shkodin, Soviet Electrochem, (Elektrokhimiya) **6**, 1757 (1970).
4. H. P. Bennetto and J. J. Spitzer, Chem. Comm., 990 (1971).
5. W. L. Argo and G. N. Lewis, J. Amer. Chem. Soc., **37**, 1983 (1915).
6. W. A. Cannon and M. S. Toy, Electrochem. Technol., **4**, 520 (1966).
7. Z. N. Timofeeva and A. M. Sukhotin, Zhur. Fiz. Khim., **33**, 137 (1959).
8. L. I. Kozeniuk, A. M. Shkodin and T. P. Soyoyan, Ukrain. Khim. Zhur., **34**, 135 (1968).
9. L. J. Kozeniuk, L. I. Karkuzaki and T. P. Sogoyan; Ukrain. Khim. Zhur., **30**, 237 (1964).
10. L. Ya. Shapovalova and A. M. Shkodin, Elektrokhimiya, **4**, 113 (1968).
11. K. K. Kundu, M. N. Das and A. L. De, J. Chem. Soc., **A**, 373 (1972).
12. A. L. De, M. N. Das and K. K. Kundu, J. Chem. Soc., **A**, 378 (1972).
13. A. K. Rakshit, M. N. Das and K. K. Kundu, J. Chem. Soc., **A**, 381 (1972).

(d) Potentiometric Titrations in Protic Solvents

Potentiometric titrations in nonaqueous solvents can be divided into two types, the "thermodynamic titration," where a plot of emf versus quantity of titrant added is used to measure equilibrium constants, and the "analytical titration" in which interest is focussed on the position of the quantitative end point. A representative sample of the "thermodynamic titrations" have been tabulated in (a)–(c) of this section.

The area of purely analytical potentiometric titrations is vast and the intent here is to cite a selection of methods and interesting advances in each of the major nonaqueous solvents and which will hopefully indicate a fairly representative overview of the problems and methods encountered in this field.

For more detailed accounts of potentiometric titrations the reader should consult several authoritative texts and reviews on this subject (see e.g.[a,b]).

The information on potentiometric titrations in protic solvents in the following tables is presented alphabetically by solvent and includes details on indicator and reference electrodes, titrand and titrant, solute and supporting electrolytes and additional comments of interest. A complete bibliography is given at the end of this section.

Symbols used:

Me	Methyl
Et	Ethyl
Pr	Propyl
Bu	Butyl
ϕ	Phenyl and in cases of higher substitution (3-BrϕOH = 3-bromophenol)
py	Pyridine and for substituted pyridines (2-Mepy = α-picoline)
Tos	4-MeϕSO$_3^-$
X	Halogen

[a] "Titrations in Nonaqueous Solvents," W. Huber, Academic Press, (1967).
[b] Analytical Chemistry, Analytical Reviews (see e.g., April 1970).

ACETAMIDE, (CH$_3$CONH$_2$)

Indicator electrode	Reference electrode	Titrand	Titrant	Comments	Ref.
"Gebremste"	"Gebremste"	HCl	NaNHAc		1
		KNHAc	HCl		
Mo	Mo	KNHAc	HBr	$K_s = 3 \times 10^{-11}$	2
		HgNHAc	HClO$_4$	$K_s = 3 \times 10^{-11}$	
		HBr }	NaNHAc		
		ϕCO$_2$H }			
		Ac$_2$NH			

ACETIC ACID HOAc, (CH$_3$CO$_2$H)

Indicator electrode	Reference electrode	Solute	Titrant	Supporting electrolyte	Comments	Ref.
Ag	Calomel	AcCl	AgOAc (in py)			14
Au	Au	HClO$_4$	ϕNEt$_2$			8
Chloranil (Pt)	Chloranil (Pt)	ϕNH$_2$ and other bases	HClO$_4$	LiCl	LiCl bridge	3
Glass		Urea H$_2$SO$_4$	NaOAc		Ion exchange electrodes	11
		Ac$_2$O	ϕNH$_2$ (in HOAc)			14
Glass	AgCl	Quinoline	HClO$_4$			4
		o-AcOϕCO$_2$(CH$_2$)$_2$NEt$_2\cdot$HCl	HClO$_4$ (in dioxane)			19

II. EMF DATA

Glass	HOAc	Bu$_3$MeNOH		5
	φ Guanidine imidazole	2-Naphthalene-sulfonic acid	Potential breaks in various solvents compared	9
	p-ClφNH$_2$	p-MeφSO$_3$H		
	p-N$_2$NφCOMe	2,4-(NO$_2$)$_2$φSO$_3$H		
	R py	HClO$_4$	Cf. mixed solvents	10
	Choline (i.e., HOCH$_2$CH$_2$NMe$_3$OH)	HClO$_4$	As impurities in CH$_2$=CHCN	12
			Also by visual indicator	
	R$_3$N	HClO$_4$ (in HOAc or dioxane)	Solvent dried by Ac$_2$O excess	13
	LiNO$_3$			
	H$_2$N(R)CO$_2$H	HClO$_4$	(20°C) K_a of amino acids	15
	Pb(OAc)$_2$	HCl	Complex formation	17
	Hg(OAc)$_2$	KCl		
	Cd(OAc)$_2$	KBr		
	Zn(OAc)$_2$	KI		
	(CH$_2$CH$_2$OCONH$_2$)$_2$	HClO$_4$		18
	MOAc	HClO$_4$ p-MeφSO$_3$H		
Calomel (in HOAc)	NaOAc py	HClO$_4$	Ion exchange electrodes	11
Polyethylene				
Pt	Calomel	TiCl$_3$		7
	TiCl$_3$	CrCl$_2$		
	CrCl$_2$	CrO$_3$		
	SbCl$_3$	Br$_2$		
	As$_2$O$_3$	NaMnO$_4$		
	SeO$_2$			
	Hg			

ACETIC ACID HOAc, (CH$_3$CO$_2$H) (Continued)

Indicator electrode	Reference electrode	Solute	Titrant	Supporting electrolyte	Comments	Ref.
Pt	Calomel	FeCl$_2$ Quinol Cl$_4$ quinol Quinone p-NH$_2\phi$OH ϕ_2NH				6
QH	Calomel	HClO$_4$	NaOAc			6
QH(Pt), chloranil, and glass	Calomel	Sulfonamides (NaOAc)	HClO$_4$		Also calorimetric indicators	16
Sb	Calomel	HClO$_4$	NaOAc			6
Te	Calomel	HClO$_4$	NaOAc			6

AMMONIA

Indicator electrode	Reference electrode	Solute	Titrant	Comments	Ref.
Ag	Pt	(NH$_4$)$_2$S	AgI	$-50°$C	20
Glass		ROH ROK	KNH$_2$ NH$_4$Cl		21
Pt	Pt	(NH$_4$)$_2$S	KNH$_2$	Apparatus design	20

II. EMF DATA

BUTANOL (n-C_4H_9OH)

Indicator electrode	Reference electrode	Solute	Titrant	Comments	Ref.
Glass	Calomel	H_3PO_4 $RPO(OH)_2$ $R_2PO(OH)$	NaOBu	K_a of acids	22

t-BUTANOL (Me_3COH)

Indicator electrode	Reference electrode	Solute	Titrant	Ref.
Glass	Calomel	RSO_3H RSO_2NH_2 RSO_2NHR	Bu_3MeNOH	23

n-BUTYLAMINE, (n-$BuNH_2$)

Indicator electrode	Reference electrode	Solute	Titrant	Ref.
Glass	Calomel	HOAc Cetyl ϕOH HCl_4	Bu_3MeNOH in py Bu_3MeNOH in py	24 5

280 NONAQUEOUS ELECTROLYTES

DICHLOROACETIC ACID, Cl_2CHCO_2H

Indicator electrode	Reference electrode	Solute	Titrant	Supporting electrolyte	Comments	Ref.
Chloranil	Calomel	Urea MeCONHϕ (Meϕ)$_3$N H$_2$O	HClO$_4$	LiClO$_4$	LiClO$_4$ bridge pK weak bases	25
Chloranil (Pt)	Chloranil (Pt)	ϕNH$_2$ and other bases	HClO$_4$	LiCl	LiCl bridge	3

ETHANOL, (CH_3CH_2OH)

Indicator electrode	Reference electrode	Solute	Titrant	Supporting electrolyte	Comments	Ref.
Glass	Calomel	HBr ZnBr$_2$ CuBr$_2$ NiBr$_2$ CdI$_2$ MnCl$_2$	NaOEt	NaClO$_4$, with *acac*	K (chelation) (with *acac*)	26
		Carvone (ex spearmint oil) +H$_2$NOH·HCl	KOH		analytical	27

II. EMF DATA

Indicator electrode	Reference electrode	Solute	Titrant	Comments	Ref.
Glass	Calomel	Aspirin (commercial)	KOH	or in 95% ethanol	28
		acac $\phi COCH_2COMe$ $\phi COCH_2CO\phi$ Dimedone-2-Thenoyl- CH_2COCF_3	Bu$_4$NOH	K_a of diketones; complexes	29
		$Me_2\overset{+}{N}H_2O_2CH$ HCO_2H Me_2NH	KOEt HClO$_4$	Bu$_4$NClO$_4$ LiClO$_4$ NaClO$_4$ KI KSCN as impurities in DMF	30
		H_3PO_4 $RPO(OH)_2$ $R_2PO(OH)$	NaOEt	K_a of acids	22
		$R\phi CO_2H$ $X\ CO_2H$	Bu$_4$NOEt	K_a of acids	31

ETHYL CELLOSOLVE, (EtOCH$_2$CH$_2$OH)

Indicator electrode	Reference electrode	Solute	Titrant	Comments	Ref.
Al, stainless steel	Calomel	SbCl$_5$	TiCl$_3$		33
Glass	Calomel	$R_2C{=}N{-}NH\phi(3){-}CO_2H(4,6)(NO_2)_2$	NaOH	Equiv. wt. detn.	32

ETHYLENE GLYCOL [$(CH_2OH)_2$]

Indicator electrode	Reference electrode	Solute	Titrant	Ref.
Glass	Calomel (in MeOH)	$RNHCH_2CH_2S_2O_3Na$	$HClO_4$	34

ETHYLENEDIAMINE, (en) ($NH_2CH_2CH_2NH_2$)

Indicator electrode	Reference electrode	Solute	Titrant	Ref.
Glass	Calomel	ϕOH	KOH in iPrOH	35
		HOAc		
		HCl		
		ϕOH	Bu_3MeNOH in py	24
		$HClO_4$	Bu_3MeNOH (in py)	5
Pt	Calomel	ϕCO_2H	$NaO(CH_2)_2NH_2$	37
Pt	Sb	$(\phi CO)_2CH_2$	Bu_4NOMe (in ϕH—MeOH)	39
		$AcOCH_2CO_2Et$		
		$(RCO)_2NH$		
		ϕSO_2NH_2		
		$R\ CONH_2$		
		$RNHCONHR'$		
		$iPrNO_2$		

II. EMF DATA

Indicator electrode	Reference electrode	Solute	Titrant	Comments	Ref.
Sb	Sb	φOH HOAc HCl	NaO(CH$_2$)$_2$NH$_2$		36
		ArOH	NaO(CH$_2$)$_2$NH$_2$		38

FORMAMIDE, (HCONH$_2$)

Indicator electrode	Reference electrode	Solute	Titrant	Comments	Ref.
Ag	Ag	AgNO$_3$	Et$_4$NCl Et$_4$NBr Et$_4$NI Et$_4$NN$_3$ KSCN NaBφ$_4$	K_{sp} complexes	40
	AgCl	KCl	AgNO$_3$ (in H$_2$O) μ	K$_2$SO$_4$ bridge	42
Pt	H$_2$	FeC$_{10}$H$_{10}$ Fe(OφC=NC$_2$H$_4$-NHCH$_2$)$_2$ Co(OφC=NC$_2$H$_4$-NHCH$_2$)$_2$	Coulometric	Indirect via unknown working reference; E^0 evaluation	41

FORMIC ACID

Indicator electrode	Reference electrode	Solute	Titrant	Comments	Ref.
Quinhydrone; Glass	Calomel (aq)	$HClO_4$ H_2SO_2 HCl HNO_3 p-tolyl SO_3H HOAc Cl_3CCO_2H ClH_2CCO_2H Picric acid $HO\phi CO_2H$	py	Acid strength	43

II. EMF DATA

METHANOL, (CH₃OH)

Indicator electrode	Reference electrode	Solute	Titrant	Supporting electrolyte	Comments	Ref.
Ag	Ag	AgCl	Et$_4$NCl Et$_4$NBr Et$_4$NI Et$_4$NN$_3$ Et$_4$NSCN AgOAc		K_{sp} complexes	40
Ag	Calomel (aq.)	AgClO$_4$	LiCl LiBr NaI	LiClO$_4$	K_{sp} complexes	45
Glass	AgCl	H$_2$SO$_4$ HClO$_4$ (CO$_2$H)$_2$ ϕCO$_2$H HCO$_2$H (CHCO$_2$H)$_2$ o-HOϕCO$_2$H CH$_2$(CO$_2$H)$_2$ H$_3$BO$_3$	Et$_4$NOH (in ϕH—MeOH) KOH (in MeOH)		Complex structure stoichiometry. Constant-rate delivery (chronopotentiometric)	54
Glass	Calomel	CuCl$_2$ ZnCl$_2$	R$_2$NCH$_2$CH$_2$OH			53

METHANOL, (CH$_3$OH) (Continued)

Indicator electrode	Reference electrode	Solute	Titrant	Supporting electrolyte	Comments	Ref.
Glass	Calomel	acac	Bu$_4$NOH	Bu$_4$NClO$_4$	K_a of di-ketones; complexes	29
		ϕCOCH$_2$COMe		LiClO$_4$		
		ϕCOCH$_2$COϕ		NaClO$_4$		
		Dimedone		KI		
		2-Thenoyl-CH$_2$COCF$_3$		KSCN		
Glass	Calomel	MOAc	HClO$_4$			49
		M(OAc)$_2$	p-MeϕSO$_3$H			
Glass	Calomel	HClO	Bu$_3$MeNOH			5
Glass	Calomel	Acids	KOMe	NaClO$_4$		44
		Phenols	Bu$_4$NOMe			
Glass	Calomel	HNO$_3$	KOH			46
		(CO$_2$H)$_2$				
		CH$_2$(CO$_2$H)$_2$				
		CH(CO$_2$H)$_2$				
		cis(CHCO$_2$H)$_2$				
		MeCH(OH)CO$_2$H				
		[CH(OH)CO$_2$H]$_2$				
		HOC(CH$_2$CO$_2$N)$_2$CO$_2$H				
		KCl	AgNO$_3$		KNO$_3$ bridge combustion method for organic 95% MeOH	50

II. EMF DATA

Glass	Calomel	$R_3SiOPOF_2$ $(R_3SiO)_2POF$	NaOMC KOMC		51
Glass	Calomel	$Ca(H_2PO_4)_2$ $NH_4H_2PO_4$ NaH_2PO_4	NaOH	Cf. mixed solvents	47
		HX	Et_4NOH	From KX via ion exchange	48
		H_3PO_4 $RPO(OH)_2$ $R_2PO(OH)$ HCl	NaOMe	K_a of acids	22
Hg	Calomel	ZnX_2	EDTA	Hg-EDTA	52
(Pt)	H_2	$FeC_{10}H_{10}$ Fe—OϕC=NC$_2$H$_4$ ... C_2H_4—NH—C_2H_4—NH Co—O—ϕ—C=NC—H ... NC_2H_4—NH—C_2H_4NH	coulometric	Indirect correlation, via unknown working ref. E^0 evaluation	41

PROPANOL ALCOHOL (n-C_3H_7OH)

Indicator electrode	Reference electrode	Solute	Titrant	Comments	Ref.
Glass	Calomel	H_3PO_4	NaOPr	K_a of acids	22
		$RPO(OH)_2$			
		$R_2PO(OH)$			
		KOAc	$HClO_4$		49
		NaOHc	p-MeϕSO$_3$H		
		NH_4OAc			
		LiOAc			
		AgOAc			
		$Sr(OAc)_2$			
		$Ba(OAc)_2$			
		$Ca(OAc)_2$			
		$Mn(OAc)_2$			
		$Zn(OAc)_2$			
		$Pb(OAc)_2$			
		$Cu(OAc)_2$			
		$Mg(OAc)_2$			
		$Ni(OAc)_2$			
		$Cd(OAc)_2$			
		$Co(OAc)_2$			

II. EMF DATA

iso-PROPANOL (iPrOH)

Indicator electrode	Reference electrode	Solute	Titrant	Comments	Ref.
Glass	AgCl	H_2SO_4	Et_4NOH (in ϕH—MeOH)	Constant rate delivery; chronopotentiometry	54
		$HClO_4$	KOH (in MeOH)		
		$(CO_2H)_2$			
		ϕCO_2H			
		HCO_2H			
		$(CHCO_2H)_2$			
		$o\text{-}HO\phi CO_2H$			
		$CH_2(CO_2H)_2$			
Glass	Calomel	H_3BO_3			
		H_3PO_4	NaOMe	K_a acids	56
		$RPO(OH)_2$			
		R_2POOH			
		HOAc	Bu_3MeNOH		55
		$HClO_4$	Bu_3MeNOH		5

PROPIONIC ACID (CH$_3$CH$_2$CO$_2$H)

Indicator electrode	Reference electrode	Solute	Titrant	Comments	Ref.
Glass	AgCl	o-AcOϕCO$_2$C$_2$H$_4$NEt$_2$·HCl	HClO$_4$ (in dioxane)		19
	Calomel	py; NaOAc	HClO$_4$	Cf. acetic acid	57

TRIFLUOROACETIC ACID

Indicator electrode	Reference electrode	Solute	Comments	Ref.
Glass	Calomel	HClO$_4$	Bu$_3$MeNOH	5

II. EMF DATA

References

1. G. Jander and G. Winkler, J. Inorg. Nuclear Chem., **9**, 24 (1959).
2. G. Jander and G. Winkler, J. Inorg. Nuclear Chem., **9**, 32 (1959).
3. M. Prytz, Acta. Chem. Scand., **1**, 507 (1947).
4. R. A. Glenn, Analyt. Chem., **25**, 1916 (1953).
5. E. A. M. F. Dahmen and H. B. Van der Heijde, Anal. Chim. Acta., **16**, 378 (1957).
6. A. Heyrovsky and O. Tomicek, Coll. Czech. Chem. Comm., **15**, 984 (1950).
7. A. Heyrovsky and O. Tomicek, Coll. Czech. Chem. Comm., **15**, 997 (1950).
8. G. Jander and H. Klaus, J. Inorg. Nuclear Chem., **1**, 228 (1955).
9. J. Belisle and D. J. Pietrzyk, Analyt. Chem., **38**, 969 (1966).
10. V. K. Kondratov, N. V. Malysheva and N. D. Rus'yanova, Russ. J. Anal. Chem., **21**, 887 (1966).
11. E. L. Filippov, A. V. Gordievskii and V. S. Shterman, Russ. J. Anal. Chem., **20**, 1214 (1965).
12. R. L. Maute and M. L. Owens, Analyt. Chem., **20**, 1177 (1955).
13. J. S. Fritz and M. O. Fulda, Analyt. Chem., **25**, 1037 (1953).
14. K. Stuerzer, Z. Analyst. Chem., **216**, 409 (1966).
15. T. Jasinski and L. Kozlowska, Roczniki Chem., **39**, 1861 (1965).
16. O. Tomicek, Coll. Czech. Chem. Comm., **13**, 116 (1948).
17. O. W. Kolling, Inorg. Chem., **1**, 561 (1962).
18. Z. Hippe and T. Krzyzanouska, Chem. Analit., **10**, 179 (1965).
19. R. Clegarski, H. Elbert and A. Regosz, Chem. Analit., **10**, 1217 (1965).
20. P. W. Schenk, Angew, Chem. Inter. Edn., **5**, 554 (1966).
21. W. M. Baumann and W. Simon, Z. Analyt. Chem., **216**, 273 (1966).
22. V. A. Drozdov, N. A. Kolchina and A. P. Kreshkov, Russ. J. Phys. Chem., **40**, 1159 (1966).
23. W. Huber, Z. Analyt. Chem., **216**, 260 (1966).
24. H. B. Van der Heijde, Anal. Chim. Acta., **16**, 392 (1957).
25. R. Schall, J. Chim. Phys., **52**, 719 (1955).
26. M. C. Day and G. M. Rouayheb, J. Chem. and Eng. Data, **5**, 508 (1960).
27. R. N. Lal and T. K. Sen, Perf. Essen. oil Rec., 731 (1965).
28. A. J. Gilman and S. Y. Shen, J. Chem. Educ., **42**, 540 (1965).
29. R. T. Iwamoto, J. Kleinberg and D. C. Luehrs, Inorg. Chem., **4**, 1739 (1965).
30. V. A. Aliferova, V. G. Baranova and I. Ya Turyan, J. Analyt. Chem. (USSR), **18**, 110 (1963).
31. R. Thuaire; Bull. Soc. Chim. France, 3944 (1966).
32. J. Calderon and M. Milian, Anales De Fisica Quimica, **62B**, 1185 (1966).
33. H. Hara and C. Yoshimura, Japan Analyst, **15**, 139 (1966).
34. J. O. Macdonald, Analyt. Chem., **37**, 1170 (1965).
35. V. Z. Deal and G. E. A. Wyld, Analyt. Chem., **27**, 47 (1955).
36. J. H. Elliot, R. T. Hall and M. L. Moss, Compt. Rend., **20**, 784 (1948).
37. G. Gran, Acta. Chem. Scand., **4**, 559 (1950).
38. H. Brockmann and E. Meyer, Chem. Ber., **86**, 1514 (1953).
39. D. A. Lee, Analyt. Chem., **38**, 1168 (1966).
40. R. Alexander, E. C. F. Ko, Y. C. Mao and A. J. Parker, J. Amer. Chem. Soc., 0, 3703 (1967).
41. H. M. Koepp, H. Strehlow and H. Wendt, Z. Elektrochem., **64**, 483 (1960).
42. C. Berger and L. R. Dawson, Analyt. Chem., **24**, 994 (1952).
43. N. P. Dzyuva, N. A. Izmailov and A. M. Shkodin, Zhur. ObSchei Khim., **23**, 27 (1953).

44. M. L. Condon and J. Julliard, Bull. Soc. Chim. France, 2535 (1963).
45. D. C. Luehrs, Ph.D. Thesis, Univ. of Kansas, (1965).
46. L. N. Bykova, N. A. Kazarayan, A. P. Kreshkov, and E. S. Rubisova, Russ. J. Anal. Chem., **20**, 457 (1965).
47. A. P. Kreshkov and L. B. Kuznetsova, Russ. J. Anal. Chem., **22**, 875 (1967).
48. K. A. Komarova, E. K. Kreshkova and A. N. Yarovenko, Russ. J. Anal. Chem., **21**, 356 (1966).
49. T. Jasinski and H. Smagowski, Chem. Analit., **10**, 217 (1965).
50. H. T. Havacek and W. B. Swann, Chemist Analyst, **56**, 16 (1967).
51. V. A. Drozdov, A. P. Kreshkov and I. Yu Orlova, J. Analyt. Chem. (USSR), **21**, 186 (1966).
52. J. L. Vandenbalk, J. Pharm. Belg., **47**, 491 (1965).
53. G. Douheret, Bull. Soc. Chim. France, 2921 (1965).
54. A. P. Kreshkov, V. D. Matreev and G. P. Svistunova, J. Analyt. Chem. (USSR), **21**, 1315 (1966).
55. H. B. Van der Heijde, Anal. Chim. Acta., **16**, 392 (1957).
56. V. A. Drozdov, N. A. Kolchina and A. P. Kreshkov; Russ. J. Phys. Chem., **40**, 1159 (1966).
57. C. Hennart and E. Merlin; Chim. Anal., **40**, 20 (1958).

(e) Potentiometric Titrations in Aprotic Solvents

The following tables are intended to present a representative sample of analytical potentiometric titrations in aprotic solvents. For more detailed accounts the reader should consult several authoritative texts and reviews (see e.g.[a,b]).

The information is presented alphabetically by solvent and includes details on indicator and reference electrodes, titrand and titrant, solute and supporting electrolytes and additional comments of interest. A complete bibliography is given at the end of this section:

Symbols used:

    Me Methyl
    Et Ethyl
    Pr Propyl
    Bu Butyl
    ϕ Phenyl and in cases of higher substitution (3-BrϕOH = 3-bromophenol)
    py Pyridine and for substituted pyridines (2-Mepy = α-picoline)
    Tos 4-MeϕSO$_3^-$
    X Halogen

[a] "Titrations in Nonaqueous Solvents," W. Huber. Academic Press (1967).
[b] Analytical Chemistry, Analytical Reviews, (see e.g. April 1970).

ACETIC ANHYDRIDE (Ac$_2$O)

Indicator electrode	Reference electrode	Solute	Titrant	Comments	Ref.
Au	Au	KOAc TlOAc	AcBr AcCl	"gebremste" electrode	1
Glass		pyrazolones	HClO$_4$ (in HOAc)	Half-neutralization potentials	6
Glass	AgCl	Aspirin– CH$_2$CH$_2$– Et$_2$N·HCl	HClO$_4$ (in dioxane)		5
Glass	Calomel	R-py-urea-R' py-C$_2$H$_4$(R)-urea-R' (py=pyridyl)	HClO$_4$		2
		HClO$_4$ HI HBr H$_2$SO$_4$ HCl HNO$_3$ RCO$_2$H	KOAc	K_a of acids	3
		KOAc NH$_4$OAc NaOAc LiOAc Mn(OAc)$_2$ Ni(OAc)$_2$ CO(OAc)$_2$	HClO$_4$ p-MeφOH (in dioxane)		4

II. EMF DATA

ACETONE [$(CH_3)_2CO$]

Indicator electrode	Reference electrode	Solute	Titrant	Supporting electrolyte	Comments	Ref.
Ag	Calomel (aq)	$AgClO_4$	NaI LiCl LiBr Bu_4NI	$LiClO_4$	K_{sp}; complexes	7
Cu	Calomel	$CuCl_2$ $Cu(NO_3)_2$ $Cu(ClO_4)_2$	LiCl LiBr	$LiClO_4$ (1M)	Complex formation Hydrated salts	21
Glass	AgCl (in MeOH)	H_2SO_4 $HClO_4$ $(HCO_2)_2$ ϕCO_2H HCO_2H $HCO_2CH=CHCO_2H$ $o\text{-}\phi(OH)CO_2H$ $HCO_2CH_2CO_2H$ H_3BO_3	Et_4NOH (in ϕOH-MeOH) KOH (in MeOH)		Constant-rate method	17
	Calomel	HOAc $p\text{-}CH_3\phi SO_3H$ HCO_2H	Bu_3MeNOH (in py) Me_4NOH			8
		$HClO_4$	Bu_3MeNOH (in py)	Bu_3MeNI		9
		ϕNH_2 $\phi NHMe$	$HClO_4$		K_b of bases	10

ACETONE [$(CH_3)_2CO$] (Continued)

Indicator electrode	Reference electrode	Solute	Titrant	Supporting electrolyte	Comments	Ref.
Glass	Calomel	Urea and derivatives	Et_4NOH or $HClO_4$			12
		NH_4OAc	Et_4NOH			18
		$Cu(OAc)_2$				
		$Zn(OAc)_2$				
		$Co(OAc)_2$				
		$NH_4O_2C\phi$				
		$Mg(O_2C\phi)_2$				
		$Cu(O_2C\phi)_2$				
		$Cu(naphthionate)_2$				
		$Cu(naphthenate)_2$				
		NH_4ClO_4	KOH (in MeOH)		Propellant samples	19
			Bu_4NOH (in MeOH)			
			NaOMe (in MeOH)			
			KOH (in PrOH)			
		$HClO_4$	$R\phi NR_2$		pK of bases	20
	Glass	R_3N	HCl; NaOH		90% acetone K_b base	13
Glass and Ag	Quinhydrone	R_3N	Ag^+	$NaClO_4$ or Bu_4NClO_4	Coulometric; complex K_d	11
Glass and Sb	Calomel	$ArSO_3NH_4$	NaOMe			15
			$HClO_4$			
Graphite	Calomel	H_2SO_4	NaOH (in EtOH)			14
		H_2SiF_6				

II. EMF DATA

W	Calomel	H_2F_2 HF $F_3C(CF_2)_nCO_2H$	KOH (in EtOH)		(90% acetone) Agar bridge	16

ACETONITRILE MeCN (CH_3CN)

Indicator electrode	Reference electrode	Solute	Titrant	Supporting electrolyte	Comments	Ref.
Ag	Ag	Et_4NCl Et_4NBr Et_4NI	$AgNO_3$		Complex formation consts. K_{sp}	22
		$AgClO_4$	Et_4NCl Et_4NBr Et_4NN_3 KSCN	Et_4NBr	K_{sp}; complexes	23
	Calomel (aq)	$AgClO_4$	Et_4NCl Et_4NBr Bu_4NI	Et_4NClO_4	K_{sp}; complexes	7
Glass	Ag	$HClO_4$	Me_4NOH Et_4NOH			24
		py ϕ_2 guanidine	Picric acid	Et_4NClO_4; $LiClO_4$ 0.1–1M	Nonaqueous filling for glass electrode	25

ACETONITRILE MeCN (CH_3CN) (*Continued*)

Indicator electrode	Reference electrode	Solute	Titrant	Supporting electrolyte	Comments	Ref.
Glass	Ag	R_3N	$HClO_4$	Et_4NClO_4		31
		Weak bases	$HClO_4$ (in various solvents)		Indicator properties	38
	Calomel	$HClO_4$	Bu_3MeNOH (in py)			9
		R_3N	$H_2O/H^+(Pt)$	$LiClO_4$	Coulometric	27
		$R\phi NR_2$	$H_2O/H^+(Pt)$	$LiClO_4$	Coulometric	28
		H_2SO_4	Morpholine			29
		HNO_3				
		HCl				
		R_3N	$HClO_4$		Base strength	30
		R_3N mixtures	$HClO_4$ (in dioxane)			36
		Weak bases (mixtures)	$HClO_4$			41
Glass and H_2	AgBr	Picric acid	ϕNH_2 quinoline py ϕ_2 guanidine piperidine Et_3N Et_2NH	Et_4NBr	K_a; Et_4NBr bridge	26
Glass and Sb	Calomel	HOAc ϕSH $Me\phi SO_2H$ ϕCO_2H	Bu_3MeNOH in py Me_4NOH in py			48
$H_2(Pt)$	Ag	R_3N	$HClO_4$	Et_4NClO_4	"Polarographic"	31

II. EMF DATA

Electrode	Reference	Species	Counter-ion	Electrolyte	Notes	Ref.
Hg	Ag	Et$_4$NCl Et$_4$NBr Et$_4$NI	Hg(ClO$_4$)$_2$			22
I$_2$(Pt)	Ag	I$^-$ I$_2$ I	SnCl$_2$ FeCl$_3$	LiClO$_4$		32
Mo	Mo	R$_{4-n}$SnCl$_n$	φ$_4$AsCl φ$_4$NCl φ$_4$NBr		R=Me, Et, Pr, Bu, φ Complex formation	39
Pt	Ag	(NH$_4$)$_2$Ce(NO$_3$)$_6$	(HO)$_2$φ	Et$_4$NClO$_4$	Effect of HOAc, picrates NO$_3^-$, H$_2$O	33
		φ$_2$NH φ$_2$Benzidine Me$_4$Benzidine	Cu(ClO$_4$)$_2$			
		Cu(ClO$_4$)$_2$	Ferrocene			34
		R-Ferrocene	Cu(ClO$_4$)$_2$			35
		Bu$_4$NI	Cu(ClO$_4$)$_2$			40
		(HO)$_2$φ (H$_2$N)$_2$CS				
H$_2$		Ferrocene	(Ferrocene)$^+$	Et$_4$NClO$_4$	E^0 values; electrochem	42
Pt and Cu(Hg)	Calomel (aq)	CuCl$_2$ CuCl	Cl$^-$ Cl$^-$	Et$_4$NClO$_4$	Complex formation	37

ARSENIOUS CHLORIDE ($AsCl_3$)

Indicator electrode	Reference electrode	Solute	Titrant	Comments	Ref.
AgCl	AgCl	Et_4NCl, $FeCl_3$	py	Also conductimetric titrations	43
		Et_2NH			
		Et_4NCl			
		Et_4NCl, $SbCl_5$			

BENZENE (ϕH)

Indicator electrode	Reference electrode	Solute	Titrant	Supporting electrolyte	Comments	Ref.
QH(Pt)	QH(Pt)	Cl_3CCO_2H	Et_2NH	$i\text{-}Am_4NI$	Also ATT	44

CHLOROBENZENE (ϕCl)

Indicator electrode	Reference electrode	Solute	Titrant	Ref.
Glass	Calomel	HOAc	Bu_3MeNOH (in py)	8
		$HClO_4$	Bu_3MeNOH (in py)	45

II. EMF DATA

CHLOROFORM (CHCl₃)

Indicator electrode	Reference electrode	Solute	Titrant	Supporting electrolyte	Ref.
Glass	Calomel	RNH_2	HCl (in iPrOH)		85
Pt	Calomel	piperidine-CH_2CHMe—$CO\phi Me \cdot HCl$	$HClO_4$ (in dioxane)	$Hg(OAc)_2$	86

DIETHYL ETHER (Et₂O)

Indicator electrode	Reference electrode	Solute	Titrant	Supporting electrolyte	Comments	Ref.
Ag	AgCl AgBr AgI	$AgClO_4$	LiCl LiBr LI	$LiClO_4$	K_{sp}	46
	AgBr	$AgClO_4$	LiI	$LiClO_4$	Complex formation	47

N,N-DIMETHYLACETAMIDE (DMA)

Indicator electrode	Reference electrode	Solute	Titrant	Comments	Ref.
Ag	Ag	AgNO$_3$	Et$_4$NCl Et$_4$NBr Et$_4$NI Et$_4$NN$_3$ KSCN Bu$_4$NOAc NaBϕ_4	K_{sp} complexes	23

N,N-DIMETHYLFORMAMIDE (DMF)

Indicator electrode	Reference electrode	Solute	Titrant	Supporting electrolyte	Comments	Ref.
Ag	Ag	AgNO$_3$	Et$_4$NCl Et$_4$NBr Et$_4$NI KSCN Et$_4$NN$_3$ Bu$_4$NOAc		K_{sp}, complexes	23
Ag	AgCl	AgClO$_4$	Et$_4$NCl	Et$_4$NClO$_4$ 0.1M	complex, K_{eq}	49

II. EMF DATA

Electrode	Reference	Sample	Titrant	Notes	Ref
Chloranil	Calomel		KOH aq.		50
Glass	Calomel	Acid mixtures phthalic, succinic	NaOH alcoholic		51
		Acid groups in acrylonitrile polymers			
		R_3POOH	NaOMe		61
		ϕOH	KOH in iPrOH		48
		HOAc			
		HCl			
		$HClO_4$	Bu_3MeNOH		9
		$o\text{-}AcO\phi CO_2H$	NaOMe (in ϕH-MeOH)		57
		$Me\phi CO_2H +$ $HO_2C\phi CO_2H$ (para)	KOH		58
			Bu_4NOH (aq)		
			KOH (in EtOH)		
			KOMe (in ϕH)		
		Stearic acid	Et_4NOH		18
		$Cu(Stearate)_2$			
Glass and H_2	Calomel	RCO_2H	LiOiPr	pK_a acids	56
		ROH			
		RNH_2	HCl		
Glass, SG and Pt	Calomel, Sb Pb	RCONHOH	Bu_4NOH	Also attempted titrations in other solvents	52

N,N-DIMETHYLFORMAMIDE (DMF) (Continued)

Indicator electrode	Reference electrode	Solute	Titrant	Supporting electrolyte	Comments	Ref.
H_2	$Hg(/Hg^{II})$	py	$HClO_4$		pH scale	59
		$N(CH_2CH_2OH)_3$				
		$H_2NCH_2CH_2OH$				
		$\phi_2$guanidine				
		$BuNH_2$				
		piperdine				
		RCO_2H	LiOiPr			
		ROH				
Pt	Calomel	Glutethimide	NaOMe (in ϕH-MeOH)			55
		NH_2 Glutethimide				
		Bemegride				
	Calomel (aq)	RSH	$CuCl_2$		Bridge KCl in MeOH	53
		$CrCl_3$	$CuCl_2$		Bridge KCl in MeOH; Redox potentials	54
		$TiCl_3$				
		$FeCl_2$				
		RSH	$CuCl_2$		Bridge KCl in MeOH	60

II. EMF DATA

2,4-DIMETHYL SULFOLANE

Indicator electrode	Reference electrode	Solute	Titrant	Comments	Ref.
Glass	Calomel	R_3N	$HClO_4$ (in dioxane)	Bridge Et_4NClO_4 in 3-Me-sulfolane	62
		R_3NO	Bu_4NOH (in iPrOH)		
		RCO_2H			
		ROH			
		HCO_3			
		$HClO_4$			
		$R(CO_2H)_2$			

DIMETHYL SULFOXIDE (DMSO)

Indicator electrode	Reference electrode	Solute	Titrant	Supporting electrolyte	Comments	Ref.
Ag	Ag	$AgNO_3$	Et_4NCl		K_{sp}-complexes	23
			Et_4NBr			
			Et_4NI			
			Et_4NN_3			
			KSCN			
			Bu_4NOAc			

DIMETHYL SULFOXIDE (DMSO) (Continued)

Indicator electrode	Reference electrode	Solute	Titrant	Supporting electrolyte	Comments	Ref.
Ag	Ag	AgNO$_3$	LiCl		K_{sp}-complexes	64
			KBr			
			KI			
			KCN			
	Calomel (aq)	AgClO$_4$	KCN	Et$_4$NClO$_4$	K_{sp}-complexes	7
			Et$_4$NCl			
			Et$_4$NBr			
			Bu$_4$NI			
			KSCN			
	Calomel (non aq)	AgNO$_3$	LiCl	NH$_4$NO$_3$ 0.1M	K_{sp}-complexes	63
			LiBr			
			LiI			
Glass	AgCl (in DMSO)	CO$_2$	HClO$_4$	Et$_4$NClO$_4$		66
	Calomel	R$_3$POOH	NaOMe			61
		BuNH$_3$ClO$_4$	NaOMe			65
		BuNH$_3$Cl	KOMe			
		NH$_4$ClO$_4$	Bu$_4$NOH			
		NH$_4$Cl				
		Bu$_2$NH$_2$Cl				
		C$_9$H$_{19}$NH$_3$Cl				

II. EMF DATA

Indicator electrode	Reference electrode	Solute	Titrant	Comments	Ref.
Sb	Calomel	N$_2$H$_6$SO$_4$			
		Alanine			
		Pr$_3$NHCl			
		C$_{16}$H$_{33}$NH$_3$Cl			
		NH$_4$	NaOMe 0.1N		67
		CdX$_2$			
		ZnX$_2$			
		NiX$_2$			
		MnX$_2$			
		PbX$_2$			
		SnX$_2$			
		MgX$_2$			
		RϕCO$_2$H	NaOMe	K_a weak acids Hammett equation	68
		RϕOH			

ETHYL ACETATE (EtOAc)

Indicator electrode	Reference electrode	Solute	Titrant	Comments	Ref.
Glass	Calomel	R$_3$N	HClO$_4$ (in dioxane)	half-neutralization potentials	30
		p-MeϕSO$_3$H			
		Me-SO$_3$H			

ETHYL METHYL KETONE (EtCOMe)

Indicator electrode	Reference electrode	Solute	Titrant	Comments	Ref.
Glass	Calomel	R-py-urea-R′ py-C_2H_4-(R)-urea-R′	Et_4NOH (in $\phi H + MeOH$)		2
		RNHCSNHR py-C_2H_4NRCSNHR′ py-CH_2NRCSNHR′ py-$(CH_2)_n$-X-CY-NHR	Et_4NOH		69
		iBuOPO(OH)Me (iBuOPO(OH)Me)$_2$O	Et_4NOH HCl (in MeOH)	Back titration of Et_4NOH	70
		R-pyCO_2H	Et_4NOH		56

ETHYLENE DICHLORIDE (1,2-DICHLOROETHANE)

Indicator electrode	Reference electrode	Solute	Titrant	Ref.
Glass	Calomel	R_3N	$HClO_4$ p-MeϕSO_3H	30

II. EMF DATA

HEXAMETHYL PHOSPHORAMIDE (Me_6N_3PO)

Indicator electrode	Reference electrode	Solute	Titrant	Comments	Ref.
Ag	Ag	$AgNO_3$	Et_4NCl Et_4NBr Et_4NN_3 KSCN	K_{sp}-complexes	23

ISOBUTYL METHYL KETONE (MIBK) (iBuCOMe)

Indicator electrode	Reference electrode	Solute	Titrant	Comments	Ref.
Glass	Calomel (aq)	RSO_3H	Naphthalene SO_3H p-MeϕSO H $2,4(NO_2)_2\phi SO_3H$ $HClO_4$	Hammett functions	71
Glass and Sb	Calomel	o-AcOϕCO$_2$H Barbiturates	NaOMe (in ϕH + MeOH)		72

3-METHYL SULFOLANE

Indicator electrode	Reference electrode	Solute	Titrant	Comments	Ref.
Glass	Calomel	R_3N R_3NO RCO_2H ROH HNO_3 $HClO_4$ $R(CO_2H_2)$	$HClO_4$ (in iPrOH)	Bridge Et_4NClO_4	62

METHYL THIOCYANATE (MeSCN)

Indicator electrode	Reference electrode	Solute	Titrant	Comments	Ref.
Glass	Calomel (in MeOH)	R_3N RCO_2H	$HClO_4$ Me_4NOH (in ϕH(MeOH)	half-neutralization potential = $f(K_b^{H_2O}), f(K_a^{H_2O})$	73

II. EMF DATA

NITROBENZENE (ϕNO_2)

Indicator electrode	Reference electrode	Solute	Titrant	Supporting electrolyte	Comments	Ref.
Glass	Calomel	R_3N	$HClO_4$		Basicity studies	30
Glass	Calomel (in MeOH)	$Ar(R)(OH)(CH_2)_nNR_2$	$HClO_4$ (in dioxane)		Basicity studies	75
Quinhydrone	Quinhydrone	$MeCONH_2$	$HClO_4$	(piperidine ClO_4 bridge)	pK weak bases	74
		$MeCONH\phi$				
		iPr_2O				
		Et_2O				
		Bu_2O				
		Me_2pyrone				

NITROETHANE ($CH_3CH_2NO_2$)

Indicator electrode	Reference electrode	Solute	Titrant	Supporting electrolyte	Comments	Ref.
Ag	Calomel (aq)	$AgClO_4$	Et_4NCl	Et_4NClO_4	K_{sp}-complexes	7
			Et_4NBr			
			Bu_4NI			

NITROMETHANE (MeNO$_2$)

Indicator electrode	Reference electrode	Titrant	Supporting electrolyte	Ref.
Glass	Calomel	R$_3$N	HClO$_4$ p-MeϕSO$_3$H	30

PROPYLENE CARBONATE (PC)

Indicator electrode	Reference electrode	Solute	Titrant	Supporting electrolyte	Comments	Ref.
Ag	Ag	AgClO$_4$	Et$_4$NCl	Et$_4$NClO$_4$ 0.1N	Complex K_{eq}	76

II. EMF DATA

PYRIDINE (PY)

Indicator electrode	Reference electrode	Solute	Titrant	Ref.
C, Glass, and Sb	Calomel	$ArSO_3Na$	NaOMe (in MeOH/ϕH)	15
Glass	Calomel	ϕOH	Bu_3MeNOH	8
		4 Me-2,6-$(tBu)_2\phi OH$		
		ϕOH	Bu_3MeNOH (in iPrOH)	
		$HClO_4$	Bu_3MeNOH	9
	Calomel (in MeOH)	Me_2SO_4	Bu_3EtNOH	77
		$MeHSO_4$		
		H_2SO_4		
		$RNHCH_2CH_2S_2O_3H$	Bu_4NOH	78
Sb, Te, and Glass	Calomel	$(RCO_2)_nSnR_{(4-n)}$	NaOMe (in MeOH/ϕH)	79

PYRIDINE

Indicator electrode	Reference electrode	Solute	Titrant	Ref.
Pt	Sb, Pt	$(\phi CO)_2CH_2$	Bu_4NOMe (in ϕH-MeOH)	80
		$(RCO)_2NH$		
		$AcOCH_2CO_2Et$		
		ϕSO_2NH_2		

SULFOLANE (TETRAMETHYLENE SULFONE)

Indicator electrode	Reference electrode	Solute	Titrant	Comments	Ref.
Glass	Calomel	R_3N	$HClO_4$ (in dioxane)	Bridge: Et_4NClO_4 in 3-Me sulfolane	62
		R_3NO	Bu_4NOH (in iPrOH)		
		RCO_2H			
		ROH			
		HCO_3			
		$HClO_4$			
		$R(CO_2H)_2$			

TETRAETHOXYSILANE

Indicator electrode	Reference electrode	Solute	Comments	Ref.
QH	AgCl	$R\phi NH_2$	$HClO_4$ (in dioxane)	81

II. EMF DATA

TETRAHYDROFURAN (THF) (C_4H_8O)

Indicator electrode	Reference electrode	Solute	Titrant	Supporting electrolyte	Comments	Ref.
H_2	Ag	Bu_4NOH	$HClO_4$	Bu_4NClO_4	pH scale	83
Pt	Ag	p-$\phi(OH)_2$	$LiAlH_4$	$LiClO_4$	by back titration	82
		BuOH				
		iBuOH				
		AmOH				
		$C_{18}H_{37}OH$				
		cholesterin				
		$C_{20}H_{41}OH$				
		ϕOH				

TETRAMETHOXYSILANE

Indicator electrode	Reference electrode	Solute	Titrant	Ref.
QH	AgCl	$R\phi NH_2$	$HClO_4$ (in dioxane)	81

TETRAPROPOXYSILANE

Indicator electrode	Reference electrode	Solute	Titrant	Comments	Ref.
QH	AgCl	RφNH₂	HClO₄ (in dioxane)	end-point poor	81

TRIBUTYL PHOSPHATE (Bu₃PO₄)

Indicator electrode	Reference electrode	Solute	Titrant	Supporting electrolyte	Comments	Ref.
Glass	AgCl Calomel	Bu₃N	HClO₄ HCl HNO₃		H₂O sat.	84
Pt	AgCl	FeCl₃	U(ClO₄)₄	HClO₄		

II. EMF DATA

References

1. G. Jander and H. Surawski, Z. Elektrochem., **65**, 527 (1961).
2. V. K. Kondratov and E. G. Novikov, J. Analyt., Chem. (USSR), **22**, 1050 (1967).
3. V. V. Belozerskaya, K. I. Evstramova, and N. A. Goncharova, Elektrokhimiya, **5**, 107 (1969).
4. T. Jasinski and H. Smagowski, Chem. Analit, **10**, 217 (1965).
5. R. Clegarski, H. Ellert, and A. Regosz, Chem. Analit., **10**, 1217 (1965).
6. N. P. Dzyuba and V. P. Georgierskii, J. Analyt. Chem. (USSR), **22**, 107 (1967).
7. D. C. Luehrs, Ph.D. Thesis, Univ. of Kansas (1965).
8. H. B. Ban Der Heijde, Anal. Chim. Acta., **16**, 392 (1957).
9. E. A. M. F. Dahmen and H. B. Van Der Heijde, Anal. Chim. Acta., **16**, 378 (1957).
10. F. Aufauvre and M. L. Dondon, Bull. Soc. Chim. France, **716** (1963).
11. D. L. Maricle, K. K. Mead, and C. A. Steali, Analyt. Chem., **37**, 237 (1965).
12. A. P. Kreshkov, V. N. Nerskaya, and A. N. Yarovenko, Russ. J. Anal. Chem., **21**, 307 (1966).
13. K. I. Evstratova, N. A. Goncharova, A. I. Ivanova, V. I. Kurov, and V. Ya. Solomko, Russ. J. Anal. Chem., **22**, 978 (1967).
14. E. V. Bezrogova, Zhur. Analit. Khim, **19**, 1498 (1964).
15. T. Jasinski and R. Korewa, Chem. Analit, **13**, 1319 (1968).
16. M. N. Chelnokova and E. I. Patyukova, Russ. J. Anal. Chem., **21**, 795 (1966).
17. A. P. Kreshkov, V. D. Matreev, and G. P. Svistunova, J. Analyt. Chem. (USSR), **21**, 1315 (1966).
18. G. M. Gal'pern, V. A. Il'ina and V. V. Nozhenkina, Indust. Lab., **32**, 1065 (1966).
19. R. J. Baczuk and R. J. Dubois, Analyt. Chem., **38**, 623 (1966).
20. F. Aufauvre, M. Dantonnet and M. L. Dondon, Bull. Soc. Chim. France, 3566 (1965).
21. J. Gazo and Z. Kompisova, Chem. Zvesti., **20**, 105 (1966).
22. J. J. Campion, Ph.D. Thesis, Univ. of Pittsburgh (1966).
23. R. Alexander, E. C. F. Ko, Y. C. Mac and A. J. Parker, J. Amer. Chem. Soc., **89**, 3703 (1967).
24. M. K. Chantooni and I. M. Kolthoff, J. Amer. Chem. Soc., **87**, 4428 (1965).
25. J. Badoz-Lambling, J. Desbarres and J. Tocussel, Bull. Soc. Chim. France, 53 (1962).
26. K. Cruse and E. Tomberg, Z. Elektrochem., **63**, 404 (1959).
27. C. A. Streuli, Analyt. Chem., **28**, 130 (1965).
28. R. B. Hanselman and C. A. Streuli, Analyt. Chem., **28**, 916 (1956).
29. F. E. Critchfield and J. B. Johnson, Analyt. Chem., **26**, 1803 (1954).
30. H. K. Hall, J. Phys. Chem., **60**, 63 (1956).
31. J. Desbarres, Bull. Soc. Chim. France, 2103 (1962).
32. J. Desbarres, Bull. Soc. Chim. France, 502 (1961).
33. V. Pleskov, Acta. Physicochim (USSR), **21**, 41 (1946).
34. B. Kratochvil and P. F. Quirk, Analyt. Chem., **42**, 402 (1970).
35. B. Kratochvil and P. F. Quirk, Analyt. Chem., **42**, 535 (1970).
36. J. S. Fritz, Analyt. Chem., **25**, 407 (1953).
37. R. T. Iwamoto and S. E. Manahan, Inorg. Chem., **4**, 1409 (1965).
38. S. Bhowmik, M. K. Chantooni and I. M. Kolthoff, Analyt. Chem., **39**, 1627 (1967).
39. G. Tagliavini and P. Zanella, Anal. Chim. Acta., **40**, 33 (1968).
40. B. Kratochvil, R. Markuszewski and D. A. Zatko, Analyt. Chem., **38**, 770 (1966).
41. W. Huber, Z. Analyt. Chem., **216**, 260 (1966).
42. H. M. Koepp, H. Strehlow and H. Wendt, Z. Elektrochem., **64**, 483 (1960).

43. L. H. Anderson and I. Lindquist, Acta. Chem. Scand., **9,** 79 (1955).
44. H. C. Downes and V. K. La Mer, J. Amer. Chem. Soc., **53,** 888 (1931).
45. H. A. Laitinen and C. J. Nyman, J. Amer. Chem. Soc., **70,** 3002 (1948).
46. B. Athin, L. G. Sillen and E. Wahlin, Acta. Chem. Scand., **3,** 321 (1949).
47. B. Adin, L. Evers and L. G. Sillen, Acta. Chem. Scand., **6,** 759 (1952).
48. V. Z. Deal and G. E. A. Wyld, Analyt. Chem., **27,** 47 (1955).
49. J. N. Butler, J. Phys. Chem., **72,** 3288 (1968).
50. N. Daune-Dubois and A. Kirrman, Compt. Rend., **236,** 1361 (1953).
51. V. Groebe, B. Phillip, H. Reichert and A. Tryonadi, Faserforschung U. Textiltechnik, **15,** 304 (1964).
52. R. Christian and T. W. Stamey, Talanta, **13,** 144 (1966).
53. Z. Hladky and J. Vrestal, Coll. Czech. Chem. Comm., **34,** 1098 (1969).
54. Z. Hladky and J. Vrestal, Coll. Czech. Chem. Comm., **34,** 984 (1969).
55. S. P. Agarwal and M. I. Blake, J. Pharm. Sci., **54,** 1668 (1966).
56. R. Schaal and M. Teze, Bull. Soc. Chim. France, 1372 (1962).
57. M. I. Blake and S. L. Lin, Analyt. Chem., **38,** 549 (1966).
58. R. E. Lewis and F. C. Trussell, Anal. Chim. Acta., **34,** 243 (1966).
59. G. H. Mergele and C. D. Ritchie, J. Amer. Chem. Soc., **89,** 1447 (1967).
60. Z. Hladky, Wiss. Z., **9,** 5 (1967).
61. T. Jasinski, A. Modro, Chem. Analit., **10,** 929 (1965).
62. G. H. Harlow and D. H. Morman, Analyt. Chem., **39,** 1869 (1967).
63. H. L. Peeters and N. A. Rumbaut, Bull. Soc. Chim. Belges, **76,** 33 (1967).
64. V. Milicevic and T. Skerlak, Glasnik Drustva Hem. Nrbih, **11,** 49 (1962).
65. K. K. Barnes and C. K. Mann, Analyt. Chem., **36,** 2502 (1964).
66. L. V. Haynes and D. T. Sawyer, Analyt. Chem., **39,** 332 (1967).
67. T. Jasinski and K. Stefaniuk, Chemia Anal., **10,** 983 (1967).
68. T. Jasinski and K. Stefaniuk, Chem. Analit., **10,** 211 (1965).
69. V. K. Kondratov and E. G. Novikov, Russ. J. Anal. Chem., **23,** 542 (1968).
70. V. A. Drozdov, N. A. Kolchina and A. P. Kreshkov, Indust. Lab. (Zavodskaya Lab p. 160) **31,** 193 (1965).
71. J. Belisle and D. J. Pietrzyk, Analyt. Chem., **38,** 969 (1966).
72. M. I. Blake and S. L. Lin, J. Pharm. Sci., **55,** 781 (1966).
73. T. Jasinski, R. Korewa and H. Smagowski, Chem. Analit., **11,** 745 (1966).
74. R. Schall, J. Chim. Phys., **52,** 719 (1955).
75. S. V. Bogatkov, E. M. Cherkosova and E. Ya. Skobeleva, Russ. J. Gen. Chem., **36,** 138 (1966).
76. J. N. Butler, Analyt. Chem., **39,** 1799 (1967).
77. W. M. Banick and E. C. Francis, Talanta, **13,** 979 (1966).
78. J. O. MacDonald, Analyt. Chem., **37,** 1170 (1965).
79. A. Groagova and M. Pribyl, Fres. Z. Anal. Chem., **234,** 423 (1968).
80. D. A. Lee, Analyt. Chem., **38,** 1168 (1966).
81. Z. Kokot and W. Rodziewicz, Chem. Analit., **11,** 175 (1966).
82. T. Higuchi, C. J. Lintner and R. H. Schleif, Analyt. Chem., **22,** 534 (1950).
83. R. Buvet and J. Perichon, Bull. Soc. Chim. France, 3697 (1967).
84. H. A. C. Mckay, J. H. Miles and M. J. Waterman, J. Chem. Soc., A, 4209 (1964).
85. J. E. Jackson, Anal. Chem., **25,** 1764 (1953).
86. E. Sisman and R. Vasiliev, Farmacia, **11,** 393 (1963).

(f) Potentiometric Titrations in Mixed Solvents

The following tables are intended to present a representative sample of analytical potentiometric titrations in mixed solvent systems. For more detailed accounts the reader should consult several authoritative texts and reviews (see e.g.[a,b]).

The information is presented alphabetically by solvent and includes details on indicator and reference electrodes, titrand and titrant, solute and supporting electrolytes and additional comments of interest. A complete bibliography is given at the end of the section.

Symbols used:

- Me   Methyl
- Et   Ethyl
- Pr   Propyl
- Bu   Butyl
- ϕ Phenyl and in cases of higher substitution (3-BrϕOH = 3-bromophenol).
- py   Pyridine and for substituted pyridines (2-Mepy = α-picoline)
- tos   4-MeϕSO$_3$
- X   Halogen

[a] "Titrations in Nonaqueous Solvents," W. Huber, Academic Press, (1967).
[b] Analytical Chemistry, Analytical Reviews, (see e.g. April 1970).

ACETIC ACID-ACETIC ANHYDRIDE

Indicator electrode	Reference electrode	Solute	Titrant	Ref.
Glass	Calomel	R py R_2 py Quinoline R py CO_2H	$HClO_4$	1
		ε-Caprolactam	$HClO_4$	2

ACETIC ACID–ACETONITRILE

Indicator electrode	Reference electrode	Solute	Titrant	Comments	Ref.
Pt	Glass and Sb	$(NH_4)_2Ce(NO_2)_6$	$FeCl_2$	Under N_2	3

ACETIC ANHYDRIDE–BENZENE + CHLOROFORM

Indicator electrode	Reference electrode	Solute	Titrant	Comments	Ref.
Glass	Calomel	Acetophenetidin Caffeine	$HClO_4$ (in HOAc + Ac_2O)	A C B 1:1:9 mixture	4, 5

II. EMF DATA

ACETIC ACID–CHLOROFORM

Indicator electrode	Reference electrode	Solute	Titrant	Comments	Ref.
Glass	Calomel	R py R$_2$ py Quinoline Isoquinoline Quinaldine	HClO$_4$	50% HOAc	1
		KOAc NaOAc LiOAc	HClO$_4$ (in dioxane)	95% CHCl$_3$	6
		KOAc HaOAc LiOAc NH$_4$OAc	HClO$_4$ (in dioxane)	10% HOAc	7

ACETONE–DIETHYLAMINE

Indicator electrode	Reference electrode	Solute	Titrant	Comments	Ref.
Glass	Calomel	(XϕO)$_n$SiR$_{4-n}$	Bu$_4$NOH	1:2 mixture	8

322   NONAQUEOUS ELECTROLYTES

ACETONE–METHANOL

Indicator electrode	Reference electrode	Solute	Titrant	Comments	Ref.
Glass	Calomel (in MeOH)	LiCl, MnCl$_2$, FeCl$_2$, CrCl$_2$, ZnCl$_2$, CoCl$_2$, SnCl$_2$, SrCl$_2$, Mn(NO$_3$)$_2$, La(NO$_3$)$_3$, Hg(NO$_3$)$_2$, Co(NO$_3$)$_2$, Cd(NO$_3$)$_2$, Zn(NO$_3$)$_2$, Cu(NO$_3$)$_2$, Fe(NO$_3$)$_2$, Mg(NO$_3$)$_2$, Al(NO$_3$)$_3$, Ni(NO$_3$)$_2$	Et$_4$NOH	Differentiating titrations	9

ACETONE–WATER

Indicator electrode	Reference electrode	Solute	Titrant	Supporting electrolyte	Comments	Ref.
AgCl	AgCl	AgNO$_3$	KCl		KNO$_3$	K_{sp}	
Glass	AgCl	HOAc ϕCO$_2$H	KOH	KCl; KNO$_3$	K_a; solvent effects	10
Glass		RϕNR$_2$	HClO$_4$	KCl	0–100% Me$_2$CO; Base strength = f (solvents)	11

II. EMF DATA

Indicator electrode	Reference electrode	Solute	Titrant	Comments	Ref.
Glass	Calomel	Dibasol, Dionine, Papaverine	KOH	Differentiating titration 80% Me_2CO	12
(Pt) H_2NCSNH_2 + $(H_2NC(NH)S)_2$	Calomel (aq.)	H_2NCSNH_2	$Zn(ClO_4)_2$, $Cd(ClO_4)_2$, $Co(ClO_4)_2$, $Ni(ClO_4)_2$	$HClO_4$, $NaClO_4$, $I = 2$; 50, 80, 90, 95% acetone; Complex formation constants	13

ACETONITRILE–CHLOROFORM

Indicator electrode	Reference electrode	Solute	Titrant	Comments	Ref.
Glass	Calomel	$R(CO_2H)_2$	KOH (in EtOH), Et_4NOH (in ϕH + MeOH)	1:4 mixture; differentiating titrations	14

BENZENE–CHLOROFORM

Indicator electrode	Reference electrode	Solute	Titrant	Comments	Ref.
Glass	Calomel	Nicotine and derivatives	$HClO_4$ (in HOAc)	9:1 mixture by vol.	15

BENZENE–METHANOL

Indicator electrode	Reference electrode	Solute	Titrant	Comments	Ref.
Sb	Calomel	RCO_2H	NaOMe	3:1 mixture by vol. (wool wax acids)	16

BENZENE–PYRIDINE

Indicator electrode	Reference electrode	Solute	Titrant	Comments	Ref.
Glass	Calomel (in MeOH)	RCO_2H, $R\phi OH$, ROH, weak acids	Bu_4NOH (in MeOH)	As petroleum impurities differentiating titration 1:1 mixture	17

II. EMF DATA

CHLOROFORM–ISOPROPANOL

Indicator electrode	Reference electrode	Solute	Titrant	Comments	Ref.
Glass	Calomel	Isoquinoline (+ MeI)	KCN	20% CHCl$_3$ by vol.	18

CHLOROFORM–PROPYLENE GLYCOL

Indicator electrode	Reference electrode	Solute	Titrant	Comments	Ref.
Glass	Calomel	H$_2$NϕNH$_2$	HCl, HClO$_4$, H$_2$SO$_4$, (in glycol)	3:1 mixture	19

DIMETHYL SULFOXIDE–CHLOROBENZENE

Indicator electrode	Reference electrode	Solute	Titrant	Comments	Ref.
Glass	Calomel	Mineral oil acids	Me$_4$NOH	Analytical "acid numbers"	20

DIMETHYL SULFOXIDE–WATER

Indicator electrode	Reference electrode	Solute	Titrant	Supporting electrolyte	Comments	Ref.
Ag	Ag	LiCl	$AgNO_3$	$LiClO_4$ 0.1M Et_4NClO_4 0.1M	Complex formation constants	21
AgCl	AgCl	$AgNO_3$	KCl	KNO_3	K_{sp}	10
		HOAc	KOH	KCl; KNO_3	K_a; solvent effects	10
		ϕCO_2H				
Glass	Calomel	o-RϕCO_2H			20, 35, 50, 65, 85, 90 vol. % DMSO pK_a, ortho effect	22
		p-RϕCO_2H				

DIOXIDE–WATER

Indicator electrode	Reference electrode	Solute	Titrant	Supporting electrolyte	Comments	Ref.
Glass	AgCl	$Be(ClO_4)_2$	$HClO_4$	$LiClO_4$ 3M	0.2 mole fraction dioxane	25
			LiOH		Hydrolysis constants of Be^{++}	

II. EMF DATA

Indicator electrode	Reference electrode	Solute	Titrant	Supporting electrolyte	Comments	Ref.
Glass	Calomel	$BeSO_4$	$AcCH_2CONH\phi$		10–40°; 50% dioxane complex formation $\Delta G°, \Delta H°, \Delta S°$	23
		2-py CO_2H 2-py CH_2CO_2H 6-Me 2-py NHMe	NaOH	$(HClO_4)$ KNO_3 0.1M	50% dioxane K_a	24
		$Cu(ClO_4)_2$ $Ni(ClO_4)_2$ $Co(ClO_4)_2$ $Zn(ClO_4)_2$ $Cd(ClO_4)_2$	6-Me 2-py NHMe 2-py CH_2CO_2H 2-py CO_2H		Complex formation constants	

ETHANOL–WATER

Indicator electrode	Reference electrode	Solute	Titrant	Supporting electrolyte	Comments	Ref.
Ag	Calomel	$AgNO_3$	$H_2NCH_2CH_2OH$ $HN(CH_2CH_2OH)_2$ $N((CH_2CH_2OH)_2$	$LiNO_3$ 0.4M	0–100% EtOH	26
AgCl	AgCl	$AgNO_3$	KCl	KNO_3	K_{sp}	10
Glass	AgCl	HOAc ϕCO_2H	KOH	KCl; KNO_3	K_a; solvent effects	10
	Calomel	RR'POOH	NaOH		pK_a values 75 and 95% EtOH	27

ETHYLENE GLYCOL–ISOPROPANOL

Indicator electrode	Reference electrode	Solute	Titrant	Comments	Ref.
Glass	Calomel	$HO(CH_2)_2NMe_3OH$	$HClO_4$ (in dioxane)	As impurities in CH_3 = CHCN 50 vol. % iPrOH	1
		RCO_2H	NaOH (in MeOH)		

ETHYLENE GLYCOL–METHANOL

Indicator electrode	Reference electrode	Solute	Titrant	Ref.
Glass	Calomel	$M_nH_{3-n}PO_4$	NaOH, HCl	29

II. EMF DATA

ETHYL METHYL KETONE–METHANOL

Indicator electrode	Reference electrode	Solute	Titrant	Comments	Ref.
Glass	Calomel	$NaNO_3$	Et_4NClO_4	NO_3^- by ion exchange followed by titration	28
		KNO_3			
		NH_4NO_3			
		$Ca(NO_3)_2$			
		$Sr(NO_3)_2$			
		$Ni(NO_3)_2$			
		$Cr(NO_3)_3$			
		$Cd(NO_3)_2$			
		KNO_2	$HClO_4$		
		$NaNO_2$			
		$Rpy\ CO_2H$	Et_4NOH		1

FORMAMIDE–WATER

Indicator electrode	Reference electrode	Solute	Titrant	Comments	Ref.
Glass	Calomel	$CuCl_2$	R_2NH	Complex formation, 50, 76, 87, 100% $HCONH_2$	30
		$ZnCl_2$			

FORMIC ACID–NITROMETHANE

Indicator electrode	Reference electrode	Solute	Titrant	Comments	Ref.
Glass	Calomel	R_4NOAc	$HClO_4$ (in dioxane)	3% HCO_2H, analysis of R_4NX from CH_2CHCN copolymerization	31

METHANOL–WATER

Indicator electrode	Reference electrode	Solute	Titrant	Supporting electrolyte	Comments	Ref.
AgCl	AgCl	$AgNO_3$	KCl	KNO_3	K_{sp}	
Glass	AgCl	HOAc ϕCO_2H	KOH	KCl; KNO_3	K_a; solvent effects	10
Glass	Calomel	RCO_2H $RASO(OH)_2$ RCOSH $RB(OH)_2$ $(RCO)_2CH_2$ $RPO(OH)_2$ $R\phi OH$	KOH Bu_4NOMe		$K_a = f$ (solvent) 0, 20, 39, 55, 70, 84, 100% wt. MeOH	32

II. EMF DATA

Solute	Titrant	Supporting electrolyte	Comments	Ref.
LaCl$_3$ PrCl$_3$ NdCl$_3$ SmCl$_3$ EnCl$_3$ GdCl$_3$ TbCl$_3$ DyCl$_3$ HOCl$_3$ ErCl$_3$ YCl$_3$ HCl	Ac$_2$CH$_2$ (ϕCO)$_2$CH$_2$ 2-HOϕCHO	NaCl	Complex formation constants	33
(NO$_2$)$_2\phi$H Cl$_2\phi$NO$_2$OH Br$_2\phi$NO$_2$OH	KOH		20, 40, 60, 80% MeOH K_a of phenols, cf. spectra	34

PROPYLENE CARBONATE–WATER

Indicator electrode	Reference electrode	Solute	Titrant	Supporting electrolyte	Comments	Ref.
Ag	Ag	AgClO$_4$ Et$_4$NCl	Et$_4$NCl AgClO$_4$	Et$_4$NClO$_4$ 0.1M	0.004–3.57M H$_2$O complex formation constants	35

References

1. V. K. Kondratov, N. V. Malysheva and N. D. Rus'yanova, Russ. J. Anal. Chem., **21**, 887 (1966).
2. V. K. Akimov, S. M. Gel'fer and B. N. Kolokolov, Russ. J. Anal. Chem., **21**, 646 (1966).
3. A. R. V. Murthy and G. P. Rao, Indian J. Chem., **4**, 49 (1966).
4. M. I. Blake and S. L. Lin, J. Pharm. Sci., **54**, 1512 (1965).
5. M. I. Blake and S. L. Lin, Anal. Chem., **38**, 549 (1966).
6. M. N. Das and I. L. Shresta, Anal. Chem., **39**, 1300 (1967).
7. C. W. Pifer, M. Schmall and E. G. Wollish, Anal. Chem., **26**, 215 (1954).
8. V. A. Drozdov, V. N. Knyazev and A. P. Kreshkov, Russ. J. Anal. Chem., **22**, 373 (1967).
9. A. P. Kreshkov, E. N. Sayvshkina, A. N. Yarovenko and L. N. Zeknina, Izv. Vyssh. Vcheb. Zaved. Khim. Tekhnol., **8**, 196 (1965).
10. J. P. Morel, D.Sc. Thesis, Univ. Clermont-Ferrand (1969).
11. F. Aufauvre, D.Sc. Thesis, Univ. Clermont-Ferrand (1969).
12. K. I. Evstratova and I. M. Perelman, Aptechnoe Delo., **12**, 27 (1963).
13. E. I. Arykova, T. V. Kramareva, S. V. Larinonov, V. M. Shul'man and V. V. Yudina, Russ. J. Phys. Chem., **11**, 580 (1966).
14. A. P. Kreshkov, Wiss. Z., **6**, 255 (1964).
15. R. H. Cundiff and P. C. Markunas, Anal. Chem., **27**, 650 (1955).
16. E. T. Donahue and J. Radell, Anal. Chem., **26**, 590 (1954).
17. B. E. Buell, Anal. Chem., **30**, 762 (1967).
18. O. Gimesi and G. Rady, Acta Chim. Acad. Sci. Hung., **55**, 25 (1968).
19. C. H. Kalidas, Z. Analyt. Chem., **223**, 260 (1966).
20. R. Kahsnitz and G. Mohlmann, Erdoel. U. Kohle., **20**, 861 (1967).
21. J. N. Butler and J. C. Synnott, J. Phys. Chem., **73**, 1470 (1969).
22. M. Hojo, M. Utaka and Z. Yoshida, Kogyo Kagaku Zasshi, **69**, 885 (1966).
23. H. J. Harries and G. Wright, J. Inorg. and Nuclear Chem., **31**, 3149 (1969).
24. S. M. Rosalie and J. L. Walter, J. Inorg. and Nuclear Chem., **28**, 2969 (1966).
25. H. Ohtaki, Inorg. Chem., **6**, 808 (1967).
26. P. K. Migal and K. I. Ploae, Russ. J. Inorg. Chem., **10**, 1368 (1965).
27. C. M. Andrejasich, G. W. Mason, and D. F. Peppard, J. Inorg. and Nuclear Chem., **27**, 697 (1965).
28. A. M. Birun, K. A. Komarova, E. K. Kreshkova and A. N. Yarovenko, Izv. Vyssh. Ucheb. Zared Khim. Tekhnol., **9**, 546 (1966).
29. A. P. Kreshkov and L. B. Kuznetsova, Russ. J. Anal. Chem., **22**, 875 (1967).
30. G. Douheret, Bull. Soc. Chim. France, 2921 (1965).
31. J. Badoz-Lambling, J. Desbarres and J. Tocussel, Bull. Soc. Chim. France, 53 (1962).
32. J. Julliard, Bull. Soc. Chim. France, 3069 (1965).
33. N. K. Davidenko and A. A. Zholdakov, Russ. J. Inorg. Chem., **13**, 1662 (1968).
34. J. Julliard and O. Muthe, Compt. Rend., **263C**, 5 (1966).
35. J. N. Butler, D. R. Cogley and W. Zurosky, J. Electrochem. Soc., **115**, 445 (1968).

III. VAPOR PRESSURE

(a) Single Solvents

Vapor pressure data for pure nonaqueous solvents have been reported extensively, even before 1900. With the exception of the liquid ammonia system a limited amount of information is available relative to vapor pressures of nonaqueous electrolyte solutions.

The presentation of the information in this section follows an alphabetical solvent classification and within a particular solvent table the electrolytes are arranged alphabetically by name. Information on the temperature range, concentration range, vapor pressure data, differential pressures and solvent activities are included wherever possible. In some cases calculations have been made using the original data and this is indicated as a footnote together with other comments where appropriate. A complete bibliography is given at the end of this section.

Symbols used:

t (°C)	Temperature
m	Molality
M	Molarity
P	Vapor pressure of solution
ΔP	Differential pressure between solution and pure solvent
a_1	Solvent activity
x_1/x_2	Mole fraction ratio of solvent to solute

ACETONE $(CH_3)_2CO$
Sodium Iodide NaI
Ref. [11]

t	m^*	P_{av}	P_0	a_1
9.9	1.214	109.1	107.5	1.015
16.7	1.677	146.1	145.0	1.008
20.0	1.990	164.7	165.6	0.994
23.0	2.421	183.8	184.4	0.997
25.0	2.503	193.8	200.1	0.969
26.0	2.617	201.7	207.5	0.972
29.0	2.538	231.4	236.3	0.979
32.0	2.452	264.3	268.5	0.985
35.9	2.364	307.5	315.6	0.974
39.9	2.235	359.5	371.5	0.968
50.0	2.002	536.0	548.8	0.977

* Calculated from the data.

ACETONITRILE CH_3CN
Lithium Chloride and Aluminum Chloride $LiCl + AlCl_3$
Ref. [38]
$t = 25°C$

M_{LiCl}	M_{AlCl_3}	P	ΔP	a_1^*
0.7	1	77.8	11.2	0.874

Lithium Perchlorate $LiClO_4$
Ref. [38]
$t = 25°C$

M	P	ΔP	a_1^*
1	79.4	9.6	0.892

* Calculated from the data.

III. VAPOR PRESSURE

ACETONITRILE CH_3CN (Continued)

Sodium Iodide

Ref. [37]

$t = 5°C$

M	P^*	ΔP^*	a_1
0.041	32.93		1.000
0.142	32.90	0.03	0.999
0.344	32.57	0.36	0.989
0.415	32.47	0.46	0.986
0.445	32.87	0.86	0.974
0.597	31.88	1.05	0.968
0.745	31.48	1.45	0.956
1.040	31.02	1.91	0.942
1.127	30.59	2.34	0.929
1.333	30.39	2.54	0.923
1.657	29.64	3.29	0.900

$t = 10°C$

M	P^*	ΔP^*	a_1
0.041	42.75		1.000
0.142	42.66	0.09	0.998
0.344	42.41	0.34	0.992
0.415	42.15	0.60	0.986
0.445	41.85	0.90	0.979
0.597	41.51	1.24	0.971
0.745	40.91	1.84	0.957
1.040	40.10	2.65	0.938
1.127	39.80	2.95	0.931
1.333	39.29	3.46	0.919
1.657	38.60	4.15	0.903

ACETONITRILE CH_3CN (Continued)

Sodium Iodide (Continued)

Ref. [37]

$t = 15°C$

M	P^*	ΔP^*	a_1
0.041	55.05		1.000
0.142	54.77	0.28	0.995
0.344	54.28	0.77	0.986
0.415	54.00	1.05	0.981
0.445	54.11	0.94	0.983
0.597	53.67	1.38	0.975
0.745	52.57	2.48	0.955
1.040	51.91	3.14	0.943
1.127	51.36	3.69	0.933
1.333	50.54	4.51	0.918
1.657	49.77	5.28	0.904

$t = 20°C$

M	P^*	ΔP^*	a_1
0.041	70.26		1.000
0.142	69.91	0.35	0.995
0.344	69.14	1.12	0.984
0.415	68.85	1.41	0.980
0.445	69.00	1.26	0.982
0.597	68.43	1.83	0.974
0.745	67.10	3.16	0.955
1.040	66.18	4.08	0.942
1.127	65.55	4.71	0.933
1.333	64.64	5.62	0.920
1.657	63.52	6.74	0.904

III. VAPOR PRESSURE

ACETONITRILE CH₃CN (Continued)

Sodium Iodide (Continued)
$t = 25°C$

M	P	ΔP	a_1	Ref.
0.041	88.84*		1.000	37
0.142	88.31*	0.53*	0.994	37
0.292	87.51	1.33	0.985	42
0.344	87.33*	1.51*	0.983	37
0.415	87.15*	1.69*	0.981	37
0.445	86.97*	1.87*	0.979	37
0.477	86.59	2.25	0.975	42
0.491	86.39	2.45	0.972	42
0.597	86.35*	2.49*	0.972	37
0.745	84.84*	4.00*	0.955	37
0.878	84.46	4.38	0.951	42
1.040	83.51*	5.33*	0.940	37
1.127	82.89*	5.95*	0.933	37
1.192	82.52	6.32	0.929	42
1.333	81.91*	6.93*	0.922	37
1.542	80.99	7.85	0.912	42
1.657	80.40*	8.44*	0.905	37

Ref. [37]
$t = 30°C$

M	P^*	ΔP^*	a_1
0.041	111.22	0.11	0.999
0.142	110.66	0.67	0.994
0.344	109.55	1.78	0.984
0.415	109.10	2.23	0.980
0.445	108.88	2.45	0.978
0.597	107.77	3.56	0.968
0.745	106.43	4.90	0.956
1.040	104.76	6.57	0.941
1.127	104.32	7.01	0.937
1.333	102.53	8.80	0.921
1.657	101.20	10.13	0.909

ACETONITRILE CH$_3$CN (Continued)

Sodium Iodide (Continued)

Ref. [37]

$t = 35°C$

M	P^*	ΔP^*	a_1
0.041	138.50	0.14	0.999
0.142	137.53	1.11	0.992
0.344	136.28	2.36	0.983
0.415	135.59	3.05	0.978
0.445	135.59	3.05	0.978
0.597	134.20	4.44	0.968
0.745	132.82	5.82	0.958
1.040	130.32	8.32	0.940
1.127	129.77	8.87	0.936
1.333	127.27	11.37	0.918
1.657	126.16	12.48	0.910

$t = 40°C$

M	P^*	ΔP^*	a_1
0.041	171.07	0.17	0.999
0.142	169.87	1.37	0.992
0.344	168.16	3.08	0.982
0.415	167.30	3.94	0.977
0.445	166.79	4.45	0.974
0.597	165.25	5.99	0.965
0.745	163.36	7.88	0.954
1.040	160.97	10.27	0.940
1.127	160.28	10.96	0.936
1.333	157.54	13.70	0.920

III. VAPOR PRESSURE

ACETONITRILE CH₃CN (Continued)

Sodium Iodide (Continued)

Ref. [37]

$t = 45°C$

M	P^*	ΔP^*	a_1
0.041	209.36	0.63	0.997
0.142	207.89	2.10	0.990
0.344	206.21	3.78	0.982
0.415	204.95	5.04	0.976
0.445	204.11	5.88	0.972
0.597	202.22	7.77	0.963
0.745	199.91	10.08	0.952
1.040	197.39	12.60	0.940
1.127	196.55	13.44	0.936
1.333	192.98	17.01	0.919

$t = 50°C$

M	P^*	ΔP^*	a_1
0.041	254.42	1.28	0.995
0.142	252.73	3.07	0.988
0.344	250.68	5.12	0.980
0.415	249.41	6.39	0.975
0.445	248.13	7.67	0.970
0.597	246.59	9.21	0.964
0.745	243.78	12.02	0.953
1.040	240.96	14.84	0.942
1.127	239.43	16.37	0.936
1.333	235.08	20.72	0.919

* Calculated from the data.

AMMONIA NH₃

Ammonium Acetate NH₄Ac
Ref. [15]

$t = 14.96°C$

M	P^*	ΔP	a_1^*
1.010	5390	72	0.987

* Calculated from the data.

Ammonium Bromide NH₄Br

t	m^*	x_1/x_2	P	Solid phase	Ref.
−80	4.4	13.20	33.7	C	35
−78.3	4.7	12.40	38.1	C	35
−73.9	5.7	10.30	52.2	C	35
−69.2	6.5	9.00	72.2	C	35
−68.2	6.7	8.81	77.2	C	35
−67.5	18.9	3.11		B	44
−64.1	7.3	8.05	100.7	C	35
−58.6	7.8	7.55	142.5	B, C	35
−56.2	7.9	7.40	162.6	B	35
−50.7	8.1	7.25	223.9	B	35
−49	19.0	3.09		B	44
−46.5	8.5	6.91	281.2	B	35
−43	19.3	3.05		B	44
−42.5	8.7	6.78	347.5	B	35
−35.8	9.1	6.45	481.9	B	35
−32			540	D	47
−28.7	9.8	6.02	660.7	B	35
−27.8			655	D	47
−27			160	B	47
−27			90	A	47
−25.5	10.1	5.82	760	B	35
−25.5			690	D	47
−25.3			700	D	47
−24			110	A	47
−23			180	B	47
−22.8	10.4	5.66	854	B	35
−21.8			835	D	47
−19.9	10.7	5.47	965	B	35
−19			145	A	47

AMMONIA NH₃ (Continued)

Ammonium Bromide NH₄Br (Continued)

t	m*	x_1/x_2	P	Solid phase	Ref.
−18.8			900	D	47
−15			1045	D	47
−14.5			240	B	47
−12			1200	D	47
−10			1280	D	47
−10	19.3	3.04	500	A, B	43
−10	20.5	2.86	436	A, B	43
−10	21.2	2.77	405	A, B	43
−10	21.9	2.68	408	A, B	43
−10	24.4	2.41	408	A, B	43
−10	28.8	2.04	408	A, B	43
−9			300	B	47
−8.7	12.3	4.77	1500	B	35
−8			1420	D	47
−5			1590	D	47
−5	19.3	3.04	641	B	43
−5	20.5	2.86	562	B	43
−5	21.2	2.77	523	A, B	43
−5	21.9	2.68	511	A, B	43
−5	22.7	2.59	515	A, B	43
−5	23.5	2.41	512	A, B	43
−5	24.4	2.04	511	A, B	43
−5			405	B	47
−3			290	A	47
−2.8			1745	D	47
0	19.3	3.04	811	B	43
0	19.9	2.95	762	B	43
0	20.5	2.86	716	B	43
0	21.2	2.77	672	A, B	43
0	21.9	2.68	637	A, B	43
0	22.7	2.59	637	A, B	43
0	24.4	2.41	636	A, B	43
0	26.5	2.22	633	A, B	43
0	28.8	2.04	634	A, B	43
0			570	A	47
0			350	B	47
3.0	15.0	3.92	1500	B	35
5	19.3	3.04	1007	B	43
5	19.9	2.95	954	B	43

AMMONIA NH₃ (Continued)

Ammonium Bromide NH₄Br (Continued)

t	m^*	x_1/x_2	P	Solid phase	Ref.
5	20.5	2.86	900	B	43
5	21.2	2.77	852	B	43
5	21.9	2.68	798	A, B	43
5	22.7	2.59	780	A, B	43
5	24.4	2.41	780	A, B	43
5	28.8	2.04	778	A, B	43
6	22.7	2.59	812	A, B	43
6	24.4	2.41	812	A, B	43
6	26.5	2.22	811	A, B	43
7	21.2	2.77	925	B	43
7			840	B	47
8	19.9	2.95	1084	B	43
8	20.5	2.28	1028	B	43
9			585	A	47
9.8			910	B	47
10	19.3	3.04	1253		43
10	19.9	2.95	1185		43
10	20.5	2.86	1121		43
10	21.2	2.77	1069		43
10	21.9	2.68	999		43
10	22.7	2.59	944	A	43
10	23.5	2.50	945	A	43
10	24.4	2.41	945	A	43
10	26.5	2.22	941	A	43
10	28.8	2.04	942	A	43
14.6			1145	B	47
14.8			775	A	47
15	19.3	3.04	1538	B	43
15	19.9	2.95	1461	B	43
15	20.5	2.86	1386	B	43
15	22.7	2.59	1161	A	43
15	23.5	2.50	1134	A	43
15	24.4	2.41	1135	A	43
15	28.8	2.04	1130	A	43
19.5			1365	B	47
20	23.5	2.50	1368	A	43
20	24.4	2.41	1350	A	43
20	28.8	2.04	1347	A	43
20			1025	A	47

III. VAPOR PRESSURE

AMMONIA NH₃ (Continued)

Ammonium Bromide NH₄Br (Continued)

t	m^*	x_1/x_2	P	ΔP^*	a_1^*	Solid phase	Ref.
25	24.4	2.41	1591			A	43
25	25.64	2.29	1623	5885	0.216	A	14
25	28.8	2.04	1585			A	43
25	32.09	1.83	1623	5885	0.216	A	14
25	37.64	1.56	1623	5885	0.216	A	14
25			1600			B	47
26			1320			A	47
30.1			1845			B	47
30.1			1660			A	47

* Calculated from the data.
A(NH₄Br·NH₃)
B(NH₄Br·3NH₃)
C(NH₄Br·4NH₃)
D(NH₄Br·6NH₃)

For graphical data and equations not included above see: Refs: [35], [44], [45], [48].

Ref. [18]

$t = 0°C$

M	P	ΔP	a_1^*
0.217	3208	12.7	0.996
0.655	3187	33.9	0.989
1.301	3157	63.9	0.980
3.265	3050	170.8	0.947
5.277	2899	322.1	0.900
6.311	2798	423.4	0.869
9.956	2281	939.9	0.708
11.40	2036	1185	0.632
13.70	1637	1584	0.508
14.08	1559	1662	0.484

AMMONIA NH₃ (Continued)

Ammonium Bromide NH₄Br (Continued)
Ref. [15]
$t = 14.96°C$

M	$P*$	ΔP	a_1*
0.198	5445	17	0.997
0.377	5429	33	0.994
0.379	5429	33	0.994
0.557	5415	47	0.991
0.708	5404	58	0.989
0.926	5386	76	0.986
1.536	5342	120	0.978
2.194	5288	174	0.968
2.956	5222	240	0.956
3.507	5214	248	0.955
3.577	5164	298	0.945
4.317	5081	381	0.930

Ref. [14]
$t = 25°C$

$m*$	x_1/x_2	P	$\Delta P*$	a_1*
0.341	172	7463	45	0.994
0.369	159	7459	49	0.993
0.402	146	7455	53	0.993
0.455	129	7451	57	0.992
0.625	94.0	7432	76	0.990
0.820	71.6	7415	93	0.988
1.043	56.3	7392	116	0.985
1.335	44.0	7359	149	0.980
1.472	39.9	7349	159	0.979
1.673	35.1	7325	183	0.976
2.387	24.6	7254	254	0.966
2.718	21.6	7222	286	0.962
3.355	17.5	7135	373	0.950
4.135	14.2	7043	465	0.938
4.774	12.3	6951	557	0.926
5.488	10.7	6834	674	0.910
6.085	9.65	6711	797	0.894
6.503	9.03	6625	883	0.882
7.946	7.39	6281	1227	0.837
8.635	6.80	6090	1418	0.811
10.23	5.74	5597	1911	0.745
10.54	5.57	5507	2001	0.733

III. VAPOR PRESSURE

AMMONIA NH_3 (Continued)

Ammonium Bromide NH_4Br (Continued)
Ref. [14]
$t = 25°C$

m^*	x_1/x_2	P	ΔP^*	a_1^*
11.36	5.17	5226	2282	0.696
12.52	4.69	4808	2700	0.640
13.66	4.30	4401	3107	0.586
14.94	3.93	3962	3546	0.528
16.63	3.53	3406	4102	0.454
17.85	3.29	3055	4453	0.407
19.57	3.00	2579	4929	0.344
21.20	2.77	2260	5248	0.301
22.94	2.56	1965	5543	0.262
23.49	2.50	1842	5666	0.245

* Calculated from the data.

Ammonium Chloride NH_4Cl

t	m	x_1/x_2	P	ΔP	a_1^*	Solid phase	Ref.
−52.55	†		259	4	0.985	A	17
−52.47	†		260	4	0.985	A	17
−48.5	19.8	2.97	28			A	44
−46.54	†		368	7	0.981	A	17
−37.55	†		596	17	0.972	A	17
−37.12	†		610	17	0.973	A	17
−31.43	†		807	29	0.965	A	17
−29	21.5	2.73	136			A	44
−28.54	†		925	38	0.961	A	17
−24.48	5.0*	11.74*	1107	58	0.950	A	17
−20.00	3.66*	16.04*	1373	53*	0.963		17
−20.00	5.61*	10.47*	1332	94	0.934	A	17
−20.00	7.03*	8.35*	1303	123*	0.914		17
−9.96	3.83*	15.33*	2109	66*	0.970		17
−9.96	7.08*	8.29*	2003	172*	0.921		17
−9.96	8.60*	6.83*	1919	266	0.878	A	17
−4.97	10.4**	5.65*	2220	444	0.833	A	17
0.00	0.294‡		3206	14.6	0.995		18
0.00	0.505‡		3199	22.4	0.993		18
0.00	1.135‡		3178	42.5	0.987		18
0.00	1.706‡		3162	58.6	0.982		18
0.00	3.721‡		3106	115.4	0.964		18
0.00	4.12*	14.25*	3096	125*	0.961		17

AMMONIA NH₃ (Continued)

Ammonium Chloride NH₄Cl (Continued)

t	m	x_1/x_2*	P	ΔP	a_1*	Solid phase	Ref.
0.00	5.670‡		3033	187.6	0.942		18
0.00	7.19*	8.17	2962	259*	0.920		17
0.00	8.596‡		2862	359.2	0.889		18
0.00	9.956‡		2750	470.8	0.854		18
0.00	10.94*	5.37	2581	640*	0.801		17
0.00	11.70‡		2566	655.4	0.797		18
0.00	12.42‡		2471	749.9	0.767		18
0.00	12.37*	4.75	2469	752	0.767	A	17
0.00	21.8*	2.69	2098	1123	0.651	B	17
2.94	14.02*	4.19	2548	1042	0.710	A	17
2.95	22.0*	2.67	1293	2299	0.360	B	17
3.94	14.7*	3.99	2554	1169	0.686	A	17
4.92	11.01*	5.33	3106	751*	0.805		17
4.92	13.75*	4.27	2796	1061*	0.725		17
4.92	15.36*	3.82	2545	1312	0.660	A	17
4.93	22.1*	2.66	1412	2447	0.366	B	17
5.92	22.2*	2.64	1459	2539	0.365	A	17
6.43	16.7*	3.52	2477	1594	0.608	A	17
6.43	22.0*	2.67	1459	2539	0.365	A	17
6.93	16.9*	3.47	2402	1741	0.580	A	17
6.93	21.2*	2.77	1667	2476	0.402	A	17
6.94	22.2*	2.64	1528	2617	0.369	B	17
7.13	19*	3.1	1729	2445	0.414	A	17
9.91	4.52*	12.99	4409	181*	0.961		17
9.91	7.28*	8.07	4244	346*	0.925		17
9.91	11.10*	5.29	3719	871*	0.810		17
9.91	13.91*	4.22	3353	1237*	0.731		17
9.91	15.66*	3.75	3020	1570*	0.658		17
9.91	17.43*	3.37	2646	1944*	0.576		17
9.91	20.05*	2.93	2196	2394*	0.478		17
9.91	21.36*	2.75	1920	2670*	0.418		17
9.92	22.34*	2.63	1734	2617	0.399	B	17
14.88	22.60*	2.60	2122	3320	0.390	B	17
14.90	11.20*	5.24	4420	1035*	0.810		17
14.90	14.12*	4.16	3991	1464*	0.732		17
14.90	15.82*	3.71	3597	1868*	0.658		17
14.90	17.67*	3.32	3165	2290*	0.580		17
14.90	20.20*	2.91	2654	2801*	0.487		17
14.90	21.66*	2.71	2328	3127*	0.427		17
14.96	0.304	193	5441*	21	0.996		15
14.96	0.349	168	5436*	26	0.995		15
14.96	0.614	95.6	5416*	46	0.992		15

III. VAPOR PRESSURE

AMMONIA NH_3 (Continued)
Ammonium Chloride NH_4Cl (Continued)

t	m	x_1/x_2*	P	ΔP	a_1*	Solid phase	Ref.
14.96	0.761	77.2	54.5*	57	0.990		15
14.96	1.430	41.1	5376*	86	0.984		15
14.96	1.958	30.0	5355*	107	0.980		15
14.96	2.330	25.2	5344*	118	0.978		15
14.96	2.514	23.4	5342*	120	0.978		15
14.96	2.721	21.6	5335*	127	0.977		15
19.89	5.22*	11.25	6114	293*	0.954		17
19.89	7.44*	7.89	5923	484*	0.924		17
19.89	11.31*	5.19	5218	1189*	0.814		17
19.89	14.30*	4.11	4709	1698*	0.735		17
19.89	16.00*	3.67	4251	2156*	0.663		17
19.89	17.90*	3.28	3759	2648*	0.587		17
19.89	20.34*	2.89	3172	3235*	0.495		17
19.89	22.03*	2.67	2800	3607*	0.437		17
19.89	22.93	2.56	2580	3827	0.403	B	17
24.83	23.2*	2.5	3117	4364	0.417	B	17

$t = 25°C$

m	x_1/x_2	P	ΔP	a_1*	Ref.
0.0051		7505*	1.21	0.9999	20
0.0098		7504*	2.13	0.9997	20
0.0278		7501*	5.42	0.9993	20
0.0575		7495*	10.65	0.9985	20
0.101	581*	7489*	17.20	0.9977	20
0.141	416*	7484*	22.5	0.997	20
0.190	309*	7479*	27.4	0.996	20
0.378	155*	7466*	40.3	0.995	20
0.411*	143	7471	37*	0.995	14
0.441*	133	7468	40*	0.995	14
0.481*	122	7459	49*	0.993	14
0.529*	111	7455	53*	0.993	14
0.559*	105	7452	56*	0.993	14
0.680*	86.4	7442	66*	0.991	14
0.717	81.9*	7438*	67.7	0.991	20
0.851*	69.0	7434	74*	0.990	14
0.900	65.2*	7432*	74.3	0.990	20
0.967*	60.7	7425	83*	0.989	14
1.145*	51.3	7417	91*	0.988	14
1.692	34.7	7386	122*	0.984	14

AMMONIA NH₃ (Continued)

Ammonium Chloride NH₄Cl (Continued)

$t = 25°C$

m	x_1/x_2	P	ΔP	$a_1{}^*$	Ref.
1.835	32.0	7376	132*	0.982	14
2.358	24.9	7352	156*	0.979	14
2.936	20.0	7324	184*	0.975	14
3.280	17.9	7307	201*	0.973	14
3.670	16.0	7290	218*	0.971	14
4.255	13.8	7259	249*	0.969	14
4.813	12.2	7223	285*	0.962	14
5.896	9.96	7140	368*	0.951	14
6.240	9.41	7124	384*	0.949	14
6.868	8.55	7070	438*	0.942	14
7.850*	7.48	6965	543*	0.928	14
8.751*	6.71	6856	652*	0.913	14
9.456*	6.21	6745	763*	0.898	14
12.01*	4.89	6290	1218*	0.838	14
13.59*	4.32	5958	1550*	0.794	14
15.06*	3.90	5582	1926*	0.743	14
16.00*	3.67	5303	2205*	0.706	14
17.42*	3.37	4978	2530*	0.663	14
18.94*	3.10	4565	2943*	0.608	14
20.25*	2.90	4212	3296*	0.561	14
22.41*	2.62	3643	3865*	0.485	14
24.67*	2.38	3149	4359*	0.419	14
28.50*	2.06	3140	4368*	0.418	14

Ref. [20]

$t = 25°C$

m	$\gamma_\pm$	$-\ln \gamma_\pm$
0.0025	0.653	0.4255
0.0100	0.456	0.784
0.0225	0.341	1.077
0.040	0.275	1.290
0.16	0.139	1.974
0.360	0.0769	2.562
0.640	0.0488	3.018
1.00	0.0344	3.365

III. VAPOR PRESSURE

AMMONIA NH₃ (Continued)

Ammonium Chloride NH₄Cl (Continued)
Ref. [20]
$t = 25°C$

m	$\gamma_\pm$	$-\ln \gamma_\pm$
1.42	0.0254	
1.96	0.0201	
4.00	0.0111	
7.40	0.00718	
10.7	0.00604	
13.7	0.00587	
16.3	0.00608	
18.9	0.00659	
21.8	0.00732	
24.4†	0.00822	

For graphical data not given above, see Refs. [44], [45].
* Calculated from the data.
† Saturated.
‡ Concentration units are moles/liter, not moles/1000 g.
A(NH₄Cl·3NH₃).
B(NH₄Cl).

Ammonium Fluoride NH₄F
For graphical data at $-35°C$ see Ref. [45]

Ammonium Iodide NH₄I

t	m	x_1/x_2	P	Solid phase	Ref.
−79	11.7	5.01	10.6	E	44
−64.5	14.5	4.06	3	D	44
−48	14.8	3.98	17	D	44
−36	15.5	3.80	49	D	44
−36	19.3	3.04	37.2	C	44
−36	28.0	2.10	19.3	B	44
−35.5	9.3	6.30	414.2	D	46

AMMONIA NH$_3$ (Continued)

Ammonium Iodide NH$_4$I (Continued)

t	m	x_1/x_2	P	Solid phase	Ref.
−29			435	F	47
−28.1	9.9	5.95	514.3	D	46
−27			130	C	47
−27			10	A	47
−25.6	10.1	5.83	552.3	D	46
−25			510	F	47
−23.6	10.4	5.67	583.3	D	46
−23			135	C	47
−21.2			575	F	47
−19.8	10.7	5.50	646.1	D	46
−19			150	C	47
−17			655	F	47
−16.6	11.2	5.22	702.4	D	46
−16	19.7	2.98	160.9	B	46
−16	21.1	2.78	133.4	B	46
−16	22.2	2.65	117.5	A	46
−16			17	A	47
−14	17.8	3.30	241.4	C	46
−14			180	C	47
−14	20.3	2.89	167.1	B	46
−14	21.0	2.79	150.1	B	46
−14			20	A	47
−13	17.8	3.30	257.3	C	46
−12.8			770	F	47
−12	17.7	3.32	274.3	C	46
−12	21.1	2.78	168.9	B	46
−11.8	17.8	3.30	277.5	D	46
−10.7	21.4	2.75	117.7	B	46
−10	17.0	3.46	346.7	D	46
−10	22.4	2.62	159.8	A	46
−10			27	A	47
−9.8			235	C	47
−8	16.0	3.67	443.4	D	46
−8	22.2	2.65	186.7	B	46
−8	22.6	2.60	176.8	A	46
−6			1000	F	47
−6	15.2	3.87	562.1	D	46
−5.8			38	A	47
−5			1035	F	47
−5			290	C	47
−4	22.8	2.58	214.3	A	46
−0.1	76.3	0.77	55	A	44

III. VAPOR PRESSURE

AMMONIA NH_3 (Continued)
Ammonium Iodide NH_4I (Continued)
Ref. [18]
$t = 0°C$

M	P	ΔP	$a_1{}^*$
0.204	3208	13.2	0.996
0.405	3196	25.2	0.992
0.877	3167	54.4	0.983
1.829	3098	122.9	0.962
2.698	3020	200.5	0.938
4.499	2804	416.9	0.871
6.278	2509	711.8	0.779
7.754	2217	1004.2	0.688
10.37	1632	1589	0.507
13.88	956.0	2265	0.297
15.52	729.0	2492	0.226
17.68	521.4	2700	0.162
18.49	465.3	2756	0.144
23.08	268.9	2952	0.083

Ref. [47]
Saturated Solutions

t	P	Solid phase
0	1250	F
0	380	C
0	57	A
4.7	1415	F
9	100	A
9.4	1620	F
9.9	580	C
11.4	1735	F
14.6	700	C

Ref. [15]
$t = 14.96°C$

M	P^*	ΔP	$a_1{}^*$
0.290	5416	46	0.992
0.554	5370	92	0.983
1.064	5270	192	0.965

AMMONIA NH₃ (Continued)
Ammonium Iodide NH₄I (Continued)

t	m^*	x_1/x_2	P	Solid phase	Ref.
17.1	24.6	2.39	544.4	A	46
17.5	172.7	0.34	127.4	A	44
19			180	A	47
19.5			840	C	47
24.5	25.3	2.32	730.0	A	46

Ref. [14]
$t = 25°C$

M^*	x_1/x_2	P	ΔP^*	a_1^*
0.294	200	7471	37	0.995
0.332	177	7460	48	0.994
0.365	161	7457	51	0.993
0.399	147	7451	57	0.992
0.435	135	7449	59	0.992
0.511	115	7436	72	0.990
0.588	99.9	7426	82	0.989
0.707	83.1	7409	99	0.987
0.883	66.5	7388	120	0.984
0.979	60.0	7372	136	0.982
1.223	48.0	7334	174	0.977
1.317	44.6	7321	187	0.975
1.600	36.7	7278	230	0.969
1.944	30.2	7220	288	0.962
2.467	23.8	7124	384	0.949
2.770	21.2	7080	428	0.943
2.921	20.1	7041	467	0.938
3.191	18.4	6975	533	0.929
3.475	16.9	6911	597	0.920
3.788	15.5	6833	675	0.910
4.774	12.3	6547	961	0.872
5.290	11.1	6383	1125	0.850
5.757	10.2	6218	1290	0.828
6.091	9.64	6124	1384	0.816
6.924	8.48	5807	1701	0.773
7.586	7.74	5530	1978	0.737
8.425	6.97	5174	2334	0.689
9.020	6.51	4875	2633	0.649
9.335	6.29	4773	2735	0.636
9.986	5.88	4486	3022	0.597
10.93	5.37	4060	3448	0.541

AMMONIA NH₃ (Continued)

Ammonium Iodide NH₄I (Continued)
Ref. [14]
$t = 25°C$

M^*	x_1/x_2	P	ΔP^*	a_1^*
11.67	5.03	3755	3753	0.500
12.06	4.87	3582	3926	0.477
12.91	4.55	3167	4341	0.422
13.50	4.35	3013	4495	0.401
13.82	4.25	2823	4685	0.376
14.57	4.03	2554	4954	0.340
14.90	3.94	2411	5097	0.321
15.29	3.84	2324	5184	0.310
15.49	3.79	2213	5295	0.295
15.96	3.68	2103	5405	0.280
16.45	3.57	1957	5551	0.261
18.01	3.26	1587	5921	0.211
19.57	3.00	1328	6180	0.177
22.67	2.59	960	6548	0.128
24.47	2.40	842	6666	0.112
29.51	1.99	738	6770	0.098
29.66	1.98	736	6772	0.098

Ref. [47]
Saturated Solutions

t	P	Solid phase
25	995	C
27	280	A
30	1160	C
35	455	A
50	940	A
56	1140	A

For graphical data and equations not given above, see: Refs. [44], [45], [46], [48].
* Calculated from the data.
A(NH₄I·NH₃)
B(NH₄I·2NH₃)
C(NH₄I·3NH₃)
D(NH₄I·4NH₃)
E(NH₄I·5NH₃)
F(NH₄I·6NH₃)

AMMONIA NH$_3$ (Continued)

Ammonium Nitrate NH$_4$NO$_3$
Ref. [15]
$t = 14.96°C$

M	P^*	ΔP	a_1
0.253	5436	26	0.995
0.285	5432	30	0.994
0.476	5420	42	0.992
0.521	5417	45	0.992
0.745	5397	65	0.988
1.055	5365	97	0.982
1.270	5350	112	0.979
1.973	5288	174	0.968
2.223	5254	208	0.962

Ref. [14]
$t = 25°C$

M^*	x_1/x_2	P	ΔP^*	a_1^*
0.419	140	7450	58	0.9923
0.435	135	7448	60	0.9920
0.466	126	7440	68	0.9909
0.529	111	7437	71	0.9905
0.642	91.4	7426	82	0.989
0.728	80.6	7417	91	0.988
0.842	69.7	7404	104	0.986
0.979	60.0	7387	121	0.984
1.135	51.8	7366	142	0.981
1.241	47.3	7355	153	0.980
1.801	32.6	7279	229	0.969
2.706	21.7	7166	342	0.954
3.764	15.6	6992	516	0.931
4.813	12.2	6827	681	0.909
5.151	11.4	6726	782	0.896
6.453	9.10	6381	1127	0.850
7.331	8.01	6130	1378	0.816
8.329	7.05	5807	1701	0.773
9.321	6.30	5511	1997	0.734
11.23	5.23	4891	2617	0.651
12.91	4.55	4386	3122	0.584
14.64	40.1	3859	3649	0.514
17.02	3.45	3311	4197	0.441
19.00	3.09	2914	4594	0.388
20.68	2.84	2667	4841	0.355

III. VAPOR PRESSURE

AMMONIA NH₃ (Continued)

Ammonium Nitrate NH₄NO₃ (Continued)

Ref. [14]

$t = 25°C$

M*	x_1/x_2	P	ΔP*	a_1*
22.76	2.58	2358	5150	0.314
26.69	2.20	1926	5582	0.257
30.42	1.93	1635	5873	0.218
31.74	1.85	1529	5979	0.204
33.26	1.76	1434	6074	0.191
37.64	1.56	1230	6278	0.164
45.17	1.30	970	6538	0.129
48.13	1.22	880	6628	0.117
72.67	0.808	863	6645	0.115
78.92	0.744	863	6645	0.115

* Calculated from the data.

Barium Nitrate Ba(NO₃)₂

Ref. [18]

$t = 0°C$

M	P	ΔP	a_1*
0.288	3208	12.6	0.996
0.566	3201	20.1	0.994
0.833	3193	27.5	0.991

* Calculated from the data.

Calcium Nitrate Ca(NO₃)₂

Ref. [8]

Saturated Solutions

t	P	ΔP*	a_1
−40	306	232	0.519
−30	517	380	0.576
−25	673	464	0.592
−20	890	573	0.608
−15	1128	645	0.636
−10	1434	747	0.657

AMMONIA NH$_3$ (Continued)

Calcium Nitrate Ca(NO$_3$)$_2$ (Continued)
Ref. [18]
$t = 0°C$

M	P	ΔP	a_1*
0.143	3214	6.8	0.998
0.405	3207	13.9	0.996
0.604	3203	17.8	0.994
1.480	3183	38.1	0.988
2.254	3144	77.2	0.976
2.962	3060	161.3	0.950
3.633	2928	292.6	0.909
4.335	2696	525.6	0.837
4.780	2516	704.8	0.781
5.01	2397	823.5	0.744

* Calculated from the data.

Cesium Bromide CsBr
Ref. [18]
$t = 0°C$

M	P	ΔP	a_1*
0.215	3210	11.2	0.997

Cesium Iodide CsI
Ref. [18]
$t = 0°C$

M	P	ΔP	a_1*
0.185	3210	11.4	0.997
0.616	3186	34.5	0.989
0.941	3169	51.8	0.984
1.478	3139	81.6	0.975
2.349	3085	136.4	0.958
4.382	2922	299.2	0.907
5.342	2828	393.4	0.878
5.84	2773	448.2	0.861

* Calculated from the data.

III. VAPOR PRESSURE

AMMONIA NH₃ (Continued)

Lithium Chloride LiCl
Ref. [18]
$t = 0°C$

M	P	ΔP	a_1*
0.342	3204	17.2	0.995

* Calculated from the data.

Lithium Nitrate LiNO₃

t	m	P	Ref.
−13.0	†	239.32	2
−6.02	†	259.94	2
1.53	†	321.90	2
8.48	†	382.9	2
10	‡	102	10
15.5	25.41*	487.45	2
20	‡	155	10
22.4	25.41*	627.39	2
30	‡	230	10
30.05	25.41*	818.13	2
35.00	25.41*	998.16	2
50	‡	455	10

* Data given as 63.66% LiNO₃. It was assumed that that weight % was given.
‡ Saturated solutions.
† Saturated solutions, overall composition is 63.66% (assumed weight %) LiNO₃.

Potassium Bromide KBr
Ref. [18]
$t = 0°C$

M	P	ΔP	a_1*
0.210	3209	12.4	0.996
0.582	3190	31.4	0.990
1.097	3163	57.5	0.982
1.558	3139	81.8	0.975
1.962	3117	104.4	0.968
2.26	3100	121.2	0.962

* Calculated from the data.

AMMONIA NH_3 (Continued)

Potassium Iodide KI

Ref. [40]

m^*	x_2	$-\log \gamma_1$					
		$-25°C$	$-35°C$	$-45°C$	$-55°C$	$-65°C$	$-75°C$
0.931	0.0156	0.00321	0.00317	0.00317	0.00379	0.00449	0.00520
1.058	0.0177	0.00405	0.00395	0.00375	0.00424	0.00472	0.00552
1.168	0.0195	0.00488	0.00473	0.00413	0.00456	0.00509	0.00593
1.315	0.0219	0.00668	0.00656	0.00688	0.00717	0.00778	0.00878
1.543	0.0256	0.00848	0.00837	0.00840	0.00876	0.00938	0.01155
1.872	0.0309	0.01162	0.01148	0.01193	0.01263	0.01382	0.01560
2.806	0.0456	0.03147	0.03012	0.03046	0.03129	0.03319	0.03620
4.440	0.0703	0.0783	0.0758	0.07736	0.07901	0.08316	0.08938

m^*	x_2	a_2		
		$-75°C$	$-65°C$	$-35°C$
0.997	0.0167	1.00	1.00	1.00
1.174	0.0196	1.08	1.07	1.06
1.469	0.0244	1.26	1.23	1.19
1.960	0.0323	1.58	1.53	1.47
2.345	0.0384	1.94	1.84	1.75
2.935	0.0476	2.94	2.72	2.45
3.254	0.0525	4.01	3.58	3.19
3.662	0.0587	5.30	4.65	4.05
4.190	0.0666	6.68	5.86	4.99
4.892	0.0769	7.90	6.53	5.93
5.878	0.0910	9.65	8.28	6.74

III. VAPOR PRESSURE

AMMONIA NH$_3$ (Continued)

Potassium Iodide KI (Continued)
Ref. [18]
$t = 0°C$

M	P	ΔP	a_1*
0.242	3206	15.2	0.995
0.469	3192	29.1	0.991
1.014	3158	63.2	0.980
1.711	3105	116.4	0.964
2.674	3015	205.5	0.947
2.992	2981	239.9	0.925
4.543	2776	445.2	0.862
5.680	2585	636.4	0.803
6.962	2332	889.2	0.724
9.196	1855	1366.4	0.576
10.24	1642	1579	0.510
11.09	1474	1747	0.458

* Calculated from the data.

Potassium Nitrate KNO$_3$
Ref. [18]
$t = 0°C$

M	P	ΔP	a_1*
0.224	3209	12.4	0.996
0.626	3191	29.6	0.991
1.04	3175	46.0	0.986

Rubidium Bromide RbBr
Ref. [18]
$t = 0°C$

M	P*	ΔP*	a_1
0.237	3207	13.5	0.996
0.530	3193	27.7	0.991
0.979	3173	47.7	0.985
1.35	3159	61.9	0.981

* Calculated from the data.

AMMONIA NH₃ (Continued)

Rubidium Iodide RbI
Ref. [18]
$t = 0°C$

M	P	ΔP	$a_1{}^*$
0.174	3210	11.2	0.997
0.407	3196	24.8	0.992
0.901	3166	54.5	0.983
1.801	3104	117.2	0.964
2.792	3019	201.8	0.937
4.404	2840	380.8	0.882
6.649	2513	707.9	0.780
7.759	2326	894.9	0.722
9.139	2097	1123.5	0.651
9.589	2025	1196.3	0.629
10.08	1952	1269.4	0.606

* Calculated from the data.

Silver Bromide AgBr₂
Ref. [18]
$t = 0°C$

M	P	ΔP	$a_1{}^*$
0.128	3214	6.9	0.998

Silver Iodide AgI₂
Ref. [18]
$t = 0°C$

M	P	ΔP	$a_1{}^*$
0.274	3210	11.0	0.997
0.542	3202	19.1	0.994
0.921	3192	29.1	0.991
2.066	3164	56.9	0.982
3.313	3135	86.3	0.973
4.646	3101	119.5	0.963
7.748	2999	222.4	0.931
10.04	2903	318.3	0.901
15.68	2602	619.3	0.808
19.43	2373	847.9	0.737
21.49	2242	979.2	0.696
22.61	2166	1055	0.672

* Calculated from the data.

III. VAPOR PRESSURE

AMMONIA NH_3 (Continued)

Sodium Bromide NaBr
Ref. [18]
$t = 0°C$

M	P	ΔP	a_1*
0.311	3203	18.0	0.994
0.892	3173	48.0	0.985
1.642	3131	90.3	0.972
1.920	3114	106.8	0.967
3.602	2987	233.9	0.927
5.013	2835	385.5	0.880
6.21	2661	560.2	0.826

* Calculated from the data.

Sodium Chloride NaCl

t	m	P	ΔP	a_1	Solid phase	Ref.
−60	0.108*	164*				16
−50	0.195*	306*				16
−40	0.370*	535*				16
−30	0.719*	887*				16
−23.06	1.167*	1224*	22.2	0.982	B	13
−17.97	1.687*	1515*	40.0	0.974	B	13
−17.95	1.682*	1515*	39.4	0.975	B	13
−12.87	2.428*	1857*	71.0	0.963	B	13
−11.48	2.691*	1967*	85.1	0.959	B	13
−11.39	2.706*	1967*	85.1	0.959	B	13
−10.98	2.790*	2004*	89.7	0.957	B	13
−10.94	2.795*	2004*	90.2	0.957	B	13
−9.94	2.989*	2080*	101.0	0.954	B	13
−8.46	2.955*	2213	103.6	0.955	A	13
−8.44	2.951*	2213*	106.3	0.954	A	13
−4.97	2.634*	2557*	105.8	0.960	A	13
−4.88	2.619*	2557*	104.5	0.961	A	13
−0.01	2.191*	3117*	104.4	0.968	A	13
0.0	0.178†	3210	10.5	0.997		18
0.0	1.071†	3170	51.0	0.984		18
0.0	1.378†	3156	65.2	0.980		18
0.0	1.865†	3133	87.8	0.973		18

AMMONIA NH₃ (Continued)
Sodium Chloride NaCl (Continued)

t	m	P	ΔP	a_1	Solid phase	Ref.
0.0	2.20†	3118	103.2	0.968		18
0.06	2.188*	3117*	104.2	0.968	A	13
4.97	1.805*	3771*	94.5	0.976	A	13
4.98	1.801*	3771*	99.0	0.974	A	13
9.92	1.285		77			15
9.92	1.365		88			15
9.92	Sat'd		96		A	15
9.95	1.462*	4517*	95.4	0.979	A	13
9.96	1.462*	4517*	94.0	0.980	A	13
14.93	1.173*	5370*	93.0	0.983	A	13
14.94	1.174*	5370*	91.5	0.983	A	13
14.96	0.375	5430	32	0.994		15
14.96	0.401	5427	35	0.994		15
14.96	0.414	5426	36	0.994		15
14.96	0.426	5427	35	0.994		15
14.96	0.437	5423	39	0.993		15
14.96	0.640	5411	51	0.991		15
14.96	0.797	5400	62	0.989		15
14.96	0.835	5395	67	0.988		15
14.96	0.872	5393	69	0.987		15
14.96	1.051	5383	79	0.986		15
14.96	1.050	5380	82	0.985		15
14.96	1.071	5380	82	0.985		15
14.96	1.098	5377	85	0.984		15
14.96	Sat'd	5369	93	0.983	A	15
20.04	0.937*	6336*	92.2	0.986	A	13
24.88	0.750*	7435*	85.6	0.989	A	13
24.89	0.747*	7435*	85.0	0.989	A	13
29.86	0.596*	8666*	85.2	0.990	A	13
29.93	0.579*	8666*	82.5	0.991	A	13

The data is also presented graphically in Refs. 13, 15, 18.
* Calculated from the data. Concentrations were calculated from solubility data.
† Concentration units are moles/liter not moles/1000 g.
  A(NaCl).
  B(NaCl·5NH₃).

AMMONIA NH₃ (Continued)

Sodium Iodide NaO
Ref. [18]
$t = 0°C$

M	P	ΔP	a_1*
0.174	3209	11.5	0.996
0.655	3181	40.4	0.988
1.038	3161	59.9	0.981
1.833	3098	123.0	0.962
3.426	2938	282.7	0.912
5.657	2574	647.4	0.799
7.032	2243	977.6	0.696
8.324	1869	1352	0.580
8.80	1682	1539	0.522

* Calculated from the data.

Sodium Nitrate NaNO₃
Ref. [7]
Saturated Solutions

t	P	ΔP*	a_1
−35.0	325	375	0.464
−30.7	400	465	0.463
−26.0	500	585	0.461
−22.1	600	705	0.460
−16.7	760	890	0.460
−15.6	800	930	0.462
−10.5	1000	1140	0.467
−6.3	1200	1330	0.474

AMMONIA NH_3 (Continued)

Sodium Nitrate $NaNO_3$ (Continued)

Ref. [18]

$t = 0°C$

M	P^*	ΔP	a_1^*
0.188	3210	11.3	0.997
0.400	3198	22.6	0.993
0.778	3179	42.0	0.987
1.599	3134	86.9	0.973
3.067	3036	184.7	0.943
5.064	2853	368.0	0.886
7.156	2599	622.1	0.807
9.599	2262	958.7	0.702
11.96	1949	1272	0.605
13.34	1782	1439	0.553
14.35	1675	1546	0.520
14.43	1666	1555	0.517
15.00	1605	1616	0.498

Ref. [15]

$t = 14.96°C$

M	P^*	ΔP	a_1^*
0.222	5443	19	0.997
0.229	5440	22	0.996
0.454	5421	41	0.992
0.646	5405	57	0.990
0.907	5378	84	0.985
0.981	5378	84	0.985
1.160	5362	100	0.982
1.744	5309	153	0.972
2.267	5262	200	0.963
2.493	5233	229	0.958
2.525	5229	233	0.957

* Calculation from the data.

III. VAPOR PRESSURE

AMMONIA NH$_3$ (Continued)

Sodium Thiocyanate NaSCN

Ref. [30]

t	m^*	P^*	ΔP^*	a_1
−20	4.6	368	37	0.908
−20	5.7	353	53	0.870
−20	7.9	309	96	0.763
−20	10.4	237	168	0.585
−10	4.6	564	56	0.910
−10	5.7	514	106	0.829
−10	7.9	454	166	0.733
−10	10.4	348	272	0.561
−0	4.6	824	92	0.900
−0	5.7	743	172	0.812
−0	7.9	641	275	0.700
−0	10.4	485	430	0.530
−0	14.2	221	695	0.241
−0	19.6	110	806	0.120
−0	23.6	75	840	0.082
10	4.6	1152±1	159±1	0.879
10	5.7	1051	259	0.802
10	7.9	911	400	0.695
10	10.4	633	678	0.483
10	14.2	324	987	0.258
10	19.6	206	1105	0.171
10	23.6	111	1199	0.100
20	4.6	1538±6	284±6	0.844
20	5.7	1407	415	0.772
20	7.9	1267	556	0.695
20	10.4	835	988	0.458
20	14.2	470	1352	0.258
20	19.6	312	1511	0.171
20	23.6	182	1640	0.100
30	4.6	2054	433	0.826
30	5.7	1872	614	0.753
30	7.9	1711	776	0.688
30	10.4	1136	1350	0.457
30	14.2	671	1815	0.270
30	19.6	455	2032	0.183
30	23.6	271	2216	0.109

AMMONIA NH₃ (Continued)

Sodium Thiocyanate NaSCN (Continued)

Ref. [14]

t	m^*	P^*	ΔP^*	a_1
40	4.6	2843±7	455±7	0.862
40	5.7	2523	775	0.765
40	7.9	2153	1144	0.653
40	10.4	1510	1787	0.458
40	14.2	937	2361	0.284
40	19.6	630	2668	0.191
40	23.6	356	2942	0.108
50	4.6	38.4	529	0.878
50	5.7	3457	875	0.798
50	7.9	3041±7	1291±7	0.702
50	10.4	2023	2309	0.467
50	14.2	1235	3098	0.285
50	19.6	853	3479	0.197
50	23.6	485	3847	0.112
60	4.6	4864±13	663±35	0.880
60	5.7	4482	1106	0.802
60	7.9	3997	1593	0.715
60	10.4	2631	2943	0.472
60	14.2	1668	3910	0.299
60	19.6	1132	4471	0.202
60	23.6	661	4941	0.118
70	7.9	5114	2089±45	0.710
70	10.4	3454	3742	0.480
70	14.2	2234	5019	0.308
70	19.6	1433	5732	0.200
70	23.6	838	6324	0.117
80	7.9	6202	2533±55	0.710
80	10.4	4350	4438	0.495
80	14.2	2836	6026	0.320
80	19.6	1778	7024	0.202
80	23.6	1102	7714	0.125
90	10.4	5393	5181±65	0.510†
90	14.2	3674±13	6823±65	0.350
90	19.6	2204	8291	0.210
90	23.6	1411	9040	0.135

* Calculated from the data. P and ΔP values for $-20°C$ to $-50°C$ were calculated from P_0 and a_1 values. For $60°$ to $90°$ P and ΔP were calculated from P and P_0 values.

† There is an error in the original paper listing this value as 0.051.

III. VAPOR PRESSURE

AMMONIA NH_3 (Continued)
Strontium Nitrate $Sr(NO_3)_2$
Ref. [18]
$t = 0°C$

M	P	ΔP	a_1^*
0.367	3207	13.7	0.996
0.664	3201	20.3	0.994
0.939	3195	26.1	0.992
1.530	3178	42.8	0.987
1.736	3170	50.7	0.984
1.91	3162	59.3	0.982

* Calculated from the data.

Tin Bromide $SnBr_4$
For graphical data see Ref. [48]

Tin Iodide SnI_4
For graphical data see Ref. [48]

BENZENE C_6H_6
Tetrabutylammonium Thiocyanate n-Bu_4NCNS
Ref. [34]
$t = 15–40°C$

m	$\log m$	a_1	$-\log a_1$	γ_1
0.0420	−2.623	0.998	1.9991	1.001
0.0457	−2.660	0.999	1.9996	1.003
0.133	−1.125	0.998	1.9991	1.008
0.144	−1.160	0.998	1.9991	1.009
0.620	−1.793	0.998	1.9991	1.046
0.762	−1.882	0.997	1.9987	1.056
1.20	0.079	0.997	1.9987	1.090
1.26	0.099	0.987	1.9946	1.084
1.63	0.213	0.976	1.9897	1.100
1.77	0.247	0.967	1.9856	1.101
4.15	0.618	0.916	1.9621	1.213
7.85	0.895	0.944	1.9607	1.474

Plots also given for (1) temperature dependence of the activity m of benzene solutions and (2) dependence of the average activity of the solvent on the logarithm of the solution concentration.

* Averaged data from 15°–40°C values. There is little dependence on t in this range.

BROMINE Br_2

Trimethylammonium Chloride $(CH_3)_3NHCl$
Ref. [24]
$t = 25°C$

X_2	Vapor* density	Rel. vapor-density lowering†
0.06		0.008
0.10	1.812	0.038
0.14	1.676	0.125
0.16	1.515	0.209
0.18	1.316	0.313
0.20	1.160	0.394
0.30	0.651	0.660
0.40	0.393	0.795
0.50	0.236	0.877
0.60	0.149	0.922
0.70	0.100	0.948
0.80	0.071	0.963
0.90	0.057	0.970
1.0	0.048	0.975

Also given as a plot of the above data.
* Analogous to vapor pressure.
† Analogous to $(1 - a_1)$.

Trimethylammonium Bromide $(CH_3)_3NHBr$
Ref. [24]
$t = 25°C$

X_2	Vapor* density	Rel. vapor-density lowering†
0.04	1.900	0.008
0.06	1.877	0.020
0.08	1.835	0.042
0.10	1.769	0.076
0.14	1.490	0.222
0.16	1.327	0.307
0.18	1.166	0.391
0.20	1.011	0.472
0.30	0.474	0.753
0.40	0.224	0.883
0.50	0.124	0.935
0.60	0.059	0.969
0.70	0.050	0.974
0.80	0.034	0.982
0.90	0.023	0.988
1.0	0.017	0.991

Also given is a plot of the above data.
* Analogous to vapor pressure.
† Analogous to $(1 - a_1)$.

III. VAPOR PRESSURE

BROMINE Br$_2$ (Continued)

Triamylammonium Chloride (Am)$_3$NHCl
Ref. [24]
$t = 25°C$

X_2	Vapor* density	Rel. vapor-density lowering†
0.05	1.909	0.003
0.10	1.735	0.094
0.15	1.406	0.266
0.16	1.331	0.305
0.20	1.005	0.475
0.25	0.622	0.675
0.30	0.391	0.796
0.40	0.184	0.904
0.50	0.090	0.953
0.60	0.052	0.973
0.70	0.031	0.984
0.80	0.019	0.990
0.90	0.015	0.992
1.0	0.013	0.993

Also given is a plot of the above data.
* Analogous to vapor pressure.
† Analogous to $(1 - a_1)$.

Tetrabutylammonium Bromide (Bu)$_4$NBr
Ref. [24]
$t = 25°C$

X_2	Vapor* density	Rel. vapor-density lowering†
0.04	1.882	0.017
0.06	1.819	0.050
0.08	1.727	0.098
0.10	1.599	0.165
0.14	1.254	0.345
0.16	1.048	0.453
0.18	0.837	0.563
0.20	0.674	0.648

Also given is a plot of the above data.
* Analogous to vapor pressure.
† Analogous to $(1 - a_1)$.

CARBON TETRACHLORIDE CCl_4
Titanium Tetrachloride $TiCl_4$
Ref. [26]

X_2	30.0°C			40.0°C			50.0°C		
	P*	ΔP†	a_1†	P*	ΔP†	a_1†	P*	ΔP†	a_1†
0.2512	120.2	20.8	0.852	182.9	27.5	0.869	266.7	36.6	0.879
0.3692	105.6	35.4	0.749	165.7	44.7	0.787	239.5	63.8	0.790
0.5002	92.5	48.5	0.656	145.4	65.0	0.691	212.0	91.2	0.699

Plots also given: (1) X_1 vs P (2) γ_{CCl_4} vs X_1 and (3) γ_{TiCl_4} vs X_1.
* Values averaged from the data.
† Calculated from the average data.

DIETHYL ETHER $(C_2H_5)_2O$
Lithium Aluminium Hydride $LiAlH_4$
Ref. [32]
$t = 24.95°C$

X_1/X_2*	X_1*X_2**	P
3.19	3.25	495.7
1.89	1.92	372.6
1.65	1.68	299.3
1.49	1.51	239.7
1.38	1.40	199.9
1.12	1.14	139.9
0.89	0.90	140.0
0.55	0.56	140.0
0.20	0.20	139.9
0.003	0.003	29.7
0.00	0.00	0.0

Ref. [32]

t‡	P‡
−23.6	9.34
0.05	38.88
9.83	65.90
23.28	128.22
29.88	172.93
42.74	298.3

* Calculated from weight analysis.
** Calculated from aluminium analysis.
‡ Averaged data for pressures of saturated solutions. Concentration data was not reported.

III. VAPOR PRESSURE

DIETHYL ETHER $(C_2H_5)_2O$ *(Continued)*

Lithium Perchlorate $LiClO_4$
Ref. [23]
$t = 23.5°C$

x_2	P	f_1
0.0	503	1.000
0.047	489	1.020
0.059	484	1.023
0.101	482	1.066
0.141	470	1.088
0.177	470	1.135
0.199	466	1.157
0.221	451	1.151
0.237	454	1.183
0.241	453	1.187
0.257	433	1.159
0.280	426	1.176
0.280	432	1.193
0.284	418	1.161
0.287	398	1.110
0.317	370	1.077
0.319	341	1.019
0.321	361	1.057

Also plots given of (1) p vs x_2 and (2) f_1 vs x_2.

DIMETHYLFORMAMIDE
Lithium Chloride LiCl
Ref. [38]

M	t	P	ΔP^*	a_1^*
1	25	3.55	0.53	0.870
1	60	23.55	2.75	0.895

Lithium Perchlorate $LiClO_4$
Ref. [38]

M	t	P	ΔP^*	a_1^*
1	25	3.16	0.72	0.814
1	60	22.6	3.7	0.859

DIETHYLFORMAMIDE (Continued)

Tetramethylammonium Hexafluorophosphate $(CH_3)_4NPF_6$
Ref. [38]

M	t	P	ΔP^*	a_1^*
0.20	25	3.50	0.38	0.902
0.20	60	23.5	0.8	0.967

* Calculated from the data.

Lithium Chloride and Aluminum Chloride $LiCl + AlCl_3$
Ref. [38]

M_{LiCl}	M_{AlCl_3}	t	P	ΔP^*	a_1^*
1	0.075	25	3.32	0.56	0.856
1	0.075	60	22.80	3.50	0.867

Lithium Chloride and Cupric Chloride $LiCl + CuCl_2$
Ref. [38]

M_{LiCl}	M_{CuCl_2}	t	P	ΔP^*	a_1^*
1	0.5	25	3.31	0.57	0.853
1	0.5	50	23.4	2.9	0.890

Lithium Perchlorate and Cupric Fluoride $LiClO_4 + CuF_2$
Ref. [38]

M_{LiClO_4}	M_{CuCl_2}	t	P	ΔP^*	a_1^*
1	0.5	25	3.27	0.61	0.843
1	0.5	60	23.60	2.70	0.897

* Calculated from the data.

ETHANOL C_2H_5OH

Lithium Chloride LiCl
Ref. [1]
$t = 15°C$

m*	ΔP	a_1*
0.222	0.3397	0.989
0.280	0.4870	0.985
0.407	0.7538	0.977
0.492	1.1585	0.964
0.869	1.770	0.945
0.902	1.795	0.944

Plot also given of ΔP vs g solute/mole solvent.

Potassium Iodide KI
Ref. [1]
$t = 15°C$

m*	ΔP	a_1*
0.037	0.0628	0.998
0.060	0.1127	0.996
0.101	0.1897	0.994

Plot also given of ΔP vs g solute/mole solvent.
* Calculated from the data.

METHANOL CH₃OH

Lithium Chloride LiCl
Ref. [1]
$t = 15°C$

m^*	ΔP	a_1^*
0.058	0.1448	0.998
0.121	0.3923	0.995
0.146	0.5871	0.992
0.169	0.7627	0.990
0.217	0.7768	0.989
0.226	1.0065	0.986
0.282	1.125	0.985
0.397	1.542	0.979
0.405	1.740	0.976
0.506	2.057	0.972
0.634	2.660	0.964
0.725	2.868	0.961
0.791	3.344	0.954
0.983	4.34	0.941
1.282	5.76	0.921

Plot also given of ΔP vs solute mole solvent.
* Calculated from the data.

Ref. [41]
$t = 25°C$

m	$\gamma_\pm$ LiCl
0.3	0.361
0.5	0.331
1.0	0.336
2.0	0.458
3.0	0.695
4.0	1.18
5.0	2.03
6.0	3.30
7.0	4.31
8.0	6.09
9.0	8.20

III. VAPOR PRESSURE

METHANOL CH_3OH (Continued)

Lithium Bromide LiBr
Ref. [41]

	15°C		20°C		25°C		30°C	
m	a_1	ϕ	a_1	ϕ	a_1	ϕ	a_1	ϕ
0.2928	0.984	0.859	0.985	0.810	0.986	0.75	0.988	0.64
0.6303	0.964	0.906	0.966	0.86	0.968	0.80	0.969	0.78
1.3 4	0.909	1.091	0.910	1.08	0.911	1.07	0.912	1.00
2.211	0.831	1.31	0.833	1.29	0.834	1.28	0.835	1.27
2.439	0.804	1.39	0.806	1.38	0.808	1.36	0.810	1.35
3.318	0.697	1.70	0.699	1.68	0.701	1.67	0.702	1.66
4.467	0.525	2.25	0.528	2.23	0.530	2.22	0.534	2.19
6.641	0.233	3.42	0.239	3.36	0.244	3.32	0.249	3.27
7.464	0.159	3.84	0.166	3.75	0.172	3.68	0.179	3.60
12.184	0.039	4.15	0.042	4.06	0.046	3.94	0.049	3.86

$t = 25°C$

m	$\gamma_\pm$ LiBr
0.3	0.361
0.5	0.331
1.0	0.336
2.0	0.479
3.0	0.766
4.0	1.63
5.0	3.78
6.0	7.40
7.0	13.27
8.0	31.9
9.0	46.6
11.0	90.8

METHANOL CH$_3$OH (*Continued*)

Potassium Iodide KI
Ref. [1]
$t = 15°C$

m*	ΔP	a_1*
0.013	0.0577	0.999
0.022	0.0743	0.999
0.037	0.154	0.998
0.057	0.238	0.997
0.104	0.423	0.994
0.197	0.798	0.989
0.262	0.974	0.987
0.294	1.123	0.985
0.346	1.324	0.982
0.395	1.514	0.979
0.465	1.780	0.976
0.552	2.1265	0.971
0.570	2.1855	0.970
0.642	2.484	0.966
0.764	2.918	0.960
0.824	3.230	0.956

Plot also given of ΔP vs g solute/mole solvent.
* Calculated from the data.

Tetramethylammonium Iodide (CH$_3$)$_4$NI
Ref. [1]
$t = 15°C$

m*	ΔP	a_1*
0.014	0.0218	1.000
0.019	0.0397	0.999

Plot also given of ΔP vs g solute/mole solvent.
* Calculated from the data.

N-METHYLFORMAMIDE

Lithium Hexafluoroarsenate $LiAsF_6$
Ref. [38]
$t = 25°$

M	P	ΔP^*	a_1^*
1.1	550	40	0.93

Lithium Perchlorate $LiClO_4$
Ref. [38]
$t = 25°$

M	P	ΔP^*	a_1^*
1	561	29	0.95

* Calculated from the data.

PHOSGENE $COCl_2$

Aluminium Chloride $AlCl_3$
Ref. [6]

Wt % $AlCl_3$	0°C			25°C		
	P	ΔP^*	a_1^*	P	ΔP^*	a_1^*
0.0	555		1.00	1406		1.00
5.0	540	15	0.97	1372	34	0.98
10.0	525	30	0.95	1335	71	0.95
15.0	508	47	0.92	1293	113	0.92
20.0	490	65	0.88	1243	163	0.88
25.0	462	93	0.83	1180	226	0.84
30.0	430	125	0.77	1107	299	0.79
35.0	394	161	0.71	1015	391	0.72
40.0	354	201	0.64	897	509	0.64
45.0	308	247	0.55	748	658	0.53
50.0	260	295	0.47	551	855	0.39
55.0				328	1078	0.23

Also given: plot of P vs wt. % $AlCl_3$.
* Calculated from the data.

PROPYLENE CARBONATE
Lithium Perchlorate LiClO$_4$
Ref. [38]

M	t	P	ΔP^*	a_1^*
1	25	0.052	0.017	0.753
1	60	0.66	0.14	0.825

Tetramethylammonium Hexafluorophosphate (CH$_3$)$_4$NPF$_6$
Ref. [38]

M	t	P	ΔP^*	a_1^*
0.125	25	0.062	0.007	0.899
0.125	60	0.52	0.28	0.650

Lithium Chloride and Aluminium Chloride LiCl + AlCl$_3$
Ref. [38]

M_{LiCl}	M_{AlCl_3}	t	P	ΔP^*	a_1^*
0.7	1.0	25	0.035	0.034	0.507
0.7	1.0	60	0.42	0.038	0.917

* Calculated from the data.

III. VAPOR PRESSURE

PYRIDINE C_5H_5N

Hydrochloric Acid HCl

Ref. [33]

$t = -96°C$

X_2/X_1*	P_{HCl}*
2.00 ± 0.03	0 ± 0.5
2.18	6.0
2.50	6.5
2.56	6.5
3.06	7.0
3.41	7.0
3.75	7.5
4.03	8.0
4.12	56.0
4.25	82.5
4.78	102.5
5.44	102.5
5.87	104.0
6.06	108.5
6.31	125.0
6.56	127.5
7.00	165.0
7.19	178.5
7.44	192.0

* Values interpolated from a graph of X_2/X_1 vs P_{HCl}.

SULFUR DIOXIDE SO_2

Potassium Iodide KI

Ref. [21]

$t = 10°C \quad P_0 = 171.456$ cm

m	ΔP (cm)	a_1
0.0000	0.000	1.00000
0.0136	0.175	0.99898
0.0233	0.170	0.99909
0.0252	0.341	0.99801
0.4257	2.349	0.98630
0.5040	2.750	0.98396
0.9367	4.621	0.97305
1.0512	4.839	0.97149
1.3141	7.206	0.95846
1.3194	5.510	0.96786
2.6734	31.999	0.81337

SULFUR DIOXIDE SO_2 (Continued)

Potassium Iodide KI (Continued)
Ref. [21]

$t = 15°C \quad P_0 = 205.96$ cm

m	ΔP (cm)	a_1
0.0000	0.000	1.00000
0.0136	0.105	0.99949
0.0235	0.212	0.99879
0.0252	0.407	0.99802
0.0684	0.812	0.99606
0.1249	1.137	0.99448
0.2195	1.825	0.99114
0.4267	2.760	0.98660
0.5051	3.186	0.98453
0.9400	5.295	0.97429
1.0540	5.595	0.97284
1.3174	7.783	0.96221
1.3225	6.683	0.96755
2.6825	36.979	0.82045
2.8474	43.406	0.78925

$t = 20°C \quad P_0 = 245.33$ cm

m	ΔP (cm)	a_1
0.0000	0.000	1.00000
0.0085	0.172	0.99930
0.0137	0.182	0.99926
0.0234	0.289	0.99882
0.0253	0.466	0.99810
0.0685	0.980	0.99601
0.1254	1.483	0.99396
0.2205	2.111	0.99140
0.4278	3.218	0.98682
0.5064	3.665	0.98506
0.9738	6.049	0.97534
1.0570	6.398	0.97392
1.3214	8.550	0.96515
1.3275	7.682	0.96869
2.6921	42.580	0.82644
2.8570	49.933	0.79647

III. VAPOR PRESSURE

SULFUR DIOXIDE SO_2 (Continued)
Potassium Iodide KI (Continued)
Ref. [15]
$t = 25°C \quad P_0 = 290.63 \text{ cm}$

m	ΔP (cm)	a_1
0.0000	0.000	1.00000
0.0085	0.203	0.99930
0.0137	0.248	0.99915
0.0234	0.328	0.99887
0.0253	0.544	0.99813
0.0687	1.282	0.99559
0.1260	1.741	0.99401
0.2216	2.490	0.99143
0.4290	3.768	0.98704
0.5078	3.812	0.98688
0.9480	6.846	0.97644
1.0612	7.129	0.97547
1.3257	9.735	0.96650
1.3323	8.920	0.96931
2.7048	48.328	0.83371
2.8678	57.525	0.80207

			f_n		
m	N_2	10°	15°	20°	25°
0.0001	0.0000128	0.827	0.821	0.807	0.789
0.0005	0.0000641	0.655	0.642	0.619	0.655
0.001	0.000128	0.551	0.534	0.508	0.471
0.005	0.000640	0.265	0.248	0.221	0.186
0.01	0.001280	0.155	0.141	0.120	0.0940
0.05	0.006365	0.0800	0.0735	0.0642	0.0528
0.1	0.012650	0.0508	0.0467	0.0408	0.0335
0.2	0.024984	0.0309	0.0284	0.0248	0.0204
0.3	0.037014	0.0224	0.0206	0.0180	0.0148
0.4	0.048750	0.0177	0.0163	0.0142	0.0117
0.5	0.060203	0.0148	0.0136	0.0118	0.00973
0.6	0.071384	0.0127	0.0117	0.0102	0.00838
0.7	0.082303	0.0113	0.0104	0.00903	0.00742
0.8	0.092967	0.0102	0.00933	0.00815	0.00670
0.9	0.103387	0.00933	0.00857	0.00749	0.00615
1.0	0.113569	0.00875	0.00805	0.00703	0.00677
1.5	0.161201	0.00724	0.00665	0.00531	0.00477
2.0	0.203974	0.00709	0.00653	0.00569	0.00468
2.5	0.242597	0.00746	0.00686	0.00599	0.00492
3.0	0.277545	0.00794	0.00730	0.00638	0.00524

p-XYLENE $CH_3C_6H_4CH_3$

Tetrapentylammonium Thiocyanate $(C_5H_{11})_4NSCN$
Ref. [36]

	52°C			70°C			90°C		
X_2	P*	ΔP†	a_1†	P*	ΔP†	a_1†	P*	ΔP†	a_1†
0.3345				70.1 ± 0.1	4.5	0.94	149.9	9.3	0.94
0.4082	28.7 ± 0.1	5.5	0.84	63.6	11.0	0.85	135.6	23.6	0.85
0.4837	24.8 ± 0.1	9.4	0.73	54.5	20.1	0.73	115.5	43.7	0.73
0.6245	17.6 ± 0.3	16.6	0.51	39.7	34.9	0.53	86.5	72.7	0.54
0.8180	7.9 ± 0.1	26.3	0.23	17.5	57.1	0.23			
0.9223	3.4 ± 0.1	30.8	0.10	7.7	66.9	0.10	17.1	142.1	0.11

Also given: (1) P vs X_2 plots, (2) a_1 vs t plots, (3) a_1 vs X_2 plots for 52°, 60°, 70°, 80°, 90° (4) $P = f(t)$ equations of the pure solvent and six concentrations studied.

* Graphical results were reported only. In an attempt to recreate the original data that was shown graphically, P values were calculated from the $P = f(t)$ equations.

† Calculated from the calculated P values. a_1 values were reported at all concentrations and at 60° and 80° as well. However, these were undoubtably derived from the $P = f(t)$ equations.

(b) Mixed Solvents

The presentation of the data in this section is according to solvent which are arranged in alphabetical order by nonaqueous component. Within this classification the electrolytes are arranged alphabetically by name. Information on the temperature range, concentration range, vapor pressure data, differential vapor pressures and solvent activities are included wherever possible.

Symbols used:

- t Temperature (°C)
- m Molality
- M Molarity
- P Vapor pressure of solution
- P_0 Vapor pressure of pure solvent
- ΔP Differential pressure between solution and pure solvent
- a_1 Solvent activity
- a_2 Solute activity
- N_a Mole fraction of alcohol
- N_w Mole fraction of water
- a_a Alcohol activity
- a_w Water activity

AMMONIA–WATER NH_3H_2O

Ammonium Nitrate NH_4NO_3

Ref. [3]

N_{NH_3}*	N_{H_2O}*	m*	P	t
29.56	70.44	6.35	−12.95	53.2
			−7.25	73.5
			0.10	109.0
			7.98	159.5
			10.95	188.0
			18.80	272.5
			27.72	405.0
			35.20	555.7

Also P vs t plot given.

AMMONIA–WATER NH_3—H_2O (Continued)

Calcium Nitrate $Ca(NO_3)_2$
Ref. [3]

N_{NH_3}*	N_{H_2O}*	m	t	P
25.82	74.18	1.77	−14.5	34.8
		(1.23)**	−9.3	48.1
			−2.2	73.2
			5.3	110.2
			15.1	181.7
			21.15	242.0
			29.5	354.0
			36.35	474.0
59.69	40.31	7.69	12.1	82.6
		(5.35)**	−4.9	124.1
			+3.97	202.3
			11.0	291.1
			18.0	413.0
			25.0	575.3
			32.4	782.7

Also P vs t plots given.

Ammonium Thiocyanate NH_4CNS
Ref. [3]

N_{NH_3}†	N_{H_2O}†	M†	P	t
23.16	76.84	2.45	−15.3	39.6
			−8.03	62.0
			−0.81	94.5
			6.78	143.9
			10.16	170.8
			17.19	237.6
			23.0	314.1
			30.3	433.1
			35.7	543.1

Also P vs t plot given.

III. VAPOR PRESSURE

AMMONIA–WATER NH_3—H_2O (Continued)

Sodium Iodide NaI
Ref. [3]

N_{NH_3}*	N_{H_2O}*	m*	t	P
24.79	75.21	3.19	−14.4	40.9
			8.0	61.14
			1.2	90.4
			3.15	116.3
			10.7	170.8
			17.9	242.4
			25.8	347.5
			35.0	519.2
77.65	22.35	12.33	−14.6	87.52
			−13.0	96.6
			−9.6	120.3
			−8.0	143.0
			−0.9	190.3
			6.0	276.0
			13.07	394.8
			20.07	551.8
			26.9	753.8

Also P vs t plots given.

Lithium Nitrate $LiNO_3$
Ref. [2]

N_{NH_3}*	N_{H_2O}**	m*	t	P
29.61	70.39	3.10	−14.45	50.82
			−8.10	73.7
			0.75	110.2
			6.45	161.9
			13.25	225.23
			20.00	307.97
			28.00	441.75
			35.10	598.09

AMMONIA–WATER NH_3—H_2O (Continued)

Lithium Nitrate $LiNO_3$ (Continued)

Ref. [2]

N_{NH_3}*	N_{H_2O}**	m*	t	P
34.40	66.60	11.87	−13.5	16.7
			−12.15	19.64
			−4.15	31.9
			1.67	46.27
			8.47	66.77
			14.42	91.14
			21.56	131.0
			28.65	183.2
			35.17	246.85
38.57	61.43	10.31	−14.85	24.54
			−6.97	41.60
			0.03	63.8
			7.00	95.31
			13.95	136.4
			20.4	188.39
			27.05	256.98
			35.2	371.1
38.81	61.19	6.06	−15.4	52.80
			−7.8	84.35
			−0.10	131.81
			8.4	203.00
			15.23	283.43
			21.00	369.72
			27.5	491.54
			35.1	683.10
87.04	12.96	17.58	−11.6	121.34
			−5.5	169.64
			1.3	242.08
			8.17	345.39
			15.39	493.04
			20.45	623.82
			27.57	862.7
			34.4	1140.92

Also P vs t plots given.

III. VAPOR PRESSURE

AMMONIA–WATER NH_3–H_2O (Continued)

Calcium Chloride $CaCl_2$
Ref. [3]

N_{NH_3}	N_{H_2O}*	m	t	P
27.41	72.59	1.33	−13.4	54.0
			−5.01	88.6
			1.8	129.8
			2.45	136.5
			9.27	193.1
			16.42	272.6
			23.06	370.1
			29.75	498.2
			35.60	640.3

Also P vs t plot.

* Calculated from the data. There is some uncertainty in these values as the original paper referred to these solutions in unlabeled "%" values for the solution components. It was assumed that "%" values referred to weight % values and X and m values were thus calculated.

** It is not possible to tell from the original paper whether hydrated salts were used and if they were, whether the hydration weights were taken into consideration for the weight percent calculations. Therefore molalities are given for both cases with the hydrated salt molality in brackets.

† Calculated from the data. There is some uncertainty in these values as again the notation used in the reference is unclear. It has been assumed that "%" NH_3 refers to weight percent of NH_3 in the NH_3–H_2O solution and that the weight of the salt is not included. The salt concentration has been calculated from the value given in weight of salt per cc of solution.

DIOXANE–WATER

Lithium Chlorate LiClO$_3$
Ref. [39]
$t = 25°C$

Weight % solution	m	p	$N_{dioxane}$ in vapor	$P_{dioxane}$	a_w	$a_{dioxane}$
Lithium chlorate in 44.5% dioxane	0.0	40.57	47.23	19.16	1.0000	1.0000
	1.0863	39.58	49.50	19.59	0.9337	1.0224
	2.0172	37.14	50.46	18.74	0.8594	0.9781
	3.0094	34.35	51.33	17.63	0.7809	0.9201
	4.1095	30.77	51.98	15.99	0.6903	0.8346
	5.0434	27.78	52.37	14.55	0.6179	0.7594
	6.7324	22.50	52.94	11.91	0.4946	0.6216
	8.9653	17.28	53.40	9.23	0.3760	0.4817
	11.5036	12.65	53.68	6.79	0.2737	0.3544
	14.0631	9.96	53.79	5.36	0.2149	0.2797
Lithium chlorate in 65.5% dioxane	0.0	45.50	56.88	25.88	1.0000	1.0000
	1.0177	44.91	61.26	27.51	0.8869	1.0630
	1.9613	42.47	63.99	27.18	0.7793	1.0502
	3.0403	38.04	66.92	25.46	0.6412	0.9838
	3.8451	34.60	68.90	23.84	0.5484	0.9212
	5.0592	29.67	70.99	21.06	0.4391	0.8138
	6.2070	25.19	72.20	18.19	0.3568	0.7029
	7.6221	21.36	73.09	15.61	0.2931	0.6032

Plots also given: (1) a_w and $a_{dioxane}$ vs m, (2) $\gamma_\pm$ vs m.

III. VAPOR PRESSURE

DIOXANE–WATER (Continued)
Sodium Chlorate $NaClO_3$
Ref. [39]
$t = 25°C$

Weight % solution	m	P	$N_{dioxane}$ in vapor	$P_{dioxane}$	a_w	$a_{dioxane}$
Sodium chlorate in 44.5% dioxane	0.0	40.57	47.23	19.16	1.0000	1.0000
	1.0332	41.22	51.66	21.29	0.9309	1.1112
	1.4716	41.32	53.00	21.90	0.9071	1.1430
	2.0795	41.47	54.76	22.71	0.8762	1.1853
	2.4637	41.62	55.90	23.27	0.8571	1.2145
	3.2081	41.72	58.23	24.29	0.8141	1.2677
	3.4873	41.72	59.09	24.65	0.7973	1.2865
	4.0314	41.72	60.89	25.40	0.7623	1.3257
Sodium chlorate in 64.5% dioxane	0.0	45.50	56.88	25.88	1.0000	1.0000
	0.8301	46.20	60.75	28.07	0.9241	1.0346
	1.0200	46.23	61.35	28.36	0.9108	1.0958
	1.2800	46.26	62.23	28.79	0.8904	1.1124
	1.4665	46.25	62.83	29.06	0.8761	1.1229
	1.6866	46.30	63.59	29.44	0.8593	1.1376
	2.0000	46.27	64.64	29.91	0.8338	1.1557
	2.3311	46.25	65.82	30.44	0.8058	1.1762

Plots also given: (1) a_w and $a_{dioxane}$ vs m, (2) $\gamma_\pm$ vs m.

ETHANOL–WATER C_2H_5OH—H_2O

Acetanalide $CH_3CONHC_6H_5$

Ref. [4]

$t = 20°C$

N_A	m	P_a	P_w	a_a*	a_w*
70	1.0	25.3	13.9	0.91	1.02

Ammonium Sulfate $(NH_4)_2SO_4$

Ref. [5]

$t = 20°C$

N_A	m	P_a	P_w	a_a*	a_w*
10	0.5	14.6	16.0	1.11	0.97

Sodium Nitrate $NaNO_3$

Ref. [5]

$t = 20°C$

N_A	m	P_a	P_w	a_a*	a_w*
10	0.5	13.7	15.9	1.04	0.97

Sodium Sulfate Na_2SO_4

Ref. [5]

$t = 20°C$

N_A	m	P_a	P_w	a_a*	a_w*
10	0.5	15.1	16.2	1.15	0.99

* Calculated from the data.

ETHANOL–WATER $C_2H_5OH—H_2O$ (Continued)

Potassium Carbonate K_2CO_3
Ref. [5]
$t = 20°C$

N_A	m	P_a	P_w	a_a^*	a_w^*
10	0.5	14.7	16.5	1.23	1.01

Potassium Nitrate KNO_3
Ref. [5]
$t = 20°C$

N_A	m	P_a	P_w	a_a^*	a_w^*
10	0.5	13.6	16.7	1.04	1.02

Potassium Bromide KBr
Ref. [5]
$t = 20°C$

N_A	m	P_a	P_w	a_a^*	a_w^*
10	0.5	13.9	16.4	1.06	1.00

Ammonium Chloride NH_4Cl
Ref. [5]
$t = 20°C$

N_A	m	P_a	P_w	a_a^*	a_w^*
10	0.5	13.8	15.5	1.05	0.94

* Calculated from the data.

ETHANOL–WATER C_2H_5OH–H_2O (Continued)

Sodium Carbonate Na_2CO_3
Ref. [5]
$t = 20°C$

N_A	m	P_a	P_w	$a_a{}^*$	$a_w{}^*$
10	0.5	14.8	16.3	1.13	0.99

Lithium Chloride LiCl
$t = 25°C$

M_a	m	p_w	p_a	a_a	a_w	$a_a/a_a{}^\circ$	$a_w/a_w{}^\circ$	Ref.
100	0	0	58.98	1.000		1.000		9
	0.5	0	57.21	0.9699		0.9699		
	1.0	0	54.27	0.9202		0.9202		
	4.0	0	26.72	0.453		0.453		
98	0	1.141	57.24	0.9764	0.04806	1.000	1.000	9
	0.5	0.908	55.98	0.9497	0.0383	0.9725	0.795	
	1.0	0.788	53.78	0.9118	0.0331	0.934	0.690	
	4.0	0.277	26.96	0.4573	0.0116	0.468	0.2425	
95	0	2.286	56.03	0.9504	0.0962	1.000	1.000	9
	0.5	1.816	54.73	0.9286	0.0764	0.977	0.794	
	1.0	1.636	52.69	0.8940	0.0688	0.9405	0.7155	
	4.0	0.580	27.44	0.4654	0.0244	0.490	0.254	
90	0	5.381	52.42	0.8900	0.2264	1.000	1.000	9
	0.5	4.547	52.26	0.8874	0.1914	0.998	0.845	
	1.0	3.852	50.72	0.8611	0.1621	0.9685	0.716	
	4.0	1.062	28.21	0.4786	0.0446	0.538	0.197	
80	0	9.81	48.04	0.8155	0.4130	1.000	1.000	9
	0.5	8.49	47.91	0.8133	0.3573	0.998	0.865	
	1.0	7.33	47.00	0.7980	0.3085	0.979	0.747	
	4.0	2.534	29.65	0.5033	0.1066	0.6175	0.258	
70	0	12.95	43.62	0.7404	0.5575	1.000	1.000	9
	0.5	11.47	44.06	0.7481	0.4829	1.011	0.866	
	1.0	10.19	43.92	0.7457	0.4282	1.008	0.768	
	4.0	4.19	31.13	0.5286	0.1763	0.714	0.315	
50	0	17.24	36.65	0.6223	0.7253	1.000	1.000	9
	0.5	15.78	37.77	0.6410	0.6639	1.030	0.915	
	1.0	14.53	38.13	0.6473	0.6114	1.040	0.543	
	4.0	7.65	33.17	0.5631	0.3219	0.902	0.444	

* Calculated from the data.

III. VAPOR PRESSURE

ETHANOL–WATER C_2H_5OH – H_2O (Continued)

Lithium Chloride LiCl (Continued)
$t = 25°C$

N_a	m	p_w	p_a	a_a	a_w	a_a/a_a,	a_w/a_w,	Ref.
25	0	19.48	29.08	0.4938	0.8195	1.000	1.000	9
	0.5	18.84	30.19	0.5126	0.7928	1.038	0.967	
	1.0	18.18	30.90	0.5247	0.7649	1.063	0.933	
	4.0	12.70	32.08	0.5446	0.5344	1.132	0.652	
6.4	0	21.94	12.29	0.2083	0.9230	1.000	1.000	9
	0.5	21.65	13.08	0.2217	0.9110	1.064	0.987	
	1.0	21.07	13.87	0.2352	0.8864	1.129	0.960	
	4.0	16.66	18.01	0.3054	0.7010	1.465	0.760	
	0.0	21.91	12.32	0.2089*	0.9218*	1.000	1.000	
	4.0	16.64	18.03	0.3057*	0.7000*	1.462	0.760	
4.0	0.0	22.40	8.16	0.1384*	0.9424*	1.000	1.000	12
	0.5	22.11	8.84	0.1499*	0.9302*	1.083	0.987	
	1.0	21.58	9.36	0.1587*	0.9079*	1.148	0.964	
	2.0	20.52	10.42	0.1767*	0.8633*	1.276	0.915	
	4.0	17.71	12.11	0.2053*	0.7451*	1.485	0.791	
2.0	0.0	23.00	4.08	0.0692*	0.9676*	1.000	1.000	12
	0.5	22.73	4.48	0.0760*	0.9562*	1.098	0.988	
	1.0	22.14	4.83	0.0819*	0.9314*	1.183	0.963	
	4.0	18.40	6.45	0.1094*	0.7741*	1.581	0.800	
0	0	23.77			1.000		1.000	9
	0.5	23.45			0.9865		0.9865	
	1.0	23.05			0.9696		0.9695	
	4.0	19.21			0.8081		0.808	

* Calculated from the data.
Also given are plots: (1) p_A vs m, (2) P_w vs m, (3) a_w vs m, (4) a_a vs m, (5) $a_a/a_a°$ vs N_a, (6) $a_a/a_w°$ vs N_a.

Sodium Chloride NaCl
Ref. [5]
$t = 20°C$

N_a	m	P_a	P_w	a_a*	a_w*
10	0.6	14.2	16.3	1.08	0.99

ETHANOL–WATER C_2H_5OH—H_2O (Continued)

Sodium Bromide NaBr
Ref. [5]
$t = 20°C$

N_a	m	P_a	P_w	a_a^*	a_w^*
10	0.5	13.9	16.1	1.06	0.98

Potassium Chloride KCl
Ref. [5]
$t = 20°C$

N_a	m	P_a	P_w	a_a^*	a_w^*
10	0.5	14.2	15.5	1.09	0.95

Benzoic Acid
Ref. [4]
$t = 20°C$

N_a	m	P_a	P_w	a_a^*	a_w^*
71	1.0	25.6	13.9	0.92	1.01

Phenylacetic Acid
Ref. [4]
$t = 20°C$

N_a	m	P_a	P_w	a_a^*	a_w^*
71	1.0	26.0	13.8	0.93	1.02

Salicylic Acid
Ref. [4]
$t = 20°C$

N_a	m	P_a	P_w	a_a^*	a_w^*
71	1.0	26.0	14.4	0.93	1.06

* Calculated from data.

III. VAPOR PRESSURE

HYDROGEN PEROXIDE–WATER H_2O_2—H_2O

Lithium Nitrate $LiNO_3$

Ref. [29]

$t = 50°C$

Mole Fraction $LiNO_3 = 0.0632 \pm 0.001$

Total pressure (mm)	Mole fraction H_2O_2		Mole fraction H_2O_2	ΔP
	In solvent	In vapor		
74.3	0.0424	0.0004	0.0	−13.2
67.6	0.1167	0.0004	0.1	−12.4
58.6	0.2124	Very small	0.2	−12.2
40.5	0.3628	0.0221	0.3	−11.9
32.8	0.4469	0.0489	0.4	−11.0
25.6	0.5376	0.113	0.5	−9.6
21.6	0.6185	0.154	0.6	−7.6
16.1	0.7101	0.314	0.7	−5.3
12.4	0.8160	0.583	0.8	−3.5
10.0	0.9042	0.824	0.9	−1.4
9.1	0.9452	0.893	1.0	−1.1

Sodium Nitrate $NaNO_3$

Ref. [29]

$t = 50°C$

Mole Fraction $NaNO_3 = 0.0647 \pm 0.001$

Total pressure (mm)	Mole fraction H_2O_2		Mole fraction H_2O_2	ΔP
	In solvent	In vapor		
82.0	0.0127	Very small	0.0	−9.8
81.5	0.0228	0.0002	0.1	−6.8
79.8	0.0451	0.0002	0.2	−5.0
73.0	0.1239	0.0005	0.3	−4.4
61.2	0.2391	0.0036	0.4	−3.7
47.8	0.3717	0.0085	0.5	−3.0
37.2	0.4734	0.0305	0.6	−2.0
35.0	0.5017	0.0330	0.7	−1.4
29.1	0.5785	0.0502	0.8	−1.6
18.4	0.7391	0.1884	0.9	−0.0
13.6	0.8285	0.4430	1.0	−1.0
11.0	0.8854	0.6957		
10.5	0.9250	0.7381		
9.5	0.9638	0.8952		
9.0	0.9949	0.9702		

HYDROGEN PEROXIDE–WATER H_2O_2—H_2O (Continued)

Potassium Nitrate KNO_3
Ref. [29]
$t = 50°C$
Mole Fraction $KNO_3 = 0.0646 \pm 0.001$

Total pressure (mm)	Mole fraction H_2O_2		Mole fraction H_2O_2	ΔP
	In solvent	In vapor		
84.1	0.0217	Very small	0.0	−7.5
79.1	0.1022	Very small	0.1	−2.8
75.3	0.1527	Very small	0.2	−0.2
58.5	0.3313	0.0011	0.3	0.2
35.3	0.5617	0.0395	0.4	0.4
34.7	0.5670	0.0365	0.5	0.6
34.5	0.5712	0.0276	0.6	1.0
25.7	0.6789	0.0843	0.7	1.2
16.1	0.8243	0.236	0.8	0.6
15.2	0.8417	0.344	0.9	0.3
11.8	0.9016	0.530	1.0	−1.2
11.7	0.9246	0.452		
10.0	0.9579	0.832		
9.5	0.9669	0.916		

Rubidium Nitrate $RbNO_3$
Ref. [29]
$t = 50°C$
Mole Fraction $RbNO_3 = 0.039 \pm 0.001$

Total pressure (mm)	Mole fraction H_2O_2		Mole fraction H_2O_2	ΔP
	In solvent	In vapor		
87.1	0.0000	0.0000	0.0	−5.6
76.8	0.1293	0.0022	0.1	2.7
51.6	0.3763	0.0108	0.2	−0.2
30.6	0.5926	0.1219	0.3	0.7
24.7	0.6480	0.1509	0.4	1.3
15.6	0.8303	0.3611	0.5	1.3
10.8	0.9179	0.7235	0.6	1.1
10.2	0.9282	0.8000	0.7	0.7
9.84	0.9421	0.8601	0.8	0.4
			0.9	−0.6
			1.0	−1.4

III. VAPOR PRESSURE

References

1. O. F. Tower and A. F. O. Germann, J. Am. Chem. Soc., **36**, 2449 (1914).
2. R. O. E. Davis, L. B. Olmstead and F. O. Lundstrum, J. Am. Chem. Soc., **43**, 1575 (1921).
3. R. O. E. Davis, L. B. Olmstead and F. O. Lundstrum, J. Am. Chem. Soc., **43**, 1580 (1921).
4. R. Wright, J. Chem. Soc., **121**, 2251 (1922).
5. R. Wright, J. Chem. Soc., **124**, 2068 (1924).
6. A. F. O. Germann and G. McIntyre, J. Phys. Chem., **29**, 102 (1925).
7. N. Kameyama and S. Yagi, J. Soc. Chem. Ind., Japan **31**, 1141 (1928), Suppl. Binding **31**, 272 B (1928); C. A. **23**, 4875 (1929).
8. N. Kameyama, J. Soc. Chem. Ind., Japan, **32**, Suppl. Binding 242 B (1929); C. A. **24**, 2360 (1939).
9. R. Shaw and J. A. V. Butler, Proc. Roy. Soc., **129**, 519 (1930).
10. N. Kameyama, J. Soc. Chem. Ind., Japan **34**, Suppl. Binding 236 (1931); C. A. **25**, 5610 (1931).
11. H. Oosaka, Bull. Inst. Phys-Chem. Research (Tokyo) **10**, 466 (1931); Published with Sci. Papers Inst. Phys-Chem. Res. (Tokyo) **15**, #304; C. A. **25**, 5610 (1931).
12. J. A. V. Butler and D. W. Thomson, Proc. Roy. Soc., **141**, 86 (1933).
13. S. Abe and R. Hara, J. Soc. Chem. Ind., Japan, **36**, 1324 (1933).
14. H. Hunt and W. E. Larsen, J. Phys. Chem., **38**, 801 (1934).
15. A. I. Shattenstein and L. S. Uskova, Acta Physiochim. U.R.S.S. **11**, 5 (1935).
16. S. Abe, S. Sigetomi and R. Hara, J. Soc. Chem. Ind., Japan, **38**, Suppl. Binding 163 (1935); C. A. **29**, 5332 (1935).
17. S. Abe, K. Watanabe and R. Hara, J. Soc. Chem. Ind., Japan, **38**, Suppl. Binding 642 (1935); C. A. **30**, 2079 (1936).
18. M. Linhard, Z. Physik Chem., **A175**, 438 (1936).
19. E. L. Chernyak, Zhur. Obshei Khim., **8**, 1341 (1938).
20. H. W. Ritchey and H. Hunt, J. Phys. Chem. **43**, 407 (1939).
21. W. G. Eversole and A. L. Hanson, J. Phys. Chem., **47**, 1 (1943).
22. N. M. Baron and K. P. Mishchenko, Zhur. Obschei Khim., **18**, 2068 (1948).
23. K. Ekelin and L. G. Sillen, Acta Chem. Scand., **7**, 987 (1953).
24. P. L. Mercer and C. A. Kraus, Proc. Nat'l Acad. Sci., **42**, 487 (1956).
25. W. Hutton, Jr., Ph.D. Thesis Michigan State U. (1959).
26. G. A. Ryder, M. R. Kamal and L. N. Canjar, J. Chem. Eng. Data, **6**, 594 (1961).
27. K. P. Mishchenko and M. K. Fedorov, J. Applied Chem., **34**, 1792 (1961).
28. V. F. Sergeeva and E. S. Moiseeva, Russ. J. Gen. Chem., **32**, 2370 (1962).
29. M. E. Everhard, P. M. Gross and J. W. Turner, J. Phys. Chem., **66**, 923 (1962).
30. G. C. Blytas and F. Daniels, J. Am. Chem. Soc., **84**, 1075 (1962).
31. G. C. Blytas, D. J. Kertesz and F. Daniels, J. Am. Chem. Soc., **84**, 1083 (1962).
32. S. C. Chattoraj, C. A. Hollingsworth, D. H. McDaniel and G. B. Smith, J. Inorg. Nuclear Chem., **24**, 101 (1962).
33. P. A. Kitty and D. Nicholls, Chem. and Ind., 1123 (1963).
34. K. P. Mishchenko and G. M. Poltoraskii, J. Struct. Chem., **5**, 18 (1964).
35. W. Lindenberg and W. Ilgner, Z. Anorg. Chem., **355**, 36 (1965).
36. J. A. O'Mally, C. Owens, C. Schmid, D. Quimby and C. M. King, J. Phys. Chem., **72**, 3584 (1968).
37. V. V. Kushchenko and K. P. Mishchenko, J. Applied Chem., **41**, 620 (1968).
38. R. Keller, J. N. Foster, D. C. Hanson, J. F. Hon and J. S. Muirhead, NASA Technical Report CR-1425 (1969).

39. A. N. Campbell and B. G. Oliver, Canad. J. Chem., **47,** 2671 (1969).
40. P. Damay and G. Lepontre, J. Chim. Phys., **66,** 809 (1969).
41. P. A. Skabichevskii, Russ. J. Phys. Chem., **43,** 1432 (1969).
42. B. D. Briggs, unpublished work.
43. H. W. B. Roozeboom, Rec. Trav. Chim., **4,** 361 (1885).
44. G. Spacu and P. Voichescu, Z. Anorg. Allg. Chem., **233,** 197 (1937).
45. G. W. Watt and W. R. McBride, J. Am. Chem. Soc., **77,** 1317 (1955).
46. W. Lindenberg, Z. Anorg. Allg. Chem., **318,** 1 (1962).
47. M. L. Troost, Comptes Rend., **92,** 715 (1881).
48. E. Bannister and G. W. A. Fowles, J. Chem. Soc., 4375 (1958).

IV. CRYOSCOPY

The use of cryoscopic measurements for the determination of activity coefficients in aprotic solvents has been very limited, although much information is available on the evaluation of molecular weights and studies of solvent association. Relative to protic solvents more information is available and the investigations of Gillespie and coworkers in sulphuric acid and related compounds have been compiled to indicate the general scope of this technique in protic solvents. The information is presented in alphabetical order by solvent.

DIMETHYL SULFOXIDE

LiCl [3]
Technique: Fr. Pt. Depression

m^*	$-\log_{10}\gamma_\pm$
0.005	0.07$_6$
0.01	0.11$_4$
0.02	0.18$_8$
0.03	0.19$_0$
0.04	0.21$_6$
0.08	0.29$_0$
0.40	0.46$_9$

CsI [4]
Technique: Fr. Pt. Depression

$\log_{10}m$	$-\log_{10}\gamma_\pm$
−2.5	0.05
−2.0	0.11$_4$
−1.75	0.14$_9$
−1.5	0.19$_4$
−1.25	0.24$_4$
−1.1	0.27$_3$
−1.0	0.29$_2$
−0.85	0.32$_0$
−0.75	0.33$_8$
−0.6	0.36$_6$

* Other concentrations were investigated.

DIMETHYL SULFOXIDE (*Continued*)

RbI
Ref. [5]
Technique: Fr. Pt. Depression

$\log_{10} m$	$-\log_{10} \gamma_{\pm}$
-2.3	0.07_2
-2.1	0.09_8
-1.9	0.12_5
-1.7	0.15_0
-1.5	0.18_3
-1.3	0.21_8
-1.1	0.25_0

HYDROGEN FLUORIDE

System	Investigation	Ref.
LiF	Cryoscopic data	24
NaF		
KF		
NH_4F		

IV. CRYOSCOPY

N-METHYLACETAMIDE
Ref. [1]

Osmotic Coefficients, ϕ

n	LiCl	NaCl	KCl	CsCl	LiBr	NaBr	KBr	CsBr	NaI	KI	CsI	LiNO$_3$	NaNO$_3$	KNO$_3$
0.01	0.996	0.993	0.992	0.992	0.994	0.994	0.992	0.994	0.995	0.992	0.992	0.993	0.990	0.990
0.05	1.008	0.994	0.990	0.987	0.999	0.999	0.990	0.994	1.003	0.990	0.990	0.995	0.979	0.978
0.10	1.023	1.001	0.994	0.987	1.011	1.010	0.993	0.996	1.019	0.994	0.991	1.004	0.972	0.967
0.20	1.050	1.020			1.039	1.035	1.004	0.994	1.053	1.008	0.997	1.027	0.962	0.948
0.30	1.068	1.042			1.070	1.061	1.017		1.089	1.027	1.003	1.054	0.955	
0.40					1.101	1.086	1.030		1.124	1.048	1.009	1.082	0.948	
0.50					1.133	1.111			1.159	1.071		1.112	0.942	
0.60					1.166	1.135			1.194	1.095		1.143	0.937	
0.70						1.158			1.228	1.122			0.931	
0.80									1.262				0.926	

N-METHYLACETAMIDE (Continued)
Ref. [1]

Activity Coefficients, $\gamma_{\pm}$

m	LiCl	NaCl	KCl	CsCl	LiBr	NaBr	KBr	CsBr	NaI	KI	CsI	LiNO$_3$	NaNO$_3$	KNO$_3$
0.01	0.982	0.967	0.974	0.973	0.977	0.977	0.974	0.977	0.979	0.973	0.974	0.975	0.970	0.969
0.05	0.993	0.966	0.958	0.952	0.974	0.974	0.957	0.968	0.983	0.956	0.956	0.966	0.937	0.934
0.10	1.020	0.971	0.955	0.943	0.988	0.988	0.955	0.966	1.005	0.954	0.952	0.974	0.915	0.907
0.20	1.073	0.996			1.033	1.028	0.964	0.962	1.065	0.968	0.953	1.007	0.886	0.864
0.30	1.120	1.031			1.088	1.075	0.980		1.135	0.993	0.959	1.051	0.864	
0.40					1.151	1.126	1.000		1.213	1.025	0.966	1.102	0.847	
0.50					1.220	1.179			1.296	1.062		1.160	0.832	
0.60					1.295	1.235			1.386	1.105		1.224	0.818	
0.70						1.293			1.482	1.154			0.805	
0.80									1.584				0.794	

IV. CRYOSCOPY

Ref. [2]

Osmotic Coefficients, ϕ

m	Li formate	Na formate	Li acetate	Na acetate	K acetate	Li propionate	Na propionate	K propionate
0.01	0.985	0.990	0.988	0.989	0.991	0.989	0.990	0.992
0.05	0.959	0.975	0.969	0.972	0.984	0.976	0.980	0.990
0.10	0.936	0.959	0.952	0.957	0.981	0.964	0.972	0.991
0.20	0.903	0.923	0.923	0.931	0.978	0.943	0.956	0.995
0.30	0.881		0.897	0.906	0.976	0.922	0.940	0.996
0.40	0.867		0.872	0.882	0.974	0.902		0.996
0.50					0.973			0.994
0.60					0.970			
0.70					0.968			
0.80					0.965			

N-METHYLACETAMIDE (Continued)
Ref. [2]

Activity Coefficients, $\gamma_\pm$

m	Li formate	Na formate	Li acetate	Na acetate	K acetate	Li propionate	Na propionate	K propionate
0.01	0.961	0.969	0.966	0.967	0.972	0.969	0.971	0.974
0.05	0.898	0.931	0.919	0.924	0.946	0.931	0.939	0.958
0.10	0.847	0.896	0.879	0.888	0.932	0.901	0.917	0.953
0.20	0.776	0.831	0.819	0.833	0.916	0.855	0.881	0.951
0.30	0.726		0.769	0.787	0.906	0.815	0.848	0.951
0.40	0.691		0.726	0.745	0.898	0.779		0.950
0.50					0.891			
0.60					0.884			
0.70					0.878			
0.80					0.872			

IV. CRYOSCOPY

Osmotic Coefficients, ϕ

m	Me$_4$NCl	Et$_4$NCl	Pr$_4$NCl	Bu$_4$NCl	Me$_4$NBr	Et$_4$NBr	Pr$_4$NBr	Bu$_4$NBr	Et$_4$NI	Pr$_4$NI	Bu$_4$NI
0.01	0.991	0.992	0.992	0.993	0.988	0.990	0.989	0.989	0.986	0.985	0.985
0.05	0.981	0.988	0.989	0.992	0.969	0.977	0.975	0.977	0.959	0.957	0.956
0.10	0.974	0.988	0.991	0.997		0.966	0.963	0.967	0.930	0.929	0.928
0.20	0.961	0.990	0.997	1.008		0.949	0.946	0.954		0.883	0.883
0.30	0.947	0.993	1.004	1.018		0.932	0.932	0.944		0.844	0.846
0.40	0.931	0.994	1.010	1.028		0.917	0.920	0.935		0.810	0.814
0.50	0.914	0.995	1.016	1.037		0.901	0.909	0.929		0.779	0.786
0.60		0.995	1.022	1.044		0.885	0.900	0.923			0.763
0.70		0.995	1.027	1.051		0.870	0.891	0.918			0.743
0.80		0.993	1.031	1.057		0.853	0.883	0.913			0.726
0.90						0.837	0.876	0.910			0.712
1.00			1.034	1.061			0.869	0.907			0.701

N-METHYLACETAMIDE (Continued)
Ref. [2]

Activity Coefficients, $\gamma_\pm$

m	Me$_4$NCl	Et$_4$NCl	Pr$_4$NCl	Bu$_4$NCl	Me$_4$NBr	Et$_4$NBr	Pr$_4$NBr	Bu$_4$NBr	Et$_4$NI	Pr$_4$NI	Bu$_4$NI
0.01	0.971	0.973	0.974	0.975	0.967	0.969	0.968	0.969	0.962	0.960	0.959
0.05	0.941	0.953	0.956	0.962	0.917	0.932	0.928	0.932	0.899	0.894	0.893
0.10	0.920	0.945	0.951	0.963		0.904	0.898	0.906	0.842	0.837	0.835
0.20	0.888	0.940	0.952	0.974		0.863	0.856	0.870		0.750	0.749
0.30	0.859	0.939	0.959	0.990		0.829	0.823	0.843		0.683	0.683
0.40	0.831	0.939	0.967	1.006			0.796	0.822		0.628	0.630
0.50	0.803	0.939	0.976	1.022		0.771	0.773	0.804		0.582	0.586
0.60		0.938	0.984	1.038		0.774	0.752	0.788			0.549
0.70		0.937	0.993	1.052		0.719	0.734	0.775			0.518
0.80		0.934	1.001	1.066		0.694	0.717	0.763			0.492
0.90			1.008	1.078		0.671	0.702	0.752			0.469
1.00							0.688	0.743			0.450

See also Ref. [26].
CCl$_4$, o-dichlorobenzene, and bromobenzene; problems with decomposition of NMA.

IV. CRYOSCOPY

NITRIC ACID

System	Investigation	Ref.
N_2O_5 H_2O	Self-dissociation of HNO_3	13

SULFURIC ACID

System	Investigation	Ref.
Acetone Acetic acid Ammonium sulfate Potassium sulfate SO_2Cl_2 $Cl\ SO_3H$ $POCl_3$ $(CH_3)_2SO_4$	Principles and methods	6
H_2O SO_3	Basic strength of water and acid strength of disulfuric acid	7
HNO_3 N_2O_5 N_2O_4 N_2O_3	Cryoscopic proof of the formation of the nitronium ion	8
Ionized sulfates	Self-ionization equilibria and ionic equilibria	9
HNO_3	Ionization in sulfuric oleum	10
Nitronium perchlorate NH_4ClO_4	Basicity of perchlorate ion and acidity of perchloric acid	11
Nitro compounds Sulphonyl compounds Sulphuryl compounds	Basicities	12
Acetic anhydride Benzoic anhydride	Evidence for acetylium and benzoylium ions	14
Unsaturated aromatic Ketones and aldehydes	Cryoscopic measurements	15
Organic compounds	Review	16
Univalent metal sulfates	Cryoscopic data	17

SULFURIC ACID *(Continued)*

System	Investigation	Ref.
Aromatic sulfides, sulfoxides and sulphones	Cryoscopic data	18
Nitrobenzene and nitrotoluene	Cryoscopic data	19
Phenols	Cryoscopic data	20
General	Equilibrium method	21
Metal sulfates (in H_2SO_4–SO_3)	Cryoscopic data	22
Organic and inorganic bases	Cryoscopic data	23
Alkali metal sulfates	Cryoscopic titratium technique	25

References

1. R. H. Wood, R. K. Wicker, II and R. W. Kreis J. Phys. Chem., **75,** 2313 (1971).
2. R. W. Kreis and R. H. Wood, J. Phys. Chem., **75,** 2319 (1971).
3. J. S. Dunnett and R. P. H. Gasser, Trans. Faraday Soc., **61,** 922 (1965).
4. M. D. Archer and R. P. H. Gasser, Trans. Faraday Soc., **62,** 3451 (1966).
5. J. M. Crawford and R. P. H. Gasser, Trans. Faraday Soc., **63,** 2758 (1967).
6. R. J. Gillespie, E. D. Hughes and C. K. Ingold, J. Chem. Soc., 2473 (1950).
7. R. J. Gillespie, J. Chem. Soc., 2493 (1950).
8. R. J. Gillespie, J. Graham, E. D. Hughes, C. K. Ingold and E. R. A. Peeling, J. Chem. Soc., 2504 (1950).
9. R. J. Gillespie, J. Chem. Soc., 2516 (1950).
10. R. J. Gillespie and J. Graham, J. Chem. Soc., 2532 (1950).
11. R. J. Gillespie, J. Chem. Soc., 2527 (1950).
12. R. J. Gillespie, J. Chem. Soc., 2542 (1950).
13. R. J. Gillespie, E. D. Hughes and C. K. Ingold, J. Chem. Soc., 2552 (1950).
14. R. J. Gillespie, J. Chem. Soc., 2997 (1950).
15. R. J. Gillespie and J. A. Leisten, J. Chem. Soc., 1 (1954).
16. R. J. Gillespie and J. A. Leisten, Quart. Rev. Chem. Soc., **8,** 40 (1954).
17. R. J. Gillespie and J. V. Oubridge, J. Chem. Soc., 80 (1956).
18. R. J. Gillespie and R. C. Passerini, J. Chem. Soc., 3850 (1956).
19. R. J. Gillespie and E. A. Robinson, J. Chem. Soc., 4233 (1957).
20. R. J. Gillespie and J. V. Oubridge, J. Chem. Soc., 2804 (1959).
21. S. J. Bass and R. J. Gillespie, J. Chem. Soc., 814 (1960).
22. R. J. Gillespie and K. C. Malhotra, J. Chem. Soc., 1944 (1967).
23. R. J. Gillespie and K. C. Malhotra, J. Chem. Soc., 1933 (1968).
24. R. J. Gillespie and D. A. Humphreys, J. Chem. Soc., 2311 (1970).
25. B. Dacre and P. A. H. Wyatt, Trans. Faraday Soc., **57,** 1958 (1961).
26. O. D. Bonner and G. B. Woolsey, J. Phys. Chem., **75,** 2879 (1971).

V. HEATS OF SOLUTION CALORIMETRY

(a) Single Solvents

The heats of solution data in single solvents are arranged alphabetically by solvent. Within each solvent table the electrolytes are listed alphabetically by chemical formulas. Values of standard heats of solution at infinite dilution, heats of solution for a particular concentration and heats of dilution are included wherever possible. A complete bibliography is given at the end of section V(c).

Symbols used:

$\Delta H°$ Standard heat of solution (the reader should consult the original reference in order to find the exact extrapolation used)
$\Delta H_\infty°$ Standard heat of solution at infinite dilution
ΔH_{dil} Heat of dilution
ΔH Heat of solution (normally at a specified concentration)

ACETIC ACID CH₃COOH

Electrolyte	$\Delta H°$ (kcal/mol)	Ref.
NaOAc	−4.8	45
SbCl₅	Graphical presentation	59
SnBr₄	Graphical presentation	59
SnCl₄	Graphical presentation	59

Additional Information

Electrolyte	Molar ratio range (solute:solvent)	$-\Delta H_{range}$ (kcal/mol)	Ref.
C₅H₅N	1:282–1:41	7.11–6.97	75
H₂SO₄	1:365–1:100	6.93–3.80	75
HSO₃F	1:1265–1:105	25.04–24.52	75
KOAc	1:222–1:51	7.95–7.52	75
NaOAc	1:2573–1:1665	4.93–4.70	45
α-Picoline	1:380–1:73	8.20–8.04	75
SO₃	1:204–1:63	33.10–32.81	75

ACETIC ANHYDRIDE

Electrolyte	$-\Delta H°$ (kcal/mol)	Ref.
HSO_3F	23.1–24.6	52

ACETONE $(CH_3)_2CO$

Electrolyte	$\Delta H°$ (kcal/mol)	Ref.
C_5H_5NOAc	−0.236	41
Et_4NBr	3.3	36
Et_4NCl	0.0	36
$NaClO_4$	8.50	49
NaI	10.82 (10°C)	49
	10.54 (25°C)	49
	10.26 (40°C)	49
	Graphical presentation	24

Additional Information

Electrolyte	Conc. range (mole/1000 g)	ΔH_{range} (kcal/mol)	Ref.
KCNS	0.0499–0.1287	2.72–2.66	49
$Mg(ClO_4)_2$	0.0212	29.09	49
NaCNS	0.0496–0.0953	5.06–5.02	49
NaI	0.027	9.560	54
NH_4CNS	0.0986	1.83	49

V. HEATS OF SOLUTION CALORIMETRY

ACETONITRILE

Electrolyte	$\Delta H°$ (kcal/mol)	Ref.
Me$_4$NBr	4.21	62
Me$_4$NCl	2.96	62
Me$_4$NI	1.33	62

Additional Information

Electrolyte	Conc. range (mole/1000 g)	$-\Delta H_{range}$ (kcal/mol)	Ref.
NaI	0.11–0.02 (10)*	6.62–6.85	54

* Number of data points.

Electrolyte	Conc. (mole/liter)	$\Delta H°$ (kcal/mol)	Ref.
LiClO$_4$	1.000	−6.3	63
Me$_4$NPF$_6$	0.063	4.5	63
0.7M LiCl + 1M AlCl$_3$	0.700	−3.2	63

ACETYL CHLORIDE

Electrolyte	$\Delta H°$ (kcal/mol)	Ref.
SbCl$_5$	Graphical presentation	58
TiCl$_4$	Graphical presentation	58

ALLYL ALCOHOL

Electrolyte	Conc. range (moles/1000 g solvent)	ΔH_{range} (kcal/mol)	Ref.
NaI	0.11–0.04 (8)*	5.640–8.150	54

* Number of data points.

LIQUID AMMONIA

Electrolyte	ΔH (kcal/mol)	Ref.
CaI_2	62.8 ± 0.7 (ΔH_∞) at $-33°C$	44
$CdCl_2$	$-90.591*$	18
$CdCl_2 \cdot NH_3$	$-125.041*$	18
$CdCl_2 \cdot 2NH_3$	$-146.781*$	18
$CdCl_2 \cdot 4NH_3$	$-201.321*$	18
$CdCl_2 \cdot 6NH_3$	$-242.781*$	18
$CdCl_2 \cdot 10NH_3$	$-318.741*$	18
HF	42.48	40
NaBr	-9.73 (0.04M) at $-33°C$	37
	-7.64 (0.4M)	
NH_4Br	-9.50 (0.0438 moles) at $-33°C$	38
NH_4F	14.31	
NH_4HF_2	5.75	
NaCl	$-3.17†$ ($-40°C$)	73
	$-8.82†$ ($-30°C$)	73
	$-8.88†$ ($-23°C$)	73
	$-8.04†$ ($-18°C$)	73
	$-6.94†$ ($-13°C$)	73
	$-6.69†$ ($-11.5°C$)	73
	$-6.45†$ ($-11°C$)	73
	$-6.23†$ ($-10°C$)	73
	$-5.58†$ ($-8.5°C$)	73
	$-5.93†$ ($-5°C$)	73
	$-6.41†$ ($0°C$)	73
	$-6.91†$ ($5°C$)	73
	$-7.32†$ ($10°C$)	73
	$-7.43†$ ($15°C$)	73
	$-6.78†$ ($20°C$)	73
	-2.7 ($20°C$)	76
	$-6.00†$ ($25°C$)	73
	$-3.62†$ ($30°C$)	73

V. HEATS OF SOLUTION CALORIMETRY

LIQUID AMMONIA (Continued)

Electrolyte	ΔH (kcal/mol)	Ref.
$ZnCl_2$	−99.55*	19
$ZnCl_2 \cdot NH$	−112.080*	19
$ZnCl_2 \cdot 2NH_3$	−152.480*	19
$ZnCl_2 \cdot 4NH_3$	−198.070*	19
$ZnCl_2 \cdot 6NH_3$	−242.040*	19
$ZnCl_2 \, 10NH_3$	−314.200*	19

* From EMF data at 25°C.
† From heat of evaporation data. Values from −8.5° to 30° appear to have an incorrect sign in the original paper.

(i) Additional Information

Electrolyte	Conc. range mole ratio (solvent/solute)	ΔH_{range} (kcal/mol)	Ref.
$Ba(NO_3)_2$	94.5–15,800 (7)*	14.704–16.98	43
CsI	475–15,200 (5)	5.252–6.79	43
HgI_2	4100–15,400 (2)	21.61–22.62	43
KI	201–65,200 (11)	9.016–11.69	43
$LaI_3 \cdot 6NH_3$	1220–18,500 (6)	41.2–42.67	43
NaCl	192–15,800 (11)	6.419–8.42	43
NH_4Br	57.4–228.4 (6)	7.858–10.435	57
NH_4Cl	25.4–270.5 (11)	4.532–6.754	57
	198–16,525 (9)	7.884–9.68	43
NH_4I	194–15,840 (6)	15.641–17.72	43

LIQUID AMMONIA (Continued)

(i)   Additional Information (Continued)

Electrolyte	Conc. range (m)	$-\Delta H_{range}$ (kcal/mol)	Ref.
Guanidine HBr	0.285–0.183 (2)	10.05–10.02	56
Guanidine HCl	0.367–0.285 (2)*	8.56–8.60	56
Guanidine HNO	0.227–0.198 (2)	5.45–5.53	56
Guanidine NCNS	0.262–0.143 (2)	9.86–9.93	56
KI	0.803–0 (7)	6.47–7.82	56
KNH_2	0.733–0.276 (3)	1.48–1.97	56
KSCN	0.520–0 (4)	3.86–4.78	56
$NH_2OH \cdot HCl$	0.315–0.277 (2)	16.43–16.21	56
NH_4NO_3	0.366 (1)	6.75	56
NH_4SCN	0.745–0 (7)	9.10–9.72	56

Electrolyte	Conc. range (g salt/24.89 g NH_3)	ΔH_{range} (kcal/mol)	Ref.
HgI_2	2.0202 (1)*	20.123	57
LiBr	0.5454 (1)	19.715	57
LiI	0.9225 (1)	18.090	57
PbI_2	2.481–0.7733 (4)	26.020–27.360	57
RbBr	0.6271 (1)	0.440	57

(ii)   Heats of Solution of Metals in Liquid Ammonia

Metal	Conc. (molar)	ΔH (kcal/mol)	Ref.
K	0.228 (6)	−3.15	17
	0.0051	−2.15	17
	0.161 (5)	−12.74	17
Li	0.0065	−11.90	17
	0.81 (10)	−1.48	17
Na	0.0042	−0.65	17

For further information on heats of solution of metals in liquid NH_3 see for instance References 37, 38, 43, 47, 48.

* Number of data points.

V. HEATS OF SOLUTION CALORIMETRY

BENZENE

Electrolyte	$\Delta H°$ (kcal/mol)	Ref.
C_5H_4NOAc	−0.207	41

n-BUTANOL

Electrolyte	$\Delta H°$ (kcal/mol)	Ref.
$Ba(ClO_4)_2$	−3.6	4, 5
$Be(ClO_4)_2$	−80.5	5
$Ca(ClO_4)_2$	−16.6	4, 5
$LiClO_4$	−10.8	4
$Mg(ClO_4)_2$	−38.5	4, 5
$Pb(ClO_4)_2$	−15.7	4
$Ra(ClO_4)_2$	−2.9	5
$Sr(ClO_4)_2$	−12.4	4, 5

Additional Information

Electrolyte	Conc. range (moles × 10^5/25 ml)	$-\Delta H_{range}$ (kcal/mol)	Ref.
CCl_3COOH	22.8–76.3	2.7–2.5	65
$CHCl_2COOH$	73.0–177.5	1.7–1.6	65
HSO_3F	25.6–153.0	24.9–25.3	65
H_2S_2O	23.7–84.8	50.1–41.9	65
H_2SO_4	52.0–238.9	15.5–11.2	65

CARBON TETRACHLORIDE

Electrolyte	$\Delta H°$ (kcal/mol)	Ref.
C_5H_4NOAc	−0.1068	41

CHLOROBENZENE

Electrolyte	$\Delta H°$ (kcal/mol)	Ref.
Bu$_4$NPic	6.01 (25°C)	74
	6.36 (35°C)	74
	6.56 (45°C)	74

CHLOROFORM

Electrolyte	$\Delta H°$ (kcal/mol)	Ref.
C$_5$H$_4$NOAc	1.2059	41

DEUTROMETHANOL CH$_3$OD

Electrolyte	$\Delta H°$ (kcal/mol)	Ref.
CsOOCCF$_3$	0.39	78
KF	−5.35	78
KI	0.09	78
KOOCCF$_3$	−1.12	78
LiOOCCF$_3$	−9.65	78
NaBr	−3.97	78
NaI	−6.98	78
NaOOCCF$_3$	−4.94	78
Pent$_4$NCl	−2.07	78
Ph$_4$AsCl	−0.75	78
RbOOCCF$_3$	0.33	78

o-DICHLOROBENZENE

Electrolyte	$\Delta H°$ (kcal/mol)	Ref.
Bu$_4$NPic	5.71 (25°C)	74
	6.03 (35°C)	74
	6.35 (45°C)	74

V. HEATS OF SOLUTION CALORIMETRY

N-N-DIMETHYL FORMAMIDE

Electrolyte	$\Delta H°$ (kcal/mol)	Ref.
Am$_4$NBr	5.74	71
Am$_4$NCl	−0.32	71
Am$_4$NClO$_4$	8.57	71
Am$_4$NI	9.05	71
Ba(ClO$_3$)$_2$	−12.80 ± 0.16	68
Bu$_4$NBBu$_4$	2.1	79
	2.1	79
Bu$_4$NBr	0.0	79
Bu$_4$NCl	1.40	71
Bu$_4$NClO$_4$	0.80	71
Bu$_4$NI	4.75	71
CaBr$_2$	−39.6 ± 0.3	70
Ca(BrO$_3$)$_2$	−4.35 ± 0.05	68
CaCl$_2$	−23.8 ± 2.2	70
Ca(ClO$_3$)$_2$	−18.4 ± 0.2	68
CsBPh$_4$	−0.92	71
CsBr	−1.08 (estimated)	28, 46
CsI	−4.25	11, 28, 46
	−4.25	10
CsOOCCF$_3$	−1.10	71
Et$_4$NBr	2.24	71
Et$_4$NCl	1.66	71
	1.34	60
Et$_4$NClO$_4$	1.80	60
	1.71	71
Et$_4$NI	3.54	71
	3.420 ± 0.025	26
	2.95	60
Hex$_4$NBr	0.65	71
Hex$_4$NClO$_4$	9.65	71
KBPh$_4$	−4.68	71
KBr	−3.89	11, 28, 46
	−3.6 ± 0.1	79
KI	−8.04	10
	−8.1 ± 0.1	79
LiBr	−21.3	11, 28, 46
	−18.4 ± 0.2	79
	−18.5	10

N-N-DIMETHYL FORMAMIDE (Continued)

Electrolyte	$\Delta H°$ (kcal/mol)	Ref.
LiCl	−11.82	10
	−14.5	11, 28, 46
	−11.8 ± 0.1	79
	−10.7*	72
LiI	−19.1	10
	−26.06†	71
	−25.5	80
LiOOCCF$_3$	−7.68	71
Me$_4$NClO$_4$	2.23	71
Me$_4$NI	4.101 ± 0.014	26
	2.90	71
NaBPh$_4$	−12.05	60
	−17.30	71
	−19.8 ± 0.4	79
NaBr	−7.39	11, 28, 46
NaClO$_4$	−10.0 ± 0.2	79
NaI	−13.1	10
	−13.95	11, 46
	−12.8 ± 0.2	79
NaOOCCF$_3$	−3.60	71
Pent$_4$NBPent$_4$	8.0	79
Pent$_4$NBr	2.7	79
Pent$_4$NI	6.6	79
Ph$_4$AsCl	−2.20	71
	−3.5 ± 0.1	79
Ph$_4$AsClO$_4$	3.52	70
Ph$_4$AsI	0.69	60
Ph$_4$PCl	−1.38	70
Pr$_4$NCl	1.93	71
Pr$_4$NClO$_4$	3.35	71
Pr$_4$NI	1.996 ± 0.021	26
	1.98	71
RbBPh$_4$	−1.99	71
RbI	−6.64	10
RbOOCCF$_3$	−0.94	71
SrBr$_2$	−34.1 ± 0.4	70
SrCl$_2$	−15.8 ± 0.6	70
Sr(ClO$_3$)$_2$	−14.96 ± 0.44	68

* From EMF calculations of ΔH_t of LiCl(DMF) → LiCl(H$_2$O).
† Footnote j in Ref. 72 referring to a private communication.

V. HEATS OF SOLUTION CALORIMETRY

N-N-DIMETHYL FORMAMIDE (Continued)

(i) Additional Information

Electrolyte	Conc. range (g/25 ml)	$-\Delta H_{range}$ (kcal/mol)	T	Ref.
AlCl$_3$	0.116–0.3656 (4)*	58.74–56.70	26.9°	12
AsCl$_3$	0.101–0.3980 (5)	16.25–14.42	26.9°	12
CdCl$_2$	0.079–0.518 (4)	11.02–10.79	26.9°	12
SbCl$_3$	0.098–0.673 (5)	10.62–9.75	26.9°	12
SbCl$_5$	0.1172–0.4313 (5)	44.10–43.43	26.9°	12
SiCl$_4$	0.054–0.3192 (5)	74.97–62.65	26.9°	12
SnCl$_4$	0.117–0.5493 (4)	44.10–41.42	26.9°	12
SO$_3$	0.071–0.7161 (5)	45.60–43.84	26.9°	12
TeBr$_4$	0.200–1.221 (6)	26.5–21.0	26.9°	12
TeCl$_4$	0.1436–0.5546 (4)	20.92–17.78	26.9°	12
TiBr$_4$	0.1032–0.3560 (5)	51.52–52.99	26.9°	12
TiCl$_4$	0.104–0.424 (5)	34.45–26.15	26.9°	12

Electrolyte	Conc. range (moles/1000 g) × 10^4	$-\Delta H_{range}$ (kcal/mol)	Ref.
GdCl$_3$	0.656–31.0	22.96–37.4	11
MgCl$_2$	1.51–39.0	172.06–3711	11

(ii) Additional Information

Electrolyte	Conc. range g mole acid (×10^{-3})	$-\Delta H_{range}$ (kcal/mol)	Ref.
CH$_3$COOH	5.41–2.60 (5)	1.75–1.57	31
HCOOH	1.88–8.16 (5)*	2.55–2.75	31
HSO$_3$F	0.417–4.33 (6)	33.17–32.18	31
H$_2$S$_2$O$_7$	0.747–2.76 (5)	60.80–58.35	31
H$_2$SO$_4$	1.39–5.98 (5)	16.69–13.79	31

* Number of data points.

N-N-DIMETHYL FORMAMIDE (Continued)
(ii) Additional Information (Continued)

Electrolyte	Conc. (moles/liter)	ΔH (kcal/mol)	Ref.
$CuCl_2 \pm 1M$ LiCl	0	-13.1	63
$CuF_2 + 1M$ $LiClO_4$	0	-0.2	63
LiCl	1.000	-8.9	63
$LiClO_4$	1.000	-18.3	63
Me_4NPF_6	0.160	2.9	63
$0.04M$ $AlCl_3 + 1M$ LiCl	0.0408	-48.8	63
$0.5M$ $CuCl_2 + 1M$ LiCl	0.499	-10.9	63

DIMETHYL SULFOXIDE DMSO

Electrolyte	$\Delta H°$ (kcal/mol)	Ref.
Bu_4NBBu_4	9.9 ± 0.1	79
Bu_4NBr	5.1 ± 0.1	79
Bu_4NCl	3.0 ± 0.1	79
Bu_4NI	7.2 ± 0.2	79
CsI	-2.84	29
Et_4NBr	3.27	29
Et_4NCl	2.42	29
Et_4NClO_4	3.90	60
Et_4NI	4.86	29
KBr	-2.7 ± 0.1	51
KI	-6.5 ± 0.1	51
	-6.15	29
LiBr	-2.7 ± 0.1	51
LiCl	-10.9 ± 0.3	51
LiI	-24.2 ± 0.3	51
$NaBPh_4$	-14.2 ± 0.2	79
	-14.23	29

V. HEATS OF SOLUTION CALORIMETRY

DIMETHYL SULFOXIDE DMSO (*Continued*)

Electrolyte	$\Delta H°$ (kcal/mol)	Ref.
NaClO$_4$	-7.9 ± 0.1	79
NaI	-11.5 ± 0.1	79
	-11.53	29
NaPic	3.41	69
Pent$_4$NBPent$_4$	18.4 ± 0.3	79
Pent$_4$NBr	9.6 ± 0.1	79
Pent$_4$NI	12.1 ± 0.3	79
Ph$_4$AsCl	-1.4 ± 0.2	79
Ph$_4$AsI	3.44	29
PrOTs	1.2 ± 0.1	77

* *n*-Propyl *p*-toluenesulfonate.

ETHANOL

Electrolyte	$\Delta H°$ (kcal/mol)	Ref.
Ba(ClO$_4$)$_2$	-5.4	4, 5
Be(ClO$_4$)$_2$	-97.7	5
CaBr$_2$	-21.471†	33
CaCl$_2$	-17.555†	33
Ca(ClO$_4$)$_2$	-21.8	4, 5
CaI$_2$	-19.833†	33
Ca(NO$_3$)$_2$	-8.710†	33
Et$_4$NBr	5.26	36
Et$_4$NCl	0.52	36
Et$_4$NPic	8.9	36
FeCl$_3$	-17.4 (24°)	34
HgCl$_2$	0†	33
HSO$_3$F	-29.66–31.66	52

ETHANOL (Continued)

Electrolyte	$\Delta H°$ (kcal/mol)	Ref.
KCl	3.63*	2
KI	−0.7	25, 32
	0.7	36
LiCl	−12.93	25, 32, 36
	−11.743†	33
LiClO$_4$	−12.9	36
	−11.2	4
LiNO$_3$	−4.655†	33
LiPi	−2.9	36
Me$_4$NBr	7.16	36
Me$_4$NCl	3.52	36
Mg(ClO$_4$)$_2$	−45.5	4, 5
NaBr	−2.65	25, 32, 36
NaCl	0.48*	2
NaClO$_4$	−0.8	36
NaI	−5.80	25, 32, 36
	−4.587†	33
NaPic	0.4	36
Pb(ClO$_4$)$_2$	−18.3	4
Ra(ClO$_4$)$_2$	−4.2	5
Sr(ClO$_4$)$_2$	−15.6	4, 5
ZnCl$_2$	−9.767†	33

* Refers to $\delta(\Delta H°) = \Delta H_S° - \Delta H_{H_2O}°$, using data from references 21, 23 and 25.
† Heat of dissolution for one gram-molecular proportion of the salt.

Additional Information

Electrolyte	Conc. range (moles/1000 g solvent)	$-\Delta H_{\text{dilution}}$ (kcal/mol)	Ref.
LiCl	0.60717–0.06365	1.020	33
	0.38717–0.03827	0.878	33
	0.13140–0.01334	0.813	33
	0.06365–0.00637	0.713	33

N-ETHYLACETAMIDE NEA

Electrolyte	$\Delta H°$ (kcal/mol)	Ref.
Bu$_4$NBBu$_4$	8.5 ± 0.1	79
Bu$_4$NBr	3.8 ± 0.1	79
Bu$_4$NI	9.1 ± 0.3	79
LiBr	−12.0 ± 0.3	79
LiCl	−10.0 ± 0.2	79
LiI	−15.8 ± 0.3	79
NaBPh$_4$	−2.5	79
NaClO$_4$	−1.6 ± 0.1	79
NaI	−6.5 ± 0.1	79
Pent$_4$NBPent$_4$	13.6 ± 0.3	79
Pent$_4$NBr	6.4 ± 0.1	79
Pent$_4$NI	12.1 ± 0.2	79
Ph$_4$AsCl	2.1	79

ETHYL ACETATE

Electrolyte	$\Delta H°$ (kcal/mol)	Ref.
C$_5$H$_4$NOAc	−0.2362	41

Additional Information

Electrolyte	$-\Delta H°$ (kcal/mol)	Ref.
HSO$_3$F	23.34–24.03	52

ETHYLENE DIAMINE (EDA)

Electrolyte	Conc. range (mole ratio, solvent/solute)	$-\Delta H_{range}$ (kcal/mol)	Ref.
AgBr	2265.0–2308.0 (2)	9.80–9.97	55
AgCl	817.0–5722.0 (3)*	11.03–12.76	55
AgCN	1118.6–2152.8 (2)	9.93–9.87	55
AgI	1615.0–6540.0 (2)	8.13–8.90	55
AgNO$_3$	5871.0–5976.0 (4)	25.17–24.90	55
CsI	840.8 (1)	2.47	55
EDA·2HCl	1888.0–7290.0 (4)	19.55–20.30	55
Hg(CN)$_2$	1468.0–14,950.0 (3)	16.60–17.34	55
HgI$_2$	1927.0–10,410.0 (3)	28.60–28.62	55
KI	221.1–984.3 (5)	5.90–6.12	55
KNO$_3$	637.0–1252.4 (3)	−0.98–0.97	55
LiI	992.2–2597.0 (3)	28.6–28.8	55
LiNO$_3$	342.3–1805.0 (3)	10.25–10.35	55
NaBr	448.8–726.0 (3)	10.00–9.50	55
NaI	381.7–3279.0 (5)	14.55–15.75	55
NaNO$_3$	512.8–540.6 (3)	3.28–3.27	55
Pb(NO$_3$)$_2$	9854.0–9861.0 (2)	32.2–32.2	55
RbI	998.0 (1)	4.50	55
TlBr	1384.0–1625.0 (2)	3.90–3.85	55
TlCl	1202.0–1945.0 (2)	4.03–4.02	55
TlNO$_3$	1123.0–1322.0 (3)	11.53–11.40	55

ETHYL METHYL KETONE (MEK)

Electrolyte	Conc. range (mole/1000 g)	$-\Delta H_{range}$ (kcal/mol)	Ref.
NaI	0.08–0.02 (7)*	8.140–9.170	54

* Number of data points.

V. HEATS OF SOLUTION CALORIMETRY

FORMAMIDE

Electrolyte	$\Delta H°$ (kcal/mol)	Ref.
$BaBr_2$	-16.3 ± 0.6	70
$Ba(BrO_3)_2$	0.27 ± 0.10	68
$BaCl_2$	-10.0 ± 0.3	70
$Ba(ClO_3)_2$	-4.95 ± 0.06	68
Bu_4NBr	4.218	67
$CaBr_2$	-29.9 ± 0.8	70
$Ca(BrO_3)_2$	-4.64 ± 0.10	68
$CaCl_2$	-22.7 ± 0.8	70
$Ca(ClO_3)_2$	-11.3 ± 0.3	68
CsBr	1.81	6, 30
CsCl	0.95	6, 30
	3.35*	2
	0.90	28
CsF	-7.59	6, 30
CsI	2.22	6, 30
	-1.01	28
Et_4NBr	2.782	67
KBr	0.232	6, 30
	0.23	28
KCl	0.834	6, 30
	0.82	28
KF	-3.18	6, 30
KI	-1.0	30
	2.22	28
LiBr	-13.39	6, 30
LiCl	-9.42	6, 30
	-9.42	28
LiF	4.9	30
LiI	-18.17	6, 30
Me_4NBr	5.628	67

FORMAMIDE (Continued)

Electrolyte	$\Delta H°$ (kcal/mol)	Ref.
NaBr	−4.41	6, 30
NaCl	−2.10	6, 30
	3.04*	2
	−2.10	28
NaF	1.4	30
	1.40	28
NaI	−7.42	6, 30
Pr$_4$NBr	3.022	67
RbBr	0.750	6, 30
RbCl	0.710	6, 30
	0.71	28
RbF	−5.27	6, 30
RbI	0.23	6, 30
SrBr$_2$	−26.0 ± 0.6	70
Sr(BrO$_3$)$_2$	−3.21 ± 0.10	68
SrCl$_2$	−17.6 ± 0.5	70
Sr(ClO$_3$)$_2$	−8.28 ± 0.07	68

* Refers to $\delta(\Delta H°) = \Delta H_s° - \Delta H_{H_2O}°$, using data from references 21, 23 and 25.

Additional Information

Electrolyte	Conc. range (mole/liter)	$-\Delta H_{range}$ (kcal/mol)	Ref.
KI	0.275 –0.144 (2)*	1.071–1.053	54
NaBr	0.0861–0.0332 (4)	4.830–4.594	54
NaCl	0.0717–0.0474 (2)	2.716–2.661	54
NaI	0.117 –0.0256 (4)	1.030–7.657	54

* Number of data points.

V. HEATS OF SOLUTION CALORIMETRY

FORMIC ACID HCOOH

Electrolyte	$\Delta H°$ (kcal/mol)	Ref.
KCl	−0.302	39
RbCl	−0.912	39

Additional Information

Electrolyte	Conc. range (mole/liter)	$-\Delta H_{range}$ (kcal/mol)	Ref.
CsCl	0.0204–5.5783	1.24–0.59	53
CsI	0.0333–0.8045	2.14–2.48	53
NaCl	0.0251–0.6742	0.18–0.08	53

FURFUROL

Electrolyte	Conc. range (mole/1000 g solvent)	$-\Delta H$ (kcal/mol)	Ref.
NaI	0.0145	10.360	54

HEXAMETHYLPHOSPHORIC TRIAMIDE HMPT

Electrolyte	$\Delta H°$ (kcal/mol)	Ref.
LiBr	−19.7 ± 0.2	79
LiCl	−12.3 ± 0.2	79
LiI	−24.2 ± 0.2	79

n-HEXANOL

Electrolyte	$\Delta H°$ (kcal/mol)	Ref.
Ba(ClO$_4$)$_2$	−3.8	5
Be(ClO$_4$)$_2$	−73.1	5
Ca(ClO$_4$)$_2$	−16.5	5
Mg(ClO$_4$)$_2$	−32.9	5
Ra(ClO$_4$)$_2$	−2.4	5
Sr(ClO$_4$)$_2$	−10.2	5

METHANOL

Electrolyte	$\Delta H°$ (kcal/mol)	Ref.
AgClO$_4$	−3.3	1
AgNO$_3$	0.73	27
Am$_4$NBr	6.84	71
Am$_4$NCl	−2.37	78
Am$_4$NI	12.54	71
Ba(ClO$_4$)$_2$	−14.3	1
Bu$_4$NBBu	14.1	79
Bu$_4$NBr	4.19	71
	5.0 ± 0.2	79
Bu$_4$NClO$_4$	6.86	71
Bu$_4$NI	8.57	71
	10.4 ± 0.2	79
Ca(ClO$_4$)$_2$	−28.2	1
CsBPh$_4$	8.84	71
CsCl	2.4	1
	2.0*	2
	1.855($\Delta H_\infty°$)	42
	1.850†	42
	2.88	71

V. HEATS OF SOLUTION CALORIMETRY

METHANOL (*Continued*)

Electrolyte	$\Delta H°$ (kcal/mol)	Ref.
CsI	4.1	1
CsOOCCF$_3$	0.41	78
Et$_4$NBr	4.38	36
	4.44	62
	4.91	71
Et$_4$NCl	0.82	62
	0.96	71
	−0.31	36
Et$_4$NClO	8.24	36
	8.95	71
Et$_4$NI	8.11	62
	8.25	71
Et$_4$NPiC	8.2	36
HCl	$-19.700 \pm 0.5(\Delta H_\infty°)$	42
HexNBr	1.63	71
HexNClO$_4$	15.46	71
HSO$_3$F	−24.79–27.13	52
KBr	0.870†	3, 42
	1.0	36
	$0.685 \pm 0.05(\Delta H_\infty°)$	42
	1.34	71
KCl	$0.760 \pm 0.05(\Delta H_\infty°)$	42
	1.080†	3, 42
	0.95*	2
	1.4	36
	1.57	71
KF	−5.25	78
KI	$0.175 \pm 0.1(\Delta H_\infty°)$	42
	−0.6	36
	−0.15	71, 78
KNO$_3$	4.1	36
KOOCCF$_3$	−1.07	78

METHANOL (*Continued*)

Electrolyte	$\Delta H°$ (kcal/mol)	Ref.
LiBr	$-13.4+ \pm 0.2$	79
LiCl	-12.4	1
	-12.38	25, 32
	-12.3	36
	$-11.480 \pm 0.1 (\Delta H_\infty°)$	42
	-11.50†	42
	-12.05	71
	-11.4 ± 0.2	79
LiClO$_4$	-12.2	1
LiI	-16.9 ± 0.2	79
LiNO$_3$	-4.51	27
LiOOCCF$_3$	-9.37	78
LiPic	-1.30	36
Me$_4$NBr	6.98	36
	7.35	71
	7.24	62
	7.246 ± 0.011 ($-5.06°$)	83
	7.213 ± 0.012 ($9.81°$)	83
	6.949 ± 0.010 ($25.04°$)	83
	6.457 ± 0.014 ($40.01°$)	83
	5.873 ± 0.018 ($55.00°$)	83
Me$_4$NCl	3.18	36
	3.18	62
	3.38	71
Me$_4$NI	9.74	62
	9.74	71
Mg(ClO$_4$)$_2$	-46.6	1
NaBPh$_4$	-8.02	60
	-10.06	72
	-3.9	79
NaBr	-4.30	25, 32
	-4.30	36
	$-4.00 \pm 0.20 (\Delta H_\infty°)$	42
	-3.87	27
	-4.05	71, 78

METHANOL (Continued)

Electrolyte	$\Delta H°$ (kcal/mol)	Ref.
NaCl	$-1.450 \pm 0.1(\Delta H_\infty°)$	42
	$-2.5(\Delta H_\infty°)$	35, 36
	-2.00†	3, 42
	-2.13	2
	-2.50	25, 32
	-1.90	71
NaClO$_4$	-2.4	1
	-2.62	36
	-2.6 ± 0.1	79
NaI	-8.1	1
	-7.52	25, 32, 36
	$-7.00 \pm 0.10(\Delta H_\infty°)$	42
	-7.14	71, 78
	graphical presentation	24
NaNO$_3$	1.12	27
NaOCH$_3$	17.2	69
NaOOCCF$_3$	-4.78	78
NaPic	1.9	36
	1.77	69
NH$_4$Br	0.62	27
NH$_4$NO$_3$	2.58	27
Pb(ClO$_4$)$_2$	-24.8	1
Pent$_4$NBPent$_4$	19.5	79
Pent$_4$NBr	7.8 ± 0.1	79
Pent$_4$NI	14.1 ± 0.3	79
Ph$_4$AsCl	-0.98	78
	5.7	79
Ph$_4$PCl	-0.12	71
Pr$_4$NBr	3.88	71
	4.04	62
Pr$_4$NCl	0.46	61
Pr$_4$NI	6.07	71
RbBPh$_4$	7.48	71

METHANOL (Continued)

Electrolyte	$\Delta H°$ (kcal/mol)	Ref.
RbCl	1.275($\Delta H_\infty°$)	42
	1.280†	42
	2.20	71
RbOOCCF$_3$	0.33	78
Sr(ClO$_4$)$_2$	−24.6	1

* Refers to $\delta(\Delta H°) = \Delta H_S° - \Delta H_{H_2O}°$, using data from references 21, 23 and 25.
† Values from smoothed curve of 0, 20.86, 44.18, 76.0 and 100 wt. % methanol.

Additional Information

Electrolyte	Conc. range (moles/1000 g solvent)	$-\Delta H_{dilution}$ (kcal/mol)	Ref.
LiCl	0.63744–0.06296	0.991	33
	0.38116–0.03779	0.904	33
	0.12621–0.01358	0.559	33
	0.06296–0.006302	0.419	33

N-METHYL ACETAMIDE

Electrolyte	$\Delta H°$ (kcal/mol)	Ref.
Bu$_4$NBr	5.837 ± 0.010(35°C)	82
CsI	1.9	8
Et$_4$NBr	6.093 ± 0.012(35°C)	82
KBr	0.3	8
KCl	1.3	8
KF	−2.0	8
KI	−2.2	8
LiI	−21.3	8
Me$_4$NBr	6.942 ± 0.018(35°C)	82
NaI	−7.1	8
Pr$_4$NBr	5.675 ± 0.006(35°C)	82
RbI	−0.4	8

V. HEATS OF SOLUTION CALORIMETRY

METHYLAMINE

Electrolyte	Conc. range mole ratio (solvent/solute)	$-\Delta H_{range}$ (kcal/mol)	Ref.
KI	215–15,830 (6)*	6.635–7.18	43

* Number of data points.

N-METHYL FORMAMIDE

Electrolyte	$\Delta H°$ (kcal/mol)	Ref.
BaBr$_2$	-18.0 ± 0.4	70
Ba(ClO$_3$)$_2$	-7.68 ± 0.12	68
Bu$_4$NBr	5.365 ± 0.015	82
CaBr$_2$	-34.05 ± 0.7	70
Ca(BrO$_3$)$_2$	-4.42 ± 0.10	68
CsCl	0.89	28, 46
	0.890 ± 0.042	9
	3.41*	2
CaCl$_2$	-24.3 ± 0.9	70
Ca(ClO$_3$)$_2$	-13.8 ± 0.1	68
CsI	0.71	7
Et$_4$NBr	5.052 ± 0.018	82
KBr	-0.82	7
KCl	0.31	28, 46
	0.373	7
	0.308 ± 0.078	9
	3.81	2
KF	-2.60	7
KI	-3.222	7
LiBr	-16.2 (estimated)	46
LiCl	-13.1 (estimated)	9
	-13.1	28, 46
LiI	-21.11	7
Me$_4$NBr	6.542 ± 0.017	82

N-METHYL FORMAMIDE (Continued)

Electrolyte	$\Delta H°$ (kcal/mol)	Ref.
NaBr	−4.393 ± 0.105	9
	−4.39	28, 46
NaCl	−1.24	28, 46
	−1.244 ± 0.046	9
	2.17*	2
NaI	−8.26	7
	−8.258 ± 0.085	9, 46
Pr_4NBr	4.965 ± 0.015	82
RbI	−1.64	7
$SrBr_2$	−28.0 ± 0.3	70
$SrCl_2$	−16.2 ± 1.3	70
$Sr(ClO_3)_2$	−9.96 ± 0.12	68

* Refers to $\delta(\Delta H°) = \Delta H_S° - \Delta H_{H_2O}°$, using data from references 21, 23 and 25.

N-METHYLPYRROLIDINONE NMP

Electrolyte	$\Delta H°$ (kcal/mol)	Ref.
Bu_4NBBu_4	4.8 ± 0.1	79
Bu_4NBr	3.9 ± 0.1	79
Bu_4NI	6.8 ± 0.2	79
KI	−6.1 ± 0.1	79
LiBr	−14.5 ± 0.2	79
LiCl	−8.2 ± 0.1	79
LiI	−20.1 ± 0.3	79
$NaBPh_4$	−19.0 ± 0.4	79
$NaClO_4$	−8.8 ± 0.1	79
NaI	−11.0 ± 0.2	79
$Pent_4NBPent_4$	10.1 ± 0.1	79
$Pent_4NBr$	6.8 ± 0.3	79
$Pent_4NI$	10.4 ± 0.2	79
PhAsCl	−1.0 ± 0.1	79

V. HEATS OF SOLUTION CALORIMETRY

NITROBENZENE $C_6H_5NO_2$

Electrolyte	$\Delta H°$ (kcal/mol)	Ref.
Et_4NClO_4	+2.40	36
Et_4NPic	+5.20	36

NITROMETHANE CH_3NO_2

Electrolyte	$\Delta H°$ (kcal/mol)	Ref.
Et_4NBr	2.59	36
Et_4NCl	−0.59	36
Et_4NClO_4	2.43	36
Et_4NPic	5.7	36
Me_4NBr	4.12	36

Additional Information

Electrolyte	Conc. range wt. of acid g moles × 10^3 in 25 ml of $MeNO_2$	$-\Delta H_{range}$ (kcal/mol) (at 26.9°C)	Ref.
$AsCl_3$	4.151–8.350 (4)	1.14–0.78	52
H_2SO_4	0.496–1.939 (5)	11.12–2.13	52
HSO_3F	0.442–2.05 (5)*	29.77–21.26	52
$SbCl_5$	0.172–0.885 (5)	40.31–16.32	52
$SnCl_4$	0.197–0.848 (6)	27.99–8.59	52

* Number of data points.

n-PENTANOL

Electrolyte	$\Delta H°$ (kcal/mol)	Ref.
Ba(ClO$_4$)$_2$	-4.0	5
Be(ClO$_4$)$_2$	-75.8	5
Ca(ClO$_4$)$_2$	-17.2	5
Mg(ClO$_4$)$_2$	-35.3	5
Ra(ClO$_4$)$_2$	-2.6	5
Sr(ClO$_4$)$_2$	-10.7	5

PIPERIDINE

Electrolyte	Conc. range (mole/1000 g solvent)	$-\Delta H_{\text{range}}$ (kcal/mol)	Ref.
NaI	0.06–0.03 (4)*	9.75–10.40	54

* Number of data points.

n-PROPANOL

Electrolyte	$\Delta H°$ (kcal/mol)	Ref.
Ba(ClO$_4$)$_2$	-4.2	4, 5
Be(ClO$_4$)$_2$	-85.8	5
Ca(ClO$_4$)$_2$	-18.2	4, 5
LiClO$_4$	-10.8	4
Mg(ClO$_4$)$_2$	-40.6	4, 5
Pb(ClO$_4$)$_2$	-16.6	4
Ra(ClO$_4$)$_2$	-3.3	5
Sr(ClO$_4$)$_2$	-13.4	4, 5

V. HEATS OF SOLUTION CALORIMETRY

2-PROPANOL

Electrolyte	$\Delta H°$ (kcal/mol)	Ref.
Ba(ClO$_4$)$_2$	−4.8	5
Ca(ClO$_4$)$_2$	−20.2	5
Mg(ClO$_4$)$_2$	−45.0	5
Sr(ClO$_4$)$_2$	−15.0	5

PROPYLENE CARBONATE

Electrolyte	$\Delta H°$ (kcal/mol)	Ref.
AsPh$_4$I	5.74	60
CsBPh$_4$	2.38	16, 64
CsOOCCF$_3$	3.04	16, 64
Et$_4$NBr	4.90	15, 64
Et$_4$NCl	3.30	15, 64
Et$_4$NI	6.08	15, 64
KBPh$_4$	0.74	16, 64
KOOCCF$_3$	3.73	16, 64
LiClO$_4$	−9.51	16, 64
LiI	−15.05	16, 64
LiOOCCF$_3$	2.36	16, 64
Me$_4$NBr	5.27	15, 64
Me$_4$NI	5.34	15, 64
NaBPh$_4$	−10.72	16, 64
NaClO$_4$	−3.05	16, 64
NaI	−5.04	16, 64
NaOOCCF$_3$	3.32	16, 64
RbBPh$_4$	2.41	16, 64
RbCF$_3$COO	3.98	16, 64

Electrolyte	Conc. (mole/liter)	ΔH(kcal/mol)	Ref.
LiClO	1.000	−7.4	63
Me$_4$NPF	0.100	4.6	63
0.5M LiCl ± 1M AlCl$_3$	0.500	(−2)*	63

* Incomplete or very slow dissolution.

PYRIDINE

Electrolyte	Conc. range (mole/1000 g)	ΔH_{range} (kcal/mol)	Ref.
NaI	0.06–0.02 (5)*	12.000–12.700	54

* Number of data points.

SULFOLANE

Electrolyte	$\Delta H°$(kcal/mol) (at 30°C)	Ref.
AsPH$_4$I	3.87	60
CsClO$_4$	2.21	60
Et$_4$NBr	4.99	60
Et$_4$NCl	3.94	60
Et$_4$NClO$_4$	3.63	60
Et$_4$NI	5.53	60
KClO$_4$	1.10	60
LiClO$_4$	−5.62	60
NaBPH$_4$	−10.90	60
NaClO$_4$	−5.40	60
NaI	−7.41	60
RbClO$_4$	2.02	60

(b) Nonaqueous-Aqueous Mixed Solvents

The heats of solution data in nonaqueous-aqueous solvents are arranged alphabetically by nonaqueous component. Within each solvent table the electrolytes are listed alphabetically by chemical formulae, except for the quarternary ammonium compounds which are given according to increasing chain length. Values of standard heats of solution at infinite dilution, heats of solution for a particular concentration and heats of dilution are included wherever possible. A complete bibliography is given at the end of section V(c).

Symbols used:

$\Delta H°$ Standard heat of solution (the reader should consult the original reference in order to find the exact extrapolation used)

$\Delta H_\infty°$ Standard heat of solution at infinite dilution

ΔH_{dil} Heat of dilution

ΔH Heat of solution (normally at a specified concentration)

DIMETHYL SULFOXIDE–WATER
95 Volume % Dimethyl Sulfoxide

Electrolyte	$\Delta H°$ (kcal/mol)	Ref.
Bu$_4$NBBu$_4$	11.4 ± 0.1	77
Bu$_4$NBr	5.7 ± 0.1	77
Bu$_4$NCl	3.3 ± 0.1	77
Bu$_4$NI	7.4 ± 0.2	77
Pr$_4$OTs*	1.4 ± 0.1	77

* *n*-Propyl *p*-toluene sulfonate.

DIMETHYL SULFOXIDE–WATER (*Continued*)

90 Volume % Dimethyl Sulfoxide

Electrolyte	$\Delta H°$ (kcal/mol)	Ref.
Bu_4NBBu_4	10.9 ± 0.3	77
Bu_4NBr	6.1 ± 0.1	77
Bu_4NCl	4.1 ± 0.1	77
Bu_4NI	7.9 ± 0.1	77
Pr_4OTs*	2.4 ± 0.1	77

* *n*-Propyl *p*-toluene sulfonate.

85 Volume % Dimethyl Sulfoxide

Electrolyte	$\Delta H°$ (kcal/mol)	Ref.
Bu_4NBBu_4	13.4 ± 0.1	77
Bu_4NBr	6.6 ± 0.1	77
Bu_4NCl	4.3 ± 0.1	77
Bu_4NI	8.1 ± 0.2	77
Pr_4OTs*	2.6 ± 0.1	77

* *n*-Propyl *p*-toluene sulfonate.

80 Volume % Dimethyl Sulfoxide

Electrolyte	$\Delta H°$ (kcal/mol)	Ref.
Bu_4NBBu_4	13.1 ± 0.3	77
Bu_4NBr	7.1 ± 0.1	77
Bu_4NCl	4.4 ± 0.1	77
Bu_4NI	8.3 ± 0.1	77
Pr_4OTs*	3.1 ± 0.1	77

* *n*-Propyl *p*-toluene sulfonate.

V. HEATS OF SOLUTION CALORIMETRY

DIOXANE–WATER
20% Dioxane

Electrolyte	Conc. (molar)	$\Delta H°$ (kcal/mol)	Ref.
CsCl	0.0114 (7)*	3.794	20
	0.0576	3.829	
KBr	0.0069 (8)	4.084	20
	0.0823	4.197	
KCl	0.0084 (6)	3.904	20
	0.1169	3.929	
KI	0.0152 (5)	3.857	20
	0.0498	3.892	
LiCl	0.0169 (9)	−9.399	20
	0.1250	−9.169	
NaBr	0.0248 (7)	−0.754	20
	0.0981	−0.651	
NaCl	0.0246 (7)	0.726	20
	0.1005	0.782	
NaI	0.0119 (7)	−2.896	20
	0.0451	−2.860	
		Graphical presentation	24
RbCl	0.0131 (6)	3.763	20
	0.0424	3.801	

* Number of data points.

METHANOL–WATER

10 wt. % Methanol

Electrolyte	ΔH_∞° (kcal/mol)*	Ref.
CsCl	4.40	42
HCl	−17.55	42, 3
KBr	5.05	42, 3
KCl	4.41	42, 3
KI	5.20	42
LiCl	−8.00	42
NaBr	0.45	42
NaCl	1.40	42, 3
NaI	−1.25	42
RbCl	4.17	42

* Values from smoothed curves of 0, 20.86, 44.18, 76.0 and 100 wt. % methanol.

20 wt. % Methanol

Electrolyte	ΔH_∞° (kcal/mol)*	Ref.
CsCl	4.45	42
HCl	−17.32	42, 3
KBr	5.23	42, 3
KCl	4.61	42, 3
KI	5.30	42
LiCl	−7.60	42
NaBr	0.65	42
NaCl	1.68	42, 3
NaI	−1.00	42
RbCl	4.29	42

* Values from smoothed curves of 0, 20.86, 44.18, 76.0 and 100 wt. % methanol.

METHANOL–WATER (*Continued*)

20.86 wt. % Methanol

Electrolyte	ΔH_∞° (kcal/mol)	Ref.
HCl	-17.28 ± 0.15	42
KBr	5.275 ± 0.020	42
KCl	4.625 ± 0.025	42
KI	5.285 ± 0.025	42
LiCl	-7.585 ± 0.050	42
NaBr	0.68 ± 0.02	42
NaCl	1.69 ± 0.03	42
NaI	-0.955 ± 0.02	42

30 wt. % Methanol

Electrolyte	ΔH_∞° (kcal/mol)*	Ref.
CsCl	4.42	42
HCl	-17.2	42, 3
KBr	5.15	42, 3
KCl	4.62	42, 3
KI	5.20	42
LiCl	-7.80	42
NaBr	0.66	42
NaCl	1.75	42, 3
NaI	-1.15	42
RbCl	4.34	42

* Values from smoothed curves of 0, 20.86, 44.18, 76.0 and 100 wt. % methanol.

METHANOL–WATER (Continued)

40 wt. % Methanol

Electrolyte	ΔH_∞° (kcal/mol)*	Ref.
CsCl	4.34	42
HCl	−17.24	42, 3
KBr	4.95	42, 3
KCl	4.54	42, 3
KI	5.00	42
LiCl	−8.05	42
NaBr	0.55	42
NaCl	1.72	42, 3
NaI	−1.40	42
RbCl	4.35	42

* Values from smoothed curves of 0, 20.86, 49.18, 76.0 and 100 wt. % methanol.

40 Mole % Methanol

Electrolyte	ΔH° (kcal/mol)	Ref.
NaI	2.030 (20°)	84
	2.066	84

44.18 wt. % Methanol

Electrolyte	ΔH_∞° (kcal/mol)	Ref.
HCl	−17.00	42
KBr	4.820 ± 0.015	42
KCl	4.495 ± 0.010	42
KI	4.40 ± 0.1	42
LiCl	−8.25	42
NaBr	0.340 ± 0.010	42
NaCl	1.600 ± 0.005	42
NaI	−1.710 ± 0.010	42
RbCl	4.345	42

V. HEATS OF SOLUTION CALORIMETRY

METHANOL–WATER (*Continued*)

45 Mole % Methanol

Electrolyte	$\Delta H°$ (kcal/mol)	Ref.
NaI	2.280 (20°)	84
	2.260	84

50 Mole % Methanol

Electrolyte	$\Delta H°$ (kcal/mol)	Ref.
NaI	2.600 (20°)	84
	2.560	84

50 wt. % Methanol

Electrolyte	$\Delta H_\infty°$ (kcal/mol)*	Ref.
CsCl	4.16	42
HCl	−17.44	42, 3
KBr	4.55	42, 3
KCl	4.28	42, 3
KI	4.62	42
LiCl	−8.40	42
NaBr	0.30	42
NaCl	1.59	42, 3
NaI	−1.75	42
RbCl	4.29	42

* Values from smoothed curves of 0, 20.86, 44.18, 76.0 and 100 wt. % methanol.

METHANOL–WATER (Continued)

60 wt. % Methanol

Electrolyte	ΔH_∞° (kcal/mol)*	Ref.
CsCl	3.92	42
HCl	−17.74	42, 3
KBr	4.07	42, 3
KCl	3.87	42, 3
KI	4.10	42
LiCl	−8.80	42
NaBr	−0.04	42
NaCl	1.375	42, 3
NaI	−2.15	42
RbCl	4.15	42

* Values from smoothed curves of 0, 20.86, 44.18, 76.0 and 100 wt. % methanol.

70 wt. % Methanol

Electrolyte	ΔH_∞° (kcal/mol)*	Ref.
CsCl	3.60	42
HCl	−18.03	42, 3
KBr	3.53	42, 3
KCl	3.40	42, 3
KI	3.47	42
LiCl	−9.35	42
NaBr	−0.60	42
NaCl	1.03	42, 3
NaI	−2.80	42
RbCl	3.91	42

* Values from smoothed curves of 0, 20.86, 44.18, 76.0 and 100 wt. % methanol.

V. HEATS OF SOLUTION CALORIMETRY

METHANOL–WATER (*Continued*)

76 wt. % Methanol

Electrolyte	$\Delta H_\infty°$ (kcal/mol)	Ref.
CsCl	3.32	42
HCl	−18.40	42
KBr	3.800 ± 0.015	42
KCl	2.960 ± 0.025	42
KI	3.060 ± 0.010	42
LiCl	−9.70 ± 0.05	42
NaBr	−0.740 ± 0.020	42
NaCl	0.775 ± 0.025	42
NaI	−3.275 ± 0.010	42
RbCl	3.670 ± 0.020	42

80 wt. % Methanol

Electrolyte	$\Delta H_\infty°$ (kcal/mol)*	Ref.
CsCl	3.15	42
HCl	−18.50	42, 3
KBr	2.82	42, 3
KCl	2.79	42, 3
KI	2.70	42
LiCl	−9.95	42
NaBr	−1.32	42
NaCl	0.50	42, 3
NaI	−3.75	42
RbCl	3.15	42

* Values from smoothed curves of 0, 20.86, 44.18, 76.0 and 100 wt. % methanol.

METHANOL–WATER (*Continued*)

90 wt. % Methanol

Electrolyte	ΔH_∞° (kcal/mol)*	Ref.
CsCl	2.58	42
HCl	−19.03	42, 3
KBr	2.02	42, 3
KCl	2.06	42, 3
KI	1.70	42
LiCl	−10.65	42
NaBr	−2.40	42
NaCl	−0.35	42, 3
NaI	−5.05	42
RbCl	2.63	42

* Values from smoothed curves of 0, 20.86, 44.18, 76.0 and 100 wt. % methanol.

1-PROPANOL–WATER

Electrolyte	x_1 (mol fraction propanol)	$\Delta H_{\text{Dilution}}$ (kcal/mol)*	Ref.
HCl	0.025	−1.95 ± 0.04	66
HCl	0.05	−1.8 ± 0.1	66
HCl	0.10	−2.27 ± 0.09	66
HCl	0.15	−2.62 ± 0.02	66
HCl	0.20	−2.82 ± 0.03	66

* Initial concentration of HCl, $10.31M$; final concentration, $\sim 0.002M$.

V. HEATS OF SOLUTION CALORIMETRY

2-PROPANOL–WATER

Electrolyte	x_1 (mol fraction propanol)	$\Delta H_{\text{Dilution}}$ (kcal/mol)*	Ref.
HCl	0.025	-1.93 ± 0.03	66
HCl	0.05	-1.60 ± 0.03	66
HCl	0.10	-1.91 ± 0.05	66
HCl	0.15	-2.58 ± 0.03	66
HCl	0.20	-3.09 ± 0.04	66

PROPYLENE GLYCOL–WATER

Electrolyte	x_1 (mol fraction glycol)	$\Delta H_{\text{Dilution}}$ (kcal/mol)	Ref.
HCl	0.025	-2.19 ± 0.01	66
HCl	0.05	-1.94 ± 0.09	66
HCl	0.10	-1.8 ± 0.1	66
HCl	0.15	-2.04 ± 0.06	66
HCl	0.20	-2.6 ± 0.1	66
HCl	0.25	-2.88 ± 0.09	66

UREA–WATER

2.20M Urea

Electrolyte	ΔH° (kcal/mol)	Ref.
Bu$_4$NBr	-1.319 ± 0.029	80
	0.246 ± 0.018 (35°)	80

UREA–WATER (*Continued*)

4.89M Urea

Electrolyte	$\Delta H°$ (kcal/mol)	Ref.
Bu$_4$NBr	-0.815 ± 0.019	80
	0.593 ± 0.010 (35°)	80

8.01M Urea

Electrolyte	$\Delta H°$ (kcal/mol)	Ref.
Bu$_4$NBr	-0.427 ± 0.023	80
	0.761 ± 0.025 (35°)	80

10.20M Urea

Electrolyte	$\Delta H°$ (kcal/mol)	Ref.
Bu$_4$NBr	-0.343 ± 0.024	80
	0.796 ± 0.018 (35°)	80

1.0m Urea

Electrolyte	$\Delta H°$ (kcal/mol)	Ref.
Bu$_4$NBr	-1.663 ± 0.010	81
Et$_4$NBr	1.165 ± 0.012	81
Me$_4$NBr	5.437 ± 0.012	81

2.0m Urea

Electrolyte	$\Delta H°$ (kcal/mol)	Ref.
Bu$_4$NBr	-1.412 ± 0.010	81
Me$_4$NBr	5.167 ± 0.012	81

V. HEATS OF SOLUTION CALORIMETRY

UREA–WATER (Continued)

4.0m Urea

Electrolyte	$\Delta H°$ (kcal/mol)	Ref.
Bu$_4$NBr	-1.002 ± 0.010	81
Et$_4$NBr	0.465 ± 0.018	81
Me$_4$NBr	4.736 ± 0.012	81

6.0m Urea

Electrolyte	$\Delta H°$ (kcal/mol)	Ref.
Bu$_4$NBr	-0.694 ± 0.010	81

2.2m Urea

Electrolyte	$\Delta H°$ (kcal/mol)	Ref.
Et$_4$NBr	0.840 ± 0.012	81

8.3m Urea

Electrolyte	$\Delta H°$ (kcal/mol)	Ref.
Bu$_4$NBr	-0.440 ± 0.015	81

10.9m Urea

Electrolyte	$\Delta H°$ (kcal/mol)	Ref.
Bu$_4$NBr	-0.260 ± 0.020	81

(c) Nonaqueous-Nonaqueous Mixed Solvents

The heats of solution data in nonaqueous-nonaqueous mixed solvents are arranged alphabetically according to nonaqueous component. The complete bibliography is given at the end of this section.

Symbols used:

$\Delta H°$ Standard heat of solution (the reader should consult the original reference in order to find the exact extrapolation used
$\Delta H_\infty°$ Standard heat of solution at infinite dilution
$\Delta H_{dil}°$ Heat of dilution
ΔH Heat of solution (normally at a specified concentration)

ACETIC ACID–PERCHLORIC ACID

Electrolyte	Conc. range HClO$_4$ (mole/liter)	Mole ratio range (solute/solvent)	$-\Delta H_{range}$ (kcal/mol)	Ref.
NaOAc	0.248–0.539	1:2732–1:1699	10.69–10.38	45

BENZENE–CHLOROBENZENE

50 Volume % Benzene

Electrolyte	$\Delta H°$ (kcal/mol)	Ref.
Bu$_4$NPic	7.13 (25°C)	74
	7.45 (35°C)	74
	7.75 (45°C)	74

V. HEATS OF SOLUTION CALORIMETRY

METHANOL–DIMETHYL SULFOXIDE

10 Volume % Dimethyl Sulfoxide

Electrolyte	$\Delta H°$ (kcal/mol)	Ref.
NaOCH	1.6 ± 0.1	69
NaPic	2.65 ± 0.14	69

20 Volume % Dimethyl Sulfoxide

Electrolyte	$\Delta H°$ (kcal/mol)	Ref.
$NaOCH_3$	15.9 ± 0.3	69

30 Volume % Dimethyl Sulfoxide

Electrolyte	$\Delta H°$ (kcal/mol)	Ref.
$NaOCH_3$	15.0 ± 0.3	69
NaPic	3.18 ± 0.28	69

50 Volume % Dimethyl Sulfoxide

Electrolyte	$\Delta H°$ (kcal/mol)	Ref.
$NaOCH_3$	13.3 ± 0.7	69
NaPic	3.04 ± 0.12	69

70 Volume % Dimethyl Sulfoxide

Electrolyte	$\Delta H°$ (kcal/mol)	Ref.
$NaOCH_3$	10.2 ± 0.2	69

METHANOL–DIMETHYL SULFOXIDE (Continued)

80 Volume % Dimethyl Sulfoxide

Electrolyte	$\Delta H°$ (kcal/mol)	Ref.
NaOCH$_3$	8.96 ± 0.32	69
NaPic	3.23 ± 0.03	69

95.4 Volume % Dimethyl Sulfoxide

Electrolyte	$\Delta H°$ (kcal/mol)	Ref.
NaOCH$_3$	6.56 ± 0.32	69

References

1. S. I. Drakin and C. Yu-Min, Russ. J. Phys. Chem., **38**, 1526 (1964).
2. Y. M. Kessler, Electrokhimiya, **2**, 1338 (1966).
3. W. M. Latimer and C. M. Slansky, J. Amer. Chem. Soc., **62**, 2019 (1940).
4. L. N. Erbanova, S. I. Drakin and M. Kh. Karapet'yants, Russ. J. Phys. Chem., **38**, 1450 (1964).
5. L. N. Erbanova, M. Kh. Karapet'yants and S. I. Drakin, Russ. J. Phys. Chem., **39**, 1467 (1965).
6. G. Somsen and J. Coops, Rec. Trav. Chim., **84**, 985 (1965).
7. L. Weeda and G. Somsen, Rec. Trav. Chim., **85**, 159 (1966).
8. L. Weeda and G. Somsen, Rec. Trav. Chim., **86**, 263 (1967).
9. R. P. Held and C. M. Criss, J. Phys. Chem., **69**, 2611 (1965).
10. L. Weeda and G. Somsen, Rec. Trav. Chim., **86**, 893 (1967).
11. R. P. Held and C. M. Criss, J. Phys. Chem., **71**, 2487 (1967).
12. R. C. Paul, S. C. Ahluwalia and S. S. Pahil, Indian J. Chem., **3**, 300 (1965).
13. W. H. Smyrl and C. W. Tobias, J. Electrochem. Soc., **115**, 33 (1968).
14. W. Smyrl, Ph.D. Thesis, Univ. of California, Berkeley (1966).
15. Y. C. Wu and H. L. Friedman, J. Phys. Chem., **70**, 2020 (1966).
16. Y. C. Wu and H. L. Friedman, J. Phys. Chem., **70**, 501 (1966).
17. S. R. Gunn and L. G. Green, J. Chem. Phys., **36**, 368 (1962).
18. C. S. Gardner, E. W. Green and D. M. Yost, J. Amer. Chem. Soc., **57**, 2055 (1935).
19. N. Elliott and D. M. Yost, J. Amer. Chem. Soc., **56**, 1057 (1934).
20. D. Feakins, B. C. Smith and L. Thakur, J. Chem. Soc., (A), 714 (1966).
21. G. V. Karpenko, K. P. Mishchenko and G. M. Poltoratskii, J. Struct. Chem., **8**, 367 (1967).
22. N. A. Izmailov, Doklady Akad. Nauk SSSR, **149**, 1103 (1963).
23. N. A. Izmailov, Doklady Akad. Nauk SSSR, **149**, 1364 (1963).

24. S. V. Shadskii and K. P. Mishchenko, Doklady Akad. Nauk SSSR, **158**, 1180 (1964).
25. G. A. Krestov, J. Struct. Chem., **3**, 501 (1962).
26. O. N. Bhatnagar and C. M. Criss, J. Phys. Chem., **73**, 174 (1969).
27. B. Jakuszewski, S. Taniewska-Osinska and R. Logwinenko, Bull. Acad. Polon. Sci., **9**, 127 (1961).
28. C. M. Criss, U.S. A.E.C. Report TID-22366 (1965).
29. E. M. Arnett and D. R. McKelvey, J. Amer. Chem. Soc., **88**, 2598 (1966).
30. G. Somsen, Rec. Trav. Chim., **85**, 517 (1966).
31. R. C. Paul, S. C. Ahluwalia and S. S. Pahill, Indian J. Chem., **3**, 305 (1965).
32. K. P. Mishchenko, Acta Physiochim., **3**, 693 (1935).
33. J. N. Pearce and H. B. Hart, J. Amer. Chem. Soc., **42**, 2411 (1922).
34. M. Bodtelsky and R. D. Larisch, J. Chem. Soc., 3612 (1950).
35. R. L. Moss and J. H. Wolfenden, J. Chem. Soc., 118 (1939).
36. F. A. Askew, E. Bullock, H. T. Smith, R. K. Tinkler, O. Gatty and J. H. Wolfenden, J. Chem. Soc., 1368 (1934).
37. C. A. Kraus and F. C. Schmidt, J. Amer. Chem. Soc., **56**, 2294 (1934).
38. L. V. Coulter and R. H. Maybury, J. Amer. Chem. Soc., **71**, 3394 (1949).
39. G. P. Kotlyarova and E. F. Ivanova, Russ. J. Phys. Chem., **38**, 221 (1964).
40. T. H. Higgins and E. F. Westrum Jr., J. Phys. Chem., **65**, 830 (1961).
41. J. H. Mathews, J. Amer. Chem. Soc., **33**, 1291 (1911).
42. C. M. Slansky, J. Amer. Chem. Soc., **62**, 2430 (1940).
43. S. R. Gunn and L. G. Green, J. Phys. Chem., **64**, 1066 (1960).
44. S. P. Wolsky, E. J. Zdanuk and L. V. Coulter, J. Amer. Chem. Soc., **74**, 6196 (1952).
45. W. L. Jolly, J. Amer. Chem. Soc., **74**, 6199 (1952).
46. R. P. Held, U.S. AEC TID-22374 (1965); C. A. **64**, 15078d.
47. F. C. Schmidt, F. J. Studer and J. Sottysiak, J. Amer. Chem. Soc., **60**, 2780 (1938).
48. L. V. Coulter and L. Monchick, J. Amer. Chem. Soc., **73**, 5867 (1951).
49. K. P. Mishchenko and V. V. Sokolov, J. Struct. Chem., **5**, 760 (1964).
50. G. A. Krestov and V. I. Klopov, J. Struct. Chem., **5**, 769 (1964).
51. R. F. Rodewald, K. Mahendran, J. L. Bear and R. Fuchs, J. Amer. Chem. Soc., **90**, 6698 (1968).
52. R. C. Paul, R. Kaushal, K. S. Dhindsa, S. S. Pahil and S. C. Ahluwalia, J. Indian Chem. Soc., **44**, 964 (1967).
53. G. P. Kotlyarova and E. F. Ivanova, Russ. J. Phys. Chem., **40**, 537 (1966).
54. K. P. Mishchenko and A. M. Sukhotin, Doklady Akad. Nauk SSSR, **98**, 103 (1954).
55. F. C. Schmidt, S. Godomsky, F. K. Ault and J. C. Huffman, J. Chem. Eng. Data, **14**, 71 (1969).
56. H. D. Mulder and F. C. Schmidt, J. Amer. Chem. Soc., **73**, 5575 (1951).
57. F. C. Schmidt, J. Sottysiak and H. D. Kluge, J. Amer. Chem. Soc., **58**, 2509 (1936).
58. R. C. Paul, P. S. Gill and J. Singh, Indian J. Chem., **2**, 219 (1964).
59. R. C. Paul, S. C. Ahluwalia, S. K. Rehani and S. S. Pahil, Indian J. Chem., **3**, 207 (1965).
60. G. Choux and R. L. Benoit, J. Amer. Chem. Soc., **91**, 6221 (1969).
61. C. V. Khrishnan and H. L. Friedman, J. Phys. Chem., **73**, 3943 (1969).
62. R. H. Boyd and P. S. Wang, Abstracts, 155th National Meeting of the American Chemical Society, San Francisco, Calif. (1968).
63. R. Keller, J. N. Foster, D. C. Hanson, J. F. Hon and J. S. Muirhead, NASA Contractor Report, NASA CR-1425 (1969).
64. R. Jasinski, Scientific Report No. 2, Air Force Cambridge Research Laboratories, AFCRL-69-0381 (1969).

65. R. C. Paul, K. S. Dhindsa, S. C. Ahluwalia and S. P. Narula, J. Indian Chem. Soc., **48**, 381 (1971).
66. J. H. Stern and S. L. Hansen, J. Chem. Eng. Data, **16**, 360 (1971).
67. C. DeVisser and G. Somsen, Rec. Tran. Chim., **90**, 1129 (1971).
68. A. Finch, P. J. Gardner and C. J. Steadman, J. Phys. Chem., **71**, 2996 (1967).
69. J. W. Larsen, K. Amin and J. H. Fendler, J. Amer. Chem. Soc., **93**, 2910 (1971).
70. A. Finch, P. J. Gardner and C. J. Steadman, J. Phys. Chem., **75**, 2325 (1971).
71. C. V. Krishnan and H. L. Friedman, J. Phys. Chem., **75**, 3606 (1971).
72. J. N. Butler and J. C. Synnott, J. Amer. Chem. Soc., **92**, 2602 (1970).
73. S. Abe, S. Sigetomi and R. Hara, J. Soc. Chem. Ind. Japan, **38**, Suppl. binding, 163 (1935).
74. S. Goldman and G. C. B. Cave, Canad. J. Chem., **49**, 4096 (1971).
75. R. C. Paul, J. Singh, S. C. Ahluwalia and S. S. Pahil, Indian J. Chem., **2**, 134 (1964).
76. P. Chall and O. Doepke, Z. Elektrochem., **37**, 357 (1931).
77. R. Fuchs, D. S. Plumlee Jr. and R. F. Fodewald, Thermochim. Acta, 2, 515 (1971).
78. C. V. Krishnan and H. L. Friedman, J. Phys. Chem., **75**, 388 (1971).
79. R. Fuchs, J. L. Bear and R. F. Rodewald, J. Amer. Chem. Soc., **91**, 5797 (1969).
80. T. S. Sarma and J. C. Ahluwalia, J. Phys. Chem., **76**, 1366 (1972).
81. R. B. Cassel and W-Y. Wen, J. Phys. Chem., **76**, 1369 (1972).
82. C. DeVisser and G. Somsen, J. Chem. Thermodynamics, **4**, 313 (1972).
83. M. J. Mastroianni and C. M. Criss, J. Chem. Thermodynamics, **4**, 321 (1972).
84. S. Taniewska-Osinska, R. Logwinienko and M. Pluta, Acta Chim. Lodz., **16**, 45 (1971).

VI. POLAROGRAPHY

The use of nonaqueous solvents for electrochemical investigations of inorganic, organometallic and organic compounds, in both academic and industrial laboratories, has become widespread in recent years and is of increasing interest. Voltammetric studies in suitable aprotic solvents can be used for the study of equilibria for rates and mechanisms of organic reactions, in electroorganic synthesis and for structural problems.

The polarographic studies of a large number of metal species in aqueous media give ill-defined waves due to the occurrence of maxima, hydrolytic reactions and catalytic hydrogen waves, whereas such investigations in nonaqueous solvents do not present these problems.[1,2,3] In addition the lower oxidation states of metal ions can be studied in nonaqueous media.[4] The polarographic determinations of half-wave potentials of metal ions in different solvents indicate a measure of relative strengths of the metal ion solvent bonds.[5]

In coordination compounds, as well as organometallic compounds, polarography has been successfully used to study their stability, composition, the rates and mechanisms of reactions and to establish relationships between the structure and redox behavior. The estimation of ionization potentials from reversible redox potentials of organometallics determined in aprotic solvents has been made and calculated values agree with those obtained from other methods.[6,7]

Voltammetric information on limiting currents, half-wave potentials and the dependence of these on the concentration of depolarizer, temperature, solvent composition and free energy relationships are important for the elucidation of the mechanism of electrode processes.[8]

Electrochemical redox potentials of aromatic hydrocarbons in aprotic solvents can be correlated to such parameters as electron affinities, u.v. absorption frequencies as well as energy levels calculated from molecular orbital theory.[9]

References

1. I. Zlotowsky and I. M. Kolthoff, J. Amer. Chem. Soc. **66**, 1431 (1944).
2. I. Zlotowsky and I. M. Kolthoff, J. Phys. Chem. **49**, 386 (1945).
3. J. W. Oliver and J. W. Ross, Jr., J. Inorg. Nucl. Chem. **25**, 1515 (1963).
4. A. I. Popov and D. H. Geske, J. Amer. Chem. Soc., **79**, 2074 (1957).
5. D. L. McMasters, R. B. Dunlap, J. R. Kuempel, L. W. Kreider and T. R. Shearer, Anal. Chem., **39**, 103 (1967).
6. A. Stanienda, Z. Naturforsch, **B23**, 1235 (1968).
7. S. P. Gubin, S. A. Smirnova, L. I. Denisovich and A. A. Lubovich, J. Organometal. Chem. **30**, 243 (1971).
8. P. Zuman, Proceedings of 3rd. Int. Cong. on Polarography, Southampton (1964), Ed. G. J. Hills, Vol. II, Macmillan, London (1966).
9. A. Streitasieser, "Molecular Orbital Theory for Organic Chemists," John Wiley and Sons, Inc., New York, N. Y. (1961).

(a) Inorganic Electrolytes

The following section on nonaqueous voltammetry and polarography relative to inorganic electrolytes covers solute concentration, supporting electrolyte, temperature, working electrode, reference electrode and half-wave potentials. The tables are organized in alphabetical order by solvent and are based on the literature through 1972.

Symbols used:

DME	Dropping mercury electrode
SCE	Standard calomel electrode
Re(Pt) or RPE	Rotating platinum electrode
MPE	Mercury pool electrode
RPGE	Rotating pyrolytic graphite electrode
HMD	Hanging mercury drop
mM	Millimoles

ACETIC ACID
0.1M $NaClO_4$ at Pt-foil

Species	$E_{1/4}V$ (vs SCE)	Ref.
I^-	+0.44	1

ACETIC ANHYDRIDE
0.1M $NaClO_4$ at Pt-foil

Species	$E_{1/4}V$ (vs SCE)	Ref.
I^-	+0.20; +0.60	1

VI. POLAROGRAPHY

ACETONE
0.1M NaClO₄ at Pt-foil

Species	$E_{1/4} V$ (vs SCE)	Ref.
I⁻	+0.22; +0.62	1

0.1M Et₄NClO₄ at DME 25°C.
0.496–1.144 mM

Species	$E_{1/2} V$ (vs SCE)	Ref.
Dy(III)	−1.44	8
Er(III)	−1.43	
Eu(II)	−1.79	
Eu(III)	−0.03	
Gd(III)	−1.49	
Ho(III)	−1.44	
Nd(III)	−1.49	
Pr(III)	−1.47	
Sm(II)	−1.70	
Sm(III)	−1.13	
Tb(III)	−1.38	
Yb(II)	−1.71	
Yb(III)	−0.71	

0.001M Bu₄NI at DME
1 mM

Species	$E_{1/2} V$ (vs SCE)	Ref.
Rb(I)	−1.916	9

ACETONE–WATER (50% ACETONE)
0.1M NaClO₄ at Pt-foil

Species	$E_{1/4} V$ (vs SCE)	Ref.
I⁻	+0.42; +0.54	1

ACETONITRILE
0.1M Et₄NClO₄ at DME at 25°C

Species	Conc. ± 10³ M/l	$E_{1/2}V$ (vs SCE)	Ref.
AgClO₄	1 mM	+0.1[a,b]	10
AgClO₄	1 mM	+0.32[a]	
Al(ClO₄)₃·9H₂O	1 mM	−1.42	
Be(ClO₄)₂·4H₂O	1 mM	−1.6	
Cd(ClO₄)₂·6H₂O	0.1 mM	−0.5[a,b]	
Co(ClO₄)₂·6H₂O	1 mM	−0.65[a]	
Cr(ClO₄)₃·6H₂O	1 mM	−1.12[a]	
Cu(ClO₄)₂·6H₂O	0.2 mM	+1.0; −0.5[b]	
Cu(ClO₄)₂·6H₂O	1 mM	−0.36[a]	
Et₄NBr	0.5 mM	+0.7[b]; +1.0[b]	
Eu(II)	1 mM	−1.67	
Eu(III)	1 mM	+0.15	
Fe(ClO₄)₂·6H₂O	0.4 mM	+1.6[b,c]	
Fe(ClO₄)₃·6H₂O	1 mM	−1.00[a]	
Fe(ClO₄)₃·6H₂O	0.4 mM	+1.1[b,c]	
Gd(III)	1 mM	−1.5	
Hg(ClO₄)₂·3H₂O	0.15 mM	+0.70; +0.5[b]	
La(III)	1 mM	−1.5	
LiCl	0.5 mM	+1.1[b]; +1.7[b]	
Mg(ClO₄)₂	1 mM	−1.84	
Mn(ClO₄)₂·6H₂O	1 mM	−1.12[a]	
NaI	0.5 mM	+0.3[b]; +0.6[b]	
Ni(ClO₄)₂·6H₂O	1 mM	−0.33[a]	
Pr(III)	1 mM	−1.5	
Sm(III)	1 mM	−1.62	
Yb(II)	1 mM	−1.69	
Yb(III)	1 mM	−0.57	
Zn(ClO₄)₂·6H₂O	0.1 mM	−0.9[a,b]	

[a] 0.1N NaClO₄ as supporting electrolyte.
[b] At rotated Pt electrode.
[c] 0.2N NaClO₄ as supporting electrolyte.

VI. POLAROGRAPHY

ACETONITRILE (Continued)
0.1M NaClO₄ at Pt-foil

Species	$E_{1/4}V$ (vs SCE)	Ref.
I⁻	+0.25; 0.55	1

Pt electrode at 25.0°C ± 0.1; 1.0 mM

Species	Supporting electrolyte	$E_{1/2}V$ (vs Ag/AgNO₃)	Ref.
Cu(II) chloride	0.1M Et₄NClO₄	+0.587; −0.819	11
Cu(II) chloride	0.1M Et₄NClO₄ + 2 mM [Cl⁻]	+0.495; +0.155; −0.920; −1.141	
Cu(II) chloride	0.1M Et₄NClO₄ + 10 mM [Cl⁻]	−0.055; −1.260	
Cu(I) chloride	0.1M Et₄NClO₄ + 1 mM [Cl⁻]	+0.585; +0.248; −0.832; −1.110	
Cu(I) chloride	0.1M Et₄NClO₄ + 7 mM [Cl⁻]	+0.190; −1.262	
Cu(II) bromide	0.1M Et₄NClO₄	+0.580; −0.818	
Cu(II) bromide	0.1M Et₄NClO₄ + 1 mM [Cl⁻]	+0.488; +0.110; −0.820	

0.1M LiClO₄ at RE(Pt)
1 mM

Species	$E_{1/2}V$ (vs Ag/AgNO₃)[a]	Ref.
Br₂	0.0[c]; 0.58	12
Bu₄NBr₃	0.2[b]; 0.67; 1.43	
Et₄NBr	0.42; 0.71; 1.41	
IBr	−0.11[b]; 0.26[b]; 0.53[b]	
ICl	−0.15[c]; 0.26[c]; 0.59[c]	
Me₄NIBr₂	−0.7[bc]; −0.08[b]; 0.81	
Me₄NI₂Br	−0.12[b]; 0.34; 0.75; 1.88[d]; 1.34	
Me₄NIBrCl	−0.03[bc]; 0.97; 1.49	
Me₄NICl₂	−0.21[b]; 0.12[b]; 1.14[b]	

[a] Anodic process given negative sign.
[b] Data on forward trace.
[c] Approximate data.
[d] Data on reverse trace.

ACETONITRILE (*Continued*)

At Pt electrode
NaI 0.1 to 4.6 mM
Py 50 to 1000 mM

Species	$E_{1/2}V$ (vs Ag/AgCl)	Ref.
NaI + Py	+0.152 to +0.336	13

0.1M LiClO$_4$ at RE(Pt)
1 mM

Species	$E_{1/2}V$ (vs Ag/AgNO$_3$)[a]	Ref.
I$_2$	−0.16; 0.32; −; 1.79[b]	
I(Py)$_2$ClO$_4$	0.22; 0.11; −0.95[c]	
Me$_4$NI$_3$	−0.18; 0.32; 1.08; 1.80[b]	
NaI	−0.01; 0.31; 1.05; 1.82[b]	14

[a] Anodic waves are given negative sign.
[b] Refer to reverse trace.
[c] Apparent value.

0.1M Et$_4$NClO$_4$ at DME at 25.00°C.
0.2 to 1.5 mM

Species	$E_{1/2}V$ (vs SCE)	Ref.
Th(ClO$_4$)$_4$	−1.18 to −1.22	15

0.1M Et$_4$NClO$_4$ at DME at 25°C
16.2 mM

Species	$E_{1/2}V$ (vs SCE)	Ref.
Nickel perchlorate	−0.43	16

VI. POLAROGRAPHY

ACETONITRILE (Continued)

0.001M Bu₄NI at DME
1 mM

Species	$E_{1/2}V$ (vs Rb)	Ref.
Ag(I)	+2.40	9
Cd(II)	+1.72	
Co(II)	+1.37	
Cr(II)	+0.86	
Cu(I)	+1.65	
Mn(II)	+0.90	
Na(I)	−1.829[a]	
Ni(II)	+1.69	
Rb(I)	−1.954[a,b]	
Zn(II)	+1.36	

[a] V vs SCE.
[b] Pr₄NI as supporting electrolyte.

0.1M Et₄NClO₄ at DME at 25°C.

Species	$E_{1/2}V$ (vs SCE)	Ref.
Eu(II)	−1.67	17
Eu(III)	+0.19	
Nd(III)	−1.57	
Pr(III)	−1.50	
Sm(I)	−1.69	
Sm(III)	−1.02	
Yb(II)	−1.70	
Yb(III)	−0.58	

ACETONITRILE (Continued)

0.1M Et$_4$NClO$_4$ at 25.0° ± 0.01°C.

0.405 mM

Species	$E_{1/2}V$ (vs SCE)	Ref.
Eu(II)	−1.62	18
Eu(III)	+0.10	
Nd(III)	−1.45	
Sm(II)	−1.55	
Sm(III)	−1.04	
Yb(II)	−1.58	
Yb(III)	−0.60	

0.1M NaClO$_4$ at DME at 20°C.

0.1 mM

Species	$E_{1/2}V$ (vs SCE)	Ref.
Acetoacetylferrocene[a]	0.560	19
Acetylferrocene[a]	0.555	
Co(ClO$_4$)$_2$·6H$_2$O	~1.3	
Cu(ClO$_4$)·H$_2$O	0.96	
Cu(ClO$_4$)$_2$·6H$_2$O	0.96	
Fe(ClO$_4$)$_2$·6H$_2$O	1.55	
Fe(ClO$_4$)$_3$·6H$_2$O	1.15	
Ferrocene[a]	0.31	

[a] 0.5 mM.

VI. POLAROGRAPHY

ACETONITRILE *(Continued)*
0.1M LiClO$_4$ at DME

Species	Conc. $\times 10^3$ M/l	$E_{1/2}$V (vs MPE)[a]	Ref.
Se(ClO$_4$)$_3$	4.5	-1.825; -2.299	20
Y(III)	3.0	-1.709	

[a] Mercury pool electrode.

At DME at 25.00 ± 0.02°C.
1 mM solutions

Species	$E_{1/2}$V (vs SCE)	Ref.
GeCl$_4$	$+0.33$; -0.16; -1.10[a]	21
GeCl$_4$	-1.27[b]	
GeCl$_4$	~ -1.0[c]	
SnCl$_2$	$+0.38$; $+0.54$; -0.08; -0.72_3[a]	
SnCl$_2$	-0.85_2[b]	
SnCl$_4$	$+0.47$; $+0.3$; -1.06[a]	
SnCl$_4$	-1.05[b]	
SnCl$_4$	-1.07[c]	
SnI$_4$	$+0.25$; $+0.11$; ~ -0.40[a]	
SnI$_4$	-1.05[b]	
ZrCl$_4$	$+0.38$; -0.94; -1.71; -1.97[d]	
ZrCl$_4$	-2.28[b]	
ZrCl$_4$	-1.20; -2.28[e]	

[a] 0.1M Et$_4$NClO$_4$ as supporting electrolyte.
[b] 0.1M Et$_4$NCl as supporting electrolyte.
[c] 0.1M Et$_4$NI as supporting electrolyte.
[d] 0.1M Et$_4$NClO$_4$ at DME and RE (Pt).
[e] 0.05M Et$_4$NClO$_4$ + 0.05M Et$_4$NI.

ACETONITRILE (Continued)

DME at 25 ± 0.1°C.
0.5 mM

Species	Supporting electrolyte	$E_{1/2}V$ (vs SCE)	Ref.
Ni(II)	0.1M LiClO$_4$ + 1–0.05 mM LiCl	−0.54	22
Ni(II)	0.1M LiClO$_4$ + 1–0.05 mM LiBr	−0.48	
Ni(II)	0.1M LiClO$_4$ + 1–0.05 mM LiI	−0.48 to −0.42	
Ni(II)	0.1M Et$_4$NClO$_4$ + 20–0.01 mM LiCl	−0.54	
Ni(II)	0.1M Et$_4$NClO$_4$ + 1–0.02 mM Et$_4$NBr	−0.50	
Ni(II)	0.1M Et$_4$NClO$_4$ + 1–0.02 mM Et$_4$NI	−0.60 to −0.40	
Ni(II)	0.1M Bu$_4$NClO$_4$ + 0.2–0.02 mM LiCl	−0.60	
Ni(II)	0.1M Bu$_4$NClO$_4$ + 1–0.02 mM Bu$_4$NBr	−0.47 to −0.46	
Ni(II)	0.1M Bu$_4$NClO$_4$ + 1–0.2 mM Bu$_4$Ni	−0.40 to −0.39	

0.1M Et$_4$NClO$_4$ at DME at 25°C.
~1.0 mM

Species	$E_{1/2}V$ (vs SCE)	Ref.
Ba(ClO$_4$)$_2$	−1.63	23
Cd(ClO$_4$)$_2$	−0.26	
Co(ClO$_4$)$_2$	−0.61	
KClO$_4$	−1.96	
Mn(ClO$_4$)$_2$	−1.08	
NaClO$_4$	−1.85	
Ni(ClO$_4$)$_2$	−0.29	
(Ph–Ph)$_2$Cr(I)	−0.73	
RbClO$_4$	−1.98	
TlClO$_4$	−0.27	
Zn(ClO$_4$)$_2$	−0.62	

VI. POLAROGRAPHY

ACETONITRILE (Continued)

0.1M Bu_4NI at DME at $25 \pm 0.2°C$.
1 mM

Species	$E_{1/2}V$ (vs Hg pool)	Ref.
CaI_2	−1.24	24
$Cd(ClO_4)_2$[a]·$4H_2O$	−0.61	
CsI[c]	−1.34	
$Cu(ClO_4)_2$[a]·xH_2O	−0.14	
$KClO_4$[a,d]	−2.29	
KI	−1.34	
LiCl	−1.42	
$LiClO_4$[a]	−2.33	
MgI_2	−1.16	
NaI	−1.25	
$Pb(ClO_4)_2$[a,c]·$3H_2O$	−0.44	
RbCl[b]	−1.36	
$Zn(ClO_4)_2$[a]·$4H_2O$	−1.07	

[a] 0.1M Bu_4NClO_4 supporting electrolyte.
[b] <0.5 mM (saturated).
[c] <1.0 mM (saturated).
[d] 0.5 mM.

0.1 M Et_4NClO_4 at DME
1.0 mM

Species	$E_{1/2}V$ (vs SCE)	Ref.
$Am(ClO_4)_3$	−1.65; −2.0	25

0.1M $NaClO_4$ at DME at 30°C.
1.0 mM

Species	$E_{1/2}V$ (vs SCE)	Ref.
$Co(ClO_4)_2$	−0.545	26

ACETONITRILE (*Continued*)

0.1M Et$_4$NClO$_4$ at DME at 25°C.

Species	Conc. $\times$ 10^3 M/l	$E_{1/2}V$ (vs SCE)	Ref.
NbCl$_5$	1.14	-0.26; -0.68	27
PaCl$_5$	0.94	-0.22	27
TaCl$_5$	1.00	-1.16	27

0.1M Et$_4$NClO$_4$ at RPGE[a] at 25 $\pm$ 0.2°C.

Species	Conc. $\times$ 10^3 M/l	$E_{1/2}V$ (vs Ag/AgNO$_3$)	Ref.
Pr$_4$NI	0.578	$+0.02$; $+0.37$; $+1.00$	28a
I$_2$	0.555	$+0.30$; $+0.96$	
Pr$_4$NI	0.597	$+0.33$; $+1.11$	
Pr$_4$NI + I$_2$	0.607	$+0.28$	

[a] Rotating pyrolytic graphite electrode.

0.1M Et$_4$NClO$_4$ at DME at 25°C.
$\sim$0.37 mM to 2 mM

Species	$E_{1/2}V$ (vs SCE)	Ref.
Diethyl-4-nitrophenyl phosphate	-1.07_4	28b
O,O-Diethyl-O,4-nitrophenyl-thiophosphate	-1.11_6	
Diethyl-4-nitrobenzylphosphate	-1.17_0	
Diphenyl-4-nitrophenylphosphate	-1.09_8	
Diphenyl-2,6-dimethyl-4-nitrophenylphosphate	-1.14_5	
i(4-methylphenyl)-4-nitrophenylphosphate	-1.09_2	
Di(2-methylphenyl)-4-nitrophenylphosphate	-1.100	
Diphenyl-4-nitrobenzylphosphate	-0.865; 1.211	

VI. POLAROGRAPHY

ACETONITRILE *(Continued)*

At DME at 25 ± 0.02°C.

0.9 to 1.81 mM

Species	$E_{1/2}V$ (vs SCE)	Ref.
$TiCl_4$	+0.34[a]; −1.75[a]	29
$TiCl_4$	−0.80[b]; −1.39[b]; −1.92[b]	
$TiCl_4$	−0.77[c]; −1.20[c]; −1.86[c]	
$TiCl_4$	−1.46[d]	
TiI_4[e]	+0.22[a]; −0.34[a]	
TiI_4[e]	−0.86[b]; −1.41[b]; −1.95[b]	
TiI_4[e]	−1.38[c]	

[a] 0.1M Et_4NClO_4 as supporting electrolyte.
[b] 0.1M Et_4NCl as supporting electrolyte.
[c] 0.1M Et_4NI as supporting electrolyte.
[d] 0.05M Et_4NCNS + 0.05M Et_4ClO_4.
[e] 1 mM solution.

0.1M $NaClO_4$ at DME at 25 ± 0.2°C.

1 mM

Species	$E_{1/2}V$ (vs SCE)	Ref.
$AgClO_4$	+0.42	85
$Zn(ClO_4)_2$	−0.66	

0.1M Et_4NClO_4 at DME at 25 ± 0.5°C.

1 mM

Species	$E_{1/2}V$ (vs SCE)	Ref.
$HClO_4$	−0.70	81
Hydrobromic acid	−0.90	
Hydrochloric acid	−1.06	
Phosphoric acid	−1.75	
Sulfuric acid	−1.20	

ACETONITRILE *(Continued)*

0.1M Et_4NClO_4 at DME at 25 ± 0.5°C.

1 mM

Species	$E_{1/2}V$ (vs SCE)	Ref.
$Ba(ClO_4)_2$	−1.63	83
$Ca(ClO_4)_2 \cdot 6H_2O$	−1.82	
$Cd(ClO_4)_2 \cdot 6H_2O$[b]	−0.27	
KI[a]	−1.94	
$LiClO_4$	−1.95	
$NaClO_4$	−1.85	
NaI[a]	−1.85	
NH_4ClO_4	−1.83	
RbI[a]	−1.95	
$Sr(ClO_4)_2$	−1.76	
$Zn(NO_3)_2 \cdot 6H_2O$	−0.70	

[a] Bu_4NI supporting electrolyte.
[b] $NaClO_4$ supporting electrolyte.

0.1M Bu_4NClO_4 at DME at 25 ± 0.1°C.

Species	$E_{1/2}V$ (vs $Ag/AgNO_3$)[a]	Ref.
$AgClO_4$	+0.13	84
$Al(ClO_4)_3 \cdot 6H_2O$	−1.64	
$Ba(ClO_4)_2$	−1.92	
$Ca(ClO_4)_2$	−2.14	
$Cd(ClO_4)_2$	−0.55	
$Co(ClO_4)_2 \cdot 6H_2O$	−0.96	
$Cr(ClO_4)_3 \cdot 6H_2O$	−1.39; −0.3	
$Cu(ClO_4)_2 \cdot 6H_2O$	−0.62; >−0.36	
$Fe(ClO_4)_2 \cdot 6H_2O$	−1.33	
$Fe(ClO_4)_3 \cdot 6H_2O$	−1.33; >−0.36	
$HClO_4$	−0.78	
$Hg(ClO_4)_2 \cdot 3H_2O$	>−0.36	
$KClO_4$	−2.24	
$LiClO_4$	−2.26	
$Mn(ClO_4)_2 \cdot 6H_2O$	−1.40	
$NaClO_4$	−2.14	
$Pb(ClO_4)_2$	−0.33	
$Sn(ClO_4)_2$	−0.23	
$Sr(ClO_4)_2$	−2.08	
$TlClO_4$	−0.55	
$Zn(ClO_4)_2 \cdot 6H_2O$	−0.91	

VI. POLAROGRAPHY

ACRYLONITRILE
0.001M Bu₄NI at DME
~1 mM

Species	$E_{1/2}V$ (vs Rb)	Ref.
Ag(I)	+2.40	9
Cd(II)	+1.72	
Co(II)	+1.43	
Cr(II)	+0.86	
Cu(I)	+1.64	
Mn(II)	+0.86	
Ni(II)	+1.58	
Zn(II)	+1.35	

0.1M Et₄NClO₄ at DME at 25 ± 0.2°C.
1 mM

Species	$E_{1/2}V$ (vs SCE)	Ref.
AgClO₄	+0.49	85
Cd(ClO₄)₂	−0.19	
Co(ClO₄)₂	−0.48	
Cr(ClO₄)₃	−1.05	
Cu(ClO₄)₂	−0.27	
Mn(ClO₄)₂	−1.05	
Ni(ClO₄)₂	−0.33	
Zn(ClO₄)₂	−0.56	

ACETONITRILE–WATER
0.1M NaClO₄ at DME at 30°C.
1.0 mM

Species	V/V Acetonitrile	$E_{1/2}V$ (vs SCE)	Ref.
Co(ClO₄)₂	40	−1.092	26
Co(ClO₄)₂	60	−1.070	
Co(ClO₄)₂	80	−0.978	

ACETONITRILE–WATER (Continued)
0.1M NaClO$_4$ at DME at 20°C.
0.1 mM

Species	$E_{1/2}V$ (vs SCE)	Ref.
Copper (II)	0.700[a]	19
	0.595[b]	
	0.530[c]	
	0.440[d]	
	0.400[e]	
	0.335[f]	

[a] 5%.
[b] 10%.
[c] 16%.
[d] 35%.
[e] 50%.
[f] 70% of water.

LIQUID AMMONIA
LiClO$_4$ at DME
~1 mM

Species	$E_{1/2}V$ (vs Hg pool)	Ref.
CdCo$_3$	-0.87[a]; -0.82[b]; -0.79[c]; -0.75[d]; -0.72[e]; -0.71[f]	30
PbSO$_4$	-0.40[a]; $-0.4-$[b]; -0.39[c]; -0.38[d]; -0.37[e]; -0.37[f]	
ZnCO$_3$	-1.38[a]; -1.35[b]; -1.32[c]; -1.31[d]; -1.28[g]; -1.26[e]; -1.26[f]	

[a] 25°C.
[b] 55°C.
[c] 69°C.
[d] 80 and 99°C.
[e] 104 and 142°C.
[f] 190°C.
[g] 99°C.

VI. POLAROGRAPHY

LIQUID AMMONIA (Continued)
Saturated Bu$_4$NI at DME at −36°C.
1 mM

Species	$E_{1/2}V$ (vs Hg pool)	Ref.
NaI	~−1.65	31

0.1M NH$_4$I at DME at 25°C. at 10 atm-pressure
10 mM

Species	$E_{1/2}V$ (vs Hg pool)	Ref.
Cd(II)	−0.75	32
Cu(NO$_3$)$_2$	−0.11; −0.50; −0.31	
Pb(NO$_3$)$_2$	−0.35	
Tl(I)	−0.255	

1M NH$_3$—NH$_4$NO$_3$ at HMD[a] at 23 ± 2°C.
0.2 mM

Species	pH	E_pV (vs SCE)	Ref.
Pd(II)	7.0	−0.743	33
	8.0	−0.750	
	9.0	−0.751	
	10.0	−0.762	
Pd(II)[b]	—	−0.191	
Pd(II)[c]	—	−0.224	
Pd(II)[d]	—	−0.328	

[a] Hanging mercury drop.
[b] 0.1M KNO$_3$ supporting electrolyte + 0.05M pyridine.
[c] 0.1M KNO$_3$ supporting electrolyte + 0.20M pyridine.
[d] 0.1M KNO$_3$ supporting electrolyte + 2.0M pyridine.

NaI at DME at 24 ± 0.2°C.
1.24 mg/g NaI

Species	$E_{1/2}V$ (vs Ag/AgI)	Ref.
Lead iodide	−0.027	34

LIQUID AMMONIA (Continued)

0.1M NaClO$_4$ at RE(Pt) at -77°C.

1.78 mM

Species	$E_{1/2}V$ (vs Pt sheet)	Ref.
NH$_4$ClO$_4$	+1.0	35

1M H$_2$SO$_4$ at DME at -33°C.

0.63 mM

Species	$E_{1/2}V$ (vs Hg pool)	Ref.
Bismuth(III)	-0.42	86

BENZONITRILE

0.1M Et$_4$NClO$_4$ at DME at 25.0 ± 0.2°C.

1 mM

Species	$E_{1/2}V$ (vs SCE)	Ref.
Al(ClO$_4$)$_3$	-1.47; -1.41[b]	23
Ba(ClO$_4$)$_2$	-1.58; -1.69[b]	
Ca(ClO$_4$)$_2$	-1.73	
Cd(ClO$_4$)$_2$	-0.17; -0.24[b]	
Co(ClO$_4$)$_2$	-0.49; -0.61[b]	
Cr(ClO$_4$)$_3$	-0.40; -0.97; -1.04[b]	
Cu(ClO$_4$)$_2$	$+0.85$[a]; -0.20	
Fe(ClO$_4$)$_2$	-0.91[b]	
Fe(ClO$_4$)$_3$	$+0.7$[a]; -0.79	
LiClO$_4$	-1.82	
Mg(ClO$_4$)$_2$	-1.62	
Mn(ClO$_4$)$_2$	-0.98; -1.11[b]	
NaClO$_4$	-1.73	
Ni(ClO$_4$)$_2$	-0.29; -0.37[b]	
Sr(ClO$_4$)$_2$	-1.72	
Zn(ClO$_4$)$_2$	-0.53; -0.61[b]	

[a] Chronopotentiometric determination.
[b] 0.1M LiClO$_4$ as supporting electrolyte.

VI. POLAROGRAPHY

BENZONITRILE (Continued)
0.1M Et$_4$NClO$_4$

Species	$E_{1/2}V$ (vs SCE)	Ref.
Eu(II)	−1.68	17
Eu(III)	+0.18	
Nd(III)	−1.50	
Pr(III)	−1.54	
Sm(II)	−1.58	
Sm(III)	−1.00	
Yb(II)	−1.64	
Yb(III)	−0.63	

0.001M Bu$_4$NI at DME
~1 mM

Species	$E_{1/2}V$ (vs Rb)	Ref.
Ag(I)	+2.42	9
Cd(II)	+1.71	
Co(II)	+1.39	
Cr(II)	+0.90	
Cu(I)	+1.68	
Mn(II)	+0.90	
Na(I)	−1.640[a]	
Ni(II)	+1.59	
Rb(I)	−1.778[a]	
Zn(II)	+1.35	

[a] V SCE.

BUTANOL
0.2M LiClO$_4$ at Pt-foil

Species	$E_{1/4}V$ (vs SCE)	Ref.
I$^-$	+0.42	1

0.2M LiCl at DME at 25°C.
0.5 mM

Species	$E_{1/2}V$ (vs Ag/AgCl)	Ref.
Bi(III)	+0.040	80
Cd(II)	+0.581	
Co(II)	+1.063	
Cr(II)	+1.200	
Cr(III)	+0.471	
Fe(II)	+1.103	
Mn(II)	+1.266	
Ni(II)	+0.422	

t-BUTANOL–WATER
0.1M NaClO$_4$ + 0.05M Sodium Hippurate at DME
pH = 4; 0.9 mM

Species	% of t-Butanol	$E_{1/2}V$ (vs SCE)	Ref.
Ni(II)	10	−1.102	36
	20	−1.127	
	30	−1.130	
	40	−1.122	
	50	−1.104	
	60	−1.099	

VI. POLAROGRAPHY

t-BUTYLALCOHOL–WATER
0.1M KCl/KBr/KI at DME at 29°C.
0.9 mM

Species	$E_{1/2}V$ (vs SCE)	Ref.
Mn(II)	-1.460[a]	37
	-1.451[b]	
	-1.453[c]	
	-1.441[d]	
	-1.438[e]	
	-1.431[f]	
	-1.417[g]	

[a] 10%.
[b] 20%.
[c] 30%.
[d] 40%.
[e] 50%.
[f] 60%.
[g] 80% t-Butylalcohol.

n-BUTYL PHOSPHATE
0.1M LiClO$_4$ at DME at 25°C.
~0.5–3 mM

Species	$E_{1/2}V$ vs (Ag SCE)	Ref.
Ag(I)	$+0.57$	62
Al(III)	-1.20	
Cd(III)	-0.37	
Cr(III)	-0.61; -1.24	
Cu(II)	$+0.30$	
Fe(III)	-1.06; -1.22; -1.31	
Hg(O)	$+0.60$	
In(III)	-1.18	
Mn(II)	-1.36	
Ni(II)	-1.0	
Pb(II)	-0.31	
Tl(I)	-0.27	
Tl(III)	-0.58; -1.20	
U(VI)	-0.22; -0.50	
Zn(II)	-0.70	
Zr(IV)	-1.20	

n-BUTYL PHOSPHATE + WATER
0.5M LiClO$_4$ at DME at 25°C.
0.5–3.0 mM

Species	% Water (vol.)	$E_{1/2}V$ vs(Ag SCE)	Ref.
Pb(II)	0.3	−0.30	62
	0.9	−0.28	
	1.8	−0.26	
	2.7	−0.25	

DIETHYL SULFATE
At room temp.; 0.5 mM

Species	Supporting electrolyte	Working electrode	$E_{1/2}V$ (vs SCE)	Ref.
Cu(II)	0.1M Et$_4$NClO$_4$	DME	+0.60	70
	0.1M Bu$_4$NClO$_4$	DME	+0.63	
	0.1M Et$_4$NClO$_4$	RPE	+0.71; +0.68	
	0.1M Bu$_4$NClO$_4$	RPE	+0.74; +0.73	

DIETHYL SULFITE
0.1M Bu$_4$NClO$_4$ at RPE at room temp.
0.5 mM

Species	$E_{1/2}V$ (vs SCE)	Ref.
Cu(II)	+0.71; +0.50	70

1,2-DIMETHOXYETHANE
0.1M NaClO$_4$ at Pt-foil

Species	$E_{1/4}V$ (vs SCE)	Ref.
I$^-$	+0.39; +0.65	1

DIMETHYLACETAMIDE

0.1M Et$_4$NClO$_4$ at DME at 25 ± 0.1°C
0.1–2.0 mM

Species	$E_{1/2}V$ (vs SCE)	Ref.
Ce(III)	−2.10	2
Dy(III)	−2.24	
Er(III)	−2.27	
Eu(II)	−2.10[a]	
Eu(III)	−0.60	
Gd(III)	−2.18	
Ho(III)	−2.	
La(III)	−2.11	
Nd(III)	−2.07	
Pr(III)	−2.06	
Sm(II)	−2.08	
Sm(III)	−1.71	
Yb(II)	−2.23[a]	
Yb(III)	−1.27	

° Contain 1.5% water.

0.1M Et$_4$NClO$_4$ at DME at 25°C

Species	Conc. × 10^3 M/l	$E_{1/2}V$ (vs SCE)	Ref.
ThCl$_4$	0.1–2.0	−1.90; −2.26	5
Th(ClO$_4$)$_4$6H$_2$O	0.1–1.0	−2.07; −2.25	
Th(NO$_3$)$_4$5H$_2$O	0.15–0.5	−1.32; −1.87	

0.1M Et$_4$NClO$_4$ at DME at 25.0 ± 0.1°C

Species	$E_{1/2}V$ (vs SCE)	Ref.
CsClO$_4$	−2.03	6
KClO$_4$	−2.08	
LiClO$_4$[a]	−2.38	
NaClO$_4$	−2.06	
NH$_4$ClO$_4$	−2.08	
RbClO$_4$	−2.04	

[a] Solution contains 3% of water (by volume).

DIMETHYLACETAMIDE (Continued)
0.1M Et_4NClO_4 at DME at 25.0°C ± 0.1

Species	Conc. × 10^3 M/l	$E_{1/2}V$ (vs SCE)	Ref.
UCl_4	0.1–1.5	−0.65; −1.17; −2.05	7
$UO_2(ClO_4)_2 \cdot 2H_2O$	0.47	−0.88; −1.44	
$UO_2(ClO_4)_2$	0.04–0.6	−0.55; −1.04	
$UO_2(NO_3)_2 \cdot 6H_2O$	0.15–1.5	−0.55; −0.97	

DIMETHYLFORMAMIDE
0.1M Et_4NClO_4 at DME at 25°C
1.0 mM

Species	$E_{1/2}V$ (vs SCE)	Ref.
$Ba(ClO_4)_2$	−2.04	53
$Cd(ClO_4)_2$	−0.58	
$Co(ClO_4)_2$	−1.25	
$KClO_4$	−2.09	
$Mn(ClO_4)_2$	−1.56	
$NaClO_4$	−2.07	
$Ni(ClO_4)_2$	−0.92	
$(Ph-Ph)_2Cr(I)$	−0.70	
$RbClO_4$	−2.07	
$TlClO_4$	−0.45	
$Zn(ClO_4)_2$	−0.98	

0.2M Et_4NClO_4 at DME at 24.9 ± 0.1°C.
6.58 mM

Species	$E_{1/2}V$ (vs SCdAE)[a]	Ref.
$NbCl_5$	−0.1; −0.6; −1.1	54

[a] Saturated $CdCl_2$ (in DMF)/Cd—amalgam electrode.

VI. POLAROGRAPHY

DIMETHYLFORMAMIDE (Continued)
0.1M Bu₄NI at DME at 25 ± 0.2°C, 2 mM

Species	$E_{1/2}V$ (vs Hg pool)	Ref.
Ba(ClO₄)₂	−1.49	82
CaCl₂	−1.78	
Ca(ClO₄)₂·6H₂O	−1.84	
Co(ClO₄)₂·6H₂O	−1.32	
CsCl	−1.53	
Cu(ClO₄)₂·6H₂O[c]	−0.15	
KI	−1.55	
LiCl[a]	−1.81	
NaI[b]	−1.53	
Ni(ClO₄)₂·6H₂O	−1.06	
RbI	−1.52	
Sr(ClO₄)₂	−1.68	
Zn(ClO₄)₂·6H₂O	−1.12	

[a] 1 mM solution.
[b] 4 mM solution.
[c] 0.2M NaClO₄ supporting electrolyte.

0.2M LiCl at DME at 25°C.
0.5 mM

Species	$E_{1/2}V$ (vs Ag/AgCl)	Ref.
Cr(III)	+0.290	80
Cr(II)	+1.15	

0.1M Et₄NClO₄ at DME at 25°C.
1 mM

Species	$E_{1/2}V$ (vs DME)	Ref.
Cd(ClO₄)₂	−0.37	45
Co(ClO₄)₂	−0.96	
Cu(ClO₄)₂	> +0.3	
Et₄NClO₄	−2.56	
Fe(ClO₄)₃	> +0.3; −1.06	
KClO₄	−1.84	
LiClO₄	−1.88	
Me₄NClO₄	−2.45	
Mn(ClO₄)₂	−1.24	
NaClO₄	−1.80	
Ni(ClO₄)₂	−0.69	
Sr(ClO₄)₂	−1.88	
Zn(ClO₄)₂	−0.97	

DIMETHYLFORMAMIDE (Continued)

0.1M Et$_4$NClO$_4$ at DME at 25°C.
$\sim$0.37 to 2 mM

Species	$E_{1/2}V$ (vs SCE)	Ref.
Diethyl-4-nitrophenyl phosphate	-0.971	28b
O,O-diethyl O-4-nitrophenyl thiophosphate	-1.116	
Diethyl-4-nitrobenzyl phosphate	-1.140	
Diphenyl-4-nitrophenyl phosphate	-1.064	

DME at 25 ± 0.1°C.

Species	Conc. × 10^3 M/l	Supporting electrolyte	$E_{1/2}V$ (vs SCE)	Ref.
Et$_4$NHgCl$_3$	0.1–2.5	0.1M Et$_4$NClO$_4$ +0.005% PVC	+0.110 to +0.173 −0.153 to −0.242	49
Et$_4$NHgI$_3$	0.1–2.50	0.1M Et$_4$NClO$_4$	+0.107 to +0.169 −0.226 to −0.282	48
Et$_4$NCl	0.1–5.00	0.1M Et$_4$NClO$_4$ +0.005% PVC	+0.122 to +0.175 −0.136 to −0.233	49
Et$_4$NI	0.1–5.00	0.1M Et$_4$NClO$_4$	+0.130 to +0.175 −0.183 to −0.278	48
HgCl$_2$	0.1–2.5	0.1M Et$_4$NClO$_4$ +0.005% PVC	+0.112 to +0.151 −0.164 to −0.234	49
HgI$_2$	0.1–2.50	0.1M Et$_4$NClO$_4$	+0.121 to +0.144 −0.210 to −0.279	48

0.1M Et$_4$NClO$_4$ at DME at 25°C.

Species	Conc. × 10^3 M/l	$E_{1/2}V$ (vs SCE)	Ref.
Bromine	0.1–2.52	+0.147 to +0.222 −0.161 to −0.258 −1.332 to −1.337	50
HBr	0.26–5.25	+0.156 to +0.233 −0.155 to −0.217 −1.310 to −1.390	
HBr$_3$	0.10–2.55	+0.202 to +0.217 −0.145 to −0.244 −1.374 to −1.346	
HHgBr$_3$	0.25–2.50	+0.166 to +0.204 −0.176 to −0.240 −1.365	

VI. POLAROGRAPHY

DIMETHYLFORMAMIDE (Continued)
0.1M Et$_4$NClO$_4$ at DME at 21.0 ± 0.1°C.
0.1–1.0 mM

Species	$E_{1/2} V$ (vs SCE)	Ref.
Ce(III)	−2.07	51
Dy(III)	−2.18	
Er(III)	−2.21	
Eu(II)	−2.05	
Eu(III)	−0.70	
Ho(III)	−2.20	
La(III)	−2.07	
Nd(III)	−2.07	
Pr(III)	−2.07	
Sm(II)	−2.04	
Sm(III)	−1.87	
Tb(III)	−2.16	
Yb(II)	−2.14	
Yb(III)	−1.39	
Yd(III)	−2.16	

0.1M Bu$_4$NClO$_4$ at DME

Species	$E_{1/2} V$ (vs SCE)	Ref.
Oxygen (sat.)	−0.80; −2.8	47

0.1M Et$_4$NClO at DME
1.006 to 4.023 mM

Species	$E_{1/2} V$ (vs Cd amalgam/CdCl$_2$)	Ref.
Vanadyl sulfate	−0.439 to −0.325; −1.310 to 1.309	46

0.1M NaClO$_4$ at DME at 30°C.
1.0 mM

Species	$E_{1/2} V$ (vs SCE)	Ref.
Co(ClO$_4$)$_2$	−1.066	26

DIMETHYLFORMAMIDE (*Continued*)
0.1M Et$_4$NClO$_4$ at DME at 25°C.

Species	Conc. $\times 10^3$ M/l	$E_{1/2}V$ (vs SCE)	Ref.
Mercuric bromide	0.1–1.0	+0.142 to +0.225 −0.073 to −0.146	52
Tetraethyl ammonium bromide	0.1–5.0	+0.199 to +0.264 −0.092 to −0.184	
Tetraethyl ammonium tri-bromomercurate	0.1–2.5	+0.191 to +0.247 −0.122 to −0.196	

DIMETHYLFORMAMIDE–WATER
0.1M NaClO$_4$ at DME at 30°C.
1.0 mM

Species	V/V DMF	$E_{1/2}V$ (vs SCE)	Ref.
Co(ClO$_4$)$_2$	20	−1.163	26
	40	−1.163	
	60	−1.162	
	80	−1.150	

DIMETHYL SULFOXIDE
0.1M NaClO$_4$ at Pt-foil

Species	$E_{1/4}V$ (vs SCE)	Ref.
I$^-$	+0.48; +0.70	1

0.1M Bu$_4$NClO$_4$ at DME

Species	$E_{1/2}V$ (vs SCE)	Ref.
Oxygen (sat.)	−0.73; −2.40	47

0.1M Et$_4$NClO$_4$ at DME

Species	$E_{1/2}V$ [vs KCl(std) DMSO/AgCl—Ag]	Ref.
Oxygen (0.09 to 1.65 mM)	−1.15; −2.50	73

VI. POLAROGRAPHY

DIMETHYL SULFOXIDE (Continued)
0.1M Et$_4$NClO$_4$ at DME at 21.0 ± 0.1°C.
0.1–1.0 mM

Species	$E_{1/2}V$ (vs SCE)	Ref.
Ce(III)	−2.24	51
Dy(III)	−2.08	
Er(III)	−2.09	
Eu(II)	−2.12	
Eu(III)	−0.81	
Ho(III)	−2.09	
La(III)	−2.26	
Nd(III)	−2.20	
Pr(III)	−2.20	
Sm(II)	−2.12	
Sm(III)	−2.02	
Tb(III)	−2.19	
Yb(II)	−2.21	
Yb(III)	−1.48	
Yd(III)	−2.16	

0.1M Et$_4$NClO$_4$ at gold electrode

Species	$E_{1/4}V$ (vs SCE)	Ref.
Superoxide ion[a] (O$_2^-$)	−2.05; −0.75; +0.70	74
Hydroxide ion[b] (OH$^-$)	+0.75[c]	

[a] 1 mM conc.
[b] 4.30 mM conc.
[c] Gold or Pt electrode.

0.5 mM at room temp.

Species	Supporting electrolyte	Working electrode	$E_{1/2}V$ (vs SCE)	Ref.
Cu(II)	0.1M Et$_4$NClO$_4$	DME	−0.07	70
	0.1M Bu$_4$NClO$_4$	DME	−0.07	
	0.1M Et$_4$NClO$_4$	RPE[a]	+0.04; −0.09	
	0.1M Bu$_4$NClO$_4$	RPE	+0.04; −0.08	

[a] Rotating platinum electrode.

DIMETHYL SULFOXIDE (*Continued*)

Species	Conc. × 10³ M/l	$E_{1/2}V$ (vs SCE)	Ref.
Cobalt(II)		-1.4^b	75
		-1.39^a	
Hydrogen chloride		-1.08^c	
		-0.67 ± 0.02^d	
Nickel(II)	0.5	-1.0^b	
	0.1	-1.08^a	
Oxygen		$-0.72; -1.13^a$	
Oxygen (sat.)		$-0.65; -1.20^b$	
Perchloric acid	0.89	-0.67 ± 0.02^d	
Sulfuric acid		$-1.06^c; -0.67 \pm 0.02$	

[a] 0.1M NaClO₄ at RE(Pt) at 30°C.
[b] 0.1M NaClO₄ at DME at 25°C.
[c] 0.1M Et₄NClO₄ at 30°C at DME.
[d] 0.1M NaClO₄ at RE(Pt) at 30°C.

0.1F Et₄NClO₄ at Hg electrode at 25.0 ± 0.1°C.
0.831–23.9 mM

Species	$E_{t=0}V$ (vs NHE)	Ref.
CO_2	-1.95 to $=1.86$	76
$CO_2{}^a$	-1.60 to -1.58	

[a] Gold electrode.

1M KClO₄ at RE(Pt) at 25°C.
99.9 mM

Species	$E_{1/2}V$ (vs Ag/Ag⁺)	Ref.
NO_2^-	$\sim +0.660$	77

VI. POLAROGRAPHY

DIMETHYL SULFOXIDE (Continued)
0.1F LiClO$_4$ at Pt-electrode at 25°C.
~3 mM

Species	$E_{1/2}V$ (vs SCE)	Ref.
Hydrogen bromide	+0.90 to +1.05 −0.61	78
Hydrogen chloride (anhy)	+0.90 to +1.05 −0.61	
Hydrogen iodide	+0.65; +0.35 −0.6	

0.1M Et$_4$NClO$_4$ at DME at 25.0°C.
~1.0 mM

Species	$E_{1/2}V$ (vs SCE)	Ref.
Ba(ClO$_4$)$_2$	−2.08	53
Cd(ClO$_4$)$_2$	−0.70	
Co(ClO$_4$)$_2$	−1.43	
KClO$_4$	−2.12	
Mn(ClO$_4$)$_2$	−1.72	
NaClO$_4$	−2.09	
Ni(ClO$_4$)$_2$	−1.07	
(Ph–Ph)$_2$Cr(I)	−0.72	
RbClO$_4$	−2.09	
TlClO$_4$	−0.54	
Zn(ClO$_4$)$_2$	−1.09	

0.1M Et$_4$NClO$_4$ at DME at 21°C.

Species	$E_{1/2}V$ (vs SCE)	Ref.
BaCl$_2$	−2.09	79
Ca(NO$_3$)$_2$	−2.30	
CsCl	−2.03	
HfCl$_4$	−1.07; −1.17	
KClO$_4$	−2.11	
LiCl	−2.45	
Mg(ClO$_4$)$_2$	−2.28	
NaClO$_4$	−2.07	
NbCl$_5$	−0.92	
NH$_4$NO$_3$	−2.13	
RbCl	−2.06	
SbI$_3$	−0.36; −0.52; −1.13	
SiCl$_4$	−0.52; −1.09	
Sr(NO$_3$)$_2$	−2.10	
TiCl$_2$(C$_2$H$_5$)$_2$	−1.21	
ZrCl$_4$	−1.04	

DIOXAN–WATER
0.1N NaOH at DME
H_6TeO_6 0.5 mM

Species	% of Dioxan (vol.)	$E_{1/2}$	Ref.
Tellurate	10	−1.57	39
	20	−1.535	
	30	−1.515	
	40	−1.495	

0.1N NaOH at DME
K_2CrO_4 1 mM

Species	% of Dioxan (vol.)	$E_{1/2}$	Ref.
Chromate	10	−1.2125	39
	20	−1.23	
	30	−1.215	
	40	−1.22	
	50	−1.187	
	60	−1.155	

ETHANOL
0.1M LiCl at DME

Species	Conc. ($\times 10^3$ M/l)	$E_{1/2}V$ (vs MPE)[a]	Ref.
$Sc(ClO_4)_3$	1.5	−1.681	20
Y(III)	3.0	−1.496; −1.825	20

[a] Mercury pool electrode.

VI. POLAROGRAPHY

ETHANOL *(Continued)*
0.2M LiCl at DME at 25°C.
0.5 mM

Species	$E_{1/2}V$ (vs Ag/AgCl)	Ref.
Bi(III)	+0.056	80
Cd(II)	+0.590	
Co(II)	+1.082	
Cr(II)	+1.238	
Cr(III)	+0.508	
Cu(II)	+0.280	
Fe(II)	+1.126	
Mn(II)	+1.294	
Ni(II)	+0.478	
Zn(II)	+1.209	

LiClO$_4$ at DME at 25 ± 0.1°C.
~1 mM

Species	$E_{1/2}V$ (vs SCE)	Ref.
Pr(ClO$_4$)$_3$	−1.55 to −1.61; <−1.70; ~−1.45	38
Pr(ClO$_4$)$_3$	−1.60 to −1.73[a]; −1.78 to −1.83[a]; −1.86 to −1.91[a]	
Pr(ClO$_4$)$_3$	−1.55[b]; −1.65[b]; −1.75[b]; −1.80[b]	

[a] Contain varying amounts of LiCl.
[b] Contain LiCl 0.1 to 0.2F.

ETHANOL–WATER

0.1M NaClO$_4$ + 0.05M Sodium Hippurate at DME

0.9 mM; pH = 4

Species	% Ethanol	$E_{1/2}V$ (vs SCE)	Ref.
Ni(II)	10	−1.040	36
	20	−1.047	
	30	−1.054	
	40	−1.065	
	60	−1.049	

0.1N NaOH at DME

HbTeO$_b$ 0.5 mM

Species	% Ethanol (vol)	$E_{1/2}V$	Ref.
Tellurate	10	−1.55	39
	20	−1.5	
	30	−1.465	
	40	−1.435	
	50	−1.4	

0.1N NaOH at DME

K$_2$CrO$_4$ 1 mM

Species	% Ethanol (vol)	$E_{1/2}$	Ref.
Chromate	10	−1.12	39
	20	−1.15	
	30	−1.163	
	40	−1.17	
	50	−1.15	
	60	−1.115	
	70	−1.07	

VI. POLAROGRAPHY

ETHANOL–WATER (Continued)
DME; 1.5 mM

Species	% of Ethanol	Supporting electrolyte	pH	$E_{1/2}V$ (vs MPE)	Ref.
$ScCl_3$	25	0.1M LiCl	3.11	−1.802	20
	50	0.1M LiCl	3.06	−1.818	
	25	0.1M Me_4NCl	2.86	−1.852	
	50	0.1M Me_4NCl	3.27	−1.734	

0.1M LiCl at DME
3.0 mM

Species	% Ethanol	pH	$E_{1/2}V$ (vs MPE)	Ref.
Y(III)	25	2.78	−1.856; −1.932; −2.093	20
	50	5.11	−1.773; −2.018	
	75	2.81	−1.884; −1.990	
	75	4.76	−1.757	

0.1M KCl/KBr/KI at DME at 29°C.
0.9 mM

Species	$E_{1/2}V$ (vs SCE)	Ref.
Mn(II)	−1.462[a]	37
	−1.457[b]	
	−1.445[c]	
	−1.455[d]	
	−1.442[e]	
	−1.441[f]	
	−1.420[g]	

[a] 10%.
[b] 20%.
[c] 30%.
[d] 40%.
[e] 50%.
[f] 60%.
[g] 80%.

ETHYLENEDIAMINE
Et_4NNO_3 at DME
0.5 to 8 mM

Species	$E_{1/2}V$ (vs NCE)	Ref.
LiCl	−2.41	40
$NaClO_4$	−1.97	
$NaNO_3$	−2.15	
$KClO_4$	−2.00	
RbCl	−2.11	
$RbNO_3$	−1.86	
CsCl	−1.82	
$CsNO_3$	−2.15	
NH_4Cl	−2.05	
O_2	∼−0.20; −1.58	
$CdCl_2$	−0.46; −0.64[a]; −0.74[b]	
$Cd(NO_3)_2$	−0.60; −0.50[a]; −0.60[b]	
$Cd(ClO_4)_2$	−0.54; −0.70[a]; −0.80[b]	

[a] $NaNO_3$ supporting electrolyte.
[b] LiCl supporting electrolyte.

At DME
0.1 mM

Species	$E_{1/2}V$ (vs NCE)	Ref.
$CdCl_2$	−0.64[a]; −0.74[b]; −0.46[c]	41
$Cd(ClO_4)_2$	−0.70[a]; −0.80[b]; −0.50[c]	
$Cd(NO_3)_2$	−0.50[a]; −0.60[b]; −0.60[c]	

[a] $NaNO_3$.
[b] LiCl.
[c] Et_4NNO_3 as supporting electrolyte.

VI. POLAROGRAPHY

ETHYLENEGLYCOL
0.2M LiCl at DME at 25°C.
0.5 mM

Species	$E_{1/2}V$ (vs Ag/AgCl)	Ref.
Bi(III)	+0.088	80
Cd(II)	+0.600	
Co(II)	+1.148	
Cr(II)	+1.317	
Cu(II)	+0.218	
Fe(II)	+1.240	
Mn(II)	+1.396	
Ni(II)	+0.926	
Zn(II)	+1.028	

0.1M NaClO$_4$ at Pt-foil

Species	$E_{1/4}V$ (vs SCE)	Ref.
I⁻	+0.42; +0.52	1

ETHYLENEGLYCOL-1,2-DIMETHOXYETHANE (1:1)
0.1M NaClO$_4$ at Pt-foil

Species	$E_{1/4}V$ (vs SCE)	Ref.
I⁻	+0.37; +0.56	1

ETHYLENEGLYCOL–WATER
0.1M KCl/KBr/KI at DME at 29°C.
0.9 mM

Species	$E_{1/2}V$ (vs SCE)	Ref.
Mn(II)	-1.460[a];	37
	-1.450[b]	
	-1.448[c]	
	-1.450[d]	
	-1.432[e]	
	-1.420[f]	

[a] 10%.
[b] 30%.
[c] 40%.
[d] 50%.
[e] 60%.
[f] 80% of ethylene glycol.

ETHYLENE GLYCOL–WATER (Ethylene Glycol 10 to 50%)
0.25M K$_2$SO$_4$ at DME
1 mM

Species	$E_{1/2}V$ (vs SCE)	Ref.
Tl$_2$SO$_4$	-0.463 to -0.450	42
	-0.465 to -0.450[a]	

[a] 0.25M Potassium oxalate supporting electrolyte.

ETHYLENE SULFITE
0.1M Et$_4$NClO$_4$ at DME at 25°C.
~1.0 mM

Species	$E_{1/2}V$ (vs SCE)	Ref.
Cd(ClO$_4$)$_2$	-0.21	71
Co(ClO$_4$)$_2$	-0.67	
Mn(ClO$_4$)$_2$	-1.19	
Ni(ClO$_4$)$_2$	-0.60	
(Ph–Ph)$_2$Cr(I)	-0.73	
TlClO$_4$	-0.26	
Zn(ClO$_4$)$_2$	-0.63	

VI. POLAROGRAPHY

FORMAMIDE
0.2M NaClO$_4$ at DME 25 ± 0.5°C.
2 mM

Species	$E_{1/2}V$ (vs SCE)	Ref.
Cr(ClO$_4$)$_3$6H$_2$O	−1.56	43
Mn(ClO$_4$)$_2$6H$_2$O	−1.57	
Fe(ClO$_4$)$_3$6H$_2$O	−1.40	
Co(ClO$_4$)$_2$6H$_2$O	−1.20	
Ni(ClO$_4$)$_2$6H$_2$O	−0.98	
Cu(ClO$_4$)$_2$6H$_2$O	−0.12	
Zn(ClO$_4$)$_2$6H$_2$O	−1.07	
3CdSO$_4$8H$_2$O	−0.69	

0.2M LiCl at DME at 25°C.
0.5 mM

Species	$E_{1/2}V$ (vs Ag/AgCl)	Ref.
Bi(III)	+0.152	80
Cd(II)	+0.672	
Cr(III)	+0.830	
Zn(II)	+1.044	

0.1M NaClO$_4$ at DME at 30°C.
1.0 mM

Species	$E_{1/2}V$ (vs SCE)	Ref.
Co(ClO$_4$)$_2$	−1.110	26

FORMAMIDE–WATER (*Continued*)
1M NaClO$_4$ at 25°C.
0.5 mM

Species	% of Formamide by volume	$E_{1/2}V$ (vs SCE)	Ref.
Co(II)	7.5	−1.296	44
	15.0	−1.280	
	30.0	−1.270	
	45.0	−1.245	
	60.0	−1.227	
Mn(II)	7.5	−1.484	
	15.0	−1.492	
	30.0	−1.508	
	45.0	−1.519	
	60.0	−1.544	
Ni(II)	7.5	−1.037	
	15.0	−1.022	
	30.0	−0.980	
	45.0	−0.975	
	60.0	−0.964	
Zn(II)	7.4	−1.016	
	15.0	−1.021	
	30.0	−1.024	
	45.0	−1.047	
	60.0	−1.077	

0.1M NaClO$_4$ at DME at 30°C.
1.0 mM

Species	V/V of Formamide	$E_{1/2}V$ (vs SCE)	Ref.
Co(ClO$_4$)$_2$	20	−1.134	26
	40	−1.143	
	60	−1.145	
	80	−1.138	

FORMIC ACID

0.5M Sodium formate at DME at 25.00 ± 0.01°C.

1 mM

Species	$E_{1/2}V$ (vs SCFAE)[a]	Ref.
Bi^{+++}	+0.14	55
Cd^{++}	−0.34	
In^{+++}	−0.49	
Ni^{++}	−0.61	
Pb^{++}	−0.22	
Sb^{+++}	+0.00	
Sn^{++}	−0.28	
Tl^+	−0.29	
Zn^{++}	−0.71	

[a] Saturated calomel in formic acid electrode.

GLYCERINE

0.0005M Et_4NClO_4 + 0.0053M Phenol at DME at 20 ± 0.1°C.

5.3 mM to 98.7 mM

Species	$E_{1/2}V$ (vs SCE)	Ref.
$NaH_2PO_2 \cdot H_2O$	−0.64	56
$NaH_2PO_3 \cdot 2.5H_2O$	−0.605	
$Na_2HPO_3 \cdot 5H_2O$	−0.42	
$Na_2HPO_4 \cdot 2H_2O$	−0.225	
$Na_3PO_4 \cdot 6H_2O$	−0.31	
$Na_2H_2P_2O_5$	−0.605	
$Na_3HP_2O_5 \cdot 6H_2O$	−0.34	
$Na_2H_2P_2O_6 \cdot 6H_2O$	−0.37	
$Na_3HP_2O_6 \cdot 8H_2O$	−0.37	
$Na_2H_2P_2O_7 \cdot 6H_2O$	−0.33	

ISOBUTYRONITRILE
0.001M Bu$_4$NI at DME
1 mM

Species	$E_{1/2}V$ (vs SCE)	Ref.
Rb(I)	−1.791	9

0.05M Et$_4$NClO$_4$ at DME
0.2 to 2.1 mM

Species	$E_{1/2}V$ (vs SCE)	Ref.
CsI	−1.80[a]	57
Et$_4$NBr	−0.33; +0.30	
KI	−1.77[a]	
LiClO$_4$	−1.90	
Me$_4$NCl	−0.31; +0.20	
NaClO$_4$	−1.77[a]; −1.70[a]	
NaI	−0.42[b]; +0.17[b]	
NaSCN	−0.01 to −0.02; +0.30	
RbI	−1.79[a]	

[a] 0.05M Bu$_4$NI as supporting electrolyte.
[b] 0.05M NaClO$_4$ as supporting electrolyte.

METHANOL
0.2M LiCl at DME at 25°C.
0.5 mM

Species	$E_{1/2}V$ (vs Ag/AgCl)	Ref.
Bi(III)	+0.076	80
Cd(II)	+0.595	
Co(II)	+1.136	
Cr(II)	+1.301	
Cr(III)	+0.531	
Cu(II)	+0.246	
Fe(II)	+1.214	
Mn(II)	+1.357	
Ni(II)	+0.917	
Zn(II)	+1.140	

VI. POLAROGRAPHY

METHANOL (Continued)
0.1M LiCl at DME at 25.0 ± 0.1
0.5–5 mM

Species	$E_{1/2}V$ (vs SCE)	Ref.
$ZrOCl_2 \cdot 8H_2O$	−1.41	61
$Zr(SO_4)_2 \cdot 4H_2O$	−1.25; −1.55	

0.1M NaClO$_4$ at Pt-foil

Species	$E_{1/4}V$ (vs SCE)	Ref.
I$^-$	+0.40	1

METHANOL–WATER
0.1M KCl/KBr/KI at DME at 29°C.
0.9 mM

Species	$E_{1/2}V$ (vs SCE)	Ref.
Mn(II)	−1.468[a]	37
	−1.467[b]	
	−1.458[c]	
	−1.459[d]	
	−1.460[e]	

[a] 10%.
[b] 20%.
[c] 30%.
[d] 40%.
[e] 50% methanol.

METHANOL–WATER (*Continued*)

0.1N NaOH at DME
H_6TeO_6 0.5 mM

Species	% Methanol (vol.)	$E_{1/2}V$	Ref.
Tellurate	10	−1.56	39
	20	−1.525	
	30	−1.48	
	40	−1.47	
	50	−1.462	

0.1N NaOH at DME
K_2CrO_4 1 mM

Species	% Methanol (vol.)	$E_{1/2}V$	Ref.
Chromate	10	−1.07	39
	20	−1.085	
	30	−1.095	
	40	−1.1	
	50	−1.09	
	60	−1.085	
	70	−1.075	

0.1M $NaClO_4$ + 0.05M Sodium hippurate at DME
0.9 mM; pH = 4

Species	% Methanol	$E_{1/2}V$ (vs SCE)	Ref.
Ni(II)	10	−1.036	36
	20	−1.030	
	30	−1.022	
	40	−1.020	
	50	−1.028	
	60	−1.030	

VI. POLAROGRAPHY

N-METHYLACETAMIDE

0.12M Et$_4$NBr at DME at 35 ± 0.2°C
0.689 to 1.72 mM

Species	$E_{1/2}V$ (vs Hg pool)	Ref.
Cadmium iodide	−0.438 ± 0.003	3
Lead iodide	−0.162 ± 0.003	
Oxygen	−0.226 to 0.003	

0.1M Et$_4$NClO$_4$ at DME at 30 ± 0.1°C

Species	$E_{1/2}V$ (vs SCE)	Ref.
Copper perchlorate[a]	+0.069	4
Thallium nitrate[b]	−0.421	
Oxygen	−0.52; −1.6	

[a] 2.0 mM.
[b] 0.4 mM.

METHYL METHANE SULFONATE

0.5 mM at room temp.

Species	Supporting electrolyte	Working electrode	$E_{1/2}V$ (vs SCE)	Ref.
Cu(II)	0.1M Et$_4$NClO$_4$	DME	+0.63	70
	0.1M Bu$_4$NClO$_4$	DME	+0.64	
	0.1M Et$_4$NClO$_4$	RPE	+0.71; +0.65	
	0.1M Bu$_4$NClO$_4$	Rpe	+0.69; +0.66	

NITROMETHANE

0.1M Et$_4$NClO$_4$ at DME at 25°C.
M1.0 mM

Species	$E_{1/2}V$ (vs SCE)	Ref.
Cd(ClO$_4$)$_2$	+0.03	53
Co(ClO$_4$)$_2$	−0.26	
Mn(ClO$_4$)$_2$	−0.33	
Ni(ClO$_4$)$_2$	−0.19	
(Ph–Ph)$_2$Cr(I)	−0.76	
TlClO$_4$	−0.21	
Zn(ClO$_4$)$_2$	−0.26	

0.1M Et$_4$NClO$_4$ at DME at 25°C.
∼1.0 mM

Species	$E_{1/2}V$ (vs SCE)	Ref.
Cd(ClO$_4$)$_2$	+0.03	58
Co(ClO$_4$)$_2$	−0.26	
Mn(ClO$_4$)$_2$	−0.33	
Ni(ClO$_4$)$_2$	−0.19	
(Ph–Ph)$_2$Cr(I)	−0.76	
TlClO$_4$	−0.21	
Zn(ClO$_4$)$_2$	−0.26	

[a] $E_{1/2}$ values vs aqueous SCE are referred to the *bis*-biphenyl chromium(I) iodide-scale.

0.1M Et$_4$NClO$_4$ at R.E.(Pt) at 20 ± 2°C.
∼0.5 mM to 2 mM

Species	$E_{1/2}V$ (vs Ag/AgCl)	Ref.
Cl$^-$	+1.38	59
Cl$_2$	+1.1; +1.38[a]	
HCl	+0.13; +1.8	

[a] 0.01M HClO$_4$ as supporting electrolyte.

VI. POLAROGRAPHY

NITROMETHANE (Continued)
0.1M Bu$_4$NClO$_4$ at DME

Species	$E_{1/2}V$ (vs Ag/AgCl)	Ref.
HClO$_4$	+0.27; +0.6[a]; +0.6[b]	60
H$_2$SL$_4$	+0.1; +0.3[b]	

[a] At Pt electrode.
[b] Pt electrode with LiClO$_4$.

PENTANOL
0.2M LiCl at DME at 25°C.
0.5 mM

Species	$E_{1/2}V$ (vs Ag/AgCl)	Ref.
Co(II)	+1.030	80
Cr(II)	+1.223	
Fe(II)	+1.120	
Mn(II)	+1.265	

PHENYLACETONITRILE
0.1M Et$_4$NClO$_4$ at DME at 25 ± 0.2°C.
1 mM

Species	$E_{1/2}V$ (vs SCE)	Ref.
AgClO$_4$	+0.52	85
Ba(ClO$_4$)$_2$	−1.58	
Ca(ClO$_4$)$_2$	−1.79	
Cd(ClO$_4$)$_2$	−0.18	
Co(ClO$_4$)$_2$	−0.52	
Ct(ClO$_4$)$_3$	−0.98	
Cu(ClO$_4$)$_2$	−0.23	
LiClO$_4$	−1.83	
Mg(ClO$_4$)$_2$	−1.63	
Mn(ClO$_4$)$_2$	−1.02	
NaClO$_4$	−1.73	
Sr(ClO$_4$)$_2$	−1.73	
Zn(ClO$_4$)$_2$	−0.51	

PHENYLACETONITRILE (Continued)
0.001M Bu$_4$NI at DME
1 mM

Species	$E_{1/2}V$ (vs SCE)	Ref.
Rb(I)	−1.75	9

PROPANEDIOL-1,2-CARBONATE
0.1M Et$_4$NClO$_4$ at DME at 25.0 ± 0.1°C.

Species	Conc. × 10^3 M/l	$E_{1/2}V$ (vs SCE)	Ref.
UCl$_4$	0.2–4.0	−1.05; −1.37; ∼−1.92	7
UO$_2$–NO$_3$)$_2$	0.74	−0.41; −0.97	
UO$_2$(NO$_3$)$_2$·6H$_2$O	0.14–2	−0.44; −1.30	
UO$_2$(ClO$_4$)$_2$·2H$_2$O	0.48	−0.37; −0.55	

0.1M Et$_4$NClO$_4$ at DME at 25°C.

Species	Conc. × 10^3 M/l	$E_{1/2}V$ (vs SCE)	Ref.
ThCl$_4$	0.36	−1.64; −1.93	5
Th(ClO$_4$)$_4$·6H$_2$O	0.45	−1.13; −1.44	
Th(NO$_3$)$_4$·5H$_2$O	0.1–3.0	−1.28; −1.64	

VI. POLAROGRAPHY

PROPANEDIOL-1,2-CARBONATE (Continued)
0.1M Et$_4$NClO$_4$ at DME at 25°C.
~1.0 mM

Species	$E_{1/2}V$ (vs SCE)	Ref.
Ba(ClO$_4$)$_2$	−1.67	63
Cd(ClO$_4$)$_2$	−0.11	
Co(ClO$_4$)$_2$	−0.73	
KClO$_4$	−1.84	
Mn(ClO$_4$)$_2$	−1.08	
NaClO$_4$	−1.86	
Ni(ClO$_4$)$_2$	−0.48	
(Ph–Ph)$_2$Cr(I)	−0.76	
RbClO$_4$	−1.97	
TlClO$_4$	−0.24	
Zn(ClO$_4$)$_2$	−0.54	

0.1M Et$_4$NClO$_4$ at DME at 25 ± 0.1°C.
~1 mM

Species	$E_{1/2}V$ (vs SCE)	Ref.
Ce(III)	−1.66	64
Dy(III)	−1.63	
Er(III)	−1.63	
Eu(II)	−1.72	
Eu(III)	+0.05	
Ho(III)	−1.64	
La(III)	−1.68	
Nd(III)	−1.68	
Pr(III)	−1.66	
Sm(II)	−1.70	
Sm(III)	−1.13	
Tn(III)	−1.62	
Yb(II)	−1.76	
Yb(III)	−0.68	
Yd(III)	−1.65	

PROPANOL
0.2M LiCl at DME at 25°C.
0.5 mM

Species	$E_{1/2}V$ (vs Ag/AgCl)	Ref.
Bi(III)	+0.048	80
Cd(II)	+0.585	
Co(II)	+1.066	
Cr(III)	+0.487	
Cu(II)	+0.291	
Fe(II)	+1.112	
Mn(II)	+1.272	
Ni(II)	+0.439	

2-PROPANOL
0.2M LiClO$_4$ at Pt-foil

Species	$E_{1/4}V$ (vs SCE)	Ref.
I$^-$	+0.42	1

ISOPROPYLALCOHOL
0.2M LiCl at DME at 25°C.
0.5 mM

Species	$E_{1/2}V$ (vs Ag/AgCl)	Ref.
Cr(II)	+1.218	80
Cr(III)	+0.396	
Cu(II)	+0.284	
Fe(II)	+1.128	
Mn(II)	+1.298	

VI. POLAROGRAPHY

PROPANOL–WATER
0.1M KCl/KBr/KI at DME at 29°C.
0.9 mM

Species	$E_{1/2}V$ (vs SCE)	Ref.
Mn(II)	−1.459[a]	37
	−1.458[b]	
	−1.451[c]	
	−1.452[d]	
	−1.449[e]	
	−1.438[f]	
	−1.421[g]	

[a] 10%.
[b] 20%.
[c] 30%.
[d] 40%.
[e] 50%.
[f] 60%.
[g] 80% Propanol.

0.1M NaClO$_4$ + 0.05M Sodium Hippurate at DME
0.9 mM; pH = 4

Species	% Propanol	$E_{1/2}V$ (vs SCE)	Ref.
Ni(II)	10	−1.082	36
	20	−1.104	
	30	−1.109	
	40	−1.112	
	50	−1.104	
	60	−1.088	

0.1N NaOH at DME
H$_6$TeO$_6$ 0.5 mM

Species	% Propanol	$E_{1/2}$	Ref.
Tellurate	10	−1.55	39
	20	−1.497	
	30	−1.458	
	40	−1.438	
	50	−1.397	

PROPANOL–WATER (Continued)
0.1N NaOH at DME
K_2CrO_4 1 mM

Species	% Propanol	$E_{1/2}$	Ref.
Chromate	10	−1.213	39
	20	−1.257	
	30	−1.235	
	40	−1.215	
	50	−1.17	
	60	−1.115	
	70	−1.065	

PROPIONITRILE
0.1M Et_4NClO_4 at DME at 25 ± 0.2°C.
1 mM

Species	$E_{1/2}V$ (vs SCE)	Ref.
$AgClO_4$	+0.45	85
$Ba(ClO_4)_2$	−1.63	
$Ca(ClO_4)_2$	−1.83	
$Cd(ClO_4)_2$	−0.25	
$Co(ClO_4)_2$	−0.64	
$Cr(ClO_4)_3$	−1.07	
$Cu(ClO_4)_2$	−0.31	
$Fe(ClO_4)_3$	−0.94	
$LiClO_4$	−1.93	
$Mg(ClO_4)_2$	−1.72	
$Mn(ClO_4)_2$	−1.10	
$NaClO_4$	−1.82	
$Ni(ClO_4)_2$	−0.36	
$Sr(ClO_4)_2$	−1.78	
$Zn(ClO_4)_2$	−0.59	

0.001M Bu_4NI at DME
1 mM

Species	$E_{1/2}V$ (vs SCE)	Ref.
Rb(I)	−1.876	9

VI. POLAROGRAPHY

PROPYLENE CARBONATE
0.02M LiClO$_4$ at Pt wire

Species	Conc. $\times 10^3$M/l	E_pV (vs SCE)	Ref.
Cs(I)	4–20	-1.10 ± 0.02 to -1.09 ± 0.02	65
K(I)	4–20	-2.06 ± 0.05 to -2.06 ± 0.01	
Na(I)	4–20	-1.53 ± 0.01 to -1.49 ± 0.03	
Rb(I)	4–10	-1.15 ± 0.03 to -1.09 ± 0.05	

1M LiClO$_4$ at Pt-wire at 28 ± 0.2°C.
1 to 30 mM

Species	E_pV (vs Li/Li$^+$)	Ref.
Water	+1.4	66

PYRIDINE
0.1M NaClO$_4$ at Pt-foil

Species	$E_{1/4}V$ (vs SCE)	Ref.
I$^-$	+0.37; +0.89	1

0.1M LiClO$_4$ at DME at 25 ± 0.2°C.
0.093 to 9.25 mM

Species	$E_{1/2}V$ (vs N AgE)	Ref.
H$_2$SO$_4$	-1.11 ± 0.08; -1.35 ± 0.03	67

PYRIDINE (Continued)
0.1 to 0.5M LiNO$_3$ at DME at 25°C.
0.15 to 5 mM

Species	$E_{1/2}V$ (vs SCE)	Ref.
Al(III)	-1.25 to -1.35[c]; -1.43 to -1.49[c]	68
Pb(II)	-0.59 to -0.97; -0.54[a]	
Tl(I)	-0.48 to -0.54	
Zn(II)	-1.60; -1.68; -1.63[b]	

[a] At 3°C.
[b] 7°C.
[c] At 0.6M LiCl supporting electrolyte vs NAgE.

SULFOLANE
0.5 mM; room temp.

Species	Supporting electrolyte	Working electrode	$E_{1/2}V$ (vs SCE)	Ref.
Cu(II)	0.1M Et$_4$NClO$_4$	DME	$+0.50$	70
	0.1M Bu$_4$NClO$_4$	DME	$+0.46$	
	0.1M Et$_4$NClO$_4$	RPE	$+0.58$; $+0.48$	
	0.1M Bu$_4$NClO$_4$	RPE	$+0.56$; $+0.49$	

0.1M Et$_4$NClO$_4$ at DME at 30°C.
~2.0 mM

Species	$E_{1/2}V$ (vs AgRE)	Ref.
CsClO$_4$	-2.66	72
KClO$_4$	-2.66	
LiClO$_4$	-2.67	
NaClO$_4$	-2.56	
RbClO$_4$	-2.67	

VI. POLAROGRAPHY

3-METHYL SULFOLANE
0.5 mM; room temp.

Species	Supporting electrolyte	Working electrode	$E_{1/2}V$ (vs SCE)	Ref.
Cu(II)	0.1M Et$_4$NClO$_4$	DME	+0.50	70
Cu(II)	0.1M Bu$_4$NClO$_4$	DME	+0.52	
Cu(II)	0.1M Et$_4$NClO$_4$	RPE	+0.58; +0.48	
Cu(II)	0.1M Bu$_4$NClO$_4$	RPE	+0.58; +0.51	

2,4-DIMETHYL SULFOLANE
0.5 mM at room temp.

Species	Supporting electrolyte	Working electrode	$E_{1/2}V$ (vs SCE)	Ref.
Cu(II)	0.1M Et$_4$NClO$_4$	DME	+0.52	70
	0.1M Bu$_4$NClO$_4$	DME	+0.56	
	0.1M Et$_4$NClO$_4$	RPE	+0.60; +0.52	
	0.1M Bu$_4$NClO$_4$	RPE	+0.60; +0.52	

LIQ. SO$_2$
At DME at −25°C.
∼5 mM

Species	$E_{1/2}V$ (vs Hg-calomel pool)	Ref.
KBr	+0.09; +1.7	69
KCl	−0.02 to +0.10; −0.49 to +1.62	
KI	+0.16 to +0.30[a]	
NH$_4$SCN	−0.58 to −0.47; +1.3	

[a] At −30 to −33°C.

TRIMETHYL PHOSPHATE
0.1M Et$_4$NClO$_4$ at DME at 25°C.
~1.0 mM

Species	$E_{1/2}V$ (vs SCE)	Ref.
Ba(ClO$_4$)$_2$	−2.03	53
Cd(ClO$_4$)$_2$	−0.49	
Co(ClO$_4$)$_2$	−1.39	
KClO$_4$	−2.06	
Mn(ClO$_4$)$_2$	−1.85	
NaClO$_4$	−2.07	
Ni(ClO$_4$)$_2$	−0.90	
(Ph–Ph)$_2$Cr(I)	−0.70	
RbClO$_4$	−2.05	
TlClO$_4$	−0.39	
Zn(ClO$_4$)$_2$	−0.82	

References

1. R. T. Iwamoto, Anal. Chem., **31**, 955 (1959).
2. V. Gutmann and G. Peychal-Heiling, Monatsh. Chem., **100**, 813 (1969).
3. D. E. Sellers and G. W. Leonard, Jr., Anal. Chem., **33**, 334 (1961).
4. L. A. Knecht and I. M. Kolthoff, Inorg. Chem., **1**, 195 (1962).
5. V. Gutmann, H. Bildstein and H. Fleischer, Monatsh. Chem., **99**, 1680 (1968).
6. V. Gutmann, M. Michlmayr and G. Peychal-Heiling, Anal. Chem., **40**, 619 (1968).
7. H. Bildstein, H. Fleischer and V. Gutmann, Inorg. Chim. Acta, **2**, 347 (1968).
8. J. F. Coetzee and W. S. Siao, Inorg. Chem., **2**, 14 (1963).
9. J. F. Coetzee, D. K. McGuire and J. L. Hedrick, J. Phys. Chem., **67**, 1814 (1963).
10. I. M. Kolthoff and J. F. Coetzee, J. Am. Chem. Soc., **79**, 1852 (1957).
11. L. Sestili, C. Furlani, A. Ciana and F. Garbassi, Electrochim Acta., **15**, 225 (1970).
12. A. I. Popov and D. H. Geske, J. Am. Chem. Soc., **80**, 5346 (1958).
13. G. Pezzatini and R. Guidelli, Electrochim. Acta, **16**, 1415 (1971).
14. A. I. Popov and D. H. Geske, J. Am. Chem. Soc., **80**, 1340 (1958).
15. I. M. Kolthoff and S. Ikeda, J. Phys. Chem., **65**, 1020 (1961).
16. I. V. Nelson and R. T. Iwamoto, J. Electroanal. Chem., **6**, 234 (1963).
17. J. B. Headridge and D. Pletcher, J. Electroanal. Chem., **15**, 312 (1967).
18. E. J. Cokal and E. N. Wise, J. Electroanal. Chem., **11**, 406 (1966).
19. O. Gürtler and H. Hennig, J. Electroanal. Chem., **30**, 253 (1971).

VI. POLAROGRAPHY 521

20. J. Sancho, A. Aldaz and A. Pujante, J. Electroanal. Chem., **25**, 505 (1970).
21. F. G. Thomas and I. M. Kolthoff, J. Electroanal. Chem., **31**, 423 (1971).
22. E. Itabashi and S. Ikeda, J. Electroanal. Chem., **26**, 103 (1970).
23. R. C. Larson and R. T. Iwamoto, J. Am. Chem. Soc., **82**, 3239, 3526 (1960).
24. S. Wawzonek and M. E. Runner, J. Electrochem. Soc., **99**, 457 (1952).
25. C. Musikas and B. Mjassoedov, Radiochem. Radio and Lett. **2**, 21 (169).
26. J. N. Gaur and N. K. Goswami, Electrochim. Acta, **15**, 519 (1970).
27. K. Schwochau and L. Astheimer, J. Inorg. Nucl. Chem., **32**, 119 (1970).
28a. G. Dryhurst and P. J. Elving, Anal. Chem., **39**, 606 (1967).
28b. W. M. Gulick and D. H. Geske, J. Am. Chem. Soc., **88**, 2928 (1966).
29. I. M. Kolthoff and F. G. Thomas, J. Electrochem. Soc., **111**, 1065 (1964).
30. W. Hubicki and M. Dabkowska, Anal. Chem., **33**, 90 (1961).
31. H. A. Laitinen and C. J. Nyman, J. Am. Chem. Soc., **70**, 3002 (1948).
32. W. B. Schaap, R. F. Conley and F. C. Schmidt, Anal. Chem., **33**, 498 (1961).
33. E. P. Parry and K. B. Odham, Anal. Chem., **40**, 1031 (1968).
34. D. E. Sellers and G. W. Leonard, Jr., Anal. Chem., **34**, 1457 (1962).
35. R. N. Hammer and J. J. Lagowski, Anal. Chem., **34**, 597 (1962).
36. J. K. Gupta and C. M. Gupta, Indian J. Chem. **8**, 264 (1970).
37. J. K. Gupta and C. M. Gupta, Monatsh. Chem., **100**, 2019 (1969).
38. S. H. Cohen, R. T. Iwamoto and J. Kleinberg, J. Phys. Chem., **67**, 1275 (1963).
39. H. Sadek, R. M. Issa and B. A. Abdel-Nakey, Electrochim Acta, **16**, 401 (1971).
40. G. Schöber and V. Gutmann, Monatsh. Chem., **89**, 401 (1958).
41. G. Schöber and V. Gutmann, Z. anal. Chem., **173**, 2 (1960).
42. S. Lal and S. N. Srivastava, Indian J. Appl. Chem., **32**, 227 (1969).
43. G. H. Brown and H. S. Hsuing, J. Electrochem. Soc., **107**, 57 (1960).
44. S. K. Jha and S. N. Srivastava, J. Prakt chemie, **38**, 295 (1968).
45. G. P. Kumar and D. A. Pantony, Proc. 3rd Int. Congr. on polarogr. Southampton 1964, Ed. G. J. Hills Vol. II, Macmillan (Lond.) 1966 pp. 1035.
46. C. W. Manning and W. C. Purdy, Anal. Chim Acta, **51**, 483 (1970).
47. D. L. Maride and W. G. Hodgson, Anal. Chem., **37**, 1562 (1965).
48. Y. Matsui, Y. Kurosaki and Y. Date, Bull. Chem. Soc., Japan **43**, 1707 (1970).
49. Y. Matsui, Y. Kurosaki and Y. Date, Bull. Chem. Soc., Japan, **43**, 2046 (1970).
50. Y. Matsui and Y. Date, Bull. Chem. Soc., Japan, **43**, 2828 (1970).
51. G. Gritzner, V. Gutmann and G. Schöber, Monatsh. Chem., 1056 (1965).
52. Y. Matsui, R. Kawakado and Y. Date, Bull. Chem. Soc., Japan, **41**, 2913 (1968).
53. V. Gutmann and R. Schmid, Monatsh. Chem., **100**, 2113 (1969).
54. L. R. Sherman and V. S. Archer, Anal. Chem., **42**, 1356 (1970).
55. T. A. Pinfold and F. Sobba, J. Am. Chem. Soc., **78**, 5193 (1956).
56. V. M. Baudler and A. S. Bongardt, Z. Anorg. Allgem. Chem. **350**, 186 (1967).
57. J. F. Coetzee and J. L. Hedrick, J. Phys. Chem., **67**, 221 (1963).
58. R. Schmid and V. Gutmann, Monatsh. Chem., **100**, 1564 (1969).
59. J. C. Marchon and J. B. Lambling, Bull. Soc. Chim., **12**, 4660 (1967).
60. G. Cauquis and D. Serve, Bull. Soc. Chim., **1**, 302 (1966).
61. E. L. Colichman and W. H. Ludewig, Anal. Chem., **25**, 1909 (1953).
62. J. L. Jones, S. Adisesh, R. M. Smith and J. H. Karnes, Anal. Chim Acta, **49**, 487 (1970).
63. V. Gutmann, M. Kogelnig and M. Michlmayr, Monatsh. Chem., **99**, 693, 699 (1968).
64. G. Peychal-Heiling and V. Gutmann, Z. Anal. Chem., **248**, 6 (1969).
65. I. Fried and H. Barak, J. Electroanal. Chem., **30**, 279 (1971).
66. B. Burrows and S. Kirkland, J. Electrochem. Soc., **115**, 1164 (1968).

67. M. S. Spritzer, J. M. Costa and P. J. Elving, Anal. Chem., **37,** 211 (1965).
68. A. Cisak and P. J. Elving, J. Electrochem. Soc., **110,** 160 (1963).
69. P. J. Elving, J. M. Markowitz and I. Rosenthal, J. Phys. Chem., **65,** 680 (1961).
70. J. L. Hanley and R. T. Iwamoto, J. Electroanal. Chem., **24,** 271 (1970).
71. V. Gutmann and O. Duschek, Monatsh. Chem., **100,** 1047 (1969).
72. J. F. Coetzee, J. M. Simon and R. J. Bertozzi, Anal. Chem., **41,** 766 (1969).
73. E. L. Johnson, K. H. Pool and R. E. Hamm, Anal. Chem., **38,** 183 (1966).
74. A. D. Goolsby and D. T. Sawyer, Anal. Chem., **40,** 83 (1968).
75. I. M. Kolthoff and T. B. Reddy, J. Electrochem. Soc., **108,** 980 (1961).
76. L. V. Haynes and D. T. Sawyer, Anal. Chem., **39,** 332 (1967).
77. J. A. Wargon and A. J. Arvia, Electrochem Acta, **16,** 1619 (1971).
78. M. Michlmayr and D. T. Sawyer, J. Electroanal. Chem., **23,** 387 (1969).
79. V. Gutmann and G. Schöber, Z. Anal. Chem., **171,** 339 (1959).
80. D. B. Brus and T. DeVries, J. Am. Chem. Soc., **78,** 733 (1956).
81. J. F. Coetzee and I. M. Kolthoff, J. Am. Chem. Soc., **79,** 6110 (1957).
82. G. H. Brown and R. Al-Urfalia, J. Am. Chem. Soc., **80,** 2113 (1958).
83. I. M. Kolthoff and J. F. Coetzee, J. Am. Chem. Soc., **79,** 870 (1957).
84. A. I. Popov and D. H. Geske, J. Am. Chem. Soc., **79,** 2074 (1957).
85. R. C. Larson and R. T. Iwamoto, J. Am. Chem. Soc., **82,** 3526 (1960).
86. A. G. Smelley, F. E. Brantley and A. F. Findeis, Anal. Chem., **38,** 449 (1966).

(b) Organic Electrolytes

The following section on nonaqueous voltammetry and polarography relative to organic electrolytes covers solute concentration, supporting electrolyte, temperature, working electrode, reference electrode and half-wave potentials. The tables are organized in alphabetical order by solvent, and the survey has been limited to literature coverage from 1969 to 1972, in view of the comprehensive and recently published survey covering this aspect of the work up to 1968.[a]

Symbols used:

DME	Dropping mercury electrode
SCE	Standard calomel electrode
RE(Pt) or RPE	Rotating platinum electrode
MPE	Mercury pool electrode
RPGE	Rotating pyrolytic graphite electrode
HMD	Hanging mercury drop
mM	Millimoles

[a] C. K. Mann and K. K. Barnes; "Electrochemical Reactions in Nonaqueous Systems," Marcel Dekker Inc; New York (1970).

ACETIC ACID
0.1M LiClO$_4$ at RE(Pt) at 25°C
~0.5 mM

Species	$E_{1/2}V$ (vs SCE)	Ref.
tris(4,7-Dimethyl-1,10-phenanthroline) Ferrocene	+0.280[a]	104

[a] 0.5M LiClO$_4$.

0.1M LiClO$_4$ at Pt disk
5.0 mM

Species	$E_{1/2}V$ (vs SCE)	Ref.
CH(NO$_2$)$_3$	+1.60	23

ACETIC ANHYDRIDE

0.1M LiClO$_4$ at RE(Pt) at 25°C

~0.5 mM

Species	$E_{1/2}V$ (vs SCE)	Ref.
Benzyl ferrocene	+0.355	104
bis(2,9-Dimethyl-1,10-phenanthroline) Cu(I)	+0.715	
tris(4,7-Dimethyl-1,10-phenanthroline Fe(II)	+0.900	
Ferrocene	+0.385	

0.3M LiClO$_4$ at Pt

1 mM

Species	$E_{1/2}V$ (vs Ag/AgCl)	Ref.
I$_2$	~+1.57; ~+2.28	101

ACETONE

1M LiClO$_4$ at RE(Pt) at 25°C

~0.5 mM

Species	$E_{1/2}V$ (vs SCE)	Ref.
Benzyl ferrocene	+0.435	104
bis(2,9:Dimethyl-1,10-phenanthroline) Cu(1)	+0.760	
tris(4,7-Dimethyl-1,10-phenanthroline) Fe (11)	+0.940	
Ferrocene	+0.460	

0.05M Et$_4$NClO$_4$ at DME at 25°C

0.5 mM

Species	$E_{1/2}V$	Ref.
1,2-Naphthoquinone	−0.57; −1.01	28
1,4-Naphthoquinone	−0.69; −1.18	

VI. POLAROGRAPHY

ACETYL ACETONE

0.1M LiClO₄ at RE(Pt) at 25°C

~0.5 mM

Species	$E_{1/2}V$ (vs SCE)	Ref.
Benzyl ferrocene	+0.355	104
bis(2,9-Dimethyl-1,10-phenanthroline) Cu(I)	+0.730	
tris(4,7-Dimethyl-1,10-phenanthroline) Fe(II)	+0.905	
Ferrocene	+0.355	

ACETONITRILE

0.1M Et₄NClO₄ at DME at 25.000 ± 0.002°C

Species	Conc. × 10³ M/l	$E_{1/2}V$ (vs SCE)	Ref.
Azobenzene	0.76	−1.405; −1.755	32
4-Chloroazobenzene		−1.325; −1.660	
4-(N,N-Dimethylamino) azobenzene		−1.600; −1.820	
4-Methoxyazobenzene		−1.495; −1.823	
4-Nitroazobenzene	0.26	−0.860; −1.195	
4-Sulfonic acid derivative of azobenzene		−1.350; 1.655	
4-(Trimethylammonio) azobenzene		−1.172; −1.537	

ACETONITRILE (Continued)

0.1M Et$_4$NClO$_4$, DME at 25 ± 0.5°C

0.5 to 0.7 mM

Species	$E_{1/2}V$ (vs SCE)	Ref.
Acridine	−1.620; −1.994	31
Benzo[o] cinnoline	−1.554; −1.863; −2.396	
Benzo(f) quinoline	−2.140	
Benzo(h) quinoline	−2.208	
Benzo(f) quinoxaline	−1.744; −2.128; −2.673	
Cinnoline	−1.686; −2.134	
Isoquinoline	−2.220	
Phenanthridine	−2.118; −2.415	
Phenanthroline	−1.227; −1.681	
m-Phenanthroline	−2.092; −2.287	
o-Phenanthroline	−2.053; −2.269	
p-Phenanthroline	−2.044; −2.229	
Phthalazine	−1.976; −2.315; −2.498	
Pyrazine	−2.080	
Pyridazine	−2.120	
Pyridine	−2.622	
Pyrimidine	−2.340	
Quinazoline	−1.799; −2.478	
Quinoline	−2.105	
Quinoxaline	−1.702; −2.163	

0.1M Et$_4$NClO$_4$ at DME at 25°C

0.26 to 0.99 mM

Species	$E_{1/2}V$ (vs SCE)	Ref.
Benzoic acid	−1.87 to −1.85	131

0.1M Et$_4$NClO$_4$ at Pt microelectrode at 25°C

Species	$E_{1/2}V$ (vs Ag/0.01M AgClO$_4$)	Ref.
Anthracene	+0.91	150
1,2-Benzanthracene	+1.00	
Chrysene	+1.22	
9,10-Diphenylanthracene	+0.92	
Phenanthrene	+1.28	
Pyrene	+1.06	
Triphenylene	+1.46	

VI. POLAROGRAPHY

ACETONITRILE (Continued)
0.1M NaClO₄ at RE (Pt)
0.0396 to 0.784 mM

Species	$E_{1/2}V$ (vs SCE)	Ref.
N,N-Dimethylaniline	+0.705 to +0.680	145
Diphenylamine	+0.845 to +0.830	
Di-4-tolylamine	+0.710 to +0.700; +1.54 to +1.50	
Triphenylamine	+0.870 to +0.855	

0.1M Et₄NClO₄ at Pt electrode
0.99 mM

Species	$E_{1/2}V$ (vs SCE)	Ref.
Benzo(a)pyrene	+1.2	83

0.1M Bu₄NClO₄ at DME at 25.0 ± 0.05°C
1 mM

Species	$E_{1/2}V$ (vs Ag/AgClO₄)	Ref.
Anthracene	−2.276	86
Naphthalene	−2.856	

0.1M LiClO₄ at RE(Pt) at 25°C
∼0.5 mM

Species	$E_{1/2}V$	Ref.
Benzylferrocene	+0.325	104
bis(2,9-Dimethyl-1,10-phenanthroline) Cu(I)	+0.685	
$tris$(4,7-Dimethyl-1,10-phenanthroline) Fe(II)	+0.860	
Ferrocene	+0.350	

0.1M Et₄NClO₄ at Pt
1.52 to 1.87 mM

Species	E_pV (vs SCE)	Ref.
Quinhydrone	+1.12; to +0.13; −0.34	90

ACETONITRILE (*Continued*)
0.1M NaClO$_4$ at Pt electrode at 22°C
5 to 10 mM

Species	$E_p V$ (vs Ag/AgNO$_3$)	Ref.
n-Butyl mercaptan	+1.49	88
s-Butyl mercaptan	+1.33; +1.99	
t-Butyl mercaptan	+1.59	
Diallyl sulfide	+1.74	
Dibenzyl sulfide	+1.48	
Dibenzylthiophene	+1.35; +1.64	
Di-n-butyl sulfide	+1.45	
Di-s-butyl sulfide	+1.43	
Di-t-butyl sulfide	+1.06; +1.7	
Diethyl sulfide	+1.50	
Dimethyl sulfide	+1.41	
Diphenyl sulfide	+1.26; +1.51; +1.81	
Di-i-propyl sulfide	+1.47	
p-Dithiane	+1.46; +1.91	
Ethylene sulfide	+1.51	
Ethylene trithiocarbonate	+1.53; 0.17[a]	
Pentamethylene sulfide	+0.55; +1.42; 0.12[a]	
Propylene sulfide	+1.69	
Tetrahydrothiophene	+1.45	
Thiophene	+1.84	
sym-Trithiane	+1.30; +1.47	

[a] Refers to reduction peaks.

0.1N Et$_4$NClO$_4$ at RE (Pt) at room temp.
1 mM

Species	$E_{1/2} V$ (vs SCE)	Ref.
2-Acetylamino	1.14; 2.40; −2.01; −2.73	113
1-Aminoanthracene	0.55; 1.27; −2.10; −2.51	
2-Aminoanthracene	0.56; 1.30; −2.17; −2.55	
4-Aminoazobenzene	0.92; 1.48; −1.54; −1.74	
2-Aminobiphenyl	0.93; 1.35	
4-Aminobiphenyl	0.76; 2.51	
2-Aminofluorene	0.64; 1.35	
1-Aminonaphthalene	0.68; 1.25	
2-Aminonaphthalene	0.75; 0.94	
Aniline	0.98	
Anthracene	1.19; −2.07; −2.52	
Azobenzene	1.89; −1.45; −1.74	
1,2-Benzanthracene	1.33; 1.83; −2.11; −2.45	

VI. POLAROGRAPHY

ACETONITRILE (Continued)
0.1N Et₄NClO₄ at RE (Pt) at room temp.
1 mM

Species	$E_{1/2}V$ (vs SCE)	Ref.
Benzidine	0.52; 0.73	
1,2-Benzpyrene	1.34; −2.22; −2.58	
3,4-Benzpyrene	1.10; 2.26; −1.95; −2.24	
5,6-Benzpyrido-(2′, 3′−1,2)-carbazole	1.06; 1.41; −1.98; −2.34	
5,6-Benzpyrido-(3′, 2′−1,2)-carbazole	1.04; 1.39; −1.95; −2.37	
Biphenyl	1.82; −2.70	
m-Bromoaniline	1.08	
p-Bromoaniline	0.97	
m-Chloroaniline	1.05	
p-Chloroaniline	0.94	
1,2,3,4-Dibenzanthracene	1.34; 1.71; −1.55; −1.93	
1,2,5,6-Dibenzanthracene	1.40; 1.77; −2.12; −2.42	
3,4,9,10-Dibenzpyrene	1.17; 1.98; −1.90; −2.10	
4-Diethylaminoazobenzene	0.84; 1.30; −1.62; −1.81	
4-Dimethylaminoazobenzene	0.86; 1.40; −1.58; −1.84	
2,4-Dimethylaniline	0.70; 1.00	
2,5-Dimethylaniline	0.79	
3,5-Dimethylaniline	0.81	
N,N-Dimethylaniline	0.73	
9,10-Dimethyl-1,2-benzanthracene	1.12; −2.15; −2.42	
Diphenylamine	0.86; 1.03	
Fluorene	1.50; −2.77	
3′-Fluoro-4-dimethylaminoazobenzene	0.90; 1.27; −1.50; −1.66	
3-Fluoro-10-methyl-1,2-benzanthracene	1.27; 1.92; −2.10; −2.40	
4-Fluoro-10-methyl-1,2-benzanthracene	1.28; 1.54; −2.04; −2.38	
Hexamethylbenzene	1.58	
p-p′-Hydrazotoluene	0.34	
N-Hydroxy-2-acetylamino	1.00; 1.55; −2.12; −2.78	
1-Hydroxynaphthalene	1.14; 1.76; −2.47; −2.66	
2-Hydroxynaphthalene	1.27; 2.35; −2.42; 2.63	
m-Methylaniline	0.84	
N-Methylaniline	0.77	
o-Methylaniline	0.85	
p-Methylaniline	0.78; 1.07	
2-Methyl-4-dimethylaminoazobenzene	0.80; 1.22; −1.60; −1.88	
2′-Methyl-4-dimethylaminoazobenzene	0.90; 1.33; −1.59; −1.84	
3′-Methyl-4-dimethylaminoazobenzene	0.91; 1.32; −1.57; −1.80	
4′-Methyl-4-dimethylaminoazobenzene	0.90; 1.34; −1.58; −1.86	
Naphthalene	1.65; −2.63	
Perylene	1.04; 1.41; −1.73; −2.21	
Phenol	1.47; 1.94	
Phenothiazine	0.56; 1.00	
Pyrene	1.25; −2.19; −2.64	

ACETONITRILE (Continued)

0.2M Et$_4$NBr at Hg plated Pt wire at 22°C

5 to 10 mM

Species	E_pV (vs Ag/AgNO$_3$)	Ref.
Benzene sulfonamide	−2.82	105
p-Cyanobenzene sulfonamide	−2.21; −2.79; +2.60	
N,N-Diethyl-p-cyanobenzene sulfonamide	−1.03; +0.93; −2.07; +1.94 −2.68; +2.56	
N,N-Diethyl-p-toluene sulfonamide	−1.52; +1.34; −2.93	
N,N-Dimethyl-p-toluene sulfonamide	−1.46; +1.20; −2.85	
N-Methyl-p-toluene sulfonamide	−2.91	
n-Propyl-p-cyanobenzene sulfonamide	−2.10; +1.91; −2.62; +2.45	
N-Propyl-p-toluene sulfonamide	−2.87	
p-Toluene sulfonamide	−2.97	

0.1M Et$_4$NBr at mercury plated Pt wire at ~22°C

~5 mM Soln

Species	E_pV (vs Ag/AgNO$_3$)	Ref.
1-Bromo-2,2-diphenylcyclopropane	−2.97; −1.31; −1.69	68
1-Bromo-1-methoxy methyl	−2.95; −1.43; −1.79	
1-Bromo-1-methyl-2,2-diphenyl-cyclopropane	−3.00; −1.26; −1.58	
1,2-Dibromoethane	−2.22	
1-Iodo-1-methyl-2,2-diphenyl-cyclopropane	−2.28; −1.26; −1.58	

0.1M Et$_4$NClO$_4$ at Pt button

2 mM

Species	$E_{1/2}V$ (vs SCE)	Ref.
5,10-Dihydro-5,10-dimethylphenazine	+0.11; +0.83	99
5,10-Dihydro-5,10-diphenylphenazine	+0.20; +0.94	
5,10-Dihydro-5-methylphenazine	+0.07; +0.63	
5,10-Dihydro-5-methyl-10-phenyl-phenazine	+0.13; +0.87	
5-Methylphenazinium methyl sulfate	+0.05; −1.01	

VI. POLAROGRAPHY

ACETONITRILE (Continued)
0.1M Bu$_4$NBr at DME
1 mM

Species	$E_{1/2}V$ (vs Hg pool)	Ref.
Anthraquinone	−0.43; −1.05	100
Benzoquinone	−0.10; −0.84	
Ethylbromide[a]	−2.13	
2-Methyl-1,4-naphthaquinone	−0.27; −0.94	

[a] 4.86 mM soln.

0.1M LiClO$_4$ at RE (Pt disk) at 20°C
0.05 to 0.5 mM

Species	$E_{1/2}V$ (vs SCE)	Ref.
Diphenyldiazomethane	+0.95; +1.73	96

0.1M Bu$_4$NClO$_4$ at DME
1 mM

Series	$E_{1/2}V$ (vs Ag/AgCl)	Ref.
4,5-Methylenephenanthrene	−2.48	97

0.1M Et$_4$NClO$_4$ at Pt at 22.5 ± 0.5°C
1.78 mM to 3.58 mM

Species	E_pV (vs SCE)	$E_{1/2}V$ (vs SCE)	Ref.
2,2′-Dinitrobibenzyl	−1.23; −1.33	−1.20; −1.30	89
4,4′-Dinitrobibenzyl	−1.23	−1.20	
m-Nitrobenzyl bromide	−1.08	−1.09	
m-Nitrobenzyl chloride	−1.13	−1.09	
p-Nitrobenzyl bromide	−0.86	−1.09	
p-Nitrobenzyl chloride	−0.97	−1.09	
o-Nitrobenzyl bromide	−0.91	−1.15	
o-Nitrobenzyl chloride	−1.01	−1.15	
m-Nitrotoluene	−1.20	−1.17	
p-Nitrotoluene	−1.23	−1.20	
o-Nitrotoluene	−1.29	−1.26	

ACETONITRILE (Continued)

0.5N NaClO$_4$ at Pt electrode
10 mM

Species	$E_{1/2}V$ (vs Ag/AgClO$_4$ and NaClO$_4$)	Ref.
Anthracene	+0.88	95

0.1M Et$_4$NClO$_4$ at DME at 25°C
0.43 mM

Species	$E_{1/2}V$ (vs SCE)	Ref.
Tetrakis(dimethylamino)ethylene	−0.75; −0.61	73

Pt electrode
5 mM

Species	EV (vs Ag/Ag$^+$)	Ref.
Allylamine	+2.0	79
Benzylamine	+1.5	
n-Butylamine	+1.2	
t-Butylamine	+1.2	
Cyclohexylamine	+1.4	
Hexylamine	+1.4	
Isobutylamine	+1.2	
Methylamine	+1.2	
n-Propylamine	+1.2	

0.1M LiClO$_4$ at Pt electrode
5 mM Soln.

Species	$E_{1/2}V$ (vs Ag/Ag$^+$)	Ref.
Tetraphenyl-2,3,4,5-pyrrole	+0.61; +1.23	48

VI. POLAROGRAPHY

ACETONITRILE (Continued)

Species	$E_{1/2}V$ (vs SCE)	Ref.
Acetophenone	−1.85; −2.27	56
α-Dimethylaminoacetophenone	−1.80; −2.20	
α-Morpholinoacetophenone	−1.76; −2.17	
α-Piperidinoacetophenone	−1.81; −2.17	
α-Trimethylammoniumacetophenone	−1.06; −1.85; −2.30	

0.1M LiClO₄ at RE (Pt disk) at 20 ± 1°C
0.4 to 0.6 mM

Species	$E_{1/2}V$ (vs SCE)	Ref.
q-Diazofluoren	+1.22	55
1-Diazo-2,3,4,5-tetraphenylcyclopentadien	+1.01	
Diphenyldiazomethane	+0.95	
Diphenylpicrylhydrazyl	+0.693	
p-Nitrodiphenyldiazomethane	+1.14	
p-Nitrophenylbenzoyldiazomethane	+1.75	
p-Nitrophenyldiazomethane	+1.33	
p-Nitrophenylmethyldiazomethane	+1.16	
Phenyl-benzoyldiazomethane	+1.47	
p-Phenyldiphenyldiazomethane	+0.89	

0.1M Et₄NClO₄ at Hg drop or RE (Pt) at 20°C

Species	$E_{1/2}V$ (vs SCE)	Ref.
α,β-Dimorpholino-styrolene	+0.06; +0.35	54
1,2-Dipiperidino-ethylene	−0.20; +0.35	
α,β-Dipiperidono-styrolene	−0.06; +0.26	
1,2-Di(N-propylamino)-1-neopentyl-ethylene	−0.18	

ACETONITRILE (Continued)

0.1M LiClO$_4$
0.1 m Molar + 100 mM Ethanol

Species	$E_{1/2}V$ (vs SCE)	Ref.
Anthracene	+1.25	34
9-Methoxyanthracene	+0.96	

0.1M Et$_4$NClO$_4$

Species	$E_{1/2}V$ (vs SCE)	Ref.
9-Phenylanthracene	+1.13	35

Species	E_pV (vs SCE)	Ref.
Ubiquinone-1	−0.87; −1.74; −1.43; −0.81	36

0.1M NaClO$_4$ at RP disk at 25°C
0.266 mM

Species	$E_{1/2}V$ (vs SCE)	Ref.
Aniline	+0.860	26

0.05M Et$_4$NClO$_4$ at DME at 25°C
0.5 mM

Species	$E_{1/2}V$ (vs SCE)	Ref.
1,2-Naphthoquinone	−0.58; −1.18	28
1,4-Naphthoquinone	−0.70; −1.33	

VI. POLAROGRAPHY

ACETONITRILE (Continued)

0.1M Bu$_4$NClO$_4$ at DME at 20 ± 0.5°C

0.5 mM; pH = 7–9

Species	$E_{1/2}V$ (vs SCE)	Ref.
C$_6$H$_5$—CH=N—C$_6$H$_5$	−1.87	24
C$_6$H$_5$CHO	−1.78	
C$_6$H$_5$—NO$_2$	−1.09	

0.1M Et$_4$NClO$_4$ at GCE[a] at 25 ± 0.5°C

Species	Conc. × 10^3 M/l	EV (vs Ag–CH$_3$CN)	Ref.
BzNH[b]	1.67	0.30; 0.69; 1.22	13
ClBzNH[c]	1.05	0.30; 0.74; 1.25	
MeNH[d]	1.23	0.23; 0.73; 1.22	
PrNH[e]	1.47	0.57; 1.02; 1.56	

[a] Glassy carbon electrode.
[b] 1-*n*-Benzyl-1,4-dihydronicotinamide.
[c] 1-(2,6-dichlorobenzyl)-1,4-dihydronicotinamide.
[d] 1-Methylnicotinamide.
[e] 1-*n*-propyl-1,4-dihydronicotinamide.

Species	$E_{1/2}V$ (vs SCE)	Ref.
Acetophenone anil	−1.90	10
α-Aminoanthracene	+0.42	9
Aniline	+0.86	9
Benzaldehyde anil	−1.76	10
Benzophenone anil	−1.66	10
Benzophenone-N-methylimine	−2.04	10
2-Bromo-2-nitropropane	−0.6; −1.5	8
Cinnamaldehyde	−1.52	10
Diphenylpickrylhydrazyl	+0.68	9
Fluorenone anil	−1.29	10
α-Naphthylamin	+0.63	9
Pinacamphone anil	−2.43	10
Thioxanthene	+1.33; +2.33	7

ACETONITRILE (*Continued*)
0.1M LiClO$_4$ at RE (Pt)

Species	$E_{1/2}V$ (vs Ag/AgCl)	Ref.
Nitrosobenzene	+1.0; +1.44	53

0.1M Et$_4$NClO$_4$ at DME
1.2 mM

Species	$E_{1/2}V$ (vs SCE)	Ref.
Conjugate base of 1-hydroxy-9,10-anthraquinone[a]	−1.27	106
1-hydroxy-9,10-anthraquinone	−0.77; −1.20	

[a] 1.83 mM.

0.1M LiClO$_4$ at Pt disk
~5.0 mM

Species	$E_{1/2}V$ (vs SCE)	Ref.
CH(NO$_2$)$_3$	+1.95	23

0.1M LiClO$_4$ at Pt at 20°C
2.0 mM

	$E_{1/2}V$ (vs Ag/AgNO$_3$)	Ref.
2,4,6-*Tri-t*-Butylaniline	0.530	22

Bu$_4$NI at DME at 25 ± 0.2°C

Species	$E_{1/2}V$ (vs SCE)	Ref.
m-Nitroaniline	−1.31; −2.15	20
o-Nitroaniline	−1.24; −2.16	
p-Nitroaniline	−1.18; −2.44	
m-Nitrophenol	−1.12; −1.41	
o-Nitrophenol	−0.95; −1.95	
p-Nitrophenol	−1.19; −1.69	

VI. POLAROGRAPHY

ACETONITRILE (Continued)

0.1M Bu$_4$NClO$_4$ at DME at 25 ± 0.05°C

1.0 mM

Species	$E_{1/2}V$ (vs Ag/AgClO$_4$)	Ref.
Anthracene	−2.276	34
Naphthalene	−2.856	

Species	$E_{1/2}V$ (vs Ag/AgNO$_3$)	Ref.
Orthothio-oxalate of heterocyclic compound	+0.68[a]; +1.12[a]; ∼ −0.1[a]	15
Pyrimidine	−2.628; −2.340[a]	6

[a] $E_{1/2}V$ (vs SCE).

0.1M Et$_4$NClO$_4$ at planar Pt button at 22.5 ± 0.5°C

Species	Conc. × 10^3 M/l	$E_{1/2}V$ (vs SCE)	Ref.
2,2′-Dinitrodibenzyl	2.44	−1.20	31
4,4′-Dinitrodibenzyl	1.78	−1.20	
m-Nitrobenzyl bromide		−1.09	
m-Nitrobenzyl chloride	2.34	−1.09	
o-Nitrobenzyl bromide		−1.15	
o-Nitrobenzyl chloride	2.58	−1.15	
p-Nitrobenzyl bromide		−1.09	
p-Nitrobenzyl chloride	2.31	−1.09	
m-Nitrotoluene	4.60	−1.17	
o-Nitrotoluene	3.58	−1.26	
p-Nitrotoluene	2.68	−1.20	

0.1N Et$_4$NClO$_4$ at RE(Pt) room temp.

Species	$E_{1/2}V$ (vs SCE)	Ref.
2-Acetylaminofluorene	+1.14; +2.40; −2.01; −2.73	64
1-Aminoanthracene	+0.55; +1.27; −2.10; −2.51	
2-Aminoanthracene	+0.56; +1.30; −2.17; −2.55	
1-Aminonaphthalene	+0.68; +1.25	
2-Aminonaphthalene	+0.75; +0.94	
4-Aminoazobenzene	+0.92; +1.48; −1.54; −1.74	
2-Aminobiphenyl	+0.93; +1.35	

ACETONITRILE *(Continued)*
0.1N Et$_4$NClO$_4$ at RE(Pt) room temp.

Species	$E_{1/2}V$ (vs SCE)	Ref.
4-Aminobiphenyl	+0.76; +2.51	
2-Aminofluorene	+0.64; +1.35	
Aniline	+0.98	
Anthracene	+1.19; −2.07; −2.52	
Azobenzene	+1.89; −1.45; −1.74	
1,2-Benzanthracene	+1.33; +1.83; −2.11; −2.45	
Benzidine	+0.52; +0.73	
1,2-Benzpyrene	+1.34; −2.22; −2.58	
3,4-Benzpyrene	+1.10; +2.26; −1.95; −2.24	
5,6-Benzpyrido-(2′,3′-1,2)-carbazole	+1.06; +1.41; −1.98; −2.34	
5,6-Benzpyrido-(3′,2′-1,2)-carbazole	+1.04; +1.39; −1.95; −2.37	
Biphenyl	+1.82; −2.70	
m-Bromoaniline	+1.08	
p-Bromoaniline	+0.97	
m-Chloroaniline	+1.05	
p-Chloroaniline	+0.94	
1,2,3,4,-Dibenzanthracene	+1.34; +1.71; −1.55; −1.93	
1,2,5,6-Dibenzanthracene	+1.40; +1.77; −2.12; −2.42	
3,4,9,10-Dibenzpyrene	+1.17; +1.98; −1.90; −2.10	
4-Dimethylaminoazobenzene	+0.84; +1.30; −1.62; −1.81	
2,4-Dimethylaniline	+0.70; +1.00	
2,5-Dimethylaniline	+0.79	
3,5-Dimethylaniline	+0.81	
N,*N*-Dimethylaniline	+0.73	
9,10-Dimethylbenzanthracene	+1.12; −2.15; −2.42	
Diphenylamine	+0.86; +1.03	
Fluorene	+1.50; −2.77	
3-Fluoro-10-methylbenzanthracene	+1.27; +1.92; −2.10; −2.40	
4-Fluoro-10-methylbenzanthracene	+1.28; +1.54; −2.04; −2.38	
Hexamethylbenzene	+1.58	
p,*p*′-Hydroazotoluene	+0.34	
N-Hydroxy-2-acetylaminofluorene	+1.00; +1.55; −2.12; −2.78	
1-Hydroxynaphthalene	+1.14; +1.76; −2.47; −2.66	
2-Hydroxynaphthalene	+1.27; +2.35; −2.42; −2.63	
m-Methylaniline	+0.84	
N-Methylaniline	+0.77	
o-Methylaniline	+0.85	
p-Methylaniline	+0.78; +1.07	
Naphthalene	+1.65; −2.63	
Perylene	+1.04; +1.41; −1.73; −2.21	
Phenol	+1.47; +1.94	
Phenothiazine	+0.56; +1.00	
Pyrene	+1.25; −2.19; −2.64	

ACETONITRILE (Continued)
0.1M Bu$_4$NClO$_4$ at DME at 25 ± 0.05°C

Species	$E_{1/2}V$ (vs Ag/AgClO$_4$)	Ref.
Anthracene	−2.276	167
Naphthalene	−2.856	

ALLYL ALCOHOL
0.1M LiClO$_4$ at RE(Pt) at 25°C
∼0.5 mM

Species	$E_{1/2}V$ (vs SCE)	Ref.
Benzyl ferrocene	+0.330	104
tris(4,7-Dimethyl-1,10-phenanthroline) Fe(II)	+0.890	
Ferrocene	+0.365	

ACETONITRILE–ACETIC ACID (3:1)
0.1M LiClO$_4$
1 mM solution

Species	$E_{1/2}V$ (vs SCE)	Ref.
9-Acetoxyanthracene	1.20	46
9-Acetoxyanthracene[a]	1.22	
Anthracene	1.24	
Anthracene[a]	1.17	

[a] 0.25M sodium acetate.

ACETONITRILE–WATER
0.1M Bu$_4$NBr at DME
1 mM

Species	$E_{1/2}V$ (vs Hg pool)	Ref.
Anthraquinone[a]	−0.45; −0.89	100
Anthraquinone[b]	−0.45; −0.83	
Anthraquinone[c]	−0.48; −0.71	
Anthraquinone[d]	−0.58	
Benzoquinone[e]	−0.09; −0.46	
Benzoquinone[f]	−0.06; −0.28	
Duroquinone[g]	−0.34; −0.76	
Duroquinone[h]	−0.33; −0.67	
Duroquinone[i]	−0.35; −0.62	
Duroquinone[j]	−0.39	
2-Methyl-1,4-naphthaquinone[k]	−0.27; −0.66	
2-Methyl-1,4-naphthaquinone[l]	−0.29; −0.57	
2-Methyl-1,4-naphthaquinone[m]	−0.30; −0.50	

[a,b,c,d] correspond to 0.5, 1.0, 4.76, and 20.0% water by volume respt.
[e,f] correspond to 1.0 and 4.76% of Water volume respt.
[g,h,i,j] correspond to 0.5, 1.0, 2.44, and 4.76% water by volume respt.
[k,l,m] correspond to 1.0, 2.44, and 4.76% water by volume respt.

0.1M LiClO$_4$ at DME

Species	$E_{1/2}V$ (vs Hg pool)	Ref.
2-Methyl-1,4-naphthohydroquinone[a]	+0.713	100
2-Methyl-1,4-naphthohydroquinone[b]	+0.095	
2-Methyl-1,4-naphthohydroquinone[c]	+0.062	
2-Methyl-1,4-naphthohydroquinone[d]	+0.022	

[a,b,c,d] correspond to 1.48; 4.76, 20.00, and 42.86% water by volume respt.

LIQUID AMMONIA
1M H$_2$SO$_4$ at DME at −33°C

Species	$E_{1/2}V$ (vs Hg pool)	Ref.
Bi(III)	−0.42	103

VI. POLAROGRAPHY

LIQUID AMMONIA (Continued)
NH$_4$NO$_3$ at RE(Pt-disk) at 0.0 ± 0.2°C
2.88 mM

Species	EV (vs Pb/Pb(NO$_3$)$_2$LiNO$_3$ in NH$_3$)	Ref.
m-Dinitrobenzene	-0.08; -0.25	91

NaI at DME at 24 ± 0.2°C
0.298 to 0.336 mg/g NaI

Species	$E_{1/2}V$ (vs Ag/AgI)	Ref.
4,4'-bis(Acetamino)-azoxybenzene	-0.342	164
p-Nitroacetanilide	-0.082; -0.448	
m-Nitroaniline	-0.102; -0.404	
p-Nitroaniline	-0.25; -0.50	
p-Nitrosoacetanilide	-0.37	

n-BUTANOL
0.1M LiClO$_4$ at RE(Pt) at 25°C
$\sim$0.5 mM

Species	$E_{1/2}V$ (vs SCE)	Ref.
Benzylferrocene	$+0.400$	104
$tris$(4,7-Dimethyl-1,10-phenanthroline Fe(II)	$+0.920$	
Ferrocene	$+0.445$	

1M Et$_4$N p-toluene sulfonate at DME
0.2 to 1.0M

Species	$E_{1/2}V$ (vs SCE)	Ref.
Dimedone	-1.5 to -1.8	5

DICHLOROMETHANE

0.20M Bu₄NClO₄ at Pt disk electrode

1.5 mM

Species	$E_{1/2}V$ (vs SCE)	Ref.
Anthracene	+1.34	92
9,10-Diphenylanthracene	+1.20	
9-Phenylanthracene	+1.30	
Rubrene	+0.78	
Tetracene	+0.95	

DIGLYME

0.5M Sodium tetraphenylborate at DME at 25°C

1 mM

Species	$E_{1/2}V$ (vs SCE)	Ref.
Sodium borohydride	−0.15	105

DIMETHOXYETHANE

0.1M Bu₄NClO₄ at RE (Pt)

Species	EV (vs AgClO₄/Ag)	Ref.
Azulene	−2.3	81
Quinone	−1.2	
Trinitrobenzene	−1.3	

DIMETHYLFORMAMIDE

0.1M Bu₄NI at DME

Species	$-E_{1/2}V$ (vs Ag/AgCl)	Ref.
Acridine	1.62, 2.38	132
Azobenzene	1.39, 2.07	
2,2′-Azopyridine	1.04, 1.65	
3,3′-Azopyridine	1.21, 1.85	
Benzo[c]cinnoline	1.55, 2.40	
3,4-Benzocridine	1.73, 2.38	
5,6-Benzoquinoline	2.20, 2.72	
7,8-Benzoquinoline	2.23, 2.72	

VI. POLAROGRAPHY

DIMETHYLFORMAMIDE (Continued)
0.1M Bu$_4$NI at DME

Species	$-E_{1/2}V$ (vs Ag/AgCl)	Ref.
Benzylideneaniline	1.94, 2.43	
2,2'-Bipyridyl	2.19, 2.76	
4,4'-Bipyridyl	1.91, 2.47	
2,2'-Biquinoline	1.77, 2.20	
Cinnoline	1.68, 2.62	
Dibenzo[a, c]phenazine	1.35, 2.12	
5,6,7,8-Dibenzoquinoxaline	1.78	
Dibenzylideneazine	1.77, 2.26	
2,6-Dimethylpyrazine	2.28	
2,5-Dimethylpyridine	2.82	
2,6-Dimethylpyridine	2.85	
2,4-Dimethylquinoline	2.285	
2,6-Dimethylquinoline	2.31	
2,3-Dimethylquinoxaline	1.90	
6,7-Dimethylquinoxaline	1.82	
Diphenylmethyleneaniline	1.94	
Isoquinoline	2.22	
3-Methylisoquinoline	2.28	
2-Methylpyrazine	2.23	
2-Methylpyridine	2.80	
3-Methylpyridine	2.77	
4-Methylpyridine	2.86	
2-Methylquinoline	2.20	
4-Methylquinoline	2.25	
6-Methylquinoline	2.19	
8-Methylquinoline	2.22	
2-Methylquinoxaline	1.85	
1,5-Naphthyridine	1.86	
Phenanthridine	2.12, 2.64	
1,10-Phenanthroline	2.12, 2.70	
Phenazine	1.20, 2.01	
2-Phenylpyridine	2.30, 2.78	
4-Phenylpyridine	2.24, 2.80	
4-Phenylpyrimidine	2.00, 2.67	
Phthalazine	2.02	
Pyrazine	2.17	
Pyridazine	2.22	
Pyridine	2.76	
1,2-bis-(2-Pyridyl)ethylene	1.92, 2.32	
1,2-bis-(4-Pyridyl)ethylene	1.69, 2.11	
Pyrimidine	2.35	
Quinoline	2.175	
Quinoxaline	1.80	
2,3,5,6-Tetramethylpyrazine	2.50	
1,3,5-Triazine	2.105	
2,4,6-Trimethylpyridine	2.91	

DIMETHYLFORMAMIDE (Continued)

0.1N Et₄NBr at DME
1 mM

Species	$E_{1/2}V$ (vs SCE)	Ref.
2-Aldehydophenazine	−1.04	134
2-Chloro-6-methoxyphenazine	−1.22	
1-Chlorophenazine	−1.24	
2-Chlorophenazine	−1.11	
1-CONH₂Phenazine	−1.02	
2-Cyanophenazine	−0.93	
1,6-Diethoxyphenazine	−1.39	
1-Methoxyphenazine	−1.28	
2-Methoxyphenazine	−1.29	
2-Methylphenazine	−1.27	
2-Nitrophenazine	−0.83	
Phenazine	−1.24	
2-Phenylphenazine	−1.18	

0.1M Et₄NClO₄ at DME

Species	$E_{1/2}V$ (vs Ag/AgNO₃)	Ref.
trans-Azobenzene	−1.81; −2.29 −1.82[a]; −1.80[b]; −1.77[c]	135

[a] 0.1M KClO₄.
[b] 0.1M NaClO₄.
[c] 0.1M LiCl or LiClO₄.

0.1M Et₄NClO₄ at DME at 25 ± 0.5°C
1 mM

Species	$E_{1/2}V$ (vs SCE)	Ref.
Quinoline	∼−2.1; ∼−2.6	149

0.1M LiCl at DME
0.05 to 0.5 mM

Species	$E_{1/2}V$ (vs Hg-pool)	Ref.
2,2-Dinitropropane	−0.455 to −0.458	151
2,2-Dinitropropane[a]	−0.462[a]; −0.491[b]	
2-Nitropropane	−1.110	

[a] 1% (volume) of phenol.
[b] 20% (volume) of phenol added.

VI. POLAROGRAPHY

DIMETHYLFORMAMIDE *(Continued)*
0.1M Et_4NClO_4 at mercury HDE
0.783 to 1.2 mM

Species	$E_{1/2}V$ (vs SCE)	Ref.
Benzfurazan	−1.26	152
o-Dinitrobenzene	−0.72; −1.08; −1.66; −2.24	
p-Dinitrobenzene	−0.535; −0.84; −2.24	
o-Nitrophenol	−0.85; −1.74	

0.1N LiCl at DME at 25.0 ± 0.1°C
0.4 mM

Species	$E_{1/2}V$ (vs SCE)	Ref.
bis(2-p-Acetamidophenyl-1,3-indandione)	−0.81; −1.66	136
bis(2-Anisyl-1,3-indandione)	−0.86; −1.68	
2-Bromo-2-phenyl-1,3-indandione	−0.10; −1.67	
2-Chloro-2-phenyl-1,3-indandione	−0.10; −1.66	
2-Dimethylamino-2-phenyl-1,3-indandione	−0.53; −0.85; −0.95; −1.34	
bis(2-p-Dimethylaminophenyl-1,3-indandione)	−0.91; −1.71	
bis(2-p-Dimethylaminophenyl-1,3-indandione) methyl nitrate	−0.57; −1.11; −1.30; −1.54	
2-Methyl-2-phenyl-1,3-indandione	−0.97; −1.31; −1.45; −1.57	
bis(2-p-Nitrophenyl-1,3-indandione)	−0.38; −0.83; −1.12; −1.67	
2-Phenyl-1,3-indandione	−1.67	
bis 2-Phenyl-2-(p-Dimethylamino)-phenyl-1,3-indandione)	−0.94; −1.79	
bis(2-Phenyl-1,3-indandione)	−0.77; −1.63	
Sodium salt of 2-phenyl-1,3-indandione	−1.76	

0.1N Et_4NClO_4 at DME at 25 ± 0.5
1 mM

Species	$E_{1/2}V$ (vs SCE)	Ref.
6-Chloroquinoline	−1.85; −2.13; −2.55	139

DIMETHYLFORMAMIDE (Continued)

0.1M Et$_4$NClO$_4$ at DME at 25 ± 0.5°C
~0.8 mM

Species	$E_{1/2}V$ (vs SCE)	Ref.
8-Hydroxyquinoline	−1.82; −2.5	140

0.04M Bu$_4$NI at DME at 22°C
0.33 mM

Species	$E_{1/2}V$ (vs SCE)	Ref.
Diphenylacetylene	−2.11; −2.42	141
p,p'-Dimethyldiphenylacetylene	−2.22	
cis-sym-Diphenylethylene	−2.07; −2.36	
trans-sym-Diphenylethylene	−2.08; −2.38	
p-Methoxydiphenylacetylene	−2.24	
m-Methyldiphenylacetylene	−2.17; −2.48	
o-Methyldiphenylacetylene	−2.18; −2.49	
p-Methyldiphenylacetylene	−2.16	
1-Naphthylphenylacetylene	−1.91; −2.26	
p-Nitrodiphenylacetylene	−0.89; −1.48; −2.24	

0.1N KNO$_3$ at DME
0.95 to 9.52 mM

Species	$E_{1/2}V$ (vs SCE)	Ref.
12,15-Dihydro-12,15-dioxo-2,3,6,7-dibenzotriptycene	−0.676; −1.276	143
12,15-Dihydro-12,15-dioxotriptycene	−0.64; −1.22	

0.1M Et$_4$NClO$_4$ at DMF
0.98 mM

Species	$E_{1/2}V$ (vs Ag/Ag$^+$)	Ref.
Azobenzene	−1.81; −2.29	147

VI. POLAROGRAPHY

DIMETHYLFORMAMIDE (Continued)
0.1M Bu$_4$NClO$_4$ at DME at 25.0 ± 0.05°C
1 mM

Species	$E_{1/2}V$ (vs Ag/AgClO$_4$)	Ref.
Anthracene	−2.405	86
Naphthalene	−2.917	

0.1M Et$_4$NBr at DME
1 mM

Species	$E_{1/2}V$ (vs Ag/AgBr)	Ref.
Camphor anil	−2.14	87
α-Methylbenzylidene-α-methyl-benzylamine	−1.75	
Norcamphor anil	−1.98	

0.1M Bu$_4$NI at DME
1.5 mM

Species	$E_{1/2}V$ (vs SCE)	Ref.
Anthracene	−1.920	92
1,9-Diphenylanthracene	−1.846	
1,10-Diphenylanthracene	−1.787	
9,10-Diphenylanthracene	−1.835	
1-Phenylanthracene	−1.878	
2-Phenylanthracene	−1.872	
9-Phenylanthracene	−1.863	
Rubrene	−1.410	
Tetracne	−1.530	

DIMETHYLFORMAMIDE (*Continued*)
0.1M Bu₄NI at DME at 30°C

Species	$E_{1/2}V$ (vs SCE)	Ref.
α,α-Dimethyl-*trans*-stilbene	−2.59	93
α,α'-Diphenyl-*trans*-stilbene	−2.05	
2,4,6,2',4',6'-Hexamethyl-*trans*-stilbene	−2.46	
2-Methylnaphthalene	−2.52	
α-Methyl-*trans*-stilbene	−2.26	
Naphthalene	−2.49	
cis-stilbene	−2.18	
α,α'-D₂-*cis*-stilbene	−2.19	
trans-stilbene	−2.15	
2,4,6-Trimethyl-*trans*-stilbene	−2.28	

0.1M Bu₄NClO₄ at HMD
1.03 mM at 3.25 mM

Species	$E_{1/2}V$ (vs SCE)	Ref.
4,4'-azobispyridine-1,1'-dioxide	−0.76 to −0.80[a]; −1.40 to −1.42[a]	94
	−0.67 to −0.71[b]; −1.30 to −1.35[b]	
	−0.77 to −0.78[c]; −0.69 to −0.71[d]	

[a] E values for cathodic process.
[b] E values for anodic process.
[c] E values for cathodic process at Pt disk electrode.
[d] E values for anodic process at Pt disk electrode.

0.1M Bu₄NClO₄ at DME
2.69 mM

Species	$E_{1/2}V$ (vs Ag/AgCl)	Ref.
4,5-methylenephenanthrene	−2.39; −2.78	97

VI. POLAROGRAPHY

DIMETHYLFORMAMIDE (Continued)
0.1M Bu₄NBr at DME
0.938 to 1.45 mM

Species	$E_{1/2}V$ (vs Hg pool)	Ref.
Anthraquinone	−0.34; −1.10	100
Benzoquinone	−0.92	
2-methyl-1,4-naphthaquinone	−0.20; −1.02	
2-methyl-1,4-naphthaquinone[a]	−0.34; −1.05	

[a] 0.1M NaNO₃ Supporting electrolyte.

0.44M Bu₄NI at Pt-electrode at 25 ± 1.0°C
1.71 to 3.31 mM

Species	E_pV (vs Ag wire)	Ref.
Diethylfumarate	−0.80; −0.76[a]	98

[a] $E_{p/2}$ value of 10 mM conc.

0.1M Bu₄NI at DME at 25.0 ± 0.2°C
∼1 mM

Species	$E_{1/2}V$ (vs Hg pool)	Ref.
Acepleiadylene	−1.10	72
Anthracene	−1.405	
Azulene	−1.1	
2,3-Benzobiphenylene	−1.67	
1,2-Benzpyrene	−1.31	
4,5-Benzpyrene	−1.58	
Biphenyl	−2.05	
Biphenylene	−1.73	
Chrysene	−1.77	
1,2,3,4-Dibenzanthracene	−1.53	
1,2,5,6-Dibenzanthracene	−1.55	
1,2,4,5-Dibenzpyrene	−1.36	
Dihydropyrene	−1.946	
1,2-Dimethylnaphthalene	−2.046	
1,4-Dimethylnaphthalene	−2.023	
1,6-Dimethylnaphthalene	−2.055	
1,7-Dimethylnaphthalene	−2.056	
2,3-Dimethylnaphthalene	−2.090	
2,6-Dimethylnaphthalene	−2.070	

DIMETHYLFORMAMIDE (Continued)

0.1M Bu$_4$NI at DME at 25.0 ± 0.2°C

~1 mM

Species	$E_{1/2}V$ (vs Hg pool)	Ref.
Fluoranthene	−1.23	
Sym-Hexahydropyrene	−2.200	
2-methylanthracene	−1.450	
9-methylanthracene	−1.417	
3-Methylfluoranthene	−1.267	
7-Methylfluoranthene	−1.273	
8-Methylfluoranthene	−1.304	
1-Methylnaphthalene	−2.022	
2-Methylnaphthalene	−2.042	
2-Methylphenanthrene	−1.942	
3-Methylphenanthrene	−1.980	
9-Methylphenanthrene	−1.969	
1-Methylpyrene	−1.549	
2-Methylpyrene	−1.541	
4-Methylpyrene	−1.555	
Naphthalene	−1.994	
Perylene	−1.17	
Phenanthrene	−1.920	
Pyrene	−1.526	
Tetrahydropyrene	−2.170	
Triphenylene	−1.91	

0.1M Bu$_4$NClO$_4$ at DME at 25°C

1–3 mM

Species	$E_{1/2}V$ (vs SCE)	Ref.
Cyclooctatetraene	~−1.15; ~−1.4	69

at DME

Species	$E_{1/2}V$ (vs Hg pool)	Ref.
4-*t*-Butyl	−2.10	71
4,4′-Di-*t*-butyl	−2.14	

VI. POLAROGRAPHY

DIMETHYLFORMAMIDE (Continued)
0.1M Pr_4NClO_4 at DME at 20 ± 1°C
0.1 to 1 mM

Species	$E_{1/2}V$ (vs SCE)	Ref.
Etioprophyrin I	−1.37; −1.80; −2.67	78
Tetraphenylprophin	−1.08; −1.52; −2.38; −2.53	
[$N(n$-Pr)$_4$]$_2$tetraphenylprophin	−1.45; −1.87; −2.26; −2.44	

0.1M Pr_4NClO_4 at DME
~1 mM

Species	$E_{1/2}V$ (vs SCE)	Ref.
H_2EtioI[a].	−0.92; −1.34	77

[a] Etioprophine I.

0.05M Et_4NI at DME at 25°C
1 mM

Species	$E_{1/2}V$ (vs SCE)	Ref.
Diphenyl sulfide	−2.580	82
β,β'-Dithienyl sulfide	−2.570	

0.1M Et_4NI at DME at 25 ± 0.1°C
0.1 mM to 10 mM

Species	$E_{1/2}V$ (vs SCE)	Ref.
Acrylonitrile	−0.47	158

0.1N LiCl at DME at 25 ± 0.5°C
1 mM

Species	$E_{1/2}V$ (vs SCE)	Ref.
2-Mercapto-5-ethyl-3-thenyl-idene-cyclohexylamine	−1.69	160
2-Methylmercapto-5-methyl-3-thenylidenecyclohexylamine	>-1.9	

DIMETHYLFORMAMIDE (Continued)
0.5F Et$_4$NClO$_4$ at DME at 25°C
1.5 mM soln.

Species	$E_{1/2}V$ (vs SCE)	Ref.
4-Hydroxycoumarin	−1.95	66
3,3′-Methylene-bis-(4-hydroxycoumarin)	−1.23; −2.2	

0.1M Et$_4$NCl at DME at 25°C
1 to 10 mM soln.

Species	$E_{1/2}V$ (vs SCE)	Ref.
α-Bromobutanal	+0.9 to 1.0; +1.5	65
α-Chlorobutanal	+0.9 to 1.0; +1.5	
α-Chloroheptanal	+0.9 to 1.0; +1.5	
α-Chloroisobutanal	+1.00; +2.1	
2-Chloro-2-methylpentanal	+1.0	
α-Fluorobutanal	∼+1.6	
α-Fluoroisobutanal	∼+1.7	

0.1M Bu$_4$NClO$_4$ at DME at 25.0 ± 0.05°C
1.0 mM

Species	$E_{1/2}V$ (vs Ag/AgClO$_4$)	Ref.
Anthracene	−2.405	34
Naphthalene	−2.917	

0.1M Et$_4$NBr at DME
1.0 mM

Species	$E_{1/2}V$ (vs Ag/AgBr)	Ref.
Camphor anil	−2.14	30
α-Methylbenzylidene-α-methyl-benzylamine	−1.75	
Norcamphor anil	−1.98	

VI. POLAROGRAPHY

DIMETHYLFORMAMIDE (Continued)
0.1N Et$_4$NClO$_4$ at DME at 25°C
2 mM solutions

Species	$E_{1/2}V$ (vs SCE)	Ref.
9-Bromo-o-carborane	−2.21	51
o-Carborane	−2.51	
9-Chloro-o-carborane	−2.34	
9,12-Di-iodo-o-carborane	−1.81	
1,2-Diphenyl-o-carborane	−1.14; −1.40	
Iodobenzene	−1.76	
9-Iodo-o-carborane	−2.13	
1-(p-Iodophenyl)-m-carborane	−1.65; −2.51	
1-(p-Iodophenyl)-o-carborane	−1.60; −2.00	
1-Methyl-2-bromo-o-carborane	−0.64; −2.70	
1-Methyl-o-carborane	−2.44	
1-Phenyl-2-bromo-o-carborane	−0.60; −2.30	
1-Phenyl-m-carborane	−2.45	
1-Phenyl-o-carborane	−1.95	

0.1M Et$_4$NClO$_4$ at DME at 25 ± 0.1°C
or 20° (Ref. 10)

Species	Conc. × 10^3 M/l	$E_{1/2}V$ (vs SCE)	Ref.
Acetophenone anil	1.0	−1.90	10
Benzaldehyde anil	1.0	−1.83	10
Benzophenone anil	1.0	−1.72	10
Benzophenone-N-methylimine	1.0	−2.02	10
Cinnamaldehyde anil	1.0	−1.57	10
Deuteroporphyrin	1.2	−1.29; −1.68; −2.53	1
Fluorenone anil	1.0	−1.34	10
Mesoporphyrin	1.2	−1.34; −1.73; −2.57	1
Pinocamphone anil	1.0	−2.42	10
Tetraphenylbacteriochlorin	0.062	−1.10; −1.55	2
Tetraphenylchlorin	0.22	−1.12; −1.52; −2.43	2
Tetraphenylporphyrin	0.36	−1.08; −1.45; −2.36; −2.48	2

DIMETHYLFORMAMIDE (Continued)
DME

Species	Conc. × 10³ M/l	Supporting electrolyte	$E_{1/2}V$ (vs SCE)	Ref.
Cyclo-octatetraene	1.25	0.1M Bu$_4$NClO$_4$	−1.74; −1.97	5
3,8-Dimethyl-2-methoxyazocine	1.25	0.1M Bu$_4$NClO$_4$	−2.17	
2-Methoxyazocine	1.25	0.1M Bu$_4$NClO$_4$	−1.84	
3,5,6,8-2-Methoxyazocine	1.25	0.1M Bu$_4$NClO$_4$	−2.46	
Methoxycyclo-octatetraene	1.25	0.1M Bu$_4$NClO$_4$	−1.87; −2.05	
8-Methyl-2-methoxyazocine	1.25	0.1M Bu$_4$NClO$_4$	−1.92	
4,6,8-Trimethyl-2-methoxyazocine	1.25	0.1M Bu$_4$NClO$_4$	−2.22	

DME

Species	Conc. × 10³ M/l	Supporting electrolyte	$E_{1/2}V$ (vs Hg-pool)	Ref.
N-Methylphthalimide	1.0	0.5M Et$_4$NClO$_4$	−1.32; −2.15	11
Oxyphthalimide	1.0	0.5M Et$_4$NClO$_4$	−2.25	
Phthalimide	1.0	0.5M Et$_4$NClO$_4$	−1.42; −2.25	
Phthalylglycine	1.0	0.5M Et$_4$NClO$_4$	−1.35; −1.65; −2.20	
Phthalylglycine-methyl ester	1.0	0.5M Et$_4$NClO$_4$	−1.30; −2.15	

0.1M Et$_4$NClO$_4$ at DME at RPDE
~1.0 mM

Species	pH	$E_{1/2}V$ (vs Ferrocene)	Ref.
p-Benzoquinone	1.0	0.73	14
	7.1	0.62	
	10.2	0.47	
	12.9	0.05	

VI. POLAROGRAPHY

DIMETHYLFORMAMIDE (Continued)
0.1M Bu$_4$NClO$_4$ at DME at 20 ± 0.5°C
0.5 mM

Species	pH	$E_{1/2}V$ (vs SCE)	Ref.
C$_6$H$_5$—CH=NC$_6$H$_5$	7–9	−1.90	24
(C$_6$H$_5$—CH=N—CH$_2$)$_2$	7–9	−2.27	
C$_6$H$_5$—CH=N—CH$_3$	7–9	−2.27	
C$_6$H$_5$NO$_2$	7–9	−1.13	
C$_6$H$_5$—CHO	7–9	−1.83	

0.1F Et$_4$NClO$_4$ + 0.005% polyvinyl chloride at DME at 25 ± 0.1°C
0.01–1.0 mM

Species	$E_{1/2}V$ (vs SCE)	Ref.
Bis(2-phenyl-3-indolinone)azine	−0.34; −0.635	19

0.1M Et$_4$NClO$_4$ at DME
0.5 mMolar Solution

Species	$E_{1/2}V$ (vs SCE)	Ref.
Perlauric acid	−0.20	44

0.05M Et$_4$NClO$_4$ at DME at 25°C
0.5 mM

Species	$E_{1/2}V$ (vs SCE)	Ref.
1,2-Naphthoquinone	−0.51; −1.12	28
1,4-Naphthoquinone	−0.63; −1.35	

DIMETHYLFORMAMIDE (Continued)
Pb electrode; 1000 mM

Species	Supporting electrolyte	$E_{1/2}V$ (vs Hg/HgCl$_2$)	Ref.
EtBr	0.25M Et$_4$NClO$_4$	-1.94	27
	0.25M NaClO$_4$	-2.12	
EtI	0.25M NaClO$_4$	-1.70	
	0.25M Et$_4$NClO$_4$	-1.60	
MeBr	0.25M Et$_4$NClO$_4$	-1.69	
	0.25M NaClO$_4$	-1.95	
MeI	0.25M NaClO$_4$	-1.54	
	0.25M Et$_4$NClO$_4$	-1.50	

0.1M Pr$_4$NClO$_4$ at DME

Species	$E_{1/2}V$ (vs SCE)	Ref.
4-Bromo-2,6-dimethylbenzonitrile	$-1.849; -2.367$	39
4-Bromo-2,6-dimethylbenzonitrile N-oxide	$-1.90; -2.39$	
2,6-Dimethylbenzonitrile	-2.403	
2,6-Dimethylbenzonitrile N-oxide	$-1.95; -2.38$	
4-Methoxy-2,6-dimethylbenzonitrile	-2.592	
4-Methoxy-2,6-dimethylbenzonitrile N-oxide	$-2.24; -2.62$	
4-Nitro-2,6-dimethylbenzonitrile	$-0.879; -1.706$	
4-Nitro-2,6-dimethylbenzonitrile N-oxide	$-0.915; -1.132$	
2,4,6-Trimethylbenzonitrile	-2.505	
2,4,6-Trimethylbenzonitrile N-oxide	$-2.12; -2.52$	

0.1M Et$_4$NClO$_4$ at DME at 25°C

Species	Conc. $\times 10^3$ M/l	$E_{1/2}V$ (vs SCE)	Ref.
Benzhydryl bromide	0.1–2.5	$+0.171$ to $+0.193$	37
		-0.114 to -0.195	
		-0.977 to -1.067	
Benzhydryl chloride	0.1–0.6	-1.642 to -1.679	
4-Nitropyridine N-oxide		$-0.80; -1.69$	38

VI. POLAROGRAPHY

DIMETHYLFORMAMIDE (Continued)
0.1M Pr$_4$NClO$_4$ at DME
1 mM solutions

Species	$E_{1/2}V$ (vs SCE)	Ref.
2,2'-dinitrobiphenyl	-0.99; -1.32; -2.0 $\sim -1.00^a$; $\sim -1.4^a$; $\sim -2.0^a$	33
2-nitrobiphenyl	-1.14; -1.78 -1.17^a; -1.88^a	

[a] Planar platinum disk electrode and peak potentials.

Species	$E_{1/2}V$ (vs SCE)	Ref.
Acetophenone	-1.87; -2.32	56
α-Dimethylaminoacetophenone	-1.83; -2.28	
α-Morpholinoacetophenone	-1.81; -2.22	
α-Piperidienoacetophenone	-1.84; -2.25	
α-Trimethylammoniumacetophenone	-1.08; -1.88; -2.30	

0.1N Et$_4$NClO$_4$ at DME at 25°C

Species	Conc. $\times 10^3$ M/l	$E_{1/2}V$ (vs SCE)	Ref.

(CH$_3$)$_3$C — [ring with C=O top, C(CH$_3$)$_3$ top-right, R—C—R' bottom]

R	R'			
H	CH$_3$	1.6	-1.28; -2.01	41
CH$_3$	CH$_3$	0.9	-0.81; -1.42	
CH$_3$	C$_2$H$_5$	0.9	-1.41	
C$_6$H$_5$	C$_6$H$_5$	0.9	-1.11; -1.41	
H	CH$_2$Br	1.6	-1.52	
H	CN	1.6	-0.26; -0.50	

DIMETHYLFORMAMIDE (*Continued*)
0.1N LiClO$_4$ at Pt electrode

Species	$E_{1/2}V$ (vs Hg/Hg^{2+})	Ref.
C$_2$H$_5$OCO\\ 　　　CH—CH=CH—CHO· C$_2$H$_5$OCO/	+0.17; +0.14 to 0.20	154
C$_2$H$_5$OCO\\ 　　　CH—CH=CH·CHO· COCH$_3$/	+0.14	
COCH$_3$\\ 　　　CH—CH=CH·CHO· COCH$_3$/	+0.19	
CO(CH$_2$)$_3$CO\\ 　　　CH—CH=CH·CHO CO(CH$_2$)$_3$CO/	+0.13	
CN\\ 　　　CH—CH=CH·CHO CONH$_2$/	−0.05	

VI. POLAROGRAPHY

DIMETHYLFORMAMIDE (Continued)
0.1M Et$_4$NClO$_4$ at DME at 25°C

Species		Conc. $\times 10^3$ M/l	$E_{1/2}V$ (vs SCE)	Ref.
p-Quinolide				
4R	4R'			
CH$_3$	Br	0.9	−0.31; −0.46	42
CH$_3$	Cl	0.9	−1.13	
CH$_3$	NO$_2$	0.9	−1.30	
CH$_3$	OCH$_3$	0.9	−1.31; −2.20	
CH$_3$	OH$_3$	0.9	−1.59; −2.18	
CH$_3$	CH$_3$	0.9	−1.84; −2.27	
CH$_3$	C(CH$_3$)$_3$	1.6	−2.06; −2.35	
H$_2$C——CH$_2$		0.9	−1.98	
CH$_3$	C$_2$H$_5$	1.6	−2.22	
Quinobromide compounds				
Br	4R	1.6		42
4Br	H	1.6	−1.66	
4Br	CH$_3$	0.62	−0.31; −0.46	
4Br	CH$_2$Br	0.9	−0.36; −1.40	
4Br	Br	0.9	+0.06; −0.24	
2Br	CHO	0.9	−0.13; −0.42	
2Br	COCH$_3$	0.9	−0.11; −0.37	
2Br	CN	0.9	−0.01; −0.32	
2Br	Br	0.9	+0.05; −0.39	
2Br	NO$_2$	0.9	+0.13; −0.29	

DIMETHYLFORMAMIDE (*Continued*)
0.05M Et$_4$NI at DME
10 mM

Species	$E_p V$ (vs Ag pool)	Ref.
C$_6$H$_5$–N=CH–C$_6$H$_5$	1.40	115
(4-Cl)C$_6$H$_4$–N=CH–C$_6$H$_5$	1.19, 1.25	
C$_6$H$_5$–N=CH–(naphthyl)	1.27, 1.41	
C$_6$H$_5$–N=CH–C$_6$H$_4$(OH)	1.10	
(2-Cl)C$_6$H$_4$–N=CH–C$_6$H$_4$(OH)	1.01	
(3-Cl)C$_6$H$_4$–N=CH–C$_6$H$_4$(OH)	1.00	
C$_6$H$_5$–N=CH–(naphthyl-OH)	0.97	
(2-Cl)C$_6$H$_4$–N=CH–(naphthyl-OH)	0.92	
(3-Cl)C$_6$H$_4$–N=CH–(naphthyl-OH)	0.90	
C$_6$H$_5$–CH=N–C$_6$H$_4$–C$_6$H$_4$–N=CH–C$_6$H$_5$	1.22, 1.28, 1.36	
C$_6$H$_5$–CH=N–C$_6$H$_3$(Cl)–C$_6$H$_3$(Cl)–N=CH–C$_6$H$_5$	1.095, 1.22, 1.32	
C$_6$H$_5$–CH=N–C$_6$H$_3$(Cl)–C$_6$H$_3$(Cl)–N=CH–C$_6$H$_5$	1.19, 1.31	

VI. POLAROGRAPHY

DIMETHYLFORMAMIDE (Continued)
0.05M Et₄NI at DME
10 mM

Species	$E_p V$ (vs Ag pool)	Ref.
naphthyl-CH=N-C₆H₄-C₆H₄-N=CH-naphthyl	1.10, 1.25, 1.40	
(2-OH-phenyl)-CH=N-C₆H₄-C₆H₄-N=CH-(2-OH-phenyl)	1.00, 1.06, 1.18, 1.41	
(2-OH-phenyl)-CH=N-(3-Cl-C₆H₃)-(3-Cl-C₆H₃)-N=CH-(2-OH-phenyl)	0.90, 0.99	
(2-OH-phenyl)-CH=N-C₆H₄-(3,3'-Cl₂)-C₆H₄-N=CH-(2-OH-phenyl)	0.98, 1.01	
naphthyl(OH)-CH=N-C₆H₄-C₆H₄-N=CH-(phenyl-OH)	0.96, 1.05	
naphthyl(OH)-CH=N-(Cl-C₆H₃)-(Cl-C₆H₃)-N=CH-(phenyl-OH)	0.93	
naphthyl(OH)-CH=N-C₆H₄(Cl,Cl)-C₆H₄-N=CH-(phenyl-OH)	0.93	

0.05M Et₄NI at DME
∼0.5 mM

Species	$E_{1/2}V$ (vs SCE)	Ref.
2,4-Dimethyl-α-methyl styrene	−2.639	121
3,5-Dimethyl-α-methyl styrene	−2.503	
4-Isopropyl-α-methyl styrene	−2.522	
α-Methyl styrene	−2.473	
Styrene	−2.433	

DIMETHYLFORMAMIDE (Continued)
0.04M Et$_4$NI at DME

Species	$E_{1/2}V$ (vs SCE)	Ref.
C$_6$H$_5$CH=CH—CH$_2$OH	−2.57; −2.80; −2.49[a]; −2.70[a]	117
C$_6$H$_5$CH=CH$_2$	−2.60; −2.51[a]	
C$_6$H$_5$COOH	−2.24[b]; −2.17[a]; −2.86[a]	
C$_6$H$_5$COOCH$_3$	−2.32; −2.83; −2.29[a]; −2.61[a]	
C$_6$H$_5$CONH$_2$	−2.53; −2.76; −2.44[a]; −2.56[a]	
C$_6$H$_5$CHO	−1.93; −2.51; −1.83[a]; −2.33[a]	

[a] In presence of 5% water.
[b] Hydrogen wave.

Et$_4$NBr at DME
2.5 mM

Species	$E_{1/2}V$		Ref.
	I	II	
Acetophenone semicarbazone	1.86	DB[a]	102
Acetophenone thiosemicarbazone	1.55	2.10	
Anisylaldehyde m-nitrophenylhydrazone	2.22[b]		
Benzaldehyde benzoylhydrazone	1.50	1.93	
Benzaldehyde p-bromophenylhydrazone	1.67	2.10	
Benzaldehyde p-carboxyphenylhydrazone	1.84	DB[a]	
Benzaldehyde isonicotinoylhydrazone	1.21	1.81; 2.11	
Benzaldehyde nicotinoylhydrazone	1.44	1.89 DB	
Benzaldehyde m-nitrophenylhydrazone	2.16[b]		
Benzaldehyde phenylhydrazone	−1.79	2.15	
Benzaldehyde semicarbazone	1.66	2.14	
Benzaldehyde thiosemicarbazone	1.48	2.05	
Benzyl ethyl ketone isonicotinoylhydrazone	1.28	2.04	
p-Bromophenylacetone semicarbazone	1.86[b]		
p-Chloroacetophenone semicarbazone	1.70	2.00	
Cumyl aldehyde phenylhydrazone	1.88	DB[a]	
p-Dimethylaminobenzaldehyde m-nitrophenylhydrazone	2.21[b]		
p-Dimethylaminobenzaldehyde phenylhydrazone	2.07	DB[a]	

VI. POLAROGRAPHY

DIMETHYLFORMAMIDE (Continued)
Et₄NBr at DME
2.5 mM

Species	$E_{1/2}V$	Ref.
p-Dimethylaminobenzaldehyde thiosemicarbazone	1.81	DB[a]
2,4-Hydroxybenzaldehyde semicarbazone	1.64[c]	DB[a,c]
2-Hydroxy-5-chlorobenzaldehyde semicarbazone	1.81	DB[a]
Isonicotinamide	1.53	2.00
Salicylaldehyde semicarbazone	1.05[c]; 1.52	DB[a]

[a] Decanted from background.
[b] Summary wave.
[c] Prewave.

0.1M Et₄NClO₄ at DME
1.2 mM

Species	$E_{1/2}V$ (vs SCE)	Ref.
Conjugate base of 1-hydroxyanthraquinone	+0.575[b]	106
Conjugate base of 1-hydroxy-9-10-anthraquinone[a]	−1.365; −1.36[b]	
1-Hydroxy-9-10-anthraquinone	−0.645; −1.26	

[a] 1.83 mM.
[b] At RE(Pt).

0.1M LiClO₄ at RE(Pt) at 25°C
~0.5 mM

Species	$E_{1/2}V$ (vs SCE)	Ref.
bis(2,9-Dimethyl-1,10-phenanthroline)Cu(I)	+0.615	104
tris(4,7-Dimethyl-1,10-phenanthroline)Fe(II)	+0.945	

DIMETHYLFORMAMIDE (Continued)

0.05N Et$_4$NI at DME at 25°C
0.87 to 0.919 mM

Species	$E_{1/2}V$ (vs SCE)	Ref.
Anthracene	−1.93; −2.28	123
9-Vinylanthracene	−1.78; −2.26	
1-Vinylnaphthalene	−2.08; −2.48	
2-Vinylnaphthalene	−2.09 −2.50	

0.1M Bu$_4$NClO$_4$ at DME at 25 ± 0.05

Species	$E_{1/2}V$ (vs Ag/AgClO$_4$)	Ref.
Anthracene	−2.405	167
Naphthalene	−2.917	

0.1M Bu$_4$NClO$_4$ at DME
1.32 mM

Species	$E_{1/2}V$ (vs SCE)	Ref.
4,4′-Azopyridine-1,1′-dioxide	−0.73; −1.35	168
	−0.76 to −0.80[a]	
	−0.71 to −0.69[a]	
	−1.40[a] to −1.33[a]	
	−0.77[c]; −0.71[b]	

[a] 1.03 mM at hanging mercury drop electrode.
[b] 1.74 mM at platinum disk electrode.

DIMETHYLFORMAMIDE (Continued)
0.1M Et$_4$NI at DME
0.1 to 0.2 mM

Species	$E_{1/2}V$ (vs SCE)	Ref.
1-Aminoanthraquinone	−0.44; −0.96; −0.43[a]; −0.73[a]; −0.43[b]	125
Anthraquinone	−0.33; −0.95	
Dinitrodianthrimide	−0.04; −0.62; −0.905; −1.03; −0.08[e]; −0.41[e]; −0.59[e]	
Mononitrodianthrimide	−0.07; −0.34; −0.59; −0.78; −1.01; −0.04[e]; −0.27[e]; −0.53[e]	
Nitroanthraquinone	−0.07; −0.70; −0.975; −0.07[d]; −0.52[d]; −0.69[d]; −1.09[e]; −0.37[e]	
Nitrobenzene	−0.58; −1.13; −0.525[c]; −0.895[c]	
1-Nitronaphthalene	−0.56; −1.15	
2-Nitronaphthalene	−0.55; −1.17	

[a] 9.5% alcohol added.
[b] 54% alcohol added.
[c] 2.1% water added.
[d] 9.5% alcohol added.
[e] 0.05M acetic anhydride added.

0.1M Et$_4$NI at DME at 20°C
0.2 to 2.5 mM

Species	$E_{1/2}V$ (vs mercury base)	Ref.
CH$_2$=CHCN	+1.67	129

CH$_2$=CCN
    |
    CH$_3$
	+1.79	

CH$_3$CH=CHCN	+1.91	
CH$_3$CH$_2$CH=CHCN	+1.91	
CNCH$_2$CH$_2$CH=CHCN	+1.73	
CH$_2$=CH—CH=CHCN	−1.31; −1.91	
C$_6$H$_5$CN	−1.82	
C$_6$H$_5$CH=CHCN	−1.32; −1.81	

DIMETHYLFORMAMIDE WATER
0.1M Et$_4$NClO$_4$ at DME
0.5 mM Solution

Species	% Water	$E_{1/2}V$ (vs)	Ref.
Perlauric acid	5	−0.02	44
	10	+0.04	
	20	+0.09	

(65% DMF)
0.1M HClO$_4$ + 0.5M NaClO$_4$ at DME at 25°C
0.3 mM

Species	$E_{1/2}V$ (vs SCE)	Ref.
Mercuric diethyldithiocarbonate	−0.36; −0.52	29

0.1M NaNO$_3$ at DME

Species	$E_{1/2}V$ (vs Hg pool)	Ref.
2-Methyl-1,4-naphthohydroquinone[a]	+0.927	100
2-Methyl-1,4-naphthohydroquinone[b]	+0.768	
2-Methyl-1,4-naphthohydroquinone[c]	+0.543	
2-Methyl-1,4-naphthohydroquinone[d]	+0.380	

[a] 2.24 mM + 0.5% water by volume.
[b] 3.10 mM + 4.76% water by volume.
[c] 2.61 mM + 20.00% water by volume.
[d] 1.86 mM + 42.86% water by volume.

(95% Dimethyl formamide)
0.04M Et$_4$NI at DME
0.5 to 1 mM

Species	$E_{1/2}V$ (vs SCE)	Ref.
Acetate of cinnamic alcohol	−2.21; −2.69	142
C$_{16}$ alcohol	−2.78	
Cinnamic alcohol	−2.49; 2.70	
Retinol	−2.13; −2.32; −2.69	
Retinyl acetate	−1.87; −2.32; −2.68	

VI. POLAROGRAPHY

DIMETHYLFORMAMIDE WATER (*Continued*)
DME; 0.5 mM

Species	Supporting electrolyte	$E_{1/2}V$ (vs SCE)	Ref.
tert-Butylcyclohexane dibromide	0.03M Me$_4$NBr	-1.15 ± 0.01	25
	0.36M Bu$_4$NBr	-1.13 ± 0.02	
Decalin dibromide	0.03M Me$_4$NBr	-1.02 ± 0.01	
	0.36M Bu$_4$NBr	-0.96 ± 0.02	

DIMETHYLSULFOXIDE
0.8M NaClO$_4$ at DME
3 mM

Species	$E_{1/2}V$ (vs SCE)	Ref.
Riboflavin	$-0.71; -1.02; -1.25;$ $-0.77^a; -1.08^a; -1.30^a$	110

[a] These are peak potentials at 0.1M NaClO$_4$ at hanging mercury drop electrode at conc. 1.06 mM.

Et$_4$NClO$_4$ at DME
2 mM

Species	$E_{1/2}V$ (vs SCE)	Ref.
Ammonia	$+0.02$	100
Ammonium chloride	$-0.17; -2.04$	
Et$_4$NCl	-0.17	
Hydrogen chloride	$+0.01; -0.17; -1.14;$	
Perchloric acid	-1.14	
Potassium chloride	$+0.01; -0.17; -2.22;$	
Potassium perchlorate	-2.11	

DIMETHYLSULFOXIDE (Continued)
Et$_4$NClO$_4$ at DME
2 mM

Species	$E_{1/2}V$ (vs SCE)	Ref.
Acetic acid	-1.91; -2.22;	103
2-Aminoethanol	$+0.07$; $+0.02$;	
2-Aminoethanethiol hydrochloride	-0.17; -1.90; -2.40;	
4-Aminophenylsulfone	-0.17; -2.24	
Butylamine	$+0.00$	
Cystine dihydrochloride	-0.17; -0.90; -1.44; -2.10	
Cysteine hydrochloride	-0.17; -1.38; -1.83; -0.07; -2.14	
Diethyldisulfide	-1.77; -2.04	
2,2'-Dithiobis (ethylamine)- dihydrochloride	-0.17; -0.97; -1.32 -1.96; -2.37	
Ethanethiol	-0.17	

0.1M Bu$_4$NClO$_4$ at HMD[a] at 25°C
~1 mM

Species	$E_{1/2}V$ (vs SCE)	Ref.
C$_6$H$_5$CH—CH—COCH$_3$	-1.62; -2.70; -2.20[b] -1.56[b]; -2.14[c]	104
C$_6$H$_5$CH—CHCOt—C$_4$H$_9$	-2.24[d]	

[a] Hanging mercury drop.
[b] In presence of 0.05% water.
[c] In presence of 5% water.
[d] Peak potential at 0.84 mM in 0.2M Bu$_4$NClO$_4$.

0.1M Bu$_4$NClO$_4$ at DME at 25.0 ± 0.05°C
1 mM

Species	$E_{1/2}V$ (vs Ag/AgClO$_4$)	Ref.
Anthracene	-2.155	86
Naphthalene	-2.763	

VI. POLAROGRAPHY

DIMETHYLSULFOXIDE (Continued)
0.1F Et$_4$NOH at gold electrode
1.73 mM

Species	E_p (vs SCE)	Ref.
(Et$_4$N)$_2$C$_2$O$_4$	+0.46 to 0.20[a]; +0.72 to 0.70[a] +0.66 to 0.46[b]	137
Oxalate ion	+0.25 to +0.01; +0.75 to +0.70; +0.90 to +0.88	

[a] 10% water added.
[b] 25% water.

0.1M Et$_4$NClO$_4$ at gold electrode
~5.58 mM

Species	$E_p V$ (vs SCE)	Ref.
Formic acid	−1.25	138
Tetraethylammonium formate	+0.4 +0.30 to 0.23[a]; +0.35 to +0.50[a]; +0.42 to +0.24[b]; +0.42 to +0.62[b]	

[a] Contain 0.1% water.
[b] Contain 0.2% water.

0.1M Et$_4$NClO$_4$ at RE (Pt disk) at 22–24°C
0.01–0.04M

Species	$E_{1/2} V$ (vs sat. Ag/AgClO$_4$)	Ref.
m-Dinitrobenzene	~−1.4; ~−1.75	84

0.1M Et$_4$NClO$_4$ at Pt
1.52–1.87 mM

Species	$E_p V$ (vs SCE)	Ref.
Quinhydrone	+0.98; −0.31	90

DIMETHYLSULFOXIDE (Continued)
0.1M Pr$_4$NClO$_4$ at DME
~1 mM

Species	$E_{1/2}V$ (vs SCE)	Ref.
H$_2$TPP[a]	−0.70; −1.05; −1.47	77

[a] Tetraphenyl prophine.

0.1M Pr$_4$NClO$_4$ at DME
0.2–1 mM

Species	$E_{1/2}V$ (vs SCE)	Ref.
Chlorophyll	−1.12; −1.54 −1.05; −1.46	80
Ethyl chlorophyllide	−1.11; −1.54	
meso-Ethyl chlorophyllide	−1.17; −1.64	

Species	0.1M Et$_4$NClO$_4$ at DME		
	Conc. × 10^3 M/l	$E_{1/2}V$ (vs SCE)	Ref.
Acetic acid		~−2.3[a]	45
Benzoquinone	0.55	−0.40; −1.24[b]	
Duroquinone	0.39	−0.73; −1.53[b]	
Tetrachloroquinone	0.83	+0.08; −0.73[b]	
Tetraethylammonium bisulfate		−2.5[b]	

[a] At 30°C.
[b] At 25°C.

0.1M Bu$_4$NClO$_4$ at DME at 25 ± 0.05°C

Species	$E_{1/2}V$ (vs Ag/AgClO$_4$)	Ref.
Anthracene	−2.155	34
Naphthalene	−2.763	

VI. POLAROGRAPHY

DIMETHYLSULFOXIDE (Continued)
0.2M Bu$_4$NClO$_4$ at HMDE[a] at 25 ± 0.1°C
2.0 mM

Species	$E_{1/2}V$ (vs SCE)	Ref.
Dibenzoylmethane	−1.42; −1.72 −2.28; −1.36	3

[a] Hanging mercury drop electrode.

0.1M Et$_4$NClO$_4$ at PRDE[a] at ∼25°C
20 mM

Species	$E_{1/2}V$ (vs Ag/AgClO$_4$)	Ref.
m-Dinitrobenzene	∼−1.5; −2.0	17

[a] Platinum rotating disk electrode.

0.1M Bu$_4$NClO$_4$ at DME at 25°C
∼2.0 mM

Species	$E_{1/2}V$ (vs SCE)	Ref.
Acetophenone	−1.94	18
Benzil		
1,4-Dibenzoyl-2,3-diphenyl-2,3-butanediol	−1.70; −1.95	
p,p'-Dimethoxybenzil	−1.20, −1.96	
1,3-Diphenyl-1,3-Propanedione	−1.38; −1.69; −2.25	
Enolate of 1,3-Diphenyl-1,3-propanedione	−2.25; −1.04; −1.76	
1,3-bis(p-Fluorophenyl)-1,3-propanedione	−1.41, −1.73, −2.3	
1,3-bis(p-Methoxyphenyl)-1,3-propanedione	−1.56, −1.87, −2.5	
1-Phenyl-1,3-butanedione	−1.54, −2.41, −2.7	
1-Phenyl-1,2-propanedione	−1.10, −2.16	

DIMETHYLSULFOXIDE (Continued)

0.1F Et$_4$NClO$_4$, Pt working electrode

Species	Conc × 10^3 M/l	$E_{1/2}V$ (vs SCE)	Ref.
Dimethyl hydrazine	9.0	+0.00; −1.0	30
Hydrazine	9.9	+0.00; −1.0	30

0.05M Et$_4$NClO$_4$ at DME at 25°C
0.5 mM

Species	$E_{1/2}V$ (vs SCE)	Ref.
1,2-Naphthoquinone	−0.49; −1.18	28
1,4-Naphthoquinone	−0.58; −1.29	28

0.1M Bu$_4$NClO$_4$ at DME at 25 ± 0.05°C

Species	$E_{1/2}V$ (vs Ag/AgClO$_4$)	Ref.
Anthracene	−2.155	167
Naphthalene	−2.763	

0.1M Et$_4$NClO$_4$ at DME at 25.0 ± 0.1°C
0.4–2.0 mM

Species	$E_{1/2}V$ (vs SCE)	Ref.
Aminophenylsulfone	−2.24	29
Asparagine	−2.382 ± 0.01	
Aspartic acid	−2.141 ± 0.02; −2.618 ± 0.007	
Butyric acid	−2.36	
n-Butylamine	−2.23	

VI. POLAROGRAPHY

DIMETHYLSULFOXIDE (Continued)
0.1M Et$_4$NClO$_4$ at DME at 25.0 ± 0.1°C
0.4–2.0 mM

Species	$E_{1/2}V$ (vs SCE)	Ref.
Cysteine HCl	−0.168 ± 0.007	
	−1.321 ± 0.018	
	−1.788 ± 0.008	
	−2.014 ± 0.011	
	−2.660 ± 0.019	
Dihydroxyphenylalanine	−2.375 ± 0.016; −2.625 ± 0.026	
Glutamic acid	−2.123 ± 0.019; −2.604 ± 0.023	
Glutamine	−2.428 ± 0.021; −2.09 ± 0.02	
Glutaric acid	−2.39	
Hydronyproline	−2.317 ± 0.013	
Isolencine	−2.435 ± 0.013	
Malic acid	−2.25	
Methionine	−2.435 ± 0.011	
Methylglutaric	−2.40	
Methylsuccinic acid	−2.41	
Phenylalanine	−2.398 ± 0.007	
Proline	−2.307 ± 0.008	
Propionic acid	−2.36	
n-Propylamine	−2.20	
Succinic acid	−2.29	
Threonine	−2.368 ± 0.005	
Tryptophan	−2.388 ± 0.007	
Tyrosine	−2.379 ± 0.019	

DIOXANE–WATER
(80% by Vol. of Dioxane)
0.175M Et$_4$NBr at DME at 20°C
~2 mM

Species	E_pV (vs Hg pool)	Ref.
Dibutylfumarate	−1.13	61

DIOXANE–DIMETHYL FORMAMIDE-WATER

(65% Dioxane, 27% DMF V/V)
0.175M Et$_4$NBr at DME at 20°C
~2 mM

Species	E_pV (vs Hg pool)	Ref.
Dibutylfumarate	−1.03	61
Dibutylmaleate	−1.27	
Dioctylfumarate	−1.05	
Dioctylmaleate	−1.37	

(47.5% Dioxane; 47.5% DMF V/V)
0.175M Et$_4$NBr at DME at 20°C
~2 mM

Species	E_pV (vs Hg pool)	Ref.
Dibutylfumarate	−1.02	61
Dioctylfumarate	−1.04	

10% Dioxane, 80% DMF V/V)
0.175M Et$_4$NBr at DME at 20°C
~2 mM

Species	E_pV (vs Hg pool)	Ref.
Dibutylmaleate	−1.17	61
Dioctylmaleate	−1.22	

VI. POLAROGRAPHY

ETHANEL
0.1M LiClO$_4$ at Pt-electrode
~2.5 mM soln.

Species	$E_{1/2}V$ (vs Ag/Ag$^+$)	Ref.
Tetraphenyl-2,3,4,5-pyrrole	+0.28	48

0.1M LiClO$_4$ at RE(Pt) at 25°C
~0.5 mM

Species	$E_{1/2}V$ (vs SCE)	Ref.
Benzylferrocene	+0.385	104
bis(2,9-Dimethyl-1,10-phenanthroline) Cu(I)	+0.720	
tris(4,7-Dimethyl-1,10-phenanthroline) Fe(II)	+0.895	
Ferrocene	+0.410	

ETHANOL–WATER
0.1M LiCl at DME
0.4 mM soln.

Species	Wt. % Water	$E_{1/2}V$ (vs SCE)	Ref.
p-Nitrotoluene	>80	~−0.7	40
	80–20	~−0.8; ~−1.1	

ETHANOL–WATER (*Continued*)
(50% Ethanol)
Dil. H$_2$SO$_4$ at DME
10–50 mM

Species	$E_{1/2}V$ (vs SCE)			Ref.
X—C$_6$H$_4$CH=NNHCONH$_2$				
X	$E_{1/2}^a$	$E_{1/2}^b$	$E_{1/2}^c$	
H	−0.729	−0.791	−0.832	57
2-CH$_3$	−0.705	−0.765	−0.826	
3-CH$_3$	−0.069; −0.711	−0.773	−0.834	
4-CH$_3$	−0.17; −0.733	−0.792	−0.855	
2-C$_2$H$_5$		−0.762	−0.826	
3-C$_2$H$_5$	−0.043; −0.712	−0.764	−0.825	
4-C$_2$H$_5$	−0.151; −0.719	−0.777	−0.838	
2-i-C$_3$H$_7$	−0.712	−0.775	−0.837	
3-i-C$_3$H$_7$	−0.717	−0.778	−0.840	
4-i-C$_3$H$_7$	−0.126; −0.726	−0.784	−0.845	
2-t-C$_4$H$_9$	−0.741	−0.801	−0.871	
3-t-C$_4$H$_9$	−0.120; −0.716	−0.776	−0.835	
4-t-C$_4$H$_9$	−0.197; −0.727	−0.784	−0.843	
3-Cl	+0.373; −0.689	−0.755	−0.820	
4-Cl	+0.227; −0.700	−0.764	−0.829	
3,4-Cl$_2$	+0.60; −0.668	−0.730	−0.791	
4-OCH$_3$	−0.268		−0.876	
3,4-(OCH$_3$)$_2$	−0.153		−0.851	
3-OH	−0.002;	−0.810	−0.870	
4-CN	+0.628;	−0.720	−0.785	
3-CF$_3$	+0.53; −0.711	−0.777	−0.844	

[a] 0.5M H$_2$SO$_4$.
[b] 0.05M H$_2$SO$_4$.
[c] 0.005M H$_2$SO$_4$.

VI. POLAROGRAPHY

ETHANOL–WATER (*Continued*)

(50% Ethanol)

0.2N CH_3COONa + 0.2N CH_3COOH at DME at 20 ± 0.1°C or

0.2M H_3BO_3 + 0.05M $Na_2B_4O_7$

0.143 mM

Species	$E_{1/2}V$ (vs mercury pool)	Ref.
α-Tetraloneoxime	−1.31[a]; −1.33[b]; −1.41[c]; −1.43[d]; −1.46[e]; −1.53[f]; −1.60[g]; −1.61[h]	122

[a] pH = 3.72.
[b] pH = 4.05.
[c] pH = 4.99.
[d] pH = 5.57.
[e] pH = 6.09.
[f] pH = 7.09.
[g] pH = 9.08.
[h] pH = 8.98.

(50% Ethanol)

0.05M Et_4NI at DME at 25°C

1 mM

Species	$E_{1/2}V$ (vs SCE)	Ref.
α,α′-Dithienyl sulfide	−2.455	82
α,β′-Dithienyl sulfide	−2.535	

ETHANOL–WATER (Continued)
(50% Ethanol)
HAc–LiAc at DME

Species	$E_{1/2}V$ (vs SCE)	Ref.
β-Acetyltetralin	-1.14^b; -1.38^d; -1.67^d; -1.45^i; -1.68^i; -1.52^j; -1.70^j; -1.70^k; -1.75^{mno}	127
5-Chloroacetyltetralin	-1.30^b; -1.38^e; -1.48^e; -1.50^g; -1.70^g; -1.70^i; -1.71^j; -1.74^k; $-1.76^{l,m,n}$; -1.77^o	
5-Fluoroacetyltetralin	-1.22^a; -1.38^b; -1.42^e; -1.48^f; -1.48^g; -1.80^g; -1.80^i; $-1.82^{i,k}$; -1.84^l; -1.85^h; -1.87^o	
5-Methylacetyltetralin	-1.35^b; -1.44^e; -1.45^e; -1.47^g; -1.85^g; -1.50^h; $-1.85^{i,j}$; $-1.86^{k,l,m}$; -1.88^o	

[a] pH = 2.1.
[b] pH = 3.6.
[c] pH = 4.4.
[d] pH = 4.8.
[e] pH = 5.0.
[f] pH = 5.7.
[g] pH = 6.1.
[h] pH = 6.4.
[i] pH = 6.7.
[j] pH = 7.3.
[k] pH = 8.4.
[l] pH = 8.7.
[m] pH = 9.5.
[n] pH = 10.7.
[o] pH = 11.7.

VI. POLAROGRAPHY

ETHANOL–WATER (*Continued*)
(50% Ethanol)
At DME at 25 ± 0.1°C at pH = 2.0
~0.2 mM

Species		$E_{1/2}V$ (vs SCE)	Ref.

Sulphilimines

R$_1$
\
  S=N—(*p*-CH$_3$C$_6$H$_4$SO$_2$)
/
R$_2$

R$_1$	R$_2$		
methyl	methyl	+0.635	59
n-propyl	*n*-propyl	+0.574	
methyl	phenyl	+0.470	
methyl	*o*-anisyl	+0.375	
methyl	*m*-anisyl	+0.430	
methyl	*p*-anisyl	+0.375	
phenyl	phenyl	+0.406	
o-anisyl	*o*-anisyl	+0.400	
m-anisyl	*m*-anisyl	+0.460	
p-anisyl	*p*-anisyl	+0.450	

(50% Ethanol)
DME at 25 ± 0.05°C
1.0 mM

Species	pH	$E_{1/2}V$ (vs SCE)	Ref.
p-Dimethylaminoazobenzene	1.48	+0.13; −1.2	52
	1.90	−1.2	
	2.89	−0.11; −1.2	
	3.86	−0.21; −1.2	
	4.72	−0.34; −1.2	
	5.02	−0.43; −1.2	
	5.68	−0.46; −1.2	
	6.94	−0.54; −1.2	
	8.42	−0.62; −1.2	
	8.92	−0.64; −1.2	
	9.50	−0.67	
	13.5	−0.88	

ETHANOL–WATER (Continued)

(46% Ethanol by wt.)

DME

0.2 mM

Species	$E_{1/2}V$ (vs SCE)			Ref.
Substituted Benzophenones, substituent X	$E_{1/2}{}^a$	$E_{1/2}{}^b$	$E_{1/2}{}^c$	
H	−0.86	−1.404	−1.497	60
3-CH$_3$	−0.85$_0$	−1.416	−1.507	
4-CH$_3$	−0.86$_6$	−1.428	−1.531	
4-i-C$_3$H$_7$	−0.85$_7$	−1.436	−1.525	
3,5-(CH$_3$)$_2$	−0.86$_5$	−1.409	−1.515	
4,4′-(CH$_3$)$_2$	−0.88$_5$	−1.470	−1.567	
3,4′-(CH$_3$)$_2$	−0.85$_9$	−1.439	−1.547	
3,4-(CH$_3$)$_2$	−0.87$_5$	−1.459	−1.551	
4,4′-(C$_2$H$_5$)$_2$	−0.87$_7$	−1.465	−1.559	
4,4′-(i-C$_3$H$_7$)$_2$	−0.85$_8$	−1.472	−1.560	
4,4′(t-C$_4$H$_9$)$_2$		−1.477	−1.553	
3,4,3′,4′-(CH$_3$)$_4$	−0.87$_5$	−1.504	−1.602	
4-C$_6$H$_5$		−1.351	−1.433	
3-CH$_2$CN		−1.317	−1.417	
4-CH$_2$CN	−0.80$_7$	−1.327	−1.425	
3-CH$_2$OC$_6$H$_5$		−1.354	−1.433	
4-CH$_2$OC$_6$H$_5$		−1.298	−1.385	
3-COO$^{(-)}$	−0.79$_3$	−1.418	−1.510	
3-COOCH$_3$		−1.334	−1.421	
4-COOCH$_3$	−0.76	−1.222	−1.264	
3-CONH$_2$	−0.82	−1.332	−1.409	
4-CONH$_2$		−1.251	−1.296	
3-CN		−1.277	−1.310	
4-CN		−1.182	−1.222	
3-F		−1.346	−1.430	
4-F	−0.86$_6$	−1.410$_5$	−1.502$_5$	
3-Cl		−1.332	−1.421	
4-Cl	−0.79$_2$	−1.356	−1.440	
3-Br		−1.312	−1.406	
4-Br	−0.78$_0$	−1.338	−1.441	
4,4′-(Cl)$_2$	−0.75	−1.315	−1.3875	
2-NH$_2$	−0.86$_8$	−1.515	−1.608	
4-OH	−0.96$_8$	−1.591	−1.735	
2,4-(OH)$_2$		−1.598	−1.728	
4,4′-(OH)$_2$	−1.03	−1.705		
4,4′-(OCH$_3$)$_2$	−0.94	−1.530	−1.616	
3,4,3′,4′-(OCH$_3$)$_4$	−0.92$_6$	−1.479	−1.578	
3-SO$_2$CH$_3$		−1.212	−1.313	
4-SO$_2$CH$_3$	−0.72$_3$	−1.155	−1.223	

VI. POLAROGRAPHY

ETHANOL–WATER (*Continued*)
(40% Ethanol)
DME at 33 ± 0.2°C
1.04 mM
$\mu = 0.27M$ (Supporting electrolyte)

Species	pH	$E_{1/2}V$ (vs SCE)	Ref.
5-Nitroacenaphthene	1.7	0.385	45
	2.95	0.490	
	3.9	0.50; 0.865	
	4.8	0.560; 1.00	
	5.7	0.600; 1.180	
	7.15	0.710; 1.40	
	8.25	0.820; 1.527	
	9.55	0.880; 1.440	
	10.4	0.880; 1.440	
	10.95	0.820; 1.430	
	11.35	0.730; 1.20	

(40% Ethanol)
0.1M KNO₃ at DME at 32.2°C
1 mM soln. at pH = 7.20

Species	$E_{1/2}V$ (vs SCE)	Ref.
Ethylthioglycollate	−0.409	67

ETHANOL–WATER (*Continued*)

(30% by vol. Ethanol)

0.2M NaCl at DME at 25 ± 0.1°C

Species	pH	$E_{1/2}V$ (vs SCE)	Ref.
CH_3COCHN_2	2.95	−0.79	35
	5.10	−1.05	
	5.90	−1.0; −1.13	
	7.60	−1.0; −1.19	
	9.20	−1.04; −1.27; −1.50	
$CH_2ClCOCHN_2$	3.95	−0.91; −0.97; −1.44	
	5.25	−0.91; −1.05; −1.46	
	6.50	−0.92; −1.17; −1.47	
	7.70	−0.93; −1.23; −1.58	
	8.25	−0.92; −1.24; −1.65	
	9.10	−0.94; −1.29; −1.71	
	9.85	−0.91; −1.27; −1.75	
	10.15	−0.92; −1.26; −1.79	
	10.50	−0.95; −1.28	
	10.90	−0.92; −1.27	
CCl_3COCHN_2	2.00	−1.00	
	3.80	−1.03; −1.20	
	5.90	−1.02; −1.36	
	7.80	−1.01; −1.41	
	9.20	−1.00; −1.36	

ETHANOL–WATER (Continued)

(20% Ethanol)

0.1M KCl + 0.1M Buffer at DME at 25 ± 0.1°C

0.2–1 mM

Species		Parent Compd. $E_{1/2}V$ (vs SCE)	Reduced form $E_{1/2}V$ (vs SCE)	Ref.
Diphosphophyridine nucleotide and substituents				
R_1	R_2			
H	H	−0.91	−0.34	63
—$CH_2C_6H_5$	—$CONH_2$[a]	−1.00	−0.36	
—$CH_2C_6H_4SO_3^-$	—$CONH_2$	−0.96	−0.44	
—$CH_2CH_2SO_3^-$	—$CONH_2$	−1.02	−0.45	
—$CH_2C_6H_5$	—CN[a]	−0.78		
—CH_3	—CN[b]	−0.88	−0.25	
—CH_3	—$COCH_3$[b]	−0.87; −1.13	−0.34	
—$CH_2C_6H_5$	—$COCH_3$[a]	−0.88	−0.2	
—$CH_2C_6H_5$	—$CONH_2$[a,c]	−0.79; −1.48	−0.14	

[a] Anion Cl^-.
[b] Anion I^-.
[c] R_3 is $CONH_2$.

ETHANOL–WATER (*Continued*)
0.05M Et$_4$NI at DME
0.1 mM

Species	% Ethanol	pH	$E_{1/2}V$ (vs SCE)	Ref.
p-Di-iodobenzene	50		−1.480	12
Di-iodomethane	50		−0.92; −1.40	
Ethyl iodide	50		−1.57	
p-Iodoacetanilide	50		−1.510	
Iodoacetic acid	50	10.0	−0.853	
	50	3.8	−0.50	
p-Iodoaniline	50	8.0	−1.595	
	50	4.0	−1.42	
p-Iodoanisole	50		−1.562	
Iodobenzene	50		−1.54	
m-Iodobenzoic acid	50	8.2	−1.507	
	50	3.9	−1.443	
o-Iodobenzoic acid	50	8.2	−1.504	
	50	3.9	−1.17	
p-Iodobenzoic acid	50	8.5	−1.500	
	50	3.9	−1.437	
p-Iodochlorobenzene	50		−1.475	
p-Iodo-*N*,*N*-dimethylaniline	50	8.5	−1.615	
p-Iododiphenyl	50		−1.506	
o-Iodophenol	50	10.0	−1.560	
	50	4.0	−1.41	
p-Iodophenol	50	11.0	−1.66	
	50	4.0	−1.542	
p-Iodophenyltrimethyl-ammonium iodide	50		−1.27	
p-Iodotoluene	50		−1.580	
Methyl iodide	50		−1.39	
Methyl-*m*-iodobenzoate	50		−1.433	
Methyl-*o*-iodobenzoate	50		−1.268	
Methyl-*p*-iodobenzoate	50		−1.427	
n-Propyl iodide	50		−1.63	

VI. POLAROGRAPHY

ETHANOL–WATER (*Continued*)
(10% Ethanol)
DME at $25.0 \pm 0.1°C$
$\sim$0.2 mM

Species	$E_{1/2}V$ (vs SCE)	Ref.
Acetylacetone	> -1.85	58
Benzylacetone	~ -1.84	
Dibenzoylmethane	-1.73	
5,5-Dimethylcyclohexanedione-1,3 (dimedone)	> -1.85	
2-Phenylhexahydroindanedione-1,3	> -1.85	
2-Phenyl-5,6,7,7a-tetrahydro-indandione-1,3	> -1.85	
2-Phenylindanedione-1,3 and derivatives	-1.58	
2-Phenyl-4,5,6,7-tetrahydro-indiandione-1,3 and its derivatives	-1.47	
2,4,5-Triphenylcyclopentane-4-dione-1,3	-1.30	

[a] These values obtained in Britton–Robinson buffers $\mu = 0.25$.

ETHANOL–WATER (*Continued*)

(10% Ethanol)

0.1M KOH at DME at 25.0 ± 0.1°C

0.2 mM

Species	$E_{1/2}V$ (vs SCE)	Ref.
2-Benzyl-1,3-indandione	−1.60	124
2-Benzyl-4,5,6,7-tetrahydro-2-phenyl-1,3-indandione	−1.04[a]	
2-p-Bromophenyl-4,5,6,7-tetrahydro-1,3-indandione	−1.09; −1.46	
Dimedon	> −1.85	
2-(Diphenylacetyl)-1,3-indandione	−1.53[a]	
1,3-Diphenyl-1,3-propanedione	−1.73[a]	
Ethyl 1,3-dioxo-2-indancarboxylate	−1.53	
Hexahydro-2-phenyl-1,3-indandione	> −1.85	
1,3-Indandione	−1.57	
2-p-Methoxyphenyl-1,3-indandione	−1.59	
2-Methyl-2-phenyl-1,3-indandione	−1.23	
2-p-Nitrophenyl-1,3-indandione	−0.84; −1.59	
2,4-Pentanedione	> −1.85	
1-Phenyl-1,3-butanedione	∼ −1.84	
2-Phenyl-2,3-indandione	−1.58	
4,5,6,7-Tetrahydro-2-methyl-2-phenyl-1,3-indandione	−1.05	
4,5,6,7-Tetrahydro-2-p-nitrophenyl-1,3-indandione	−0.88; −1.51	
4,5,6,7-Tetrahydro-2-phenyl-1,3-indandione	−1.47	
5,6,7,7a-Tetrahydro-2-phenyl-1,3-indandione	> −1.85	
2,4,5-Triphenyl-4-cyclopentene-1,3-dione	−1.30[a]	

VI. POLAROGRAPHY

ETHANOL–WATER (*Continued*)

Buffer Soln. at DME at $25.0 \pm 0.1°C$

0.2 mM

Species	$E_{1/2}V$ (vs SCE)	Ref.
Ethylester of 5-nitropyromucic acid	-0.09^a; -0.53^b; -0.83^b	118
2-Nitrofuran	-0.22^a; -0.64^b; -0.92^b	
5-Nitrofurfuryl alcohol	-0.18^a; -0.63^b; -0.87^b	
5-Nitropyromucic acid	-0.10^a; -0.53^b; -0.86^b	
2-Nitrosylvan	-0.20^a; -0.61^b	

[a] pH 2.0 at 10% ethanol.
[b] pH 10 at 10% ethanol.

0.1N LiCl at DME at 25°C

0.5 mM

Species	$E_{1/2}V$ (vs SCE)	Ref.
Benzophenone	-0.865^a	116
	-0.875^b	
	-0.890^c	
	-0.900^d	
	-0.920^e	
	-0.940^f	

[a] 10% alcohol.
[b] 20%.
[c] 30%.
[d] 40%.
[e] 50%.
[f] 60% alcohol by volume.

ETHANOL–WATER (*Continued*)

(60% Ethanol)
Buffer at DME
15–20 mM

Species	$E_{1/2}V$ (vs SCE)						Ref.
	$E_{1/2}^a$	$E_{1/2}^b$	$E_{1/2}^c$	$E_{1/2}^d$	$E_{1/2}^e$	$E_{1/2}^f$	
4'-Benzylacetophenone	−1.195	−1.305	−1.345 −1.555		−1.610	−1.620	114
4',4''-Biacetophenone	−1.050	−1.150 −1.440	−1.230 −1.540			−1.360 −1.590	
p-Diacetylbenzene	−0.780	−0.995	−1.015	−1.090 −1.645		−1.152 −1.730	
4',4''-Ethylenediacetophenone	−1.150	−1.285 −1.590	−1.325 −1.600			−1.610	
4',4''-Methylenediacetophenone	−1.150	−1.220		−1.580		−1.580	
4'-Phenethylacetophenone	−1.180	−1.335	−1.360 −1.620	−1.620		−1.625	
4'-Phenoxyacetophenone	−1.210	−1.270 −1.625	−1.310 −1.625	−1.630	−1.630	−1.630	
4',4''-Oxydiacetophenone	−1.115		−1.250 −1.515	−1.550		−1.585	
4',4''-Thiodiacetophenone	−1.130	−1.175 −1.335	−1.180 −1.340	−1.350		−1.410	

[a] pH = 2.5.
[b] pH = 3.3.
[c] pH = 3.9.
[d] pH = 6.2.
[e] pH = 7.9.
[f] pH = 9.4.

(50% Ethanol)
0.2M Bu₄NClO₄ at HMD[a] electrode
0.1–0.5 mM

Species	$E_{1/2}V$ (vs SCE)	Ref.
$C_6H_5CH=CHCOCH_3$	−0.62[b]; −1.42 ± 0.04[c]; −1.36 ± 0.04[d]; −1.76[d]	104

[a] Hanging mercury drop.
[b] pH range 1–8, with $E_{1/2}$ = −0.62−0.078 pH.
[c] pH range 1–8.
[d] pH above 10.

VI. POLAROGRAPHY

ETHANOL–WATER (*Continued*)

(40% Ethanol)

Buffers at DME at 33 ± 0.2°C

$\mu = 0.27M$

Species	$E_{1/2}V$ (vs SCE)	Ref.
5-nitro-ace-naphthene	+0.385[a]; +0.490[b]; +0.500[c] +0.865[c]; +0.560[d]; +1.00[d]; +0.600[e]; +1.180[e] +0.710[f]; +1.40[f] +0.820[g]; +1.527[g] +0.880[h]; +1.440[h] +0.880[i]; +1.440[i] +0.820[j]; +1.43[j] +0.730[k]; +1.20[k]	166

[a] pH = 1.7.
[b] pH = 2.95.
[c] pH = 3.9.
[d] pH = 4.8.
[e] pH = 5.7.
[f] pH = 7.15.
[g] pH = 8.25.
[h] pH = 9.55.
[i] pH = 10.4.
[j] pH = 10.95.
[k] pH = 11.35.

(75% Ethanol)

0.05M Et$_4$NOH at DME

Species	E_pV (vs Hg pool)	Ref.
Carbontetrachloride	−0.87; −1.8	107
Hexachlorobutadiene	−1.45; −2.55	

ETHANOL–WATER *(Continued)*

(50% Ethanol)

Buffer at DME at 30 ± 0.1°C

0.212–1.538 mM)

Species	$E_{1/2}V$ (vs SCE)	Ref.
Thiosalicylic acid	+0.010[a] −0.270 to −0.275[b] −0.230[c] −0.450[d] −0.500[e] −0.495[f]	108

[a] $HClO_4$ + $NaClO_4$ pH = 1.8.
[b] Acetate buffer + KNO_3 pH = 6.0.
[c] Phosphate buffer + KNO_3 pH = 5.1.
[d] Ammonia buffer + KNO_3 pH = 9.1.
[e] Borax + KNO_3 pH = 10.45.
[f] NaOH + KNO_3 pH = 13.7.

ETHYLENEDIAMINE

0.3M LiCl at Pt-electrode at 25–27°C

17 mM

Species	$E_{1/2}V$ (vs Zn(Hg)ZnCl$_2$)	Ref.
Anthracene	−0.02	133
Biphenyl	−0.63	
Naphthalene	−0.50	

ETHYLENEGLYCOL–WATER

NH_4Br at DME at 25 ± 0.1°C

1 mM

Species	$E_{1/2}V$ (vs SCE)	Ref.
Zn(II)	−1.070[a]; −1.040[b]; −1.010[c]; −0.958[d]; −0.940[e]; −0.915[f]	102

[a,b,c] Correspond to 20, 40, and 60% Ethylene glycol at 0.5M NH_4Br.
[d,e,f] Correspond to 20, 40, and 60% Ethylene glycol at 1M NH_4Br.

VI. POLAROGRAPHY

METHANOL
0.1M LiClO$_4$ at RE (Pt) at 25°C
∼0.5 mM

Species	$E_{1/2}V$ (vs SCE)	Ref.
Benzyl ferrocene	+0.340	104
bis(2,9-Dimethyl-1,10-phenanthroline) Cu(1)	+0.660	
tris(4,7-Dimethyl-1,10-phenanthroline); Fe(11)	+0.890	
Ferrocene	+0.370	

0.1M LiCl at DME at 20–21°C
1 mM

Species	$E_{1/2}V$ (vs SCE)	Ref.
Solochromviolett RS	−0.57; −0.78; −0.98	70

0.1M LiCl at DME at 20–21°C
0.5 mM

Species	$E_{1/2}V$ (vs SCE)	Ref.
Azoxybenzene	−1.00	162

0.1M LiClO$_4$ at Pt Electrode at 20°C
4.0 mM

Species	$E_{1/2}V$ (vs Ag/AgCl)	Ref.
2,4,6-tri-Tertiary butyl aniline	0.230	22

Species	Working electrode	$E_{1/2}V$ (vs SCE)	Ref.
C(NO$_2$)$_4$	DME	+0.45; +0.29	21

METHANOL (*Continued*)
0.1M LiClO$_4$ at Pt Electrode
2.5 mM

Species	$E_{1/2}V$ (vs Ag/Ag$^+$)	Ref.
Tetraphenyl-2,3,4,5-pyrrole	+0.28	48

METHANOL–WATER
(20% Methanol)
pH = 5.0 at DME

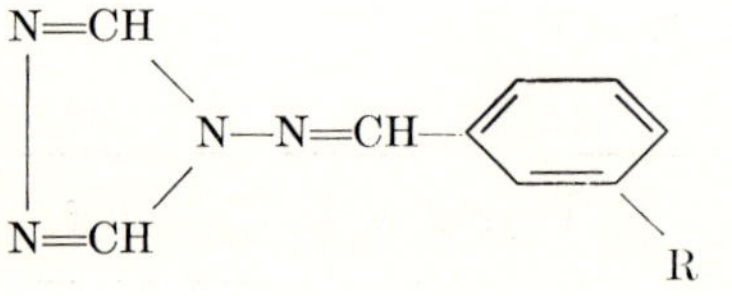

Species	$E_{1/2}V$ (vs AgCl)	Ref.

N=CH–N(–N=CH–)–N=CH–C$_6$H$_4$–R

R		
H	−0.92	43
p-CH(CH$_3$)$_2$	−1.00	
p-OH	−0.97	
p-N(CH$_3$)$_2$	−0.94	
p-Br	−0.85	
p-Cl	−0.87	
m-NO$_2$	−0.95	
m-OH	−0.85	
m-Cl	−0.85	
o-Cl	−0.81	
o-OH	−0.84	

VI. POLAROGRAPHY

METHANOL–WATER (Continued)
(20% Methanol)
pH = 5, at DME

Species	$E_{1/2}V$ (vs AgCl)	Ref.
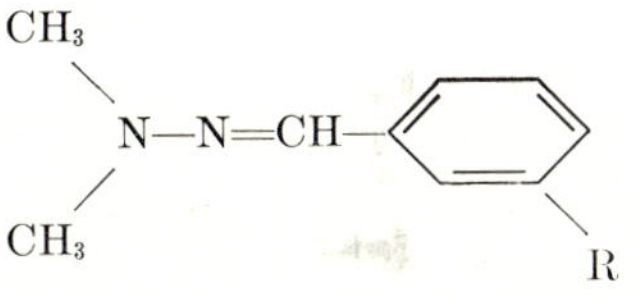 CH₃\N–N=CH–C₆H₄–R / CH₃		
R		
H	−0.92	43
p-OH	−0.95	
o-OH	−0.98	
p-N(CH₃)₂	−0.98	
p-Cl	−0.95	
o-Cl	−0.88	
m-Cl	−0.87	
p-NO₂	−0.95	
p-Br	−0.92	
m-NO₂	−1.04	

(92% Methanol)
0.02N Et₄NI at DME at 25°C
∼0.6–0.9 mM

Species	$E_{1/2}V$ (vs SCE)	Ref.
Anthracene	−1.92	123
1-Methyl-3-Vinyl naphthalene	−2.06	
9-Vinyl anthracene	+1.78; −2.25	
1-Vinyl naphthalene	−2.06	
2-Vinyl naphthalene	−2.08	

METHANOL–WATER (Continued)
(92% Methanol)
0.02N Et$_4$NI at DME at 25°C
~0.69 mM

Species	$E_{1/2}V$ (vs SCE)	Ref.
β-acetyl tetraline	−1.759	127
5-Chloroacetyl tetralin	−1.06	127
5-Fluoroacetyl tetralin	−1.756	127
2'3'4'5'-Tetramethylacetophenone	−1.90	128
2'3'5'6'-Tetramethylacetophenone	−2.06	128
2'4'5'-Trimethylacetophenone	−1.87	128
2'4'6'-Trimethylacetophenone	−2.00	128

(92% Methanol)
0.05M Et$_4$NI at DME
2.0 mM–8 mM

Species	$E_{1/2}V$ (vs SCE)	Ref.
1,3-Diphenyl propenone	−1.26; −1.80	159
1,35,-Triphenyl pyrazoline	−2.00	

(92% Methanol)
0.02N Et$_4$NI at DME
~0.25–~0.38 mM

Species	$E_{1/2}V$ (vs SCE)		Ref.
	Wave I	Wave II	
Acetophenone	−1.596		114
4'-Benzylacetophenone	−1.607		
4',4''-Biacetophenone	−1.390	−1.590	
p-Diacetylbenzene	−1.206	−1.720	
4',4''-Ethylenediacetophenone	−1.596		
4',4''-Methylenediacetophenone	−1.580		
4',4''-Oxydiacetophenone	−1.566		
4'-Phenethylacetophenone	−1.614		
4'-Phenoxyacetophenone	−1.646		
4'-Phenylacetophenone	−1.552		
4'-(Phenylthio)-acetophenone	−1.495		
4',4''-Thiodiacetophenone	−1.386		

VI. POLAROGRAPHY

METHANOL–WATER (Continued)

(92% Methanol)
0.02N Et₄NI at DME
0.738–1.64 mM

Species	$E_{1/2}V$ (vs SCE)	Ref.
2-Ethyl pyridine	−1.72	120
2-Methyl-5-ethynyl pyridine	−2.06	

(92% Methanol)
0.02N Et₄NI at DME
∼0.6 mM

Species	$E_{1/2}V$ (vs SCE)	Ref.
1-Methyl-3-vinyl naphthalene	−2.06	123
1-Vinyl naphthalene	−2.06	
2-Vinyl naphthalene	−2.08	

0.02N Et₄NI at DME at 25°C
0.45–0.88 mM

Species	$E_{1/2}V$ (vs SCE)	Ref.
4′-(4-Biphenylyl)acetophenone	−1.532	126
4′-Cyclohexyl acetophenone	−1.718	
Fluoren-2-yl methyl ketone	−1.567	
2′-Methyl-4′-m-tolylacetophenone	−1.658	
4′-Phenylacetophenone	−1.552	

0.1M LiCl at DME
0.4 mM

Species	Wt. % water	$E_{1/2}V$ (vs SCE)	Ref.
p-Nitrotoluene	>80	∼−0.7	40
	80–20	∼−0.7	

METHANOL–WATER (Continued)

(50% by Volume of Methanol)
Buffer Soln at DME at 25 ± 0.5°C
~0.96 mM

Species	pH	$E_{1/2}V$ (vs SCE)	Ref.
Benzil	8.5	−0.688	94
	9.9	−0.734	
	11.0	−0.780	
	11.8	−0.788	
	12.0	−0.800	
	12.2	−0.802	
	12.4	−0.808	
	12.6	−0.813	
	12.8	−0.819	
	13.0	−0.815	

At DME
0.218 mM

Species	$E_{1/2}V$ (vs SCE)	Ref.
Benzanthrone	−0.46[a]; −0.50[b]; −0.56[c]; −0.62[d]; −0.64[e]; −0.73[f]; −0.80[g]; −0.84[h]; −0.98[i]; −1.07[jkl]; −0.43[m]; −1.03[n]; −1.05[o]; −1.07[p]	130

70% Methanol:
[a] pH = 0.65.
[b] pH = 1.30.
[c] pH = 2.25.
[d] pH = 3.25.
[e] pH = 4.00.
[f] pH = 5.00.
[g] pH = 6.20.
[h] pH = 7.00.
[i] pH = 9.00.
[j] pH = 10.60.
[k] pH = 11.15.
[l] pH = 11.75.

99% Methanol:
[m] 0.75N to 0.18N H_2SO_4.
[n] 0.1N LiCl.
[o] 0.01N KOH.
[p] 0.1N KOH.

VI. POLAROGRAPHY

METHANOL–GLACIAL ACETIC ACID
(98% Methanol)
1.4M Sodium acetate at DME at 25 ± 0.1°C
0.01–0.5 mM

Species	$E_p V$ (vs SCE)	Ref.
Disulfide	−0.35; −0.53	144
Free sulfur	−0.54	
Mercapton compound	−0.35	

METHANOL–BENZENE
0.1N LiCl + 0.5 NCH₃COOH at DME at 25 ± 0.1°C
0.5 mM

Species	$E_{1/2} V$ (vs SCE)	Ref.
Dithiodiglycolic acid dianelide	−0.267[a]; −0.670[a] −0.258[b]; −0.763[b] −0.228[c]; −0.795[c] −0.202[d]; −0.814[d] −0.820[e] −0.245[f]; −0.649[f]	155

[a] 90% Methanol.
[b] 80%.
[c] 70%.
[d] 60%.
[e] 50%.
[f] 100% Methanol.

METHANOL–BENZENE, 4:1
0.25M NH₄NO₃ at DME
0.5 mM solutions

Species	$E_{1/2} V$ (vs SCE)	Ref.
Percapric acid	+0.22	44
Percaprylic acid	+0.23	
Perlauric acid	+0.23	
Permyristic acid	+0.23	

2-METHOXY-ETHANOL
0.2M Bu$_4$NBr at DME

Species	$E_{1/2}V$ (vs Hg pool)	Ref.
3,4-Benz-pyrene	−1.6; −1.9	62
3,4,5,8,9,10-Hexahydro-perylene	−2.165; −2.075	
Naphthalene	−1.98	
Perylene	−1.75	
Tetracene	−1.94	

N-METHYLACETAMIDE
0.12M Et$_4$NBr at DME at 35 ± 0.2°C
1 mM

Species	$E_{1/2}V$ (vs Hg pool)	Ref.
4,4'-bis(Acetamino)azoxybenzene	−0.88 ± 0.003; −0.96 ± 0.003	163
Azobenzene	−0.858 ± 0.003	
Benzil	−0.724 ± 0.003	
p-Nitroacetanilide	−0.693 ± 0.003; −0.994 ± 0.003	
p-Nitroaniline	−0.854 ± 0.003; −1.144 ± 0.003	

METHYLENE CHLORIDE
0.5M Bu$_4$NClO$_4$ at DME at 25°C
~1 mM

Species	$E_{1/2}V$ (vs Ag/AgI)	Ref.
Chloranil	+0.35[a]	153
2,3-Dichloro-5,6-dicyanoquinone	+0.78[b]	
1,2,4,5-Tetracyanobenzene	−0.35[a]	
Tetracyanoethylene	+0.54[b]	
7,7',8,8'-Tetracyanoquinodimethane	+0.47[a]	

[a] Contains hexamethylbenzene or pyrene.
[b] Contains hexamethylbenzene.

VI. POLAROGRAPHY

METHYLENE CHLORIDE (*Continued*)
0.2M Bu₄NClO₄ at Pt-disk electrode
0.5–1.2 mM

Species	E_p (vs SCE)	Ref.
9,10-Diphenylanthracene	+1.22; +1.16	148
Rubrene	+0.82; +0.74	
Tetracene	+0.98 to +1.02	
1,3,6,8-Tetraphenylpyrene	+1.17; +1.09	

NITROMETHANE
0.1M Bu₄NClO₄ at DME

Species	$E_{1/2}V$ (vs Ag/AgCl)	Ref.
CH₃COOH	−0.65; −0.8[a]	161

[a] At Pt electrode.

0.1M Et₄NClO₄ at Pt
1.52–1.87 mM

Species	$E_p V$ (vs SCE)	Ref.
Quinhydrone	+0.86; −0.23; −0.61	90

0.1M Me₄NClO₄ at Pt electrode
~0.7 mM

Species	$E_{1/2}V$ (vs Ag/AgCl)	Ref.
N,N,N',N'-Tetramethylbenzidine	+0.70; +0.91; ±0.01; ±0.01	50

0.1M LiClO₄ at RE (Pt)

Species	$E_{1/2}V$ (vs Ag/AgCl)	Ref.
Nitrosobenzene	+2.0	53

NITROMETHANE (*Continued*)
0.1M Et₄NClO₄ at Pt electrode at 20°C
1.5 mM Soln

Species	$E_{1/2}V$ (vs Ag/AgCl)	Ref.
Tetraphenyl-2,3,4,5-pyrrole	+1.17; +1.77	49

0.1M Et₄NClO₄ at RE (Pt) at 25°C
~0.5 mM

Species	$E_{1/2}V$ (vs SCE)	Ref.
Benzyl ferrocene	+0.300	104
bis(2,9-Dimethyl-1,10-phenanthroline) Cu(I)	+0.670	
tris(4,7-Dimethyl-1,10-phenanthroline) Fe(II)	+0.880	
Ferrocene	+0.335	

1-PROPANOL
0.1M LiClO₄ at RE (Pt) at 25°C
~0.5 mM

Species	$E_{1/2}V$ (vs SCE)	Ref.
Benzyl ferrocene	+0.385	104
tris(4,7-Dimethyl-1,10-phenanthroline Fe(II)	+0.915	
Ferrocene	+0.420	

2-PROPANOL
0.1M LiClO₄ at RE (Pt) at 25°C
~0.5 mM

Species	$E_{1/2}V$ (vs SCE)	Ref.
Benzyl ferrocene	+0.405	104
tris(4,7-Dimethyl-1,10 phenanthroline) Fe(II)	+0.925	
Ferrocene	+0.445	

VI. POLAROGRAPHY

PROPANOL–WATER
NH_4Br at DME at $25 \pm 0.1°C$
1 mM

Species	$E_{1/2}V$ (vs SCE)	Ref.
Zn(II)	-1.080^a; -1.1005^b; -1.085^c -1.01^d; -1.025^e; -1.085^f	102

[a,b,c] Correspond to 20, 40 and 60% propanol at 0.5M NH_4Br.
[d,e,f] Correspond to 20, 40 and 60% propanol at 1M NH_4Br.

i-PROPANOL–WATER
0.1M LiCl at DME 0.4 mM solution

Species	Wt. % of water	$E_{1/2}V$ (vs SCE)	Ref.
p-Nitrotoluene	>80	~−0.7	40
	80–20	~−0.8; ~−1.1	

PROPYLENE CARBONATE
0.1M Et_4NClO_4 at DME
1.2 mM

Species	$E_{1/2}V$ (vs SCE)	Ref.
1-Hydroxy-9,10-anthraquinone	−0.74; −1.20	106
Conjugate base of 1-hydroxy-9,10-anthraquinone[a]	−1.20	
Conjugate base of 1-hydroxy-anthraquinone[b]	+0.51	

[a] 1.83 mM.
[b] RE (Pt).

Conc. 0.5×10^{-3} M/l; temp 25°C
Supporting electrolyte 0.05M Et_4NClO_4

Species	Working electrode	$E_{1/2}V$ (vs SCE)	Ref.
1,2-Naphthoquinone	DME	−0.55; −0.92	28
1,4-Naphthoquinone	DME	−0.67; −1.05	28

PYRIDINE

0.5M LiClO$_4$ at Graphite Electrode at 25.0 ± 0.1°C

Species	$E_{p/2}V$ (vs Ag+/Ag)	Ref.
Catechol	+0.29	101
2,2-Diphenyl-1-pioxylhydrazyl	+0.29	101
Ferrocene	+0.43; +0.55	28
Hydroquinone	+0.31	101
Phenol	+0.75	101
Quinone	−0.19	101
Resorcinol	+0.61; +0.91	101

0.1M LiClO$_4$ at RE (Pt) at 25°C
∼0.5 mM

Species	$E_{1/2}V$ (vs SCE)	Ref.
Benzylferrocene	+0.490	104
bis(2,9-Dimethyl 1,10-phenanthroline) (Cu(I))	+0.335	
tris(4,7-Dimethyl 1,10-phenanthroline) Fe(II)	+0.985	
Ferrocene	+0.505	

VI. POLAROGRAPHY

PYRIDINE (Continued)
0.1M Et$_4$NClO$_4$ at DME at 25 ± 0.2°C
Precursor Conc. 0.1–32 mM

Species pyridinium ion precursor	$E_{1/2}V$ (vs NAgE)	Ref.
Acetic acid	−1.72 ± 0.02[a]	112
	−1.72 to −1.58[b]	
Benzoic acid	−1.62 ± 0.01; −1.62 to −1.41[c]	
	−1.67 ± 0.02[d]	
	−1.62 ± 0.01[f]; −1.60 ± 0.08[g]	
	−1.65 ± 0.08[h]	
	−1.20 to 1.16[i]; −1.31 to −1.38[j]	
	−1.09 to −1.12[k]; −1.30 to −1.33[l]	
	−1.09 ± 0.02[m]; −1.35 ± −0.01[n]	
None	−1.67 ± 0.02[e]	

[a] Plus 0–6 mM φ COONEt$_4$.
[b] Plus 0–6.7 mM LiOAc.
[c] Plus 1–10 mM LiClO$_4$.
[d] Plus 0–8 mM φ COONEt$_4$.
[e] Plus 0–8 mM φ COONEt$_4$.
[f] Plus 0–5 mM φ COONa.
[g] Plus 0–4 mM φ COONEt$_4$.
[h] Plus 0–6.25 mM φ COONEt$_4$ at 40°C.
[i,j] 0.1M LiClO$_4$ + 0–8 mM Et$_4$NClO$_4$.
[k,l] 0.1M LiClO$_4$ + 0–10 mM Et$_4$NClO$_4$.
[m,n] 0.1M LiClO$_4$ + 0–7 mM φ COONEt$_4$.

0.5M LiClO$_4$ at Graphite Electrode at 25.0 ± 0.1°C
4 mM

Species	$E_{p/2}V$ (vs Ag/AgNO$_3$)	Ref.
Hydroquinone	+0.30; −0.12	119
Quinone	−0.18	

PYRIDINE (Continued)

0.1M LiClO$_4$ at DME at 25°C
0.25–1.3 mM

Species	$E_{1/2}V$ (vs Hg pool)	Ref.
Diethyl fumerate	−1.25; −1.43; −1.70	146
Diethyl maleate	−1.16; −1.35; −1.70	
Fumaric acid	−1.12; −1.33	
Maleic acid	−1.10; −1.30	
Maleic anhydride	−0.70; −0.87; −1.13; −1.32	

0.1M LiClO$_4$ at DME at 25 ± 0.2°C
0.071–15.7 mM

Species	$E_{1/2}V$ (vs NAgE)	Ref.
CF$_3$COOH	−1.11 ± 0.17; −1.30 ± 0.02	165
CH$_3$COOH	−1.12 ± 0.04; −1.36 ± 0.07	
C$_6$H$_5$COOH	−1.10 ± 0.03; −1.36 ± 0.04	
2,4-Cl$_2$C$_6$H$_3$OH	−1.36 ± 0.04	
n-BuBr	−1.38 ± 0.04	
HPyrNO$_3$	−1.13 ± 0.05; −1.39 ± 0.03	
EtPyrBr	−1.34 ± 0.03	
Phthalic acid	−1.09 ± 0.03; −1.36 ± 0.01	
Salicylic acid	−1.05 ± 0.02; −1.22 ± 0.04	

PYRIDINE–WATER

0.1M LiClO$_4$ at DME at 25.0 ± 0.2°C

Species	$E_{1/2}V$ (vs Hg pool)	Ref.
Pyridine	−1.41 ± 0.03[a]	111
	∼−1.47[b]	
	∼−1.52[c]	
	∼−1.57[d]	

[a] 0–10%.
[b] 20%.
[c] 40%.
[d] 60% water by volume.

VI. POLAROGRAPHY

TETRAHYDROFURAN

0.1M Bu₄NClO₄ at Pt wire electrode at 25°C

1 mM

Species	$E_p V$ (vs Ag wire)	Ref.
Diphenylpioxylhydrazyl	+0.500; −0.122; −1.714; −2.030 +0.436; −0.183; −1.788; −2.110	109

0.1M Bu₄NClO₄ at DME

0.1–1.0 mM

Species	$E_{1/2} V$ (vs SCE)	Ref.
3,8-Dimethyl-2-methoxyazocine	−2.28	5
2-Methoxyazocine	−1.94	

TETRAHYDROFURAN–WATER

(30% Tetrahydrofuran)

0.1M NaClO₄ at DME

1 mM

Species	$E_{1/2} V$ (vs SCE)	Ref.
(Acetyl-2)-thiophene	−1.07[a]; −1.09[b]; −1.13[c] −1.18[d]; −1.30[e]	156

[a] pH = 2.28.
[b] pH = 2.97.
[c] pH = 3.52.
[d] pH = 4.32.
[e] pH = 6.10.

TETRAHYDROFURAN–WATER (*Continued*)
(30% Tetrahydrofuran)
0.1M NaClO$_4$ at DME at 25°C
1 mM

Species	$E_{1/2}V$ (vs SCE)	Ref.
(Difuroyl-2) methane	−0.93[a]; −1.10[b]; −1.30[c] −1.44[b]; −1.50[c] −1.37[d]; −1.52[d] −1.63[e]	157

[a] pH = 2.05.
[b] pH = 4.70.
[c] pH = 8.29.
[d] pH = 9.54.
[e] pH = 11.50.

References

1. G. Peychal-Heiling and G. S. Wilson, Anal. Chem., **43**, 545 (1971).
2. G. Peychal-Heiling and G. S. Wilson, Anal. Chem., **43**, 550 (1971).
3. R. C. Buchta and D. H. Evans, J. Electrochem. Soc., **117**, 1494 (1970).
4. J. L. Huntington and D. G. Davis, J. Electrochem. Soc., **118**, 57 (1971).
5. L. B. Anderson, J. F. Hansen, T. Kakihana and L. A. Paquette, J. Am. Chem. Soc., **93**, 161 (1971).
6. J. E. O'Reilly and P. J. Elving, J. Am. Chem. Soc., **93**, 1871 (1971).
7. P. T. Kissinger, P. T. Holt and C. N. Reilley; J. Electroanal. Chem., **33**, 1 (1971).
8. D. E. Bartak and M. D. Hawley, J. Electroanal. Chem. **33**, 13 (1971).
9. M. Breitenbach and K. H. Heckner; J. Electroanal. Chem., **33**, 45 (1971).
10. C. P. Andrieux and J. M. Saveant, J. Electroanal. Chem., **33**, 453 (1971).
11. G. Farnia, A. Romanin, G. Capobianco and F. Torzo, J. Electroanal. Chem., **33**, 31 (1971).
12. R. A. Caldwell and S. Hacobian, Aust. J. Chem., **21**, 1 (1968).
13. W. J. Blaedel and R. G. Haas, Anal. Chem., **42**, 918 (1970).
14. J. Badoz-Lambling and G. Demange-Guerin, Anal. Lett., **2**, 123 (1969).
15. N. D. Canfield, J. Q. Chambers and D. L. Coffen, J. Electroanal. Chem., **24**, 7 (1970).
16. E. Kariv, J. Hevmolin and E. Gileadi, J. Electrochem. Soc., **117**, 342 (1970).
17. J. S. Dunning and D. N. Bennion, J. Electrochem. Soc., **117**, 485 (1970).
18. R. C. Buchta and D. H. Evans, Anal. Chem., **40**, 2181 (1868).
19. R. Andruzzi, M. E. Cardinali, I. Carelli and A. Trazza, J. Electroanal. Chem., **26**, 211 (1970).
20. E. S. Jayadenappa, Indian J. Chem., **7**, 1146 (1969).
21. V. A. Petrosyan, V. I. Slovetskii, S. G. Mairanovskii and A. A. Fainzilberg, Soviet Elektrokhimiya, **6**, 1539 (1970).
22. G. Canquis, G. Fauvelot and J. Rigandy, Bull. Soc. Chim., **12**, 4928 (1968).
23. V. A. Kokorekina, L. G. Feoktistov, S. A. Sheveler and A. A. Fainzilberg, Soviet Elektrokhimiya, **6**, 1770 (1970).

24. M. Uehara and J. Nakaya, Bull. Chem. Soc. Japan, **43**, 3136 (1970).
25. H. R. Koch and M. G. McKeon, J. Electroanal. Chem., **30**, 331 (1971).
26. M. Breitenbach and K. H. Heckner, J. Electroanal. Chem., **29**, 309 (1971).
27. M. Fleischmann, D. Pletcher and C. J. Vance, J. Electroanal. Chem., **29**, 325 (1971).
28. T. Fujinaga, K. Izutsu and T. Nomura, J. Electroanal. Chem., **29**, 203 (1971).
29. T. R. Koch and W. C. Purdy, Anal. Chim. Acta. **54**, 271 (1971).
30. M. Michlmayr and D. T. Sawyer, J. Electroanal. Chem., **23**, 375 (1969).
31. S. Millefiori, J. Heterocyclic Chem., **7**, 145 (1970).
32. K. G. Boto and F. G. Thomas, Aust. J. Chem., **24**, 975 (1971).
33. J. W. Rogers and Wm. H. Watson, Anal. Chim. Acta, **54**, 41 (1971).
34. V. D. Parker, Acta Chem. Scand. **24**, 3162 (1970).
35. V. D. Parker, Acta Chem. Scand. **24**, 3171 (1970).
36. M. F. Marcus and M. D. Hawley, Biochim. Biophys. Acta **226**, 234 (1971).
37. Y. Matsui, T. Soga and Y. Date, Bull. Chem. Soc., Japan, **44**, 513 (1971).
38. H. Miyazaki and T. Kubota, Bull. Chem. Soc., Japan, **44**, 279 (1971).
39. H. Miyazaki, K. Nichikida and T. Kubota, Bull. Chem. Soc., Japan, **44**, 277 (1971).
40. H. Sadek and B. A. Abd-El-Nakey, Electrochimica, Acta, **116**, 739 (1971).
41. A. I. Prokofev, S. P. Solodovnikov, D. Kh. Rasuleva, A. A. Volod'kin and V. V. Ershov, Izv. Akad. Nauk SSSR Ser. Khim, **7**, 1566 (1970).
42. D. Kh. Rasuleva, A. A. Volod'kin, V. V. Ershov, A. I. Prikof'ev and S. P. Solodovnikov, Izv. Akad. Nauk. SSSR. Ser. Khim, **7**, 1569 (1970).
43. Yu. P. Kitaev, I. M. Skrebkova and L. I. Maslova, Izv. Akad. Nauk. SSSR. Ser. Khim., **10**, 2068 (1970).
44. I. W. Siddiqi and R. M. Johnson, J. Electroanal. Chem., **31**, 211 (1971).
45. I. M. Kolthoff and T. B. Reddy, J. Electrochem. Soc., **108**, 980 (1961).
46. V. D. Parker, Acta. Chem. Scand., **24**, 3151 (1970).
47. C. P. Andrieux and J. M. Saveant, J. Electroanal. Chem., **26**, 223 (1970).
48. M. Libert, C. Caullet and S. Longchamp, Bull. Soc., Chim., **6**, 2367 (1971).
49. M. Libert and C. Caullet, Bull. Soc. Chim., **5**, 1947 (1971).
50. D. Delahaye, G. Barkey and C. Caullet, Bull. Soc., Chim., **8**, 3082 (1971).
51. L. I. Zakharkin, V. N. Kalinin and A. P. Snyakin, Dokl. Akad. Nauk. SSSR, **195**, 970 (1970).
52. H. A. Laitinen and T. J. Kneip, J. Am. Chem. Soc., **78**, 736 (1956).
53. G. Canquis, M. Genies, H. Lemaire, A. Rassat and J. P. Ravet, J. Chem. Phys., **47**, 4642 (1967).
54. C. P. Andrieux and J. M. Saveant, J. Electroanal. Chem., **28**, 339 (1970).
55. F. Pragst and W. Jugelt, Electrochim. Acta, **15**, 1769 (1970).
56. C. P. Andrieux and J. M. Saveant, J. Electroanal. Chem., **28**, 12 (1970).
57. B. Fleet and P. Zuman, Collection Czechoslovak. Chem. Commun., **32**, 2066 (1967).
58. J. Stradins, I. Tutane and G. Vanegs, Proc. 3rd Int. Congress on Polarography 1964, Southampton, Ed. by J. Hills, Vol. II, pp. 731, Macmillan (Lond.) 1966.
59. M. Vajda and F. Ruff, Proc. 3rd Int. Congress on Polarography 1964, Southampton, Ed. by J. Hills, Vol. II, pp. 759, Macmillan (Lond.) 1966.
60. P. Zuman, O. Exner, R. F. Rekker and W. Th. Nauta, Collection of Czechoslovak. Chem. Communs., **33**, 323 (1968).
61. J. S. Double, C. E. R. Jones and G. E. F. Reynolds, Proc. 3rd Int. Congress on Polarography, Southampton 1964, Ed. by J. Hills, Vol. II, pp. 885, Macmillan (Lond.) 1966.
62. I. Bergman, Proc. 3rd Int. Congress on Polarography, Southampton 1964, Ed. by J. Hills, Vol. II, pp. 925, Macmillan (Lond.) 1966.

63. I. Bergman, Proc. 3rd Int. Congress on Polarography, Southampton, 1964, Ed. by J. Hills, Vol. II, pp. 985, Macmillan (Lond.) 1966.
64. T. A. Gough and M. E. Peover, ibid., pp. 1017.
65. N. S. Moe, ibid., pp. 1077.
66. A. Inesi and L. Rampazzo, Electrochim. Acta, **16,** 1469 (1971).
67. R. S. Saxena and U. S. Chaturvedi, Electrochim. Acta, **16,** 1107 (1971).
68. J. L. Webb, C. K. Mann and H. M. Walborsky, J. Am. Chem. Soc., **92,** 2042 (1970).
69. B. J. Huebert and D. E. Smith, J. Electroanal. Chem., **31,** 333 (1971).
70. L. Holleck, J. M. Abd el Kadz and A. M. Shams el din, J. Electroanal. Chem., **20,** 287 (1969).
71. M. D. Curtis and A. L. Allred, J. Am. Chem. Soc., **87,** 2554 (1965).
72. A. Streitwieser, Jr. and I. Schwager, J. Phys. Chem., **66,** 2316 (1962).
73. K. Kuwata and D. H. Geske, J. Am. Chem. Soc., **86,** 2100 (1964).
74. W. F. Little, C. N. Reilley, J. D. Johnson, K. N. Lynn and A. P. Sanders, J. Am. Chem. Soc., **86,** 1376 (1964).
75. W. F. Little, C. N. Reilley, J. D. Johnson and A. P. Sanders, J. Am. Chem. Soc., **86,** 1382 (196).
76. D. W. Hall and C. D. Russell, J. Am. Chem. Soc., **89,** 2316 (1967).
77. R. H. Felton and H. Linschitz, J. Am. Chem. Soc., **88,** 1113 (1966).
78. D. W. Clack and N. S. Hush, J. Am. Chem. Soc., **87,** 4238 (1965).
79. K. K. Barnes and C. K. Mann, J. Org. Chem., **32,** 1474 (1967).
80. R. H. Felton, G. M. Sherman and H. Linschitz, Nature **203,** 637 (1964).
81. R. E. Dessy and R. L. Pohl, J. Am. Chem. Soc., **90,** 2005 (1968).
82. L. K. Gladkova, S. G. Mairanovskii and F. M. Stoyanovich, Elektrokhimiya, **7,** 325 (1971).
83. L. Jeftic and R. N. Adams, J. Am. Chem. Soc., **92,** 1332 (1970).
84. J. S. Dunning and D. N. Bennion, J. Electrochem. Soc., **117,** 485 (1970).
85. E. Kariv, J. Hermolin and E. Gileadi, J. Electrochem. Soc., **117,** 342 (1970).
86. J. R. Jezorek and H. B. Mark, Jr., J. Phys. Chem., **74,** 1627 (1970).
87. A. J. Fry and R. G. Reed, J. Am. Chem. Soc., **91,** 6448 (1969).
88. P. T. Cottrell and C. K. Mann, J. Electrochem. Soc., **116,** 1499 (1969).
89. J. G. Lawless, D. E. Bartak and M. D. Hausley, J. Am. Chem. Soc., **91,** 7121 (1969).
90. B. R. Eggins and J. Q. Chambers, J. Electrochem. Soc., **117,** 186 (1970).
91. W. H. Tiedemann and D. N. Bennion, J. Electrochem. Soc., **117,** 203 (1970).
92. A. J. Bard, K. S. V. Santhanam, J. T. Maloy, J. Phelps and L. O. Wheeler, Discussions, Faraday Soc., **45,** 167 (1968).
93. R. Dietz and M. E. Peover, Diss. Faraday Soc., **45,** 154 (1968).
94. H. E. Stapelfeldt and S. P. Perone, Anal. Chem., **41,** 623 (1969).
95. E. J. Majeski, J. D. Stuart and W. E. Ohnesovge, J. Am. Chem. Soc., **90,** 633 (1968).
96. W. Jufelt and F. Pragst. Angew. Chem. Internat. Edit, **7,** 290 (1968).
97. J. Janata, J. Gendell, R. G. Lawton and H. B. Mark, Jr., J. Am. Chem. Soc., **90,** 5226 (1968).
98. W. V. Childs, J. T. Maloy, C. P. Keszthelyi and A. J. Bard, J. Electrochem. Soc., **118,** 874 (1971).
99. R. F. Nelson, D. W. Leedy, E. T. Seo and R. N. Adams, Z. Analyt. Chem., **224,** 184 (1967).
100. S. Wawzonek, R. Berkey, E. W. Blaha and M. E. Runner, J. Electrochem. Soc., **103,** 456 (1956).
101. W. R. Turner and P. J. Elving, Anal. Chem., **37,** 467 (1965).

VI. POLAROGRAPHY 609

102. Yu. P. Kitaev and I. M. Skrebkova, Izv. Akad. Nauk. SSSR., **4,** 727 (1968).
103. B. Y. Giang, G. D. Christian and W. C. Purdy, J. Polarog. Soc., **13,** 17 (1967).
104. J. P. Zimmer, J. A. Richards, J. C. Turner and D. H. Evans, Anal. Chem., **43,** 1000 (1971).
105. P. T. Cottrell and C. K. Mann, J. Am. Chem. Soc., **93,** 3579 (1971).
106. I. Piljac and R. W. Murray, J. Electrochem. Soc., **118,** 1758 (1971).
107. L. K. Kutanina, K. G. Berezina and Zh. D. Gudzenko, Industrial Lab., **32,** 344 (1966).
108. A. N. Kuman, H. L. Nigam and T. D. Seth, J. Polarog. Soc., **12,** 93 (1966).
109. B. L. Funt and D. G. Gray, Can. J. Chem., **46,** 1337 (1968).
110. S. V. Tatwawadi, K. S. V. Santhanam and A. J. Bard, J. Electroanal. Chem., **17,** 411 (1968).
111. L. Floch, M. S. Spritzer and P. J. Elving, Anal. Chem., **38,** 1074 (1966).
112. J. E. Hickey, M. S. Spritzer and P. J. Elving, Anal. Chim. Acta. **35,** 277 (1966).
113. T. A. Gough and M. E. Peover, Proc. 3rd Int. Congr. Polarogr. Southampton 1964, Ed. by J. Hills, Vol. II Macmillan (Lond.) pp. 1017 (1966).
114. V. D. Bezuglyi, L. A. Melnik, E. M. Shamis and M. M. Dashevskii, J. Gen. Chem., USSR, **36,** 1878 (1966).
115. V. N. Dmitrieva, V. B. Smelyakova, B. M. Krasovitskii and V. D. Bezuglyi, J. Gen. Chem. USSR, **36,** 421 (1966).
116. K. Wettig, Sov. Electrochem., **3,** 235 (1967).
117. V. G. Mairanovskii and G. I. Samokhvalov. Sov. Electrochem., **2,** 663 (1966).
118. Ya. P. Stradyn, G. O. Reikhmanis and R. A. Gavar, Sov. Electrochem. **1,** 852 (1965).
119. W. R. Turner and P. J. Elving, J. Electrochem. Soc., **112,** 1215 (1965).
120. V. D. Bezuglyi and T. A. Alekseeva, J. Am. Chem. USSR, **20,** 220 (1965).
121. V. D. Bezuglyi and Yu. P. Ponomarev, J. Anal. Chem., USSR, **22,** 369 (1967).
122. N. A. Lezhneva and E. A. Kruglov, J. Anal. Chem. USSR **21,** 1208 (1966).
123. T. A. Alekseeva and V. D. Bezuglyi, J. Gen. Chem. USSR, **36,** 2046 (1966).
124. Ya. Stradyn, E. Gren, V. Kampar and G. Vanag, J. Gen. Chem. USSR, **35,** 223 (1965).
125. E. S. Levin and Z. I. Fodiman, J. Gen. Chem. USSR, **34,** 1055 (1964).
126. V. D. Bezuglyi and L. A. Mel'nik, J. Gen. Chem. USSR, **34,** 2117 (1964).
127. V. D. Bezuglyi, L. A. Melnik and V. N. Dmitrieva, J. Gen. Chem. USSR, **34,** 1041 (1964).
128. V. D. Bezuglyi and L. A. Mel'nik, J. Gen. Chem. USSR, **34,** 2495 (1964).
129. I. G. Sevast'yanova and A. P. Tomilov, J. Gen. Chem. USSR, **33,** 2741 (1963).
130. L. Ya. Kheifets, E. A. Preobrazhenskaya and V. D. Bezuglyi, J. Gen. Chem. USSR, **35,** 1704 (1965).
131. R. Takahashi and P. J. Elving, Israel J. Chem. **4,** 195 (1966).
132. B. J. Tabner and J. R. Yandle, J. Chem. Soc., A. 381 (1968).
133. H. W. Sternberg, R. E. Markloyand and I. Wender, J. Electrochem. Soc., **113,** 1060 (1966).
134. L. L. Gordienko, Sov. Electrochem., **1,** 1351 (1965).
135. G. H. Aylward, J. L. Garnett and J. H. Sharp, Chem. Commun., **5,** 137 (1966).
136. J. Stradins, I. Tutane, O. Nieland, Doklady, Chem. Proceed., **166,** 140 (1966).
137. E. Jacobsen and D. T. Sawyer, J. Electroanal. Chem., **16,** 361 (1968).
138. E. Jacobsen, J. L. Roberts and D. T. Sawyer, J. Electroanal. Chem., **16,** 351 (1968).
139. T. Fujinaga and K. Takoaka; J. Electroanal. Chem., **16,** 99 (1968).
140. T. Fujinaga, K. Izutsu and K. Takoaka, J. Electroanal. Chem., **16,** 89 (1968).

141. R. E. Soida, D. O. Cowan and W. S. Koski, J. Am. Chem. Soc., **89,** 230 (1967).
142. V. G. Mairanovskii, L. A. Vakulova and G. I. Samokhvalov, Sov. Elektrochem., **3,** 18 (1967).
143. R. Yu. Mamedzade, E. I. Klabunovskii and A. A. Balandin, Bull. Acad. Sci. USSR, **5,** 976 (1967).
144. M. Kashiki and K. Ishida, Bull. Chem. Soc., Japan, **40,** 97 (1967).
145. V. Dvorak, I. Nemec and J. Zyka, Microchem. J. **12,** 99 (1967).
146. R. Takahashi and P. J. Elving, Electrochim Acta, **12,** 213 (1967).
147. G. H. Aylward, J. L. Garnett and J. H. Sharp, Anal. Chem., **39,** 457 (1967).
148. J. Phelps, K. S. V. Santhanam and A. J. Bard, J. Am. Chem. Soc., **89,** 1752 (1967).
149. T. Fujinega, K. Izutsu and K. Takaoka, J. Electroanal. Chem., **12,** 203 (1966).
150. M. E. Peover and B. S. White, J. Electroanal. Chem., **13,** 93 (1967).
151. R. A. Wasserman and W. C. Purdy, J. Electroanal. Chem., **9,** 51 (1965).
152. J. Q. Chambers III and R. N. Adams, J. Electroanal. Chem., **9,** 400 (1965).
153. M. E. Peover and J. D. Davies, Proc. 3rd Int. Congr. Southampton, Vol. 2, (1964), pp. 1003.
154. R. Gelin, M. Bréant and D. Makula, C. R. Acad. Sc., **260,** 5767 (1965).
155. G. Faraone and M. Trozzi, Annali di chimica, **54,** 1207 (1964).
156. C. Caullet, J. M. Bessin and J. C. Bodard, C. R. Acad. Sc., **261,** 1848 (1965).
157. C. Caullet, G. Laur and A. Nonat, C. R. Acad. Sc., **261,** 1974 (1965).
158. S. Lazarov, A. Trifonov and T. Vitanov, Z. Physik Chem., (Leipzig) **226,** 221 (1964).
159. V. F. Lavsushin, V. D. Bezuglyi, G. G. Belous and V. G. Tishchenko, J. Gen. Chem., USSR, **34,** 5 (1964).
160. S. G. Mairanovskii, I. A. D'yachenko and Ya. L. Gol'dfarb, Sov. Elektrokhimiya, **7,** 33 (1970).
161. G. Cauquis and D. Serve, Bull. Soc. Chim., **1,** 302 (1966).
162. L. Holleck and A. M. S. Eldin, Electrochim Acta, **13,** 199 (1968).
163. D. E. Sellers and G. W. Leonard, Jr., Anal. Chem., **33,** 334 (1961).
164. D. E. Sellers and G. W. Leonard, Jr., Anal. Chem., **34,** 334 (1961).
165. M. S. Spritzer, J. M. Costa and P. J. Elving, Anal. Chem., **37,** 211 (1965).
166. S. L. Gupta and N. Kishore, Electrochim. Acta, **15,** 1367 (1970).
167. J. R. Jezorek and H. B. Mark, Jr., J. Phys. Chem., **74,** 1627 (1970).
168. J. L. Sadler and A. J. Bard, J. Electrochem. Soc., **115,** 343 (1968).

(c) Organometallic Compounds

The following section on nonaqueous voltammetry and polarography relative to organometallic compounds covers solute concentration, supporting electrolyte, temperature, working electrode, reference electrode and half-wave potentials. The tables are organized in alphabetical order by solvent and the survey covers the literature to up to 1972.

Symbols used:

DME	Dropping mercury electrode
SCE	Standard calomel electrode
RE(Pt) or RPE	Rotating platinum electrode
MPE	Mercury pool electrode
RPGE	Rotating pyrolytic graphite electrode
HMD	Hanging mercury drop
mM	Millimoles

ACETIC ACID

0.1M LiClO at RE (Pt) at 25°C

0.5 mM

Species	$E_{1/2}V$ (vs SCE)	Ref.
tris(4,7-Dimethyl 1,10-phenanthroline) Fe(II)	+0.815	74
Ferrocene	+0.280	

ACETIC ANHYDRIDE

0.1M LiClO$_4$ at RE (Pt) at 25°C

0.5 mM

Species	$E_{1/2}V$ (vs SCE)	Ref.
Benzyl ferrocene	+0.355	74
bis(2,9-Dimethyl-1,10-phenanthroline) Cu(I)	+0.715	
tris(4,7-Dimethyl-1,10-phenanthroline) Fe(II)	+0.900	
Ferrocene	+0.385	

ACETONE
0.1M LiClO$_4$ at RE (Pt)

Species	$E_{1/2}V$ (vs SCE)	Ref.
4,7-Dimethyl-1,10-phenanthroline ferrous	+0.940	4

0.1M Et$_4$NClO$_4$ at DME at 20 ± 0.1°C
0.065 mM

Species	$E_{1/2}V$ (vs Ag/AgCl) 10.1M LiCl	Ref.
Fe(*SacSac)$_3^{3+}$	−0.21	3
Rh(SacSac)$_3^{3+}$	+1.05	
Os(SacSac)$_3^{3+}$	−0.05	
Ru(SacSac)$_3^{3+}$	+0.04	

Species	Supporting electrolyte	Working electrode	$E_{1/2}V$ (vs SCE)	Ref.
(π-C$_5$H$_5$)Fe(CO)$_2$	0.2M NH$_4$PF$_6$	carbon	+0.48	2
	0.1M Bu$_4$NCl	carbon	>0.30	
	0.2M NH$_4$PF$_6$	RE (Pt)	~0.55	

0.1M Et$_4$NClO$_4$ at DME at 20 ± 1°C
0.01 to 1 mM

Species	$E_{1/2}V$ (vs Ag/AgCl)	Ref.
Co(SacSac)$_2$[a]	−0.66; −1.46	1
Co(SacSac)$_3$	~−0.7; −1.45	
Fe(SacSac)$_3$	−0.21	
Ir(SacSac)$_3$	−1.165; −1.170	
Ni(SacSac)$_2$	−0.940; −1.16; −1.47	
Os(SacSac)$_3$	−0.05	
Pd(SacSac)$_2$	−0.956; −1.40; −1.54	
Pt(SacSac)$_2$	−1.000; −1.504	
Rh(SacSac)$_3$	−1.05	
Ru(NO)(SacSac)$_2$Cl		
Ru(SacSac)$_3$	+0.04	

[a] Dithioacetylacetone.

VI. POLAROGRAPHY

ACETONE (Continued)
0.1M LiClO$_4$ at RE (Pt) at 25°C
0.5 mM

Species	$E_{1/2}V$ (vs SCE)	Ref.
Benzyl ferrocene	+0.435	74
bis(2,9-Dimethyl-1,10-phenanthroline) Cu(I)	+0.760	
tris(4,7-Dimethyl-1,10-phenanthroline Fe(II)	+0.940	
Ferrocene	+0.460	

ACETONITRILE
0.1M Bu$_4$NPF$_6$ at DME
~1 mM

Species	$E_{1/2}V$ (vs SCE)	Ref.
[π-(3)-1,2-B$_9$C$_2$H$_{11}$]$_2$Ni	−2.10	20
[Et$_4$N][(π-(3)-1,7-B$_9$C$_2$H$_{11}$)$_2$Ni]	−2.09	
(π-C$_5$H$_5$)Co(π-(3)-1,2-B$_9$C$_2$H$_{11}$)	−2.11	
[Et$_4$N][(π-(3)-1,2-B$_9$C$_2$H$_{11}$)$_2$Co	−2.24	
[Et$_4$N][(π-(3)-1,7-B$_9$C$_2$H$_{11}$)$_2$Co]	−2.52	
[Me$_4$N]$_2$[(B$_9$C$_2$H$_{11}$)Co(B$_8$C$_2$H$_{10}$)-Co(B$_9$C$_2$H$_{11}$)]	−2.36	

ACETONITRILE (Continued)

0.2M LiClO$_4$ at Pt-foil

~0.818 to 3.25 mM

Species	$E_{1/2}V$ (vs SCE)	Ref.
—CH=CH$_2$	+0.325 ± 0.002	26
—COCH$_3$	+0.573 ± 0.004	
1,1'-(COCH$_3$)	+0.796 ± 0.004	
—COC$_6$H$_5$	+0.571 ± 0.004; +0.907 ± 0.008	
—COC$_6$H$_5$ osmocene	+0.866 ± 0.010	
1,1'-(COC$_6$H$_5$)	+1.089 ± 0.018	
—COOH	+0.550 ± 0.002	
1,1'-Diethyl ferrocene	+0.194 ± 0.001	
Ethyl ferrocene	+0.245 ± 0.003	
Ferrocene	+0.307 ± 0.002	
Methyl ferrocenyl carbinol	+0.298 ± 0.003	
Phenyl ferrocenyl carbinol	+0.318 ± 0.003	
Phenyl ruthenocenyl carbinol	+0.755 ± 0.004	
Osmocene	+0.633 ± 0.005; +1.50 ± 0.01	
Ruthenocene	+0.693 ± 0.003	

0.2M LiClO$_4$ at Pt foil

0.1 mM

Species (ferrocene and substituted ferrocenes)	$E_{1/4}V$ (vs SCE)	Ref.
PhOCH$_2$	+0.378 ± 0.001	25
CH$_3$CO	+0.587 ± 0.003	
(CH$_3$)$_3$Si	+0.346 ± 0.002	
bis-(CH$_3$)$_3$Si	+0.346 ± 0.001	
p-OCH$_3$C$_6$H$_4$	+0.323 ± 0.001	
p-ClC$_6$H$_4$	+0.387 ± 0.001	
p-BrC$_6$O$_4$	+0.396 ± 0.001	
p-CH$_3$COC$_6$H$_4$	+0.426 ± 0.001	
p-NO$_2$C$_6$H$_4$	+0.464 ± 0.001	
Dioctyl	+0.228 ± 0.001	
Didenyl	+0.222 ± 0.001	
Cinnamoyl	+0.594 ± 0.001	

VI. POLAROGRAPHY

ACETONITRILE (Continued)
0.2M LiClO₄ at Pt foil
0.1 mM

Species (ferrocene and substituted ferrocenes)	$E_{1/4}V$ (vs SCE)	Ref.
Formyl	$+0.624 \pm 0.002$	25
Ferrocene substituent	$+0.341 \pm 0.002$	
Di-C_2H_5	$+0.224 \pm 0.001$	
Di-CH_3	$+0.241 \pm 0.004$	
$CH_3CHC(CH_3)_3$	$+0.258 \pm 0.002$	
Sec-C_4H_9	$+0.288 \pm 0.001$	
i-C_3H_7	$+0.286 \pm 0.001$	
n-C_3H_7	$+0.284 \pm 0.001$	
C_2H_5	$+0.281 \pm 0.002$	
Di-$PhCH_2$	$+0.296 \pm 0.001$	
CH_3	$+0.281 \pm 0.002$	
α-$PhCH_3H_6$	$+0.305 \pm 0.002$	
β-PhC_2H_4	$+0.296 \pm 0.001$	
α-PhC_2H_4	$+0.315 \pm 0.002$	
$PhCH_2$	$+0.314 \pm 0.001$	
$CH_3CH{=}CH$	$+0.316 \pm 0.002$	
$PhCH{=}CH$	$+0.328 \pm 0.003$	
CH_3OCH_3	$+0.340 \pm 0.003$	
Ph	$+0.366 \pm 0.001$	
Di-Ph	$+0.370 \pm 0.001$	

0.1M Et₄NClO₄ at DME at 25 ± 0.02°C
0.2 to 1 mM

Species	$E_{1/2}V$ (vs SCE)	Ref.
Ferrocene	$+0.379 \pm 0.001$	24

0.1M Et₄NBr at Mercury plated Pt-wire at ~22°C
~5 mM

Species	E_pV (vs Ag/AgNO₃)	Ref.
1-Methyl-2, 2-diphenylcyclo propyl mercuric bromide	$-2.57; -0.62; -1.22$	23

ACETONITRILE (Continued)

0.1M Et_4NClO_4 at 25°C

~2.5 mM

Species	$E_{1/2}V$ (vs SCE)	Ref.
Co(III)[trans(14)diene]$^+$2NO$_2$	-0.40^a; -0.38^a; -0.48^b; -1.66^b; -1.66^b; -1.64^b	22
Co(III)[trans(14)diene]$^+$2CN	-1.10^a; -1.10^a; -1.10^b; -1.10^b; -1.78^b; -1.68^b; -1.09^c; -1.80^c	
Co(III)[trans(14)diene]$_3$$^+$2NH$_3$	-0.91^b; -0.91^b; -1.45^b; -1.43^b	
Ni(II)[trans(14)diene]$^{2+}$	-1.26^a; -1.24^a; -1.22^b; -1.21^b; $+1.76^b$; $+1.03^b$; -1.23^c; $+2.06^c$	
Ni(II)[cis(14)diene]$^{2+}$	-1.25^a; -1.24^a; -1.25^b; -1.25^b; $+1.73^b$; $+1.12^b$; -1.25^c	
Ni(II)[(CH$_3$)$_2$trans(14)diene]$^{2+}$	-1.21^a; -1.19^a; -1.17^b; -1.16^b; $+1.92^b$; $+1.22^b$; -1.20^c	
Ni(II)[A(13)T$^w-$]$^{2+}$	-2.15^a; $+0.69^b$; $+0.73^c$	
Ni(II)[A(14)T$^z-$]$^{2+}$	-1.99^a; -2.00^b; $+0.61^b$	
Ni(II)[teta]$^{2+}$	-1.25^a; -1.22^a; -1.24^b; -1.23^b; $+1.25^b$; $+1.21^b$; -1.29^c	
Ni(II)[tetb^y]$^{2+}$	-1.30^a; -1.17^a; -1.25^b; -1.24^b; $+1.80^b$; $+1.28^b$; -1.25^c	
Ni(II)[CH$_3$)$_2$teta]$^{2+}$	-1.21^a; -1.20^a; -1.21^b; -1.19^b; $+1.42^b$; $+1.36^b$; -1.22^c	
Cu(II)[2-trans(14)diene]$^{2+}$	-0.65^a; -0.61^a; -0.73^b; -0.53^b; $+1.60^b$; $+1.52^b$	
Cu(II)[β-trans(14)diene]$^{2+}$	-0.65^a; -0.59^a; -0.70^b; -0.55^b; $+1.64^b$; $+1.52^b$; -0.77^c	
Fe(II)[trans(14)diene^x]$_2$$^{2+}$	-1.90^a; -1.86^b; $+0.16^b$; $+0.15^b$	
Fe(II)[teta^y]$_2$$^{2+}$	-1.92^a; $+0.18^b$; $+0.18^b$	

VI. POLAROGRAPHY

ACETONITRILE *(Continued)*
0.1M Et_4NClO_4 at 25°C
~2.5 mM

Species	$E_{1/2}V$ (vs SCE)	Ref.
Co(II)[*trans*(14)diene₂]²⁺	−1.59[z]; −1.19[a]; −1.39[b]; −1.36[b]; +0.72[b]; +0.40[b]	22
Co(II)[*trans*(14)diene 2H₂O]₂²⁺	−1.55[a]; −1.29[a]; −1.40[b]; −1.39[b]; +0.94[b]; +0.31[b]; −1.42[c]; +1.21[c]	
Co(II)[*trans*(14)diene 2Py]₂²⁺	−1.46[a]; −1.25[a]; −1.42[b]; −1.41[b]; +1.06[b]; +0.25[b]; +1.43[c]; +1.44[c]	
Co(II)[teta]²⁺	−1.90[a]; −1.66[a]; −1.76[b]; −1.63[b]; +0.51[b]; +0.37[b]	
Co(II)[(CH₃)₂-*trans*(14)diene]²⁺	−1.37[a]; −1.24[a]; −1.36[b]; −1.36[b]; +0.75[b]; +0.53[b]; −1.33[c]	
Co(III)[*trans*(14)diene]⁺Br₂	+0.17[b]; +0.17[b]; −1.45[b]; −1.45[b]	
Co(III)[*trans*(14)diene]⁺Cl₂	−0.20[b]; −0.17[b]; −1.55[b]; −1.47[b]; −0.20[c]; −1.80[c]	
Co(III)[*trans*(14)diene]⁺2N₃	−0.42[a]; −0.42[a]; −0.45[b]; −0.42[b]; −1.73[b]; −1.64[b]	
Co(III)[*trans*(14)diene]³⁺2H₂O	+0.32[b]; −0.23[b]; −1.53[b]; −1.50[b]; −0.27[c]	
Co(III)[*trans*(14)diene]⁺2NCS	+0.44[b]; +0.59[b]; −0.40[b]; +0.62[b]; −1.46[b]; −1.45[b]; −0.45[c]; −1.48[c]	
Co(III)[*trans*(14)diene](OH)O₂CCH₃	−0.58[a]; −0.44[a]; −0.60[b]; −0.55[b]; −1.78[b]; −1.70[b]	
Cu(II)[teta]²⁺	−0.82[b]; +1.68[b]; +1.54[b]; −0.83[c]	
Zn(II)[*trans*(14)diene]²⁺	−1.64[a]	

[a] Hanging drop mercury.
[b] Pt electrode (stationary).
[c] Pt electrode rotating).
[w] 11, 13-Dimethyl 1, 4, 7, 10-tetraazacyclotrideca-10, 12-diene [A(13)T⁻].
[x] 5, 7, 7, 12, 14, 14 Hexamethyl-1, 4, 8, 11-tetraazacyclotetradeca-4, 11-diene (*trans*-(14)diene).
[y] 5, 7, 7, 12, 14, 14 Hexamethyl-1, 4, 8, 11-tetraazacyclotetradecane (teta or tetb).
[z] 12, 14-Dimethyl-1, 4, 8, 11-tetraazacyclotetradeca-11, 13-diene [A(14)T⁻].

ACETONITRILE (Continued)
[0.1M LiClO$_4$ at RE (Pt)]

Species	$E_{1/2}V$ (vs SCE)	Ref
4,7-Dimethyl 1,10-phenanthroline ferrous	+0.860	4

Et$_4$NClO$_4$ at RPE at 25°C
0.5 mM

Species	$E_{1/2}V$ (vs Ag/Ag$^+$)[b] 0.1M in CH$_3$CN	Ref.
Ni(TAAB)$^{+a}$	−0.640; −0.930	19
Ni(TAAB)$^{2+}$	−0.640; −0.930; −2.035[b]	

[a] Tetrabenzo [b, f, j, n][1, 5, 9, 13] tetraaza cyclohexadecine.
[b] Ligand reduction.

DME; 0.1 mM

Species	Supporting electrolyte	$E_{1/2}V$ (vs AgNaCl SCE)	Ref.
Cu(acac)$_2$	0.1M Et$_4$NClO$_4$ + LiClO$_4$ (0.1 mM)	−0.58	18
	0.1M Et$_4$NClO$_4$ + LiClO$_4$ (0.5 mM)	−0.62	

0.1N Et$_4$NClO$_4$ at RPE

Species	Temp.	$E_{1/2}V$ (vs SCE)	Ref.
(π-C$_5$H$_5$)$_2$Ni	−40	+0.77; −0.09	17
(C$_5$H$_5$)Ni(B$_9$C$_2$H$_{11}$)	25	+0.46; −0.52	

ACETONITRILE (Continued)
0.1M Et$_4$NClO$_4$ at DME

Species	$E_{1/2}V$ (vs SCE)	Ref.
MOCl$_3$(Py)$_3$	+0.94; −1.47	16
MOCl$_3$(Py)bipy	+0.83; −1.00	
MOCl$_2$(bipy)$_2^+$	+0.92; −0.63; −1.08	

0.1M Et$_4$NClO$_4$ at DME
0.5–7.5 mM

Species	$E_{1/2}V$ (vs Ag/AgCl)	Ref.
Co(CO)$_3$NO	−1.07; ∼−1.6; −2.3	14
Co(CO)$_3$NO[a]	−1.07	15
Co(CO)$_2$NOSbϕ_3[a]	−1.18	15
Co(CO)$_2$NOAsϕ_3[a]	−1.27	15
Co(CO)$_2$NOPϕ_3[a]	−1.46	15
[Co(CO)$_2$NO]$_2$diphos[a]	−1.53	15
CoCONO(Sbϕ_3)$_2$[a]	−1.65	15
CoCONO(Asϕ_3)$_2$[a]	−1.70	15
CoCONO(Pϕ_3)$_2$[a]	−1.89	15
CoCONO diphos[a]	−2.02	15
NO[a]	−0.96	15

[a] 0.1 to 1 mM.

Species	Supporting electrolyte	Working electrode	$E_{1/2}V$ (vs SCE)	Ref.
[(π-C$_5$H$_5$)Fe(CO)]$_2^2$	0.2M NH$_4$PF$_6$	carbon	+0.19	2
	0.1M Et$_4$NCl	carbon	+0.42	
	0.2M NH$_4$PF$_6$	RE (Pt)	∼1.0	
	0.1M Et$_4$NCl	RE (Pt)	>0.88	

ACETONITRILE (Continued)

0.1M NaClO$_4$ at Re (Pt disk) at 20°C

0.107 mM Solutions

Species	$E_{1/2}V$ (vs SCE)	Ref.
Co(dipy*)$_3{}^{2+}$	+0.220; +0.212	13
Co(dipy)$_3{}^{3+}$	+0.222; +0.214	
Fe(dipy)$_3{}^{2+}$	+0.980; +0.975	
Fe(dipy)$_3{}^{3+}$	+0.978; +0.973	
Fe(phen*)$_3{}^{2+}$	+0.997; +0.992	
Fe(phen)$_3{}^{3+}$	+0.996; +0.990	
Fe(Cpdien*)$_2{}^a$	+0.308; +0.303	
Fe(Cpdien^a)$_2{}^+$	+0.305; +0.300	

* dipy = 2,2′-dipyridyl.

phen = 1,10-phenanthroline.

*^{a} Cpdien = cyclopentadienyl 0.0741 mM.

0.1M Et$_4$NClO$_4$ at DME at 20 ± 1°C

~1–12 mM

Species	$E_{1/2}V$ (vs SCE)	Ref.
π-C$_3$H$_5$Fe(CO)$_2$NO	−1.43	12
π-(1-CH$_3$C$_3$H$_4$)Fe(CO)$_2$NO	−1.50	
π-(2-CH$_3$C$_3$H$_4$)Fe(CO)$_2$NO	−1.63	
π-(1-ClC$_3$H$_4$)Fe(CO)$_2$NO	−1.16	
π-(2-ClC$_3$H$_4$)Fe(CO)$_2$NO	−1.10	
π-(2-BrC$_3$H$_4$)Fe(CO)$_2$NO	−1.31	

VI. POLAROGRAPHY

ACETONITRILE *(Continued)*
0.1M Bu$_4$NBF$_4$ at 25 ± 0.1°C
1 mM

Species	$E_{1/2}V$ (vs SCE)	Ref.
CoBr$_3$(C$_5$H$_5$)$_2$	−0.98,[a] −1.92[a]	11
Fe(C$_5$H$_5$)$_2$	+0.42,[a] +0.48[b]	
Ni(C$_5$H$_5$)$_2$	+0.01,[a] −1.74[a]	
	+0.01,[b] +0.75[b]	
Ni(t-Bu$_2$)(C$_5$H$_4$)$_2$	−0.11,[a] −1.86[a]	
(C$_5$H$_5$)$_2$Ni P(C$_6$H$_5$)$_3$	+0.24,[a] +0.01,[a] −1.78[a]	
C$_5$H$_5$Ni P(C$_6$H$_5$)$_3$Cl[c]	+0.24,[a] −1.03,[a] −1.79[a]	
NiCl$_2$2(C$_6$H$_5$)$_3$P	+0.32, −1.03	

[a] At DME.
[b] At platinum electrode.
[c] 2 mM.

0.1M Bu$_4$NBF$_4$ at 25 ± 0.1°C
1 mM

Species	$E_{1/2}V$ (vs SCE)	Ref.
Fe(C$_5$H$_5$)$_2$	+0.42,[a] +0.48[b]	10
Fe(C$_5$H$_5$)(p-FC$_6$H$_4$)C$_5$H$_4$	+0.47[a]	
Fe(C$_5$H$_5$)C$_5$H$_4$CH$_2$C$_5$H$_4$OsC$_5$H$_5$	+0.14,[a] +0.41[b]	
Fe(C$_5$H$_5$)(C$_5$H$_4$CH$_2$C$_5$H$_4$)RuC$_5$H$_5$	+0.34,[a] +0.54,[a] +0.40,[b] +0.88[b]	
Fe(C$_5$H$_5$)(C$_5$H$_4$COC$_5$H$_4$)RuC$_5$H$_5$	+0.57[b]	
Os(C$_5$H$_5$)$_2$	+0.16,[a] +0.75,[b] +1.37[b]	
Ru(C$_5$H$_5$)$_2$	+0.36,[a] +0.78[b]	
Ru(C$_5$H$_5$)C$_5$H$_4$C$_2$H$_5$	+0.32,[a] +0.66[b]	
Ru(C$_5$H$_5$)C$_5$H$_4$C$_6$H$_5$	+0.37[a]; +0.75[b]	
Ru(C$_5$H$_5$)C$_5$H$_4$CN	+1.05[b]	
Ru(C$_5$H$_5$)C$_5$H$_4$OCH$_3$	+0.31,[a] +0.65[b]	
Ru(C$_5$H$_5$)C$_5$H$_4$COCH$_3$	+0.57,[a] +1.01[b]	
Ru(C$_5$H$_5$)C$_5$H$_4$OCOCH$_3$	+0.39,[a] +0.84[b]	
Ru(C$_5$H$_5$)C$_5$H$_4$COOCH$_3$	+0.57,[a] +0.99[b]	

[a] At DME.
[b] At platinum electrode.

ACETONITRILE (Continued)

0.2M LiClO$_4$ at Pt-foil at 25°C

1 mM

Species	$E_{1/2}V$ (vs SCE)	Ref.
m-Amino phenyl ferrocene	+0.328	5, 9
p-Amino phenyl ferrocene	+0.243	
p-Bromo ferrocenyl azobenzene	+0.392	
o-Bromo phenyl ferrocene	+0.388 ± 0.002	
m-Bromo phenyl ferrocene	+0.390	
p-Bromo phenyl ferrocene	+0.377	
m-Carbobenzhydroxy phenyl ferrocene	+0.385	
p-Carbobenzhydroxy phenyl ferrocene	+0.409	
Carboethoxy ferrocene	+0.556	
m-Carboethoxy phenyl ferrocene	+0.390	
p-Carboethoxy phenyl ferrocene	+0.402	
m-Carbomethoxy phenyl ferrocene	+0.385	
o-Carbomethoxy phenyl ferrocene	+0.380 ± 0.001	
p-Carbomethoxy phenyl ferrocene	+0.396	
Carboxyl ferrocene	+0.552	
m-Carboxyl phenyl ferrocene	+0.381	
o-Carboxy phenyl ferrocene	+0.373 ± 0.002	
p-Carboxy phenyl ferrocene	+0.397	
C(C$_6$H$_5$)$_3$ ferrocene	+0.372	
o-Chloro phenyl ferrocene	+0.386 ± 0.002	
CH(OCH$_3$)C$_6$H$_5$ ferrocene	+0.332	
CH(OH)C$_3$H$_5$ ferrocene	+0.324	
o-CH$_2$OH-phenyl ferrocene	+0.352 ± 0.001	
COCH$_3$ ferrocene	+0.567	
CO$_2$CH(C$_6$H$_5$)$_2$ ferrocene	+0.587	
CO$_2$CH$_3$ ferrocene	+0.561	
COC$_6$H$_5$ ferrocene	+0.571	
CON(C$_6$H$_5$)$_2$ ferrocene	+0.496	
CONHC$_6$H$_5$ ferrocene	+0.522	

VI. POLAROGRAPHY

ACETONITRILE (Continued)
0.2M LiClO$_4$ at Pt-foil at 25°C
1 mM

Species	$E_{1/2}V$ (vs SCE)	Ref.
CONH$_2$ ferrocene	+0.531	5, 9
p-Cyano phenyl ferrocene	+0.430	
2,6-Dimethyl-4-nitro phenyl ferrocene	+0.415 ± 0.001	
o-Ethoxy phenyl ferrocene	+0.295 ± 0.002	
Ferrocene	+0.315	
Ferrocenyl azobenzene	+0.389	
o-Fluoro phenyl ferrocene	+0.359 ± 0.003	
p-Hydroxy phenyl ferrocene	+0.293	
Iodo ferrocene	+0.457	
p-Iodo ferrocenyl azobenzene	+0.390	
o-Iodo phenyl ferrocene	+0.389 ± 0.002	
p-Methoxy ferrocenyl azobenzene	+0.388	
o-Methoxy phenyl ferrocene	+0.292 ± 0.003	
p-Methyl ferrocenyl azobenzene	+0.390	
2-Methyl-4-nitro phenyl ferrocene	+0.434 ± 0.001	
2-Methyl-5-nitro phenyl ferrocene	+0.424 ± 0.003	
2-Methyl-6-nitro phenyl ferrocene	+0.436 ± 0.001	
m-Methyl phenyl ferrocene	+0.392	
o-Methyl phenyl ferrocene	+0.340 ± 0.001	
p-Methyl phenyl ferrocene	+0.325	
m-N$_2$C$_6$H$_5$-phenyl ferrocene	+0.380	
p-N$_2$C$_6$H$_5$-phenyl ferrocene	+0.389	
m-NHCOC$_6$H$_5$-phenyl ferrocene	+0.356	
p-NHCOC$_6$H$_5$-phenyl ferrocene	+0.330	
m-Nitro phenyl ferrocene	+0.423	
o-Nitro phenyl ferrocene	+0.444 ± 0.004	
p-Nitro phenyl ferrocene	+0.447	
Phenyl ferrocene	+0.346	
o-Phenyl phenyl ferrocene	+0.340 ± 0.001	
p-Phenyl phenyl ferrocene	+0.346	

ACETONITRILE (Continued)
0.2M NaClO$_4$ at Pt-disk at 25°C
~2 mN

Species	$E_{1/4}V$ (vs Ag/AgClO$_4$)	Ref.
Acetamido ferrocene	−0.005	8
1′-Acetyl-1-acetamido ferrocene	+0.234	
2-Acetyl-1-acetamido ferrocene	+0.313	
1′-Acetyl-1-bromo ferrocene	+0.469	
1′-Acetyl-1-chloro ferrocene	+0.490	
1′-Acetyl-1-cyano ferrocene	+0.661	
3-Acetyl-1, 1′-dimethyl ferrocene	+0.211	
1′-Acetyl-1-methoxy carbonylamino ferrocene	+0.220	
Cyano ferrocene	+0.438	
1, 1′-Diacetyl ferrocene	+0.551	
1, 1′-Dibromo ferrocene	+0.380	
1, 1′-Di(ethoxy carbonylamino)ferrocene	−0.076	
1, 1′-Di(methoxy carbonylamino)-ferrocene	−0.074	
Ferrocene	+0.063	
Methoxy carbonylamino ferrocene	−0.007	
Methoxy ferrocene	+0.005	

0.05M Et$_4$NClO$_4$ at DME

Species	$E_{1/2}V$ (vs Ag/AgNO$_3$)	Ref.
Cyclopentadiene carbonylhydridobis (trichlosilyl) iron	−1.5; −2.4	7
Cyclopentadiene dicarbonyltrichlorosilyl iron	−1.93; −2.4	
bis(Dicarbonyl cyclopentadienyl iron)	−1.84	

VI. POLAROGRAPHY

ACETONITRILE (Continued)
0.1M NaClO$_4$ at Pt wire electrode at 22°C
5–10 mM

Species	$E_p V$ [vs Ag/AgNO$_3$ (0.1M CH$_3$CN)]	Ref.
n-Butyl mercaptan	+1.49	6
s-Butyl mercaptan	+1.33	
t-Butyl mercaptan	+1.59	
Diallyl sulfide	+1.74	
Dibenzyl sulfide	+1.48	
Dibenzyl thiophene	+1.35	
Di-n-butyl sulfide	+1.45	
Di-s-butyl sulfide	+1.43	
Di-t-butyl sulfide	+1.06	
Diethyl sulfide	+1.50	
Dimethyl sulfide	+1.41	
Diphenyl sulfide	+1.26	
Di-i-propyl sulfide	+1.47	
p-Dithiane	+1.46	
Ethylene sulfide	+1.51	
Ethylene trithiocarbonate	+1.53	
Tetrahydrothiophene	+1.45	
Pentamethylene sulfide	+0.55	
Propylene sulfide	+1.69	
Thiophene	+1.84	
sym-Trithiane	+1.30	

At DME at 25 ± 1°C
1.04–1.62 mM

Species	$E_{1/2} V$ (vs SCE)	Ref.
CpVTe[a]	+0.184 to +0.196	21
	+0.168[b]	
	+0.192 to +0.196[c]	
	+0.839[c]	

[a] π-Cyclopentadienyl-π-cycloheptatrienylvanadium (0).
[b] At −20°C.
[c] At rotating Pt. electrode.

ACETONITRILE (Continued)
0.1M LiClO$_4$ at Re (Pt) at 25°C
0.5 mM

Species	$E_{1/2}V$ (vs SCE)	Ref.
Benzyl ferrocene	+0.325	74
bis(2,9-Dimethyl-10-phenanthroline) Cu(I)	+0.685	
tris(4,7-Dimethyl-1,10-phenanthroline) Fe(II)	+0.860	
Ferrocene	+0.350	

ACETYLACETONE
0.1M LiClO$_4$ at RE (Pt) at 25°C
0.5 mM

Species	$E_{1/2}V$ (vs SCE)	Ref.
Benzyl ferrocene	+0.355	74
bis(2,9-Dimethyl-1,10-phenanthroline) Cu(I)	+0.730	
tris(4,7-Dimethyl-1,10-phenanthroline) Fe(II)	+0.905	
Ferrocene	+0.355	

ALLYL ALCOHOL
0.1M LiClO$_4$ at RE (Pt) at 25°C
0.5 mM

Species	$E_{1/2}V$ (vs SCE)	Ref.
Benzyl ferrocene	+0.330	74
tris(4,7-Dimethyl-1,10-phenanthroline) Fe(II)	+0.890	
Ferrocene	+0.365	

VI. POLAROGRAPHY

BENZENE–METHANOL (1:1)

0.3M Sodium Acetate at DME at 30 ± 0.1°C

0.1 to 1.5 mM

Species	$E_{1/2}V$ (vs Hg pool)	Ref.
Cobalt naphthenate	−1.32	27
Cobalt octoate 6.06*	−1.42; −0.23	
Cobalt tallate	−1.45	
Copper octoate 8.0 ± 0.2	−0.23; −0.77	
Iron octoate 6.0 ± 0.2	−0.75; −1.58	
Lead naphthenate	−0.68	
Lead octoate 24.0 ± 0.2	−0.75	
Lead tallate	−0.69	
Manganese octoate 6.0 ± 0.2*	−1.68	
Manganese tallate	−1.64	
Nickel octoate 6.0 ± 0.2	−1.36; −1.70	
Zinc naphthenate	−1.32	

BENZENE–METHANOL (1:4) by VOL.

0.5N LiCl at DME at 20 ± 1°C

Species	$E_{1/2}V$ (vs calomel electrode)	Ref.
$[(C_5H_5)Cr(1)(C_7H_7)]I$	−0.720; −0.695[a]	28
$(C_5H_5)Cr(0)(C_7H_7)$	−0.720; −0.696[a]	
$[(C_6H_6)_2Cr(1)]I$	−0.835; −0.807[b]	
$(C_6H_6)_2Cr(0)$	−0.835; −0.807[b]	

[a] 0.5M NaOH.
[b] 0.35M NaOH.

1-BUTANOL

0.1M LiClO$_4$ at RE (Pt)

Species	$E_{1/2}V$ (vs SCE)	Ref.
4,7-Dimethyl-1,10-phenanthroline ferrous	+0.920	4

1-BUTANOL (Continued)
0.1M LiClO$_4$ at RE (Pt) at 25°C
0.5 mM

Species	$E_{1/2}V$ (vs SCE)	Ref.
Benzyl ferrocene	+0.400	74
tris(4,7-Dimethyl-1,10-phenanthroline) Fe(II)	+0.920	
Ferrocene	+0.445	

CHLOROFORM
0.5M LiCl in MEOH at DME
5.0 mM

Species	pH	$E_{1/2}^a V$ (vs SCE)	Ref.
Fe(III)BPHA[b]	1.0	−0.56	29
In(III)BPHA	5.4	−0.79	
Pb(II)BPHA	9.0	−0.62	
Sb(III)BPHA	1.0 (0.1MHCl)	−0.27	
V(V)BPHA	7.0	−0.20	

[a] Correction to $E_{1/2}$ values due to 1R drop should be made.
[b] N-Benzoyl-N-phenyl hydroxylamine.

DICHLOROMETHANE
0.05M Et$_4$NClO$_4$ at RPE at 20 ± 2.0°C
1.0 mM

Species	$E_{1/2}^a V$ (vs SCE)	Ref.
[Fe(NO)(S$_2$C$_2$Ph$_2$)$_2$]^x	+0.71; −0.02; −0.83	66
{Fe(NO)[S$_2$C$_2$(4-MeC$_6$H$_4$)$_2$]}	+0.54; −0.15; −0.85	
{Fe(NO)[S$_2$C$_2$(4-MeO·C$_6$H$_4$)$_2$]}	+0.46; −0.17; −0.90	
{Fe(NO)[S$_2$C$_2$(3-MeO·C$_6$H$_4$)$_2$]}	+0.62; −0.10; −	
{Fe(NO)[S$_2$C$_2$(2-MeO·C$_6$H$_4$)$_2$]}	+0.95; +0.45; −0.22; −0.42[a]	
{Fe(NO)[S$_2$C$_2$(3,4-CH$_2$O$_2$C$_6$H$_3$)$_2$]}	+0.88; −0.23; −	
{Fe(NO)[S$_2$C$_2$(2,5-(MeO)$_2$C$_6$H$_3$)$_2$]}	+1.31; −0.16; −0.48	

[a] Value using DME.

VI. POLAROGRAPHY

DICHLOROMETHANE (Continued)
0.05 Et$_4$NClO$_4$ at 20 ± 2.0°C
1.0 mM

Species		Working electrode	$E_{1/2} V$ (vs SCE)	Ref.
\[Fe(L^a)(S$_2$C$_2$Ar$_2$^{b})$_2$\]				
L	Ar			
PEt$_3$	Ph	DME	+0.48; −0.71	66
PEt$_3$	4-Me·C$_6$H$_4$	DME	+0.42; −0.77	
PEt$_3$	4-MeO·C$_6$H$_4$	DME	+0.38; −0.78	
PEt$_3$	3-MeO·C$_6$H$_4$	DME	+0.51; −0.69	
PEt$_3$	2-MeO·C$_6$H$_4$	DME	+0.36; −0.82; −1.17	
PEt$_3$	3,4-CH$_2$O$_2$C$_6$H$_3$	DME	+0.46; −0.70	
PEt$_3$	2,5-(MeO)$_2$C$_6$H$_3$	DME	+0.39; −0.79; −1.26	
P(CH$_2$CH$_2$CN)$_3$	Ph	DME	+0.63; −0.43	
PPh$_3$	Ph	DME	+0.53; −0.47	
PPh$_3$	4-Me·C$_6$H$_4$	RPE	+0.48; −0.53	
PPh$_3$	4-MeO·C$_6$H$_4$	DME	+0.43; −0.48	
PPh$_3$	3-MeO·C$_6$H$_4$	DME	+0.57; −0.48	
P(OPh)$_3$	Ph	RPE	+0.78; −0.39	
P(OPh)$_3$	4-Me·C$_6$H$_4$	RPE	+0.63; −0.46	
P(OPh)$_3$	4-MeO·C$_6$H$_4$	DME	+0.64; −0.47	
P(NMe$_2$)$_3$	Ph	DME	+0.42; −0.60	
P(NMe$_2$)$_3$	4-Me·C$_6$H$_4$	DME	+0.37; −0.66	
P(NMe$_2$)$_3$	4-MeO·C$_6$H$_4$	DME	+0.34; −0.65	
P(NMe$_2$)$_3$	3-MeO·C$_6$H$_4$	DME	+0.65; −0.59	
P(NMe$_2$)$_3$	2-MeO·C$_6$H$_4$	DME	+0.36; −0.8	
P(NMe$_2$)$_3$	3,4-CH$_2$O$_2$C$_6$H$_3$	DME	+0.62; −0.57	

DICHLOROMETHANE (Continued)
0.05 Et$_4$NClO$_4$ at 20 ± 2.0°C
1.0 mM

Species [Co(L^a)(S$_2$C$_2$Ar$_2^b$)$_2$]		Working electrode	$E_{1/2}V$ (vs SCE)	Ref.
L	Ar			
PEt$_3$	Ph	DME	+0.25; −0.55	
PEt$_3$	4-Me·C$_6$H$_4$	DME	+0.20; −0.56	
PEt$_3$	4-MeO·C$_6$H$_4$	DME	+0.10; −0.66	
PEt$_3$	3-MeO·C$_6$H$_4$	DME	+0.27; −0.55	
PEt$_3$	2-MeO·C$_6$H$_4$	DME	+0.10; −0.68	
PEt$_3$	3,4-CH$_2$O$_2$C$_6$H$_3$	DME	+0.20; −0.59; −0.97	
PEt$_3$	2,5-(MeO)$_2$C$_6$H$_3$	DME	+0.13; −0.64	
P(CH$_2$CH$_2$CN)$_3$	Ph	DME	+0.44; −0.20; −1.24	
PPh$_3$	Ph	RPE	+0.30; −0.42; −1.05	
PPh$_3$	4-Me·C$_6$H$_4$	DME	+0.21; −0.44; −1.25	
PPh$_3$	4-MeO·C$_6$H$_4$	DME	+0.13; −0.49; −1.36	
P(OPh)$_3$	Ph	DME	+0.33; −0.44; −1.15	
P(OPh)$_3$	4-Me·C$_6$H$_4$	DME	+0.22; −0.38; −1.18	
P(OPh)$_3$	4-MeO·C$_6$H$_4$	DME	+0.22; −0.45; −	
P(NMe$_2$)$_3$	Ph	DME	+0.16; −0.57; −	

[a] Ligand.
[b] Aromatic ring.

VI. POLAROGRAPHY

DIMETHOXYETHANE
0.1M Bu$_4$NClO$_4$ at DME at 22°C
2 mM

Species	$E_{1/2}V$ (vs AgClO$_4$/Ag)	Ref.
(π-C$_5$H$_5$)$_2$TiCl$_2$	-1.4; -2.7	35
(π-C$_5$H$_5$)$_2$Ti(SPh)$_2$	-1.9; Ill-def	
(π-C$_5$H$_5$)$_2$Ti(S$_2$C$_6$H$_3$CH$_3$)	-1.7; -3.4	
(π-C$_5$H$_5$)$_2$ZrCl$_2$	-2.3; -2.7	
(π-C$_5$H$_5$)$_2$VCl$_2$	-1.0; -2.3	
[(π-C$_5$H$_5$)VS$_2$C$_4$F$_6$]2	-1.4	
(π-C$_5$H$_5$)V(CO)$_4$	-2.4	
(π-C$_4$H$_4$S)Cr(CO)$_3$	-2.5	
[(π-C$_5$H$_5$)Cr(CO)$_3$]$_2$Hg	-1.2	
(π-C$_6$H$_5$C$_6$H$_5$)Cr(C$_6$H$_6$)$^+$, Ph$_4$B$^-$	-1.3; -3.3	
(π-C$_{14}$H$_{10}$)$_2$Cr$^+$, PH$_4$B$^-$ (anthracene)	-1.4	
Cr(CO)$_5$CN$^-$, Na$^+$	-0.2; -2.6	
Ph$_3$PMo(CO)$_5$	-2.7; -3.7	
[(CH$_3$)$_2$N]$_3$PMo(CO)$_5$	-2.9	
Ph$_2$P(CH$_2$)$_2$N(Et)$_2$ ↘ ↙ Mo(CO)$_4$	-2.9	
(Me)$_2$N(CH$_2$)$_2$N(Me)$_2$ ↘ ↙ Mo(CO)$_4$	-3.0	
(C$_5$H$_5$N)$_3$Mo(CO)$_3$	-2.8	
(π-C$_5$H$_5$)Mo(CO)$_3$Cl	-1.4	
[(π-C$_5$H$_5$)MoS$_2$C$_4$F$_6$]$_2$	-1.7; -2.3	
[(π-C$_5$H$_5$)MoNO(S$_2$C$_4$F$_6$)]$_2$	-1.3; -1.9	
(π-C$_{10}$H$_8$)(Mo(CO)$_3$)$_2$ (azulene)	-1.5 -2.3; -2.7	
(π-C$_5$H$_5$)(π-C$_3$H$_5$)Mo(CO)$_2$	-2.9	
p-CH$_3$C$_6$H$_4$N$_2$Mo(CO)$_2$(π-C$_5$H$_5$)	-2.3	
(π-C$_5$H$_5$)MoCO(π-C$_6$H$_6$)$^+$, Ph$_4$B$^-$	-1.4	
[π-C$_5$H$_5$]Mo(P(Me)$_2$CO]$_3$	-2.0; -2.4	
(π-C$_5$H$_5$)W(CO)$_3$Cl	-1.5	
(π-C$_5$H$_5$)W(CO)$_3$H	-2.5	
(π-C$_7$H$_8$)W(CO)$_4$ (bicyclo[2.2.1]heptadiene)	-2.8 -2.8	
(π-C$_8$H$_{12}$)W(CO)$_4$ (cyclooctadiene)	-2.9	
(CH$_3$COCH=CH$_2$)$_3$W	-3.0	

DIMETHOXYETHANE (Continued)
0.1M Bu$_4$NClO$_4$ at DME at 22°C
2 mM

Species	$E_{1/2}V$ (vs AgClO$_4$/Ag)	Ref.
(C$_2$H$_5$C≡CC$_2$H$_5$)WCO	−2.7	
(π-C$_5$H$_5$)W(CO)$_2$NO	−2.4	
CpMo(CO)$_3$⟨H/P⟩Mo(CO)$_3$Cp (↙Me ↘Me)	−2.4	
W$_2$(CO)$_{10}$H$^-$, Et$_4$N$^+$	−2.7	
(OC)$_5$MnBr	−1.4	
(OC)$_5$MnCl	−2.2; −2.4	
(OC)$_5$MnH		
(OC)$_4$MnPPh$_3$ (may be dimer)	−2.0; −3.5	
CH$_3$SCH$_2$CH$_2$COMn(CO)$_4$	−2.2	
(CH$_2$)$_2$NCH$_2$CH$_2$COMn(CO)$_4$	−2.3	
(π-C$_3$H$_5$)[Mn(CO)$_5$]$_2$	−1.8	
(π-C$_5$H$_5$)$_2$Mn$_2$(NO)$_3$	−1.6	
(π-C$_5$H$_5$)$_6$Mn$_6$(NO)$_8$	−1.9	
(π-C$_5$H$_5$)Re(CO)$_3$	−3.3	
(π-C$_5$H$_5$)Fe(CO)$_2$(NCCH$_3$)$^+$, PF$_6^-$	−0.8 −2.2	
(π-C$_5$H$_5$)Fe(CO)$_2$PPh$^+$, PF$_6^-$	−1.4 −2.2	
(π-C$_5$H$_5$)Fe(CO)$_3^+$, PF$_6^-$	−1.4; −2.2	
(π-C$_5$H$_5$)Fe(CO)(PPh$_3$)$_2^+$, PF$_6$	−1.4	
Fe$_3$(CO)$_{12}$	−1.6; −2.4	
HF$_3$(CO)$_{11}^-$, Et$_3$NH$^+$	−0.2 −1.9	
CpFeS$_2$C$_4$F$_6$	−2.0	
CpFe(CO)S$_2$CN(CH$_3$)$_2$	−2.3; −2.9	
(π-C$_8$H$_8$O)Fe(CO)$_3$ (cyclo-octatrienone)	−1.7; −2.2	
(π-C$_5$H$_5$)Fe(π-C$_6$H$_7$)	−0.6	
C$_6$H$_{10}$COFe(CO)$_3$	−2.1	
(π-C$_8$H$_{10}$)Fe$_2$(CO)$_6$	−1.8	

VI. POLAROGRAPHY

DIMETHOXYETHANE (Continued)
0.1M Bu$_4$NClO$_4$ at DME at 22°C
2 mM

Species	$E_{1/2}V$ (vs AgClO$_4$/Ag)	Ref.
((OC)$_3$FeNH)$_2$	−2.2	
(π-C$_{12}$H$_8$)Fe$_2$(CO)$_6$ (acenaphthylene)	−1.8 −2.4; −3.5	
Biferrocenyl	0 −0.2	
[(π-C$_5$H$_5$)Co(CO)]$_3$	−1.6	
Co$_2$(CO)$_6$·PhC≡CPh	−1.6	
ClC(Co(CO)$_3$)$_3$	−1.1; −2.0	
Co$_2$(CO)$_7$(C$_4$HO$_2$)	−1.2	
(π-C$_2$H$_4$)(π-C$_2$H$_4$)Rh(O$_2$C$_5$H$_7$)	−1.9	
(π-C$_5$H$_5$)IrS$_2$C$_4$F$_6$	−1.7	
(Ph$_3$P)$_2$Pt·PhC≡CPh	−3.4	
(π-C$_5$H$_5$)$_3$Ni$_3$(CO)$_2$	−1.5; −2.7	
Anthracene	−2.6; −3.1	
Acenaphthylene	−2.3; −3.0	
Azulene	−2.3; −3.0	
Cycloheptatriene	−3.2	
Cyclooctatetraene	−2.5	
Biphenyl	−3.3	
Bipyridyl	−2.9; −3.2	
Bicyclo[2.2.1]heptadiene	−3.1	
PhC≡CPh	−3.0	

0.1M Bu$_4$NClO$_4$ at DME
1 mM

Species	$E_{1/2}V$ (vs SCE)	Ref.
Cp$_2$[a] TiCl$_2$	−1.1; −1.4	31

[a] π-Cyclopentadienyl.

DIMETHOXYETHANE (*Continued*)
0.1M Bu$_4$NClO$_4$ at DME
1 mM

Species	$E_{1/2}V$ (vs AgClO$_4$/Ag)	Ref.
(CH$_3$)$_2$Mg	−1.2	32
(C$_2$H$_5$)$_2$Mg	−1.2	
(*i*-C$_3$H$_7$)$_2$Mg	−1.2	
(*i*-C$_4$H$_9$)$_2$Mg	−1.2	
(C$_6$H$_5$)$_2$Mg	−1.2	
(C$_6$H$_5$CH$_2$)$_2$Mg	−1.2; −2.74	
(CH$_2$=CH CH$_2$)$_2$Mg	−1.2; −2.65	
(C$_5$H$_5$)$_2$Mg	−1.2; −2.50	
MgBr$_2$	−0.6; −2.47	
Mg(ClO$_4$)	−2.30	
(CH$_3$)$_2$Mg + MgBr$_2$	−2.49	
(C$_2$H$_5$)$_2$Mg + MgBr$_2$	−2.43; −2.66	
(*i*-C$_3$H$_7$)$_2$Mg + MgBr$_2$	−2.46; −2.74	
(*i*-C$_4$H$_9$)$_2$Mg + MgBr$_2$	−2.50; −2.78	
(C$_6$H$_5$)$_2$Mg + MgBr$_2$	−2.50; −2.83	
CH$_3$MgBr	−2.49	
C$_2$H$_5$MgBr	−2.44; −2.70	
i-C$_3$H$_7$MgBr	−2.44; −2.75	
i-C$_4$H$_9$MgBr	−2.46; −2.75	
C$_6$H$_5$MgBr	−2.46; −2.80	

DIMETHOXYETHANE (Continued)
0.1M Bu$_4$NClO$_4$ at DME at 25°C
~1 mM

Species	$E_{1/2}V$ (vs AgClO$_4$/Ag)		Ref.
	$E_{1/2}V^a$	$E_{1/2}V^b$	
Ph$_2$Bi:$^-$	−2.3	−2.1	33
Ph$_2$Sb:$^-$	−2.5	−2.0	
Ph$_2$As:$^-$	−2.7	−2.0	
Ph$_3$Ge:$^-$	−3.5	−1.2	
Ph$_3$Sn:$^-$	−2.9	−0.95	
CpFe(CO)$_2$:$^-$	−2.2	−1.6	
CpRu(CO)$_2$:$^-$	−2.6	−1.5	
CpNi(CO):$^-$	−2.4	−1.4	
PhSe:$^-$	~−1.5; ~−0.9	−1.5	
PhS:$^-$	−1.7	−1.4	
Ph$_3$Pb:$^-$	−2.0	−0.90	
Re(CO)$_5$:$^-$	−2.3	−0.8 to −1.0	
CpW(CO)$_3$:$^-$		−1.0	
Mn(CO)$_5$:$^-$	−1.7	−0.55	
CpMo(CO)$_3$:$^-$	−1.4	−0.55	
CpCr(CO)$_3$:$^-$	−1.3	−0.80	
Co(CO)$_4$:$^-$	−1.0	−0.20	
Cr(CN)(CO)$_5$:$^-$		~−0.2	
Mo(CN)(CO)$_5$:$^-$		~−0.2	
W(CN)(CO)$_5$:$^-$		~−0.2	

[a] Uncharged complex.
[b] Oxdn. of anion at Pt.E (rotating).

DIMETHOXYETHANE (Continued)
0.1M Bu$_4$NClO$_4$ at DME
1 mM

Species	$E_{1/2}V$ (vs Ag/ClO$_4$/Ag)	Ref.
Ph$_3$SiCl	-3.1	39
Ph$_3$GeCl	-2.8	
Ph$_3$SnCl	-1.6; -2.9	
Ph$_3$PbOAc	-1.4; -2.2	
Ph$_2$SiCl$_2$	-1.9	
Ph$_2$GeCl$_2$	-2.6	
Ph$_2$SnCl$_2$	-1.6; -2.7	
Ph$_2$Pb(OAc)$_2$	-1.1; -1.6	
Ph$_3$GeGePh$_3$	-3.5	
Ph$_3$SnSnPh$_3$	-2.9	
Ph$_3$SnSiPh$_3$	-3.1	
Ph$_3$PbPbPh$_3$	-2.0	
Ph$_3$SnSeSnPh$_3$	-2.4; -2.9	
Ph$_3$SnSeGePh$_3$	-2.4; -2.9	

0.15M Bu$_4$NPF$_6$ at DNE
$\sim$1 mM

Species	$E_{1/2}V$ (vs SCE)	Ref.
[π-(3)-1,2-B$_9$C$_2$H$_{11}$]$_2$Ni	$+0.28$; -0.62; -2.21	20
[Et$_4$N][(π-(3)-1,7-B$_9$C$_2$H$_{11}$)$_2$Ni]	-0.94; -2.25	
(π-C$_5$H$_5$)Co(π-(3)-1,2-B$_9$C$_2$H$_{11}$)	-1.19; -2.21	
[Et$_4$N][π-(3)-1,2-B$_9$C$_2$H$_{11}$)$_2$Co]	-1.37; -2.37	
[Et$_4$N][(π-(3)-1,7-B$_9$C$_2$H$_{11}$—$_2$Co]	-1.21; -2.70	
[Me$_4$N$_2$][(B$_9$C$_2$H$_{11}$)Co(B$_8$C$_2$H$_{10}$)-Co(B$_9$C$_2$H$_{11}$)]	-1.49; -2.43	

VI. POLAROGRAPHY

DIMETHOXYETHANE (Continued)
0.1M Bu$_4$NClO$_4$ at DME
2–4 mM

Species	$E_{1/2}V$ (vs AgClO$_4$/Ag)	Ref.
Fe$_2$(CO)$_7$(ClC$_6$H$_4$C)$_4$	−1.12	40
Fe$_3$(CO)$_8$(C$_6$H$_5$C)$_4$	−1.3	
Fe$_2$(CO)$_7$(PhC$_2$Me)$_2$	−1.5	
F$_7$B(PhCO)$_2$CH	−1.6	
Ph$_2$B(PhCO)$_2$CH	−1.7	
COT—Fe(CO)$_3$	−2.0	
Mo(bipy)(CO)$_4$	−2.2	

0.1M Bu$_4$NClO$_4$ at RE (Pt)

Species	$E_{1/2}V$ (vs AgClO$_4$/Ag)	Ref.
Cr(C$_6$H$_6$)$_2$$^+$	−1.2	40
Cr(acac)$_3$	−2.5	
CpCr(CO)$_3$	−1.1	
Fe(acac)$_3$	−1.4	
CpFe(CO)$_2$	−1.9	
Co(acac)$_3$	−1.2	
Mn(acac)$_3$	−0.8	
Mo(bipy)(CO)$_4$	−2.2	
CpMo(CO)$_3$	−1.1	
PhS	−1.4	
PhSe	−1.6	
Ph$_3$Sn	−1.9	
Ph$_3$Pb	−1.4	

DIMETHOXYETHANE (Continued)

0.1M Bu$_4$NClO$_4$ at DME at 22°C

2 mM

Species	$E_{1/2}V$ (vs AgClO$_4$/Ag)	Ref.
$(\pi$-C$_6$H$_6)_2$Cr$^+$B(C$_6$H$_5)_4^-$	-1.3	38
$(\pi$-C$_5$H$_5)(\pi$-C$_7$H$_7)$Cr$^+$PF$_6^-$	-1.2	
Cr(CO)$_6$	-2.7	
$(\pi$-C$_6$H$_6)$Cr(CO)$_3$	-3.0	
$(\pi$-C$_7$H$_8)$Cr(CO)$_3$(cycloheptatriene)	$-2.4; -2.3$[a]; -1.1[a]	
$(\pi$-C$_7$H$_8)$Cr(CO)$_4$(bicyclo[2, 2, 1]-heptadiene)	-2.8	
[(CH$_3)_2$N)$_3$P]Cr(CO)$_5$	-2.9	
[(CH$_3)_2$N)$_3$P]Cr(CO)$_4$	> -3.6	
Mo(CO)$_6$	-2.7	
$(\pi$-C$_7$H$_8)$Mo(CO)$_3$(cycloheptatriene)	-2.3	
$(\pi$-C$_7$H$_8)$Mo(CO)$_4$(bicyclo[2, 2, 1]-heptadiene)	-2.3	
Bipyridyl Mo(CO)$_4$	$-2.2; -2.8$	
$(\pi$-C$_5$H$_5)$Mo(CO)$_2(\pi$-C$_3$H$_5)$	-2.9	
$(\pi$-C$_5$H$_5)$Mo(CO)$_2(\pi$-CH$_2$=SCH$_2)$	$-0.2; -2.7$	
$(\pi$-C$_5$H$_5)$Mo(CO)$_2[\pi$-C$_6$H$_5$CH$_2^-]$	-2.3	
$(\pi$-C$_5$H$_5)$Mo(CO)$_2[\sigma$-C$_6$H$_5$CH$_2^-]$	$-2.1; -1.3$[a]	
$(\pi$-C$_5$H$_5)$Mo(CO)$_3$(CH$_2$CO$_2$CH$_3$)	$-2.0; -1.3$[a]	
$(\pi$-C$_5$H$_5)$Mo(CO)$_3$CF$_3$	$-2.1; -1.3$[a]	
[$(\pi$-C$_5$H$_5)$Mo(CO)$_3]_2$	-1.4	
[$(\pi$-C$_5$H$_5)$Mo(CO)$_3]_2$Hg	-1.3	
$(\pi$-C$_5$H$_5)$Mo(CO)$_3$Fe(CO)$_2(\pi$-C$_5$H$_5)$	-1.4	
W(CO)$_6$	-2.6	
$(\pi$-1, 3, 5-(CH$_3)_2$C$_6$H$_3)$Mn(CO)$_3^+$I$^-$ (mesitylene)	-0.9	
$(\pi$-C$_5$H$_5)$Mn(CO)$_3$	-3.0	
$(\pi$-C$_5$H$_5)$Mn(CO$_2)$NO$^+$PF$_6^-$	-0.8	
[π-C$_5$H$_5$Mn(CO)NO]$_2$	-2.0	
$(\pi$-C$_5$H$_5)_3$Mn$_2$(NO)$_3$	$-1.5; -2.5$	
[Mn(CO)$_5]_2$	-1.8	
[Re(CO)$_5]_2$	-2.3	
Fe(CO)$_5$	-2.4	
$(\pi$-C$_8$H$_8)$Fe(CO)$_3$	$-2.0; -2.5$	
$(\pi$-C$_6$H$_8)$Fe(CO)$_3$	-2.7	

VI. POLAROGRAPHY

DIMETHOXYETHANE (Continued)
0.1M Bu$_4$NClO$_4$ at DME at 22°C
2 mM

Species	$E_{1/2}V$ (vs AgClO$_4$/Ag)	Ref.
(π-C$_4$H$_6$)Fe(CO)$_3$	−2.6	
π-n-C$_6$H$_9$Fe(CO)$_3$$^+$(C$_6H_5$)$_4$$^−$	−2.6	
(n-cyclo-C$_7$H$_9$)Fe(CO)$_3$$^+BF_4$$^−$	−0.7	
(π-1,3,5-(CH$_3$)$_3$C$_6$H$_3$)Fe(π-C$_5$H$_5$)$^+$I$^−$ (mesitylene)	−2.0	
(π-C$_6$H$_6$)Fe(π-C$_5$H$_5$)$^+$PF$_6$$^−$	−1.9; −2.9	
(π-C$_9$H$_7$)$_2$Fe(indenyl)	−0.4; −2.7	
(π-C$_5$H$_5$)Fe(CO)$_2$SCOC$_6$H$_5$	−1.8; −1.2[a]	
(π-C$_5$H$_5$)Fe(CO)$_2$COCH$_3$	−2.5	
(π-C$_5$H$_5$)Fe(CO)$_2$CH$_2$C$_6$H$_5$	−2.5	
(π-C$_5$H$_5$)Fe(CO)$_2$(C$_6$H$_5$)	−2.3; −3.0	
(π-C$_5$H$_5$)Fe(CO)$_2$COCH=CH C$_6$H$_5$	−2.4 to −2.7	
[(π-C$_5$H$_5$)Fe(CO)$_2$]$_2$	−2.2	
[(π-C$_5$H$_5$)Fe(CO)$_2$]$_2$Jg	−2.0; −2.4	
(π-C$_5$H$_5$)Fe(CO)$_2$I	−1.2	
(π-C$_5$H$_5$)Fe(CO)$_2$SMe	−0.4; −2.1	
[(π-C$_5$H$_5$)Fe(CO)SCH$_3$]$_2$	−0.6; −2.4	
[CH$_3$SFe(CO)$_3$]$_2$	−1.9; −2.5	
Fe$_2$(CO)$_9$	−2.4	
(π-C$_{10}$H$_8$)Fe$_2$(CO)$_5$	−2.2	
(π-C$_5$H$_5$)Co(CO)$_2$	−2.5; −3.2	
(π-C$_5$H$_5$)CoSC$_4$F$_6$	−1.1; −2.9	
Co(S$_2$C$_4$F$_6$)$_2$$^{2−}$2Et$_4N^+$	−0.84; ∼−3.0	
Co(acac*)$_3$	−1.2; −2.6	
Co(acac)$_2$	−2.6	
[(π-C$_5$H$_5$)Ni(CO)]$_2$	−0.55; −2.4; −1.5[a]	
[(π-C$_5$H$_5$)NiS$_2$C$_4$F$_6$	−0.96; −2.4	
Ni(S$_2$C$_4$F$_6$)$_2$$^{2−}$2Et$_4N^+$	−0.84	
Ni(acac)$_2$	−2.2	
[(π-C$_5$H$_5$)Ni]$_2$CH=CH	−2.2	
Ni(CO)$_4$	−2.9	
(π-C$_5$H$_5$)RhS$_2$C$_4$F$_6$	−1.4	
(π-C$_5$H$_5$)Rh(π-CH$_2$=CH$_2$)$_2$	−0.25; −3.2	
(π-CH$_2$=CH$_2$)$_2$Rh(O$_2$C$_5$H$_7$)	−2.8	

[a] New waves; acac* = acetylacetonate.

DIMETHOXYETHANE (Continued)
0.1M Bu$_4$NClO$_4$ at DME at 22°C

Species	$E_{1/2}V$ (vs Ag$^+$/Ag)	Ref.
[CpCr(CO)$_3$]$_2$	−1.3	37
[CpMo(CO)$_3$]$_2$	−1.4	
CpMo(CO)$_3$—CH$_2$Ph	−2.1	
CpMo(CO)$_3$—CH$_2$COCH$_3$	−2.0	
CpMo(CO)$_3$—CF$_3$	−2.1	
CpMo(CO)$_3$—Fe(CO)$_2$Cp	−1.4	
CpMo(CO)$_2$—SnMe$_3$	−1.9	
CpMo(CO)$_3$—SnPh$_3$	−2.4	
CpMo(CO)$_3$—PbPh$_3$	−2.2	
[(OC)$_5$Mn—]$_2$	−1.7	
[(OC)$_5$Re—]$_2$	−2.3	
(OC)$_5$Mn—Fe(CO)$_2$Cp	−1.6	
(OC)$_5$Mn—SnMe$_3$	−1.9	
(OC)$_5$Mn—SnPh$_3$	−2.5	
(OC)$_5$Re—SnPh$_3$	−2.5	
(OC)$_5$Mn—PbEt$_3$	−1.8	
(OC)$_5$Mn—PbPh$_3$	−2.1	
(OC)$_5$Re—PbPh$_3$	−2.4	
[CpFe(CO)$_2$]$_2$	−2.2	
[CpRu(CO)$_2$]$_2$	−2.6	
CpFe(CO)$_2$—C	−2.3–2.5	
CpFe(CO)$_2$—SnPh$_3$	−2.6	
CpFe(CO)$_2$PhPb$_3$	−2.1	
[(OC)$_4$Co]$_2$[b]	−0.9	
(OC)$_4$Co—SnMe$_3$	−1.6	
(OC)$_4$Co—SnPh$_3$	−1.6	
[CpNi(CO)]$_2$	−2.4	
PhS—SPh	−1.6	
PhSe—SePh	−0.9	
Ph$_3$Si—SnPh$_3$	−3.1	
Ph$_3$Ge—GePh$_3$	−3.5	

VI. POLAROGRAPHY

DIMETHOXYETHANE (Continued)
0.1M Bu₄NClO₄ at DME at 22°C

Species	$E_{1/2}V$ (vs Ag$^+$/Ag)	Ref.
Ph$_3$Sn—SnPh$_3$	−2.9	
Ph$_3$Pb—PbPh$_3$	−2.0	
Ph$_2$As—AsPh$_2$	−2.7	
Ph$_2$Sb—SbPh$_2$	−2.5	
Ph$_2$Bi—BiPh$_2$	−2.3	
CpMo(CO)$_2$—Sn(Me)$_2$—Mo(CO)$_3$Cp	−1.8	
CpFe(CO)$_2$—Sn(Me)$_2$—Ge(CO)$_2$Cp	−2.7	
[CpMo(CO)$_3$]$_2$Sn[Fe(CO)$_2$Cp]$_2$	−1.8	
	−2.0	
	−2.5	
[CpCr(CO)$_3$]Hg	−1.3	
[CpMo(CO)$_3$]$_2$Hg	−1.3	
[(OC)$_5$Mn]$_2$Hg	−1.1	
[CpFe(CO)$_2$]$_2$Hg	−2.0	
Ph$_3$Sn—Se—SnPh$_3$	−2.4	
Ph$_3$Sn—Se—GePh$_3$	−2.4	
[CpMo(CO)PPh$_2$]$_3$	−2.0	
[(OC)$_4$CrPMe$_2$]$_2$	−1.8	
[(OC)$_3$WAsMe$_2$]$_2$	−1.8	
[(OC)$_3$FePMe$_2$]$_2$	−2.0	
[(OC)$_3$FeAsMe$_2$]$_2$	−1.9	
Fe$_2$(CO)$_9$	−2.4	
[CpNi]$_2$HC≡CH	−2.2	
[CpMo(CO)P(PH)$_2$]$_3$	−2.0	
[CpFe(CO)]$_4$	−1.9	
[CpCo(CO)]$_3$	−1.6	
Cp$_3$Ni$_3$(CO)$_2$	−1.5	

DIMETHOXYETHANE (Continued)
0.1M Bu$_4$NClO$_4$ at Pt or RE (Pt) at 22°C

No.	Compound	$-E_{1/2}^1$	$-E_{1/2}^2$	$E_{1/2}V$ (vs AgClO$_4$/Ag)		Ref.
1	[(π-C$_5$H$_5$)V(SMe)$_2$]$_2$	0.2	2.6	0.3	0.1	36
2	[(π-C$_5$H$_5$)VS$_2$C$_4$F$_6$]$_2$	1.3	2.3	1.35	1.25	
3	[Cr(CO)$_4$PMe$_2$]$_2$	1.85		1.9	1.7	
4	[W(CO)$_4$PMe$_2$]$_2$	1.9		1.9	1.6	
5	[W(CO)$_4$AsMe$_2$]$_2$	1.8		1.9	1.4	
6	[Cr$_2$(CO)$_{10}$As$_2$Me$_4$]$_n$	2.1		2.1	1.4	
7	[C$_5$H$_5$MoS$_2$Me$_2$]$_2$	0.6	3.1	0.65	0.55	
8	[(C$_5$H$_5$)MoPPh$_2$(CO)]$_3$	Ill-def	Ill-def			
9	(OC)$_4$MnS$_2$CHEt$_2$	2.0		2.2	1.8	
10a	[Mn(CO)$_4$PPh$_2$]$_2$	2.2		Irrev		
10b	[Mn(CO)$_3$SPh]$_2$	1.6	2.5	1.75	1.5	
11	[(π-C$_5$H$_5$)Fe(CO)PPh$_2$]$_2$	0.2 (Pt)	0.5	0.1 / 0.7	0.2 / 0.5	
12	[(OC)$_3$FePMe$_2$]$_2$	2.1		2.2	1.9	
13	[(OC)$_3$FeAsMe$_2$]$_2$	0.2	1.9	0.2 / 2.0	0.1 / 1.8	
14	[(ON)$_2$FePPh$_2$]$_2$	1.7	2.1	1.75 / 2.15	1.65 / 2.05	
15	[(π-C$_5$H$_5$)Fe(CO)SCH$_3$]$_2$	0.6		0.7	0.5	
16	(π-C$_5$H$_5$)Fe(CO)$_2$SCH$_3$	0.4		0.6	0.3	
17	(π-C$_5$H$_5$)FeS$_2$C$_4$F$_6$	2.1		2.2	1.8	
18	SbFe$_2$(CO)$_8$	2.3		Irrev		
19	[(π-C$_5$H$_5$)CoPPh$_2$]$_2$	0.3	2.6	Irrev / 2.65	2.55	
20	[(π-C$_5$H$_5$)CoSMe]$_2$	0.5	2.4	0.5 / 2.45	0.4 / 2.35	
21	(π-C$_5$H$_5$)CoS$_2$C$_4$F$_6$	1.1		1.2	0.9	
22	(π-C$_5$H$_5$)RhS$_2$C$_4$F$_6$	1.4		1.5	1.2	
23	[(OC)$_3$Ni←PPh$_2$—]$_2$	2.4		Irrev		
24	[(OC)$_2$NiPPh$_2$]$_2$	1.7	2.3	1.75 / 2.35	1.65 / 2.25	
25	[(π-C$_5$H$_5$)NiPPh$_2$]$_2$	2.3		2.35	2.25	
26	[(π-C$_5$H$_5$)NiSCH$_3$]$_2$	0.5	1.8	0.6 / 2.0	0.4 / 1.7	

VI. POLAROGRAPHY

DIMETHYL FORMAMIDE
0.2M NaClO$_4$ at DME at 25.0 ± 0.5°C
2 mM

Species	$E_{1/2}V$ (vs SCE)	Ref.
(C$_6$H$_6$)$_2$CrI	−0.81	47
(C$_6$H$_5$·C$_6$H$_5$)CrI	−0.75	
(C$_9$H$_7$)$_2$CoClO$_4$[a]	−0.53	

[a] Conc. was 0.47 mM.

0.5M Et$_4$NClO$_4$ at DME
5 mM

Species	$E_{1/2}V$ (vs SCE)	Ref.
TiCp$_2$[a]Cl$_2$	−0.63; −0.22; +0.19	54

[a] Cyclopentadiene.

0.1M Pr$_4$NClO$_4$ at DME at 20 ± 1°C
0.1–1 mM

Species	$E_{1/2}V$ (vs SCE)	Ref.
Cu Etioporphyrin IV	−1.48; −1.99; −2.70	55
Zn Etioporphyrin I	−1.62; −2.00; −2.77	
Zn tetrabenzo porphin	−1.47; −1.84; −2.49; −2.70	
Zn tetraphenyl porphin	−1.32; −1.73; −2.45; −2.67	
Mg octaphenyl tetraza porphin	−0.68; −1.11; −1.81; −2.18	

DIMETHYL FORMAMIDE (Continued)
0.1M Pr$_4$NClO$_4$ at DME
~1 mM

Species	$E_{1/2}V$ (vs SCE)	Ref.
Cu Etio[a]I	−0.95; −0.46; −2.05	57
Cu TPP[b,c]	−0.72; −1.20; −1.68	
Co EtioI	−1.04; −1.57	
NiTPP[d]	−1.18; −1.75	
Zn EtioI	−0.89; −1.60; −1.95	

[a] Etioporphyrin I.
[b] Tetraphenyl porphyrin.
[c] 1:3 Tetrahydrofuran: DMF.
[d] Benzene: DMF.

0.1M Bu$_4$NClO$_4$ at DME
~1 mM

Species	$E_{1/2}V$ (vs Ag/AgNO$_3$)	Ref.
(CH$_3$)$_3$Si(C$_6$H$_4$)Si(CH$_3$)$_3$	−3.45[a]	59
(CH$_3$)$_3$Ge(C$_6$H$_4$)Ge(CH$_3$)$_3$	−<3.50[a]	
(CH$_3$)$_3$Si(C$_6$H$_4$)$_2$Si(CH$_3$)$_3$	−2.97[a]; ~−3.3	
(CH$_3$)$_3$Ge(C$_6$H$_4$)$_2$Ge(CH$_3$)$_3$	−3.04[a]; ~−3.3	
(CH$_3$)$_3$Si(C$_6$H$_4$)$_3$Si(CH$_3$)$_3$	−2.79[a]; −3.07	
(CH$_3$)$_3$Ge(C$_6$H$_4$)$_3$Ge(CH$_3$)$_3$	−2.83[a]; −3.14	
(CH$_3$)$_3$Si(C$_6$H$_4$)$_4$Si(CH$_3$)$_3$	−2.71[a]; −2.90	
(CH$_3$)$_3$Ge(C$_6$H$_4$)$_4$Ge(CH$_3$)$_3$	−2.71[a]; −2.92	
((CH$_3$)$_3$Si)$_2$C$_{10}$H$_6$	−2.90[a]	
((CH$_3$)$_3$Ge)$_2$C$_{10}$H$_6$	−2.96[a]	

[a] E peak(ac).

VI. POLAROGRAPHY

DIMETHYL FORMAMIDE (Continued)
0.1M Bu$_4$NClO$_4$ at DME

Species	$E_{1/2}V$ (vs Hg pool)	Ref.
4-Trimethyl germyl	−2.05	60
4,4′-bis(Trimethyl germyl)	−2.03	
4-Trimethyl silyl	−2.03	
4,4′-bis(Trimethyl silyl)	−1.94	
4-Trimethyl stannyl	−2.07	

0.2M Bu$_4$NI at DME at 25 ± 0.1°C
0.90–1.44 mM

Species	$E_{1/2}V$ (vs Hg pool)	Ref.
Diphenyl sulfide	−2.03	48
Tetraphenyl arsonium chloride	−1.01; −2.20	
Tetraphenyl phosphonium chloride	−1.20; −1.87; −2.08	
Triphenyl arsine	−2.19	
Triphenyl arsine oxide	−1.74	
Triphenyl bismuthine	−1.90	
Triphenyl-n-butyl-phosphonium iodide	−1.24; −1.60; −2.02	
Triphenyl phosphine	−2.08	
Triphenyl phosphine oxide	−1.91	
Triphenyl stibine	−2.03	

0.1M LiClO$_4$ at Re (Pt) at 25°C
0.5 mM

Species	$E_{1/2}V$ (vs SCE)	Ref.
bis(2,9-Dimethyl-1,10-phenanthroline) Cu(I)	+0.615	74
tris(4,7-Dimethyl-1,10-phenanthroline) Fe(II)	+0.945	

DIMETHYL FORMAMIDE (Continued)
DME at 25 ± 0.1°C
1.0 mM

Species	Supporting electrolyte	$E_{1/2}V$ (vs SCE)	Ref.
[Fe(acac)$_3$]	0.1M Et$_4$NClO$_4$	−0.585	49
	0.1M Et$_4$NClO$_4$ + 30 mM LiClO$_4$	−0.482	
	0.1M Et$_4$NClO$_4$ + 30 mM Mg(ClO$_4$)$_2$	−0.507	
[Eu(acac)$_3$]	0.1M Et$_4$NClO$_4$	−1.495	
	+2 mM LiClO$_4$	−2.087	
	0.1M Et$_4$NClO$_4$	−1.339	
	+30 mM LiClO$_4$	−2.033	

0.1M Et$_4$NClO$_4$ at DME
0.5 to 7.5 mM

Species	$E_{1/2}V$ (vs Ag/AgCl)	Ref.
Co(CO)$_3$NO	−0.37; −0.96; −2.3	14

0.1M Et$_4$NClO$_4$ at DME

Species	$E_{1/2}V$ (vs SCE)	Ref.
MoCl$_3$Py$_3$	+0.94; −1.47	16
MoCl$_3$(Py)(bipy)	+0.83; −1.00	
MoCl$_3$(Py)$_3$		

VI. POLAROGRAPHY

DIMETHYL FORMAMIDE (Continued)
DME at 25 ± 0.1°C
2.0 mM

Species	Supporting electrolyte	$E_{1/2}V$ (vs SCE)	Ref.
$Cp_2{}^aTiF_2$	0.1N Et_4NClO_4	−1.21; −2.17	30
Cp_2TiCl_2		+0.18; −0.63; −1.94	
Cp_2TiBr_2		0.22^c; −1.95	
Cp_2TiI_2		0.26^c; −1.99	
$(CH_3-Cp)_2TiCl_2$		−0.73; −2.04	
Cp_2TiF_2	0.1N LiCl	−0.84	
Cp_2TiCl_2		−0.63	
Cp_2TiBr_2		−0.65	
Cp_2TiI_2		−0.68	
$TiCl_4$	0.1N Et_4NClO_4	−0.73; −1.58; −2.34	
$CpTiCl_3$		+0.16; −0.87; −1.60; −2.04	
$(CH_3)_5C_5TiCl_3$		+0.14; −1.06; −1.56; −2.34	

^{a} Cyclopentadienyl.
^{b} Pentamethyl cyclopentadienyl.
^{c} Anodic—cathodic wave.

0.1M Et_4NClO_4
0.1–5.0 mM

Species	Working electrode	$E_{1/2}V$ [vs Ag/AgCl(s)]	Ref.
C_5H_5NiNO	DME	−1.45; −2.03 to −1.78; −2.1	50
C_5H_5NiNO	HMDE^a	−1.54; −1.83; −2.10	

^{a} Hanging mercury drop electrode.

Species	Supporting electrolyte	Working electrode	$E_{1/2}V$ (vs SCE)	Ref.
$MoCl_3(tripy)$	0.1M Et_4NClO_4	DME	+0.60; −0.75; −1.57	16
$Mo(bipy)_3$	0.1M Et_4NBF_4	RPE	−0.01; −0.42; −1.13; −1.77	
$Mo(tripy)_2$		RPE	+0.68; +0.28; −0.63	

DIMETHYL FORMAMIDE (Continued)
0.1M Et$_4$NClO$_4$ at DME at 25 ± 0.2°C

Species	$E_{1/2}V$ (vs SCE)	Ref.
(Cl$_2$CHCOO)$_2$Hg	0.33	51
[3,5-(NO$_2$)$_2$C$_6$H$_3$COO]$_2$Hg	0.29	
(ClCH$_2$COO)$_2$Hg	0.25	
(O-NO$_2$C$_6$H$_4$COO)$_2$Hg	0.20	
(p-NO$_2$C$_6$H$_4$COO)$_2$Hg	0.27	
(m-NO$_2$C$_6$H$_4$COO)$_2$Hg	0.25	
(ClCH$_2$CH$_2$COO)$_2$Hg	0.18	
(C$_6$H$_5$COO)$_2$Hg	0.17	
(CH$_3$COO)$_2$Hg	0.14	
(CH$_3$CH$_2$COO)$_2$Hg	0.11	
(CH$_3$CH$_2$CH$_2$COO)$_2$Hg	0.14	
[CH$_3$(CH$_2$)$_3$COO]$_2$Hg	0.11	
[CH$_3$(CH$_2$)$_5$COO]$_2$Hg	0.13	
[ClCH$_2$(CH$_2$)$_9$COO]$_2$Hg	0.12	
[(CH$_3$)$_2$CH$_2$COO]$_2$Hg	0.12	

0.1M NaClO$_4$ at DME at 25°C
1.0 mM

Species	$E_{1/2}V$ (vs Ag/AgCl)	Ref.
cis(Pt(NH$_3$)$_2$Cl$_2$	~−0.4	52
trans-Pt(NH$_3$)$_2$Cl$_2$		

0.1M Pr$_4$NClO$_4$ at Pt disk electrode
1 mM

Species	E_pV (vs SCE)	Ref.
bis(2,5-Diphenyl-1,3-oxazole-4-yl) mercury	−1.96 to −2.01; −2.06; −2.45	53

0.1M Et$_4$NBF$_4$ at HMDE or Pt

Species	$E_{1/2}V$ (vs SCE)	Ref.
[Mo(bipy)$_3$]^0	−1.77; −2.15; −2.7 −1.13; −0.42; −0.01	56

VI. POLAROGRAPHY

DIMETHYL FORMAMIDE *(Continued)*
0.1M Bu$_4$NI at DME
2.8 mM

Species	$E_{1/2}V$ (vs SCE)	Ref.
Tris(*p*-nitrophenyl)phosphate	−0.83; −1.07; −1.9 −2.2; −2.5 −1.13 to −1.16[a]; −1.96 to −2.01[a] −2.24 to −2.49[a]; −2.51 to −2.61[a]	58

[a] E peak in volts at 2.85 mM at Pt disk electrode.

0.1M Et$_4$NClO$_4$
∼2.5 mM

Species	$E_{1/2}V$ (vs SCE)	Ref.
Ni(II)[A(13)T^{w-}]$^{2+}$	−2.26[a]	22
Ni(II)[A(14)T^{z-}]$^{2+}$	−2.02[a]; −2.00[b]	

[w] 11,13-Dimethyl 1,4,7,10-tetraaza cyclotridecer-10,12-diene.
[z] 12,14-Dimethyl 1,4,8,11-tetraaza cyclotetradeca-11,13-diene.
[a] Hanging mercury drop.
[b] Pt electrode stationary.

0.1M Et$_4$NClO$_4$ at DME
0.16–0.38 mM

Species	$E_{1/2}V$ (vs SCE)	Ref.
n-Diamyl thallium bromide	−1.093; −1.76; −2.05; −2.35	61
n-Dibutyl thallium bromide	−1.093; −1.76; −2.03; −2.33	
Diethyl thallium bromide	−1.106; −1.76; −2.03; −2.33	
Dimethyl thallium bromide	−1.09; −1.75; −2.04; −2.31	
Diphenyl thallium bromide	−0.80; −2.55	
n-Dipropyl thallium bromide	−1.12; −1.76; −2.04; −2.33	
Phenyl thallium chloride	−1.24; −2.55	
Triphenyl thallium	−1.76; −2.55	

DIMETHYL FORMAMIDE–WATER

0.1M NaClO$_4$ at DME at 28 ± 1°C

0.9 mM Ni(II) + 0.05M Malonic Acid

Species	% DME (Vol)	$E_{1/2}V$ (vs SCE)	Ref.
Ni(II) Malonic Acid	5	−1.025	63
	10	−1.033	
	15	−1.015	
	20	−1.005	

DIMETHYL FORMAMIDE–WATER (3:7)

1N NaClO$_4$ at DME

Species	$E_{1/2}V$ (vs SCE)	Ref.
Acetyl bromomercuri thiophene	−0.06; −0.73; −1.52	62
bis(2-Acetyl-5-thienyl)mercury	−1.29; −1.55	

DIMETHYLSULFOXIDE

0.1M Pr$_4$NClO$_4$ at DME

∼1 mM

Species	$E_{1/2}V$ (vs SCE)	Ref.
Mg TPP[a]	−0.67; −1.35; −1.80	57
Zn TPP	−0.71; −1.31; −1.72	
Cd TPP	−0.67; −1.25; −1.70	
Co TPP	−0.82; −1.87	
Zn TPC[b]	−0.72; −1.33; −1.70; −2.07	
Sn TPP(OAC)$_2$	−0.50; −0.81; −1.26	
Sn TPP	−2.14	
Pb TPP	−2.34	
Co EtioI[c]	−1.57	
Pb TPP[d]	−0.83; −1.10; −1.52	

[a] Tetraphenyl porphyrin.
[b] Tetraphenyl chlorin.
[c] Etioporphyrin 1.
[d] 1:3 Tetrahyorofuran—DMSO.

VI. POLAROGRAPHY

DIMETHYLSULFOXIDE (Continued)
0.5M Et$_4$NClO$_4$ at DME at 25°C
1 mM

Species	$E_{1/2}V$ (vs SCE)	Ref.
cis-[Co(CN)$_2$en$_2$]NO$_3$	−0.95; −1.48; −1.99	70
cis-Na[Co(CN)$_4$en]7/2 H$_2$O	−1.59; −1.78	
fac-[Co(CN)$_3$dien]H$_2$O	−1.37; −1.83	
mer-[Co(CN)$_3$dien]3H$_2$O	−1.09; −1.74	
trans-[Co(CN)$_2$en$_2$]NO$_3$	−0.97; −1.52; −1.99	

0.1M Et$_4$NClO$_4$ at DME at 25°C
1 mM

Species	$E_{1/2}V$ (vs SCE)	Ref.
trans-[CoF(dgH)$_2$(NH$_3$)]	−0.59; −0.81	71
trans-[CoCl(dgH)$_2$(NH$_3$)]	−0.58; −0.81	
trans-H[CoCl$_2$(dgH)$_2$]	−0.81	
trans-[Co(dgH)$_2$(NH$_3$)$_2$]ClO$_4$	−0.74	
[Co(dgH)$_3$]5/2 H$_2$O	−0.81; −2.11	
cis-[Co(dgH)$_2$en]ClO$_4$ 1/3 H$_2$O	−0.87; −2.29	
cis-[Co(dgH)$_2$(NH$_3$)$_2$]ClO$_4$·7H$_2$O	−0.82; −2.04	

0.1N Et$_4$NClO$_4$ at DME at 23–25°C
0.1–1 mM

Species	$E_{1/2}V$ (vs SCE)	Ref.
Na$_4$CuPTS[a]	−0.727; −1.111; −1.895	72
Na$_4$CuPTS[a,b]	−0.735; −1.113; −1.88	
Na$_4$CoPTS	−0.547; −1.346	
Na$_4$CoPTS[b]	−1.355; +0.455; +1.09	
Na$_4$NiPTS	−0.672; −1.165; −1.933	
Na$_4$NiPTS[b]	−0.682; −1.171; −1.925; +0.98	
Na$_4$H$_2$PTS[a]	−0.525; −0.970; −1.810	
Na$_4$H$_2$PTS[b]	−0.530; −0.980; +0.90	

[a] Metal-free phthalocyanine.
[b] $E_{1/2}$ values obtained using RE (Pt).

DIMETHYLSULFOXIDE (*Continued*)
0.1M Et$_4$NClO$_4$ at DME at 25°C
0.1–10 mM

Species	$E_{1/2}V$ (vs SCE)	Ref.
trans-[Co(CN)$_2$(NH$_3$)$_4$]NO$_3$·H$_2$O	-0.71; -1.06; -1.57	68
trans-[Co(CN)$_2$en$_2$]NO$_3$	-0.97; -1.52; -1.99	
cis-[Co(CN)$_2$en$_2$]NO$_3$	-0.95; -1.48; -1.99	
cis-[Co(CN)$_2$dip$_2$]NO$_3$·7H$_2$O	-0.48; -1.16; -1.96	
cis-[Co(CN)$_2$phen$_2$]NO$_3$·6H$_2$O	-0.39; -1.28; -1.94	

0.1F Et$_4$NClO$_4$ at Pt electrode

Species	Conc $\times 10_3$ M/l	E_pV (vs Ag/AgCl)	Ref.
Dioxobis (8-quinolinilato) Molybdenum(VI)	1.0	-1.15; -1.86; -0.98	69
Oxo-dioxotetrakis (8-quinolinolato) diMolybdenum(V)	0.98	-1.21; -1.57; -2.0	

0.1M Et$_4$NClO$_4$ at DME
0.5–7.5 mM

Species	$E_{1/2}V$ (vs Hg pool)	Ref.
Co(CO)$_3$NO	-0.4; -1.87	14

VI. POLAROGRAPHY

DIMETHYLSULFOXIDE (Continued)
0.02M LiClO$_4$ at DME at 25 ± 0.2°C

Species	$E_{1/2}V$ (vs SCE)	Ref.
(Cl$_2$CHCOO)$_2$Hg	0.41	51
[3,5-(NO$_2$)$_2$C$_6$H$_3$COO]$_2$Hg	0.26	
(ClCH$_2$COO)$_2$Hg	0.24	
(O-NO$_2$C$_6$H$_4$COO)$_2$Hg	0.25	
(p-NO$_2$C$_6$H$_4$COO)$_2$Hg	0.23	
(m-NO$_2$C$_6$H$_4$COO)$_2$Hg	0.18	
(ClCH$_2$CH$_2$COO)$_2$Hg	0.15	
(C$_6$H$_5$COO)$_2$Hg	0.09	
(CH$_3$COO)$_2$Hg	0.03	
(CH$_3$CH$_2$COO)$_2$Hg	0.05	
(CH$_3$CH$_2$CH$_2$COO)$_2$Hg	0.03	
[CH$_3$(CH$_2$)$_3$COO]$_2$Hg	0.01	
[CH$_3$(CH$_2$)$_5$COO]$_2$Hg	0.02	
[ClCH$_2$(CH$_2$)$_9$COO]$_2$Hg	0.00	
[(CH$_3$)$_2$CHCOO]$_2$Hg	0.03	

DIOXANE–WATER (90% Water)
0.1N Et$_4$NClO$_4$ at DME at 25 ± 0.1°C
2.0 mM

Species	$E_{1/2}V$ (vs SCE)	Ref.
Cp$_2$TiCl$_2$	−0.52; −1.70	30

ETHANOL
0.1M LiClO$_4$ at RE (Pt) at 25°C
0.5 mM

Species	$E_{1/2}V$ (vs SCE)	Ref.
Benzyl ferrocene	+0.385	74
bis(2,9-Dimethyl-1,10-phenanthroline) Cu(I)	+0.720	
tris(4,7-Dimethyl-1,10-phenanthroline) Fe(II)	+0.985	
Ferrocene	+0.410	

ETHANOL (*Continued*)

[0.1M LiClO$_4$ at RE (Pt)]

Species	$E_{1/2}V$ (vs SCE)	Ref.
4,7-Dimethyl-1,10-phenanthroline ferrous	+0.895	4

ETHANOL–WATER (90% Ethanol)

[0.1M NaClO$_4$ + 0.01M HClO$_4$ at DME]

Species	$E_{1/2}V$ (vs SCE)	Ref.
Ferrocene	+0.31	45
Ruthenocene	+0.26	

0.2 NaClO$_4$ at DME

~4 mM

Species	$E_{1/2}V$ (vs SCE)	Ref.
(C$_5$H$_5$)$_2$Ni	−0.08	43

ETHANOL–WATER (30% V/V Ethanol)

1M NaNO$_3$ at DME at 25.0 ± 0.1°C

~1 mM

Species	$E_{1/2}V$ (vs SCE)	Ref.
Iron(III) protoporphyrin cyanide[a]	−0.499	44
Iron(III) protoporphyrin pyridine[b]	−0.128	
Iron(III) protoporphyrin cyanopyridine[c]	−0.360	
Iron(III) protoporphyrin β or γ picoline[d]	−0.147	

[a] 0.5M NaCN pH = 10.5.
[b] 2.0M Pyridine pH = 7.9.
[c] 2.0M Pyridine, 0.0003–0.01M NaCN pH = 10.5 at RE (Pt).
[d] 2.0M Picoline pH = 8.

VI. POLAROGRAPHY

ETHANOL–WATER (at 50% Ethanol)
0.1M $NaCl_4$ at pH 11.5 and 0.5 mM conc.

Species	$E_{1/2}V$ (vs SCE)	Ref.
[Fe arene complex]$^+$ X^- with R		41
H	1.59	
C_2H_5	1.62	
n-C_3H_5	1.58	
i-C_3H_7	1.55	
$CH_2C_6H_5$	1.56	
$COCH_3$	1.28; 1.04; 1.55	
[Fe arene complex with R', –R'] X^-		
H	1.59	
CH_3	1.63	
C_2H_5	1.57	
n-C_3H_7	1.54	
i-C_3H_7	1.56	
$C(CH_3)_3$	1.56	
C_6H_5	1.44	

ETHANOL–WATER
DME; 0.125 mM

Species	Supporting electrolyte	$E_{1/2}V$ (vs SCE)	Ref.
Triphenyltinfluoride	0.1F HCl 0.1F KCl	−0.65; −0.92	42

ETHANOL–WATER (1:1)
DME at 25 ± 0.1°C
2.0 mM

Species	Supporting electrolyte	$E_{1/2}V$ (vs SCE)	Ref.
$Cp_2{}^a TiCl_2$	0.1N KCl	−0.50; −1.52	30
$CH_3-Cp-TiCpCl_2$	0.1N KCl	−0.55; −1.58	
$(CH_3-Cp)_2TiCl_2$	0.1N KCl	−0.58; −1.60	
Cp_2TiF_2	0.1N KCl	−0.78	
Cp_2TiBr_2	0.1N KCl	−0.50; −1.48	
Cp_2TiI_2	0.1N KCl	−0.52; −1.52	
Cp_2TiF_2	0.1N KF	−0.90	
Cp_2TiBr_2	0.1N KBr	−0.50; −1.48	
Cp_2TiI_2	0.1N KI	−0.52; −1.50	

[a] Cyclopentadienyl.

ETHER
At 22°C

Species	$E_{1/2}V$	Ref.
C_6H_5MgBr	2.17	46
CH_3MgBr	1.94	
C_3H_7MgBr	1.42	
C_4H_9MgBr	1.32	
C_2H_5MgBr	1.28	
$C_2H_5(CH_3)CHMgBr$	1.24	
$(CH_3)_2CHMgBr$	1.07	
$(CH_3)_3CMgBr$	0.97	
$CH_2=CHCH\,MgBr$	0.86	

VI. POLAROGRAPHY

FORMAMIDE
0.2M NaClO$_4$ at DME at 25.0 ± 0.5°C
2 mM

Species	$E_{1/2}V$ (vs SCE)	Ref.
(C$_6$H$_6$)$_2$CrI	−1.04	47
(C$_6$H$_5$·C$_6$H$_5$)$_2$CrI	−0.89	
(C$_5$H$_5$)$_2$CoI$_3$	−1.11	
(C$_9$H$_7$)$_2$CoClO$_4$	−0.71	
(C$_5$H$_5$)$_2$TiBr$_2$	−0.62	

[a] Supporting electrolyte Et$_4$NClO$_4$.

GLYME
Bu$_4$N ClO$_4$ at DME
~0.05 mM

Species	$E_{1/2}V$ (vs Ag/AgClO$_4$)	Ref.
PPh$_3$	−3.5	64
Ph$_2$PCl	−3.3	
PhPCl$_2$	−2.5	
Ph$_4$PCl	−3.1	
Ph$_3$As	−3.4	
Ph$_3$AsBr$_2$	−0.9; −3.4	
Ph$_2$AsBr	−0.9; −2.7	
Ph$_2$AsI	−1.0; −2.7	
Ph$_2$AsClO$_4$	−1.0; −2.7	
Ph$_2$AsAsPh$_2$	−2.7	
PhAsCl$_2$	−0.9	
Ph$_4$AsCl	−3.0	
Ph$_2$AsOAsPh$_2$	−2.7	
Pb$_3$Sb	−3.3	
Ph$_3$SbCl$_2$	−2.0; −3.3	
Ph$_3$SbBr$_2$	−1.4; −3.3	
Ph$_2$SbI	−1.0; −2.5	
Ph$_2$SbOAC	−1.4; −2.5	
Ph$_2$SbClO$_4$	−0.9; −2.5	
Ph$_2$SbSbPh$_2$	−2.5	
Ph$_2$SbOSbPh$_2$	−2.5	
Ph$_3$Bi	−3.1	
Ph$_3$BiBr$_2$	~−1.1; −3.1	
Ph$_2$BiCl	−1.0; −2.3	
Ph$_2$BiBiPh$_2$	−2.3	
Ph$_4$BiClO$_4$	−3.1	

GLYME (Continued)
0.1M Bu₄NClO₄ at DME at 22°C

Species	$E_{1/2}V$ (vs Ag/AgClO₄)	Ref.
i-C₃H₇HgBr	−3.27	65
C₂H₅HgBr	−3.25	
CH₃HgBr	−3.10	
C₆H₅CH₂CH₂HgBr	−3.04	
C—C₃H₅HgBr	−3.01	
CH₂=CHHgBr	−2.94	
C₆H₅HgBr	−2.92	
CH₂=CH CH₂HgBr	−2.32	
C₆Cl₅HgBr	−2.36	
C₆H₅CH₂HgBr	−2.08	
OHCCH₂HgBr	−1.85	
CH₃COCH₂HgBr	−1.86	
CH₃O₂CHgBr	−1.75	
(CH₂=CH)₂Hg	−3.34	
(C₆H₅)₂Hg	−3.32	
(C₃H₇C≡C)₂Hg	−2.90	
(C₆H₅CH=CH)₂Hg	−2.75	
(C₆Cl₅)₂Hg	−2.63	
(C₆H₅C≡C)₂Hg	−2.25	
(C₆F₅)₂Hg	−1.81	
(CCl₃)₂Hg	−1.43	

METHANOL
0.1M LiClO₄ at RE (Pt) at 25°C
0.5 mM

Species	$E_{1/2}V$ (vs SCE)	Ref.
Benzyl ferrocene	+0.340	74
bis(2, 9-Dimethyl-1, 10-phenanthroline) Cu(I)	+0.660	
$tris$(4, 7-Dimethyl-1, 10-phenanthroline) Fe(II)	+0.890	
Ferrocene	+0.370	

VI. POLAROGRAPHY

METHANOL (*Continued*)
DME at 25°C)

Species	Conc. $\times 10^3$ M/l	Supporting electrolyte	$E_{1/2}V$ (vs SCE)	Ref.
Cu(TAAB)$^{2+}$	0.5	KNO$_3$	+0.057; −1.004[b]; −1.138[b]	19
Cu(TAAB)$^+$	0.5	KNO$_3$	+0.052; −0.971[b] −1.533	
Ni(TAAB)$^{2+}$	0.5	KCl	−0.432; −0.624; −1.568[b]	
Ni(H$_8$TAAB)$^{2+}$	0.5	KCl	−0.973; −1.004[b]	
Ni[(TAAB)(OMe)$_2$]	0.1	KCl	−0.496; −1.543[b]	
Cu[(TAAB)(OMe)$_2$]	0.1	KCl	−0.195; −1.284[b]	
Co(TAAB)$^{2+}$	0.5	BU$_4$NI	−0.497; −0.840$_i$; −1.025; −1.300[b]	
Co(TAAB)(NO$_3$)$_2{}^+$	0.5	Et$_4$NClO$_4$	+0.508; −0.550; −0.887; −1.066; −1.478; −1.650[b]	

[b] $E_{1/2}$ = of ligand reduction.
TAAB = Tetrabenzo[*b, f, j, n*][1, 5, 9, 13] tetrazacyclohexadecine.

0.02M LiClO$_4$ at DME at 25 ± 0.2°C

Species	$E_{1/2}V$ (vs SCE)	Ref.
(Cl$_2$CHCOO)$_2$Hg	0.57	51
(3, 5(NO$_2$)$_2$C$_6$H$_3$COO)$_2$Hg	0.49	
(ClCH$_2$COO)$_2$Hg	0.47	
(*o*-NO$_2$C$_6$H$_4$COO)$_2$Hg	0.46	
(*p*-NO$_2$C$_6$H$_4$COO)$_2$Hg	0.43	
(*m*-NO$_2$C$_6$H$_4$COO)$_2$Hg	0.41	
(ClCH$_2$CH$_2$COO)$_2$Hg	0.38	
(C$_6$H$_5$COO)$_2$Hg	0.37	
(CH$_3$COO)$_2$Hg	0.34	
(CH$_3$CH$_2$COO)$_2$Hg	0.34	
(CH$_3$CH$_2$CH$_2$COO)$_2$Hg	0.34	
[CH$_3$(CH$_2$)$_3$COO]$_2$Hg	0.32	
[CH$_3$(CH$_2$)$_5$COO]$_2$Hg	0.32	
[ClCH$_2$(CH$_2$)$_9$COO]$_2$Hg	0.32	
[(CH$_3$)$_2$CHCOO]$_2$Hg	0.34	

METHANOL (Continued)
0.1M LiClO$_4$ at RE (Pt)

Species	$E_{1/2}V$ (vs SCE)	Ref.
4,7-Dimethyl-1,10-phenanthroline ferrous	+0.890	4

METHANOL–WATER
0.1M NaClO$_4$ at DME at 28 ± 1°C
Ni(II) 0.9 mM

Species	% Methanol	$E_{1/2}V$ (vs SCE)	Ref.
Ni(II) malonic acid 0.052M malonic acid	5	−1.07	63
	10	−1.068	
	20	−1.065	
	30	−1.065	
	40	−1.060	
	50	−1.070	
	60	−1.075	
	80	−1.072	
Ni(II) succinic acid 0.05M succinic acid	10	−1.01	
	20	−1.02	
	30	−1.0275	
	40	−1.03	
	60	−1.0325	
	65	−1.0375	

VI. POLAROGRAPHY

METHANOL–WATER (Continued)

Species	Supporting electrolyte	pH	$E_{1/2} V$ (vs SCE)	Ref.
Olefin–Hg(II) acetate				
Olefin				
Allyl acetate	Borate	9.55	−0.39; −1.09	67
Allyl acetone			−0.37; −0.73	
Allyl alcohol			−0.41; −1.02	
Crotyl alcohol			−0.33; −0.65	
Cyclohexene			−0.28; −0.80	
2,5-Dimethyl-1,5-hexadiene			−0.28; −0.92	
Ethene			−0.34; −1.00	
4-Methyl-1-pentene			−0.31; −0.99	
α-Methyl styrene			−0.28; −0.51	
Propene			−0.35; −0.90	
Styrene			−0.30; −0.73	
Vinyl acetate			−0.38; −0.88	
Vinyl-n-butylether			−0.38; −0.91	
Vinyl-isobutylether			−0.39; −0.92	
N-Vinylcarbazole			−0.27; −0.84	
Vinyl-2-chloroethylether			−0.39; −0.92	
N-Vinyl-2-pyrrolidone			−0.37; −0.73	
Allyl acetate	NaOH	12.9	−0.60; −0.99	
Allyl acetone			−0.59; −0.70	
Allyl alcohol			−0.58; −0.98	
Crotyl alcohol			−0.51; −0.63	
Cyclohexene			−0.48; −0.77	
2,5-Dimethyl-1,5-hexadiene			−0.45; −0.65 −0.95	
Ethene			−0.52; −0.95	
4-Methyl-1-pentene			−0.50; −0.98	
α-Methyl styrene			−0.47	
Propene			−0.53; −0.87	
Styrene			−0.49; −0.71	
Vinyl acetate			−0.56; −0.86	
Vinyl-n-butylether			−0.58; −0.88	
Vinyl-isobutylether			−0.58; −0.89	
N-Vinylcarbazole			−0.45; −0.84	
Vinyl-2-chloroethylether			−0.58; −0.89	
N-Vinyl-2-pyrrolidone			−0.53; −0.69	

METHANOL–WATER (*Continued*)

(50% Methanol)

Acetic acid/sodium acetate buffer at DME at 25°C and 0.2 mM soln. at pH = 5.55

Species	$E_{1/2}V$ (vs SCE)	Ref.
Olefin-Hg(II) acetate		
Olefin		
Allylacetate	−0.27; −0.65	67
Allylacetone	−0.27; −0.43	
Allylalcohol	−0.28; −0.56	
Crotyl alcohol	−0.21; −0.33	
Cyclohexene	−0.16; −0.43	
2,5-Dimethyl-1,5-hexadiene	−0.15; −0.22	
Ethene	−0.26; −0.54	
4-Methyl-1-pentene	−0.20; −0.46	
α-Methyl styrene	−0.14; −0.20	
Prepene	−0.23; −0.41	
Styrene	−0.17; −0.40	
Vinylacetate	−0.26	
Vinyl-*n*-butylether	−0.26; −0.35	
Vinyl-isobutylether	−0.26; −0.36	
N-Vinyl carbazole	−0.11; −0.62	
Vinyl-2-chloroethylether	−0.26; −0.37	
N-Vinyl-2-pyrrolidone	−0.23; −0.33	

NITROMETHANE

0.1M Et$_4$NClO$_4$ at RE (Pt) at 25°C

0.5 mM

Species	$E_{1/2}V$ (vs SCE)	Ref.
Benzyl ferrocene	+0.300	74
bis(2,9-Dimethyl-1,10-phenanthroline) Cu(I)	+0.670	
tris(4,7-Dimethyl-1,10-phenanthroline) Fe(II)	+0.880	
Ferrocene	+0.335	

VI. POLAROGRAPHY

NITROMETHANE (Continued)
0.1M Et₄NClO₄ at RE (Pt)

Species	$E_{1/2}V$ (vs SCE)	Ref.
4,7-Dimethyl-1,10-phenanthroline ferrous	+0.880	4

1-PROPANOL
0.1M LiClO₄ at RE (Pt) at 25°C
0.5 mM

Species	$E_{1/2}V$ (vs SCE)	Ref.
Benzyl ferrocene	+0.385	74
tris(4,7-Dimethyl-1,10-phenanthroline Fe(II)	+0.915	
Ferocene	+0.420	

0.1M LiClO₄ at RE (Pt)

Species	$E_{1/2}V$ (vs SCE)	Ref.
4,7-Dimethyl-1,10-phenanthroline ferrous	+0.915	4

2-PROPANOL
0.1M LiClO₄ at RE (Pt)

Species	$E_{1/2}V$ (vs SCE)	Ref.
4,7-Dimethyl-1,10-phenanthrolic ferrous	+0.925	4

2-PROPANOL (Continued)

0.1M LiClO$_4$ at RE (Pt) at 25°C

0.5 mM

Species	$E_{1/2}V$ (vs SCE)	Ref.
Benzyl ferrocene	+0.405	74
tris(4,7-Dimethyl-1,10-phenanthroline) Fe(II)	+0.925	
Ferrocene	+0.445	

PROPANOL–WATER

0.1M NaClO$_4$ at DME at 28 ± 1°C

Ni(II) 0.9 mM + 0.05M succinic acid

Species	% Propanol (vol)	$E_{1/2}V$ (vs SCE)	Ref.
Ni(II)-succinic acid	5	−1.062	63
	10	−1.102	
	20	−1.135	
	30	−1.140	
	40	−1.137	
	50	−1.132	
	60	−1.125	
	70	−1.1	

PYRIDINE

0.1M LiClO$_4$ at RE (Pt)

Species	$E_{1/2}V$ (vs SCE)	Ref.
4,7-Dimethyl-1,10-phenanthroline ferrous	+0.985	4

PYRIDINE (Continued)
0.1M LiClO₄ at RE (Pt) at 25°C
0.5 mM

Species	$E_{1/2}V$ (vs SCE)	Ref.
Benzyl ferrocene	+0.490	74
bis(2,9-Dimethyl-1,10-phenanthroline) Cu(I)	+0.335	
tris(4,7-Dimethyl-1,10-phenanthroline) Fe(II)	+0.985	
Ferrocene	+0.505	

TOLUENE
0.1M LiCl at DME
~1 mM

Species	$E_{1/2}V$ (vs Hg pool)	Ref.
Chromium(III) acetylacetonate	−1.34	73
Cobalt(III) acetylacetonate	−0.60	
Copper(II) acetylacetonate	−0.46	
Indium acetylacetonate	−0.92	
Iron(III) acetylacetonate	−0.48	
Molybdenum(VI) acetylacetonate	−0.82	
Palladium acetylacetonate	−0.74	
Titanium(IV) acetylacetonate	−0.84	

References

1. A. M. Bond, Y. A. Heath and R. L. Martin, Inorg. Chem. **10**, 2026 (1971).
2. J. A. Ferguson and T. J. Meyer, Inorg. Chem. **10**, 1025 (1971).
3. A. M. Bond, Y. A. Heath and R. L. Martin, J. Electrochem. Soc. **117**, 1362 (1970).
4. I. V. Nelson and R. T. Iwamoto, Anal. Chem. **33**, 1795 (1961).
5. H. A. Laitinen and C. J. Nyman, J. Am. Chem. Soc. **70**, 3002 (1948).
6. P. T. Cottrell and C. K. Mann, J. Electrochem. Soc. **116**, 1499 (1969).
7. J. H. Breckenridge, H. W. Vanden Born and W. E. Harris, Can. J. Chem. **49**, 398 (1971).
8. A. I. Popov and D. H. Yeske, J. Am. Chem. Soc. **80**, 1340 (1958).
9. I. M. Kolthoff and S. Ikeda, J. Phys. Chem. **65**, 1020 (1961).

10. S. P. Yubin, S. A. Smirnova, L. I. Denisovich and A. A. Lubovich, J. Organometal. Chem. **30**, 243 (1971).
11. S. P. Yubin, S. A. Smirnova and L. I. Denisovic, J. Organometal. Chem. **30**, 257 (1971).
12. Y. Paliani, S. M. Murgia and Y. Cardaci, J. Organometal. Chem. **30**, 221 (1971).
13. Z. Samec and I. Nemec, J. Electroanal. Chem. **31**, 161 (1971).
14. Y. Piazza, A. Foffani and Y. Paliani, Z. Phys. Chem. (Frankfurt) **60**, 167 (1968).
15. Y. Piazza, A. Foffani and Y. Paliani, Z. Phys. Chem. (Frankfurt) **60**, 177 (1968).
16. D. W. Dubois, R. T. Iwamoto and J. Kleinberg, Inorg. Chem. **9**, 968 (1970).
17. R. J. Wilson, L. F. Warren Jr. and M. F. Hawthorne, J. Am. Chem. Soc. **91**, 758 (1969).
18. J. N. Burnett, L. K. Hiller Jr. and R. W. Murray, J. Electrochem. Soc. **117**, 1028 (1970).
19. N. E. Tokel, V. Katovic, K. Farmery, L. B. Anderson and D. H. Busch, J. Am. Chem. Soc. **92**, 400 (1970).
20 W. E. Yeiger Jun and D. E. Smith, J. Chem. Soc. D. **1**, 8 (1971).
21. W. M. Yulick, Jr. and D. H. Yeske, Inorg. Chem. **6**, 1320 (1967).
22. D. P. Rillema, J. F. Endicott and E. Papaconstantinou, Inorg. Chem. **10**, 1739 (1971).
23. J. L. Webb, C. K. Mann and H. M. Walborsky, J. Am. Chem. Soc. **92**, 2042 (1970).
24. I. M. Kolthoff and F. Y. Thomas, J. Phys. Chem. **69**, 3049 (1965).
25. Y. L. K. Hoh, W. E. McEwen and J. Kleinberg, J. Am. Chem. Soc. **83**, 3949 (1961).
26. T. Kuwana, D. E. Bublitz and Y. Hoh, J. Am. Chem. Soc. **82**, 5811 (1960).
27. E. S. Kuta, Anal. Chem. **32**, 1065 (1960).
28. C. Furlani, A. Furlani and L. Sestili, J. Electroanal. Chem. **9**, 140 (1965).
29. D. P. Hubbard and F. Vernon, Anal. Lett. **2**, 657 (1969).
30. S. P. Yubin and S. A. Smirnova, J. Organometal. Ch m. **20**, 229 (1969).
31. R. Y. Doisneau and J. C. Marchon, J. Electroanal. Chem. **30**, 487 (1971).
32. T. Psarras and R. E. Dessy, J. Am. Chem. Soc. **88**, 5132 (1966)
33. R. E. Dessy, R. L. Pohl and R. ɔ. King, J. Am. Chem. Soc. **88**, 5121 (1966).
34. R. E. Dessy and R. L. Pohl, J. Am. Chem. Soc. **90**, 1995 (1968)
35. R. E. Dessy, R. B. King and M. Waldrop, J. Am. Chem. Soc. **88**, 5112 (1966).
36. R. E. Dessy, R. Kornmann, C. Smith and R. Haytor, J. Am. Chem. Soc. **90**, 2001 (1968).
37. R. E. Dessy, P. M. Weissman and R. L. Pohl, J. Am. Chem. Soc. **88**, 5117 (1966).
38. R. E. Dessy, F. E. Stary, R. B. King and M. Waldrop, J. Am. Chem. Soc. **88**, 471 (1966).
39. R. E. Dessy, W. Kitching and T. Chivers, J Am. Chem. Soc. **88**, 453 (1966).
40. R. E. Dessy and R. L. Pohl, J. Am. Chem. Soc. **90**, 2005 (1968).
41. D. Astruc, R. Dabard and E. Laviron, C. R. Acad. Sc. Paris **269**, 608 (1969).
42. A. Vanachayan kul a d M. D. Morris, Anal. Lett. **1**, 885 (1968).
43. Y. Wilkinson, P. L. Pauson and F. A. Cotton, J. Am. Chem. Soc. **76**, 1970 (1954).
44. D. Y. Davis and R. F. Martin, J. Am. Chem. Soc. **88**, 1365 (1966).
45. J. A. Page and Y. Wilkinson, J. Am. Chem. Soc. **74**, 61 9 (1952).
46. W. V. Evans, F. H. Lee and C. H. Lee, J. Am. Chem. Soc. **57**, 489 (1935).
47. H. S. Hsiung and Y. H. Brown, J. Electrochem. Soc. **110**, 1085 (1963).
48. S. Wawzonek and J. H. Wagen Knecht. Proc. 3rd Int. Congr. on Polarogr. Southampton 1964, Ed. Y. J. Hills, Vol. II, Macmillan (Lond) 1966, pp. 1035.
49. S. Misumi, M. Aihara and Y. Nonaka, Bull. Chem. Soc. Japan **43**, 774 (1970).
50. Y. Paliani, Z. Naturforsch B, **25**, 786 (1970).

VI. POLAROGRAPHY

51. K. P. Butin, I. P. Beletskaya, P. N. Belik, A. N. Ryabtsev and O. A. Reutov, J. Organometal. Chem. **20,** 11 (1969).
52. Y. Sundholm, Acta. Chem. Scand. **24,** 335 (1970).
53. Y. L. Smith, P. Zuniga and J. W. Rogers, Anal. Chim. Acta **56,** 312 (1971).
54. S. Valcher and M. Mastragostino, J. Electroanal. Chem. **14,** 219 (1967).
55. D. W. Clack and N. S. Hush, J. Am. Chem. Soc. **87,** 4238 (1965).
56. D. W. Dubois, R. T. Iwamoto and J. Kleinberg, Inorg. Nucl. Chem. Letters **6,** 53 (1970).
57. R. H. Felton and H. Linschitz, J. Am. Chem. Soc. **88,** 113 (1966).
58. K. S. V. Santhanam, L. O. Wheeler and A. J. Bard, J. Am. Chem. Soc. **89,** 3386 (1967).
59. A. L. Allred and L. W. Bush, J. Am. Chem. Soc. **90,** 3352 (1968).
60. M. D. Curtis and A. L. Allred, J. Am. Chem. Soc. **87,** 2554 (1965).
61. V. K. Issleib, S. Naumann, H. Matschiner and B. Walther, Z. Anorg. Allg. Chem. **381,** 226 (1971).
62. A. N. Kashin, I. M. Levinson, K. P. Butin and A. B. Ershler, Elektrokhimiya **7,** 981 (1971).
63. O. N. Srivastava, J. K. Yupta and C. M. Yupta, Electrochim. Acta **16,** 585 (1971).
64. R. E. Dessy, T. Chivers and M. Kitching, J. Am. Chem. Soc. **88,** 467 (1966).
65. R. E. Dessy, W. Kitching, T. Psarras, R. Salinger, A. Chen and T. Chivers, J. Am. Chem. Soc. **88,** 460 (1966).
66. J. A. McCleverty and B. Ratcliff, J. Chem. Soc., 1631 (1970).
67. B. Fleet and R. D. Jee, J. Electroanal. Chem. **25,** 397 (1970).
68. N. Maki, Bull. Chem. Soc., Japan **42,** 3617 (1969).
69. A. F. Isbell, Jr. and D. T. Sawyer, Inorg. Chem. **10,** 2449 (1971).
70. N. Maki and K. Ohkawa, Bull. Chem. Soc., Japan **44,** 2005 (1971).
71. N. Maki, Bull. Chem. Soc., Japan **44,** 1447 (1971).
72. L. D. Rollmann and R. T. Iwamoto, J. Am. Chem. Soc. **90,** 1455 (1968).
73. B. K. Afghan, R. M. Dagnall and K. C. Thompson, Talanta **14,** 715 (1967).
74. I. V. Nelson and R. T. Iwamoto, Anal. Chem. **35,** 867 (1963).

VII. LIGAND EXCHANGE RATES AND ELECTRODE REACTIONS

(a) Ligand Exchange Rates

The following information is a selection of ligand exchange rate data for nonaqueous systems. Information is given relative to pseudo first order rate constants for the exchange of one molecule of coordinated solvent for another, the energy of activation for the exchange and the enthalpy and entropy of activation for the exchange.

Symbols and Units

k	pseudo first order rate constant [sec^{-1}] for exchange of one molecule of coordinated solvent.
E_A	kcal/mol; measured from slope of ln k vs $1/T$.
T	25°C, unless otherwise stated.
ΔH^*	kcal/mol
ΔS^*	cal/mol. deg.

ΔH^* and ΔS^* were measured from

$$k = (KT/n) \exp [\Delta S^*/R - \Delta H^*/RT]$$

ACETONITRILE

Species	$k(\text{sec}^{-1})$	ΔH^* (kcal/mole)	ΔS^* (cal/mole-deg)	Method	Ref.
$\text{Co}(\text{CH}_3\text{CN})_6^{+2}$	1.4×10^5	8.1 ± 0.5	-7.5 ± 2	PMR	13
$\text{Co}(\text{CH}_3\text{CN})_6^{+2}$	$(320 \pm 30) \times 10^3$	11.4 ± 0.5	5.0 ± 0.2	N^{14} NMR	81
$\text{Co}(\text{CH}_3\text{CN})^{+2}$	15×10^4	8.4 ± 0.6	-7 ± 2	PMR	85
$\text{Fe}(\text{CH}_3\text{CN})_6^{+3}$	<40			PMR	11
$\text{Fe}(\text{CH}_3\text{CN})_6^{+2}$	$(550 \pm 80) \times 10^3$	9.7 ± 0.7	0.3 ± 2.2	N^{14} NMR	81
$\text{Mn}(\text{CH}_3\text{CN})_6^{+2}$	$(1.2 \pm 0.3) \times 10^7$	7.25 ± 25	-1.8 ± 8	N^{14} NMR	14
$\text{Ni}(\text{CH}_3\text{CN})_6^{+2}$	$(2.5 \pm 0.5) \times 10^3$	11.7 ± 1.0	-3.6 ± 1.0	PMR	15
$\text{Ni}(\text{CH}_3\text{CN})_6^{+2}$	3.9×10^3	10.9 ± 0.5	-8.8 ± 2	PMR	13
$\text{Ni}(\text{CH}_3\text{CN})^{+2}$	1.24×10^4	11.8 ± 0.8	-0.2 ± 2.5	PMR	85
$\text{Ni}(\text{CH}_3\text{CN})_6^{+2}$	$(2 \pm 0.3) \times 10^3$	16.4 ± 0.5	12.0 ± 2.0	N^{14} NMR	81
$\text{Vo}(\text{CH}_3\text{CN})_4^{+2}$	2.85×10^3	7.05	-20	PMR	16

VII. LIGAND EXCHANGE RATES AND ELECTRODE REACTIONS

AMMONIA (NH_3)

Species	$k(\text{sec}^{-1})$	ΔH^* (kcal/mole)	ΔS^* (cal/mole-deg)	Method	Ref.
$Co(NH_3)_6^{+2}$	$7.2 \pm 1.4 \times 10^6$	11.2 ± 0.4	10.2 ± 0.4	N^{14} NMR	2
$Co(NH_3)_6^{+2}$	1×10^7			N^{14} NMR	6
$Cr(NH_3)_6^{+2}$	6×10^{-5}			isotopic dilution N^{15}	3
$Mn(NH_3)_6^{+2}$	$(3.6 \pm 0.3) \times 10^7$	8 ± 0.5	5 ± 3	N^{14} NMR	4
$Ni(NH_3)_6^{+2}$	$(4.7 \pm 0.5) \times 10^4$	10 ± 1	-3 ± 4	N^{14} NMR	5
$Ni(NH_3)_6^{+2}$	$(1.0 \pm 0.1) \times 10^5$	11 ± 1	$+2 \pm 3$	N^{14} NMR	6
$Ni(NH_3)_6^{+2}$	$(1.9 \pm 0.1) \times 10^5$	10.1 ± 0.5	-0.5 ± 1	PMR	7
$Ni(NH_3)_6^{+2}$	$(100 \pm 10) \times 10^3$	9.9 ± 0.5	-2 ± 2	N^{14} NMR	87

AMMONIA (AQUEOUS) (NH_3)

Species	k (sec^{-1})	ΔH^* (kcal/mole)	ΔS^* (cal/mole-deg)	Method	Ref.
$Co(NH_3)_5H_2O^{+3}$	5.9×10^{-6}	26.6 ± 0.3	6.7	Isotopic dilution	78
$Cu(NH_3)_4^{+2}$	$7 \times 10^6 (32°C)$			N^{14} NMR	8
$Ir(NH_3)_5H_2O^{+3}$	6.5×10^{-8}	28.2 ± 0.2	3.0	O^{18}-isotopic dilution	80
$Ni(NH_3)_6^{+2}$	$8 \times 10^4 (32°C)$			N^{14} NMR	8
$Ni(NH_3)_6^{+2}$	$(5.6 \pm 0.5) \times 10^4$	9.5 ± 1.1	-5 ± 4	N^{14} NMR	5
$Rh(NH_3)_5H_2O^{+3}$	1.07×10^{-5}	24.1 ± 0.3	-0.65	O^{18} isotopic dilution	79

VII. LIGAND EXCHANGE RATES AND ELECTRODE REACTIONS

DIMETHYL FORMAMIDE (DMF)

Species	k (sec^{-1})	ΔH^* (kcal/mole)	ΔS^* (cal/mole-deg)	Method	Ref.
Al(DMF)$_6^{+3}$	0.15	18	5	PMR	27
Al(acetyl acetonate)(DMF)$_4^{+2}$	30			PMR	28
Al(acetylacetonate)$_2$(DMF)$_2^+$	30			PMR	28
Be(DMF)$_4^{+2}$	310	14.6 ± 0.3	2.6 ± 1	PMR	29
Be(acetylacetonate)(DMF)$_2^{+2}$	22	13.9 ± 0.8	−6 ± 2	PMR	29
Co(DMF)$_6^{+2}$	3.9 × 10^5	13.6 ± 0.5	12.6 ± 2	PMR	30
Co(DMF)$_6^{+2}$	2.3 × 10^5	7.1 ± 0.5	−10 ± 2	O^{17} NMR	31
Fe(DMF)$_6^{+2}$	5.3 × 10^5	5.3	−15	O^{17} NMR	10
Fe(DMF)$_6^{+3}$	33	12.5 ± 1.5	−10 ± 5	PMR	11
Ga(DMF)$_6^{+3}$	1.7	14.6 ± 2	−8 ± 2	PMR	32
Mg(DMF)$_6^{+2}$	>1 × 10^5	<10		PMR	29
Mn(DMF)$_6^{+2}$	4.8 × 10^6	1.2	−23	O^{17} NMR	10
Ni(DMF)$_6^{+2}$	7.7 × 10^3	9.4 ± 0.5	−9 ± 2	O^{17} NMR	31
Ni(DMF)$_6^{+2}$	3.8 × 10^3	15 ± 0.5	8 ± 2	PMR	30
Ni(DMF)$_6^{+2}$	6.9 × 10^3	14.0	6	PMR	82
VO(DMF)$_4^{+2}$	5.75 × 10^2	7.25	−16	PMR	35
VO(DMF)$_4^{+2}$	5.8 × 10^2	7.2	−16	PMR	16

DIMETHYL METHYL PHOSPHONATE (DMMP)

Species	k (sec^{-1})	ΔH^* (kcal/mole)	ΔS^* (cal/mole-deg)	Method	Ref.
Ni(dimethyl methyl phosphonate)$_6^{+2}$	2.4 × 10^4	7.5 ± 1	−11	PMR	9

DIMETHYL SULFOXIDE (DMSO)

Species	$k(\sec^{-1})$	ΔH^* (kcal/mole)	E_A (kcal/mole)	ΔS^* (cal/mole-deg)	Method	Ref.
Al(DMSO)$_6^{+3}$	$0.289 \pm 0.016 (40°C)$	20 ± 1		3.7 ± 2.5	PMR	36
Co(DMSO)$_6^{+2}$	3.1×10^5	12.2		9.8	PMR	37
Co(DMSO)$_6^{+2}$	$>1.5 \times 10^4$				PMR	39
Cu(DMSO)$_2^{+2}$	7.95×10^3	7.25		-3.6	PMR	76
Fe(DMSO)$_6^{+3}$	~ 50		10 ± 2		PMR	38
Fe(DMSO)$_6^{+2}$	1×10^4	4.16		-28.8	PMR	76
Ni(DMSO)$_6^{+2}$	4.2×10^3	12.1 ± 0.3		1.3	PMR	39
Ni(DMSO)$_6^{+2}$	7.5×10^3	8 ± 0.7		-16 ± 7	PMR	40
Ni(DMSO)$_6^{+2}$	9.3×10^3	9.3 ± 0.5		-16 ± 2	PMR	41
Ni(DMSO)$_6^{+2}$	5.57×10^3	8.05		-14	PMR	76
Ni(DMSO)$_6^{+2}$	3.2×10^3	13.0		1.4	PMR	37
VO(DMSO)$_4^{+2}$	$>1.52 \times 10^3$				PMR	16

VII. LIGAND EXCHANGE RATES AND ELECTRODE REACTIONS

ETHANOL

Species	$k(\text{sec}^{-1})$	ΔH^* (kcal/mole)	ΔS^* (cal/mole-deg)	Method	Ref.
$\text{Fe}(\text{C}_2\text{H}_5\text{OH})_6^{+3}$	2.0×10^4	6.2 ± 1.5	18 ± 5	PMR	11
$\text{Mg}(\text{C}_2\text{H}_5\text{OH})_6^{+2}$	2.8×10^6	17.7	30	PMR	12
$\text{Ni}(\text{C}_2\text{H}_5\text{OH})_6^{+2}$	1.1×10^4	10.8	4	PMR	11

ETHYLENE GLYCOL

Species	$k(\text{sec}^{-1})$	Method	Ref.
Ni^{+2}	4.4×10^3 (at 27°C)	PMR	77

METHANOL (CH_3OH)

Species	$k(sec^{-1})$	ΔH^* (kcal/mole)	E_A (kcal/mole)	ΔS^* (cal/mole-deg)	Method	Ref.
$Co(CH_3OH)_6^{+2}$	1.8×10^4	13		5.2	PMR	17
$Co(CH_3OH)_6^{+2}$	1.8×10^4	13.8		7.2	PMR	18
$Co(CH_3OH)_6^{+2}$	4×10^3	11.5 ± 0.7		-4	Isotopic dilution	86
$cis\text{-}CoCl(CH_3OH)_5^+$	1.6×10^3				PMR	19
$trans\text{-}CoCl(CH_3OH)_5^+$	8×10^2				PMR	19
$Co(H_2O)(CH_3OH)_5^{+2}$	6.3×10^5				PMR	17
$Co(H_2O)_2(CH_3OH)_4^{+2}$	2×10^6				PMR	17
$Cu(CH_3OH)_6^{+2}$	10^8	10		13	PMR	20
$Cu(CH_3OH)_6^{+2}$	8.4×10^3	10.0		13	PMR	23
$Fe(CH_3OH)_6^{+2}$	5×10^4	12		3	PMR	20
$Fe(CH_3OH)_6^{+3}$	5.1×10^3	10.1		-8	PMR	20
$Fe(CH_3OH)_6^{+3}$	2.4×10^3	10.7		-8	PMR	11
$Mg(CH_3OH)_6^{+2}$	4.7×10^3	16.7		14	PMR	21
$Mg(CH_3OH)_6^{+2}$	8.4×10^3	13 ± 2		4 ± 4	Isotopic dilution	22
$Mn(CH_3OH)_6^{+2}$	3.7×10^5	6.2		12	PMR	20
$Mn(CH_3OH)_6^{+2}$	9.5×10^5		7.4		PMR	23
$Ni(CH_3OH)_6^{+2}$	1×10^3	15.8		8	PMR	18

VII. LIGAND EXCHANGE RATES AND ELECTRODE REACTIONS

Species	$k\,(\text{sec}^{-1})$				Method	Ref.
$Ni(CH_3OH)_6^{+2}$	1.1×10^3	10.6		-9	PMR	17
$Ni(CH_3OH)_5Cl^-$	2.3×10^5		15.8		PMR	24
$Ni(CH_3OH)_6^{+2}$	0.2×10	13.1 ± 0.7		-4	Isotopic dilution	86
$NpO_2(CH_3OH)_4^{+2}$	$(9.1 \pm 0.3) \times 10^4 (0°C)$	7.5 ± 0.4	7.9 ± 0.4	8.8 ± 0.6	PMR 0.2M HCl	25
$Ti(CH_3OH)_4^{+3}$	1.9×10^5	3.3		-24	O^{17} NMR	26
$VO(CH_3OH)_4^{+2}$	3.3×10^2	12		-1.3	PMR	16

TRIMETHYL PHOSPHATE

Species	$k\,(\text{sec}^{-1})$	Method	Ref.
$Co(TMPA)_4^{+2}$	$>6.3 \times 10^3$	PMR	39
$Ni(TMPA)_6^{+2}$	$>1.8 \times 10^2$	PMR	39
$VO(TMPA)_4^{+2}$	$>0.84 \times 10^3$	PMR	16

TRIMETHYL PHOSPHITE

Species	$k\,(\text{sec}^{-1})$	Method	Ref.
$VO(TMPi)_4^{+2}$	$<0.35 \times 10^3$	PMR	16

WATER

Species	$k(\text{sec}^{-1})$	ΔH^* (kcal/mole)	E_A (kcal/mole)	ΔS^* (cal/mole-deg)	Method	Ref.
$Al(H_2O)_6^{+3}$	0.15	27		28	O^1 NMR	42
$Be(H_2O)_4^{+2}$	$>3.2 \times 10^3$				O^{17} NMR	43
$Co(H_2O)_6^{+2}$	3.1×10^5				O^{17} NMR	44
$Co(H_2O)_6^{+2}$	$(2.24 \pm 0.05) \times 10^5$	10.3 ± 0.2		5.1 ± 0.6	O^{17} NMR pH 4.2 $2M\ NH_4NO_3$	45
$Co(H_2O)_6^{+2}$	$(2.35 \pm 0.2) \times 10^6$	11.0 ± 0.7		10.6	O^{17} NMR $0.2M\ HClO_4$	46
$Co(H_2O)_6^{+2}$	2.15×10^6	10.4		5.3	O^{17} NMR $0.35M\ HClO_4$	47
$Co(H_2O)_6^{+2}$	$(2.45 \pm 0.1) \times 10^6$		11.4 ± 0.4		O^{17} NMR	48
$Co(H_2O)_6^{+2}$	1.35×10^6	8		-4	O^{17} NMR	49
$CoCl(H_2O)_5^+$	1.7×10^7	13.8 ± 0.7		21 ± 3	O^{17} NMR	46
$Co(\text{malonate})(H_2O)_4$	2.2×10^7	17.9 ± 0.8		18 ± 4	O^{17} NMR $0.1M\ HClO_4$	45
$Co(NCS)(H_2O)_5^+$	$(9.46 \pm 0.95) \times 10$ (300°K)		10.7 ± 1.1		O^{17} NMR	48
$Co(NCS)_2(H_2O)_2$	$(2-4) \times 10^8$ (300°K)		11–14 $n = 4$		O^{17} NMR	48
$Co(NCS)_3(H_2O)^-$	$>5 \times 10^8$ (300°K)				O^{17} NMR	48

VII. LIGAND EXCHANGE RATES AND ELECTRODE REACTIONS

Complex	Rate			Method	Ref
$CoNH_3(H_2O)_5^{+2}$	$(1.55 \pm 0.2) \times 10^7$	12.6 ± 0.6	17 ± 3	O^{17} NMR	45
$Co(NH_3)_2(H_2O)_4^{+2}$	$(6.5 \pm 1) \times 10^7$	9.4 ± 1.5	7 ± 5	O^{17} NMR	45
$Co(H_2O)_6^{+3}$	2×10^{-5}	27	10	Isotopic dilution O^{18}	50
$Cr(H_2O)_6^{+2}$	8×10^9				44
$Cu(H_2O)_6^{+2}$	3.3×10^6	3	-3	O^{17} NMR	44
$Cu(H_2O)_4(H_2O)_2$	Axial $\sim 2 \times 10^8$ Equit 1×10^4	Axial 5 Equit 11	Axial -4 Equit -4	O^{17} NMR	49
Diaquo triethylene tetramine nickel(II)	3.8×10^6	6		O^{17} NMR	52
$Dy(H_2O)_9^{+3}$	6.3×10^7			O^{17} NMR	51
$Fe(H_2O)_6^{+2}$	3.2×10^6	7.7	-3	O^{17} NMR	49
$Fe(H_2O)_6^{+3}$	2.4×10^4			O^{17} NMR	44
$Fe(H_2O)_6^{+3}$	$(3 \pm 0.2) \times 10^3$	8.9 ± 1	13 ± 4	O^{17} NMR 1M HNO_3	53
$Ga(H_2O)_6^{+3}$	1.8×10^3	6.3	-22	O^{17} NMR	42
$Gd(H_2O)_9^{+3}$	9×10^8	3.2	-7	O^{17} NMR	54
$Mg(H_2O)_6^{+2}$	5.3×10^5	10.2	2	O^{17} NMR	55
$Mg(H_2O)_6^{+2}$	$\sim 10^6$	~ 14		PMR	56
$Mn(H_2O)_6^{+2}$	3.6×10^7	7.8	2	PMR	57
$Mn(H_2O)_6^{+2}$	3.1×10^7	8.1	3	O^{17} NMR 0.1M $HClO_4$	49
$Mn(H_2O)_6^{+2}$	2.2×10^7			O^{17} NMR	44
$Mn(H_2O)_6^{+2}$	4×10^7	7.5	1	PMR	58

WATER (Continu

Species	$k(\text{sec}^{-1})$	ΔH^* (kcal/mole)	E_A (kcal/mole)	ΔS^* (cal/mole-deg)	Method	Ref.
$Mn(H_2O)_6^{+2}$	5.9×10^6	8.8 ± 1		5 ± 3	O^{17} NMR	34
$Mn(1,10\text{-phenantholine})(H_2O)_4^{+2}$	$(1.3 \pm 0.2) \times 10^7$	9 ± 2		6 ± 7	O^{17} NMR	34
$Mn(1,10\text{-phenantholine})_2(H_2O_2)^{+2}$	$(3.1 \pm 0.3) \times 10^7$	9 ± 2		8 ± 9	O^{17} NMR	34
$Ni(H_2O)_6^{+2}$	$(3.4 \pm 0.1) \times 10^4$	12.1 ± 0.3		2.9 ± 0.9	O^{17} NMR	59
$Ni(H_2O)_6^{+2}$	3.2×10^4				O^{17} NMR 0.1M $HClO_4$	44
$Ni(H_2O)_6^{+2}$	$(4.4 \pm 0.2) \times 10^4$	10.3 ± 0.5		-5.2 ± 2	O^{17} NMR	60
$Ni(H_2O)_6^{+2}$	$(3.6 \pm 0.2) \times 10^4$	12.3 ± 0.5		3.6 ± 1.5	O^{17} NMR 2M NH_4NO_3	61
$Ni(H_2O)_6^{+2}$	6.3×10^4	11		1	PMR	62
$Ni(H_2O)_6^{+2}$	3.2×10^4	12.1 ± 0.5		2.6 ± 2	O^{17} NMR	63
$Ni(H_2O)_6^{+2}$	$(3.0 \pm 0.3) \times 10^4$	10.8 ± 0.5		4	O^{17} NMR 10^{-3}M HCl	64
$Ni(H_2O)_6^{+2}$	2.7×10^4	11.6		1	O^{17} NMR	49
$Ni(2,2'\text{-bipyridyl})(H_2O)_4^{+2}$	4.9×10^4	12.6 ± 0.5		5.1 ± 2	O^{17} NMR 10^{-3}M HCl	63
$Ni(2,2'\text{-bipyridyl})_2(H_2O)_2^{+2}$	6.6×10^4	13.7 ± 0.5		9.2 ± 4	O^{17} NMR 10^{-3}M HCl	63
$Ni(en)(H_2O)_4^{+2}$	$(4.4 \pm 0.2) \times 10^5$	10.0 ± 0.5		1.0 ± 2	O^{17} NMR	60
$Ni(en)_2(H_2O)_2^{+2}$	$(5.4 \pm 0.3) \times 10^6$	9.1 ± 0.5		2.6 ± 2	O^{17} NMR	60

VII. LIGAND EXCHANGE RATES AND ELECTRODE REACTIONS

Complex				Method	
$Ni(NCS)_4(H_2O)_2^{-2}$	1×10^6	6		O^{17} NMR	66
$NiNH_3(H_2O)_5^{+2}$	$(2.5 \pm 0.2) \times 10^5$	10.6 ± 0.5	1.8 ± 1.5	O^{17} NMR 2M NH_4NO_3	61
$Ni(NH_3)_2(H_2O)_4^{+2}$	$(6.1 \pm 0.2) \times 10^5$	7.8 ± 0.3	-6 ± 2	O^{17} NMR	61
$Ni(NH_3)_3(H_2O)_3^{+2}$	$(2.5 \pm 0.2) \times 10^6$	10.2 ± 0.5	5.0 ± 1.5	O^{17} NMR	61
Triaquo diethylene amine nickel(II)	1.2×10^6	8		O^{17} NMR	52
Ni(tripyridine) $(H_2O)_3^{+2}$	$(5.2 \pm 0.4) \times 10^4$	10.7 ± 0.4	-1.0 ± 1	O^{17} NMR	59
$Ti(H_2O)_6^{+3}$	1×10^5	6.1	-15	O^{17} NMR	26
$V(H_2O)_6^{+2}$	9×10^1	16.4	6	O^{17} NMR	68
$V(H_2O)_6^{+3}$	1×10^3	9	-15	O^{17} NMR	69
$VO(H_2O)_4^{+2}$	5×10^2	13.7	-1	O^{17} NMR	70
$VO(H_2O)_4^{+2}$	8×10^2	13.3	-2	O^{17} NMR	72
VO(1, 2-dehydroxybenzene-3, 5-disulfonic acid)$(H_2O)_2^{-2}$	5.3×10^5	11.8	7.0	O^{17} NMR	71
VO(iminediacetate)$(H_2O)_2$	1.2×10^5	11.7	3.9	O^{17} NMR	71
VO(5-sulfoxalecylate)$(H_2O)_2^-$	1.5×10^5	10.8	1.2	O^{17} NMR	71

ACETONE (AQUEOUS)

Species	k (sec^{-1})	ΔH^* (kcal/mole)	ΔS^* (cal/mole-deg)	Method	Ref.
$Mg(H_2O)_6^{+2}$ nitrate	5×10^6	12.7	14.3	PMR	56
$Mg(H_2O)_6^{+2}$ perchlorate	4×10^6	14.5	19.8	PMR	56
$Mg(H_2O)_6^{+2}$	1.3×10^5	8.4		PMR	1

METHANOLLIC ACETONE

Species	k (sec^{-1})	ΔH^* (kcal/mole)	Method	Ref.
$Mg(CH_3OH)_6^{+2}$	2.5×10^4	12	PMR	1

References

1. W. A. Matwiyoff and H. Taube, J. Amer. Chem. Soc., **90**, 2796 (1968).
2. H. H. Glaeser, H. W. Dodgen and J. P. Hunt, Inorg. Chem., **4**, 1061 (1965).
3. H. U. D. Wiesendanger, et al., J. Chem. Phys., **27**, 668 (1957).
4. M. Grant, J. Amer. Chem. Soc., **91**, 638 (1969).
5. J. P. Hunt, H. W. Dodgen and F. Klanberg, Inorg. Chem., **2**, 478 (1963).
6. H. H. Glaeser, G. A. Lo, H. W. Dodgen and J. P. Hunt, Inorg. Chem., **4**, 206 (1964).
7. T. J. Swift and H. H. Lo, J. Amer. Chem. Soc., **88**, 2994 (1966).
8. J. P. Hunt, J. Chem. Phys., **35**, 2261 (1961).
9. L. S. Frankel, J. Molecular Spec., **29**, 273 (1969).
10. J. Baleic, Ph.D. Thesis, Univ. of Mass., July 1966, p. 60.
11. F. Breivogel, J. Phys. Chem., **73**, 4203 (1969).
12. T. Alger, J. Amer. Chem. Soc., **91**, 2220 (1969).
13. N. A. Matwiyoff and S. V. Hooker, Inorg. Chem., **6**, 1127 (1967).
14. W. L. Purcell and R. S. Mairanelli, Inorg. Chem., **9**, 1724 (1970).
15. O. Ravage, J. R. Stengle and C. H. Langford, Inorg. Chem., **6**, 1252 (1967).
16. N. S. Augermann and R. B. Jordan, Inorg. Chem., **8**, 65 (1969).
17. Z. Luz and S. Meiboom, J. Chem. Phys., **40**, 1058 (1964); ibid, p. 1066.
18. Z. Luz and S. Meiboom, J. Chem. Phys., **40**, 2686 (1964).
19. Z. Luz, J. Chem. Phys., **41**, 1748 (1964).
20. F. Breivogel, J. Chem. Phys., **51**, 445 (1969).
21. S. Nakamuna and S. Meiboom, J. Amer. Chem. Soc., **89**, 1765 (1967).
22. J. H. Swinehart and H. Taube, J. Chem. Phys., **37**, 1579 (1962); ibid, **38**, 398 (1963).
23. H. L avanon and Z. Luz, J. Chem. Phys., **49**, 2031 (1968).

VII. LIGAND EXCHANGE RATES AND ELECTRODE REACTIONS 685

24. Z. Luz, J. Chem. Phys., **51**, 1206 (1969).
25. J. C. Sheppard and J. L. Burdett, Inorg. Chem., **5**, 921 (1966).
26. A. M. Chmelnick and D. Fiat, J. Chem. Phys., **51**, 4238 (1969).
27. W. G. Movius and N. A. Matwiyoff, Inorg. Chem., **6**, 847 (1967).
28. N. A. Matwiyoff and W. G. Movius, J. Amer. Chem. Soc., **90**, 5452 (1968).
29. N. A. Matwiyoff and W. G. Movius, J. Amer. Chem. Soc., **89**, 6077 (1967).
30. N. A. Matwiyoff, Inorg. Chem., **5**, 788 (1966).
31. J. S. Babiec, C. H. Langford and T. R. Strenge, Inorg. Chem., **5**, 1362 (1966).
32. W. G. Movius and N. A. Matwiyoff, Inorg. Chem., **8**, 925 (1969).
33. J. Babiec, Thesis, Univ. of Mass., 1966, as reported in T. R. Strenge and C. H. Langford, Coord. Chem. Rev., **2**, 349 (1967).
34. M. Grant, H. Dodgen and J. P. Hunt, Inorg. Chem., **10**, 71 (1971).
35. N. S. Angermann and R. B. Jordan, J. Chem. Phys., **48**, 3983 (1968).
36. S. Thomas and W. L. Reynolds, J. Chem. Phys., **44**, 3148 (1966).
37. L. S. Frankel, Inorg. Chem., **10**, 814 (1971).
38. C. H. Langford and Fine Man Chung, J. Amer. Chem. Soc., **90**, 4485 (1968).
39. N. S. Angermann, Inorg. Chem., **8**, 2579 (1969).
40. S. Thomas and W. L. Reynolds, J. Chem. Phys., **46**, 4164 (1967).
41. S. Blacktoffe and R. Dwek, Molecular Physics, **15**, 279 (1968).
42. D. Fiat and R. E. Connick, J. Amer. Chem. Soc., **90**, 608 (1968).
43. R. E. Connick and D. Fiat, J. Chem. Phys., **39**, 1349 (1963).
44. R. E. Connick and E. D. Stover, J. Phys. Chem., **65**, 2075 (1961).
45. P. E. Hoggard, H. W. Dodgen and J. P. Hunt, Inorg. Chem., **10**, 959 (1971).
46. A. H. Zeltmann, N. A. Matwiyoff and L. O. Morgan, J. Phys. Chem., **73**, 2689 (1969).
47. A. M. Chmelnick and D. Fiat, J. Chem. Phys., **47**, 3986 (1967).
48. A. H. Zeltmann, Inorg. Chem., **9**, 2522 (1970).
49. T. J. Swift and R. E. Connick, J. Chem. Phys., **37**, 307 (1962); corrected in ibid, **41**, 2553 (1964).
50. J. P. Hunt and R. A. Plane, J. Amer. Chem. Soc., **76**, 5960 (1954).
51. J. Reuben and D. Fiat, Chem. Comm., 729 (1967).
52. D. Rablen, paper presented at the 160th A.C.S. National Meeting, Chicago, Illinois, 1970.
53. E. E. Geuser, Univ. of Calif., Lawrence Radiation Lab. Report, UCRL 9846, (1962).
54. R. Marianelli, Univ. of Calif., Lawrence Radiation Lab. Report, UCRL-17069 (1966).
55. R. E. Connick and J. Neely, J. Amer. Chem. Soc., **92**, 3476 (1970).
56. R. G. Wawro and T. J. Swift, J. Amer. Chem. Soc., **90**, 2792 (1968).
57. R. A. Bernheim, T. H. Brown, H. S. Gatowsky and D. E. Waessner, J. Chem. Phys., **30**, 950 (1959).
58. N. Bloemberger and L. O. Morgan, J. Chem. Phys., **34**, 842 (1961).
59. D. Rablen and G. Gordon, Inorg. Chem., **8**, 395 (1969).
60. A. G. Desai, H. W. Dodgen and J. P. Hunt, J. Amer. Chem. Soc., **91**, 5001 (1969).
61. A. G. Desai, et al., J. Amer. Chem. Soc., **92**, 798 (1970).
62. T. J. Swift and T. A. Stephenson, Inorg. Chem., **5**, 1100 (1966).
63. M. Grant, H. W. Dodgen and J. P. Hunt, J. Amer. Chem. Soc., **92**, 2321 (1970).
64. R. E. Connick and D. Fiat, J. Chem. Phys., **44**, 4103 (1966).
65. T. J. Swift and R. E. Connick, J. Chem. Phys., **37**, 307 (1962).
66. R. B. Jordan, H. W. Dodgen and J. P. Hunt, Inorg. Chem., **5**, 1906 (1966).
67. A. M. Chmelnick and D. Fiat, J. Chem. Phys., **51**, 4238 (1969).

68. M. Olson, Y. Ranazawa and H. Taube, J. Chem. Phys., **51,** 289 (1969).
69. D. Donham and H. Taube, reported in K. Kustin and J. Swinehart, "Fast-Metal Complex Reactions" in *Inorganic Reaction Mechanics*, J. Edwards, ed. Interscience Publishers, New York, 1970, p. 114.
70. K. Wutrich and R. E. Connick, Inorg. Chem., **6,** 583 (1967).
71. K. Withrich and R. E. Connick, Inorg. Chem., **7,** 1377 (1968).
72. J. Reuben and D. Fiat, Inorg. Chem., **6,** 579 (1967).
73. No reference.
74. N. S. Angermann and R. B. Jordan, Inorg. Chem., **8,** 65 (1969).
75. N. S. Angermann and R. B. Jordan, J. Chem. Phys., **48,** 3983 (1968).
76. G. S. Vigee and P. Ng, J. Inorg. Nucl. Chem., **33,** 2477 (1971).
77. R. Pearson and R. Lanier, J. Amer. Chem. Soc., **86,** 765 (1964).
78. H. R. Hunt and H. Taube, J. Amer. Chem. Soc., **80,** 2642 (1958).
79. F. Monacelli and E. Viel, Inorg. Chim. Acta., **1,** 467 (1967). H. L. Bott and A. J. Poe, J. Chem. Soc. (A), 1745 (1969).
80. E. Barglie and F. Monacelli, Inorg. Chim. Acta., **5,** 211 (1971).
81. R. J. West and S. F. Lincoln, Austr. J. Chem., **24,** 1169 (1971).
82. L. S. Frankel, Inorg. Chem., **10,** 2360 (1971).
83. G. Glass and R. S. Tobias, J. Amer. Chem. Soc., **89,** 6371 (1967).
84. I. R. Lautzka and D. W. Watts, Austr. J. Chem., **20,** 173 (1962).
85. J. F. O'Brian and W. L. Reynolds, Inorg. Chem., **6,** 2110 (1967).
86. T. E. Rogers, J. H. Swinehart, and H. Taube, J. Phys. Chem., **69,** 134 (1965).
87. W. L. Rice and B. B. Mayland; Inorg. Chem., **7,** 1040 (1968).

(b) Electrode Reactions in Single Solvents

The following section contains information relative to the study of electrode reactions in some nonaqueous single solvents. The organization is alphabetical by solvent and within each table alphabetical by electrolyte. The information includes techniques used, temperature, supporting electrolyte, cathodic and anodic transfer coefficients, standard rate constants, diffusion coefficients and standard exchange current densities.

Symbols and Units

α	cathodic transfer coefficient
β	anodic transfer coefficient
I_a^0	apparent exchange current density
k_S	standard rate constant
$I_0 = I_a^0 = i_0$	
k_f^0	forward rate constant at 0 V (usually vs NHE)
αn	product of α and n, the number of electrons in the rate determining step
$k_{E_{1/2}}$	forward rate constant at the half wave potential
k	forward rate constant at a given potential
k_0	forward rate constant at the equilibrium potential
i_S^0	standard exchange current density $= nFk_S$
βn_0	for oxidation reactions this has the same significance as αn
k_B^0	backward rate constant at 0 V vs NHE

ACETONITRILE[a]

Electrolyte	Technique	Temp.	Supporting electrolyte	αn_a	k_f^0 (cm/sec)	k_s (cm/sec)	D (cm²/sec)	Ref.
CO(II)	Polarography	30°C	0.1M NaClO$_4$ 0.005% gelatin	0.66	2.8×10^{-6}		3.00×10^{-5}	1
Mn(II)	Polarography (Matsuda)		0.1M NaClO$_4$ 0.005% gelatin	1.5		3.21×10^{-4}		2

[a] DME used in each case.

ACETONITRILE

Electrolyte	Technique	Temp. (°C)	Supporting electrolyte	α	k_f^0 (cm/sec)	k_s (cm,sec)	Comments	Ref.
Yb(III)	Polarography		0.1M NaClO$_4$	0.15	5.5×10^{-4}		Calculated from the first polarographic wave	4
Cd(II)	Faradaic impedance	25	0.9M NaClO$_4$	0.55		0.19		5

VII. LIGAND EXCHANGE RATES AND ELECTRODE REACTIONS

Cu(II)[a]	Chronopotentiometry	25.0 ± 0.1	0.1M Et$_4$NClO$_4$	0.12	(3.5 ± 0.3) × 10^{-3}	Results are dependent on the pretreatment of the electrode cycling between +1.5 and −1.5 V ending on a reduction and allowing electrode to set at open circuit for a few minutes Cu(II) + e$^-$ ⇌ Cu(I) results are also dependent on the method of measuring transition time.	6

[a] Pt electrode used. DME used in all other cases.

AMMONIA[a]

Electrolyte	Technique	Temp. (°C)	α	I_a^0 (mA/cm^2)	μ	Ref.
Pb(II)	Linear sweep voltametry	−45	0.56	20.0	2.0 (NO$_3^-$)	3

[a] Pb electrode used in each case.

DIMETHYL

Electrolyte	Technique	Temp. (°C)	Supporting Electrolyte	α	αn_a
Cd(II)	Faradaic impedance	25	0.9M NaClO_4	0.24	
Co(II)	Polarography	30	0.1M NaClO_4 0.005% gelatin		0.70
Mn(II)	Polarography (Gellings)		0.1M NaClO_4 0.005% gelatin		1.15
Nb(V)	Polarography	24.9 ± 0.1	0.2M $(C_2H_5)_4NClO_4$	0.596 ±0.054	
Nb(V)[a]	Chronopotentiometry	24.9 ± 0.1	0.2M $(C_2H_5)_4NClO_4$	0.618 ±0.080	
Ni(II)	Polarography (Kautecky)	25 ± 0.01	0.1M NaClO_4 Triton X-100		0.56
Yb(III)	Polarography		0.1M NaClO_4		

[a] Hg electrode used. DME used in all other cases.

DIMETHYL
Technique:
Temp.:

Electrolyte	Supporting Electrolyte	α	k_s (cm/sec)
Li(I)	0.5M LiCl	0.5	
Li(I)	2.104M LiCl	0.763 ± 0.003	$(8.5 ± 0.6) \times 10^{-5}$
Li(I)	0.210M LiCl	0.74 ± 0.02	$(9.2 ± 0.3) \times 10^{-5}$
Li(I)	0.063M LiCl	0.72 ± 0.03	$(11.9 ± 0.3) \times 10^{-5}$
Tl(I)[a]	0.5M LiCl		

[a] 40 mole percent Tl(Hg) electrode used. Li(Hg) electrode used in all other cases.

VII. LIGAND EXCHANGE RATES AND ELECTRODE REACTIONS

FORMAMIDE

k (cm/sec)	k_f^0 (cm/sec)	k_s (cm/sec)	D (cm/sec)	Comments	Ref.
		0.15			5
	5.0×10^{-13}		0.52×10^{-15}		1
		1.11×10^{-3}			2
$(3.56 \pm 3.61) \times 10^{-4}$				Rate constant at 0 V vs solid $CdCl_2$ (DMF)/Cd(Hg) reference electrode	7
$(1.40 \pm 0.65) \times 10^{-4}$					7
		1.4×10^{-9}	1.05×10^{-6}		8
		3.16×10^{-9}			4

SULFOXIDE[a]
Galvanostatic
25°

I_a^0 (A/cm²)	Comments	Ref.
5×10^{-4}		10
	correcting for effects of electrode double layer as $\alpha = 0.75$ $K_s = (4.6 \pm 0.6) \times 10^{-5}$ cm/sec	9
		9
		9
7×10^{-4}	correcting for effects of electrode double layer assumes $\alpha = 0.75$ $K_s = (2.9 \pm 0.2) \times 10^{-5}$ cm/sec correcting for effects of electrode double layer assumes $\alpha = 0.75$ $K_s = (2.8 \pm 0.4) \times 10^{-5}$ cm/sec	10

ETHANOL

Technique: Polarography

Electrolyte	Temp. (°C)	Supporting electrolyte	k (cm/sec)	Comments	Ref.
Cr(III)	20	1M KCl	2.34×10^{-13}	Solution was freshly prepared and had a red color rate constant at equilibrium potential	11
Cr(III)	30	1M KCl	5.68×10^{-11}		11
Cr(III)	40	1M KCl	7.5×10^{-10}		11
Cr(III)	50	1M KCl	4.64×10^{-10}		11
Cr(III)	60	1M KCl	5.12×10^{-10}		11
Cr(III)	70	1M KCl	2.25×10^{-10}		11

FORMAMIDE[a]

Electrolyte	Technique	Temp. (°C)	Supporting electrolyte	αn_a	k_f^0 (cm/sec)	$k_{E1/2}$ (cm/sec)	k_s (cm/sec)	D (cm²/sec)	Ref.
Co(II)	Polarography	30	0.1M NaClO$_4$ 0.005% gelatin	0.84	2.8×10^{-15}			0.18×10^{-5}	1
Eu(III)	Polarography (Kautecky)	25 ± 0.2	0.1M NaClO$_4$	0.280	1.023×10^{-5}	6.310×10^{-4}			12
			0.2M NaClO$_4$	0.265	8.318×10^{-5}	1.698×10^{-4}			12
Mn(II)	Polarography (Matsuda)		0.1M NaClO$_4$ 0.005% gelatin	1.2			4.5×10^{-4}		2

VII. LIGAND EXCHANGE RATES AND ELECTRODE REACTIONS

Electrolyte	Technique	Temp.	Supporting electrolyte	α			Ref.
Ni(II)	Polarography (Kautecky)	25 ± 0.01	0.1M NaClO₄ Triton X-100	0.62	7.1 × 10⁻¹⁰		8
Yb(III)	Polarography		0.1M NaClO₄		3.16 × 10⁻¹⁵	0.58 × 10⁻⁶	4

[a] DME used in each case.

METHANOL[a]

Electrolyte	Technique	Temp.	Supporting electrolyte	α_1	k_s	Ref.
Cd(II)	Faradaic	25°C	0.9M NaClO₄	0.27	0.013	5

[a] DME used in each case.

N-METHYL FORMAMIDE[a]

Electrolyte	Technique	Supporting electrolyte	α	k_s (cm/sec)	Ref.
Cd(II)	Faradaic impedence	0.9M NaClO₄	0.40	0.12	5
V(III)	Polarography (Kautecky)	1M NH₂SO₃H	0.20	0.00116	5

[a] DME used in each case; temp. 25°C.

PROPYLENE CARBONATE[a]

Maximum water content	Electrolyte	Technique	Temp. (°C)	Supporting electrolyte	α	I_a^0 (mA/cm²)	i_0 (mA/cm²)	Comments	Ref.
0	Cd(II)[b]	Faradaic impedence	25	0.9M NaClO₄					5
0.001 m	Li(I)	Galvanostatic		0.257M LiClO₄		>12		$k_s \approx 0.01$ cm/sec	13
0.001 m	Li(I)	Galvanostatic		0.257M LiClO₄		10.2		α was assumed to equal 0.72	13
0.001 m	Li(I)	Galvanostatic		0.257M LiClO₄		7.8			13
0.001 m	Li(I)	Galvanostatic		0.257M LiClO₄		5.4			13
0.001 m	Li(I)	Galvanostatic		0.257M LiClO₄		2.8			13
0.001 m	Li(I)	Galvanostatic		0.257M LiClO₄		1.6			13
0.02 m	Li(I)	Galvanostatic		0.229M LiClO₄		>10			13
0.02 m	Li(I)	Galvanostatic		0.229M LiClO₄		8			13
0.02 m	Li(I)	Galvanostatic		0.229M LiClO₄		3.1			13
0.02 m	Li(I)	Galvanostatic		0.229M LiClO₄		1.5			13
0.02 m	Li(I)	Galvanostatic		0.229M LiClO₄		0.30			13
0.02 m	Li(I)	Galvanostatic		0.229M LiClO₄		0.026			13
0.54 m	Li(I)	Galvanostatic		0.257M LiClO₄			75		13
0.54 m	Li(I)	Galvanostatic		0.257M LiClO₄		≈3			13
0.54 m	Li(I)	Galvanostatic		0.257M LiClO₄		≈1			13
0.54 m	Li(I)	Galvanostatic		0.257M LiClO₄		≈0.1			13

VII. LIGAND EXCHANGE RATES AND ELECTRODE REACTIONS

15 ppm	Li(I)	Current interrupter method	23	0.1M LiClO$_4$	0.64 ±0.03	0.49 ±0.09	For a film covered Li surface: $\alpha = 0.64 \pm 0.06$, $i_0 = 0.30 \pm 0.02$ mA/cm^2 values given are for a clean Li surface	14
15 ppm	Li(I)	Current interrupter method	23	0.2M LiClO	0.62 ±0.01	0.76 ±0.07	$\alpha = 0.60 \pm 0.02$, $i_0 = 0.53 \pm 0.05$ mA/cm^2	14
15 ppm	Li(I)	Current interrupter method	23	0.5M LiClO$_4$	0.59 ±0.08	1.22 ±0.03	$\alpha = 0.61 \pm 0.03$, $i_0 = 0.69 \pm 0.13$ mA/cm^2	14
15 ppm	Li(I)	Current interrupter method	23	1M LiClO$_4$	0.64 ±0.01	1.76 ±0.10	$\alpha = 0.61 \pm 0.03$, $i_0 = 0.69 \pm 0.13$ mA/cm^2	14
20 ppm	Li(I)	Galvanostatic	55	1M LiAlCl$_4$		2.35	$\partial \log I_a^0 / \partial \log \mathrm{Li}^+$ gives $\alpha = 0.8$ for LiAlCl$_4$	15
20 ppm	Li(I)	Galvanostatic	46	1M LiClCl$_4$		1.49		15
20 ppm	Li(I)	Galvanostatic	35	1M LiAlCl$_4$		1.05		15
20 ppm	Li(I)	Galvanostatic	19	1M LiAlCl$_4$		0.40		15
20 ppm	Li(I)	Galvanostatic	19	0.5M LiAlCl$_4$		0.45		15
20 ppm	Li(I)	Galvanostatic	29	0.5M LiAlCl$_4$		0.86		15
20 ppm	Li(I)	Galvanostatic	45	0.5M LiAlCl$_4$		1.65		15
20 ppm	Li(I)	Galvanostatic	56	0.5M LiAlCl$_4$		2.90		15
20 ppm	Li(I)	Galvanostatic	19	0.1M LiAlCl$_4$		0.37		15
20 ppm	Li(I)	Galvanostatic	40	0.1M LiAlCl$_4$		0.79		15
20 ppm	Li(I)	Galvanostatic	51	0.1M LiAlCl$_4$		1.10		15

PROPYLENE CARBONATE[a] (*Continued*)

Maximum water content	Electrolyte	Technique	Temp. (°C)	Supporting electrolyte	α	I_a^0 (mA/cm²)	i_0 (mA/cm²)	Comments	Ref.
20 ppm	Li(I)	Galvanostatic	19	0.05M LiAlCl₄		0.29			15
20 ppm	Li(I)	Galvanostatic	41	0.05M LiAlCl₄		0.40			15
20 ppm	Li(I)	Galvanostatic	20	1M LiPF₆		0.29		Salt solution contained $\approx$ 60 ppm H_2O	15
20 ppm	Li(I)	Galvanostatic	31	1M LiPF₆		0.52			15
20 ppm	Li(I)	Galvanostatic	46	1M LiPF₆		0.53			15
20 ppm	Li(I)	Galvanostatic	57	1M LiPF₆		0.65			15
20 ppm	Li(I)	Galvanostatic	20	1M LiBF₄		0.50		Salt solution contained $\approx$ 180 ppm H_2O	15
20 ppm	Li(I)	Galvanostatic	48	1M LiBF₄		0.72			15
20 ppm	Li(I)	Galvanostatic	55	1M LiBF₄		0.88			15
20 ppm	Li(I)	Galvanostatic	65	1M LiBF₄		0.97			15
30 ppm	Na(I)[c]	Galvanostatic	21	$\approx$ 0.3M NaPF₆ (sat.)		0.002		Na/Na(I) electrode requires brief pre-treatment to give reproducible I_a^0 values: electrode was anodized 10 minutes at 2 mA/cm²	16
30 ppm	Na(I)[c]	Galvanostatic	19	1M NaClO₄		0.21			16
30 ppm	Na(I)[c]	Galvanostatic	37	1M NaClO₄		0.90			16

VII. LIGAND EXCHANGE RATES AND ELECTRODE REACTIONS

30 ppm	Na(I)[c]	Galvanostatic	52	1M NaClO$_4$	2.6		16
30 ppm	Na(I)[c]	Galvanostatic	66	1M NaClO$_4$	6.77		16
30 ppm	Na(I)[c]	Galvanostatic	19	0.1M NaClO$_4$	0.0005		16
30 ppm	Na(I)[c]	Galvanostatic	19	0.517M NaClO$_4$	0.012		16
35 ppm	Li(I)	Galvanostatic	28	0.1M LiClO$_4$	0.114	$\partial \log I_a{}^0/\partial \log C_{Li(I)}$ gives $\alpha = 2/3$; α is temperature independent	17
35 ppm	Li(I)	Galvanostatic	40	0.1M LiClO$_4$	0.24		17
35 ppm	Li(I)	Galvanostatic	49	0.1M LiClO$_4$	0.3		17
35 ppm	Li(I)	Galvanostatic	57	0.1M LiClO$_4$	0.42		17
35 ppm	Li(I)	Galvanostatic	29	0.5M LiClO$_4$	0.4		17
35 ppm	Li(I)	Galvanostatic	49	0.5M LiClO$_4$	0.75		17
35 ppm	Li(I)	Galvanostatic	61	0.5M LiClO$_4$	1.05		17
35 ppm	Li(I)	Galvanostatic	69	0.5M LiClO$_4$	1.25		17
35 ppm	Li(I)	Galvanostatic	28	1M LiClO$_4$	0.95		17
35 ppm	Li(I)	Galvanostatic	43	1M LiClO$_4$	1.8		17
35 ppm	Li(I)	Galvanostatic	58	1M LiClO$_4$	3.4		17
35 ppm	Li(I)	Galvanostatic	67.5	1M LiClO$_4$	5.25		17
35 ppm	Li(I)	Galvanostatic	28.5	2M LiClO$_4$	0.56		17
35 ppm	Li(I)	Galvanostatic	41	2M LiClO$_4$	1.25		17
35 ppm	Li(I)	Galvanostatic	53	2M LiClO$_4$	1.85		17
35 ppm	Li(I)	Galvanostatic	61.5	2M LiClO$_4$	2.35		17

[a] Li electrode used unless otherwise indicated.
[b] DME used.
[c] Na electrode used.

TETRAHYDROFURAN[a]

Technique: Galvanostatic
Supporting Electrolyte: 0.7M NaClO$_4$
Temp.: 25°

Mole fraction tetrahydrofuran	Electrolyte	α	β	I_a^0 (mA/cm^2)	Ref.
0	Zn(II)	0.30	0.70	8.0	18
0.01	Zn(II)	0.37	0.70	0.3	
0.09	Zn(II)	0.34	0.74	0.05	
0.5	Zn(II)	0.31	0.78	0.13	
0.9	Zn(II)	0.30	0.72	0.6	
1.0	Zn(II)	0.29	0.70	4.4	

[a] Zn(Hg) electrode used in each case.

(c) Electrode Reactions in Mixed Solvents

The following section contains information relative to the study of electrode reactions in some nonaqueous mixed solvents. The organization is alphabetical by solvent and within each table alphabetical by electrolyte. The information includes techniques used, temperature, supporting electrolyte, cathodic and anodic transfer coefficients, standard rate constants, diffusion coefficients and standard exchange current densities.

Symbols and Units

α	cathodic transfer coefficient
β	anodic transfer coefficient
I_a^0	apparent exchange current density
k_s	standard rate constant
$I_0 =$	$I_a^0 = i_0$
k_f^0	forward rate constant at 0 V (usually vs NHE)
αn	product of α and n, the number of electrons in the rate determining step
$k_{E_{1/2}}$	forward rate constant at the half wave potential
k	forward rate constant at a given potential
k_0	forward rate constant at the equilibrium potential
i_s^0	standard exchange current density $= nFk_s$
βn_0	for oxidation reactions this has the same significance as αn
k_B^0	backward rate constant at 0 V vs NHE

ACETAMIDE–WATER[a]

Technique: Polarography (Meites–Israel)

Temp.: 25 ± 0.1°C

V/V	Electrolyte	Supporting electrolyte	αn_a	k_f^0 (cm/sec)	Comments	Ref.
5%	Mn(II)	1M KCNS 0.01% gelatin	0.6237	5.924×10^{-23}	A single irreversible wave is observed	19
10%	"	"	0.5475	7.991×10^{-21}		
15%	"	"	0.4985	5.7989×10^{-20}		
20%	"	"	0.4555	4.598×10^{-19}		
25%	"	"	0.4065	3.256×10^{-18}		
30%	"	"	0.3853	9.607×10^{-17}		

VII. LIGAND EXCHANGE RATES AND ELECTRODE REACTIONS

5%	Ni(II)	0.5202	2.201×10^{-20}	Two polarographic waves are observed
10%	"	0.4878	4.927×10^{-19}	
15%	"	0.4663	1.788×10^{-19}	
20%	"	0.4235	1.917×10^{-19}	
25%	"	0.4153	1.296×10^{-18}	
30%	"	0.4065	8.762×10^{-17}	
5%	Zn(II)	0.4535	3.653×10^{-18}	Two polarographic waves are observed
10%	"	0.3942	1.421×10^{-17}	
15%	"	0.3888	1.326×10^{-17}	
20%	"	0.3796	8.966×10^{-16}	
25%	"	0.3697	6.960×10^{-16}	
30%	"	0.3596	6.203×10^{-16}	

[a] DME used in each case.

ACETONITRILE–WATER[a]

V/V	Electrolyte	Technique	Temp. (°C)	Supporting electrolyte	αn_a	k_f^0 (cm/sec)	k_s (cm/sec)	D (cm²/sec)	Ref.
40%	Co(II)	Polarography	30	0.1M NaClO₄ 0.005% gelatin	0.54	1.4×10^{-10}		1.43×10^{-5}	1
60%	CO(II)	Polarography	30	0.1M NaClO₄ 0.005% gelatin	0.42	4.5×10^{-9}		1.36×10^{-5}	1
80%	Co(II)	Polarography	30	0.1M NaClO₄ 0.005% gelatin	0.81	7.0×10^{-13}		1.44×10^{-5}	1
40%	Mn(II)	Polarography (Matsuda)		0.1M NaClO₄ 0.005% gelatin	0.96		14.1×10^{-4}		2
60%	Mn(II)	Polarography (Matsuda)		0.1M NaClO 0.005% gelatin	0.75		12.75×10^{-4}		2
80%	Mn(II)	Polarography (Matsuda)		0.1M NaClO₄ 0.005% gelatin	1.45		6.0×10^{-4}		2

[a] DME used in each case.

Mole fraction Acetonitrile	Mole fraction water	Electrolyte	Technique	Temp.	Supporting electrolyte	k_s (cm/sec)	Ref.
0.5	0.5	Cd(II)	Faradaic impedence	25°C.	0.9M NaClO₄	0.0145	5

[a] DME used in each case.

VII. LIGAND EXCHANGE RATES AND ELECTRODE REACTIONS

t-BUTYL ALCOHOL–WATER[a]

Technique: Polarography (Matsuda–Ayabe)
Temp: 29°C

V/V	Electrolyte	Supporting electrolyte	α	k_s (cm/sec)	Ref.
0%	Mn(II)	0.004% gelatin	0.745	6.457×10^{-4}	20
10%	"	"	0.75	8.726×10^{-4}	
20%	"	"	0.71	5.024×10^{-4}	
30%	"	"	0.77	6.022×10^{-4}	
40%	"	"	0.81	5.320×10^{-4}	
50%	"	"	0.80	2.981×10^{-4}	
60%	"	"	0.81	4.280×10^{-4}	
80%	"	"	0.73	1.744×10^{-4}	

[a] DME used in each case.

DIMETHYL FORMAMIDE–WATER[a]

V/V	Electrolyte	Technique	Temp. (°C)	Supporting electrolyte	αn_a	k_f^0 (cm/sec)	k_s (cm/sec)	D (cm²/sec)	Comments	Ref.
0%	Co(II)	Polarography (Kautecky)	30	0.1M NaClO₄ 0.005% gelatin	0.9	2.5×10^{-16}		1.36×10^{-5}		1
20%	"	Polarography	"	"	0.81	1×10^{-14}		0.86×10^{-5}		1
40%	"	"	"	"	0.66	3.5×10^{-17}		0.60×10^{-5}		1
60%	"	"	"	"	0.81	3.2×10^{-16}		0.42×10^{-5}		1
80%	"	"	"	"	0.87	3.5×10^{-15}		0.40×10^{-5}		1

DIMETHYL FORMAMIDE–WATER[a] (Continued)

V/V	Electrolyte	Technique	Temp. (°C)	Supporting electrolyte	αn_a	k_f^0 (cm/sec)	k_s (cm/sec)	D (cm²/sec)	Comments	Ref.
20%	Mn(II)	Polarography (Gellings)		"	1.42		1.35×10^{-3}		Presence of maximum suppressor inferred from other studies by these authors	2
40%	"	"		"	1.25		1.07×10^{-3}			2
60%	"	"		"	1.24		0.9×10^{-3}			2
80%	"	"		"	1.20		0.8×10^{-3}			2
0%	Ni(II)	Polarography (Kautecky)	25 ± 0.01	0.1M NaClO₄ Triton X-100	0.737	2×10^{-13}		2.45×10^{-6}	k_f^0 is rate constant of electrode reaction at 0.1V vs. NHE uncorrected for liquid junction potentials	8
40%	"	"	"	"	0.59	1.0×10^{-10}		1.04×10^{-6}		8
60%	"	"	"	"	0.53	5.6×10^{-10}		0.57×10^{-6}		8
80%	"	"	"	"	0.53	6×10^{-10}		0.56×10^{-6}		8

[a] DME used in each case.

DIOXANE–WATER[a]

Technique: Polarography (Kautecky)
Supporting Electrolyte: 0.1M NaOH

V/V	Electrolyte	αn_a	k_f^0 (cm/sec)	D (cm^2/sec)	Ref.
0%	Cr(VI)	0.42	1.8×10^{-9}	1.57×10^{-5}	21
10%	Cr(VI)	0.88	5.3×10^{-18}	1.3×10^{-5}	
20%	Cr(VI)	0.77	2.2×10^{-16}	1.0×10^{-5}	
30%	Cr(VI)	0.87	8.4×10^{-18}	8.6×10^{-6}	
40%	Cr(VI)	0.87	8.4×10^{-18}	7.2×10^{-6}	
50%	Cr(VI)	0.87	1.8×10^{-17}	6.0×10^{-6}	
60%	Cr(VI)	0.83	4.6×10^{-17}	5.2×10^{-6}	
0%	Te(VI)	0.49	5.0×10^{-15}	1.3×10^{-5}	
10%	Te(VI)	0.61	5.5×10^{-17}	1.3×10^{-5}	
20%	Te(VI)	0.79	7.2×10^{-21}	1.0×10^{-5}	
30%	Te(VI)	0.86	5.0×10^{-22}	9.4×10^{-6}	
40%	Te(VI)	1.1	8.3×10^{-26}	5.5×10^{-6}	

[a] DME used in each case.

ETHANOL

V/V	Electrolyte	Technique	Temp.	Supporting electrolyte	α	αn_a
0%	Cr(VI)	Polarography (Kautecky)		0.1M NaOH		0.42
10%	"	"		"		0.59
20%	"	"		"		0.7
30%	"	"		"		0.71
40%	"	"		"		0.76
50%	"	"		"		0.7
60%	"	"		"		0.7
70%	"	"		"		0.78
0%	Te(VI)	"		"		0.49
10%	"	"		"		0.61
20%	"	"		"		0.69
30%	"	"		"		0.82
40%	"	"		"		0.96
50%	"	"		"		1.1
50%	Ce(III)	"	27°C	0.1M LiCl 0.01% gelatin		
50%	"	"	"	0.1M KCl	0.17	
50%	"	"	"	0.1M LiCl	0.12	
80%	"	"	"	"	0.10	
0%	Mn(II)	Polarography (Matsuda-Ayabe)	29°C	0.004% gelatin	0.745	
10%	"	"	"	"	0.75	
20%	"	"	"	"	0.81	
30%	"	"	"	"	0.77	
40%	"	"	"	"	0.75	
50%	"	"	"	"	0.73	

VII. LIGAND EXCHANGE RATES AND ELECTRODE REACTIONS 707

—WATER[a]

k_s	k_f^0 (cm/sec)	k_{eq}	$k_{E1/2}$	D (cm²/sec)	Ref.
	1.8×10^{-9}			1.57×10^{-5}	21
	3.5×10^{-13}			1.1×10^{-5}	21
	2.4×10^{-15}			0.4×10^{-6}	21
	8.5×10^{-16}			7.5×10^{-6}	21
	7.4×10^{-17}			6.5×10^{-6}	21
	2.2×10^{-15}			6.1×10^{-6}	21
	6.9×10^{-15}			5.6×10^{-6}	21
	7.2×10^{-16}			5.3×10^{-6}	21
	5.0×10^{-15}			1.3×10^{-5}	21
	6.2×10^{-17}			9.0×10^{-6}	21
	3.2×10^{-18}			7.3×10^{-6}	21
	1.15×10^{-20}			6.5×10^{-6}	21
	3.16×10^{-23}			5.3×10^{-6}	21
	1.8×10^{-25}			3.3×10^{-6}	21
		7.47×10^{-5}	6.5×10^{-4}		22
		2.22×10^{-5}	6.0×10^{-4}		22
		1.06×10^{-4}	8.5×10^{-4}		22
		1.08×10^{-4}	7.3×10^{-4}		22
6.457×10^{-4}					20
7.358×10^{-4}					20
6.505×10^{-4}					20
4.799×10^{-4}					20
7.914×10^{-4}					20
3.968×10^{-4}					20

ETHANOL

V/V	Electrolyte	Technique	Temp.	Supporting electrolyte	α	αn_a
60%	"	"	"	"	0.71	
80%	"	"	"	"	0.64	
50%	La(III)	"	27°C	0.1M LiCl	0.18	
80%	"	"	"	"	0.09	
50%	"	Polarography	"	0.1M LiCl 0.01% gelatin		
50%	"	"	"	0.1M KCl	0.14	

[a] DME used in each case

VII. LIGAND EXCHANGE RATES AND ELECTRODE REACTIONS

–WATER (Continued)

k_s	k_f^0 (cm/sec)	k_f^0	$k_{E_{1/2}}$	D (cm²/sec)	Ref.
4.785×10^{-4}					20
9.463×10^{-4}					20
		7.69×10^{-5}	7.60×10^{-4}		22
		1.19×10^{-4}	7.45×10^{-4}		22
		3.14×10^{-5}	7.60×10^{-4}		22
		1.73×10^{-5}	7.45×10^{-4}		22

ETHYLENE GLYCOL–WATER[a]

Technique: Polarography (Matsuda–Ayabe)

Temp.: 29°C

V/V	Electrolyte	Supporting electrolyte	α	k_s (cm/sec)	Ref.
0%	Mn(II)	0.1M KCl 0.004% gelatin	0.745	6.457×10^{-4}	20
10%	"	"	0.81	4.637×10^{-4}	
30%	"	"	0.75	4.060×10^{-4}	
40%	"	"	0.81	8.308×10^{-4}	
50%	"	"	0.75	4.060×10^{-4}	
60%	"	"	0.88	2.334×10^{-4}	
80%	"	"	0.81	1.874×10^{-4}	

[a] DME used in each case.

FORMAMIDE

V/V	Electrolyte	Technique	Temp. (°C)	Supporting electrolyte	αn_a
0%	Co(II)	Polarography (Kautecky)	30	0.1M NaClO$_4$ 0.005% gelatin	0.9
20%	"	Polarography	"	"	0.84
40%	"	"	"	"	0.84
60%	"	"	"	"	0.80
80%	"	"	"	"	0.79
0%	"	Polarography (Meites-Israel)	25	1M NaClO$_4$	0.5665
7.5%	"	"	"	"	0.5003
15%	"	"	"	"	0.4435
30%	"	"	"	"	0.3997
45%	"	"	"	"	0.3613
60%	"	"	"	"	0.336
0%	Eu(III)	Polarography (Kautecky)	25±0.2	0.1M NaClO$_4$	0.206
20%	"	"	"	"	0.183
40%	"	"	"	"	0.283
60%	"	"	"	"	0.371
80%	"	"	"	"	0.298
0%	"	"	"	0.2M NaClO$_4$	0.258
20%	"	"	"	"	0.230
40%	"	"	"	"	0.255
60%	"	"	"	"	0.33
80%	"	"	"	"	0.317
20%	Mn(II)	Polarography (Matsuda)		0.1M NaClO$_4$ 0.005% gelatin	1.4
40%	"	"		"	1.05
60%	"	"		"	1
80%	"	"		"	1.04

VII. LIGAND EXCHANGE RATES AND ELECTRODE REACTIONS

−WATER[a]

k_f^0 (cm/sec)	$k_{E_{1/2}}$ (cm/sec)	k_s	D (cm²/sec)	Comments	Ref.
2.5×10^{-16}			1.36×10^{-5}		1
1.0×10^{-14}			0.95×10^{-5}		1
5.6×10^{-15}			0.80×10^{-5}		1
4.5×10^{-15}			0.65×10^{-5}		1
1.4×10^{-14}			0.48×10^{-5}		1
9.770×10^{-16}			1.737×10^{-5}	D calculated from Ilkovic equation	23
2.468×10^{-14}			9.217×10^{-5}		23
6.981×10^{-13}			7.652×10^{-6}		23
6.488×10^{-12}			6.141×10^{-6}		23
5.351×10^{-11}			4.643×10^{-6}		23
1.751×10^{-10}			2.822×10^{-6}		23
6.166×10^{-5}	1.603×10^{-3}				12
9.772×10^{-5}	1.429×10^{-3}				12
1.66×10^{-5}	9.55×10^{-4}				12
5.623×10^{-6}	6.607×10^{-4}				12
1.259×10^{-5}	8.590×10^{-4}				12
2.344×10^{-5}	1.738×10^{-3}				12
3.388×10^{-5}	7.762×10^{-4}				12
2.818×10^{-5}	6.109×10^{-4}				12
2.818×10^{-6}	4.634×10^{-4}				12
1.622×10^{-5}	2.754×10^{-4}				12
		9.9×10^{-4}		presence of max. suppressor inferred from other studies by these authors	2
		15.4×10^{-4}			2
		8.4×10^{-4}			2
		7.0×10^{-4}			2

FORMAMIDE

V/V	Electrolyte	Technique	Temp. (°C)	Supporting electrolyte	αn_a
0%	Ni(II)	Polarography (Meites-Israel)	25	1M NaClO$_4$	0.6498
7.5%	"	"	"	"	0.6273
15%	"	"	"	"	0.5884
30%	"	"	"	"	0.5895
60%	"	"	"	"	0.6124
0%	"	Polarography (Kautecky)	25±0.01	0.1M NaClO$_4$ Triton X-100	0.737
40%	"	"	"	"	0.6785
60%	"	"	"	"	0.65
80%	"	"	"	"	0.65
0%	Zn(II)	Polarography (Meites-Israel)	"	1M NaClO$_4$	1.009
7.5%	"	"	"	"	1.084
15%	"	"	"	"	1.266
30%	"	"	"	"	1.237
45%	"	"	"	"	1.366
60%	"	"	"	"	1.580

[a] DME used in each case.

VII. LIGAND EXCHANGE RATES AND ELECTRODE REACTIONS

–WATER[a] (Continued)

k_f^0 (cm/sec)	$k_{E_{1/2}}$ (cm/sec)	k_s	D (cm²/sec)	Comments	Ref.
6.748×10^{-15}			9.086×10^{-6}		23
1.898×10^{-14}			6.555×10^{-6}		23
1.249×10^{-14}			5.937×10^{-6}		23
2.885×10^{-13}			5.157×10^{-6}		23
1.340×10^{-13}			3.007×10^{-6}		23
2.5×10^{-13}			2.96×10^{-6}	k_f^0-Rate constant of electrode reaction, at zero V vs NHE uncorrected for liquid junction potential	8
4.5×10^{-12}			1.96×10^{-6}		8
4.0×10^{-11}			1.44×10^{-6}		8
3.2×10^{-10}			0.845×10^{-6}		8
1.920×10^{-15}			1.443×10^{-5}		23
1.656×10^{-15}			1.091×10^{-5}		23
1.522×10^{-15}			7.889×10^{-6}		23
1.308×10^{-15}			5.567×10^{-6}		23
1.086×10^{-15}			3.763×10^{-6}		23
85.09×10^{-14}			1.336×10^{-6}		23

METHANOL–

V/V	Electrolyte	Technique	Temp. (°C)	Supporting electrolyte	α	αn_a
	Zn(II)	Galvanostatic	25.0±0.2	1M LiClO$_4$ 0.02M HClO$_4$	0.3	
0%	Cr(VI)	Polarography (Kautecky)		0.1M NaOH		0.42
10%	Cr(VI)	Polarography (Kautecky)		0.1M NaOH		0.51
20%	Cr(VI)	Polarography (Kautecky)		0.1M NaOH		0.63
30%	Cr(VI)	Polarography (Kautecky)		0.1M NaOH		0.57
40%	Cr(VI)	Polarography (Kautecky)		0.1M NaOH		0.55
50%	Cr(VI)	Polarography (Kautecky)		0.1M NaOH		0.53
60%	Cr(VI)	Polarography (Kautecky)		0.1M NaOH		0.53
70%	Cr(VI)	Polarography (Kautecky)		0.1M NaOH		0.59
25%	Eu(III)	Polarography (Kautecky)	30±0.5	0.05M Cl$^-$ 2mM HClO$_4$		
25%	Eu(III)	Polarography (Kautecky)	30±0.05	0.50M Cl$^-$ 2mM HClO$_4$		
25%	Eu(III)	Polarography (Kautecky)	30±0.5	0.20M Cl$^-$ 2mM HClO$_4$		
25%	Eu(III)	Polarography (Kautecky)	30±0.5	0.05M ClO$_4^-$ 2mM ClO$_4^-$		
25%	Eu(III)	Polarography (Kautecky)	30±0.5	0.20M ClO$_4^-$ 2mM HClO$_4$		
25%	Eu(III)	Polarography (Kautecky)	30±0.5	0.50M ClO$_4^-$ 2mM HClO$_4$		
25%	Eu(III)	Polarography (Kautecky)	30±0.5	0.05M Acetate		
25%	Eu(III)	Polarography (Kautecky)	30±0.5	0.20M Acetate		
25%	Eu(III)	Polarography (Kautecky)	3.±0.5	0.50M Acetate		

VII. LIGAND EXCHANGE RATES AND ELECTRODE REACTIONS

WATER[a]

k_s (cm/sec)	k_f^0 (cm/sec)	D (cm^2/sec)	Comments	Ref.
				24
	1.8×10^{-9}	1.57×10^{-5}	I_a^0 decreases as mole fraction of CH$_3$OH increases	21
	1.9×10^{-11}	1.4×10^{-5}		21
	1.6×10^{-12}	1.1×10^{-5}		21
	1.1×10^{-11}	1.0×10^{-5}		21
	2.1×10^{-11}	9.7×10^{-6}		21
	4.1×10^{-11}	7.7×10^{-6}		21
	4.1×10^{-11}	7.7×10^{-6}		21
	1.1×10^{-11}	7.6×10^{-6}		21
1.9×10^{-4}			Rate constant evaluated at standard potential	25
0.6×10^{-4}				25
0.8×10^{-4}				25
2.9×10^{-4}				25
1.8×10^{-4}				25
1.3×10^{-4}				25
4.5×10^{-4}				25
2.8×10^{-4}				25
2.0×10^{-4}				25

METHANOL–

V/V	Electrolyte	Technique	Temp. (°C)	Supporting electrolyte	α	αn_a
50%	Eu(III)	Polarography (Kautecky)	30±0.5	0.05M Cl⁻ 2mM HClO₄		
50%	Eu(III)	Polarography (Kautecky)	30±0.5	0.20M Cl⁻ 2mM HClO₄		
50%	Eu(III)	Polarography (Kautecky)	30±0.5	0.50M Cl⁻ 2mM HClO₄		
50%	Eu(III)	Polarography (Kantecky)	30±0.5	0.05M ClO₄⁻ 2mM HClO₄		
50%	Eu(III)	Polarography (Kautecky)	30±0.5	0.20M ClO₄⁻ 2mM HClO₄		
50%	Eu(III)	Polarography (Kautecky)	30±0.5	0.50M ClO₄⁻ 2mM HClO₄		
50%	Eu(III)	Polarography (Kautecky)	30±0.5	0.05M Acetate		
50%	Eu(III)	Polarography (Kautecky)	30±0.5	0.20M Acetate		
50%	Eu(III)	Polarography (Kautecky)	30±0.5	0.50M Acetate		
75%	Eu(III)	Polarography (Kautecky)	30±0.5	0.05M Cl⁻ 2mM HClO₄		
75%	Eu(III)	Polarography (Kautecky)	30±0.5	0.20M Cl⁻ 2mM HClO₄		
75%	Eu(III)	Polarography (Kautecky)	30±0.5	0.05M ClO₄⁻ 2mM HClO₄		
75%	Eu(III)	Polarography (Kautecky)	30±0.5	0.20M ClO₄⁻ 2mM HClO₄		
75%	Eu(III)	Polarography (Kautecky)	30±0.5	0.50M ClO₄⁻ 2mM HClO₄		
75%	Eu(III)	Polarography (Kautecky)	30±0.5	0.05M Acetate		
75%	Eu(III)	Polarography (Kautecky)	30±0.5	0.20M Acetate		
75%	Eu(III)	Polarography (Kautecky)	30±0.5	0.50M Acetate		

VII. LIGAND EXCHANGE RATES AND ELECTRODE REACTIONS 717

WATER (*Continued*)

k_s (cm/sec)	k_f^0 (cm/sec)	D (cm²/sec)	Comments	Ref.
1.9×10^{-4}				25
1.1×10^{-4}				25
0.6×10^{-4}				25
2.1×10^{-4}				25
1.5×10^{-4}				25
0.8×10^{-4}				25
3.4×10^{-4}				25
1.0×10^{-4}				25
0.9×10^{-4}				25
3.0×10^{-4}				25
1.7×10^{-4}				25
12.5×10^{-4}				25
3.3×10^{-4}				25
1.6×10^{-4}				25
3.2×10^{-4}				25
1.9×10^{-4}				25
1.3×10^{-4}				25

METHANOL–

V/V	Electrolyte	Technique	Temp. (°C)	Supporting electrolyte	α	αn_a
50%	Mn(II)	Polarography (Matsuda-Ayabe)	29	0.1M KCl 0.004% gelatin	0.76	
40%	Mn(II)	Polarography (Matsuda-Ayabe)	29	0.1M KCl 0.004% Gelatin	0.75	
30%	Mn(II)	Polarography (Matsuda-Ayabe)	29	0.1M KCl 0.004% Gelatin	0.75	
20%	Mn(II)	Polarography (Matsuda-Ayabe)	29	0.1M KCl 0.004% Gelatin		
10%	Mn(II)	Polarography (Matsuda-Ayabe)	29	0.1M KCl 0.004% Gelatin	0.75	
0%	Mn(II)	Polarography (Matsuda-Ayabe)	29	0.1M KCl 0.004% Gelatin	0.745	
0%	Te(VI)	Polarography (Kautecky)		0.1M NaOH		0.49
10%	Te(VI)	Polarography (Kautecky)		0.1M NaOH		0.5
20%	Te(VI)	Polarography (Kautecky)		0.1M NaOH		0.62
30%	Te(VI)	Polarography (Kautecky)		0.1M NaOH		0.83
40%	Te(VI)	Polarography (Kautecky)		0.1M NaOH		0.87
50%	Te(VI)	Polarography (Kautecky)		0.1M NaOH		0.98
10%	V(IV)	Polarography	35±0.1	0.1M NaClO$_4$		
20%	V(IV)	Polarography	35±0.1	0.1M NaClO$_4$		
30%	V(IV)	Polarography	35±0.1	0.1M NaClO$_4$		
40%	V(IV)	Polarography	35±0.1	0.1M NaClO$_4$		
50%	V(IV)	Polarography	35±0.1	0.1M NaClO$_4$		

[a] DME used in each case.

VII. LIGAND EXCHANGE RATES AND ELECTRODE REACTIONS 719

WATER[a] (Continued)

k_s (cm/sec)	$k_f{}^0$ (cm/sec)	D (cm^2/sec)	Comments	Ref.
6.258×10^{-4}				20
9.612×10^{-4}				20
4.118×10^{-4}				20
4.922×10^{-4}				20
5.364×10^{-4}				20
6.457×10^{-4}				20
	5.0×10^{-5}	1.3×10^{-5}		21
	1.2×10^{-14}	1.2×10^{-5}		21
	6.9×10^{-17}	1.1×10^{-5}		21
	6.3×10^{-21}	1.0×10^{-5}		21
	1.25×10^{-21}	7.3×10^{-6}		21
	7.9×10^{-24}	4.3×10^{-6}		21
	4.06×10^{-6}	0.31×10^{-5}		26
	2.52×10^{-6}	0.28×10^{-5}		26
	1.87×10^{-6}	0.26×10^{-5}		26
	0.84×10^{-6}	0.24×10^{-5}		26
	0.01×10^{-6}	0.22×10^{-5}		26

PROPANOL–WATER

V/V	Electrolyte	Technique	Supporting electrolyte	α	αn_a	k_f^0 (cm/sec)	D (cm²/sec)	I_a^0 mA/cm²	Ref.
0%	Cr(VI)	Polarography (Kautecky)	0.1M NaOH		0.42	1.8×10^{-9}	1.57×10^{-5}		21
10%	"	"	"		0.79	1.1×10^{-16}	1.1×10^{-5}		21
20%	"	"	"		0.88	1.1×10^{-18}	9.0×10^{-6}		21
30%	"	"	"		0.8	1.6×10^{-17}	7.5×10^{-6}		21
40%	"	"	"		0.79	3.0×10^{-15}	6.3×10^{-6}		21
50%	"	"	"		0.68	2.7×10^{-14}	4.9×10^{-6}		21
60%	"	"	"		0.82	5.8×10^{-16}	4.0×10^{-6}		21
70%	"	"	"		0.88	1.1×10^{-15}	3.06×10^{-6}		21
0%	Te(VI)	"	"		0.49	5.0×10^{-15}	1.3×10^{-5}		21
10%	"	"	"		0.64	1.6×10^{-17}	1.0×10^{-5}		21
20%	"	"	"		0.87	4×10^{-22}	5.3×10^{-6}		21
30%	"	"	"		1.01	2×10^{-24}	5.2×10^{-6}		21
40%	"	"	"		1.23	8.7×10^{-29}	4.5×10^{-6}		21
50%	"	"	"		1.4	4.5×10^{-31}	3.0×10^{-6}		21
	Zn(II)[a]	Galvanostatic	1M LiClO₄ 0.02M HClO₄	0.25				18	24

[a] Zn(Hg) electrode used. DME used in all other cases.

VII. LIGAND EXCHANGE RATES AND ELECTRODE REACTIONS 721

V/V	Electrolyte	Technique	Temp.	Supporting electrolyte	α	k_s (cm/sec)	Ref.
0%	Mn(II)	Polarography (Matsuda–Ayabe)	29°C	0.004% gelatin	0.745	6.457×10^{-4}	20
10%	"	"	"	"		4.790×10^{-4}	
20%	"	"	"	"	0.81	7.226×10^{-4}	
30%	"	"	"	"	0.81	4.021×10^{-4}	
40%	"	"	"	"	0.83	7.447×10^{-4}	
50%	"	"	"	"	0.79	8.190×10^{-4}	
60%	"	"	"	"	0.82	1.936×10^{-4}	
80%	"	"	"	"	0.73	3.701×10^{-4}	

ACETONITRILE-METHANOL[a]

Mole fraction Acetonitrile	Mole fraction methanol	Electrolyte	Technique	Temp.	Supporting electrolyte	α	k_s (cm/sec)	Ref.
0.8	0.2	Cd(II)	Faradaic impedence	25°C	0.9M NaClO$_4$	0.14	0.080	5
0.5	0.5	Cd(II)	Faradaic impedence	25°C	0.9M NaClO$_4$	0.14	0.028	5

[a] DME used in each case.

DIMETHYL FORMAMIDE-N-METHYL FORMAMIDE[a]

Mole fraction dimethyl formamide	Mole fraction N-methyl formamide	Electrolyte	Technique	Temp.	Supporting electrolyte	α	k_s (cm/sec)	Ref.
0.2	0.8	Cd(II)	Faradaic impedence	25°C	0.9M NaClO$_4$	0.18	0.108	5
0.5	0.5	Cd(II)	Faradaic impedence	25°C	0.9M NaClO$_4$	0.20	0.089	5
0.8	0.2	Cd(II)	Faradaic impedence	25°C	0.9M NaClO$_4$	0.14	0.081	5

[a] DME used in each case.

VII. LIGAND EXCHANGE RATES AND ELECTRODE REACTIONS

References

1. J. N. Gaur and N. K. Goswaine, Electrochim. Acta, **15,** 519 (1970).
2. J. N. Gaur and N. K. Goswaine, Electrochim. Acta, **12,** 1489 (1967).
3. D. Larkin, J. Electrochem. Soc., **119,** 189 (1972).
4. J. N. Gaur and N. K. Goswaine, Proc. Symp. Electrode Processes (1966); ed. by R. C. Kapoon, Univ. Jodbypur, India.
5. T. Biegler, Collect Czech Chem. Comm., **36,** 414 (1971).
6. T. A. Kowofohis and P. J. Lingane, J. Electroanal. Chem., **31,** 1 (1971).
7. L. R. Sherman and V. S. Archer, Anal. Chem., **42,** 1356 (1970).
8. J. N. Gaur, Electrochim. Acta, **11,** 939 (1966).
9. D. R. Cogley and J. N. Butler, J. Phys. Chem., **72,** 4568 (1968).
10. D. R. Cogley and J. N. Butler, J. Electrochem. Soc., **113,** 1074 (1966).
11. A. V. Pamfilov, A. I. Lopvskampkaya and T. S. Zueva, Ukr. Khim. Zhur., **33,** 17 (1967).
12. J. N. Gaur and K. Zutski, J. Electroanal. Chem., **11,** 390 (1966).
13. J. N. Butler, J. Phys. Chem., **73,** 4026 (1969).
14. R. Scan, J. Electrochem. Soc., **117,** 295 (1970).
15. S. G. Meibuki, J. Electrochem. Soc., **118,** 1320 (1971).
16. S. G. Meibuki, J. Electrochem. Soc., **118,** 708 (1971).
17. S. G. Meibuki, J. Electrochem. Soc., **117,** 56 (1970).
18. W. Jaenicke and P. H. Schweitzer, Z. Phys. Chem., **52,** 104 (1967).
19. S. K. Joa, S. N. Srivastova, Bul'. Chem. Soc. Japan, **40,** 810 (1967).
20. J. R. Gupta and G. M. Gupta, Monatsh. Chem., **100,** 2018 (1969).
21. H. Sadek, *et al.*, Electrochim. Acta, **16,** 401 (1971).
22. J. Sandcho and U. Almagno, Anal. De Quimica, **67,** 129 (1971).
23. S. K. Joa and S. N. Srivastova, J. Prakt. Chem., **38,** 295 (1968).
24. M. H. Miles and H. Gerischer, J. Electrochem. Soc., **118,** 837 (1971).
25. V. R. Chandraskaran and A. K. Sundarain, Proc. Anal. Acad. Sci., **74,** 133 (1971).
26. S. Lal, *et al.*, Review of Polarography, **16,** 1 (1969).

VIII. ELECTRICAL DOUBLE LAYER

A general introduction to the field of the electrical double layer may be found in references 1–5 and recent reviews on this subject have been published by Damaskin (6) and Parsons (7). Payne (8) has published the only review dealing with the electrical double layer in nonaqueous solvents, including mixtures with water. In this review he also mentions early work in the field (59, 79) which is not included in the following tables.

In all the work cited below pure, liquid mercury is the electrode material unless stated otherwise in the column of comments.

Abbreviations used:

γ	Interfacial tension as a function of the electrical variable
DC	Differential capacitance as a function of the electrical variable
DL	Electrical double layer
ads	Specific adsorption or specifically adsorbed.

ACETIC ACID

Solute	Technique	Ref.
NH_4OCOCH_3	γ	9

ACETONE

Solute	Technique	Comments	Ref.
$CsNO_3$	DC	Effect of solvents on partial charge transmission coefficient	10
K_2SO_4	DC	Effect of solvents on partial charge transmission coefficient	10
LiCl	DC	DC in alcohols, ketones, amines and amides compared	11
	γ	Cl^- more strongly ads. than from water	12
$LiNO_3$	γ	NO_3^- less strongly ads. than from water	12
NH_4CNS	γ		12
$NaClO_4$	DC	Effect of solvents on partial charge transmission coefficient	10
$NaNO_3$	DC	Effect of solvents on partial charge transmission coefficient	10

ACETONITRILE

Solute	Technique	Comments	Ref.
$(C_4H_9)_4NClO_4$	DC	Study of O_2 ads. on amalgams	13
CsI	DC	Ads. of Cs^+ and I^- detected	14
HBr	γ	Ads. of anions studied	15
HCl	γ	Ads. of anions studied	15
H_3PO_4	γ	Ads. of anions studied	15
H_2SO_4	γ	Ads. of anions studied	15
KPF_6	DC		16
Li^+		Fundamental study of ads.	17
LiBr	DC	Br^- ads.	14
$NaClO_4$	DC, γ	Comparison with aqueous slns.	18
	DC, γ	No ads. found	42
NaI	γ	Ads. of anions studied	15

AMMONIA

Solute	Technique	Comments	Ref.
KI	DC	$-36.5°$ K^+ and I^- ads.	19
KNO_3	DC	$-36.5°$ K^+ ads.	19
NH_4Cl	DC	$-36.5°$ NH_4^+ ads.	19
NH_4NO_3	DC	$-36.5°$ NH_4^+ ads.	19
NH_4NO_3 + several cyclic organics	γ	$0°$, under pressure	20
NH_4NO_3, $+Cl^-$, Br^-, I^-	γ	study of anion ads.	21
NaCl	DC	$-36.5°$	22

1-BUTANOL

Solute	Technique	Comments	Ref.
LiCl	γ	Alcohols and water compared	23
	DC	Alcohols compared	24
	DC	Effect of added H_2O investigated	25

VIII. ELECTRICAL DOUBLE LAYER 729

2-BUTANONE

Solute	Technique	Comments	Ref.
LiCl	DC	DC of LiCl in alcohols, ketones, amines and amides compared	11

N-BUTYLACETAMIDE

Solute	Technique	Comments	Ref.
KPF_6	DC, γ	Comparison of amides	26

BUTYLAMINE

Solute	Technique	Comments	Ref.
LiCl	DC	DC in alcohols, ketones, amines and amides compared	11

N-tert-BUTYLFORMAMIDE

Solute	Technique	Comments	Ref.
KPF_6	DC, γ	DL for amides compared	26

4-BUTYROLACTONE

Solute	Technique	Comments	Ref.
KPF_6	DC		16
	DC	Temperature effects	8

DIMETHYLACETAMIDE

Solute	Technique	Comments	Ref.
KPF_6	DC, γ	DL for amides compared	26

DIMETHYLFORMAMIDE

Solute	Technique	Comments	Ref.
Br^-	DC, γ	Adsorption stronger than from water, formamide and N-methyl–formamide	27, 28 29, 30
$(C_4H_9)_4N\ ClO_4$	γ	0.1M Na–amalgam	29
$(C_4H_9)_4N\ ClO_4 + O_2$	DC	Ads. of O_2 on noble metal amalgams	13
Cl^-	DC, γ	Ads. stronger than from water, formamide, and N-methylformamide	27, 28 29, 30
ClO_4^-	DC, γ		27, 28, 29
CsCl	Vibrating interface	Points of zero charge determined	31
CsI	DC	Cs^+ and I^- ads.	14
HCl + acetone	γ	DL structure related to acetone reduction	32
I^-	DC, γ	Ads. stronger than from water, formamide, and N-methylformamide	27, 28 29, 30
KCl	Vibrating interface	Point of zero charge determined	31
KI	Vibrating interface	Point of zero charge determined	31
KPF_6	DC		16
LiCl	DC		14
	Vibrating interface	Point of zero charge determined	31
$LiClO_4$	γ	DL structure related to acetone reduction	32
$LiClO_4$ + arom. hydrocarbons	γ		33
$LiClO_4$ + arom. amines	γ		68
NH_4NO_3	DC		19
NH_4SCN, + org. surfactants	DC		34
NO_3^-	DC, γ		27, 28, 29
$NaClO_4$	DC		35
SCN^-	DC, γ	Surprisingly low ads.	27, 28, 29, 30

DIMETHYLSULFOXIDE

Solute	Technique	Comments	Ref.
$(C_4H_9)_4NClO_4$	DC	O_2 ads. on noble metal amalgams	13
KBr	DC, γ	Br^- ads. compared with ads. of I^-, Cl^-, ClO_4^-, PF_6^-, NO_3^-	36
KI	DC, γ	I^- ads. compared with ads. of Br^-, Cl^-, ClO_4^-, PF_6^-, NO_3^-	36
KNO_3	DC, γ	NO_3^- ads. compared with ads. of I^-, Br^-, Cl^-, ClO_4^-, PF_6^-	36
KPF_6	DC, γ	PF_6^- ads. compared with ads. of I^-, Br^-, Cl^-, NO_3, ClO_4^-	36
LiCl	γ	ads. of Cl^- studied	37
$LiClO_4$	DC, γ	ClO_4^- ads. compared with ads. of I^-, Br^-, Cl^-, NO_3, PF_6^-	36
NH_4Cl	DC, γ	Cl^- ads. compared with ads. of I^-, Br^-, NO_3^-, PF_6^-	36
$NaNO_3$	DC, γ	NO_3^- ads. compared with ads. of I^-, Br^-, Cl^-, ClO_4^-, NO_3, PF_6^-	36

DIOXANE

Solute	Technique	Comments	Ref.
HCL + Acetone	γ	DL structure related to acetone reduction	32
$LiClO_4$ + Acetone	γ	DL structure related to acetone reduction	32

ETHANOL

Solute	Technique	Comments	Ref.
$Ca(ClO_4)_2$	DC		38
HCl	DC, γ		39
HCl + Acetone	γ	DL structure related to acetone reduction	32
K_2CO_3	DC	solid mercury electrode	40
KF	DC	K^+ ads. found	41
LiCl	γ	Cl^- ads. less than in methanol	12
	DC		41
	DC	alcohols compared	24
$LiClO_4$ + Acetone	γ	DL structure related to acetone reduction	32
$LiNO_3$	DC, γ		43
NH_4NO_3	γ		12
NH_4NO_3 + long chain org. acids	γ	ads. of stearic acid, oleic acid, and elaidic acid studied	44
NH_4NO_3 + cyclic org. compounds	γ	ads. of unsaturated alicyclic and aromatic hydrocarbons studied	45
NH_4NO_3 + unsat. org. acids	γ	comparison of ads. of cistrans isomers	46
NaI	DC, γ	I^- less ads. than in methanol	39, 12
$StBr_2$	DC		38

N-ETHYLACETAMIDE

Solute	Technique	Comments	Ref.
KPF_6	DC, γ	DL for amides compared	26

VIII. ELECTRICAL DOUBLE LAYER

ETHYLENE CARBONATE

Solute	Technique	Ref.
KPF_6	DC	16

ETHYLENE GLYCOL

Solute	Technique	Ref.
$(CH_3)_4NOH$	DC	47
$(C_4H_9)_4NOH$	DC	47
CsCl	DC	47
KCl	DC	47
LiCl	DC	47
NaCl	DC	47
RbCl	DC	47

N-ETHYLFORMAMIDE

Solute	Technique	Comments	Ref.
KPF_6	DC, γ	DL for amides compared	26

FORMAMIDE

Solute	Technique	Comments	Ref.
CsCl	DC	Cs^+ not ads.	14, 48
	DC	temperature dependence of DC measured	49
	DC, γ	Cs^+ ads. detected	50
KCl	DC, γ	temperature dependence of DC measured	49
	DC, γ	K^+ not ads.	50, 51
KF	DC	F^- not ads.	50, 52
	DC, γ	comparison of measured and calculated (from DC) γ	53
KF + thiurea	DC, γ	ads. of thiurea studied	54
KI	DC	I^- ads. studied	50, 51
KI + KF (const. ionic strength)	γ	I^- ads. studied	55
KNO_3	DC, γ	no K^+ ads. detected	51
LiCl	DC, γ		49
NH_4F + diethyl ether	DC	origin of DC hump discussed	54
NaCl	DC, γ		50
$NaNO_3$	DC	temperature dependence of DC determined	19
RbCl	DC		49
$SrCl_2$	DC	Sr^{++} not ads.	14

FORMIC ACID

Solute	Technique	Comments	Ref.
CsH_2PO_4	DC, γ	Cs^+ ads.	56
KH_2PO_4	DC, γ		56
LiH_2PO_4	DC, γ		56
NaHCOO	DC, γ	ads. of di-n-butyl ether studied	56
$NaHSO_4$	DC		56
NaH_2PO_4	DC, γ		56
Na_2SO_4	DC, γ		56

VIII. ELECTRICAL DOUBLE LAYER

1-HEPTANOL

Solute	Technique	Comments	Ref.
LiCl	DC	Alcohols compared	24

1-HEXANOL

Solute	Technique	Ref.
LiCl	DC	24

2-HEXANONE

Solute	Technique	Comments	Ref.
LiCl	DC	Alcohols, ketones, amines amides compared	11

HEXYLAMINE

Solute	Technique	Comments	Ref.
LiCl	DC	Alcohols, ketones, amines amides compared	11

METHANOL

Solute	Technique	Comments	Ref.
$(CH_3)_3NHCl$	γ		57
$(CH_3)_4NI$	γ		57
$(C_2H_5)_4NI$	γ		57
$CaCl_2$	DC		58
$CaClO_4$	DC		38
CsCl	DC		58
$CsNO_3$	DC	Effect of solvents on partial charge transmission coefficient	10
HCl	γ	Cl^- ads. found	60
HCl + Acetone	γ	DL structure related to acetone reduction	32
KCl	DC		61
KCl + bromothymol blue	DC	DL structure related to electrode reaction	62
KF	DC, γ		63, 41, 61
	DC, γ	Comparison of experimental and calculated (from DC) γ	53
	DC	Theory of the DL	64
KF + thiurea	DC	ads. of thiurea studied	65
KI + KF (const. ionic strength)	DC	ads. of I studied	65
KI + bromothymol blue	DC	DL structure related to electrode reaction	62
K_2SO_4	DC	Effect of solvents on partial charge transmission coefficient	10
$LaCl_3$	DC		9
LiCl	DC, γ		41, 61, 24, 66
LiCl + di-n-butyl ether	γ		65
LiCl + dimethyl-formamide	DC		67
$LiClO_4$	γ	DL structure related to acetone reduction	32

VIII. ELECTRICAL DOUBLE LAYER

METHANOL (Continued)

Solute	Technique	Comments	Ref.
$LiClO_4$ + aromatic amines	γ		68
$LiNO_3$	DC, γ		43
$MgCl_2$	DC		41, 58
NH_4CL	DC		63
	DC	comparison of liquid and solid Hg electrode (-35 and $-41°$)	69
NH_4F	DC	NH_4^+ ads. detected	63
NH_4NO_3	DC, γ		63, 57, 12
NH_4NO_3 + unsaturated hydrocarbons	DC, γ		45
NaBr	DC, γ		42, 12
NaCl	DC		61
$NaClO_4$	DC		35
	DC	Effect of solvents on partial charge transmission coefficient	10
NaF + thiurea	DC, γ	Comparison of thiurea ads. from aqueous and methanol solutions	71
NaI	γ		42, 12
$NaNO_3$	DC	Effect of solvents on partial charge transmission coefficient	10
$SrBr_2$	DC		38
$SrCl_2$	DC		41, 58

N-METHYLACETAMIDE

Solute	Technique	Comments	Ref.
KPF_6	DC, γ	DL for amides compared	26

N-METHYLFORMAMIDE

Solute	Technique	Comments	Ref.
CsCl	DC	Cs^+ ads. found	72, 73
KBr	DC	Br^+ ads. found	72
KCl	DC		72
KF	DC		52
	DC, γ	Comparison of measured and calculated (from DC) γ	53
KI	DC	I^- strongly ads.	72
LiCl	DC		72
NaCl	DC		72
NaCl + $(C_4H_9)_4$NBr	DC		72
RbCl	DC		72

N-METHYLPROPIONAMIDE

Solute	Technique	Comments	Ref.
KPF_6	DC, γ	DL for amides compared	26

1-PENTANOL

Solute	Technique	Ref.
LiCl	DC	24

1-PROPANOL

Solute	Technique	Ref.
LiCl	DC	24
NH_4NO_3	γ	23

VIII. ELECTRICAL DOUBLE LAYER

2-PROPANOL

Solute	Technique	Comments	Ref.
HCl	γ	DL structure related to acetone reduction	32
LiClO$_4$	γ	DL structure related to acetone reduction	32

PROPYLAMINE

Solute	Technique	Comments	Ref.
LiCl	DC	DC in alcohols, ketones, amines, amides compared	11

PROPYLENE CARBONATE

Solute	Technique	Ref.
$(C_2H_5)_4NBr$	DC	74
$(C_2H_5)_4NCl$	DC	74
$(C_2H_5)_4NClO_4$	DC	74
$(C_2H_5)_4NI$	DC	74
$(C_4H_9)_4NClO_4$	DC	74
KPF$_6$	DC	16
LiClO$_4$	DC	74
NaClO$_4$	DC	70, 74

PYRIDINE

Solute	Technique	Ref.
LiNO$_3$	DC, γ	75
NH$_4$CNS	γ	12
NaI	γ	12

SULFOLANE

Solute	Technique	Ref.
KPF_6	DC, γ	76
$NaClO_4$	DC	76

SULFURIC ACID

Solute	Technique	Comments	Ref.
Pure	DC	DC higher than for aqueous solutions	77
Rubeanic acid	DC		78

4-VALEROLACTONE

Solute	Technique	Ref.
KPF_6	DC	16

References

1. P. Delahay, *Double Layer and Electrode Kinetics*, Interscience, New York (1965).
2. D. C. Grahame, Chem. Rev., **41**, 441 (1947).
3. R. Parsons, "Equilibrium Properties of Electrified Interfaces," in *Modern Aspects of Electrochemistry*, No. 1, J. O'M. Bockris, Ed., Butterworths, London, pp. 103–179 (1966).
4. D. M. Mohilner, "The Electrical Double Layer, Part I. Elements of Double Layer Theory," in *Electroanalytical Chemistry*, Vol. 1, A. J. Bard, Ed., Marcel Dekker, Inc., New York, pp. 241–409 (1966).
5. A. N. Frumkin and B. B. Damaskin, "Adsorption of Organic Compounds at Electrodes," in *Modern Aspects of Electrochemistry*, No. 3, J. O'M Bockris and B. E. Conway, Ed., Butterworths, London, pp. 149–223 (1964).
6. B. B. Damaskin, Elektrokhimiya, **5**, 771 (1969).
7. R. Parsons, Rev. Pure Appl. Chem., **18**, 91 (1968).

VIII. ELECTRICAL DOUBLE LAYER

8. R. Payne, "The Electrical Double Layer in Nonaqueous Solutions," in *Advances in Electrochemistry and Electrochemical Engineering*, Vol. 7, P. Delahay, Ed., Interscience, New York, pp. 1–76 (1970).
9. G. R. Bachman, J. Am. Chem. Soc., **64**, 2177 (1942).
10. W. Lorenz, Z. Physik. Chem. (Leipzig), **232**, 176 (1966).
11. R. R. Salem, Zh. Fiz. Khim, **43**, 2876 (1969).
12. A. N. Frumkin, Z. Physik. Chem., **103**, 43 (1922).
13. W. Lund and M. E. Peover, J. Electroanal. Chem., **25**, 19 (1970).
14. S. Mink, J. Jastrzebska, and M. Brzostowska, J. Electrochem. Soc., **108**, 1160 (1961).
15. G. A. Korchinskii, Ukr. Khim. Zh., **28**, 693 (1962).
16. R. Payne, J. Phys. Chem., **71**, 1548 (1967).
17. R. W. Murray and C. N. Reilley, "Fundamental Electronic Studies of Adsorption Kinetics and Excited States," 1969 AD-694 575 (Available from CFSTI).
18. P. Champion, C. R. Acad. Sci., Ser. C, **269**, 1159 (1969).
19. R. Payne, Thesis, University of London, 1962; cf. ref. 8.
20. A. M. Murtazaev and I. Igamberdyev, Zh. Fiz. Chim., **14**, 217 (1940).
21. A. M. Murtazaev, Acta Physicochim. URSS, **12**, 225 (1940).
22. S. Minc and A. Muszalska, Roczniki Chem., **38**, 903 (1964).
23. C. Ockrent, J. Phys. Chem., **35**, 3354 (1931).
24. R. R. Salem, Zh. Fiz. Chim., **43**, 1839 (1969).
25. P. A. Kirkov, Zh. Fiz. Chim., **34**, 2375 (1960).
26. R. Payne, J. Phys. Chem., **73**, 3598 (1969).
27. V. D. Bezuglyi and L. A. Korshikov, Elektrokhimiya, **1**, 1422 (1965).
28. V. D. Bezuglyi and L. A. Korshikov, Elektrokhimiya, **3**, 390 (1967).
29. L. A. Demchuk, V. A. Smirnov, and D. P. Semchenko, Tr. Novocherk. Polytekhn. Inst., **141**, 23 (1965).
30. B. B. Damaskin, J. M. Ganzhina, and R. V. Ivanova, Elektrokhimiya, **6**, 1540 (1970).
31. S. Minc and M. Brzostowska, Rocz. Chem., **42**, 369 (1968).
32. V. A. Smirnov, L. A. Demchuk, and L. I. Antropov, Zh. Fiz. Chim., **42**, 1716 (1968).
33. V. D. Bezuglyi and L. A. Korshikov, Elektrokhimiya, **4**, 454 (1968).
34. W. Kemula and B. Behr, Coll. Czech. Chem. Comm., **30**, 4050 (1965).
35. J. Dojlido, R. V. Ivanova, and B. B. Damaskin, Elektrokhimiya, **4**, 567 (1968).
36. R. Payne, J. Am. Chem. Soc., **89**, 489 (1967).
37. San Hyung Kim, T. N. Andersen, and H. Eyring, J. Phys. Chem., **74**, 4555 (1970).
38. V. F. Ivanov, B. B. Damaskin, N. I. Peshkova, A. A. Ivashchenko and V. F. Balashov, Elektrokhimiya, **4**, 851 (1968).
39. G. A. Korchinskii, Zh. Fiz. Chim., **34**, 2759 (1960).
40. A. Gorodetskaya and M. A. Proscurnin, Zh. Fiz. Chim., **12**, 411 (1938).
41. S. Minc and J. Jastrzebska, J. Electrochem. Soc., **107**, 135 (1960).
42. P. Champion, C. R. Acad. Sci., Ser. C **272**, 1090 (1971).
43. V. F. Ivanov, B. B. Damaskin, and L. F. Maiorova, Elektrokhimiya, **6**, 382 (1970).
44. G. A. Korchinskii, Ukr. Khim. Zh., **28**, 473 (1962).
45. M. A. Gerovich and G. F. Rybal'chenko, Zh. Fix. Chim., **32**, 108 (1958).
46. G. A. Korchinskii Ukr. Khim. Zh., **29**, 1031 (1963).
47. D. J. Dzhaparidze, G. Tedoradze and Sh. S. Dzhaparidze, Elektrokhimiya, **5**, 955 (1969).
48. S. Minc and J. Jastrzebska, Roczniki Chem., **36**, 1901 (1962).
49. G. H. Nancollas, D. S. Reid and C. A. Vincent, J. Phys. Chem., **70**, 3300 (1966).
50. B. B. Damaskin, R. V. Ivanova and A. A. Survila, Elektrokhimiya, **1**, 767 (1965).

51. R. Payne, J. Chem. Phys., **42,** 3371 (1965).
52. S. Minc and M. Brzostowska, Roczniki Chem., **38,** 301 (1964).
53. S. Minc and M. Brzostowska, Roczniki Chem., **40,** 1759 (1966).
54. E. Dutkiewicz and R. Parsons, J. Electroanal. Chem., **11,** 196 (1966).
55. E. Dutkiewicz, Roczniki Chem., **41,** 1965 (1967).
56. J. Lawrence and R. Parsons, Trans. Faraday Soc., **64,** 1656 (1968).
57. A. M. Murtazaev and M. Abramov, Zh. Fiz. Chim., **13,** 350 (1939).
58. S. Minc and M. Brzostowska, Roczniki Chem., **34,** 1109 (1960).
59. G. Gouy, Ann. Ch'm. Phys., **9**(8), 75 (1906).
60. R. Parsons and M. A. V. Devanathan, Trans. Faraday Soc., **49,** 673 (1953).
61. S. Minc and J. Jastrzebska, Dokl. Acad. Nauk SSSR, **120,** 114 (1958).
62. J. Gupta, J. Indian Chem. Soc., **41,** 668 (1964).
63. D. C. Grahame, Z. Elektrochem., **59,** 740 (1955); **60,** 101 (1956).
64. J. R. Macdonald and C. A. Barlow, J. Chem. Phys., **36,** 3062 (1962).
65. J. D. Garnish and R. Parsons, Trans. Faraday Soc., **63,** 1754 (1967).
66. S. Minc, J. Jastrzebska and J. Adrezejczak, Electrochim. Acta, **14,** 821 (1969).
67. S. Minc and M. Brzostowska, Roczniki Chem., **36,** 1909 (1962).
68. V. D. Bezuglyi, L. A. Korshikov, and V. B. Titova, Elektrokhimiya, **6,** 1150 (1970).
69. E. S. Sevast'yanov and D. J. Leikis, Elektrokhimiya, **1,** 239 (1965).
70. T. Biegler and R. Parsons, J. Electroanal. Chem., **22,** 4 (1969).
71. F. W. Shapnik, M. Oudemann, K. W. Leu and J. N. Helle, Trans. Faraday Soc., **56,** 415 (1960).
72. B. B. Damaskin and Yu. M. Povarov, Dokl. Akad. Nauk SSSR, **140,** 394 (1961).
73. B. B. Damaskin and R. V. Ivanova, Zh. Fiz. Chim. **38,** 176 (1964).
74. V. A. Kuznitsov, N. G. Vasil'kevich and B. B. Damaskin, Elektrokhimiya, **6,** 1339 (1970).
75. B. B. Damaskin, V. F. Ivanov, V. F. Balashov and V. F. Khomina, Elektrokhimiya **7,** 127 (1971).
76. J. Lawrence and R. Parsons, Trans. Faraday Soc., **64,** 751 (1968).
77. A. A. Vlcek, Chem. Listy, **45,** 377 (1951).
78. E. Dutkiewics, Poznanskie Towarzestwo Przyjacio Nauk Wydzia Matematyczno—Przyrodniczy Prace Komisju Matematyczno—Przyrodniczej, **12,** 57 (1967).
79. H. Wild, Z. Physik. Chem., **103,** 1 (1922).

IX. NONAQUEOUS SPECTROSCOPY AND STRUCTURE OF ELECTROLYTES

The tables in this section illustrate the type of studies and areas receiving attention and is not intended to give a complete overview of all the work that has been published in the past decade.

Some work relative to aqueous electrolytes (structure of the solvent, ionic interactions and hydration) is included in view of its importance with reference to practical and theoretical considerations of electrolyte solutions.

The organization of the material is indicated below.

1. *Raman and Infrared Spectroscopy*
   A. Mainly Directed to the Structure of Water
   B. Ionic Interactions and Ionic Solvation
      a. Aqueous Electrolytes
      b. Nonaqueous Electrolytes
   C. Crystalling Hydrates

2. *Infrared and Raman Spectroscopy: Additional Studies*

3. *NMR Measurements*
   A. Aqueous Electrolytes
   B. Nonaqueous Electrolytes

4. *NMR Additional Studies*

5. *Far-Infrared Studies*

6. *UV and Visible Spectroscopy*

7. *Various Methods*
   A. Mass Spectral Studies
   B. Mössbauer Spectroscopy
   C. Sound Velocity Measurements
   D. ESR Studies
   E. Molar Refraction Studies
   F. Quadrupole Relaxation Studies
   G. X-Ray Studies

8. *Water Activity—Concentrated Salt Solution*

9. *General Articles and Reviews*

1. Raman and Infrared Spectroscopy

A. Mainly directed to the structure of water

System	Investigation	Ref.
Aqueous electrolyte solutions; KCl, 3.6M; KBr, 4.2M; NaCl, 5.2M; NaBr, 7.03M; LiCl, 5.02M; LiBr → 10.8M	All regions of the water spectrum studied	1
Aqueous electrolyte solutions; LiCl, ~14M; LiBr, ~11M; NaCl, ~5M; NaBr, ~7M; KCl, ~4M; KBr, ~5M; NH_4Cl, ~6M; NH_4Br, ~7M; $Ca(NO_3)_2$, ~6M; $LiNO_3$, ~7M	Water band region at <1000 cm^{-1}	2
Aqueous electrolyte solutions; Li, Na, and K perchlorates in D_2O/H_2O mixtures	OH, OD, and HDO stretching modes (2800–3800 and 2000–2900 cm^{-1})	3
Aqueous electrolyte solutions; KBr, 3.5M; KOH, 14.2M; LiCl, 13.7M; LiI, 3M; KF, 10.5M; KI, 6M; CsF, 10M; CsI, 2.6M; and moderately dilute solutions of various K halides and chlorides of Be, Cu, Sn, Zn, Al, La and Pr	HOH bending and OH stretching regions (1600–1700 and 3000–3800 cm^{-1}	4, 5, 6
HDO in water at high pressures (50–4000 bars) and temperatures (30–400°C) using cell with sapphire window	OD stretching of HDO in H_2O (2200–2900 cm^-); rotational structure of vibration band of free water molecules observed at supercritical temp. of 400°C and pr. below 200 bars; support for continuum model of water in the liquid and in dense supercritical state	7

B. Ionic interactions and ionic solvation

(a) Aqueous Electrolytes

System	Investigation	Ref.
Aqueous alkali metal nitrates	NO_3^- bands	8
Aqueous $Ca(NO_3)_2$	NO_3^- bands; concn. range 0.5–7.2M	9, 10
Aqueous $M(NO_3)_2$	NO_3^- bands; concn. range up to saturation solubility	11, 12
Aqueous $Cd(NO_3)_2$	NO_3^- bands; concn. range up to 5M	13
Aqueous $Hg(NO_3)_2$	NO_3^- bands; concn. range up to 5M	14

IX. NONAQUEOUS SPECTROSCOPY OF ELECTROLYTES

1. Raman and Infrared Spectroscopy (*Continued*)

B. Ionic interactions and ionic solvation (*Continued*)

(b) *Nonaqueous Electrolytes*

System	Investigation	Ref.
Alkali metal salts in DMSO, dipropyl sulfoxide and dibutyl sulfoxide	5000–100 cm^{-1} region; in all cases band obtained in 500–100 cm^{-1} region whose frequency was characteristic of cation and solvent but not of anion (vibrations of cations in solvent cage); study of ν_{s-0} indicated solvent dipoles oriented with oxygen atom in direct proximity to metal	15
Na$^+$ in THF	Unperturbed bands at 1071 and 913 cm^{-1}. In presence of Na$^+$ new bands at $\sim$1053 and $\sim$895 cm^{-1} indicating complexation of Na$^+$; band intensities dependent on ratio of ether; salt; used to approx. stability constants for stepwise complexation of Na$^+$ by THF.	16
Li, Na and NH$_4$ salts in pyridine	2000–100 cm^{-1} region; in all cases a low-frequency band was observed whose position was dependent on nature of cation, but with exception of NaI, not on anion; 8 and maxima full at 418, 199 and 180 cm^{-1} for Li, NH$_4$ and Na salts, respectively; evidence indicates all alkali cations are solvated in pyridine solutions	17
AgNO$_3$ in CH$_3$CN	Raman spectra; concn. range 0.1–9 mol/l; two frequencies 1036 and 1041 cm^{-1} observed for NO$_3^-$ symmetrical stretching; concn. dependence of relative intensities of those two frequencies is examined from viewpoint of ion–ion interactions; band contours of C$\equiv$N and C—C stretching frequencies of CH$_3$CN examined	18

1. Raman and Infrared Spectroscopy (*Continued*)

B. Ionic interactions and ionic solvation (*Continued*)

(b) *Nonaqueous Electrolytes*

System	Investigation	Ref.
Li^+ and Na^+ in DMSO, and 1-methyl-2-pyrrolidons	Far infrared and NMR; solvation numbers of Na^+ in DMSO and 1M2PY are 6 and 4 respect.; anions do not enter into inner solvation shell of alkali cations	19
Li^+ and Na^+ salts and KCNS in acetone and acetone-nitromethane solutions	Infrared and proton NMR; cation-acetone vibrational bands identified at $\sim$425 cm^{-1} for Li^+, $\sim$195 cm^{-1} for Na^+ and $\sim$148 cm^{-1} for K^+; dependence of these frequencies on isotopic substitution of the solvent (d_6-acetone) and of the solute (^6Li) indicate observed vibrations clearly involve both the cation and the solvent; observed frequencies showed anion dependence in the case of LiCl and LiBr indicating the respective halide ions participate in the observed vibration; solvation number of Li^+ in acetone is 4.	20
$NaCO(CO)_4$ in THF	Infrared and photoelectric laser Raman in CO stretching region; temp. variation of complex infrared band envelope reveals presence of two kinds of anion environment and permits spectra to be divided into four band components; study made of the variation of these band components with concn. [0.03–0.0007M] and temp. [29 to $-42°$]; anion sites identified as solvent-separated ion pairs and contact ion pairs; CO stretching frequency assignments are $F_2 = 1886$, $A_1 = 2005$ for the solvent separated ion pair and $A_1 = 1855, 2005$, $E = 1898$ cm^{-1} for the contact ion pair; equilibrium constant for conversion of contact pairs to solvent separated pairs at 29° was $K = 0.45$ with $\Delta H = 3.7$ Kcal and $\Delta S = 14$ eu	21

IX. NONAQUEOUS SPECTROSCOPY OF ELECTROLYTES

1. Raman and Infrared Spectroscopy (*Continued*)
B. Ionic interactions and ionic solvation (*Continued*)
(b) *Nonaqueous Electrolytes*

System	Investigation	Ref.
Molten $AgNO_3$, $AgNO_3$—CH_3CN mixtures; molten $LiClO_3$, and $LiClO_3$—H_2O mixtures molten $Ca(NO_3)_2 \cdot 4H_2O$ and its mixtures with KNO_3	Laser Raman studies; component in excess is a molten salt and second component is a molecular solvent; results examined from the viewpoint of the structural changes in the transition from molten salt to an ultraconcentrated solution	22

C. Crystalline Hydrates

System	Investigation	Ref.
Salt hydrates of $Zn(NO_3)_2$, KSCN KSCN and mixtures	Water bands; ambiguity arises in O—H stretching region	23
Crystalline $CaSO_4 \cdot 2H_2O$	Water bands; similar ambiguity as above	24

2. Infrared and Raman Spectroscopy: Additional Studies

System	Investigation	Ref.
Carbon deuterated glycine	Raman	25
Thiocyanate and isothiocyanates	Solvent effects	26
Metal nitrates in organic solvents	Infrared	27
Metal halides with esters of amino-acids	Infrared study of complex formation	28
Diglycine $BaCl_2 \cdot H_2O$	Raman	29
Phosphorus oxide halides and halides of Group IIIA	Structure of addition compounds	30
$HgCl_2$, $HgBr_2$, $ZnCl_2$ and $ZnBr_2$ in acetone	Raman	31
$AlBr_3$ in CH_3Br and C_2H_5Br	Raman; addition compounds	32
Cr(III) complexes with hexa-ammine and pentos-ammine	Infrared	33
H-bond strengths	Solvent effects	34

2. Infrared and Raman Spectroscopy: Additional Studies *(Continued)*

System	Investigation	Ref.
Ion-pairing effects	Spectroscopic and configurational assignments	35
Metal diasine chlorocomplexes	Low frequency infrared	36
Aqueous $HClO_6$	Degree of dissociation by Raman	37
Trans-tetrahydroxidodimethyl-stannate(IV) ion	Raman and EMF studies; four acid dissociation consts. of the aquodimethyltin(IV) ion	38
$HClO_6$	Dissociation measured by Raman and NMR	39
Solvent effects	O—H and S—H stretching vibrations	40
Ionic solutions in CH_3CN in the solid phase	Effect of temp. on infrared absorption bands	41
Sodium tetracarbonylcobaltate(I) in various solvents	Nature of solutions	42
Zinc chloride in acetonitrile	Raman and infrared	43
HBr in weakly polar solvents	Rotation in solution	44
Ionic solutions of acetonitrile	Temp. effect; solid phase	45
Ionic solutions of acrylonitrile	Temp. effect	46
Aqueous ions	Studies of hydration	47
Lithium, sodium and potassium alcoholates	Structures	48
Alkali ions	Infrared bands from alkali motion in solution	49
Copper and zinc acetates and their hydrates and ammines	Infrared study	50
Compounds of Ni(II) with nitrites	Infrared study	51
HF solutions of xenon oxide tetrafluoride	Raman spectrum and hydrolysis studies	52
$BaCl_2, 2H_2O$ and $BaCl_2, 2D_2O$	Ir and Raman spectra	53
Concentrated solutions	Quantitative study of donor–acceptor complexes by Raman spectroscopy	54
Solutions in dimethoxyethane and T.H.F.	Studies of solvation phenomena of ions and ion pairs	55
H^+ in DMSO	Structure and ir spectrum	56
$AlCl_4^-$ ion	Fundamental vibrations	57
Ions in solution	Interaction of ions with neighboring water molecules in an organic medium	58

IX. NONAQUEOUS SPECTROSCOPY OF ELECTROLYTES

2. Infrared and Raman Spectroscopy: Additional Studies (*Continued*)

System	Investigation	Ref.
Formic and acetic acids	Low frequency Raman spectra and molecular association	59
Orthophosphoric acid in nonaqueous solvents	Vibrational spectra	60
Compounds of methylforamide and dimethylformamide with some strong acids	Vibrational spectra	61
$AgNO_3$, MeCN and $AgNO_3$, 2MeCN	ir spectra of complexes in MeCN solutions	62
Tetrahalide complexes	Raman study	63
Complex anions of Au, Pt, Pd	Comparison of Raman spectra in crystalline phase and in solution	64
Cation-acetonitrile complexes	Raman studies; normal coordinate analysis	65
Alcohol-water mixtures	ir and near ir	66
Liquid and solid $AsCl_3$ and $AsBr_3$ and their solutions in tri-*n*-butyl phosphate As III	Vibrational spectra	67
CdI_3^-, $CdBr_2$, $CdBr_3^-$, $CdCl_2$, $CdCl_3^-$, $CdCl_4^{2-}$, and $MgCl_4^{2-}$	Vibrational spectra	68
Zinc chloride in some organic solvents	Spectroscopic study	69
Tri-*n*-propylamine and some carboxylic acids	Acid-base interactions in aprotic solvents	70
Water and aqueous solutions of acetonitrile	Ir spectrum	71
Electrolytes in nonaqueous solvents	Interionic vibrational spectra	72
CO(II) halide complexes with aniline, and *o*-, *m*-, and *p*-chloro-aniline	Thermochemical and spectroscopic properties	73
Alkali metal ions in dialkyl sulphoxide solutions	Solvation studies by ir	74
$AsCl_3$ in water and alcohols	Raman spectra	75
Re(III) chloride in pyridine	Spectrophotometric investigation	76
NO_3^- in chloroform and water	Solvation studies by ir	77
$AlCl_3 \cdot THF$, $AlCl_3 \cdot 2THF$, $AlBr_3 \cdot THF$, $AlBr_3 \cdot 2THF$ and $BF_3 \cdot 2THF$	Vibrational spectra between 1500 and 250 cm^{-1}	78
HF solutions	Ir study of hydrogen bonding	79

3. NMR Measurements
A. Aqueous electrolytes

System	Investigation	Ref.
Electrolyte solutions (review)	Chemical shift of NMR–lines of proton in water in electrolytic solutions in connection with hydration of ions; chemical shift of O^{17}-resonance; measurements of degree of dissociation of acids	80
Alkali halides	Investigation of times of reorientation of water molecules in the hydration shell; activation energies for reorientation process	81
Substances with alkyl groups in water and nonaqueous solvents	Molecular orientation times; comparison between pure water and electrolyte solutions; concepts of structure making and structure breaking	82, 83
Diamagnetic electrolyte solutions	Intermolecular nuclear magnetic relaxation rate of proton in water molecules; correlation times for molecular rotation in free water and hydrated water; self diffusion coefficients of water molecules	84, 85
$(CH_3)_4NCl$, $(C_2H_5)_4NCl$, $(C_3H_7)_4NCl$, and $(C_4H_9)NCl$	Self-diffusion coefficients compared with self-diffusion coefficient of H_2O in same solution	86
NaF, KF, RbF, CsF	F^{19} nuclear magnetic relaxation rates as a function of concentration; models for hydration spheres; self-diffusion of F^- in aqueous solution	87
Alkali halides and nitrates	Chemical shifts of the nuclear resonances of ^{23}Na, ^{39}K, ^{87}Rb and ^{133}CS from concns. of 0.2M up to saturation; shielding effects discussed.	88

IX. NONAQUEOUS SPECTROSCOPY OF ELECTROLYTES

3. NMR Measurements (Continued)

B. Nonaqueous electrolytes

System	Investigation	Ref.
Co(acetylacetonate)$_3$ and Cr(acetylacetonate)$_3$ in CHCl$_3$—CCl$_4$, CHCl$_3$—C$_6$H$_6$, CHCl$_3$—dioxan, CHCl$_3$-acetone and CHCl$_3$-methanol	Summary of equisolvation points; preferential solvation	89
Li$^+$ in aqueous, nonaqueous and mixed solvents	Solvation studies by ^{7}Li NMR	90
Li$^+$ in nitrilotriacetic acid (NTA)	Study of complex formation, Li(NTA)$_2^{5-}$ suggested in equilibrium with LiNTA^{2-}, Li$^+$(aq) and NTA^{3-}	91
Na$^+$ in cyclohexane	Solvation studies	92
NaI in 14 different O or N donor organic solvents	^{23}Na chemical shifts; observed range related to changes in paramagnetic term of general nuclear screening equation	93
LiCl and LiBr in mixed aqueous–methanol solutions	Chemical shifts and transverse nuclear relaxation times; correlation with viscosity data	94
Co(III) (acetylacetonate) and Cr(III) (acetylacetonate) in several solvent mixtures	Preferential solvation studies; thermodynamic model devised to interpret the results	95
Onium salts in a variety of solvents	Chemical shifts terms of ion size, ion shape, solvent polarity and nature of counterion	96
Sodium tetraphenylborate, NaClO$_4$, NaI and NaCNS in a number of nonaqueous solvents	Chemical shifts for ^{23}Na studied as a function of concentration; results discussed in terms of contact ion pairs and electron donor abilities of solvents	97
NaI and sodium tetraphenylborate in liquid NH$_3$, ethylenediamine, ethylamine, isopropylamine, tert-butyl amine, hydrazine and 1,1,3,3-tetramethylguanidine	Chemical shifts for ^{23}Na discussed in terms of solvating ability of solvent	98

4. NMR Additional Studies

System	Investigation	Ref.
Ni^{2+} in methanol and methanol–water	Solvation studies	99
Co^{2+} in methanol and methanol–water	Solvation studies	100
Weak acids in liquid NH_3	NMR study	101
Lewis acid adducts with N,N-dimethylformamide	Relative strength of Lewis acids	102
Ga(III) ions in aqueous solution	Coordination number	103
Aluminium perchlorate in dioxane and dioxane–water	Solvation studies	104
Co(II) and Ni(II) in N,N-dimethylformamide	Solvation studies and direct determination of solvation number	105
Boron trihalides in acetonitrile	Relative acceptor power	106
NaCl and $NaClO_4$ in H_2O	Relaxation studies	107
Binary solvent mixtures; 10:1 aqueous mixtures of acetone, N,N-dimethylformamide, DMSO, dioxane, ethanol and N-methylformamide	NMR studies of electrolytes	108
Tetrabutylammonium $tris$(acetylacetonato)cobalt(II) and Ni(II)	Ion pairing and magnetic ansiotropy	109, 111
Hexafluoroasenate ion in aqueous solutions	^{75}As nuclear magnetic spin-lattice relaxation times; temp. dependence	110, 120
Aluminium perchlorate in water and acetonitrile	Solvation study	112
Alkyl lithiums in ethers	Solvation of ion pairs	113
Ions in pure and mixed solvents	NMR studies	114
Co(II) and Ni(II) in acetonitrile solutions	^1H NMR studies	115
Cis- and $trans$-bisethylenediamine cobalt(III) isomers in N,N-dimethylformamide, N,N-dimethylacetamide, and DMSO	Octahedral complexes in dipolar aprotic solvents	116
BeF_4^{2-} and HF_2^-	Dissociation studies	117
General	Preferential solvation by NMR	118
$Mg(ClO_4)_2$ in water–acetone mixtures	Coordination number study	119
Benzene solutions of polar molecules	Solvent effects on chemical shifts	121

IX. NONAQUEOUS SPECTROSCOPY OF ELECTROLYTES

4. NMR Additional Studies (Continued)

System	Investigation	Ref.
Halide ion–trihalogenomethane	NMR and ir studies of association	122
Aromatic amines in nitroalkanes	Solvent effects	123
Alkali metal ions	Solvation studies	124
Aluminium chloride in acetonitrile	Study of complexes	125
Aluminium(III) halides in N,N-dimethylformamide	Outer-sphere ion pair formation	126
Palladium(II) in nitrolotriacetic acid, N-methylaminodiacetic acid, and iminodiacetic acid	NMR study of complexes	127
Diammineplatinum(II) perchlorate in acetonitrile and propionitrile	NMR study	128
Thorium oxalates and formates in water	Study in hydration	129
Cations in liquid NH_3	Cationic solvation	130
Solutions in methanol	Cation and anion shifts in the hydroxy proton resonance	131
LiCl and LiBr in methanol–water mixtures	NMR study	132
$Mg(ClO_4)_2$ in acetone–water mixtures	NMR study of solvation sphere of Mg^{2+}	133
Diamagnetic cations in aqueous mixtures of organo-phosphorus solvents	NMR study of solvation shells	134
Al(III) in n-propanol	PMR study of solvation	135
Alkylated bases in DMSO	NMR study of hydrogen bonding	136
Polar solutes	NMR proton shifts	137
Co(II) halides in nonaqueous solvents	Solvation study	138
Aluminium complexes in aqueous solution	NMR and Raman studies	139
Sodium ions	Studies of solvation shell changes from NMR linewidths	140
Al(III) in ethanol	Solvation studies	141
Alkylammonium ions	Chemical shifts and interactions with solvent in aqueous acid solutions	142
Solutions of some 1:1 electrolytes	Temp. dependence of proton relaxation rate	143
Fe(II) and Ni(II) in water	^{17}O NMR studies of hydration	144

5. Far-Infrared Studies

System	Investigation	Ref.
Li^+, Na^+, K^+, Cs^+ salts and acid forms of polyelectrolytic ethylene methacrylic acid copolymers	800–33 cm^{-1} at ambient and low temps.; models are proposed and analyzed for the vibrational modes involving ion motion	145
Quaternary ammonium halides in benzene	Ion-aggregate modes	146
$NaAlBu_4$ in cyclohexane and THF	Laser Raman and far ir; problem of quantized ion motion in solution	147
Zinc and cadmium halides with dialkylamine	Study of complexes	148
Mercuric halides and dioxane	Study of complexes	149
Water and some aqueous solutions	Spectral study	150
Hydrogen bonding	Far ir and Raman studies	151
$MCl_4{}^{x-}$ in benzene	Spectral studies of tetrahedral and octahedral complexes	152

6. UV and Visible Spectroscopy

System	Investigation	Ref.
Polyvalent anions	Change-transfer to solvent spectra	153
Bu_4NBr, ϕ_4AsCl and $NaB\phi_4$ in water	Near ir spectra of water	154
Br^- and Cl^- in various solvents	Solvation spectra	155
Ti(IV) and Zr(IV) halides with alkyl cyanides	UV spectra	156
$PbCl_4$ in nonaqueous solvents	UV spectra	157
$NiCl_2$ in different organic solvents	Absorption spectra of Ni^{2+}	158
$FeCl_3$ in nonaqueous solvents	UV spectra	159
SO_2 in water–methanol mixtures	Spectrophotometric study	160
o- and m-nitroaniline in acetone	Determination of pK_a	161
Nonpolar solvents	Charge transfer spectra	162
Trinitrotriaquochromium(III)	Spectrometric study of formation	163
Copper acetate and copper propionate in solution	Spectral study	164

IX. NONAQUEOUS SPECTROSCOPY OF ELECTROLYTES

6. UV and Visible Spectroscopy (Continued)

System	Investigation	Ref.
$CoBr_2$ in nitriles	Electronic absorption spectra	165
Tetra-n-heptylammonium iodide with hydroxy compounds	Solvation spectra of iodide ion; role of hydrogen bonding	166
SnCl with ethylbenzene, m-xylene and o-xylene	Spectroscopic investigation of complex formation	167
Manganose(II) chromate with amines	Molar conductance, ir, magnetic susceptibility, UV and visible spectra of complexes	168
Charge transfer complexes	Polarographic acid spectral studies	169
Aquomanganese(III) in perchlorate	Spectral study	170
Uranyl ion in perchlorate media	Absorption spectra; excited state UO_2^{2+}	171
SO_2 in nonpolar solvents	UV study of autodissociation	172
Ti(IV) with oxalic, malonic and chromotropic acids	Spectral study	173
Cu(II) and Ag(II) in perchloric, nitric, sulphuric and phosphoric acid media	Spectral study	174
I_2 and iodine halide solutions in SO_3	Spectral study	175
Sodium tetraphenylborate in polar organic solvents	Contact ion pair and solvent separated ion pairs	176
I^- in nonaqueous solvents	Effect of pressure on UV absorption spectra	177
Cu^{2+} and Cl^- in propylene carbonate	Complexation studies	178

7. Various Methods
A. Mass spectral studies

System	Investigation	Ref.
Halide ions in water and acetonitrile	Gas phase solvation	179
H^+ in water and methanol	Competitive solvation	180
General	Heats of hydration and solvation	181
General	Mass spectrometric measurements of diffusion coefficients	182

7. Various Methods (*Continued*)
B. Mössbauer spectroscopy

System	Investigation	Ref.
Dibutyltin dichloride	Effect of dipolar aprotic solvents on quadrupole splitting	183
Fe(II) salt solutions in mixtures of methanol and formamide	Hydration and solvation	184, 185

C. Sound velocity measurements

System	Investigation	Ref.
Bu_4NBr in acetone and in acetone-p-nitroaniline mixtures	Ionic association	186
Alkali halides in DMSO	Ion pair formation by ultrasonic absorption studies	187

D. ESR studies

System	Investigation	Ref.
Cu^{2+} in aqueous solution	Structural study	188
Mn(II) and Cu(II) with pentamethylenetetrazole	Complex formation	189
Nitro-derivative anions in mixed solvents	Spectral study	190
Mn(II) in acetonitrile	Complex studies	191
2,6-Dimethyl-p-benzosemiquinone and its sodium salt	Solvation spectra	192
Anion radicals in liquid NH_3	Spectral study	193
Alkali metals with m-dinitrobenzene	Ion association; new observations on complexes	194
General	E.S.R. review	195
Cu(II) in ammonia solution	Spectral study	196
Vanadyl acetylacetonate	Correlation between E.S.R. spectral parameters and optional spectra	197

IX. NONAQUEOUS SPECTROSCOPY OF ELECTROLYTES

7. Various Methods (*Continued*)

E. Molar refraction studies

System	Investigation	Ref.
Electrolytes in mixed solvents	Ion-solvent interactions	198

F. Quadrupole relaxation studies

System	Investigation	Ref.
Electrolyte solutions	Quadrupole relaxation	199

G. X-ray studies

System	Investigation	Ref.
Cu(II) chloride in methanol	Solubility; characteristic X-ray spectra of $CuCl_2, CH_3OH$ and $CuCl_2, 2CH_3OH$	200
Tin(IV) chloride glutaronitrile	Crystal and molecular structure	201

8. Concentrated Aqueous Electrolytes and Water Activity from Isopiestic Vapor Pressure Data

Concentration range fitted for water activity on the application of the modified BET physical adsorption isotherms,[202] solid phases in equilibrium with saturated solutions and transition temperatures

Solute	BET Conc. range fitted	Solid phases and transition temperatures (°C)[203]							
LiCl	8–20M	−80° .5H$_2$O	−68° .3H$_2$O	−20° .2H$_2$O	18.5° H$_2$O				
LiBr	8–20M		−45° .3H$_2$O	4° .2H$_2$O	44° H$_2$O				
HCl	7–16M		−86° 3H$_2$O	−27.5° 2H$_2$O	−23.5° H$_2$O				
HClO$_4$	9–16M	−50.5° 3.5H$_2$O	−40.5° 3H$_2$Oα	−47.8° 3H$_2$Oβ	−37.5° 2.5H$_2$O	−44° 2H$_2$O	−12.5° H$_2$O		

IX. NONAQUEOUS SPECTROSCOPY OF ELECTROLYTES

Ca(NO$_3$)$_2$	9–20M	–28.7° .4H$_2$O	42.7° .3H$_2$O	49° .2H$_2$O			
ZnCl$_2$	10–22M	–62° 4H$_2$O	–30° 3H$_2$O	6.5° 2.5H$_2$O	11.5° 1.5H$_2$O		
ZnBr$_2$	11–20M		–15° 3H$_2$O	–8° 2H$_2$O			
CaCl$_2$	4–10.5M	–55° 6H$_2$O	29.8° 4H$_2$Oα	20° 4H$_2$Oβ	38.4° 2H$_2$O	175.5° H$_2$O	
NaOH	14–29M	–28° 7H$_2$O	–24° 5H$_2$O	–17.7° 4H$_2$O	5.2° 3.5H$_2$O	5° 2H$_2$O	12.3° H$_2$O

9. General Articles and Reviews

Investigation	Ref.
Hydrogen bonding and solubility in nonaqueous systems	204
Interaction of complexes formed by hydrogen bonding with a solvent	205
Lewis acidity of polar organic solvents from thermodynamic measurements	206
Ionic character and molecular properties	207
Molecular motion in infrared and Raman spectra	208
Ionic association in solutions of thionine; fluorescence and solvent effects	209
Solvation of coablt(II) and nickel(II) ions in acetone–water and ethanol–water solutions	210
The association of water with bases and anions in an inert solvent	211
Existence of complex cations of definite structure in electrolyte solutions	212
Ion solvation and solvent basicity	213
Cryoscopic, spectroscopic, and conductimetric study of phosphorus pentachloride in selected nonaqueous solvent systems	214
Molar volumes of ions [structure of water in vicinity of ions]	215
Method for predicting the effect of solvation on hydrogen-bonding association equilibria	216
Solutions of lithium bromide in diethyl ether: auto-association and solvation	217
Spectroscopy of metal-ammonia solutions	218
Effect of anion solvation on acid dissociation constants in methanol, water, dimethylformamide, and dimethyl sulphoxide	219
Short-range interaction of ions from thermodynamic and electrokinetic data	220
Thermochemical study of the solvation of alkali– and alkali–earth–metal nitrates in aqueous alcohol solutions	221
Variation of solvation of ions with temperature in aqueous solutions from conductance data	222
Nature of solutions of the alkali and alkaline–earth iodides	223
Selective solvation of ions in solvents	224
Fluorosulphuric acid solvent system. The solutes water and potassium nitrate	225
Solvation of anilinium salts	226
Specific ion-solvent interactions in the mixed solvent nitromethane–water	227

IX. NONAQUEOUS SPECTROSCOPY OF ELECTROLYTES

9. General Articles and Reviews (*Continued*)

Investigation	Ref.
Hydrogen bonds in crystalline hydrates	228
Selective solvation of ions in solvents	229
Equilibria in solution. Part 1. Ion solvation and mixed solvent interaction. Part 2. Evaluation of pK and solvation numbers	230
Molecular complexity of water in organic solvents	231
Bonding in acetonitrile adducts	232
Specificity of ionic hydration and the evaluation of individual ionic properties	233
Solvent-electrolyte interaction. Some salts in water-glycine mixtures	234
Strongly basic systems. The H_2 acidity scale	235
Solvent levelling effect upon Lewis acidity	236
Solubility of silver halides and stability of silver halide complexes in selected nonaqueous media	237
Acid-base reactions in nitromethane and nitromethane acetic acid mixtures	238
Solubility products and instability constants [of silver, caesium, and potassium salts] in water, methanol, formamide, dimethylformamide, dimethylacetamide, dimethyl sulphoxide, acetonitrile, and hexamethylphosphorotriamide	239
Changes in the standard chemical potential of anions on transfer from protic to dipolar aprotic solvents	240
Influence of the structure of solvent on the solvation of ion pairs	241
Contact and solvent-separated ion pairs of carbanions. Specific solvation of alkali ions by polyglycol dimethyl ethers	242
Primary solvation number of magnesium(II) in liquid ammonia	243
Ionic reactions in acetonitrile	244
Liquid ammonia solutions. Sulphur and tetrasulphur tetranitride (S_4N_4)	245
Acetic acid solvates of copper(II) chloride	246
Ionization of sulphur chlorides in acetonitrile	247
Liquid ammonia solutions. Solutions of hydrogen sulphide	248
Acid-base reactions in liquid ammonia	249
Acidities of weak acids in dimethyl sulphoxide solutions. The H-acidity scales	250
Acidity in nonaqueous solvents. Acidity scales in dimethyl sulphoxide solution	251

9. General Articles and Reviews (*Continued*)

Investigation	Ref.
Acid properties of solutions of lithium chloride in acetic acid	252
Donor strengths in 1, 2-dichloroethane	253
Hydration of ions in acetonitrile	254
Ion pairs. Chlorocuprate equilibria in acetic acid solutions	255
Ion-solvent interaction: importance of the dipole moment and basicity of the ligand	256
Universal pH scale for solutions at different temperatures and in different solvents	257
Universal pH scale in methanol and methanol-water mixtures	258
Comparison of interactions of ions with acetonitrile and water	259
Ion pairs: isomerization of ion pairs possessing a solvated agent on their periphery into the agent separated pairs	260
Solvent effects on acidity	261
Rotational motion of solvent molecules and solvation of ions in nonaqueous electrolyte solutions	262
Stability of the mixed compounds of nickel with glycine, ethylenediamine and oxalate in solution	263
Correlation between the acidity scales of various solvents	264
Ion-pair dissociation of solvated bases and base perchlorates in anhydrous acetic acid	265
Hydrogen bonds and other ion-molecule interactions	266
Electrochemical study of the acid-base properties of a solution in molten acetamide	267
Electronic structure and donor properties of cyanamides	268
Solvation coefficients of ions [such as H, Ag etc.] in aqueous–organic mixtures: evaluation by means of the ferrocene-ferricinium couple	269
Interaction of lanthanide and nitrate ions in solution. Relative interaction strength with water and nitrate in organic solvents	270
Contact and solvent-separated ion pairs of carbanions. Role of solvent structure in alkali ion solvation	271
Co-ordination number of ions in solution	272

IX. NONAQUEOUS SPECTROSCOPY OF ELECTROLYTES

9. General Articles and Reviews (Continued)

Investigation	Ref.
Thermodynamics of ion association. Copper complexes of diglycine and triglycine	273
State of iron(III) chloride in nonaqueous solvents	274
Water retention by anhydrous zinc oxalate	275
Solvation of cations of hydrogen and the alkali metals	276
Structure of liquid formic acid and its aqueous solutions	277
Formation of some solvates of copper(II) chloride in the presence of dioxan	278
Intramolecular solvation	279
Proton-transfer complexes. Preferential solvation of p-nitrophenol-amine complexes in nonaqueous solvent mixtures	280
Solubility of complex formation equilibria of silver chloride in anhydrous dimethylformamide	281
Chemical and electrochemical properties in solvents	282
Chemical and electrochemical properties in solvents. Acid-base equilibria in dimethyl sulphoxide	283
Solvation of halide and silver ions in six aprotic polar solvents	284
Ionization of strong electrolytes. Molecular states of nitric acid and perchloric acid	285
Nature of carbon-lithium bonding in benzyl-lithium and its variation with solvent	286
Significance of the solubility of hydrogen halides	287
Stability of complexes with different ligands in nonaqueous solutions	288
Ionic equilibria in mixed solvents. Solvent effect on dissociation constants of acids	289
Structural chemistry of some solid solvates of the nonaqueous ionizing solvent seleninyldichloride	290
Formation constants of iodine–iodide complexes in water-acetonitrile and water–ethanol mixtures: solvation coefficients of I_3^- anion	291
Spectroscopic studies of alkali-metal salt solvation in dialkylsulphoxides	292
Thallium(I) halide complexes in dimethyl sulphoxide	293
Antimony halides as solvents. Polycyclic aromatic hydrocarbons as Lewis bases in antimony trichloride solution	294

9. General Articles and Reviews (*Continued*)

Investigation	Ref.
A relative acidity scale for mixed solvents	295
Chemical and electrochemical properties in nonaqueous solvents, 1, 2- and 1, 1- dichloroethane	296
Solvates of Mg, Ca, Sr, and Ba perchlorate with methanol	297
Cyclic carbonates as ionizing solvents	298
A review of electrochemistry in nonaqueous solvents	299
Solvation of ions in pure and mixed solvents	300
A laser temperature-jump apparatus and fast conductometric detection for relaxation measurements	301
Dimer/monomer equilibrium of Copper(II) acetate in mixtures of acetic acid, ethanol and water from optical and magnetic measurements	302
Alkali metal solution effects of two cyclic polyethers on solubility and spectra	303
Relaxation of charge transport and dielectric properties in electrolyte solutions	304

References

1. G. E. Walrafen, J. Chem. Phys., **36**, 1035 (1962).
2. G. E. Walrafen, J. Chem. Phys., **44**, 1546 (1966).
3. G. E. Walrafen, J. Chem. Phys., **52**, 4176 (1970).
4. W. R. Busing and D. F. Hornig, J. Phys. Chem., **65**, 284 (1961).
5. J. W. Schultz and D. F. Hornig, J. Phys. Chem., **65**, 2131 (1961).
6. T. T. Wall and D. F. Hornig, J. Chem. Phys., **47**, 784 (1967).
7. E. U. Franck and K. Roth, Discussions Faraday Soc., **43**, 108 (1967).
8. D. E. Irish and A. R. Davis, Canad. J. Chem., **46**, 943 (1968).
9. D. E. Irish and G. E. Walrafen, J. Chem. Phys., **46**, 378 (1967).
10. R. E. Hester and R. A. Plane, J. Chem. Phys., **40**, 41 (1964).
11. R. E. Hester and R. A. Plane, J. Chem. Phys., **45**, 4588 (1966).
12. R. E. Hester and C. W. J. Scaife, J. Chem. Phys., **47**, 5253 (1967).
13. A. R. Davis and R. A. Plane, Inorg. Chem., **7**, 2565 (1968).
14. A. R. Davis and D. E. Irish, Inorg. Chem., **7**, 1699 (1968).
15. B. Wm. Maxey and A. I. Popov, J. Amer. Chem. Soc., **91**, 20 (1969).
16. E. G. Höhn, J. A. Olander and M. C. Day, J. Phys. Chem., **73**, 3880 (1969).
17. Wm. J. McKenney and A. I. Popov, J. Phys. Chem., **74**, 535 (1970).
18. K. Balasubrahmanyam and G. J. Janz, J. Amer. Chem. Soc., **92**, 4189 (1970).
19. J. L. Wuepper and A. I. Popov, J. Amer. Chem. Soc., **92**, 1493 (1970).
20. M. K. Wong, Wm. J. McKenny and A. I. Popov, J. Phys. Chem., **75**, 56 (1971).
21. W. F. Edgell and J. Lyford IV, J. Amer. Chem. Soc., **93**, 6407 (1971).
22. G. J. Janz and K. Balasubrahmanyam, Revue Roumaine Chemie, **17**, 187 (1972).
23. R. E. Hester, K. Krishman and C. W. J. Scaife, J. Chem. Phys., **49**, 1100 (1968).

IX. NONAQUEOUS SPECTROSCOPY OF ELECTROLYTES

24. V. Seidl, O. Knop and M. Falk, Canad. J. Chem., **47**, 1361 (1969).
25. S. A. S. Ghazunfar, D. V. Myers and J. T. Edsall, J. Amer. Chem. Soc., **86**, 3439 (1964).
26. R. A. Cummins, Austral. J. Chem., **17**, 838 (1964).
27. J. M. P. J. Verstegen, J. Inorg. Nuclear Chem., **26**, 25 (1964).
28. M. P. Springer and C. Curran, Inorg. Chem., **2**, 1270 (1963).
29. R. S. Krishnan and K. Balasubrahmanyam, Proc. Indian Acad. Sci., **59**, 14 (1964).
30. E. W. Wartenberg and J. Goubeau, Z. Anorg. Chem., **329**, 269 (1964).
31. Z. Kecki and T. Gulik-Krzywicki, Roczniki Chem., **38**, 277 (1964).
32. B. Rice and K. C. Bald, Spectrochim. Acta, **20**, 721 (1964).
33. N. Tanaka, M. Kamada, J. Fufita and E. Kyuno, Bull. Chem. Soc. Japan, **37**, 222 (1964).
34. A. R. H. Cole and A. J. Michell, Austral. J. Chem., **18**, 102 (1965).
35. S. F. Mason and B. J. Norman, Chem. Comm., **73** (1965).
36. J. Lewis, R. S. Nyholm and G. A. Rodley, J. Chem. Soc., 1483 (1965).
37. K. Heinzinger and R. E. Weston, Jr., J. Chem. Phys., **42**, 272 (1965).
38. R. S. Tobias and C. E. Freidline, Inorg. Chem., **4**, 215 (1965).
39. J. W. Akitt, A. K. Covington, J. G. Freeman and T. H. Lilley, Chem. Comm., 349 (1965).
40. A. R. H. Cole, L. H. Little and A. J. Michell, Spectrochim. Acta, **21**, 1169 (1965).
41. A. Z. Gadzhiev and I. S. Pominov, Optics and Spectroscopy, **17**, 468 (1964).
42. W. F. Edgell, M. T. Yang and N. Koizumi, J. Amer. Chem. Soc., **87**, 2563 (1965).
43. J. C. Evans and G. Y. S. Lo, Spectrochim. Acta, **21**, 1033 (1965).
44. Pham Van Huong and J. Lascombe, Compt. Rend., **260**, 6572 (1965).
45. A. Z. Gadzhiev and I. S. Pominov, Optics and Spectroscopy, **17**, 468 (1964).
46. A. Z. Gadzhiev and I. S. Pominov, Optics and Spectroscopy, **18**, 84 (1965).
47. A. V. Karyakin, A. V. Petrov, Yu. B. Gerlit, and M. E. Zubrilina, Teor. i. eksp. Khim., **2**, 494 (1966).
48. A. P. Simonov, D. N. Shigorin, G. V. Tsareva, T. V. Talalaeva and K. A. Kocheshkov, Zhur. priklod. Spektroskopii, **3**, 531 (1965).
49. W. F. Edgell, A. T. Watts, J. Lyford and W. M. Risen, Jr., J. Amer. Chem. Soc., **88**, 1815 (1966).
50. A. I. Grigor'ev, Zhur. Neorg. Khim., **10**, 2499 (1965).
51. Yu. Ya. Kharitonov, Kh. U. Ikramov and A. V. Babaeva, Zhur. Neorg. Khim., **10**, 2424 (1965).
52. H. H. Selig, L. A. Quarterman, and H. H. Hyman, J. Inorg. Nuclear Chem., **28**, 2063 (1966).
53. G. M. Venkatesh and P. Neelakantam, Proc. Indian Acad. Sci. Sect. A, **64**, 36 (1966).
54. G. Leclere and G. Duyckaerts, Spectrochim. Acta, **22**, 403 (1966).
55. C. Carvafal, K. J. Tölle, J. Smid and M. Szarvc, J. Amer. Chem. Soc., **87**, 5548 (1965).
56. J. M. Williams and M. Kreevoy, J. Amer. Chem. Soc., **89**, 5499 (1967).
57. D. E. H. Jones and J. L. Wood, Spectrochim. Acta, **23**, 2695 (1967).
58. I. S. Perelygin and N. R. Safiullina, Zhur. Strukt. Khim., **8**, 205 (1967).
59. P. Waldstein and L. A. Blatz, J. Phys. Chem., **71**, 2271 (1967).
60. R. J. Levene, D. B. Powell and D. Steele, Spectrochim. Acta, **22**, 2033 (1966).
61. E. Spinner, Austral. J. Chem., **19**, 2091 (1966).
62. G. J. Janz, M. J. Tait and J. Meier, J. Phys. Chem., **71**, 963 (1967).
63. M. P. Hanson, Diss. Abs., **27**, B, 4342 (1967).
64. P. J. Hendra, Spectrochim. Acta, **23A**, 2871 (1967).

65. Z. Kecki and J. Golaszewska, Roczniki Chem., **41**, 1817 (1967).
66. M. Kikuchi and E. Oikawa, J. Chem. Soc. Japan, **89**, 129 (1968).
67. J. E. D. Davies and D. A. Long, J. Chem. Soc. (A), 1757 and 1761 (1968).
68. J. E. D. Davies and D. A. Long, J. Chem. Soc. (A), 2054 (1968).
69. K. Schaarschmidt, Z. Chem., **8**, 343 (1968).
70. J. W. Smith and M. C. Vitoria, J. Chem. Soc. (A), 2468 (1968).
71. Yu. V. Gurikov, L. V. Moiseeva and A. I. Sidorova, Doklady. Akad. Nauk SSSR, **182**, 1044 (1968).
72. M. J. French and J. L. Wood, J. Chem. Phys., **49**, 2358 (1968).
73. G. Beech, G. Marr and B. W. Rockett, J. Chem. Soc. (A), 629 (1969).
74. B. W. Maxey and A. I. Popov, J. Amer. Chem. Soc., **91**, 20 (1969).
75. T. M. Loehr and R. A. Plane, Inorg. Chem., **8**, 73 (1969).
76. H. Petersen, Jr. and R. S. Drago, Inorg. Chim. Acta, **3**, 155 (1969).
77. A. R. Davis, J. W. Macklin and R. A. Plane, J. Chem. Phys., **50**, 1478 (1969).
78. M. T. Forel, J. Derouault, J. Le Calve and M. Rey-Lafon, J. Chim. Phys., **66**, 1232 (1969).
79. H. Touhara, H. Shimoda, N. Nakanoski and N. Watanabe, J. Phys. Chem., **75**, 2222 (1971).
80. H. G. Hertz, Ber. Bunsengesellschaft, **67**, 311 (1963).
81. H. G. Hertz and M. D. Zeidler, Ber. Bunsengesellschaft, **67**, 774 (1963).
82. H. G. Hertz and M. D. Zeidler, Ber. Bunsengesellschaft, **68**, 821 (1964).
83. H. G. Hertz, Ber. Bunsengesellschaft, **68**, 907 (1964).
84. H. G. Hertz, Ber. Bunsengesellschaft, **71**, 999 (1967).
85. H. G. Hertz, Ber. Bunsengesellschaft, **71**, 979 (1967).
86. H. G. Hertz, B. Lindman and V. Siepe, Ber. Bunsengesellschaft, **73**, 542 (1969).
87. H. G. Hertz, G. Keller and H. Versmold, Ber. Bunsengesellschaft, **73**, 549 (1969).
88. C. Deverell and R. E. Richards, Mol. Phys., **10**, 551 (1966).
89. L. S. Frankel, T. R. Stengle and H. Langford, Chem. Comm., **17**, 363 (1965).
90. J. W. Akitt and A. J. Downs; Alkali Metal Symposium, (Chem. Soc. London). p. 199 (1967).
91. J. W. Akitt and M. Parekh, J. Chem. Soc. (A), 2195 (1968).
92. E. Schaschel and M. C. Day, J. Amer. Chem. Soc., **90**, 503 (1968).
93. E. G. Bloor and R. G. Kidd, Canad. J. Chem., **46**, 3425 (1968).
94. C. Hall, G. L. Haller and R. E. Richards, Mol. Phys., **16**, 377 (1969).
95. L. S. Frankel, C. H. Langford and T. R. Stengle, J. Phys. Chem., **74**, 1376 (1970).
96. R. A. Taylor and I. D. Kuntz, J. Amer. Chem. Soc., **92**, 4813 (1970).
97. R. H. Erlich and A. I. Popov, J. Amer. Chem. Soc., **93**, 5620 (1971).
98. M. Herlem and A. I. Popov, J. Amer. Chem. Soc., **94**, 1431 (1972).
99. Z. Luz and S. Meiboom, J. Chem. Phys., **40**, 1066 (1964).
100. Z. Luz and S. Meiboom, J. Chem. Phys., **40**, 1058 (1964).
101. T. Birchall and W. L. Jolly, J. Amer. Chem. Soc., **87**, 3007 (1965).
102. S. J. Kuhn and J. S. McIntyre, Canad. J. Chem., **43**, 375 (1965).
103. D. Fiat and R. E. Connick, J. Amer. Chem. Soc., **88**, 4754 (1966).
104. J. F. Hinton, L. S. McDowell and E. S. Amis, Chem. Comm., **776** (1966).
105. N. A. Matwtyoff, Inorg. Chem., **5**, 788 (1966).
106. J. M. Miller and M. Onyszchuk, Canad. J. Chem., **44**, 899 (1964).
107. M. Eisenstadt and H. L. Friedman, J. Chem. Phys., **44**, 1407 (1966).
108. A. Fratiello and D. P. Miller, Mol. Phys., **11**, 37 (1966).
109. W. D. W. Horroch, Jr., R. H. Fischer, J. R. Hutchison and G. N. La Mar, J. Amer. Chem. Soc., **88**, 2436 (1966).
110. M. St. J. Arnold and K. J. Packer, Mol. Phys., **10**, 141 (1966).

IX. NONAQUEOUS SPECTROSCOPY OF ELECTROLYTES 769

111. G. N. La Mar, R. H. Fischer and W. D. Horrocks, Jr., Inorg. Chem., **6,** 1798 (1967),
112. L. D. Supran and N. Sheppard, Chem. Comm., 832 (1967).
113. D. Nicholls and M. Szwarc, J. Phys. Chem., **71,** 2727 (1967).
114. J. F. Hinton and E. S. Amis, Chem. Rev., **67,** 367 (1967).
115. N. A. Matwtyoff and S. V. Hooker, Inorg. Chem., **6,** 1127 (1967).
116. I. R. Lantzke and D. W. Watts, Austral. J. Chem., **20,** 35 (1967).
117. R. Haque, Diss. Abs., **27B,** 2312 (1967).
118. L. S. Frankel, Diss. Abs., **27B,** 4340 (1967).
119. A. Fratiello, R. E. Lee, V. M. Nishida and R. E. Schuster, Chem. Comm., 173 (1968).
120. M. S. J. Arnold and K. J. Packer, Mol. Phys., **14,** 249 (1968).
121. T. Matsuo, J. Phys. Chem., **72,** 1819 (1968).
122. R. D. Green and J. S. Martin, J. Amer. Chem. Soc. **90,** 3659 (1968).
123. N. E. Alexandrou and D. Jannakoudakis, Tetrahedron Letters, 3841 (1968).
124. B. W. Maxley and A. I. Popov, J. Amer. Chem. Soc., **90,** 4470 (1968).
125. J. F. Hon, Mol. Phys., **15,** 57 (1968).
126. W. G. Movines and N. A. Matwtyoff, J. Phys. Chem., **72,** 3063 (1968).
127. B. B. Smith and D. T. Sawyer, Inorg. Chem., **7,** 1526 (1968).
128. J. F. O'Brien, G. E. Glass and W. L. Reynolds, Inorg. Chem., **7,** 1664 (1968).
129. J. Demarquay, Pham-Quang-Tho, B. Mentzen and B. Claudel, J. Chim. Phys., **65,** 1380 (1968).
130. Hang Hsin Lo, Diss. Abs., **29B,** 967 (1968).
131. R. N. Butler and M. C. R. Symons, Trans. Faraday Soc., **65,** 945 (1969).
132. C. Hall, G. L. Haller and R. E. Richards, Mol. Phys., **16,** 377 (1969).
133. R. D. Green and N. Sheppard, J. Chem. Soc., Faraday Trans. II, **68,** 821 (1972).
134. J. J. Delpuech, A. Peguy and M. R. Khaddar, J. Magnetic Res., **6,** 325 (1972).
135. H. Grasdalen, J. Magnetic Res., **6,** 336 (1972).
136. R. C. Ford and T. R. Lindstrom, J. Phys. Chem., **75,** 3963 (1971).
137. P. H. Weiner and E. R. Malinowski, J. Phys. Chem., **75,** 3971 (1971).
138. G. Beech and K. Miller, J. Chem. Soc., Dalton Trans., **801,** (1972).
139. J. W. Akitt, N. N. Greenwood and G. D. Lester, J. Chem. Soc., 2450 (1971).
140. A. M. Grotens, J. Smid and E. DeBoer, J. Chem. Soc., 759 (1971).
141. H. Grasdalen, J. Magnetic Resonance, **5,** 84 (1971).
142. J. T. Edward, Canad. J. Chem., **49,** 2364 (1971).
143. A. N. Voronovich, L. S. Lilich, S. V. Petukhov and M. K. Khripun, Dokl. Akad. Nauk SSSR, **198,** 865 (1971).
144. A. M. Chmelrick and D. Fiat, J. Amer. Chem. Soc., **93,** 2875 (1971).
145. A. T. Tsatsas, J. W. Reed and Wm. M. Risen, Jr., J. Chem. Phys., **55,** 3260 (1971).
146. J. R. Evans and G. Y. S. Lo, J. Phys. Chem., **69,** 3223 (1965).
147. A. Tsatsas and Wm. J. Risen, Jr., J. Amer. Chem. Soc., **92,** 1789 (1970).
148. M. Goldstein and E. F. Mooney, J. Inorg. Nuclear Chem., **27,** 1601 (1965).
149. Y. Mikawa, R. J. Jakobsen and J. W. Brasch, J. Chem. Phys., **45,** 4528 (1966).
150. D. A. Draegert, Diss. Abs., **28B,** 1080 (1967).
151. W. J. Hurley, Diss. Abs., **28B,** 2373 (1967).
152. M. L. Good, C. Chary, D. W. Wertz and J. R. Durig, Spectrochim. Acta., **25A,** 1303 (1969).
153. R. Sperling and A. Treinin, J. Phys. Chem., **68,** 893, 896 (1964).
154. C. Jolicoeur, The Dinh Nguyen and A. Cabana, Canad. J. Chem., **49,** 2008 (1971).
155. M. J. Blandamer, T. R. Griffiths, L. Shields and M. C. R. Symons, Trans. Faraday Soc., **60,** 1524 (1964).
156. G. W. A. Fowles and R. A. Walton, J. Chem. Soc., 2840 (1964).

157. J. Szychlinski, E. Latowska and W. Moska, Roczniki Chem., **38,** 1427 (1964).
158. N. S. Chhonkar, J. Chem. Phys., **41,** 3683 (1964).
159. R. S. Drago, D. M. Hart and R. L. Carson, J. Amer. Chem. Soc., **87,** 1900 (1965).
160. D. Deveze, Compt. Rend., **263,** 392 (1966).
161. F. Aufauvre and F. Riva, Compt. Rend., **263,** 565 (1966).
162. E. M. Voigt, J. Phys. Chem., **70,** 598 (1966).
163. A. Garnier, Compt. Rend., **265,** 198 (1967).
164. S. Aditya, J. Inorg. Nuclear Chem., **29,** 1901 (1967).
165. V. V. Zamkova and Yu. A. Kushnikov, Zhur. Neorg. Khim., **12,** 1809 (1967).
166. S. Singh and C. N. R. Rao, Trans. Faraday Soc., **62,** 3310 (1966).
167. Z. Kecki and B. Izdebska, Roczniki Chem., **40,** 1529 (1966).
168. G. Narain and P. Shukla, J. Indian Chem. Soc., **43,** 694 (1966).
169. R. D. Hohn, W. R. Carper and J. A. Blancher, J. Phys. Chem., **71,** 3960 (1967).
170. C. F. Wells and G. Davies, J. Chem. Soc., 1858 (1967).
171. J. T. Bell and R. E. Biggers, J. Mol. Spectroscopy, **25,** 312 (1968).
172. D. Deveze and P. Rumpf, Compt. Rend., **266,** 1001 (1968).
173. V. T. Athavak, N. Mahadevan and R. M. Sathe, Indian J. Chem., **6,** 462 (1968).
174. J. E. Spessard, Spectrochim. Acta, **25,** 731 (1969).
175. C. Tiar, R. Mercier and M. Camelot, Compt. Rend., **268,** 1826 (1969).
176. L. L. Böhm and G. V. Schulz, Ber. Bunsengesellschaft, **73,** 260 (1969).
177. M. J. Blandamer and T. R. Burdett, J. Chem. Soc., Faraday Trans. II, **68,** 577 (1971).
178. J. P. Scharff, Bull. Chem. Soc. France, 413 (1972).
179. R. Yamdagri and P. Kebarle, J. Amer. Chem. Soc., **94,** 2940 (1972).
180. P. Kebarle, R. N. Haynes and J. G. Collins, J. Amer. Chem. Soc., **89,** 5753 (1967).
181. P. Kebarle and A. M. Hogg, J. Chem. Phys., **42,** 798 (1965).
182. C. R. Mueller and R. W. Cahill, J. Chem. Phys., **40,** 651 (1964).
183. V. I. Gol'danskii, O. Yu. Okhlobystin, V. Ya. Rochev and V. V. Khrapov. J. Organometallic Chem., **4,** 160 (1965).
184. A. Vertes, Magyar Kem Folyoirat, **75,** 175 (1969).
185. A. Vertes, K. Burger and M. Suba, Magyar Kem Folyoirat, **75,** 317 (1969).
186. S. Petrucci and F. Fittipaldi, J. Phys. Chem., **71,** 3087 (1967).
187. D. R. Dickson and P. Kruus, Can. J. Chem., **49,** 3107 (1971).
188. S. Fujiwara and H. Hayashi, J. Chem. Phys., **43,** 23 (1965).
189. H. A. Kuska, F. M. D'tri and A. I. Popov, Inorg. Chem., **5,** 1272 (1966).
190. C. Corvafa and G. Giacometti, Ricerca Sci. Rend., **8,** 1038 (1965).
191. S. I. Chen, B. M. Fung and H. Lütfe, J. Chem. Phys., **47,** 2121 (1967).
192. T. A. Claxton and J. Oakes, Trans. Faraday Soc., **64,** 596 (1968).
193. F. J. Smentoswski and G. R. Stevenson, J. Amer. Chem. Soc., **90,** 4661 (1968).
194. R. F. Adams and N. M. Atherton, Trans. Faraday Soc., **65,** 649 (1969).
195. M. C. R. Symons, Ann. Rev. Phys. Chem., **20,** 219 (1969).
196. G. Nyberg, Mol. Phys., **17,** 87 (1969).
197. C. M. Guzy, J. B. Raynor and M. C. R. Symons, J. Chem. Soc., 2791 (1969).
198. J. Padova, Canad. J. Chem., **43,** 458 (1965).
199. C. Deverell, D. J. Frost and R. E. Richards, Mol. Phys., **9,** 565 (1965).
200. J. J. P. Martin, Bull. Soc. Chim. France, 1237 (1966).
201. D. M. Barnhart, C. N. Caughlan and M. Ul-Haque, Inorg. Chem., **7,** 1135 (1968).
202. R. A. Robinson and R. H. Stokes, J. Amer. Chem. Soc., **70,** 1870 (1948).
203. H. Stephen and T. Stephen (ed.); Solubility of Inorganic Compounds, Vol. II, Pergamon Press (1963).
204. S. Chulkaratana, Diss. Abs., **25,** 2251 (1964).

X. ORGANIC ELECTROLYTE BATTERY SYSTEMS

The following information is a selection of significant contributions in the area of nonaqueous battery investigations. An attempt has been made to summarize the solvents and electrolytes that have been studied, in addition to the anode and cathode materials.

For a recent review of organic electrolyte battery systems the reader is referred to:

1. *"Organic Electrolyte Battery Systems"* by J. T. Nelson and C. F. Green, Harry Diamond Laboratories, U. S. Army Material Command, Washington, D. C., HDL-TR-1588 (March 1972).

This report presents a very extensive tabulation of the solvents and solute materials that have been investigated relative to battery systems, and contains a useful bibliography of technical reports in this field.

Solvents

The dozen most widely investigated solvents are:

(1) Propylene carbonate (PC)
(2) Dimethylformamide (DMF)
(3) Butyrolactone (BL)
(4) Dimethylsulfoxide (DMSO)
(5) Acetonitrile (DMSO)
(6) Nitromethane (NM)
(7) Tetrahydrofuran (THF)
(8) Acetone (A)
(9) Nitrosodimethylamine (NDA)
(10) Ethylene carbonate (EC)
(11) Methyl formate (MF)
(12) Ethyl acetate (EA)

Solutes

The majority of the solute materials fall into five general classes of compounds:

(1) Simple salts, e.g., alkali and alkaline earth halides;
(2) Lewis acids (electron pair acceptors), e.g., $AlCl_3$ and BF_3;
(3) Lewis acids combined with alkali halides, e.g., $LiAlCl_4$ ($AlCl_3$ and $LiCl$);
(4) complex fluorides, e.g., $NaPF_6$ and KBF_4; and

(5) organically substituted ammonium salts, e.g., tetraethylammonium perchlorate. The first four classes are represented in the ten most commonly studied solutes:

(1) $LiClO_4$
(2) $LiCl$
(3) $AlCl_3$
(4) KPF_6
(5) $LiPF_6$
(6) $LiAlCl_4$
(7) $LiBF_4$
(8) $NaPF_6$
(9) LiF
(10) $KSCN$

The requirements governing the choice of solute materials are compatibility with and solubility in the solvent, low cost, and ease of handling.

Systems

The following are some common systems that have been employed.

Anodes	Solvents	Solutes	Cathodes
Li	MF	$LiAlCl_4$	$CuCl_2$
Mg	PC, NM	$LiClO_4$	CuF_2
Ca	DMF	$LiCl$	CuS
	DMSI	$LiPF_6$	AgF_2
	PC	KPF_6	$AgCl$
	BL	$Mg(ClO_4)_2$	AgS
	AN	$TPA\ BF_4$	ACL-70
	ACN		

Successful Cell Systems

Cell System

1. Li/PC—$LiAlCl_4$/AgCl
2. Li/PC—$NaPF_6$/CuF
3. Li/PC—KPF_6/NiX_2
4. Li/PC—$LiClO_4$/CuF_2
5. Li/PC,NM—$LiAlCl_4$/$CuCl_2$
6. Li/BL—$LiClO_4$/$CuCl_2$
7. Li/BL—KPF_6/AgF_2
8. Li/MF—$LiClO_4$/CuF_2
9. Li/MF—$LiClO_4$/ACL-70
10. Li/IPA—$LiClO_4$/CuS
11. Li/THF,DME—$LiClO_4$/CuS
12. Li/MCC—$LiAlCl_4$/CuCl

X. ORGANIC ELECTROLYTE BATTERY SYSTEMS

NONAQUEOUS LITHIUM BATTERY:

Lithium-Nickel Thiocyanate Cells

2. "Lithium Battery Development" by D. E. Semones and J. McCallum, BATTELLE Columbus Laboratories, Tech. Rpt. AFAPL-TR-71-82, Air Force Aero Propulsion Laboratory, Air Force Systems Command, Wright-Patterson Air Force Base, Ohio (September 1971).

Nickel thiocyanate positive electrodes in propylene carbonate were developed by press-forming the active metal onto a stainless steel screen. Evaluation of electrode materials included studies of cyclic voltammetry at solid electrodes, and measurements of the solubility of the electrode materials in the organic electrolytes. Charge and discharge data are given as are data on separator materials. [Polypropylene or Polyethylene]; some metal halide-lithium couples are reviewed.

3. *Electrochemical Characterization of Nonaqueous Systems for Secondary Battery Application* by M. Shaw, O. A. Paez and F. A. Ludwig, Whittaker Corp., Research and Development, San Diego, Calif. Tech. Report prepared for NASA (Contract No. NAS-3-8509). Final Report, Jan. 19, 1969. See also Tech. Rept. NASA CR-1434 (September 1969).

Cyclic voltammetric measurements were applied to the screening of 940 individual electrochemical systems in organic electrolytes. A limited number of these systems were recommended for further characterization and cell development as practical high energy density rechargeable battery systems.

Recommended Positive-Electrolyte Systems (recommended on the basis of high discharge peaks and low peak separation)

AgO/BL—LiCl + AlCl$_3$*
AgF$_2$/PC—LiBF$_4$*
Cu/AN—LiPF$_6$ + KPF$_6$
Cu/AN—LiPF$_6$ + KPF$_6$
Cu/DMF—LiPF$_6$
CuCl$_2$/AN—LiPF$_6$
CuCl$_2$/BL—AlCl$_3$
CuCl$_2$/DMF—LiCl + LiClO$_4$
CuCl$_2$/DMF—LiPF$_6$
CuCl$_2$/PC—LiClO$_4$*
CuF$_2$/DMF—LiPF$_6$*
CuF$_2$/PC—LiPF$_6$*
CuF$_2$/PC—LiClO$_4$

Zn/AN—LiClO$_4$
Zn/BL—KPF$_6$
Zn/DMF—KPF$_6$*
Zn/DMF—LiPF$_6$
Zn/DMF—LiClO$_4$
Zn/PC—KPF$_6$
ZnF$_2$/DMF—KPF$_6$
ZnF$_2$/DMF—LiClO$_4$
Cd/BL—KPF$_6$
Cd/DMF—KPF$_6$
Cd/DMF—LiBF$_4$
Cd/DMF—LiClO$_4$*

* These systems were recommended for further characterization and cell development based on the charge-discharge data for sintered electrodes.

Chronopotentiometric data is given for some of the recommended systems.

Cyclic Voltammetry of Negative-electrolyte Systems

    Lithium systems   Calcium systems
    Magnesium systems

Solvent purification discussed for

    Butyrolactone (BL)   Acetonitrile (AN)
    Dimethylformamide (DMF)   Propylene carbonate (PC)

Solute purification discussed for

    Lithium chloride   Calcium fluoride
    Lithium fluoride   Aluminum chloride
    Magnesium fluoride   Potassium hexafluorophosphate
    Magnesium chloride   Phosphorous pentachloride
    Lithium perchlorate   Boron trifluoride
    Magnesium perchlorate   Lithium hexafluorophosphate
    Calcium chloride

Electrode preparation and characterization

    Metal wire electrodes   Sintered electrodes
    Fluoride, chloride and oxide wire
      electrodes

4. *Properties of Nonaqueous Electrolytes* by R. Keller, J. N. Foster, D. C. Hanson, J. F. Hon and J. S. Muirhead. (Rocketdyne a division of North American Rockwell Corp., Canoga Park, Calif.) Tech. Report NASA CR-1425 prepared under Contract No. NASA 3-8521 August (1969)

                           Nonaqueous lithium battery

Aprotic solvents investigated

    Propylene carbonate (PC)   Acetonitrile (AN)
    Dimethylformamide (DMF)   Methyl formate (MF)

X. ORGANIC ELECTROLYTE BATTERY SYSTEMS

Solutes studied

    Lithium perchlorate
    Lithium chloride (with and without AlCl$_3$)
    Tetramethylammonium hexafluorophosphate
    Lithium hexafluoroarsenate

Electroactive battery materials

    Copper fluoride
    Copper chloride

Types of investigations

    Solubilities
    Heats of solution
    Vapor pressures
    Viscosities and densities
    Ionic velocities
    Conductance measurements
    Conductometric titrations
    Hittorf experiments
    Diffusion coefficients
    Dielectric constants

Structural Studies

    NMR
    EPR

5. *High Energy Batteries* by R. Jasinski, Plenum Press (1967)

This text contains chapters on nonaqueous inorganic electrolytes and organic electrolytes and surveys information from 1945 including government reports and the patent literature. The following is a list of the various systems that are discussed in this text.

Inorganic electrolyte systems

    Liquid NH$_3$
    Liquid SO$_2$
    Anhydrous HF

Organic electrolyte systems

    Acetonitrile
    Butyrolactone
    Dimethylformamide
    Dimethylsulfoxide
    Ethylene carbonate
    Ethyl ether
    Methyl formate
    Nitromethane
    Propylene carbonate

Negative materials

Aluminum	Lithium
Magnesium	

Positive materials

CuF_2	AgO
$CuCl_2$	Carbon black
CoF_3	AgCl
$NiCl_2$	

6. *Lithium battery development* by D. E. Semones and J. McCallum, BATTELLE Columbus Laboratories, Technical Report AFAPL-TR-71-82 September (1971), Air Force Aero Propulsion Laboratory, Air Force Systems Command, Wright-Patterson Air Force Base, Ohio

Nonaqueous lithium batteries investigated

Solvent: Propylene carbonate

Electrode system: Lithium-Nickel thiocyanate

Also a large number of compounds were considered as positive electrode materials, including thirty solid metal halide compounds.

7. *Chemoelectric Energy Conversion for Nonaqueous Reserve Batteries* 14th Quarterly report October–December 1966, Naval Ordnance Laboratory, Corona, California, NOLC Report 737. Contains the following articles:

"Nonaqueous electrolyte Studies for high energy batteries and Fuel cells" by D. N. Bennion.

Conductivity of m-dinitrobenzene in liquid NH_3, solubilities and conductances of m-dinitrobenzene in DMSO.

"Identification of Organic reduction products from IR Spectra" by M· J. Schaer.

Reduction of nitrobenzene and o, m and p-dinitrobenzene and dinitroaniline in acid ammonia solutions.

"Information Storage and Retrieval" by W. C. Spindler

Provides indexes of both reports and PIC briefs in the chemoelectric energy-conversion field.

X. ORGANIC ELECTROLYTE BATTERY SYSTEMS

8. *Study of the Composition of Nonaqueous Solutions of Potential use in High Energy Density Batteries* by J. N. Butler, D. R. Cogley, J. C. Synnott, and G. Holleck, Final Report, September 1969, Contract No. AF 19 [628] 6131, prepared for Air Force Cambridge Research Laboratories, Office of Aerospace Research, United States Air Force, Bedford, Mass, [AFCRL-69-0470].

This report discusses electrochemical studies relative to the solution thermodynamics of potential high energy battery electrolytes and cathode materials in the aprotic organic solvents, propylene carbonate, DMF and DMSO. The study includes a comprehensive survey of reversible reference electrode systems and a specific experimental review of chloride reversible electrodes based on silver, thallium, lead and cadmium. A very useful bibliography is included which covers much of the recent technical report literature relative to high energy density batteries.

9. *Investigation of Electrolyte Systems for Lithium Batteries* by R. Keller and J. F. Hon, Rocketdyne prepared for NASA, Contract NAS3-12969 (May 1969–July 1970).

Aprotic electrolytes based upon methyl formate, propylene carbonate and dimethyl formamide as solvents and lithium chloride plus aluminum chloride, lithium hexafluoroarsenate and lithium perchlorate as solutes are discussed. The effect of additives, other aprotic solvents, other salts and trace amounts of water on species in solution and other physical properties are described.

Physical property determinations

    Electrolyte stability studies
    Solubility studies
    Viscosities and densities
    Conductance measurements
    Transference experiments
    Measurement of diffusion coefficients
    NMR studies

10. *Lithium-Anode Limited Cycle Battery Investigation* by H. F. Bauman, J. E. Chilton, and A. E. Hultquist, Tech. Report AFAPL-TR-67-104, July 1967, Air Force Aero Propulsion Laboratory, Research and Technology Division Air Force Systems Command, Wright-Patterson Air Force Base, Ohio.

Lithium–cupric fluoride couple with a lithium perchlorate–propylene carbonate electrolyte.

11. *High Energy Battery Systems based on Propylene Carbonate* by R. Jasinski, Scientific Report No. 2 prepared for Air Force Cambridge Research Laboratories, Office of Aerospace Research, United States Air Force, Bedford, Mass., AFCRL-69-0381, July 1969.

Topics discussed

Electrolyte stability
Purity requirements
Purification procedures
Solvent-ion interaction phenomena
Electrode reactions of inorganic positives, organic positives and active metal negatives.

12. *Design Studies for High Rate, High Energy, Nonaqueous Electrochemical Energy Conversion Systems* by D. N. Bennion, J. S. Dunning, and W. H. Tiedemann, UCLA-ENG-7078, December 1970 (final report), prepared for U. S. Naval Weapons, Corona Lab, Corona, Calif., Contract No. N00123-70-C-0188.

Reduction of *m*-dinitrobenzene in liquid NH_3 and in DMSO.
Density, viscosity and conductivity of lithium-trifluoromethane sulfonate in Dimethylsulfite.

XI. ADDITIONAL REFERENCES AND DATA SOURCES

This section contains additional references relative to the various sections in this Volume as well as updated material for Nonaqueous Electrolytes Handbook, Volume I.

(a) Antimony Trichloride

1. Solvent Physical Properties

Property	Data	Reference and Comments
Melting point	73.2°C	[1]
	73.17°C	[2]
Boiling point	222.6°C	
Viscosity	1.84 cP at 99°C	[3] Range covered 79–192°C
Cryoscopic constant	20.3°C Mol^{-1} kg	[1b] No correction for non-ideality of solutes
	15.6°C Mol^{-1} kg	[2] With correction for non-ideality of solutes
	14.7°C Mol^{-1} kg	[4] Correction for non-ideality using true non-electrolyte solutes
	15.1°C Mol^{-1} kg	[4] Calc. from latent heat of fusion [5]
Latent heat of fusion	15.1 kJ Mol^{-1}	[5]
Enthalpy of vaporization	11.16 kcal Mol^{-1}	[6]

XI. ADDITIONAL REFERENCES AND DATA SOURCES

Heat capacity of liquid	33.3 cal Mol^{-1} °C^{-1} (mean value over range 76–98°C)	[2] Method of mixtures
Heat capacity of solid	25.8 cal. Mol^{-1} °C^{-1} (mean value over range 28–69°C)	[2] Method of mixtures
Dielectric constant	33.2 at 75°C to 30.4 at 99°C	[7]
	36.2 at 75°C to 22.7 at 205°C	[8] Real part of DC at 4.2 MHz
Density (liquid)	$2.622 - 0.002268\,(T - 100) - 0.32 \times 10^{-8}(T - 100)^3$ g. cm^{-3}	[9]
Density (solid)	3.14 g cm^{-3} at 20°C	[10]
Dipole moment	4.3 D	[2] Calc. using Onsager's "Reaction Field" eq.
	3.9 D (vapor)	[11]
	3.12 to 4.11 (various solvents)	[12]
Refractive index	1.46 (solid)	[10]
Vapour pressure of pure solvent	13.7 mm at 100°C to 100.3 mm at 155°C	[10]
Other data also Ref. [10]		
Self conductivity *see* Purification section		

2. Purification

All methods consist of further purification of the best Analytical Grade Reagent, which usually contains a lot of water. Criterion for purity is obtaining the lowest possible value of self conductivity.

Method	Conductivity ohm^{-1} cm^{-1}	Reference and Comments
1. Multiple distillation under reduced pressure of dry N_2 gas	3.17×10^{-6} at 75°C	[2], [13]. Usual value 6×10^{-6}
2. As method 1 plus final vacuum sublimation.	2.45×10^{-6} at 75°C	[14], [15]
3. Single distillation under N_2 at 1 atmosphere	$\simeq 10^{-5}$ at 99°C	[16]
4. Multiple distillation + freezing in vacuum	0.85×10^{-6} at 99°C	[17] Lowest value ever obtained. After equilibrium value rose to $2-4 \times 10^{-6}$ ohm^{-1} cm^{-1}
5. Distillation (twice) under 1 atmosphere CO_2 over metallic Sb	6.7×10^{-6} at 99°C	[18] Usual value $8-12 \times 10^{-6}$
6. Dry under vacuum + distillation (twice) + vac sublimation	7.7×10^{-6} at 76°C to 11.3×10^{-5} at 231°C	[19]

General Comments

Addition of metallic Sb has been recommended [13], [16], [18] to reduce any $SbCl_5$ present. Z. Klemensiewicz [17] in detailed study stated that multiple distillation by itself is not sufficient to produce $SbCl_5$ of lowest conductivity. He recommended freezing in vacuum.

Preliminary drying and/or distillation, followed by vacuum sublimation is the most convenient method [14], [15], [19] today.

XI. ADDITIONAL REFERENCES AND DATA SOURCES

3. Electrical Conductance of Systems in $SbCl_3$

Inorganic Halides (Halide donors)

General Reaction: $X\ Hal + SbCl_3 \rightleftarrows X^+ + SbCl_3Hal^-$

Compound (Solute)	Data	Reference and Comments
KBr	More conducting than pure $SbCl_3$	[20] Earliest ref. to $SbCl_3$ being used as a solvent.
Rb, K, NH_4 and Tl chlorides	Strong 1:1 electrolytes (99°C)	[21a] Debye Hückel slope approached as $C \rightarrow 0$
K, NH_4 and Tl bromides	Strong 1:1 electrolytes (99°C)	[21b, c]
	λ_0 Range 150–160	Slope X given by $X = 0.661\lambda_0 + 29.74$ at 75°C. See Ref. [13]
$HgCl_2$ and $HgBr$	Weak electrolytes	[21c]
HCl, SbOCl	very slightly dissociated	[21d]

Inorganic halides (Halide acceptors)

General Reaction: $X\ Hal_n + SbCl_3 \rightleftarrows X\ Hal_nCl + SbCl_2^+$

Compound (Solute)	Data	Reference and Comments
$AlCl_3$	1:1 strong electrolyte	[16]
	$\lambda_0 = 48$ and eq. is:	
	$\lambda = \lambda_0 - (0.682\ \lambda_0 + 42.16)C^{1/2}$ (99°C)	
$GaCl_3$		[16] Similar to $AlCl_3$
$AlCl_3$	Weak electrolyte	[18]

3. Electrical Conductance of Systems in SbCl$_3$ (*Continued*)

Other Inorganic Halides

Compound (Solute)	Data	Reference and Comment
SeCl$_4$, KI, HgI$_2$, I$_2$, KF, SbF$_3$	All soluble, but no conductivity data given	[18]
TeCl$_4$	Weak electrolyte, solubility 0.75 Mol l^{-1}	
Water	Water very poorly conducting and hence hydrolysis reaction: SbCl$_3$ + H$_2$O $\rightleftarrows$ SbOCl + 2HCl is well over to left	[21d]

The following compounds have been investigated as potential electrolytes but were found to have limited solubility as indicated:

Inorganic Halides

Compound	Data	Reference and Comments
LiCl, NaCl, SnCl$_2$, BiCl$_3$	Slightly soluble	[18]
FeCl$_3$	0.06 Mol l^{-1} at 100°C	
CuCl, AgCl, MgCl$_2$, SrCl$_2$, BaCl$_2$, ZnCl$_2$, CdCl$_2$, PbCl$_2$, MnCl$_2$, CoCl$_2$, NiCl$_2$, CrCl$_3$, HgCl$_2$	Insoluble	[18]

Other Inorganic Compounds

Sb$_2$O$_3$, Sb$_2$S$_3$, As$_2$S$_3$, S.	Soluble, no conductivity	[18]

XI. ADDITIONAL REFERENCES AND DATA SOURCES

Compound	Data	Reference and Comments
Na_2SO_4, K_2SO_4, $(NH_4)_2SO_4$, $BaSO_4$, $ZnSO_4$, $Sb_2(SO_4)_3$, ZnS, CuS, MgS, PbS, $KClO_4$, NH_4ClO_4, KCN, K_2CrO_4, $KMnO_4$	Insoluble	[18]
Nitrates and Carbonates	decompose in $SbCl_3$	[18]
$AgClO_4$	Dissolves due to reaction $AgClO_4 + SbCl_3 \rightarrow SbCl_2ClO_4 + AgCl \downarrow$	

Organic Halides

Compound	Data	Reference and Comments
$(CH_3)_4NCl$ and Ph_3CCl	1:1 strong electrolytes	[13] Temp. 75°C
	Approaching Debye–Hückel–Onsager slope $C \rightarrow 0$	Reaction as for inorganic halide donors
	$\lambda = \lambda_0 - X(C)^{1/2}/[1 + Y(C)^{1/2}]$ where $X = 0.661$ and $\lambda_0 = +29.74$	
	$(CH_3)_4NCl$, $\lambda_0 = 96.9$ and $Y = 2.4$	
	Ph_3CCl, $\Lambda_0 = 89.4$ and $y = 1.95$	See also {18}
$PhNH_3Cl$	Strong electrolyte	[18]
Ph_4AsCl, $\{C_2H_5\}_4NCl$	Strong electrolyte	[16]
Certain chlorides ionise mainly in mode:	$2RCl + SbCl_3 \rightleftarrows R_2Cl^+ + SbCl_4^-$	
	$K = \{R_2Cl^+\}\{SbCl_4^-\}f^2/\{RCl\}$	
and to minor extent in the mode:	$RCl + SbCl_3 \rightleftarrows R^+ + SbCl_4^-$	[13]
	$K^1 = \{R^+\}\{Cl^-\}f^2/\{RCl\}$	

3. Electrical Conductance of Systems in SbCl$_3$ (*Continued*)

Organic Halides

Compound	K	Data		K^1	Reference and Comments
Chloride					Temp. 75°C
Bornyl		strong electrolyte			
n-Decyl		strong electrolyte			
1-Ethylcyclohexyl	$2.2 \pm 0.1 \times 10^{-1}$			$2.5 \pm 0.6 \times 10^{-4}$	
Cyclohexyl	$9.7 \pm 0.2 \times 10^{-2}$			$4.1 \pm 1.2 \times 10^{-4}$	
Diphenylmethyl	$5.8 \pm 0.3 \times 10^{-2}$			$2.6 \pm 1.5 \times 10^{-5}$	
Cinnamyl	$7.5 \pm 0.5 \times 10^{-3}$			$2.1 \pm 0.8 \times 10^{-5}$	
Benzyl	$3.7 \pm 0.4 \times 10^{-5}$			$1.0 \pm 0.5 \times 10^{-7}$	

Other Organic Compounds

General Reaction:

$$Ar + SbCl_3 \rightleftarrows Ar:SbCl_2^+ + SbCl_4^-$$

$$K_{th} = f^2\{R:SbCl_2^+\}\{SbCl_4^-\}/\{C - R:SbCl_2^+\}$$

C = Total concentration of Ar

f = mean ionic activity coeff.

	$-\log K_{th}$				
Aromatic Hydrocarbons					
Pentacene	2.66				[15] Results obtained in absence of oxygen or protonating agents. K
Naphthacene	3.33				

XI. ADDITIONAL REFERENCES AND DATA SOURCES

Anthracene	4.30	related to Nr (Dewar's Reactivity index) by eq. $-\log K_{th} = 0.15 + 3.22$Nr. (See also electrochemistry section).
Perylene	4.43	
Benzo{a}anthracene	4.61	
Pyrene	5.09	
Phenanthrene	5.83	
Naphthalene, triphenylene biphenyl, benzene	Nonconductors	[15]
Pentacene, naphthacene, Perylene + oxygen	Well resolved ESR spectra. Perylene ESR similar to that in 98 % H_2SO_4. Napthacene splitting constants 5.20, 1.78 and 1.04 g cm in ratio 1.000:0.343:0.200	[22] For perylene reaction $O_2 + 4Pn + 6SbCl_3 \rightarrow 4Pn^+ + 4SbCl_4^- + 2SbOCl$ is quantitative. [14] [23]
Perylene + HCl	Complete protonation occurs $Pn + HCl + SbCl_3 \rightleftharpoons PnH^+ + SbCl_4$ General Reaction as for aromatic hydrocarbons	[14]

Amines

	$-\log K_{th}$	
Amylamine	1.4	[24]
Cyclohexylamine	1.5	Absence of oxygen.
p-Toluidine	1.8	Temp. = 75°C.
Aniline	2.05	(See also electrochemistry section).
p-Nitroaniline	2.3	
m-Nitroaniline	2.6	
Diphenylamine	2.6	
Acridine	1.1	
Indole	2.3	
Triphenylamine	Weak electrolyte—no K found	
Tribenzylamine	1:1 strong electrolyte $\lambda_0 = 92$	[24]
Pyridine	1:1 strong electrolyte $\lambda_0 = 92$	[16]

3. Electrical Conductance of Systems in $SbCl_3$ (*Continued*)

Other Organic Compounds

General Reaction as for aromatic hydrocarbons

Compound	Data	Reference and Comments
Triphenyl derivatives of group V elements		
Ph_3N	Weak electrolyte	[24] For Ph_3P eq. $Ph_3P + 2SbCl_3 \rightleftarrows Ph_3P \cdot + SbCl_2^+ + SbCl_4^-$ strong interaction of filled Sb '4d' orbitals with vacant '3d' orbitals of P is suggested.
Ph_3P	Strong electrolyte	
Ph_3As	Weak electrolyte	
Ph_3Sb	non electrolyte	
	General reaction as for Amines	[24]
Ketones	$-\log K$	
4-Methoxybenzophenone	1.82	
Anthrone	1.92	
Benzophenone	3.00	
Acetophenone		
Fluorenone	4.52	
Anthraquinone	4.96	
Dibenzylketone	5.25	

XI. ADDITIONAL REFERENCES AND DATA SOURCES

General reaction as for ketones

	$-\log K$	
Nitriles		
Benzonitrile	4.95	[24]
p-Chlorobenzonitrile	5.0	
p-Nitrobenzonitrile	5.05	
Phenols and Ethers		
Phenol	Nonelectrolyte	[24]
1-Naphthol	3.91	
2-Naphthol	4.23	
Ethyl-Phenyl Ether	5.35	
Diphenyl Ether	Nonelectrolyte	

4. Transference Numbers

Compound	Data	Reference and Comments
KCl	Concentration 0.15–0.31M Transport No. of Cl⁻ is 0.88–0.90	[25] Hitorf method. Sb anode and Pt cathode
NH₄Cl	Concentration 0.25–0.66M Transport No. of Cl⁻ is 0.87–0.91 Extrapolates to 0.81 as C→0.	These high mobilities infer affinity for reaction: $Cl^- + SbCl_3 \rightleftharpoons SbCl_3 + Cl^-$ [13]

5. Voltammetry

Type of Investigation	Data	Reference and Comments
Range of electroactivity	About 1.5 V depending on Cl$^-$ conc. (Sb/SbCl$_3$, SatKCl reference electrode)	[16], [26]. Temp. = 99°C.
Oxidation (Vitreous C working electrode)		
a. Neutral unbuffered solution (1M KCl & 1M AlCl$_3$)	3SbCl$_3$ →SbCl$_5$ + 2SbCl$_2^+$ + 2e eq. 1	
b. Solutions of high Cl$^-$ conc.	SbCl$_3$ → 2Cl$^-$ → SbCl$_5$ + 2e eq. 2	
c. Solutions of low Cl$^-$ conc.	3SbCl$_3$ → SbCl$_5$ + 2SbCl$_2^+$ + 2e eq. 3	
Reduction		
a. Neutral solution	SbCl → Sb ↓ + 3Cl$^-$ − 3e eq. 4	
b. Solutions of high Cl$^-$ conc.	SbCl$_3$ → Sb ↓ + 3Cl$^-$ − 3e Range 0.7V eq. 5	
c. Solutions of low Cl$^-$ conc.	3SbCl$_2^+$ → Sb ↓ + 2SbCl$_3$ − 3e Range 1.0V eq. 6	
Redox System SbV/SbCl$_3$	SbV/SbCl$_3$ is reversible. Eqn. is:	[16]
a. In presence of excess Cl$^-$ at vitreous C electrode.	$E = E_0 + (RT/2F) \log\{SbCl_5\}/\{Cl\}^2$ $= E_0 + 0.074\ pCl + 0.037 \log\{SbCl_5\}$ Normal potential is given by $E_0' = E_0 + 0.074\ pCl$	See eq. 2 $pCl = -\log\{Cl^-\}$
b. In presence of excess SbCl$_2^+$	System not reversible. Oxidized species very short lifetime. (Triangular wave cyclic voltammetry)	

XI. ADDITIONAL REFERENCES AND DATA SOURCES 795

Redox System $SbCl_3/Sb$ ↓

a. In presence of high Cl^- conc. System is reversible [16]. See eq. 5
$E = E_0 + (RT/3F) \log\{Cl^-\}^3$
$= E_0 + 0.074\ pCl$

b. Solution of low Cl^- conc. (excess $SbCl_2^+$) System is not reversible.

Ionic Product Determination

SbCl/Sb electrode system as indicator of Cl^- For Equilibrium $SbCl_3 \rightleftarrows Cl^- + SbCl_2^+$ [16], [27]
$K_i = \{Cl^-\}\{SbCl_2^+\}$
$pK_i = 10^{-7.8\pm0.5}$

Oxidation Studies

Ref. electrode $Sb/SbCl_3$, Sat.KCl

Aromatic hydrocarbons (Ar) $Ar/Ar^+ E_{1/2}$ Volts

Naphthacene	0.21	[26]
Perylene	0.24	
9,10-Dimethylanthracene	0.31	
3 Methylanthracene	0.34	Temp. 99°C.
9,10-Diphenylanthracene	0.39	
3,4-Benzpyrene	0.44	
Anthracene	0.51	
Pyrene	0.59	
1,2-Benzanthracene	0.63	
Acenaphthene	0.64	
Coronene	0.64	
1,2-Benzpyrene	0.66	

Aromatic Amines

$Sb/SbCl_3$. Sab. KCl electrode $E_{1/2}$ Volts at $pCl = 0$ Ref {28} {29}. $pCl = 0 \equiv$ 1M KCl soln.

5. Voltammetry (Continued)

Type of Investigation	Data	Reference and Comments
Class A Amines		
1-Aminoanthracene	0.05	Temp = 100°C.
2-Aminoanthracene	0.20	
Aniline	0.52	Electrochem. oxidation proceeds through reversible one-step oxidation. Cation radical is not stable
p-Chloroaniline	0.51	
m-Chloroaniline	0.50	
Pyridine ⎫		
Diphenylguanidine ⎬	Not oxidizable	
Antipyrine ⎭		
Class B Amines		
1-Naphthylamine	0.18	
2-Naphthylamine	0.34	
Diphenylamine	0.27	

For Class A Amines $E_{1/2}$ increases by 0.074 V for each pCl unit suggesting following reaction:

$$A: SbCl_2^+ + SbCl_4^- \rightarrow A^+ + 2SbCl_3$$

For Class B Amines $E_{1/2}$ is independent of pCl of solution. Eqn. for reaction is:

$$B - e \rightarrow B^+.$$

The Reaction: Amine + $2SbCl_3 \rightleftarrows$ Amine: $SbCl_2^+ + SbCl_4^-$ is therefore to right for Class A Amines and to left for those of Class B.

XI. ADDITIONAL REFERENCES AND DATA SOURCES

6. Cryoscopy

Compound (Solute)	Data	Reference and Comments
Non Electrolytes		
Xylene, anthracene, diphenyl-methane, acetophenone and benzophenone	$K_f = 18.4$ deg Mol^{-1} Kg	[1a] no non ideality correction
Anthracene, fluorene, benzophenone and dibenzyl	Normal solutes but have intense colors and marked $-ve$ deviation from Raoult's Law $\rightarrow K_f = 15.6$ deg Mol^{-1} Kg	[2] full correction for non ideality. *See* Solvent Physical Properties Section
Stilbene	Dimerises	[2]
	Product is mixture of 3 cyclic dimers. One has indane structure and other two are isomeric tetralin derivatives	[30]
Inorganic halides		
KCl, HgCl$_2$	Dissociation	[16] See conductivity section
AsCl$_3$	Association	
AsBr$_3$, AsI$_3$	Halogen exchange e.g., $AsBr_3 + SbCl_3 \rightleftarrows AsBr_2Cl + SbCl_2Br$.	
SbBr$_3$, SnCl$_4$, BiCl$_2$	Solid solutions formed (BiCl$_3$ elevated f.p.)	
Others		
Sulphur	Molecular S$_8$ in dilute solution	
Iodine	Molecular I$_2$ in dilute solution	

6. Cryoscopy (*Continued*)

Compound (Solute)	Data	Reference and Comments
Electrolytes		
Me$_4$NBr, Ph$_3$CCl, Me$_4$NCl, CsCl, KCl	i factors tend to 2 as C $\to$ 0 Molal osmotic coeff. ϕ is given by: $\phi = 1-1.86m_2^{1/2}\sigma(0.768\require{mediawiki-texvc}\require{}å m_2^{1/2} + bm_2)$ m_2 = molality of solute å = distance of closest approach of ions (Angstrom) b is a constant. This gives for å KCl 2.1 CsCl 4.5 Me$_4$NCl 13.0 Ph$_3$CCl 15 KCl and CsCl are associated (Bjerrum critical distance = 7.3 Å) Me$_4$NCl solvated by at least one complete solvation shell	[2] Ph$_3$C$^+$ (not spherical)

XI. ADDITIONAL REFERENCES AND DATA SOURCES 799

References

1. S. Tolloczko, (a) Z. Phys. Chem. (1899), **30**, 705; (b) Bull. Int. Acad. Sci. Cracovie, (1901), **1**, 1.
2. G. B. Porter and E. C. Baughan, J. Chem. Soc. (1958), 744.
3. Z. Klemenziewicz, Bull. Int. Acad. Sci. Cracovie (1908), **6**, 485.
4. J. R. Atkinson, E. C. Baughan and B. Dacre, J. Chem. Soc. (A), (1970), 1378.
5. J. E. Martin, Division of Chemical Standards N.P.L., Personal Communication in Ref. 4.
6. H. Oppermann, Z. Anorg. Allgem. Chem. (1967), **356**, 1.
7. Schlundt, J. Phys. Chem. (1901), **5**, 503.
8. Patzitka and R. Bertrum, J. Electroanal. Chem. (1970), **28**, 119.
9. Zhuravlev, J. Phys. Chem. U.S.S.R. (1939), **13**, 684.
10. J. W. Mellor, Comprehensive Treatise on Inorganic and Theoretical Chemistry, Longmans, (1929), **9**, 471–2.
11. J. W. Smith, "Electric Dipole Moments," Butterworths, London (1955), p. 118.
12. Gmelin, "Handbuch der Anorg. Chemie," 8 Auflage (1949), **18B2**, 408.
13. A. G. Davies and E. C. Baughan, J. Chem. Soc. (1961), 1711.
14. J. R. Atkinson, T. P. Jones and E. C. Baughan, J. Chem. Soc. (1964 Suppl.), 5808.
15. P. V. Johnson and E. C. Baughan, J. Chem. Soc. (A), (1969), 2686.
16. P. Texier, Ph.D. Thesis, Paris 1970.
17. Z. Klemenziewicz and Z. Balowna, Roczn. Chem. (1930), **10**, 481.
18. G. Jander and K. H. Swart, Z. Anorg. Allg. Chem. (1959), **299**, 252.
19. K. Saito, K. Ichikawa and M. Shimoji, Bull. Chem. Soc. Japan (1968), **41**, 1104.
20. Walden, Z. Anorg. Chem. (1900), **25**, 209.
21. (a) Z. Klemenziewicz, Z. Phys. Chem. (1924), **113**, 28. (b) Z. Klemenziewicz and Z. Balowna, Roczn. Chem. (1930), **10**, 481. (c) Z. Klemenziewicz and Z. Balowna, Roczn. Chem. (1931), **11**, 683. (d) Z. Klemenziewicz and A. Zebrowska, Roczn. Chem. (1934), **14**, 14.
22. E. C. Baughan, T. P. Jones and L. G. Stoodley, Proc. Chem. Soc. (1963), 274.
23. G. B. Porter, J. Simpson and E. C. Baughan, J. Chem. Soc. (A), (1970), 2806.
24. P. V. Johnson, Ph.D. Thesis, C.N.A.A. London, 1969.
25. Frycz and Tolloczko, Ksiega Pamiatkowa (Lwow), (1912), **1**, 1, Chem. Zentra, (1913), **1**, 91.
26. D. Bauer, J. P. Beck and P. Texier, Coll. Czech. Chem. Comm. (1971), **36**, 940.
27. D. Bauer and P. Texier, Compt. Rend. (1968), **266C**, 602.
28. D. Bauer and J. P. Beck, Coll. Czech. Chem. Comm. (1971), **36**, 323.
29. D. Bauer and J. P. Beck, J. Electro. Anal. Chem. (1971), **32**, 21.
30. M. Hiscock and G. B. Porter, J. Chem. Soc., Perkin II, (1972), 79.

(b) Electrical Conductance

ACETIC ACID

R. C. Paul, S. C. Ahluwalia, S. K. Rehari, and S. Singh Pahil, Indian J. Chem., **3**, 297 (1965). Nature of solutions of Lewis acids in acetic acid systems: antimony (V) chloride, tin (IV) chlorine, tin (IV) bromide and boron (III) fluoride.

M. M. Jones and E. Griswold, J. Amer. Chem. Soc., **76**, 3247 (1954). Conductances of some salts and ion-pair equilibria in acetic acid at 30° (bromides, acetates and formates of potassium, sodium and lithium).

ACETIC ANHYDRIDE

G. Jander and H. Surawski, Z. Elektrochem. Angew Physik Chem., **65**, 384 (1961). Weak and strong electrolytes in acetic anhydride.

ACETONE

A. M. Golub and V. A. Kalibabchuk, Russ. J. Inorg. Chem., **11**, 320 (1966). The behavior of $ThCl_4$ and the formation of thiocyanato complexes of thorium in nonaqueous solutions.

F. Accascina and S. Schiavo, Annali di Chimica., **42**, 695 (1953). Conductances of alkali metal perchlorates in acetone.

A. H. Ewald and J. A. Scudder, Austr. J. Chem., **23**, 695 (1970). The effect of pressure on the conductance of some iodides in acetone and 2-methylpropan-1-ol.

J. Thomas, Ph.D. Thesis, Case Western Reserve (1970). Conductance measurements of electrolytes in formamide, acetone and acetone-n-propanol mixtures.

G. Wikander, A. M. Nilsson, A. Holmgren, and P. Beronius, Acta. Chem. Scand., **25**, 1468 (1971). Association of ammonium bromide in acetone.

V. M. Tsentovskii, V. P. Barabanov, F. M. Kharrasava, and T. A. Busygina, Zh. Obshch. Khim., **41**, 1659 (1971). Conductance of halides of tetraalkyl (aryl) phosphonium in acetone, DMF and nitromethane.

XI. ADDITIONAL REFERENCES AND DATA SOURCES

ACETONITRILE

E. G. Hackenberg and H. Ulich, Z. Anorg. Chem., **243**, 99 (1939). Conductance of $Al(ClO_4)_3$ in anhydrous acetonitrile.

D. S. Payne, J. Chem. Soc., 1052 (1953). The electrolytic conductance of solutions of phosphorus pentachloride in acetonitrile.

R. P. T. Tomkins, E. Andalaft, and G. J. Janz, Trans. Faraday Soc., **65**, 1906 (1969). Conductance, density and viscosity of NaI in anhydrous acetonitrile at 25°C.

M. Graulier, Compt. Rend., **247**, 2139 (1958). Conductivities of ICl_3 and PCl_5 in acetonitrile.

BENZONITRILE

D. S. Payne, J. Chem. Soc., 1052 (1953). The electrolytic conductance of solutions of phosphorus pentachloride in benzonitrile.

BENZOYL CHLORIDE

D. S. Payne, J. Chem. Soc., 1052 (1953). The electrolytic conductance of solutions of phosphorus pentachloride in benzoyl chloride.

BUTANOL

O. A. Osipov, O. E. Koshireninov, and A. V. Leshchenko, Russ. J. Organ. Chem., **9**, 406 (1964). Electrical conductivity of niobium oxide chloride in organic solvents.

Golik, Allanazarev, and Cholpan, Izvestiya Akad. Nauk. Turkmenskoi SSR. Seriya Khim., Geologicheskikh. Nauk., **4**, 112 (1969) Electrical conductance of LiCl in n-butanol.

CARBON TETRACHLORIDE

K. F. Denning and J. A. Plambeck, Can. J. Chem., **50**, 1600 (1972). Conductance and viscosity of the tetrabutylammonium bromide-carbon tetrachloride system.

DIETHYL ETHER

Yu. A. Kopylov, T. N. Trofimova, and Yu. M. Stolovitskii, Russ. J. Phys. Chem., **39**, 262 (1965). Temperature dependence of electrical conductivity of monohalogene-benzenes and their solutions in diethyl ether.

N,N-DIMETHYLACETAMIDE

G. R. Lester, J. N. Vaughn, and P. G. Sears, Trans. Kentucky Acad. Sci., **19**, 28 (1958). The conductance behavior of some lithium, strontium and barium salts in N,N-Dimethylacetamide at 25°C.

G. Vicentini and C. Airoldi, J. Inorg. Nucl. Chem., **33**, 1733 (1971). Addition compounds between lanthanide bromides and N,N-dimethylacetamide.

DIMETHYLFORMAMIDE

A. M. Golub and V. A. Kalibabchuk, Russ. J. Inorg. Chem., **11**, 320 (1966). The behavior of $ThCl_4$ and the formation of thiocyanato complexes of thorium in nonaqueous solutions.

R. C. Paul and B. R. Sreenathan, Indian J. Chem., **4**, 348 (1966). Dimethylformamide as a polar solvent: electrochemical studies of weak acids.

R. C. Paul, S. Sharda, and B. R. Sreenathan, Indian J. Chem., **2**, 97 (1964). Electrochemical studies of Lewis acids in dimethylformamide. Complexes studied: SO_3, tin (IV) halides, aluminum chloride, cadmium chloride and antimony (V) chloride.

P. Bruno and M. Della Monica, J. Phys. Chem., **76**, 1049 (1972). Conductance behavior of tetraamylammonium bromide and potassium picrate in some nonaqueous solvents.

DIMETHYLSULFOXIDE

J. A. Bolzan and A. J. Aruia, Electrochim. Acta, **15**, 39 (1970). Electrical conductance of hydrogen chloride in dimethylsulfoxide at 25°C.

A. Buckingham and R. P. H. Gasser, J. Chem. Soc., 1964 (1967). Complex formation in DMSO: Mercuric iodide and potassium iodide.

D. L. Venlzky and J. E. Quick, J. Chem. Eng. Data, **17**, 23 (1972). Molecular weights and conductivities of diphenylphosphinic acid, diphenylphosphinic anhydride and acetic diphenylphosphinic anhydride in DMSO.

P. Bruno and M. Della Monica, J. Phys. Chem., **76**, 1049 (1972). Conductometric behavior of tetraamylammonium bromide and potassium picrate in some nonaqueous solvents.

XI. ADDITIONAL REFERENCES AND DATA SOURCES

DIOXANE

R. Bury and J. C. Justice, Compt. Rend., **260**, 6089 (1965). Conductivity of $KClO_4$ in dioxane at 25°C.

T. Erdey-Gruz and I. Czako Nagyne, Magy. Kem. Foly., **46**, 583 (1970). Conductance of HCl, LiCl, KCl and KF in dioxane from 5 to 25°C.

DIPHENYL SULFOXIDE

S. K. Ramalingam and S. Soundararajan, Bull. Chem. Soc. (Japan), **41**, 106 (1968). Diphenyl sulphoxide complexes of lathanide and yttrium perchlorates.

ETHANOL

S. Ernst and B. Jezowska-Trzebiatowska, J. Inorg. Nucl. Chem., **28**, 2885 (1966). Investigations of uranyl nitrate dihydrate in alcohols by conductivity.

ETHYLENE GLYCOL

M. D. Carmo and M. N. Dos Santos, Rev. Port. Quim., **11**, 50 (1969). Conductances in ethylene glycol.

FORMAMIDE

M. L. Berardelli, G. Pistoia, and A. M. Polcaro, Ric. Sci., **38**, 814 (1968). Conductances and solubilities of several halides in formamide and N-methylformamide at 25°C.

J. Thomas, Ph.D. Thesis, Case Western Reserve (1970). Conductance measurements of electrolytes in formamide, acetone and acetone-n-propanol mixtures.

P. Bruno and M. Della Monica, J. Phys. Chem., **76**, 1049 (1972). Conductometric behavior of tetraamylammonium bromide and potassium picarate in some nonaqueous solvents.

FORMIC ACID

H. I. Schlesinger and F. H. Reed, J. Amer. Chem. Soc., **41**, 1921 (1919). The behavior of mixtures of two salts containing a common ion in anhydrous formic acid solution.

H. I. Schlesinger and R. D. Mullinix, J. Amer. Chem. Soc., **41**, 72 (1919). The conductivity of alkaline earth formates in anhydrous formic acid.

HEPTANOL

O. A. Osipov, O. E. Kashireninov, and A. V. Leshchenko, Russ. J. Inorg. Chem., **9**, 406 (1964). Electrical conductivity of niobium oxide chloride in organic solvents.

HYDROGEN FLUORIDE

M. L. Kilpatrick, N. Kilpatrick, and J. G. Jones, J. Amer. Chem. Soc., **87**, 2806 (1965). The conductance of nitronium fluoroborate in anhydrous hydrogen fluoride.

R. P. Clark and J. R. Moser, J. Electrochem. Soc. **118**, (1971). Electrical conductivity measurements in anhydrous hydrogen fluoride.

METHANOL

O. A. Osipov, O. E. Kashireninov, and A. V. Leshchenko, Russ. J. Inorg. Chem., **9**, 406 (1964). Electrical conductivity of niobium oxide chloride in organic solvents.

R. E. Busby and V. S. Griffiths, J. Chem. Eng. Data, **10**, 29 (1965). Conductivity of silver nitrate in nonaqueous and mixed solvents. (nitromethane, pyridine, benzene).

D. O. Johnston, Diss, Abs., **25**, 127 (1964). Conductances, viscosities and spectra of some rare earth and cadmium salts in methanol.

S. Ernst and B. Jezowska-Trzebiatowska, J. Inorg. Nucl. Chem., **28**, 2885 (1966). Investigations of uranyl nitrate dihydrate in alcohols by conductivity.

P. A. D. DeMaine and G. E. McAlonie, J. Inorg. Nucl. Chem., **18**, 286 (1961). Conductance studies of iron-group chlorides dissolved in methanol.

S. Taniewska-O. Sinska, Lodzkie Towarzyorwe Nauk. Wydzial III. Acta Chim., **12**, 51 (1967). Specific electrical conductivity of benzoic acid solutions in methanol.

A. G. Ogston, Trans. Faraday Soc., **32**, 1679 (1936). Temperature coefficients of electrical conductivity of electrolytes in methyl and ethyl alcohols.

A. Z. Golik, G. Allanazarov, and P. F. Cholpan, Izv. Akad. Nauk., **5**, 116 (1969). Viscosity and conductance of $MgCl_2$ in alcoholic solutions.

P. A. D. De Maine and E. J. Walsh, J. Inorg. Nucl. Chem., **19**, 156 (1961). Conductance studies at 20 and 45°C for hydrated tin chlorides in methanol and in methanol–carbon tetrachloride.

XI. ADDITIONAL REFERENCES AND DATA SOURCES

METHANOL (Continued)

R. L. Kay and J. L. Hawes, J. Phys. Chem., **69**, 2787 (1965). The association of cesium chloride in anhydrous methanol at 25°C.

R. Lovas, G. Macri, and S. Petrucci, J. Amer. Chem. Soc., **92**, 6502 (1970). Electrical conductance of anhydrous potassium, magnesium, cobalt, nickel, copper and zinc m-Benzenedisulfonates in methanol at 25°.

T. H. Mead, O. L. Hughes, and H. Hartley, J. Chem. Soc., 1207 (1933) The conductivity of tetramethylammonium salts in methanol and ethanol.

N-METHYLACETAMIDE

C. M. French and K. H. Glover, Trans. Faraday Soc., **51**, 1427 (1955). Electrical conductance of solutions in N-methylacetamide. Systems studied: seven alkali metal halides and two tetra-alkylammonium salts.

N-METHYLBUTYRAMIDE

L. R. Dawson, R. H. Graves, and P. G. Sears, J. Amer. Chem. Soc., **79**, 298 (1957). Solutions of sodium and potassium halides in N-methylpropionamide and in N-methylbutyramide from 30 to 60°.

N-METHYLFORMAMIDE

G. P. Johari and P. H. Tewari, J. Phys. Chem., **69**, 3167 (1965). Magnesium sulfate as an unassociated salt in N-methylformamide.

M. L. Berardelli, G. Pistoia, and A. M. Polcaro, Ric. Sci., **38**, 814 (1968). Conductances and solubilities of several halides in formamide and N-methylformamide at 25°C.

R. D. Singh and R. Gopal, Bull. Chem. Soc. Japan, **45**, 2088 (1972). Conductance and Walden product of tetraalkylammonium ions in N-methylformamide and N-methylpropionamide at different temperatures.

R. I. Mostkova, Y. M. Kessler, and V. N. Semenova, Elektrokhimiya, **7**, 642 (1971). Conductance, viscosity and density of NaCl and CsCl in solutions in N-methylformamide at 5° and 25°C.

N-METHYLPROPIONAMIDE

L. R. Dawson, R. H. Graves, and P. G. Sears, J. Amer. Chem. Soc., **79**, 298 (1957). Solutions of sodium and potassium halides in N-methylpropionamide and in N-methylbutyramide from 30 to 60°.

R. D. Singh and R. Gopal, Bull. Chem. Soc. Japan, **45**, 2088 (1972). Conductance and Walden product of tetraalkylammonium ions in N-methylformamide and N-methylpropionamide at different temperatures.

NITROBENZENE

E. G. Hackenberg and H. Ulich, Z. Anorg. Chem., **243**, 99 (1939). Conductance of $Al(ClO_4)_3$ in anhydrous nitrobenzene.

D. S. Payne, J. Chem. Soc., 1052 (1953). The electrolytic conductance of phosphorus pentachloride in nitrobenzene.

O. A. Osipov, O. E. Kashireninov, and A. V. Leshchenko, Russ. J. Inorg. Chem. **9**, 406 (1964). Electrical conductivity of niobium oxide chloride in organic solvents.

E. Ya. Gorenbein, G. G. Rusin, and A. T. Beznis, Russ. J. Inorg. Chem., **11**, 310 (1966). Complexes of aluminum halides with tetrahydrofuran and their electrical conductivity in nitrobenzene.

E. Ya. Gorenbein and A. A. Fominskaya, Russ. J. Inorg. Chem., **12**, 1103 (1967). Electrical conductivity of complexes of aluminum bromide with organic substances in nitrobenzene.

W. Reed, D. W. Secret, R. C. Thompson, and P. A. Yeats, Can. J. Chem., **47**, 4275 (1969). Electrical conductivity studies of fluorosulfuric acid and tetramethylammonium fluorosulfate in nitrobenzene.

M. Graulier, Compt. Rend., **247**, 2139 (1958). Conductances of ICl_3, ICl, PCl_5, $AlBr_3$ and $AlCl_3$ in nitrobenzene.

F. Barreira and G. J. Hills, Trans. Faraday Soc., **64**, 1359 (1968). Pressure and temperature coefficients of conductances in nitrobenzene.

NITROMETHANE

W. E. Bull and R. G. Ziegler, Inorg. Chem., **5**, 689 (1966). Metal complexes of N,N,N',N'-tetramethylmalonamide.

N-(ρ-METHOXYBENZYLIDENE)-ρ-n-BUTYLANILINE (MBBA)

G. Heppke and F. Schneider, Ber. Bunsenges. Phys. Chem., **75**, 1231 (1971). The equivalent conductance of tetrapropylammonium picrate in NBBA.

PHOSPHORYL BROMIDE

R. C. Paul and S. K. Vasisht, J. Indian Chem. Soc., **43**, 141 (1966). Phosphoryl bromide as a polar solvent.

XI. ADDITIONAL REFERENCES AND DATA SOURCES

1-PROPANOL

T. A. Gover and P. G. Sears, J. Amer. Chem. Soc., **60**, 330 (1956). Conductance of NaI, KI, NaSCN, KSCN and $(Et)_4NBr$, $(Et)_4NI$, $(Pr)_4NBr$ and $(Pr)_4NI$ in 1-propanol at 25°.

2-PROPANOL

M. A. Matesich, J. A. Nadas, and D. Fennell Evans, J. Phys. Chem., **74**, 4568 (1970). Conductance of tetraalkylammonium salts in 2-propanol.

PROPYLENE CARBONATE

L. M. Mukherjee, D. P. Boden, and R. Lindauer, J. Phys. Chem., **74**, 1942 (1970). Behavior of electrolytes in propylene carbonate. Conductance and viscosity properties. Evaluation of ion conductances. Electrolytes studied: KI, $KClO_4$, $(i\text{-}Am)_4N(i\text{-}Am)_4B$, and $(i\text{-}Am)_4NI$.

PYRIDINE

O. A. Osipov, O. E. Kashireninov, and A. V. Leshchenko, Russ. J. Inorg. Chem., **9**, 406 (1964). Electrical conductivity of niobium oxide chloride in organic solvents.

SULFOLANE

P. Bruno and M. Della Monica, J. Phys. Chem., **76**, 1049 (1972). Conductometric behavior of tetraamylammonium bromide and potassium picrate in some nonaqueous solvents.

TETRAMETHYLUREA

B. J. Barker and J. A. Caruso, J. Amer. Chem. Soc., **93**, 1341 (1971). Solvation and chemical equilibrium studies of alkali metal salts in 1,1,3,3-tetramethylurea.

XYLENE

L. E. Simanavichus and A. M. Levinskene, Sov. Electrochem., **2**, 324 (1966). Some properties of solutions of alummium bromide in xylene.

B. L. Solnick, Diss. Abs., **30B**, 5001 (1970). A conductance-viscosity study of tri-n-amylammonium picrate in p-xylene.

ACETONE–WATER

A. Ya Deich, Russ. J. Phys. Chem., **36**, 1344 (1962). Density, viscosity and specific conductance of cadmium bromide solutions in acetone–water mixtures.

t-BUTANOL–WATER

F. Accascina, R. De Lise, and M. Goffredi, Electrochim. Acta, **15**, 1209 (1970). The conductance behavior of some alkaline perchlorates in water-*t*-Butanol mixtures at 25°C.

T. L. Broadwater and R. L. Kay, J. Phys. Chem., **74**, 3802 (1970). Ionic mobilities in tert–butyl alcohol–water mixtures at 25°. (KCl, CsCl, LiBr, NaBr, KBr, Me$_4$NBr, Bu$_4$NBr and Me$_4$NI).

J. Julliard, J. P. Morel, and L. Avedikian, J. Chem. Phys., **69**, 787 (1972). Conductivities of potassium formate, acetate, isobutyrate, cyclohexanecarboxylate and benzoate in water–tert butyl alcohol mixtures.

DIMETHYL SULFOXIDE–WATER

B. J. Yager and T. W. Cowley, Texas J. Sci., **23**, 89 (1971). The effects of ethanol–water and DMSO-water solvent systems on the conductivity of HCl, NaOH and KCl.

DIOXANE–WATER

E. M. Khairy, A. El-Said Mahgoub, and A. I. Mosaad, J. Electroanal. and Interfacial Chem., **23**, 115 (1969). The electrical conductivity of uranyl acetate in water and water–dioxane mixtures.

P. B. Das, P. K. Das, and D. Patnaik, J. Indian Chem. Soc., **36**, 410 (1959). Conductance of MgSO$_4$ in dioxane–water mixtures at 35°C.

T. Erdey-Gruz, E. Kugler, I. Nagy-Czako, and K. Balthazar-Vass, Acta Chim. (Budapest), **71**, 353 (1972). Anomalous temperature coefficient of the conductance of electrolytes in dioxane–water mixtures.

M. V. Ramana-Murti and R. C. Yadav, Electrochim. Acta, **17**, 643 (1972). Conductance of NaNO$_3$ in dioxane–water mixtures at 25°C.

L. B. Yeatts and W. L. Marshall, J. Phys. Chem., **76**, 1053 (1972). Electrical conductance and ionization behavior of NaCl in dioxane–water solutions at 300° and pressures to 4000 bars.

P. B. Das, J. Instr. Chem. India, **40**, 205 (1968). Conductance of NaCl in dioxane-H$_2$O mixtures at 40°C.

L. B. Yeatts, L. A. Dunn, and W. L. Marshall, J. Phys. Chem., **75**, 1099 (1971). Electrical conductances and ionization behavior of NaCl dioxane–water solutions at 100°C and pressures to 4000 bars.

XI. ADDITIONAL REFERENCES AND DATA SOURCES

ETHANOL–WATER

E. S. Amis and J. F. Casteel, J. Electrochem. Soc., **117** 213 (1970). The equivalent conductance of magnesium sulfate in water–ethanol solvents.

A. Than and E. S. Amis, J. Inorg. Nucl. Chem., **31**, 1685 (1969). The specific conductance of concentrated solutions of magnesium chloride in the water–ethanol system.

A. R. Tourky and A. A. Abdel-Hamid, Z. Phys. Chemie, Leipzig, **247**, 289 (1971). Studies of proton conductance in mixed solvents (HCl and KCl).

B. J. Yager and T. W. Cowley, Texas J. Sci., **23**, (1971). The effects of ethanol–water and DMSO-water solvent systems on the conductivity of HCl, NaOH and KCl.

J. P. Demey, G. Delesalle and P. Devrainne, C. R. Acad. Sc. Paris, **273**, 1677 (1971). Conductance of tetrabutylammonium iodide in ethanol-water mixtures.

G. Delesalle, J. P. Demey, P. Devrainne and J. Heubel, C. R. Acad. Sc. Paris, **274**, 455 (1972). Conductance of potassium iodide in ethanol-water mixtures.

J. F. Castrell and E. S. Amis, J. Chem. Eng. Data, **17**, 55 (1972). Specific conductance of concentrated solutions of magnesium salts in water–ethanol system at 25, 35 and 45°C.

R. Whorton and E. S. Amis, Z. Physik. Chemie, **17**, 300 (1958). The equivalent conductance of tetraethylammonium picrate in the water–ethanol system.

ETHYLENE GLYCOL–WATER

F. Accascina and A. D'Aprano, Ric. Sci., **37**, 257 (1966). Conductance of NaBr in ethylene glycol–water mixtures.

B. Sesta and M. L. Berardelli, Electrochim. Acta., **17**, 915 (1972). Conductances of $LiNO_3$, $NaNO_3$ and KNO_3 in ethylene–glycol water mixtures at 25°C.

GLYCEROL–WATER

H. Sadek, A. M. Hafez, and F. Y. Khalil, Electrochim. Acta, **14**, 1089 (1969). Conductance of KIO_3 in glycerol–water mixtures.

METHANOL–WATER

H. Tsubota and G. Atkinson, J. Phys. Chem., **71**, 1131 (1967). The association of $MnSO_4$ in methanol–water mixtures of high methanol content.

F. Barbulescu, I. St. Popescu, and I. Sass, Rev. Rou. de Chim., **11**, 903 (1966). Electrical conductivity of zinc chloride in methanol–water mixtures.

B. Sesta and M. Ludovica Berardelli, Ric. Sci., **38**, 1199 (1968). Electrical conductivity of alkali metal mitrates in methanol–water mixtures at 25°C.

F. Butera, N. Goffredi, and R. Triol, Atti. Accad. Sci. Lett. Arti. Palermo, Parte, **27**, 213 (1968). Conductances of alkali metal perchlorates in alcohol-water mixtures.

D. Hartmann and E. U. Franck, Ber. Bunsengesellschaft, 514, (1969). Conductance of KCl in alcohol–water mixtures at high temperatures and pressures.

L. G. Longsworth and D. A. MacInnes, J. Phys. Chem., **43**, 239 (1938). Ion conductances in water–methanol mixtures (NaCl and LiCl).

A. R. Tourky and A. A. Abdel-Hamid, Z. Phys. Chemie, Leipzig, **247**, 289 (1971). Studies of proton conductance in mixed solvents (HCl and KCl).

J. P. Demey, G. Delesalle, and P. Devrainne, C. R. H. Acad. Sci. Ser., **C273**, 1677 (1971). Conductance of tetrabutylammonium iodide in water–methanol mixtures at 25°C.

n-PROPANOL–WATER

A. R. Tourky and A. A. Abdel-Hamid, Z. Phys. Chemie, Leipzig, **247**, 289 (1971). Studies of proton conductance in mixed solvents (HCl and KCl) also in iso-propanol-water mixtures.

M. Kotaka, E. Kubota, J. Mori, and M. Yokoi, Nippon Kagaku Zasshi, **92**, 18 (1971). The conductance of NH_4OH, NaOH and NH_4Cl in 1-propanol water mixtures.

XI. ADDITIONAL REFERENCES AND DATA SOURCES

SULFOLANE–WATER

A. Sacco, G. Petrella, and M. Castagnolo, Z. Naturforsch, **A26,** 1306 (1971). The conductance of KCl in sulfolane–water mixtures at 35°C.

ACETONE–BENZENE

E. Ya. Gorenbein and A. I. Karpovich, Izv. Vyssh. Ucheb. Zaved., Khim. Khim. Technol., **13,** 1588 (1970). Effect of the dielectric constant of a mixture of solvents (acetone–benzene and acetone–CCl_4) on the viscosity and electrical conductance of tetrabutylammonium iodide solutions.

ACETONE–CHLOROFORM

A. M. Shkodin, and I. A. Sergeeva, Elektrokhimiya, **7,** 552 (1971). Investigation of conductance of $LiClO_4$, $NaClO_4$ solutions in mixtures of acetone with chloroform.

ACETONE–ETHANOL

G. Pistoia and G. Pecci, J. Phys. Chem., **74,** 1450 (1970). Ion-pair association of cesium and tetraethylammonium perchlorates in ethanol–acetone mixtures at 25°C.

ACETONITRILE–NITROBENZENE

G. Kortum, S. D. Gokhale, and H. Wilski, Z. Physik. Chemie, **4,** 286 (1955). Conductance of tetraethylammonium iodide in acetonitrile–nitrobenzene mixtures.

BENZENE–NITROBENZENE

J. Macau and L. Lamberts, Bull Soc. Chim. Belg., **80,** 551 (1971). Conductance of n-butylammonium picrate in alcohol–dioxane and benzene–nitrobenzene mixtures.

BUTANOL–HEXANE

A. M. Sukhotin, Russ. J. Phys. Chem., **33,** 450 (1959). Electrical conductance of NaI, LiI, LiBr, LiCl and $(iso\text{-}C_5H_{11})_4$ NI in butanol–hexane mixtures.

E. M. Ryzhkay and A. M. Sukhotin, Russ. J. Phys. Chem., **34,** 983 (1960). Electrochemistry of solutions of hydrogen chloride in solvents of low dielectric constant (butanol–hexane mixtures).

BUTANOL–METHANOL

F. Accascina, M. Battistini, and S. Schiavo, Scienza Tecnica, **5,** 107 (1961). Conductance of lithium perchlorate in butanol–water mixtures.

F. Accascina and S. Petrucci, La Ricerca Sci., **29,** 1383 (1959). The conductance of tetraethylammonium picrate in methanol–butanol mixtures at 25°.

D. Singh and I. P. Aggarwal, Z. Phys. Chem. (Frankfurt), **73,** 144 (1970). Conductance of KCl, RbCl, CsCl in methanol–butanol mixtures at 25°C.

CARBON TETRACHLORIDE–ETHANOL

A. M. Sukhotin, Russ. J. Phys. Chem., **33,** 450 (1959). Electrical conductance of solutions of NaI in ethanol–carbon tetra-chloride mixtures.

CARBON TETRACHLORIDE–METHANOL

P. A. D. De Maine and E. J. Walsh, J. Inorg. Nucl. Chem., **19,** 156 (1961). Conductance studies at 20 and 452°C for hydrated tin chlorides in methanol and in methanol–carbon tetrachloride.

DIOXANE–ETHANOL

J. Macau and L. Lamberts, Bull. Soc. Chim. Belg., **80,** 551 (1971). Conductance of n-butylammonium picrate in alcohol–dioxane and benzene–nitrobenzene mixtures.

DIOXANE–METHANOL

D. Singh and I. P. Aggarwal, Z. Physik. Chemie, **76,** 50 (1971). Conductance of tetraethylammonium bromide in methanol–dioxane mixtures at 25°C.

N. K. Levitskaya, L. I. Tkachenko, and A. M. Shkodin, Izv. Vyssh. Ucheb. Zaved. Khim. Khim. Tekhnol., **13,** 1586 (1970). Electrical conductance of tetrabutylammonium iodide in methanol–dioxane mixtures.

ETHER–PHOSPHORIC ACID

A. V. Solomin, E. I. Kryuchkova, and K. I. Omarova, Z. Prik. Khim., **42,** 1673 (1968). Solubility of certain monoalkylbenzenes in ether–phosphoric acid mixtures and electrical conductivity of the solutions.

XI. ADDITIONAL REFERENCES AND DATA SOURCES

GENERAL

A. M. Shkodin, L. P. Sadovnichaya, Popolidy Akad. Nauk. Ukr. SSR, **30,** 357 (1968). The anomalous properties of some alkali metal halides in aliphatic alcohols.

D. N. Bennion, J. S. Dunning, and W. H. Tiedemann, Technical Report, UCLA-ENG-7078 December (1970) prepared for U.S. Naval Weapons, Corona Lab, Calif. under Contract No. N00123-70-C-0188. Design studies for high rate, high energy, nonaqueous electrochemical energy conversions systems. Contains information on density, viscosity and conductivity of lithium-trifluoromethanesulfonate in dimethylsulfite and conductance of m-dinitrobenzene in liquid ammonia solutions.

R. J. Jasinski, "High Energy Battery Systems based on Propylene Carbonate," Scientific Report No. 2 prepared for Air Force Cambridge Research Labs, Office of Aerospace Research, U.S. Air Force, Bedford, Mass. (July 1969), Contract AFCRL-69-0381. Conductivities of various salt solutions including $LiClO_4$.

R. M. Fuoss, Proc. Natl. Acad. Sci., **45,** 807 (1959). Dependence of the Walden product on dielectric constant.

R. Fernandez Prini, Trans. Faraday Soc., **65,** 3311 (1969). A modified expression for the concentration dependence for the conductance of electrolyte solutions.

L. Fischer, G. Winkler, and G. Jander, Z. Electrochem., **62,** 1 (1958). General review.

G. Jander and G. Moass, Fortschr. Chem. Forsch., **2,** 619 (1953). General review.

(c) Viscosity

CARBON TETRACHLORIDE

K. F. Denning and J. A. Plambeck, Can. J. Chem., **50,** 1600 (1972). Conductivity and viscosity of the tetrabutylammonium bromide–carbon-tetrachloride system.

FORMAMIDE

P. Bruno and M. Della Monica, J. Phys. Chem., **76,** 3034 (1972). Density, viscosity, and electrical conductance of sodium salts in formamide solutions at 25°.

GLYCINE

S. Phang, Austr. J. Chem., **25,** 1575 (1972). The viscosity and conductivity of KCl and KI in glycine and urea solutions.

DIOXAN–WATER

E. M. Kartzmark, Can. J. Chem., **50,** 2845 (1972). Conductances, densities, and viscosities of solutions of sodium nitrate in water and in dioxane–water at 20°C.

B. Das, K. Singh, and P. K. Das, J. Ind. Chem. Soc., **49,** 561 (1972). Viscosities of solutions of KCl, NaCl, KBr, and NaBr in dioxan–water mixtures at 30 and 40°C and activation energies and entropies for viscous flow.

GENERAL

J. P. Bare and J. F. Skinner, J. Phys. Chem., **76,** 434 (1972). Electrolyte viscosities in associated solvents.

(d) Transference Numbers

AMMONIA (LIQ)

J. Baldwin, J. Evans, and J. B. Gill, J. Chem. Soc., **A,** 3389 (1971). The mean molal ion activity coefficients and ion transference numbers of NH_4I in liquid NH_3.

METHANOL

G. A. Vidulich, Diss. Abs., **25,** 4432 (1965). The transference numbers of lithium chloride and cadmium chloride in anhydrous methanol at 25°C.

(e) Solubility

ACETONITRILE

Solute	Solubility (M)	Temp.	Ref.
LiF	2.20×10^{-5}	25.00°	1
	3.20×10^{-5}	60.00°	
LiCl	2.70×10^{-2}	R.T.	2
	2.60×10^{-2}	25.00°	1
	1.40×10^{-2}	60.00°	1
LiClO$_4$	1.06	25.00°	1
KPF$_6$	7.50×10^{-1}	R.T.	3
CuF$_2$	7.50×10^{-4}	R.T.	2
	2.70×10^{-4}	R.T.	4
N Me$_4$PF$_6$	1.00×10^{-1}	25.00°	1
	1.80×10^{-1}	60.00°	1
N Pr$_4$PF$_6$	1.10	R.T.	3

1. Rocketdyne, "Properties of Nonaqueous Electrolytes," Final Report No. R-7703, Contract No. NAS3-8521 (1968).

2. P. R. Mallory, "Research and Development of a High Capacity Nonaqueous Battery," Final Report No. NASA CR-54969 Contract No. NAS3-6017 (1965).

3. Globe Union, "A Program to Develop a High Energy Density Primary Battery with a minimum of 200 watt hours per pound of total battery weight." Final Report No. NASA CR-72364, Contract No. NAS3-6015 (1967).

4. Electric Storage Battery, Inc., High Energy System (Organic Electrolytes), Final Report No. ECOM-01394-F, Contract No. 04-28043-AMO-01394(F) (1966).

XI. ADDITIONAL REFERENCES AND DATA SOURCES

LIQUID AMMONIA[a]

Solute	Solubility (moles/mole NH_3)
NH_4SCN	0.861
NH_4NO_3	0.829
$LiNO_3$	0.602
$NaSCN$	0.432
NH_4I	0.432
NH_4Br	0.413
NH_4Cl	0.326
$NH_4(CH_3COO)$	0.332
$NaBr$	0.228
NH_4ClO_4	0.200
$NaNO_2$	0.196
KI	0.186
RbF	0.172
NaI	0.161
$(NH_4)_2S$	0.139
CsI	0.100

[a] J. M. Freund and R. W. Graham, "Basic Research for Ammonia Vapor Activated Batteries," Report No. 1, Contract DA 36-039-SC-72306 July (1956).

BUTYROLACTONE

Solute	Solubility	Temp.	Ref.
$LiBF_4$	2.50×10^{-1}	24.00°	1
$LiPF_6$	2.20×10^{-1}	24.00°	1
NaBr	2.20×10^{-2} mole kg^{-1}	25.00°	2
NaI	3.88 mole kg^{-1}	25.00°	2
$NaPF_6$	1.14 molar	24.00°	1
KCl	6.30×10^{-4} molar	R.T.	3
KI	7.55×10^{-1} mole kg^{-1}	25.00°	2
CsF	1.96×10^{-3} molar	R.T.	4
$CaCl_2$	4.00×10^{-3} mole kg^{-1}	25.00°	2
CoF_3	2.50×10^{-4} molar	R.T.	1
Hg_2Cl_2	1.00×10^{-4} molar	25.00°	5
$HgCl_2$	1.00×10^{-1} molar	25.00°	5
CuF_2	2.20×10^{-4} molar	R.T.	3
	1.50×10^{-3} molar	R.T.	1
	2.00×10^{-4} molar	25.00°	5
CuCl	3.00×10^{-4} molar	25.00°	5
$CuCl_2$	8.00×10^{-2} molar	R.T.	6
	3.00×10^{-3} molar	25.00°	5
AgCl	1.00×10^{-4} molar	25.00°	5
	1.00×10^{-3} molar	R.T.	6
$PbCl_2$	5.00×10^{-4} molar	R.T.	6
	5.00×10^{-4} molar	25.00°	5

XI. ADDITIONAL REFERENCES AND DATA SOURCES

1. Lockheed Missiles and Space Co., "Lithium Anode Limited Cycle Secondary Battery," Final Report No. APL-TDR-64-59, Contract No. AF 33(657)-11709 (1964).
2. University of California Radiation Laboratory "Electrochemical Studies in Cyclic Esters," Report No. UCRL-8781, Contract No. W-7405-ENG-48 (1958).
3. P. R. Mallory, "Research and Development of a High Capacity Nonaqueous Secondary Battery," Final Report No. NASA CR-54969, Contract No. NAS 3-6017 (1965).
4. Battelle Memorial Institute, "Lithium Battery Research," Report No. AFAPL-TR-69-48, Contract No. 33615-68-C-1282.
5. M. L. B. Rao and R. W. Holmes, Electrochem. Technol., **6,** 105 (1968).
6. P. R. Mallory, "Evaluation of Rechargeable Lithium-Copper Chloride Organic Electrolyte Battery System," Contract No. DA-44-009-AMC-1537(T) (1967).

DIETHYL ETHER[a]

Solute	Solubility	Temp.	Ref.
$TiBr_4$	1.20×10^{-1} molar	R.T.	9
$TiBr_2(cptn)_2$	1.30×10^{-3} molar	R.T.	9
$ZrCl_4$	5.00×10^{-3} molar	R.T.	9
$ZrBr_4$	3.00×10^{-3} molar	R.T.	9

[a] W. E. Reid, J. M. Bish and A. Brenner, J. Electrochem. Soc., **104,** 21 (1957).

DIMETHYL FORMAMIDE

Solute	Solubility	Temp.	Ref.
LiF	3.20×10^{-5} molar	25.00°	1
	5.30×10^{-5}	60.00°	1
	5.30×10^{-5} mole kg^{-1}	25.00°	2
LiCl	2.42 molar	25.00°	1
	3.72 molar	60.00°	1
	2.67 mole kg^{-1}	25.00°	3
LiBr	1.91 mole kg^{-1}	25.00°	3
LiClO$_4$	4.40 molar	25.00°	1
	4.80 molar	60.00°	1
LiBF$_4$	1.60 molar	25.00°	1
LiPF$_6$	1.60 molar	25.00°	1
NaF	4.70×10^{-5} mole kg^{-1}	25.00°	2
NaCl	6.10×10^{-3} mole kg^{-1}	25.00°	2
	7.20×10^{-3} mole kg^{-1}	25.00°	3
NaBr	1.00 mole kg^{-1}	25.00°	3
NaI	4.24×10^{-1} mole kg^{-1}	25.00	3
NaSCN	3.69 mole kg^{-1}	25.00°	3
KCl	1.80×10^{-3}	R.T.	6
	2.30×10^{-3}	25.00°	2
	2.70×10^{-3}	25.00°	3
KBr	6.90×10^{-2} mole kg^{-1}	25.00°	3
KI	2.51 mole kg^{-1}	25.00°	3
KPF$_6$	≥ 2.33 molar	R.T.	4
CsCl	3.06×10^{-3} mole kg^{-1}	25.00°	2
CsBr	2.62×10^{-3} mole kg^{-1}	25.00°	2
MgCl$_2$	1.31 mole kg^{-1}	25.00°	3
MgI$_2$	4.28 mole kg^{-1}	25.00°	3

XI. ADDITIONAL REFERENCES AND DATA SOURCES

DIMETHYL FORMAMIDE (Continued)

Solute	Solubility	Temp.	Ref.
$CaCl_2$	5.41×10^{-1} mole kg^{-1}	25.00°	3
$CaBr_2$	9.60×10^{-1} mole kg^{-1}	25.00°	3
$ZnCl_2$	1.70 mole kg^{-1}	25.00°	3
$ZnBr_2$	2.46 mole kg^{-1}	25.00°	3
ZnI_2	3.79 mole kg^{-1}	25.00°	3
$CdCl_2$	4.80×10^{-2} mole kg^{-1}	25.00°	3
$CdBr_2$	2.93 mole kg^{-1}	25.00°	3
CdI_2	4.01 mole kg^{-1}	25.00°	3
Hg_2Cl_2	3.20×10^{-3} molar	25.00°	5
$HgCl_2$	$\geq 1.00 \times 10^{-1}$ molar	25.00°	5
CuF_2	3.40×10^{-4} molar	R.T.	6
	1.00×10^{-4} molar	25.00°	1
	2.00×10^{-4} molar	60.00°	1
	3.00×10^{-4} molar	25.00°	5
$CuF_2 \cdot 2H_2O$	1.00×10^{-1} molar	35.00°	7

1. Rocketdyne, "Properties of Nonaqueous Electrolytes," Final Report No. R-7703, Contract No. NAS 3-8521 (1968).
2. C. M. Criss and E. Luksha, J. Phys. Chem., **72**, 2966 (1968).
3. G. Pistoia, G. Pecci, and B. Scrosati, Ric. Sci., **37**, 1167 (1967).
4. Globe-Union, "A Program to Develop a High-Energy Density Primary Battery with a minimum of 200 watt-hours per pound of total battery," Final Report No. N68-17299 (1967).
5. M. L. B. Rao and R. W. Holmes, Electrochem. Technd., **6**, 105 (1968).
6. P. R. Mallory, "Research and Development of a High Capacity Nonaqueous Secondary Battery," Final Report No NASA CR-54969, Contract No. NAS 3-6017 (1965).
7. Livingston Electronic (Honeywell), "Development of High Energy Density Primary Batteries," Final Report No. NASA CR-72535, Contract No. NAS 3-10613 (1968).

FORMAMIDE[a]

Solute	Solubility	Temp.
NaBr	3.30 molar	25.00°C
KCl	8.90×10^{-1} molar	18.00°C
	9.00×10^{-1} molar	25.00°C
KI	3.71 molar	25.00°C
CsI	1.66 molar	25.00°C
RbBr	1.87 molar	25.00°C

[a] T. Pavlopoulos and H. Strehlow, Z. Phys. Chem. **2**, 89 (1954).

PROPYLENE CARBONATE

Solute	Solubility	Temp.	Ref.
LiF	$<5.0 \times 10^{-6}$ molar	25.00°	1
	$<5.0 \times 10^{-6}$ molar	60.00°	1
LiCl	6.1×10^{-2} molar	25.00°	2
	5.5×10^{-2} molar	R.T.	3
	3.8×10^{-2} molar	25.00°	1
	3.1×10^{-2} molar	60.00°	1
LiBr	2.43 mole/kg	25.00°	4
LiI	1.365 mole/kg	25.00°	4
LiClO$_4$	2.1 molar	25.00°	1
	3.1 molar	60.00°	1
LiAlCl$_4$	6.0×10^{-3} molar	25.00°	5
LiBF$_4$	4.2×10^{-1} molar	24.00°	6
LiPF$_6$	5.5×10^{-1} molar	24.00°	6
NaF	5.0×10^{-1} mole/kg	25.00°	4
NaCl	3.0×10^{-6} mole/kg	25.00°	4

XI. ADDITIONAL REFERENCES AND DATA SOURCES

PROPYLENE CARBONATE (*Continued*)

Solute	Solubility	Temp.	Ref.
NaBr	8.00×10^{-2} mole/kg	25.00°	4
NaI	1.11 mole/kg	25.00°	4
	1.10 mole/kg	25.00°	7
$NaBF_4$	9.35×10^{-2} mole/kg	24.00°	6
$NaPF_6$	8.60×10^{-1} molar	24.00°	6
	6.96×10^{-1} molar	R.T.	7
KCl	5.80×10^{-4} molar	R.T.	3
	4.00×10^{-4} mole/kg^{-1}	25.00°	4
KBr	6.00×10^{-33} mole/kg^{-1}	25.00°	4
KI	2.23×10^{-1} mole/kg^{-1}	25.00°	4
	2.06×10^{-1} mole/kg^{-1}	40.00°	4
	2.30×10^{-1} mole/kg^{-1}	25.00°	7
KBF_4	1.20×10^{-2} mole/kg^{-1}	25.00°	4
KPF_6	1.20 molar (?)	R.T.	8
$CaCl_2$	5.00×10^{-3} mole/kg^{-1}	25.00°	4
$CaBr_2$	7.45×10^{-1} mole/kg^{-1}	25.00°	4
$Ca(BF_4)_2$	6.78×10^{-2} mole/kg^{-1}	25.00°	4
$BaBr_2$	7.80×10^{-3} mole/kg^{-1}	25.00°	4
$AlCl_3$	3.00 molar	25.00°	2
$ZnCl_2$	5.28 mole/kg^{-1}	25.00°	4
$CdBr_2$	4.50×10^{-2} mole/kg^{-1}	25.00°	4
CdI_2	4.54×10^{-2} mole/kg^{-1}	25.00°	4
$MnCl_2$	2.00×10^{-3} mole/kg^{-1}	25.00°	4
$MnBr_2$	1.50 mole/kg^{-1}	25.00°	4

PROPYLENE CARBONATE (*Continued*)

Solute	Solubility	Temp.	Ref.
CoF_3	6.00×10^{-4} molar	R.T.	6
$CoCl_2$	6.50×10^{-3} mole/kg^{-1}	25.00°	4
$NiBr_2$	5.00×10^{-3} mole/kg^{-1}	25.00°	4
Hg_2Cl_2	1.00×10^{-4} molar	25.00°	5
$HgCl_2$	$\geq 1.00 \times 10^{-1}$ molar	25.00°	5
CuF_2	1.60×10^{-4} molar	R.T.	3
	1.90×10^{-4} molar	R.T.	9
	2.40×10^{-2} molar	R.T.	10
	5.00×10^{-3} molar	R.T.	6
	2.00×10^{-4} molar	25.00°	1
	4.00×10^{-4} molar	60.00°	1
	2.00×10^{-4} molar	25.00°	5
$CuF_2 \cdot 2H_2O$	1.20×10^{-2} molar	25.00°	5
$CuCl$	4.00×10^{-3} molar	25.00°	5
$CuCl_2$	1.21×10^{-2} molar	25.00°	10
	8.70×10^{-3} mole/kg	25.00°	4
	4.90×10^{-3} molar	25.00°	1
	4.20×10^{-3} molar	60.00°	1
	8.00×10^{-3} molar	R.T.	11
	8.00×10^{-3} molar	25.00°	5
$CuCl_2 \cdot 2H_2O$	1.23×10^{-3} mole/kg^{-1}	R.T.	11
$CuBr_2$	6.20×10^{-3} mole/kg^{-1}	25.00°	4
$AgCl$	1.00×10^{-4} molar	R.T.	11
	1.00×10^{-4} molar	25.00°	5
$PbCl_2$	1.00×10^{-4} molar	R.T.	11
	2.00×10^{-5} mole/kg	25.00°	4
	1.00×10^{-4} molar	25.00°	5

XI. ADDITIONAL REFERENCES AND DATA SOURCES

PROPYLENE CARBONATE (*Continued*)

Solute	Solubility	Temp.	Ref.
$PbBr_2$	1.00×10^{-3} mole/kg	25.00°	4
PbI_2	2.00×10^{-7} mole/kg^{-1}	25.00°	4
$BiCl_3$	7.74×10^{-1} mole/kg^{-1}	25.00°	4
$SnCl_4$	1.02 mole/kg^{-1}	25.00°	4
$NbCl_5$	3.15 mole/kg^{-1}	25.00°	4
$ThCl_4$	3.34×10^{-2} mole/kg^{-1}	25.00°	4
UCl_4	1.02×10^{-1} mole/kg^{-1}	25.00°	4
N Me$_4$F	6.50×10^{-2} molar	25.00°	1
N Me$_4$PF$_6$	1.15×10^{-1} molar	R.T.	12
	1.50×10^{-1} molar	25.00°	1
	2.20×10^{-1} molar	60.00°	1

1. Rocketdyne, "Properties of Nonaqueous Electrolytes," Final Report No. R-7703, Contract No. NAS 3-8521 (1968).

2. Lockheed Missiles and Space Co., "New Cathode-Anode Couples using Nonaqueous Electrolytes," Report No. ASD-TDR-62-837, Contract No. AF-33(616)-7957 (1962).

3. P. R. Mallory, "Research and Development of a High Capacity Nonaqueous Secondary Battery," Final Report No. NASA CR-54969, Contract No. NAS 3-6017 (1965).

4. University of California Radiation Laboratory, "Electrochemical Studies in Cyclic Ester," Report No. UCRL-8381, Contract No. W-7405-ENG-48 (1958).

5. M. L. B. Rao and R. W. Holmes, Electrochem. Technol., **6**, 105 (1968).

6. Lockheed Missiles and Space Co., "Lithium Anode Limited Cycle Secondary Battery," Final Report No. APL-TDR-64-59, Contract No. AF 33(657)-11709 (1964).

7. Y. C. Wu and H. L. Friedman, J. Phys. Chem., **70,** 501 (1966).

8. Globe Union, "A Program to Develop a High Energy Density Primary Battery with a minimum of 200 watt-hours per pound of total battery weight."

9. Electric Storage Battery Inc., "High Energy System (Organic Electrolyte)," Final Report No. ECOM-01394-F, Contract No. DA-28-043-AMO-01394(P).

10. Electrochimica Corp., "Research and Development of a High Energy Nonaqueous Battery," Final Contract No. NOW-63-0618-C (1965).

11. P. R. Mallory, "Evaluation of Rechargeable Lithium-Copper Chloride Organic Electrolyte Battery System," Contract No. DA-44-009-AMC-1537(T).

12. Gulton Industries, "Lithium-Nickel Fluoride Secondary Battery Investigation," Final Report No. AFAPL-TR-68-71, Contract No. AF 33(615)-3488 (1968).

TETRAHYDROFURAN

Solute	Solubility	Temp.	Ref.
CuF_2	2.40×10^{-3} molar	R.T.	1
$CuCl_2$	3.35×10^{-2} molar	25.00	1
$TiCl_3$	1.60×10^{-2} molar	R.T.	2
$Ti(cpdn)_2Br_2$	9.20×10^{-3} molar	R.T.	2
$ZrCl_4$	6.00×10^{-2} molar	R.T.	2
$ZrBr_4$	1.30×10^{-1} molar	R.T.	2

1. Electrochimiea Corp, "Research and Development of a High Energy Nonaqueous Battery," Final Contract No. NOW-63-0618-C (1965).

2. W. E. Reid, Jr., J. M. Bish, and A. Brenner, J. Electrochem. Soc., **104,** 21 (1957).

XI. ADDITIONAL REFERENCES AND DATA SOURCES

TETRAMETHYL UREA[a]

Solute	Solubility (g/100 g solvent)	Temp.
NaCl	0.14	75°C
NaBr	5.8	22°C
NaI	92	
NaCN	0.27	
$NaNO_3$	3.8	
$Na_2S_2O_3 \cdot 5H_2O$	0.47	
$Na(CH_3COO)_2 \cdot 3H_2O$	0.26	
KCl	0.15	75°C
KBr	0.11	22°C
LI	14.2	
KCN	0.25	
KOCN	0.16	
KSCN	29.4	
$K_2S_2O_5$	0.45	
$K(CH_3COO)$	0.21	
NH_4Cl	0.72	
NH_4NO_3	32	
$CuSO_4 \cdot (5H_2O)$	0.22	
$AgNO_3$	"good"	
$CaCl_2$	2.0	
$ZnSO_4 \cdot 7H_2O$	0.39	
$Ca(NO_3)_2 \cdot 4H_2O$	>60	
H_3BO_3	12	
$MnCl_2 \cdot 4H_2O$	85	
$FeCl_3$	6.9	
$Cu(NO_3)_2 \cdot 6H_2O$	>60	
$NI(NO_3)_2 \cdot 6H_2O$	74	
$NbCl_5 \cdot 2DMF$	good	
HCl[b]	good	
HBr[b]	moderate	

[a] A. Luttninghaus and H. W. Dirksen, Angew Chem. Int. Edn. **3**, 260 (1964).
[b] Compound formed.

ACETONITRILE

D. H. Berne and O. Popovych, Anal. Chem., **44**, 817 (1972). Solubilities and medium effects of tetraphenylgermame, tetraphenylmethane, and tetraphenylsilane in acetonitrile, methanol and some ethanol–water solvents.

DIPROPYLENE GLYCOL

B. Jeandu, J. Biais, and B. Lemanceau, J. Chim. Phys., **68**, 1472 (1971). Solubility of ammonium perchlorate in dipropylene glycol and various polyoxypropylene glycols.

FORMAMIDE

E. Cotton and R. E. Brooker, J. Phys. Chem., **62**, 1595 (1958). The solubility of some salts of sodium, potassium, magnesium and calcium in formamide.

METHANOL

O. Popovych and R. M. Friedman, J. Phys. Chem., **70**, 1671 (1966). Solubilities of potassium triisoamylbutylammonium and tetrabutylammonium tetraphenylborides and picrates in water and methanol and their medium effects at 25°.

SULFOLANE

J. A. Starkovich and M. Jarghorbani, J. Inorg. Nucl. Chem., **34**, 789 (1972). Solubilities of some chloride and perchlorate salts in sulfolane.

ETHANOL–WATER

D. H. Berne and O. Popovych, J. Chem. Eng. Data, **17**, 180 (1972). Solubility products of tetraphenylarsonium and tetraphenylphosphonium picrates in ethanol–water solvents at 25°C.

G. Delesalle and J. Heurel, Bull. Soc. Chim. France, 2626 (1972). Relation between the solubility of electrolytes and the dielectric constant of the solution in water–ethanol mixed solvents.

METHANOL–WATER

C. L. De Ligny and N. G. Van der Veen, Rec. Trav. Chim., **90**, 984 (1971). Solubilities of some tetra-alkyl carbon, -silicon, -germanium, and -tin compounds in mixtures of water with methanol, ethanol, dioxane, acetone and acetic acid and differences between the standard chemical potentials of these solutes in their solutions in water and in the mixed or nonaqueous solvents at 25°C.

XI. ADDITIONAL REFERENCES AND DATA SOURCES

R. Thuaire, Bull. Soc. Chim. (France), 3815 (1971). Study of the solubility of substituted benzoic acids in water–methanol mixtures.

I. M. Kolthoff and M. K. Chantooni, Jr., Anal. Chem., **44**, 194 (1972). Solubility product of silver tetraphenylborate in water and methanol.

iso-PROPANOL–WATER

J. Nedoma, Chem. Listy., **66**, 772 (1972). Solubility products of silver halides in isopropyl alcohol and its mixtures with water.

PYRIDINE–WATER

R. J. Raridon, W. H. Baldwin, and K. A. Kraus, J. Phys. Chem., **72**, 925 (1968). Activity coefficients of sodium chloride in saturated water–pyridine mixtures at 5 and 25°, using packed column techniques to determine solubilities.

(f) EMF

ACETIC ANHYDRIDE

G. Durand, Bull. Soc. Chim. (France), 2091 (1972). Electrochemical study of the functioning of quinhydrone and choranil electrodes in acetic anhydride.

ACETONE

D. Bax, C. L. de Ligny, and A. G. Remijnse, Rec. Trav. Chim., **91**, 1225 (1972). Activity coefficients of single ions in acetone and acetone–water mixtures.

ACETONITRILE

F. G. K. Baucke, R. Bertram, and K. Cruse, J. Electroanalyt. Chem. Interfacial Electrochem., **32**, 247 (1971). The iodine-iodide system in acetonitrile using the cell: Pt, I_2 | NaI || NaI | I_2, Pt

J. K. Senne and B. Kratochvil, Anal. Chem., **44**, 585 (1972). Standard potential of the Copper (11)–(1) Couple in acetonitrile.

Cells: Ag | $AgNO_3$ | $LiClO_4$ or Et_4NClO_4 | $CuClO_4$ $Cu(ClO_4)_2$ Et_4NClO_4 | Pt

Ag(Hg) | $AgClO_4$ | Et_4NClO_4 | $CuClO_4$ $Cu(ClO_4)_2$ Et_4NClO_4 | Pt

Liquid junction potentials were eliminated by an extrapolation procedure.

O. Bravo and R. T. Iwamoto, J. Electroanal. Chem., **23**, 419 (1969). A low-chloride calomel reference electrode of low polarizability in acetonitrile.

J. K. Senne and B. Kratochvil, Anal. Chem., **43**, 79 (1971). Potentiometric study of copper (1) complexes with univalent anions in acetonitrile. The stepwise formation constants of several complexes with Cu (1) in acetonitrile were determined by potentiometric titration of Cu (1) solutions with salts of these anions using a saturated copper amalgam indicating electrode.

XI. ADDITIONAL REFERENCES AND DATA SOURCES

ACETONITRILE (*Continued*)

O. Popovych, A. Gibofsky, and D. H. Berne, Anal. Chem., **44**, 811 (1972). Medium effects for single ions in acetonitrile and ethanol–water. Solvents based on reference electrolyte assumptions.

H. P. Bennetto and J. J. Spitzer, Chem. Comm., 990 (1971). Behavior of hydrogen and silver–silver chloride electrodes in acetonitrile solutions.

L. F. Heerman and G. A. Rechnitz, Anal. Chem., **44**, 1655 (1972). Ion selective electrode study of copper (1) complexes in acetonitrile.

I. M. Kolthoff and M. K. Chantooni, J. Phys. Chem., **76**, 2024 (1972). A critical study involving water, methanol, acetonitrile, N,N-dimethylformamide, and dimethylsulfoxide of medium ion activity coefficients, γ, on the basis of the $\gamma_{AsPH_4^+} = \gamma_{BPh_4^+}$ assumption.

AMMONIA (liq.)

J. Baldwin and J. B. Gill, J. Chem. Soc., 2040 (1971). Solutions of electrolytes in liquid NH_3. The mean molal ion activity coefficients in NH_4NO_3 solutions at $-40°C$.

J. Baldwin, J. B. Gill, and A. Prescott, J. Inorg. Nucl. Chem., **33**, 2103 (1971). The use of hydrogen electrodes in liquid ammonia.

n-BUTANOL

R. N. Roy, A. L. M. Bothwell, J. Gibbons, and W. Vernon, J. Chem. Soc., Dalton Trans., **A**, 530 (1972). Standard potential of the Ag | AgCl electrode and related thermodynamic quantities in Butanol at different temperatures. Cell: Pt, H_2 | HCl | AgCl | Ag. Temp. range: 5–45°C.

DIMETHYLFORMAMIDE

P. A. Rock, J. J. Kim, and L. F. Sylvester, J. Chem. Phys., **56**, 1863 (1972). Thermodynamics of hydrogen—isotope-exchange reactions. Electrochemistry in dimethylformamide.

Cell used: Pt, D_2 | DCl | TlCl | Tl(Hg) | TlCl | HCl | H_2, Pt
(12%)

$E^0 = 7.64$ mV and 8.12 mV; conc. range 0.01–0.24m

C. Lassigne and P. Baine, J. Phys. Chem., **75**, 3188 (1971). Solvation studies of lithium salts in dimethylformamide.

DIMETHYL SULFATE

A. Caiola, H. Guy, and J. C. Sohm, Electrochim. Acta, **15**, 1733 (1970). The behavior of the electrodes Li | LiCH$_3$CO$_2$, Hg | Hg$_2$(CH$_3$CO$_2$)$_2$ and Cu | Cu(CH$_3$CO$_2$) in propylene carbonate and dimethyl sulfate.

DIMETHYL SULFOXIDE

I. M. Kolthoff and M. K. Chantooni, J. Phys. Chem., **76**, 2039 (1972). A critical study involving water, methanol, acetonitrile, N,N-dimethylformamide, and dimethylsulfoxide of medium ion activity coefficients, γ, on the basis of the $\gamma_{\text{AsPh}_4}{}^+ = \gamma_{\text{BPh}_4}{}^+$ assumption.

ETHANOL

D. Bax, C. L. de Ligny, and A. G. Remijnse, Rec. Trav. Chim., **91**, 965 (1972). The difference between the standard chemical potentials of single ions in ethanol and ethanol–water mixtures and in water at 25°, and some related quantities.

FORMAMIDE

R. K. Agarwal, D. K. Sahu, and B. Nayak, Ind. J. Chem., **9**, 978 (1971). Relative partial molal enthalpy and relative partial molal heat capacity of HCl from EMF measurements.

B. Nayak and D. K. Sahu, Electrochim. Acta **16**, 1757 (1971). Thermodynamic properties from electric tension of the cell: Pt, H$_2$ | HCl | Hg$_2$Cl$_2$ | Hg at 25°C in formamide.

GLYCEROL

W. Vernon, A. L. M. Bothwell, J. J. Gibbons, and R. N. Roy, J. Electroanalyt. Chem. Interfacial Electrochem., **34**, 101 (1972). Stanffard potential of silver–silver chloride electrode and related thermodynamic quantities in glycerol at 5–45°C. using the cell: Pt, H$_2$ | HCl | AgCl | Ag.

METHANOL

J. Nedoma, Chem. Listy., **65**, 71 (1971). Solubility products of silver halides in water, methanol, n-propanol and their mixtures with water.

D. Bax, C. L. de Ligny, and M. Alfenaar, Rec. Trav. Chim., **91**, 452 (1972). The difference between the standard chemical potentials of single ions in methanol and methanol–water mixtures and in water at 25°C and some related quantities.

M. Mastroianni and C. M. Criss, J. Chem. Eng. Data, **17**, 222 (1972). Standard partial molal heat capacities of sodium perchlorate in water from 0–90°C and in anhydrous methanol from −5 to 55°C.

XI. ADDITIONAL REFERENCES AND DATA SOURCES

METHANOL (*Continued*)

I. M. Kolthoff and M. K. Chantooni, J. Phys. Chem., **76**, 2024 (1972). A critical study involving water, methanol, acetonitrile, N,N-dimethylformamide, and dimethylsulfoxide of medium ion activity coefficients, γ, on the basis of the $\gamma_{AsPh_4^+} = \gamma_{BPh_4^+}$ assumption.

NITROMETHANE

J.-Claude Bardin, J. Electroanalyt. Chem., **28**, 157 (1970). A study of the complexes of Ag^+ with various anions and molecules in nitromethane.

J. C. Fischer and M. Wartel, Bull. Soc. Chim. (France), 3302 (1972). Study of the Hammett acidity function for several mineral acids in nitromethane.

iso-PROPANOL

J. Nedoma, Chem. Listy., **66**, 772 (1972). Solubility products of silver halogenides in *iso*-propanol and *iso*-propanol water mixtures.

n-PROPANOL

J. Nedoma, Chem. Listy., **65**, 71 (1971). Solubility products of silver halides in water, methanol, *n*-propanol and their mixtures with water.

R. N. Roy, W. Vernon, J. J. Gibbons, and A. L. M. Bothwell, J. Chem. Thermodyn., **3**, 883 (1971). Thermodynamics of HCl in *n*-propanol from potentiometric measurements.

R. N. Roy, W. Vernon, and A. L. M. Bothwell, Electrochim. Acta, **17**, 5 (1972). S.E.T. of the Ag | AgCl electrode in 95 wt.% *n*-propanol and related thermodynamic functions.

PROPYLENE CARBONATE

G. Singh, P. A. Rock, and J. C. Hall, J. Chem. Phys., **56**, 1855 (1972). Thermodynamics of lithium isotope exchange reactions.

Cell used: Li | LiBr | TlBr | Tl(Hg) | TlBr | LiBr | Li

$E^0 = 0.76$ mV; Conc. Range 0.03–0.05m

L. M. Mukherjee and D. P. Boden, Electrochim. Acta **17**, 965 (1972). Glass electrode response to lithium ion activity in propylene carbonate solutions. The effects of the ion K^+, NH_4^+ and $(Et)_4N^+$ were studied at an ionic strength of 0.25M.

I. Fried and H. Barak, J. Electroanal. Chem., **27**, 167 (1970). A study of the calomel reference electrode in propylene carbonate.

PROPYLENE CARBONATE (Continued)

E. Kirowa-Eisner and E. Gileadi, J. Electroanal. Chem., **25**, 481 (1970). The silver–silver perchlorate reference electrode in propylene carbonate.

I. Piljac and R. T. Iwamoto, J. Electroanal. Chem., **23**, 484 (1969). The Hg | Hg$_2$Cl$_2$, KCl(sat.), Et$_4$NClO$_4$ reference electrode in propylene carbonate.

A. Caiola, H. Guy, and J. C. Sohm, Electrochim. Acta, **15**, 1733 (1970). Hg | Hg$_2$(CH$_3$CO$_2$)$_2$ and Cu | Cu(CH$_3$CO$_2$)$_2$ in propylene carbonate and dimethyl sulfate.

E. Sutzkover, Y. Nemirovsky, and M. Ariel, J. Electroanal. Chem., **38**, 107 (1972). The I$_3^-$ | I$^-$ Reference electrode in propylene carbonate.

J. Jansta, F. P. Dousek, and J. Riha, J. Electroanal. Chem., **38**, 445 (1972). Electrochemical systems for galvanic cells in aprotic solvents. Trace water elimination in propylene carbonate electrolytes using sodium–potassium alloy.

M. L'Her and J. Courtot-Coupez, Bull. Soc. Chim. (France). Electrochemistry in propylene carbonate. Determination of normal potentials of the alkali-metals and calcium. Cation solvation.

PYRIDINE

L. M. Mukherjee, J. Phys. Chem., **76**, 243 (1972). Standard potential of the ferrocene–ferricinium electrode in pyridine. Evaluation of proton medium effect.

Cells: Ag | AgPi || Fc(m) | Pt
         (m)   FcPi
                    (m)

Zn(Hg) | ZnCl$_2$(s) || Fc(m) | Pt
                       FcPi
                       (m)

Conc. $m = 1.0 \times 10^{-3}$; 2.5×10^{-3}; 3.0×10^{-3} M.

ACETONE–WATER

A. M. Nilsson and P. Beronius, Z. Phys. Chem. (Frankfurt), **79**, 83 (1972). Solvation phenomena of lithium bromide in acetone–water mixtures.

D. Bax, C. L. de Ligny, and A. G. Remijnse, Rec. Trav. Chim., **91**, 1225 (1972). Activity coefficients of single ions in acetone and acetone–water mixtures.

XI. ADDITIONAL REFERENCES AND DATA SOURCES

ACETONITRILE–WATER

T. Mussini, P. Longhi, and P. Giammario, Chimica Et Industria (Milan), **54**, 3 (1972). The cell $H_2 \mid HX \mid AgX \mid Ag$ in Acetonitrile–water mixtures. Primary medium effect and primary hydration numbers of the hydrohalic acids. 0, 5, 10 and 15%-wt acetonitrile studied.

P. Giammario, P. Longhi, and T. Mussini, Chim. Ind. (Milan), **53**, 347 (1971). H^+ sensitive electrodes in Acetonitrile–water media.

T. Mussini, P. Longhi, and P. Giammario, Chim. Ind. (Milan), **53**, 1124 (1971). The cell $H_2 \mid HX \mid AgX \mid Ag$ in AN–water mixtures: thermodynamics of HCl.

n-BUTANOL–WATER

R. N. Roy, W. Vernon, and A. L. M. Bothwell, Electrochim. Acta, **17**, 1057 (1972). The activity coefficient and related thermodynamic quantities of HCl in 5 wt% n-butanol.

Cell: Pt, $H_2 \mid HCl \mid AgCl \mid Ag$

Temp. Range: 5–45°C.

tert-BUTANOL–WATER

J. P. Morel and J. Morin, J. De Chimie Phy., **67**, 2018 (1970). Standard electrode potentials of Ag | AgCl in water–isopropanol and water–tert-butanol mixtures.

R. N. Roy, W. Vernon, A. Bothwell, and J. Gibbons, J. Chem. and Eng. Data, **17**, 79 (1972). Standard EMF of silver–silver chloride electrode and related thermodynamic quantities in tert-butanol–water mixtures (2, 6, and 8 wt%) between 5–45°C.

Cell: Pt, $H_2 \mid HCl \parallel HCl \mid AgCl \mid Ag$
        (H_2O)   (t-BuOH-H_2O)

Standard potentials calculated by a polynomial curve fitting program.

R. N. Roy, W. Vernon, and A. M. Bothwell, J. Chem. Soc., 1242 (1971). S.E.T. of the Ag | AgCl electrode in tert-butyl alcohol–water mixtures and the thermodynamics of solutions of HCl at different temperatures.

DIMETHYLFORMAMIDE–WATER

B. Jakuszewski and H. Scholl, Electrochim. Acta, **17**, 1105 (1972). Shift of the standard potential scale and changes of surface potentials between water and some nonaqueous solvents.

Cell: Hg | Hg_2Cl_2 | KCl(aq) | air | LiCl | AgCl | Ag
         0.1m   (m)
                        (mixed
                        solvent)

Conc. Range: 10^{-3}–0.13m.

R. N. Roy, W. Vernon, and A. L. M. Bothwell, J. Chem. Soc., **B**, 2320 (1971). Thermodynamic properties of HCl in DMF–water mixtures at different temperatures.

DIMETHYLSULFOXIDE–WATER

K. Yates and G. Welch, Can. J. Chem., **50**, 1513 (1972). Acidity function behavior in the 70% DMSO—H_2O system.

R. N. Roy, W. Vernon, A. Bothwell, and J. Gibbons, J. Electrochem. Soc., **119**, 694 (1972). Standard potentials of Ag | AgCl electrode and related thermodynamic quantities in dimethylsulfoxide–water mixtures from 5 to 45°C.

Cell: Pt, H_2 | HCl | AgCl | Ag

Comp. Range: 5, 10, and 20 wt% DMSO.

K. H. Khoo, J. Chem. Soc., 2932 (1971). The standard potentials of silver–silver halide electrodes and ion solvation in dimethylsulfoxide–water mixtures at 25°C.

J. Coupez-Courtot, and C. Madec, Bull. Soc. Chim. France, 4621 (1971). Stability of Ag(l) complexes. Solvation of anions.

J. Coupez-Courtot, and C. Madec, Bull. Soc. Chim. France, 4626 (1971). Solvation of the iodine molecule and the triiodide anion.

R. G. Bates, L. Johnson, and R. A. Robinson, Chem. Anal. Warsaw, **17**, 479 (1972). Solvent effects on the acid dissociation of *m*-nitroanilinium ion in water–DMSO.

XI. ADDITIONAL REFERENCES AND DATA SOURCES

DIOXAN–WATER

D. Feakins and J. P. Lorimer, Chem. Comm., 646 (1971). The EMF method for studying the transport of non-electrolytes in electrolytic solutions; results for the methanol– and dioxane–water systems.

Cells: Ag | AgCl | MCl(aq) | MCl | AgCl | Ag
(mixed solvent)

$M = H^+, Li^+, Na^+, K^+, Rb^+$ and Cs^+

and

Ag | AgCl | MCl(aq) | M(Hg) | MCl | Ag | AgCl
(mixed solvent)

S. Chakrabarti and S. Aditya, J. Chem. and Eng. Data, **17**, 46 (1972). EMF Studies of the cell:

$$Cd_xHg_y \mid CdSO_4 \mid Hg_2SO_4 \mid Hg$$
$$(m)$$

in dioxane–water media. Temp. Range: 25°, 30°, 35°, $-\Delta G°$, $-\Delta H°$ and $-\Delta S°$ obtained for the reaction,

$$Cd + H_2SO_4 \rightarrow CdSO_4 + 2Hg$$

K_d of $CdSO_4$ in 20, 40 and 60 wt% dioxane–water reported.

D. Pax, M. Alfenaar, and C. L. de Ligny, Rec. Trav. Chim., **90**, 1002 (1971). The difference of the standard chemical potentials of KCl in water and in mixtures of water with ethanol, dioxane and acetone.

T. Mussini, C. Massarani-Formaro, and P. Andrigo, J. Electroanal. Chem., **33**, 177 (1971). The cell H_2 | HX | AgX | Ag where X = Br in dioxane–water mixtures.

T. Mussini, C. Massarani-Formaro, and P. Andrigo, J. Electroanal. Chem., **33**, 189 (1971). The cell H_2 | HX | AgX | Ag in dioxane–water mixtures. Primary hydration numbers of HCl and HI.

A. K. Basu and S. Aditya, J. Indian Chem. Soc., **48**, 129 (1971). The Hg | Hg(I) acetate electrode: S.E.T. in dioxane–water media at 25°C.

A. K. Basu and S. Aditya, J. Indian Chem. Soc., **48**, 155 (1971). The Hg | Hg(I) acetate electrode: S.E.T. and related thermodynamic quantities in dioxane–water at different temperatures.

ETHANOL–WATER

B. Jakuszewski and H. Scholl, Electrochim. Acta, **17**, 1105 (1972). Shift of the standard potential scale and changes of surface potentials between water and some nonaqueous solvents.

Cell: Hg | Hg_2Cl_2 | KCl(aq) | air | LiCl | AgCl | Ag
          0.1m   m
                                    (mixed solvent)

Conc. Range: 10^{-3}–0.13m.

D. Pax, M. Alfenaar, and C. L. de Ligny, Rec. Trav. Chim., **90**, 1002 (1971). The difference of standard chemical potentials of KCl in water and in mixtures of water with ethanol, dioxane and acetone.

O. Popovych, A. Gibofsky, and D. H. Berne, Anal. Chem., **44**, 811 (1972). Medium effects for single ions in acetonitrile and ethanol–water mixtures. Solvents based on reference electrolyte assumptions.

K. H. Pool and R. G. Bates, J. Chem. Thermodyn, **1**, 21 (1969). Thermodynamics of HCl in 95 volume per cent ethanol from potentiometric measurements from 5 to 50°C.

ETHYLENE GLYCOL–WATER

K. K. Kundu, Ind. J. Chem., **10**, 303 (1972). Standard free energy and entropy of chloride ion relative to bromide ion in aquo-glycollic solvents.

K. K. Kundu, A. K. Rakshit, and M. N. Das, Electrochim. Acta, **17**, 1921 (1972). Standard potentials of Li/Li$^+$, Na/Na$^+$ and K/K$^+$ electrodes in ethylene glycol and its aqueous mixtures at 25°C and the related thermodynamic behavior of the alkali halides.

K. K. Kundu and A. K. Rakshit, Indian. J. Chem., **9**, 439 (1971). S.E.T. of Cs | Cs$^+$ electrode in ethylene glycol and 50% H_2O–ethylene glycol mixtures. A test of the Strehlow's method for free energies of transfer of some individual ions from H_2O to these solvents.

GLYCEROL–WATER

K. H. Khoo, J. Chem. Soc., Faraday Transactions I, **68A**, 554 (1972). Standard potentials of silver–silver halide electrodes and ion solvation in glycerol–water mixtures at 25°C, using the cell: Pt, H_2 | HX | AgX | Ag. Where X = Cl or Br and the cells:

H_2 | HBO_2, $NaBO_2$, KCl | AgCl—Ag

H_2 | HBO_2, $NaBO_2$, KI | AgI—Ag

for the iodide system. Composition Range: 10, 20, 30, 40, 50 and 70% (w/w) glycerol.

XI. ADDITIONAL REFERENCES AND DATA SOURCES

GLYCEROL–WATER (Continued)

R. N. Roy, W. Vernon, and A. L. M. Bothwell, J. Electroanal. Chem. Interfacial Electrochem., **30**, 335 (1971). Thermodynamic studies of HCl in 95 wt% glycerol from potentiometric measurements at 5 to 45°C.

R. N. Roy, W. Vernon, and A. L. M. Bothwell, J. Electrochem. Soc., **118**, 1302 (1971). Thermodynamics of HCl in glycerol–water mixtures from potentiometric measurements between 5 and 45°C.

METHANOL–WATER

J. Nedoma, Chem. Listy., **65**, 71 (1971). Solubility products of silver halides in water, methanol, n-propanol and their mixtures with water.

B. Jakuszewski and H. Scholl, Electrochim. Acta, **17**, 1105 (1972). Shift of the standard potential scale and changes of surface potentials between water and some nonaqueous solvents.

Cell: Hg | Hg$_2$Cl$_2$ | KCl(aq) | air | LiCl | AgCl | Ag
                0.1m   (m)
                                    (mixed solvent)

Conc. Range: 10^{-3}–0.13m.

D. Feakins and J. P. Lorimer, Chem. Comm., 646 (1971). The EMF method for studying the transport of non-electrolytes in electrolytic solutions; results for the methanol– and dioxan–water systems.

Cells: Ag | AgCl | MCl(aq) MCl | AgCl | Ag
                    (mixed solvent)

M = H$^+$, Li$^+$, Na$^+$, K$^+$, Rb$^+$ and Cs$^+$

and

Ag | AgCl | MCl(aq) | M(Hg) | MCl | AgCl | Ag.
                    (mixed solvent)

K. H. Khoo, J. Chem. and Eng. Data, **17**, 82 (1972). Redetermination of standard potential of silver–silver bromide electrode in methanol–water mixtures at 25°C.

Cell: Pt, H$_2$ | HBr | AgBr | Ag and Owen buffered cells.

Composition Range: 10, 20.22, 30, 33.4, 43.12, 50, 68.33 and 90 wt% methanol.

G. Douheret, Bull. Soc. Chim. (France), **7**, 2393 (1971). A study of the cell: Pt, H$_2$ | HCl | Hg$_2$Cl$_2$ | Hg in methanol–water mixtures.

METHANOL–WATER (*Continued*)

K. Jackowska, Rocz. Chem., **45**, 87 (1971). Determination of mean activity coefficients of CsCl in methanol–water mixtures using cells with liquid junction.

Cells: Ag | AgCl | NaCl || CsCl | AgCl | Ag

Ag | AgCl | CsCl || CsCl | AgCl | Ag

Comp. Range: 10, 20, 40 wt% methanol.

D. Feakins and A. R. Willmott, J. Chem. Soc., 3121 (1970). Free energies of transfer of $BaCl_2$ from water to 10, 20, and 40% (W W) Methanol–water mixtures, and of $SrCl_2$ to the 20% mixture at 25°C.

Cell: Ag | AgCl | MCl_2 | M(Hg) | MCl_2 | AgCl | Ag

J. H. Stern and S. L. Lansen, J. Chem. Eng. Data, **16**, 360 (1971). Thermodynamics of transfer of HCl from water to aqueous alcohols at 25°C.

D. Feakins and P. J. Voice, J. Chem. Soc. Faraday Trans., I, **68**, 1390 (1972). Free energies of transfer of the alkali metal chlorides from water to 10–99% (W/W) Methanol–water mixtures at 25°C.

R. N. Roy, W. Vernon, and A. L. M. Bothwell, J. Chem. Thermodyn., **3**, 769 (1971). S.E.T. of the Ag | AgCl electrode from 5 to 45°C and the thermodynamic properties of HCl in 95 mass per cent *iso*-propanol.

2-METHOXYETHANOL–WATER

H. P. Thun, B. R. Staples, and R. G. Bates, J. Res. Nat. Bur. Std., **74A**, 641 (1970). Thermodynamics of HCl in 80 wt% 2-methoxyethanol and 20 wt% water from 10 to 50°C. Cell: Pt, H_2 | HCl | AgCl | Ag.

METHYLCELLOSOLVE–WATER

H. Sadek, T. F. Tadros, and A. E. El-Harakany, Electrochim. Acta, **16**, 339 (1971). Thermodynamics of HCl in H_2O–methylcellosolve mixtures: S.E.T. Ag | AgCl electrode.

H. Sadek, T. F. Tadros, and A. A. El-Harakany, Electrochim. Acta, **16**, 353 (1971). Mean activity coefficients and medium effects of HCl in H_2O–methylcellosolve mixtures.

PHENOL–WATER

H. Sadek, A. El-Harakany, and N. A. El-Nadory, Electrochim. Acta, **17,** 1745 (1972). Thermodynamics of HCl in water–phenol mixtures. Standard potentials of the silver/silver chloride electrode and medium effects.

iso-PROPANOL–WATER

K. Schwabe and R. Müller, Ber. Bunsen Gesellschaft, **74,** 1248 (1970). A Study of the cell: Pt, H_2 | HBr, Hg_2Br_2 | Hg in isopropanol–water mixtures.

J. P. Morel and J. Morin, J. De Chimie Phy., **67,** 2018 (1970). Standard electrode potentials of Ag | AgCl in water–iso propanol and water–tert-butanol mixtures.

J. Nedoma, Chem. Listy., **66,** 772 (1972). Solubility products of silver halogenides in iso propanol and isopropanol–water mixtures.

R. N. Roy and A. L. M. Bothwell, J. Chem. Eng. Data, **16,** 347 (1971). Thermodynamic quantities of HCl in iso propanol–water mixtures.

V. V. Aleksandrov and Y. V. Sych, Zh. Fiz. Khim., **46,** 812 (1972). Thermodynamic properties of alkali metal iodides in iso propyl alcohol and its mixtures with water. Standard electromotive forces and activity coefficients.

R. N. Roy, W. Vernon, and A. L. M. Bothwell, J. Chem. Thermodyn., **3,** 769 (1971). S.E.T. of the Ag | AgCl electrode from 5 to 45°C and the thermodynamic properties of HCl in 95 mass per cent *iso*-propanol.

n-PROPANOL–WATER

J. Nedoma, Chem. Listy., **65,** 71 (1971). Solubility products of silver halides in water, methanol, *n*-propanol and their mixtures with water.

PROPYLENE CARBONATE–WATER

R. N. Roy, W. Vernon, J. J. Gibbons, and A. L. M. Bothwell, J. Chem. Soc., 3589 (1971). Thermodynamics of HCl in propylene carbonate–water mixtures between 5 and 45°C.

TETRAHYDROFURAN–WATER

A. Braibanti, E. Leporati, and F. Dallavalle, Inorg. Chim. Acta, **4**, 529 (1970). The protonation constant of oxamic acid in aqueous solution and in water–tetrahydrofuran mixtures, at different temperatures and ionic strengths.

R. N. Roy and A. L. M. Bothwell, J. Chem. Eng. Data, **16**, 347 (1971). Thermodynamic quantities of HCl in tetrahydrofuran–water mixtures.

BUTANOL–METHANOL

A. M. Shkodin and T. P. Sogoyan, Elektrokhimiya, **6**, 1643 (1970). Thermodynamic characteristics of hydrogen chloride in isodielectric mixed solvents.

Cell: Pt, H_2 | HCl | AgCl | Ag.

Systems: butanol–methanol,   dioxane–methanol,   dioxane–ethanol, ethanol–hexane.

DIOXANE–METHANOL

A. M. Shkodin and T. P. Sogoyan, Elektrokhimiya, **6**, 1643 (1970). Thermodynamic characteristics of hydrogen chloride in isodielectric mixed solvents.

Cell: Pt, H_2 | HCl | AgCl | Ag.

Systems: butanol–methanol,   dioxane–methanol,   dioxane–ethanol, ethanol–hexane.

METHANOL–PROPYLENE GLYCOL

K. K. Kundu, A. L. De, and N. N. Das, J. Chem. Soc., Dalton Trans., 343 (1972). Standard potentials of Ag—AgX (X = Br or Cl) electrodes in methanol–propylene glycol at 25°C.

K. K. Kundu, A. K. Rakshit, and M. N. Das, J. Chem. Soc., Dalton Trans., 381 (1972). Standard potentials of M | M$^+$ (M = Li, Na and K) electrodes in methanol–propylene glycol.

GENERAL

R. P. Buck, Anal. Chem., **44**, 270R (1972). Ion selective electrodes, potentiometry and potentiometric titrations.

F. G. K. Baucke, Electrochim. Acta, **17**, 845 (1972). Potentials of electrodes of the second kind at low concentrations of common ion electrolyte. General Discussion.

XI. ADDITIONAL REFERENCES AND DATA SOURCES

GENERAL (*Continued*)

F. G. K. Baucke, Electrochim. Acta, **17**, 851 (1972). Potentials of electrodes of the second kind at low concentrations of common ion electrolyte. Quantitative treatment of electrodes with salts with negligible complex formation.

J. F. Coetzee and J. J. Campion, J. Amer. Chem. Soc., **89**, 2513 (1967). Evaluations of relative activities of reference cations in acetonitrile and water.

O. Popovych, Crit. Rev. Anal. Chem., **1**, 73 (1970). Estimation of medium effects for single ions in nonaqueous solvents.

J. Padova, J. Chem. Phys., **56**, 1606 (1972). Thermodynamic properties of ions in methanol solution.

D. G. Hall, J. Chem. Soc. Faraday Trans. II, **68**, 25 (1972). A definition of the primary medium effect for individual ionic species in solution.

I. M. Ganzhina, B. B. Damaskin, and R. I. Kaganovich, Elektrokhimiya, **8**, 93 (1972). Determination of real chemical energies and evaluations of Na^+ and I^- solvation in DMF.

R. Alexander, A. J. Parker, J. H. Sharp, and W. E. Waghorne, J. Amer. Chem. Soc., **94**, 1148 (1972). Solvent activity coefficients of single ions; a recommended extrathermodynamic assumption.

S. Goldman and R. G. Bates, J. Amer. Chem. Soc., **94**, 1476 (1972). Calculation of thermodynamic functions for ionic hydration.

D. Bax, C. L. de Ligny, M. Alfenear, and N. J. Mohr, Rec. Trav. Chim., **91**, 604 (1972). The standard potentials of the silver–silver bromide and of the silver–silver iodide electrode in some organic solvents and their mixtures with water.

L. U. Thulin, Acta. Chem. Scand., **26**, 225 (1972). Partial Gibbs energy for alkali metal systems from emf measurements.

N. L. Weinberg, J. Chem. Educ., **49**, 120 (1972). Simplified construction of electrochemical cells.

N. A. Kazarjan and E. Pungor, Anal. Chim. Acta, **60**, 193 (1972). Further investigations of ion-selective electrodes in nonaqueous solvents.

H. Widmer, Chimia, **26**, 229 (1972). Solvation of ions and metal complexes in organic solvents.

GENERAL (*Continued*)

J. Juillard, J. P. Morel, and L. Avedikian, J. Chim. Phys. Phys. Chim. Biol., **69**, 787 (1972). Solute–solvent interactions in t-butanol–water media.

J. I. Padova, Modern Aspects of Electrochemistry, **7**, 1 (1972). Ionic solvation in nonaqueous and mixed solvents.

K. K. Kundu, Israel J. Chem., **10**, 548 (1972). Appraisal of free energy of transfer of proton from one solvent to another.

G. Gammons, B. H. Robinson, and M. J. Stern, J. Chem. Soc. Chem. Comm., 1157 (1972). Thermodynamics and kinetics of ion pair formation in aprotic solvents.

B. S. Krumgalz, Zh. Strukt. Khim., **13**, 592 (1972). Relation between solvation of ions and the structural property of the solvent.

S. Goldman, P. Sagner, and R. G. Bates, J. Phys. Chem., **75**, 826 (1971). Solute–solvent effects in the ionization of tris(hydroxymethyl) acetic acid (THAA) and related acids in water and aqueous methanol.

M. Salomon, J. Electrochem. Soc., **118**, 1609 (1971). Energetics of single ion solvation in nonaqueous solvents and the effects on electrode kinetics.

(g) Potentiometric Titrations

ACETIC ACID
W. D. Johnson and J. E. Sherwood, Aust. J. Chem., **25**, 81 (1972). Relative basicities of substituted acetanilides in acetic acid and acetic anhydride. A potentiometric and spectrophotometric investigation.

ACETONITRILE
I. M. Kolthoff and M. K. Chantooni, Jr., J. Amer. Chem. Soc., **90**, 5961 (1968). The second dissociation constant of sulfuric acid in acetonitrile and in dimethyl sulfoxide.

CHLOROSULPHONIC ACID
K. C. Mohan Rao and P. R. Naidu, J. Electroanal. Chem., **35**, 429 (1972). Monaqueous titrations with chlorosulphonic acid as titrant.

DIMETHYL SULFOXIDE
I. M. Kolthoff and M. K. Chantooni, Jr., J. Amer. Chem. Soc., **90**, 5961 (1968). The second dissociation constant of sulfuric acid in Acetonitrile and in Dimethyl Sulfoxide.

I. H. Kolthoff, M. K. Chantooni, Jr., and S. Bhowmik, J. Amer. Chem. Soc., **90**, 23 (1968). Dissociation constants of uncharged and monovalent cation acids in dimethyl sulfoxide.

METHYL *iso* BUTYL KETONE
J. Juillard and I. M. Kolthoff, J. Phys. Chem., **75**, 2496 (1971). Dissociation and homoconjugation constants of some acids in methyl *iso* butyl ketone.

N-METHYL PYRROLIDONE
A. P. Kreshkov, A. P. Gurvich, and G. M. Galperin, Zh. Anal. Khim., **27**, 1166 (1972). Acid-base titration in N-methyl pyrrolidone.

PROPYLENE CARBONATE
N. A. Baranov, N. A. Vlasov, L. P. Potekhina, and O. F. Shepot'ko, Zh. Anal. Khim., **25**, 1777 (1970). Titration of acids and bases in propylene carbonate media.

DIOXAN–WATER

J. P. Shukla and S. G. Tandon, J. Electroanal. Chem., **35**, 423 (1972). Corrections to pH measurements for titrations in dioxane–water mixtures.

J. P. Shukla and S. G. Tandon, J. Inorg. Nucl. Chem., **33**, 1681 (1971). Ionization constants of some N-aryl-hydroxamic acids in dioxane–water mixtures.

ETHANOL–WATER

P. K. Migal and N. G. Chebotar, Russ. J. Inorg. Chem., **15**, 625 (1970). Cerium, Praeseodymium and Neodymium Acetate complexes in aqueous ethanol solutions.

I. M. Bhatt and K. P. Soni, Indian J. Chem., **9**, 172 (1971). Stability constants of Ag(I) diethyldithiocarbamate in aqueous ethanol.

U. P. Kamat, P. V. Kamat, and M. G. Datar, J. Indian Chem. Soc., **48**, 302 (1971). Determination of pK_a values of hydrochlorides of organic bases in aqueous ethanol.

METHANOL–WATER

D. B. Rorabacher, W. J. MacKellar, F. R. Shu, and S. M. Bonavita, Anal. Chem., **43**, 561 (1971). Solvent effects on protonation constants, NH_4OH, acetate, polyamine, and polyaminocarboxylate ligands in CH_3OH—H_2O mixtures.

ACETIC ACID–CHLOROFORM

O. W. Kolling and K. I. Laws: Anal. Chem., **43**, 986 (1971). Potentiometric study of base strengths in the binary solvent, Acetic acid–chloroform, using the glass-saturated calomel electrode pair.

GENERAL

T. A. Khudyakova and A. P. Kreshkov, J. Electroanal. Chem., **29**, 181 (1971). The theory of conductimetric and acid-base titrations in nonaqueous solutions.

J. J. Lagowski, Anal. Chem., **42**, 305R (1970). Titrations in nonaqueous solvents.

A. Saito, Review of Polarography (Japan), **13**, 29 (1965). Acid-Base equilibrium in nonaqueous solvents.

E. Clifford Toren, Jr., P. M. Gross, and R. P. Buck, Anal. Chem., **42**, 284R (1970). Potentiometric titrations.

XI. ADDITIONAL REFERENCES AND DATA SOURCES

GENERAL (*Continued*)

J. T. Stock, Anal. Chem., **42,** 276R (1970). Amperometric titrations.

J. J. Lagowski, Anal. Chem., **44,** 524R (1972). Titrations in Nonaqueous solvents.

W. Lynn Schertz and G. D. Christian, Anal. Chem., **44,** 755 (1972). Study of effects of neutral inorganic salts on potentiometric titration curves of weak bases in nonaqueous solvents. Weak bases used: urea, caffeine, m-nitroaniline, N-dimethylaniline, aniline and pyridine. Solvents: acetone, methyl isobutyl ketone, 2-butanone. Solutes: $LiClO_4$, LiI, $Mg(ClO_4)_2$.

R. P. Buck, Anal. Chem., **44,** 270R (1972). Ion selective electrodes, potentiometry and potentiometric titrations.

N. H. Furman, Anal. Chem., **26,** 84 (1954). Potentiometric Titrations. Methods and Applications.

J. A. Riddick, Anal. Chem., **24,** 41 (1952). Acid-base Titrations in Nonaqueous Solvents. Extensive survey of theory and practice.

E. O. Toren, Anal. Chem., **40,** 402R (1968). Potentiometric Titrations. Extensive review dealing mainly with theory, nonequilibrium methods and ion-selective electrodes.

J. A. Riddick, Anal. Chem., **26,** 77 (1954). Acid-base Titrations in Nonaqueous Solvents. Discussion of acid-base concepts is included.

E. F. C. Cain, Diss. Abs., **27B,** 67 (1966). The Study of Acid-base Equilibria in Mixed Aqueous Solvents. Theoretical equations are developed to predict changes in acid and base strength as a function of dielectric constant of the solvent.

V. E. Petrakovich, O. M. Podurovskaya, and I. Ya. Turyan, Russ. J. Anal. Chem., **20,** 863 (1965). Comparison of oxidized Platinum and Glass Indicator Electrodes for Acid-base Titrations. The Effect of Nonaqueous Solvents, the Nature of the Titrant and of Different Additives.

G. A. Harlow and D. H. Morman, Anal. Chem., **38,** 485R (1966). Titrations in Nonaqueous Solvents. Covers all nonaqueous titration methods up to 1965 and combinations of methods.

N. H. Furman, Anal. Chem., **23,** 21 (1951). Potentiometric Titrations. Analytical methods including amperometric, coulometric and nonaqueous methods.

C. N. Reilley, Anal. Chem., **28,** 671 (1956). Potentiometric Titrations. Development of potentiometric methods and apparatus.

GENERAL (*Continued*)

E. C. Toren and R. P. Buck, Anal. Chem., **42**, 284R (1970). Potentiometric Titrations. Covers development in theory, ion-selective electrodes, other indicators and reference electrodes, and nonaqueous titrimetry.

G. A. Harlow and D. H. Morman, Anal. Chem., **40**, 418R (1968). Titrations in Nonaqueous Solvents. Discussion especially of theoretical developments, methods and applications.

W. L. Schertz and G. D. Christian, Anal. Chem., **44**, 75 (1972). Study of effects of neutral inorganic salts on potentiometric titration curves of weak bases in nonaqueous solvents.

K. C. Nohan-Rao and P. R. Naidu, J. Electroanal. Chem., **35**, 429 (1972). Nonaqueous titrations with chlorosulfonic acid as titrant.

V. K. Akimov, A. I. Busev, I. A. Emel'yanova, S. I. Bragina, and S. M. Gel'fer, Zh. Anal. Khim., **26**, 864 (1971). Titrimetric determination of some elements by means of nonaqueous solvents.

E. M. Semenova, M. A. Trofimov, A. A. Pendin, and B. P. Nikol'skii, Elektrokhimiya, **7**, 1557 (1971). Determination of acidity of some nonaqueous and mixed solvents (water–carboxylic acids) using a redox system of ferrocene–ferricinium cation.

(h) Vapor Pressure

ETHANOL

I. Palogoric and I. J. Gal, J. Chem. Soc. Faraday Trans. I., **68,** 1093 (1972). Ionic association of some tetrapentylammonium and tetraoctylphosphonium salts in ethanol studied by vapor pressure osmometry.

(i) Cryoscopy

FORMAMIDE

R. C. Paul, J. A. Singla, and D. S. Gill, J. Chem. Soc. Faraday Trans. I, **68,** 1894 (1972). Cryoscopic studies of 1:1 electrolytes in formamide.

(j) Heats of Solution Calorimetry

ACETONITRILE

M. F. Stennikova, G. M. Poltoratskii, and K. P. Mishchenko, Zh. Obshch. Khim., **41**, 2588 (1971). Use of the method of comparative partial molal enthalpies for the water–acetonitrile–sodium iodide system.

M. H. Abraham, J. Chem. Soc. Chem. Comm., 888 (1972). Entropies of transfer of tetraalkylammonium ions from water to Methanol, Dimethylformamide and Acetonitrile.

N,N-DIMETHYLFORMAMIDE

C. De Visser and G. Somsen, Rec. Trav. Chim., **91**, 942 (1972). A thermochemical study of some tetralkylammonium bromides in N,N-dimethylformamide at 25°.

DIMETHYL SULFOXIDE

R. E. Dodd and R. P. H. Gasser, Proc. Chem. Soc., 415 (1964). Complex formation in dimethyl sulfoxide. Heats of solution of mixtures of potassium cyanide with either silver (1) cyanide or copper (1) cyanide.

FORMAMIDE

D. S. Gill, J. P. Singla, R. C. Paul, and S. P. Narula, J. Chem. Soc. Dalton Trans., 522 (1972). Thermochemical studies and ion solvation enthalpies in formamide, N-methylformamide and N,N-dimethylformamide.

METHANOL

M. J. Mastroianni and C. H. Criss, J. Chem. Thermodyn., **4**, 321 (1972). Heat capacities of tetraalkylammonium bromides in water and in anhydrous methanol at various temperatures.

N-METHYLFORMAMIDE

C. De Visser and G. Somsen, J. Chem. Thermodyn., **4**, 313 (1972). Solvation of tetraalkylammonium bromides in N-methylformamide and N-methylacetamide.

t-BUTANOL–WATER

R. K. Mohanty, S. Sunder, and T. C. Ahluwalia, J. Phys. Chem., **76,** 2577 (1972). Excess partial molal heat capacities of n-tetraamylammonium bromide in water from 10 to 80° and in aqueous $tert$-butyl alcohol solvent system at 30° and the effects on the water structure.

DIOXAN–WATER

A. N. Campbell and O. N. Bhatnagar, Can. J. Chem., **50,** 1627 (1972). Some thermodynamic properties of lithium chlorate in water and water–dioxane mixtures. Heats of solution and dilution.

FORMAMIDE–WATER

V. P. Emelin and U. M. Kessler, Zh. Strukt. Khim., **13,** 323 (1972). Structural effects in potassium chloride, heats of dissolution in water–formamide mixed solvent.

METHANOL–WATER

S. Taniewsk, R. Logawinie, and M. Pluta, Soc. Sci. Lod., **16,** 45 (1971). Thermochemical properties of NaI solutions in water-methanol mixtures.

M. H. Abraham, J. Chem. Soc. Chem. Comm., 888 (1972). Entropies of transfer of tetraalkylammonium ions from water to methanol, dimethylformamide and acetonitrile.

PROPANOL–WATER

V. I. Klopov, A. I. Pirogov, and G. A. Krestov, Zh. Strukt. Khim., **13,** 141 (1972). Heats of dissolution of KCl and structural properties of propyl alcohol aqueous solutions at 10 to 60 degrees.

UREA–WATER

T. S. Sarma and J. C. Ahluwalia, J. Phys. Chem., **76,** 1366 (1972). Thermodynamics of transfer of tetrabutylammonium bromide from water to aqueous urea solutions and the effects on the water structure.

R. B. Cassel and W. Y. Wen, J. Phys. Chem., **76,** 1369 (1972). Thermodynamics of transfer of three tetraalkylammonium bromides from water to aqueous urea solutions.

B. Chawla, S. Subramanian, and J. C. Ahluwalia, J. Chem. Thermodyn., **4,** 575 (1972). Enthalpies and heat capacities of transfer of $NaBPh_4$ from water to aqueous urea and the effects on the water structure.

XI. ADDITIONAL REFERENCES AND DATA SOURCES

DIMETHYLFORMAMIDE–DIMETHYLSULFOXIDE

F. Rallo and F. Rodante, Gaz. Chim. Ital., **102**, 56 (1972). Ionic enthalpy of transfer of picrate ion between DMSO and DMF.

GENERAL

G. Somsen and L. Weeda, Rec. Trav. Chim., **90**, 81 (1971). Enthalpies of transfer of alkali halides between different solvents. Enthalpies of solution of alkali halides in formamide, N-methylformamide, N,N-dimethylformamide, N-methylacetamide, dimethyl sulfoxide, water, methanol and ammonia are discussed.

G. A. Krestov, Proc. 1st Intern. Conf. Calorimetry and Thermodynamics, Warsaw, August 31–Sept. 4, 1969, p. 949. Thermodynamics of dilute solutions.

C. V. Krishnan and H. L. Friedman, J. Phys. Chem., **75**, 3598 (1971). Solvation enthalpies of Hydrocarbons and normal alcohols in highly polar solvents.

G. Somsen, L. Weeda, and J. H. Los, J. Electroanal. Chem., **31**, 9 (1971)· Discontinuous-model calculations of ionic enthalpies of solvation.

W. L. Jolly, Chem. Rev., **50**, 351 (1952). Heats, Free Energies and Entropies in Liquid Ammonia.

M. A. Bernard, F. Busnot, and J. F. Le Querler, Bull. Soc. Chim. (France), 3136 (1971). Solvation of cadmium chloride by several amines.

A. N. Campbell and O. N. Bhatnagar, Can. J. Chem., **50**, 1627 (1972)· Some thermodynamic properties of lithium chlorate in water and dioxane–water mixtures.

T. Kenjo and R. M. Diamond, J. Phys. Chem., **76**, 2454 (1972). Solvation of perchlorate, tetraphenylboride and nitrate ions in some organic solvents.

M. J. Mastroianni, M. J. Pikal, and S. Lindenbaum, J. Phys. Chem., **76**, 3050 (1972). Heats of dilution of tetrabutylammonium bromide and lithium bromide in mixed aqueous solvent systems.

(k) Polarography

(l) Inorganic Compounds

ACETONITRILE
0.4M LiClO$_4$ at RDE (Pt)

Species	$E_{1/2}V$ (vs Ag/AgCl)	Ref.
Cl$^-$	+1.06[a]	2
Cl$_2$	+0.90[b]	
Cl$^-$ + Cl$_2$	+1.02; +0.95[c]	

[a] 18.9 mM at 0°C.
[b] 21.3 mM at 0°C vs. Pt.
[c] Cl$^-$ (15.1 mM) + Cl$_2$ (21.5 mM) at 0°C vs Pt.

BENZONITRILE
0.1M Bu$_4$NClO$_4$ at DME at 25 ± 0.1°C
0.5 mM

Species	$E_{1/2}V$ (vs SCE)	Ref.
Ni(11)	−0.40; −1.20	8

DIMETHYL FORMAMIDE
0.1F Et$_4$NClO$_4$ at Pt
5 mM

Species	$E_p V$ (vs SCE)	Ref.
Sulfurdioxide	−0.84[a]; −0.74[b]; −0.25[b]	4

[a] Cathodic peak.
[b] Anodic peak.

XI. ADDITIONAL REFERENCES AND DATA SOURCES

DIMETHYL SULFOXIDE
1M $KClO_4$ at RDE[a] (Pt) at 36°C
0.02547M

Species	$E_{1/2}V$	Ref.
Nitrite ion	+0.670 to +0.723[b]	1

[a] Rotating disk electrode.
[b] Stirring rate varying from 476 to 2202 rev/min.

0.1M $BuNClO_4$ at Pt

Species	$E_{1/2}V$ (vs Ag/AgCl)	Ref.
Oxygen	−0.90; −0.85	3

METHANOL
0.01M HCl + 0.2M LiCl at DME
0.2 mM

Species	$E_{1/2}V$ (vs SCE)	Ref.
$TiCl_4$	−0.35	6

PROPIONITRILE
0.1M Bu_4NClO_4 at DME at 25 ± 0.1°C
0.5 mM

Species	$E_{1/2}V$ (vs SCE)	Ref.
Ni(11)	−0.50; −1.20	8

DIMETHYL SULFOXIDE–WATER
0.15M NaClO$_4$ at DME at 30 ± 0.1°C
1 mM

Species	Vol % DMSO	$E_{1/2}V$ (vs SCE)	Ref.
UO$_2$(ClO$_4$)$_2$	40	−0.313[a], −0.312[b], −0.312[c], −0.312[d]	7
	60	−0.366[e], −0.378[f], −0.378[g], −0.381[h]	
	80	−0.396[i], −0.420[j], −0.430[k], −0.430[l]	
	99	−0.458[m], −0.474[n], −0.498[o], −0.492[p]	

[a] pH 2.40.
[b] pH 3.30.
[c] pH 4.00.
[d] pH 5.00.
[e] pH 2.50.
[f] pH 3.00.
[g] pH 4.30.
[h] pH 5.25.
[i] pH 2.60.
[j] pH 3.70.
[k] pH 4.20.
[l] pH 4.78.
[m] pH 2.95.
[n] pH 3.50.
[o] pH 4.60.
[p] pH 4.95.

ACETONITRILE–TOLUENE (50% V/V)
0.1M Et$_4$NClO$_4$ at DME at 25 ± 0.1°C
1.73 mM

Species	$E_{1/2}V$ (vs SCE)	Ref.
trans-Chlorocarbonyl bis(tris-Phenyl phosphine) rhodium (1)	−1.75[a] −1.10,[a,b] ∼ −0.23,[a,b] −2.05[a,c]	5

[a] 1.73 mM triphenylphosphine present.
[b] Anodic peaks after electrolysis.
[c] Cathodic wave after slow electrolysis.

XI. ADDITIONAL REFERENCES AND DATA SOURCES

References

1. J. A. Wargon and A. J. Arvia, Electrochim. Acta **17,** 649 (1972).
2. L. Sereno, V. A. Macagno and M. C. Giordano, Electrochim. Acta **17,** 561 (1972).
3. K. D. Legg, D. W. Shive and D. M. Hercules, Anal. Chem., **44,** 1650 (1972).
4. R. P. Martin and D. T. Sawyer, Inorg. Chem., **11,** 2644 (1972).
5. G. Pilloni, S. Valcher and M. Martelli, J. Electroanal. Chem., **40,** 63 (1972).
6. E. P. Parry and D. H. Hern, J. Electrochem. Soc., **119,** 1141 (1972).
7. T. T. Lai and J. C. Yang Bull. Chem. Soc. Japan, **45,** 1683 (1972).
8. E. Itabashi, Bull. Chem. Soc., Japan, **45,** 2455 (1972).

(2) ORGANIC COMPOUNDS

ACETIC ACID
0.3M Et$_4$NTS[a] at carbon rod electrode at room temp.

Species	E_pV (vs SCE)[b]	Ref.
Cyclohexene	2 to ~2.5	28
Cyclopentene	2.15 to 2.35	
1-Methyl cyclohexene	2.1 to 2.3	
3-Methyl cyclohexene	2.15 to 2.4	
4-Methyl cyclohexene	2.1 to 2.35	
Norbornene	2.1 to 2.3	
2-Octene	2.1 to 2.25	
1-Octene	2.4 to 2.55	

[a] Tetraethyl ammonium p-toluene sulfonate.
[b] Oxidation potentials.

ACETONE
0.1M Et$_4$NClO$_4$ at DME at 20°C
0.15 mM to 3 mM

Species	$E_{1/2}V$ (vs Ag/AgCl)	Ref.
o-Ethyl thioacetate	+0.7; −2.08	10
o-Ethyl thioacetothioacetate	+0.226; −1.428	

XI. ADDITIONAL REFERENCES AND DATA SOURCES

ACETONITRILE
0.1N LiClO$_4$ at Pt electrode
1 mM to 5 mM

Species	$E_p V$ (vs Ag/AgNO$_3$)	Ref.
Benzyl acetate	1.90	27
Benzyl alcohol	1.90	
Benzyl *tert*-butyl ether	1.95	
Benzyl cyclohexyl ether	1.80	
Benzyl ether	1.90	
Benzyl methyl ether	1.90	
o-Benzyl vanillyl alcohol	1.10	
Benzhydryl dodeyl ether	1.90	
Benzhydryl methyl ether	1.70	
Cumyl alcohol	1.90	
m-Methoxy benzyl alcohol	1.40	
o-Methoxy benzyl alcohol	1.30	
p-Methoxy benzyl alcohol	1.30	
p-Methoxybenzyl methyl ether	1.30	
2-Octyl benzyl ether	1.90	
1-Phenyl ethanol	1.70	
1-Phenyl ethyl benzoate	1.95	
1-Phenyl ethyl ether	1.90	
trans-Stilbene oxide	1.70	
Tetraphenyl ethylene oxide	1.70	

0.05M Et$_4$NClO$_4$ at DME at 25°C
~0.5 mM

Species	$E_{1/2} V$ (vs SCE)	Ref.
TPS[a]	−2.18; −1.94	30
DPS[b]	−2.26; −1.94	

[a] 1,1-Dimethyl-2,3,4,5-tetraphenyl-1-sila cyclopentadiene.
[b] 1,1-Dimethyl-2,5-di-phenyl-1-sila cyclopentadiene.

ACETONITRILE (Continued)
LiClO$_4$ at Carbon rod electrode at room temp.

Species	E_pV (vs SCE)[a]	Ref.
Cyclohexene	2.14	28
Cyclopentene	1.96	
1-Methyl cyclohexene	1.70	
Norbornene	2.02	
1-Octene	2.8[a,b]	
2-Octene	2.3[a,b]	
α-Pinene	1.41	
β-Pinene	1.89	

[a] Oxidation potentials.
[b] Estimated values.

0.1M Bu$_4$N hexafluorophosphate at DME
$\sim$0.16 mM

Species	$E_{1/2}V$ (vs SCE)	Ref.
Cycloactatraene	$-1.87; -2.65;$ -1.80[a]; -1.97[a]; -2.60[a] -1.81[b]; -2.62[b]	26

[a] 0.16M Bu$_4$N hexafluorophosphate.
[b] Solvent containing 10^{-4}M water.

0.1M Bu$_4$NBF$_4$ at RPt disk electrode at 25 $\pm$ 0.05
$\sim$0.926 mM

Species	$E_{1/2}V$ (vs Ag/Ag$^+$)	Ref.
2,2'-(3-Ethyl benzothioazolinone)-azine	+0.210; 0.723	25
2,2'-(3-Ethyl-6-sulfonato benzothiazoline)-azine	+0.232	
2,2'-(3-Methyl benzothiazolinone)-azine	+0.225; 0.742	

XI. ADDITIONAL REFERENCES AND DATA SOURCES 861

ACETONITRILE (*Continued*)
0.1M Bu₄NClO₄ at Pt disk electrode
2.24 mM

Species	$E_p V$ (vs SCE)	Ref.
N-(p-methoxybenzylidene)-p-n-butylaniline	−2.15[a]; +1.45[b]; +2.0[b]	18

[a] Cathodic peak.
[b] Anodic peak.

0.1M Et₄NClO₄ at Pt disk
1 mM

Species	$E_{1/2} V$ (vs Ag/Ag⁺)	Ref.
Dimethyl 1,1-hydrazine	−0.08; +0.16	13
Trimethyl hydrazine	−0.4; +0.74	

0.1M NaClO₄ at Pt and carbon electrodes at 25°C
1 mM

Species	$E_{1/2} V$ (vs SCE)	Ref.
Styrene	+1.9[a]; +1.54[b]	12
α-Methylstyrene	+1.76[a]; +1.48[b]	
Ethyl vinyl ether	+1.90[a]; +1.52[b]	
DPDM[c] Hene-1	+1.08[a]; +1.05[b]	
DP Benzene-1	+1.14[a]; +1.14[b]	
Phenylacetylene	+2.25[a]	
DP Byne[e]	+1.85[a]	
Ferrocene	+0.325[a]; +0.325[b]	

[a] At Pt electrode.
[b] At carbon electrode.
[c] 2,5-Diphenyl-2,5-dimethyl-2,4-hexadiene.
[d] 1,4-Diphenyl-1,3-butadiene.
[e] 1,4-Diphenyl-1,3-butadiyne.

ACETONITRILE (Continued)

0.1M TBAH[a] at Pt at 22 ± 2°C

Species	$E_{1/2}V$ (vs SCE)[b]	Ref.
Porphyrin	+0.80; +1.42	7

[a] *tetra-n*-Butylammonium hexafluorophosphate.
[b] Saturated sodium chloride calomel electrode.

0.1M Et$_4$NClO$_4$ at Pt electrode
1 mM

Species	E_pV (vs SCE)	Ref.
Phenazine	−1.290 to −1.204[a]; −1.048 to −1.128[b]; +0.33[c]; +0.06[c]; +0.43[d]; +0.12[d]	6

[a] Cathodic peak in neutral solution.
[b] Anodic peak in neutral solution.
[c] Cathodic peak in acidic solution.
[d] Anodic peak in acidic solution.

0.5M LiClO$_4$ at Pt at 30°C
0.01M to 0.1M

Species	$E_{1/2}V$ (Ag 0.1M AgNO$_3$)	Ref.
2,6-Xylenol	+1.1; +1.5 +1.1; 1.3[a]	2

[a] Values obtained with rotating Pt electrode.

XI. ADDITIONAL REFERENCES AND DATA SOURCES

ACETONITRILE (*Continued*)
0.1M NaClO₄ at gold electrode
0.2 mM to 0.9 mM

Species (tetraalkyl hydrazines)	$E_{1/2}V$ (vs SCE)	Ref.
	−0.28	20
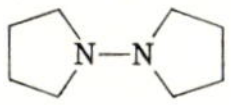	−0.04	
	0.00	
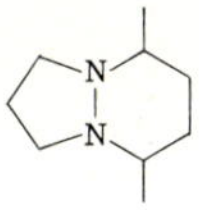	0.00	
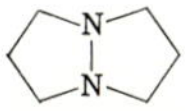	0.02	
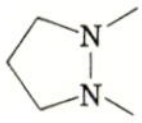	0.06	
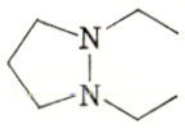	0.06	
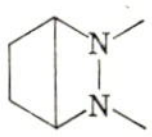	0.10	
	0.10	
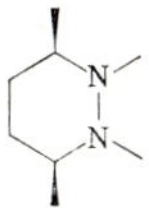	0.12	
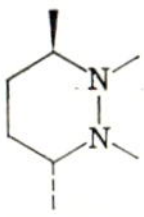	0.13[a]	
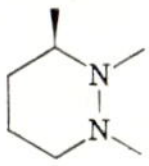	0.14[b]	

ACETONITRILE (*Continued*)

Species (tetraalkyl hydrazines)	$E_{1/2}V$ (vs SCE)	Ref.
(iPr)N-CH₂-N(iPr)-N(iPr)-CH₂-N(iPr) ring	0.18	
1,2-dimethylhexahydropyridazine	0.18	
Me₂N-NMe₂	0.22	
1,2-dimethyl-1,2-dihydrophthalazine-like	0.24	
bicyclic hexahydropyridazine	0.24	
1,2,3-trimethyl-1,2,3,6-tetrahydropyridazine	0.24	
(Et)N-CH₂-N(Et)-N(Et)-CH₂-N(Et) ring	0.25	
1,2-dimethyl-1,2,3,6-tetrahydropyridazine	0.28	
bicyclic tetrahydropyridazine	0.30	
(Me)N-CH₂-N(Me)-N(Me)-CH₂-N(Me) ring	0.32	
bicyclic N,N,O oxadiazine	0.42^c	
(tBu)N-N(tBu) oxadiazolidine	0.56	
(Et)N-N(Et) oxadiazolidine	0.56	
Ph₂NNPh₂	0.76	

XI. ADDITIONAL REFERENCES AND DATA SOURCES

ACETONITRILE (Continued)

Species (tetraalkyl hydrazines)	$E_{1/2}V$ (vs SCE)	Ref.
[structure]	0.96[d]	
[structure]	0.63[e]	
[structure]	1.07[f]	
[structure]	1.15[g]	
[structure]	>2[h]	

[a] 0.206 mM.
[b] 2.81 mM.
[c] 3.05 mM.
[d] 1.89 mM.
[e] 0.301 mM and E_p oxidation peak.
[f] 0.418 mM and E_p oxidation peak.
[g] 1.32 mM and E_p oxidation peak.
[h] 0.5 mM and E_p oxidation peak.

ACETONITRILE *(Continued)*
0.1M NaClO$_4$ at gold electrode

Species (Tertiary amines & Polyamines)	Conc. in mM	E_pV (vs SCE)	Ref.
Monoamines			
(triethylamine)	2.77	0.78	21
(quinuclidine)	4.50	1.10	
1,3-Diamines			
(N,N,N',N'-tetramethylmethylenediamine)	3.26	0.87	
(cyclohexane-spiro-imidazolidine)	1.64	0.85	
(bicyclic diamine)	4.64	1.20	
(bicyclic diamine)	3.40	0.84	
(bicyclic diamine)	1.79	0.69	
(bicyclic diamine)	1.35	0.59	

XI. ADDITIONAL REFERENCES AND DATA SOURCES

ACETONITRILE (Continued)

Species (Tertiary amines & Polyamines)	Conc. in mM	$E_p V$ (vs SCE)	Ref.
[structure: diazaadamantane]	0.58	0.86	
[structure: diphenyl diazaadamantane]	0.54	0.85	
[structure: diphenyl diazaadamantanone]	1.26	1.15	
[structure: cyclohexyl diazaadamantane]	0.39	0.70[a]	
1,4-Diamines			
[structure: TMEDA]	4.20	0.67	
[structure: N,N'-dimethylpiperazine]	3.16	0.75	
[structure: DABCO]	3.11	0.60[a]	

ACETONITRILE (Continued)

Species (Tertiary amines & Polyamines)	Conc. in mM	$E_p V$ (vs SCE)	Ref.
Tri- and Tetramines			
[structure]	2.32	0.94	
[structure]	2.94	1.03	
[structure]	1.38	1.02	
[structure]	1.14	1.37	
[structure]	0.88	0.58[a]	

[a] $E_{1/2}$ values in V vs SCE.

n-BUTYRONITRILE
0.1M TBAH[a] at Pt at 22 ± 2°C

Species	$E_{1/2} V$ (vs SCE)[b]	Ref.
Porphyrin	+0.86; +1.35[b]	7

[a] Tetra-n-butylammonium hexafluorophosphate.
[b] Saturated sodium chloride calomel electrode.

XI. ADDITIONAL REFERENCES AND DATA SOURCES

CHLOROFORM
0.25M Bu$_4$NClO$_4$ at RDE (Pt)
10^{-3}M to 5×10^{-4}M

Species	$E_{1/2}V$ (vs Ag 10^{-3}M Ag$^+$)	Ref.
N,N'-dimethyl N,N'-diphenyl-p-phenylene diamine	-0.43; $+0.065$	3
N,N'-diphenyl-p-phenylene diamine	-0.43; $+0.11$	
4-Hydroxy diphenylamine	-0.26; ~-0.10; ~-0.20	
N-(p-methoxy phenyl)-p-p-phenylenediamine	-0.51; $+0.02$	
N,N,N',N'-tetramethyl-p-phenylene diamine	-0.64; -0.11	
N,N,N',N'-tetraphenyl-p-phenylene diamine	-0.25; $+0.17$	

DICHLOROMETHANE
0.1M TBAH[a] at Pt at $22 \pm 2°$C

Species	$E_{1/2}V$ (vs SCE)[b]	Ref.
Porphyrin	$+0.83$; $+1.39$	7

[a] *tetra-n*-Butylammonium hexafluorophosphate.
[b] Saturated sodium chloride calomel electrode.

DIMETHOXYETHANE
0.5M Et$_4$NClO$_4$ at DME
0.5 mM

Species	$E_{1/2}V$ (vs SCE)	Ref.
DPS[a]	-1.94; -2.26	11
TPS[b]	-1.94; 2.18	

[a] 1,1-Dimethyl-2,5-diphenyl-1-salacylopentadiene.
[b] 1,1-Dimethyl-2,3,4,5-tetraphenyl-1-sila cyclopentadiene.

1, 2-DIMETHOXYETHANE
0.1M Bu$_4$NClO$_4$ at DME

Species	$E_{1/2}V$ (vs Ag/Ag$^+$)	Ref.
2, 2'-Bipyridyl	−2.80; −3.11	19
Di-2-Pyridyl ketone	−2.31; −2.92	
o-Phenanthroline	−2.72; −3.24	

DIMETHYLFORMAMIDE
0.1M Bu$_4$NI at Pt or DME

Species	E_pV (vs SCE)	Ref.
Decafluorobiphenyl	−1.73	5
1, 3-Difluorobenzene[a]	−2.82	
4, 4'-Difluorobiphenyl	−2.66	
4-Fluorobiphenyl	−2.64; −2.64; to −2.71[d]; −2.69 to −2.83[d]	
2-Fluoronaphthalene[e]	−2.41 to −2.425; −2.325; −2.298 to −2.307	
Pentafluorobenzene[b,c]	−2.31 to −2.39; −2.23 to −2.29; −2.48 to −2.52; −2.62 to −2.70; −2.76 to −2.82; −2.91 to 2.93	
1, 2, 3, 5-Tetrafluorobenzene[b]	−2.65 ± 0.02; −2.77 ± 0.02	
1, 2, 4, 5-Tetrafluorobenzene[b]	−2.40 ± 0.03; −2.53 ± 0.04; −2.65 ± 0.05	

[a] At DME.
[b] At Pt electrode.
[c] 2.7 mM.
[d] 1.61 mM + 18 mM hydroquinone in 0.1M Bu$_4$NClO$_4$.
[e] 1.66 mM.

XI. ADDITIONAL REFERENCES AND DATA SOURCES

DIMETHYLFORMAMIDE (Continued)
0.1F Et$_4$NClO$_4$ at Pt electrode
1 mM

Species	$E_p V$ (vs SCE)	Ref.
Phenazine	-1.17^a; -1.85^a; -1.11^b -0.29^c; $+0.03^d$	6

[a] Cathodic peak in neutral solution.
[b] Anodic peak in neutral solution.
[c] Cathodic peak in acidic solution.
[d] Anodic peak in acidic solution.

0.1M Bu$_4$NI at RDE (Pt) at 24°C

Species	$E_{1/2} V$ (vs Ag)[a]	Ref.
Cinnamonitrile[b]	-1.21; -1.83	8
Dimethylfurmarate[d]	-0.75	8
Fumaronitrile[c]	-0.68	8

[a] Silver wire spiral.
[b] 1.25 to 11.3 mM.
[c] 0.42 to 3.94 mM.
[d] 3.0 to 6.20 mM.

DIMETHYLFORMAMIDE (Continued)
0.1M Pr$_4$NClO$_4$ at DME
1 mM

Species	$E_{1/2}V$ (vs SCE)	Ref.
2,5-Diphenyloxazole	−2.16; 2.52	11
2-(3-Methoxyphenyl)-5-phenyloxazole	−2.06; −2.37	
2-(4-Fluorophenyl)-5-phenyloxazole	−2.13; −2.48	
2-(1-Naphthyl)-5-phenyloxazole	−1.86; −2.24	
2,5-Diphenyloxadiazole	−1.95; −2.34; −2.63	
2-(4-Fluorophenyl)-5-phenyloxadiazole	−1.96; −2.36; −2.64	
2-(4-Fluorophenyl)-5-(2-naphthyl)-oxadiazole	−1.81; −2.16; −2.48	
2-(4-Methoxyphenyl)-5-(2-naphthyl)-oxadiazole	−1.78; −2.22	
2-(4-Methoxyphenyl)-5-(2-naphthyl)-oxadiazole	−1.75; −2.18	
2,5-Diphenyloxazole	−2.19[a]; −2.15[b]	
2-(1-Naphthyl)-5-phenyloxazole	−1.93[a]; −1.90[b]	
2,5-Diphenyloxadiazole	−2.03[a]; −2.02[b]	

[a] At hanging mercury drop electrode.
[b] At hanging mercury drop electrode in presence of 105 mM hydraquinone.

XI. ADDITIONAL REFERENCES AND DATA SOURCES

DIMETHYLFORMAMIDE (Continued)
0.1M Et$_4$NClO$_4$ at DME
0.5 mM

Species	$E_{1/2}V$ (vs SCE)	Ref.
9-Nitroanthracene	−0.995; −0.940[a] −1.575; −1.430[a]	14
1-Nitronaphthalene	−1.030; −1.010[a] −1.700; −1.500[a]	
2-Nitronaphthalene	−1.050; −1.040[a] −1.720; −1.465[a]	
Nitrobenzene	−1.10; −1.065[a] −1.770; −1.380[a]	
7-Nitro-1,2-benzanthracene	−1.045; −1.025[a] −1.507; −1.375[a]	
2-Nitrobiphenyl	−1.190; −1.175[a] −1.720; −1.500[a]	
4-Nitrobiphenyl	−1.045; −1.020[a] −1.715; −1.535[a]	
6-Nitrochrysene	−0.985; −0.945[a] −1.580; −1.390[a]	
9-Nitrophenanthrene	−1.020; −0.980[a] −1.685; −1.460[a]	
1-Nitropyrene	−0.985; −0.925[a] −1.605; −1.375[a]	

[a] 0.1M Potassium perchlorate supporting electrolyte.

0.1M Bu$_4$NClO$_4$ at Pt disk electrode
6.65 mM

Species	E_pV (vs SCE)	Ref.
N-(p-Methoxybenzylidene)-p-n-butylaniline	−2.06[a]; 2.60[a]; −0.54[b]	18

[a] Cathodic peaks.
[b] Anodic peak.

DIMETHYL SULFOXIDE
0.1M Bu$_4$NClO$_4$ at Pt
2.14 × 10^{-3}M to 1.07 × 10^{-2}M

Species	$E_{1/2}V$ (vs Ag/AgCl)	Ref.
Lucigenin	−0.30; +0.45	4

0.1F Et$_4$NClO$_4$ at Pt electrode
1 mM

Species	E_pV (vs SCE)	Ref.
Phenazine	−1.280 to −1.190^a; −1.040 to −1.080^b −0.400 to −0.392^c; +0.056 to −0.040^d	6

^{a} Cathodic peak in neutral solution.
^{b} Anodic peak in neutral solution.
^{c} Cathodic peak in acidic solution.
^{d} Anodic peak in acidic solution.

0.1F Et$_4$NClO$_4$ at Pt electrode
1 mM

Species	E_pV (vs SCE)	Ref.
Riboflavin	−0.82^a; −0.55^b; −1.3^a	22

HEXAMETHYL PHOSPHORIC TRIAMIDE
0.2M Bu$_4$NClO$_4$ at DME

Species	$E_{1/2}V$ (vs Ag wire)	Ref.
Dipyridyl	−2.37; −2.80	29

XI. ADDITIONAL REFERENCES AND DATA SOURCES

METHANOL

0.3M Et$_4$NTS[a] at carbon rod electrode at room temp.

Species	$E_p V$ (vs SCE)[b]	Ref.
Cyclohexene	1.6 to ~1.7	28
Cyclopentene	1.55 to ~1.7	
Norbornene	1.6 to ~1.7	

[a] Tetraethylammonium p-toluene sulfonate.
[b] Oxidation potentials.

0.1M NaClO$_4$ 050.1M Na-tosylate at Carbon electrode at 25°C
1 mM

Species	$E_{1/2} V$ (vs SCE)	Ref.
Styrene	+1.425[a]; +1.480[b]	12
α-Methyl styrene	+1.350; +1.410[b]	
Ethyl vinyl ether	+1.460[a]; 1.510[b]	
DP Dene[c] 1	+1.040[a]; +1.050[b]	
DPDM[d] Hene 1	+0.925[a]; +0.950[b]	
Ferrocene	+0.370; +0.960	

[a] 0.1M NaClO$_4$.
[b] 0.1M Na-tosylate.
[c] 1.4 Diphenyl-1, 3-butadiyne.
[d] 2.5 Diphenyl-2, 5-dimethyl-2, 4-hexadiene.

NITROMETHANE
AlCl₃ at Pt electrode
3 mM

Species	$E_{1/2}V$ (vs ferracene ferricinium)	Ref.
Anthracene	+0.83	15
Dimethoxy-1,4-benzene	+0.93	
Diphenyl-9,10-anthracene	+0.88[a]; +1.28[b]	
Naphthalene	+1.30	
Perylene	+0.65[a]	

[a] 0 to +1 state.
[b] +1 to +2 state.

PYRIDINE
0.1M Et₄NClO₄ at pyrolytic graphite electrode at 25°C
1 mM

Species	E_pV (vs SCE)	Ref.
Quinone	∼−0.56[a]; ∼−0.42[b]	9

[a] Cathodic peak.
[b] Anodic peak.

TETRAHYDROFURAN
0.19M Bu₄-N-hexafluorophosphate at DME
0.24 mM

Species	$E_{1/2}V$ (vs SCE)	Ref.
Cyclooctatetraene	−1.96; −2.16; −2.78	26

XI. ADDITIONAL REFERENCES AND DATA SOURCES

ACETONE–WATER
0.5M LiCl at DME at 30 ± 0.1°C

Species	% Conc. of solvent	$-E_{1/2}V$ (vs SCE)	Ref.
Nitrosobenzene	10	0.898[a]	1
	10	1.06[b]	
	20	0.839[c]	
	40	0.934[a]	
	40	1.14[b]	
	40	0.866[c]	
	60	0.950[a]	
	70	1.15[b]	
	70	0.854[c]	

[a] In presence of 0.01M CCl.
[b] In ammonia buffer at pH = 9.
[c] In presence of 0.1M HCl.

ACETONITRILE–WATER (3:1)
0.1M Bu NBF at RPt disk electrode at 25 ± 0.05
~0.926 mM

Species	$E_{1/2}V$ (vs Ag/AgCl)	Ref.
2,2'-(3-Ethyl benzothiazolinone)-azine	+0.463; +0.950	25
2,2'-(3-Ethyl-6-sulfonatobenzo-thiazolinone)-azine	+0.550; ~+0.985	
2,2'-(3-Methyl benzothiazolinone)-azine	+0.465; +0.945	

DIOXAN–WATER

0.5M LiCl at DME at 30 ± 0.1°C

Species	% Conc. of solvent	$-E_{1/2}V$ (vs SCE)	Ref.
Nitrosobenzene	10	0.197[a]	1
	40	0.185[a]	
	70	0.177[a]	

[a] In ammonia buffer.

ETHANOL–WATER

0.1M KNO$_3$ at DME at 30 ± 0.1°C

1 mM at pH 11.66

Species	% Ethanol	$E_{1/2}V$ (vs SCE)	Ref.
Glycol dimercaptopropionate	10	−0.541	23
	20	−0.584	
	30	−0.593	
	40	−0.613	
	50	−0.620	
	60	−0.642	
	70	−0.640	

XI. ADDITIONAL REFERENCES AND DATA SOURCES

ETHANOL–WATER (Continued)

0.5M LiCl at DME at 30 ± 0.1°C

Species	% Conc. of solvent	$-E_{1/2}V$ (vs SCE)	Ref.
Nitrosobenzene	10	0.680[a]	1
	10	0.206[b]	
	20	0.633[c]	
	40	0.750[a]	
	40	0.205[b]	
	40	0.675[c]	
	60	0.792[a]	
	70	0.208[b]	
	70	0.720[c]	

[a] In presence of 0.01M HCl.
[b] In ammonia buffer at pH = 9.
[c] In presence of 0.1M HCl.

ETHANOL–WATER (50 Vol % of ethanol)
0.2M KNO₃ at DME at 30 ± 0.2°C
0.6504 mM

Species		$E_{1/2}V$ (vs SCE)	Ref.
Thiobenzoic acid	1.3	−0.090; −0.265	24
	3.5	−0.220; −0.405	
	5.30	−0.265; −0.450	
	6.30	−0.270; −0.440; −0.565	
	10.50	−0.270; −0.450; −0.680	
	11.70	−0.270; −0.445; −0.710	
	13.70	−0.270; −0.450; −0.770	

ETHANOL–WATER (50% ethanol)
Britton–Robinson Buffer at DME at 22 ± 1°C
0.05 mM

$E_{1/2}V$ (vs SCE) Ref. 16

Species	PH = 2.6	3.6	4.7	5.6	6.9	8.0	9.2	10.0	11.2	12.15	A
HPy-2[b]	−0.35	−0.42	−0.49	−0.56	−0.67	−0.76	−0.83	−0.86	−0.93	−1.05	−0.90
	−1.14	−1.20	−1.27	−1.32	−1.42	−1.48	−1.53	−1.53	−1.53	—	−1.63
						−1.79					−2.03
HPy-3[c]	−0.57	−0.62	−0.67	−0.75	−0.88	−0.93	−0.96	−0.98	−1.03	−1.15	−0.97
	−1.13	−1.18	−1.23	−1.28	−1.38	−1.47	−1.50	−1.50	−1.53	−1.32	−1.33
						−1.75					−1.66
HPy-4[d]	−0.33	−0.39	−0.45	−0.51	−0.62	−0.71	−0.77	−0.83	−0.90	−1.01	−0.30
	−1.18	−1.22	−1.28	−1.33	−1.42	−1.50	−1.50	−1.50	−1.13	—	−1.55
						−1.88			−1.10		−2.00

XI. ADDITIONAL REFERENCES AND DATA SOURCES

MPy-2[e]	−0.39	−0.47	−0.54	−0.61	−0.72	−0.80	−0.86	−0.90	−0.96	−1.08	−0.95
	−1.24	−1.31	−1.37	−1.43	−1.53	−1.60	−1.68	−1.68	−1.70	—	−0.85
						−1.80					
MPy-3[f]	−0.62	−0.68	−0.75	−0.82	−0.92	−1.00	−1.01	−1.03	−1.06	−1.25	−1.07
	−1.20	−1.26	−1.31	−1.35	−1.45	−1.55	−1.62	−1.63	−1.65	—	−1.35
						−1.74					−1.90
MPy-4[g]	−0.37	−0.43	−0.49	−0.56	−0.64	−0.72	−0.80	−0.85	−0.93	−1.05	−0.88
	−1.28	−1.33	−1.37	−1.42	−1.52	−1.59	−1.65	−1.60	−1.60	—	−1.72
						−1.78					−2.05

A = 0.175M Bu$_4$NI in 75% dioxane.

[b] 1-(2-Hydroxyphenyl)3-(2-pyridyl)2-propen-1-one.
[c] 1-(2-Hydroxyphenyl)3-(3-pyridyl)-2-propen-1-one.
[d] 1-(2-Hydroxyphenyl)-3-(4-pyridyl)-2-propen-1-one.
[e] 1-(2-Hydroxy-4-methoxyphenyl)3-(2-pyridyl)-2-propen-1-one.
[f] 1-(2-Hydroxy-4-methoxyphenyl)-3-(3-pyridyl)-2-propen-1-one.
[g] 1-(2-Hydroxy-4-methoxyphenyl)-3-(4-pyridyl)-2-propen-1-one.

ETHANOL–WATER (50% ethanol) (Continued)
0.5 mM

$E_{1/2} V$ (vs SCE) Ref. 17

Species	PH = 2.6	3.6	4.6	5.4	6.05	6.9	8.0	9.0	10.0	11.0	12.6	A
3,2-Py[b]	−0.25	−0.32	−0.39	−0.45	−0.51	−0.53	−0.63	−0.69	−0.78	−0.83	−0.90	−0.93
						−0.62	−0.71	−0.80	−0.86	−0.93	−1.00	−1.47
	−0.65	−0.72	−0.90	−0.96	−1.02	−1.08	−1.18	−1.26	−1.35	−1.45	−1.15	−2.03
							−1.75[a]	−1.77[a]	−1.80[a]			
3,3-Py[c]	−0.41	−0.48	−0.56	−0.64	−0.67	−0.71	−0.78	−0.83	−0.90	−0.95	−0.95	−0.96
					−0.74	−0.82	−0.87	−0.95	−1.02	−1.11	−1.12	−1.33
	−0.78	−0.87	−0.93	−0.99	−1.05	−1.11	−1.20	−1.20	−1.40	−1.50	−1.30	−2.02
		−1.50[a]	−1.55[a]	−1.60[a]	−1.67[a]	−1.75[a]	−1.80[a]	−1.85[a]			−1.48	
3,4-Py[d]	−0.23	−0.32	−0.39	−0.45	−0.48	−0.52	−0.58	−0.65	−0.71	−0.83	−1.06	−0.82
					−0.58	−0.67	−0.73	−0.79	−0.84			−1.45
	−0.83	−0.90	−0.97	−1.04	−1.09	−1.14	−1.20	−1.26	−1.36	−1.42	−1.45	−2.02

XI. ADDITIONAL REFERENCES AND DATA SOURCES

2,3-Py[e]	−0.30	−0.38	−0.47	−0.53	−0.58	−0.64	−0.69	−0.76	−0.80	−0.90	−0.90	−0.92
	−0.56	−0.62	−0.68	−0.71	−0.75	−0.78	−0.85	−0.95	−1.03	−1.12	−1.12	−1.33
	−1.23	−1.27	−1.30	−1.35	−1.41	−1.48	−1.58	−1.58	−1.75		−1.23	−2.00
						−1.65[a]	−1.75[a]	−1.85[a]	−1.85[a]			
2,4-Py[f]	−0.20	−0.27	−0.35	−0.40	−0.46	−0.54	−0.60	−0.67	−0.72	−0.82	−0.88	−0.80
							−0.70	−0.78	−0.83	−0.88	−0.98	−1.35
	−0.55	−0.62	−0.70	−0.75	−0.80	−0.86	−0.93	−1.00	−1.07	−1.15	−1.15	−1.92
	−1.23	−1.26	−1.29	−1.32	−1.35	−1.42					−1.28	
							−1.75[a]	−1.78[a]	−1.80[a]	−1.83[a]		

A = 0.175M Bu$_4$NI in 75% dioxane.
[b] 1-(3-Pyridyl)-3-(2-pyridyl)-2-propen-1-one.
[c] 1-(3-Pyridyl)-3-(3-pyridyl)-2-propen-1-one.
[d] 1-(3-Pyridyl)-3-(4-pyridyl)-2-propen-1-one.
[e] 1-(2-Pyridyl)-3-(3-pyridyl)-2-propen-1-one.
[f] 1-(2-Pyridyl)-3-(4-pyridyl)-2-propen-1-one.

References

1. S. K. Vijayalekshamma and R. S. Subrahmanya, Electrochim. Acta, **17**, 471 (1972).
2. C. Iwakura, M. Tsunaga and H. Tamura, Electrochim. Acta, **17**, 1391 (1972).
3. G. Cauquis and D. Serve, Anal. Chem., **44**, 2222 (1972).
4. K. D. Legg, D. W. Shive and D. M. Hercules, Anal. Chem., **44**, 1650 (1972).
5. B. H. Campbell, Anal. Chem., **44**, 1659 (1972).
6. D. T. Sawyer and R. Y. Komai, Anal. Chem., **44**, 715 (1972).
7. J. A. Ferguson, T. J. Meyer and D. G. Whitten, Inorg. Chem., **11**, 2767 (1972).
8. V. J. Puglisi and A. J. Bard, J. Electrochem. Soc., **119**, 829 (1972).
9. R. E. Panzer and P. J. Elving, J. Electrochem. Soc., **119**, 864 (1972).
10. A. M. Bond, A. R. Hendrickson and R. L. Martin, J. Electrochem. Soc., **119**, 1325 (1972).
11. S. L. Smith, L. D. Cook and J. W. Rogers, J. Electrochem. Soc., **119**, 1332 (1972).
12. M. Katz, P. Riemenschneider and H. Wendt, Electrochimica Acta, **17**, 1595 (1972).
13. G. Canquis, B. Chabaud and M. Genies, J. Electroanal. Chem., **40**, 6 (1972).
14. T. M. Krygowski, M. Stencel and Z. Galus, J. Electroanal. Chem., **39**, 395 (1972).
15. D. Bauer and A. Foucault, J. Electroanal. Chem. **39**, 385 (1972).
16. K. Butkiewicz, J. Electroanal. Chem., **39**, 407 (1972).
17. K. Butkiewicz, J. Electroanal. Chem., **39**, 419 (1972).
18. A. Lomax, R. Hirasawa and A. J. Bard, J. Electrochem. Soc., **119**, 1679 (1972).
19. R. E. Dessy, J. C. Charkoudian and A. L. Rheingold, J. Am. Chem. Soc., **94**, 738 (1972).
20. S. F. Nelsen and P. J. Hintz, J. Am. Chem. Soc., **94**, 7108 (1972).
21. S. F. Nelsen and P. J. Hintz, J. Am. Chem. Soc., **94**, 7114 (1972).
22. D. T. Sawyer and R. L. McCreery, Inorg. Chem., **11**, 779 (1972).
23. R. S. Saxena and U. S. Chaturvedi, Electrochimica Acta, **17**, 2009 (1972).
24. S. K. Tiawari and A. Kumar, Electrochimica Acta, **17**, 2085 (1972).
25. J. Janata and M. B. Williams, J. Phys. Chem., **76**, 1178 (1972).
26. D. R. Thielen and L. B. Anderson, J. Am. Chem. Soc., **94**, 2521 (1972).
27. E. A. Mayeda, L. L. Miller and J. F. Wolf, J. Am. Chem. Soc., **94**, 6812 (1972).
28. T. Shono and A. Ikeda, J. Am. Chem. Soc., **94**, 7892 (1972).
29. A. Misono, Y. Uchida, T. Yamagishi and H. Kageyama, Bull. Chem. Soc., Japan, **45**, 1438 (1972).
30. N. Tamaka, T. Ogata and Y. Uratani, Inorg. Nucl. Chem. Letters **8**, 1041 (1972).

(3) ORGANOMETALLIC COMPOUNDS

ACETIC ACID
0.1M LiClO$_4$ at carbon electrode at 25°C
1 mM

Species	$E_p V$ (vs Ag/AgCl)	Ref.
Ferrocene	∼+0.2[a]; ∼+0.27[b] ∼+0.1[c]; ∼+0.2[d]	1

[a] Cathodic peak with pyrolytic graphite electrode.
[b] Anodic peak with pyrolytic graphite electrode.
[c] Cathodic peak with glassy carbon electrode.
[d] Anodic peak with glassy carbon electrode.

ACETONE
0.1M Et$_4$NClO$_4$ at DME at 20°C
0.15 mM to 3 mM

Species	$E_{1/2} V$ (vs Ag/AgCl)	Ref.
Mercury di(o-ethylthioacetothioacetate)	+0.145; −0.41	3

0.1M Bu$_4$Nhexafluorophosphate Pt electrode at 22 ± 2°C

Species	$E_p V$ (vs SCE)	Ref.
[(π-C$_5$H$_5$)Fe(CO)]$_2$(pH$_2$PCH$_2$CH$_2$PPh$_2$)	+0.08; +0.88	6

ACETONITRILE

0.1M LiClO$_4$ at glassy carbon electrode at 25°C

1 mM

Species	$E_p V$ (vs SCE)	Ref.
Ferrocene	~+0.08[a]; ~+0.10[b] ~+0.06[c] ~+0.07[d]	1

[a] Cathodic peak.
[b] Anodic peak.
[c] Cathodic peak in presence of 0.85 mM benzoic acid vs Ag/Ag$^+$.
[d] Anodic peak in presence of 0.85 mM benzoic acid vs Ag/Ag$^+$.

0.1M TBAH[a] at Pt at 22 ± 2°C

1 mM

Species	$E_{1/2} V$ (vs SCE)[b]	Ref.
Lead(11) octaethyl porphyrin	+0.80; +1.42	17

[a] Tetra-n-butylammonium hexafluorophosphate.
[b] Saturated sodium chloride calomel electrode.

0.05M Et$_4$NClO$_4$ at DME at 25°C

~0.5 mM

Species	$E_{1/2} V$ (vs SCE)	Ref.
[Ni(bipy)$_3$](ClO$_4$)$_2$	−1.7	10

0.2M Bu$_4$NBF$_4$ at Pt electrode

1 mM

Species	$E_p V$ (vs Ag wire)	Ref.
Ru(bipy)$_3$Cl$_2$	+1.63[a], −1.09[a], −1.27[a], −1.53[a], −2.22[a] +1.70[b], −1.03[b], −1.22[b], −1.46[b]	12

[a] Reduction peaks.
[b] Oxidation peaks.

XI. ADDITIONAL REFERENCES AND DATA SOURCES

ACETONITRILE (*Continued*)
0.1M Bu$_4$N-hexafluorophosphate at Pt electrode at 22 ± 2°C

Species	E_pV (vs SCE)	Ref.
[(π-C$_5$H$_5$)Fe(CO)]$_2$(Ph$_2$PCH$_2$CH$_2$PPh$_2$)	−0.05; +0.75	6

BENZONITRILE
0.1M Bu$_4$NClO$_4$ at DME at 25 ± 0.1°C
0.5 mM Ni(11) + 0.1 mM to 1 mM LiSCN

Species	$E_{1/2}V$ (vs SCE)	Ref.
Ni(11) + LiSCN	−0.40; −0.7 to −1.76	16

BUTYRONITRILE
0.1M TBAH[a] at Pt at 22 ± 2°C
1 mM

Species	$E_{1/2}V$ (vs SCE)[b]	Ref.
Lead(11) octaethylporphyrin	+0.86; +1.35	17

[a] Tetra-*n*-butyl ammonium hexafluorophosphate.
[b] Saturated sodium chloride calomel electrode.

0.1M Bu$_4$Nhexafluorophosphate at Pt electrode at 22 ± 2°C

Species	E_pV (vs SEC)	Ref.
[(π-C$_5$H$_5$)Fe(CO)]$_2$(Ph$_2$PCH$_2$CH$_2$PPh$_2$)	+0.82	6

DICHLOROMETHANE
0.1M TBAH[a] at Pt at 22 ± 2°C
1 mM

Species	$E_{1/2}V$ (vs SCE)[b]	Ref.
Lead(11) octaethyl porphyrin	+0.68; +1.03	17

[a] Tetra-n-butyl ammonium hexafluorophosphate.
[b] Saturated sodium chloride calomel electrode.

0.1M Bu$_4$N hexafluorophosphate at Pt electrode at 22 ± 2°C

Species	E_pV (vs SCE)[a]	Ref.
[(π-C$_5$H$_5$)Fe(CO)]$_2$(Ph$_2$PCH$_2$CH$_2$PPh$_2$)	+0.10; +0.95 +0.02[b]; +0.86[b]	6

[a] Saturated sodium chloride calomel electrode.
[b] In presence of 0.1M 1,10-phenanthroline.

DIMETHOXYETHANE
0.1M Bu$_4$NClO$_4$ at DME

Species	$E_{1/2}V$ (vs Ag/Ag$^+$)	Ref.
Bipyridyl-Co(CO)(NO)	−0.50[a], −2.26[b], −2.81[b]	5
Bipyridyl-Fe(NO)$_2$	−0.56[a], −2.14[b]; −2.78[b]	
(Di-2-Pyridyl Ketone)-Fe(NO)$_2$	−0.30[a], −1.68[b], −2.40[b]	
o-Phenanthroline-Co(CO)(NO)	−0.36[a], −2.19[b], −2.82[b]	
o-Phenanthroline-Fe(NO)$_2$	−0.60[a], −2.16[b], −2.80[b]	
o-Phenanthroline-Ni(CO)$_2$	−0.74[a], −2.42[b], −2.76[b]	

[a] Oxidation potentials.
[b] Reduction potentials.

0.1M Bu$_4$NClO$_4$ at Pt disk electrode
~1 mM

Species	E_pV (vs Ag/AgClO$_4$)	Ref.
Ferrocene	−0.144 ± 0.005[a] −0.03[a]	4

[a] Oxidation potentials.

XI. ADDITIONAL REFERENCES AND DATA SOURCES

DIMETHYL FORMAMIDE
0.15M Et$_4$NClO$_4$ at DME at 20°C
1 mM to 1.75 mM

Species	$E_{1/2}V$ (vs SCE)	Ref.
Cu(baen)[a]	−1.541	15
Cu(baen-Cl$_2$)	−1.306	
Cu(baen-Br$_2$)	−1.289	
Ni(baen)	−2.004	
Ni(baen-Cl$_2$)	−1.760	
Co(acac)$_3$[b]	−0.410	
Et$_4$NCoy·2H$_2$O	−0.518	
Co(baen)(H$_2$O)$_2$	−1.845	
K[Co(baen)(ala)$_2$]·7H$_2$O[c]	−1.020; −1.845	
K[Co(baen)(Val)$_2$]·8H$_2$O[d]	−0.972; −1.834	
[Co(baen)(NH$_3$)$_2$]ClO$_4$	−0.750; −1.852	
Co(baen)(acac)	−0.945; −1.826	
[Co(baen)(dip)]ClO$_4$[e]	−0.452; −1.832	
Co(bisaen)(H$_2$O)$_2$[f]	−1.280	
K[Co(bisaen)(ala)$_2$]·12H$_2$O	−0.665; −1.279	
[Co(bisaen)(NH$_3$)$_2$]ClO$_4$	−0.590; −1.287	
Co(bisaen)(acac)	−0.400; −1.315	
[Co(bisaen)(dip)]ClO$_4$	−0.104; −1.315	

[a] N,N'-ethylene bis(acetyl acetoniminato).
[b] Acetyl acetonato.
[c] Alanine.
[d] Valine.
[e] 2,2-Dipyridyl.
[f] N,N'-ethylene bis(salicylaldehydeiminato).

0.1M LiClO at pyrolytic graphite electrode at 25°C
1 mM

Species	E_pV (vs Ag/AgCl)	Ref.
Ferrocene	+0.76[a]; +0.84[b]	1

[a] Cathodic peak.
[b] Anodic peak.

0.1M Et$_4$NClO$_4$ at HMDE[a] at 25 ± 0.1°C
0.05 mM to 1 mM

Species	$E_{1/2}V$ (vs SCE)	Ref.
Zinc tetraphenyl prophin	−1.31, −1.71; −2.32; −2.53; −0.5[b]; −1.9[c] −1.3[d], −1.7[d], −0.45[e]	2

[a] Hanging mercury drop electrode.
[b] Anodic peak at low scan rate.
[c] Cathodic peak at low scan rate.
[d] Cathodic peak in thin layer voltammetry.
[e] Anodic peak in thin layer voltammetry.

DIMETHYL SULFOXIDE

0.1 F Et$_4$NClO$_4$ at Pt electrode
0.42 mM Riboflavin + 4.2 mM Metal ion
$\Delta E_p mV$ (vs SCE)

Species	$\Delta E_p mV$ (vs SCE)	Ref.
Riboflavin radical[a] + Fe^{2+}	177[b]; −83[c]	7
Riboflavin radical + Ni^{2+}	209[b]; −52[c]	
Riboflavin radical + Y^{3+} [d]	~370[b]; −80[c]	
Riboflavin radical + La^{3+}	~400; −84[c]	
Riboflavin radical + Th^{4+} [d]	306[b]; −88[c]	
Riboflavin radical + Ca^{2+}	17[b]; −15[c]	
Riboflavin radical + Na$^+$	−5[b]; 0[c]	

[a] Riboflavin radical produced by reducing it at −0.9 V vs SCE.
[b] Anodic shift.
[c] Cathodic shift.
[d] Evidence of disproportionation.

0.1M Bu$_4$N-Hexafluorophosphate at Pt electrode at 22 ± 2°C

Species	$E_p V$ (vs SCE)	Ref.
[(π-C$_5$H$_5$)Fe(CO)]$_2$(PhPCH$_2$CH$_2$PPh$_2$)	+0.10; +0.80	6

XI. ADDITIONAL REFERENCES AND DATA SOURCES

HEXAMETHYLPHOSPHORIC TRIAMIDE
0.2M Bu$_4$NClO$_4$ at DME
5 mM

Species	$E_{1/2}V$ (vs Ag wire)	Ref.
Ni(dipy[a])$_2$	−0.73, −1.00, −1.25; −1.60; −2.05	13
Ni(AN[b])$_2$(dipy)	−0.67; −1.32, −1.94	
Ni(FN[c])(dipy)	−0.87, −1.68, −2.25	
Ni(AN)$_2$	−1.74	
Ni(FN)$_2$	−1.93	

[a] Dipyridyl (dipy).
[b] Acrylonitrile.
[c] Fumaronitrile.

γ-PICOLINE
0.1M LiClO$_4$ at DME at 25°C
1.06 mM

Species	$E_{1/2}V$ (vs Ag/Ag$^+$)	Ref.
Cobalticinium perchlorate	−0.839	9

0.1M LiClO$_4$ at DME at 25°C
1.03 mM

Species	$E_{1/2}V$ (Ag/Ag$^+$)	Ref.
Cobalticinium perchlorate	−0.870	9

PROPIONITRILE
0.1M Bu$_4$NClO$_4$ at DME at 25 ± 0.1°C
0.5 mM Ni + 0.1 mM to 1 mM LiSCN

Species	$E_{1/2}V$ (vs SCE)	Ref.
Ni(11) + LiSCN	−0.48; −1.00 to −1.46	16

PYRIDINE

0.1M LiClO$_4$ at DME at 25°C

0.28 mM to 5.10 mM

Species	$E_{1/2}V$ (vs Ag/Ag$^+$)	Ref.
Cobalticinium perchlorate	-0.821	9

0.1M LiClO$_4$ at pyrolytic graphite electrode at 25°C

1 mM

Species	E_pV (vs Ag/Ag$^+$)	Ref.
Ferrocene	$\sim+0.4$[a]; $\sim+0.45$[b]	1

[a] Cathodic peak in presence of 0.5 mM pyridinium nitrate.
[b] Anodic peak in presence of 0.5 mM pyridinium nitrate.

TETRAHYDROFURAN

0.25M Bu$_4$NClO$_4$ at lead electrode

0.5 mM

Species	$E_{p/2}V$ (vs Ag/Ag$^+$)	Ref.
AlEt$_3$	>-0.6	8
EtMgBr	-1.73	
MgEt$_2$	-1.72	
NaAlEt$_4$	-1.25	

0.1M Bu$_4$NClO$_4$ at Pt disk electrode

$\sim$1 mM

Species	E_pV (vs Ag/AgClO$_4$)	Ref.
Ferrocene	-0.123 ± 0.005[a] -0.04[a]	4

[a] Oxidation potentials.

XI. ADDITIONAL REFERENCES AND DATA SOURCES

DIMETHYL SULFOXIDE–WATER
0.15M NaClO$_4$ at DME at 30 ± 0.1°C
1 mM

Species	Vol % DMSO	$E_{1/2}V$ (vs SCE)	Ref.
UO$_2$(ClO$_4$)$_2$-p-ATPAA[a]	80	−0.370; −0.440	14
	90	−0.360; −0.460	
	95	−0.355; −0.465	

[a] p-Aminothiophenocyacetic acid.

DIOXANE–WATER (50% dioxane)
0.1M KCl at DME at 30 ± 0.1°C
0.216 mM

Species	pH	$E_{1/2}V$ (vs SCE)	Ref.
Trans-2-chloromercury-3-ethoxy cyclotridecene	1.4	−0.075; −0.750	11
	2.2	−0.050; −0.837	
	2.6	−0.050; −0.937	
	3.4	−0.050; −1.025	
	5.4	−0.050; −1.200	
	6.6	−0.080; −1.325	
	9.8	−0.125; −1.380	
	11.3	−0.212; −1.380	

References

1. R. E. Panzer and P. J. Elving, J. Electrochem. Soc., **119**, 864 (1972).
2. T. G. Lanese and G. S. Wilson, J. Electrochem. Soc., **119**, 1039 (1972).
3. A. M. Bond, A. R. Hendrickson and R. L. Martin, J. Electrochem. Soc., **119**, 1325 (1972).
4. A. Caillet and G. Demange-Guerin, J. Electroanal. Chem., **40**, 69 (1972).
5. R. E. Dessy, J. C. Charkoudian and A. L. Rheingold, J. Am. Chem. Soc., **94**, 738 (1972).
6. J. A. Ferguson and T. J. Meyer, Inorg. Chem., **11**, 631 (1972).
7. D. T. Sawyer and R. L. McCreery, Inorg. Chem., **11**, 779 (1972).
8. M. Fleishmann, D. Pletcher and C. J. Vance, J. Organomet. Chem., **40**, (1972);

9. G. H. Aylward, E. C. Watton, G. S. Buchanan and R. W. Lee, J. Electroanal. Chem., **34,** 521 (1972).
10. N. Tanaka, T. Ogata and S. Niizuma, Inorg. Nucl. Chem., Letters, **8,** 965 (1972).
11. S. S. Katiyar, M. Lalithambika and D. Devaprakhakara, Electrochimica Acta, **17,** 2077 (1972).
12. N. E. Tokel and A. J. Bard, J. Am. Chem. Soc., **94,** 2862 (1972).
13. A. Misono, Y. Uchida, T. Yamagishi and H. Kageyama, Bull. Chem. Soc., Japan, **45,** 1438 (1972).
14. T. T. Lai and J. C. Yang, Bull. Chem. Soc., Japan, **45,** 1683 (1972).
15. M. Kodama, Y. Fujii and M. Sakurai, Bull. Chem. Soc., Japan, **45,** 1629 (1972).
16. E. Itabashi, Bull. Chem. Soc., Japan, **45,** 2455 (1972).
17. J. A. Ferguson, T. J. Meyer and D. G. Whitten, Inorg. Chem., **11,** 2767 (1972).

XII. COMPOUND INDEX

(a) Solvent

The solvent index is arranged according to single solvents, nonaqueous-aqueous mixtures and nonaqueous-nonaqueous mixtures. Individual solvent systems are indexed according to solubilities, pKsp, emf, potentiometric titrations, vapor pressure, cryoscopy, heats of solution, polarography, ligand exchange rates, electrode reactions, electrical double layer, spectroscopy, and organic electrolyte batteries.

The notation used in the index for these properties is as follows:

solubility	Sl	polarography	Pg
pKsp:pKsp		pKsp:pKsp	
emf	EMF	ligand exchange rates and electrode reaction	Le
potentiometric titration	Pt	electrical double layer	Edl
vapor pressure	Vp		
cryoscopy	Cr	spectroscopy	Sp
heats of solution	Hs	organic electrolyte batteries	Ob

SINGLE SOLVENTS

A

Acetamide
 Sl: 13, 49, 69, 79
 EMF: 130, 131
 Pt: 276
Acetic acid
 Sl: 5–7, 13, 14, 16, 22, 23, 24, 28, 29, 35, 39, 40, 43, 62, 67, 72, 82, 93, 94, 96–98, 100–106
 EMF: 130–135
 Pt: 276–278, 845
 Hs: 415
 Pg: 466, 523, 611, 885
 Edl: 727
 Sp: 751
Acetic anhydride
 EMF: 830
 Pt: 295
 Hs: 416
 Pg: 466, 524, 611, 858
Acetone
 Sl: 3, 9, 11, 16, 18–24, 29, 30, 32, 34–38, 41, 43, 47, 49, 52, 54–59, 62, 66, 67, 69, 71, 72, 76, 79, 80, 84, 88, 97, 98, 100–105
 EMF: 136, 139, 830
 Pt: 295–297
 Vp: 336
 Hs: 416
 Pg: 467, 524, 612, 613, 858, 885
 EdL: 727
 Sp: 748, 729, 756, 758
 Ob: 773

Acetonitrile
 Sl: 18–20, 24, 27, 29, 31, 34, 35, 37, 41, 44, 47, 49, 54–56, 62, 64, 72, 76, 80, 102, 816, 828
 pKsp: 110–112
 EMF: 138–149, 830, 831
 Pt: 297–299, 845
 Vp: 336–341
 Hs: 417, 851
 Pg: 468–478, 525–539, 613–626, 854, 859–868, 886, 887
 Le: 672, 688, 689
 Edl: 728
 Sp: 747, 750, 751, 754, 755, 757, 758
 Ob: 773, 775–777
Acetophenone
 Sl: 62, 94, 95
Acetyl acetone
 Pg: 525, 626
Acetyl chloride
 Hs: 417
Acrylonitrile
 Pg: 479
 Sp: 750
Allyl alcohol
 Hs: 417
 Pg: 539, 626
Ammonia
 Sl: 3–11, 16, 17, 19, 31, 37, 40, 43, 44, 46, 49, 52, 55, 56, 59, 61, 62, 67, 69, 72, 76, 80, 103, 817
 EMF: 148–155, 831
 Pt: 278
 Vp: 342–369
 Hs: (liq. amm.) 418–420
 Pg: (liq. amm.) 480–482, 540, 541
 Le: 673, 674, 659
 Edl: 728
 Sp: 753, 754, 755, 758, 762
 Ob: 778

898   NONAQUEOUS ELECTROLYTES

iso-Amyl alcohol
  EMF: 154, 155
Amyl benzene
  Sl: 62, 93, 102
  Sp: 751
Antimony trichloride
  Sl: 97
Arsenous chloride
  Pt: 300

B

Benzaldehyde
  Sl: 29, 49, 94–95
Benzene
  Sl: 41, 46, 62, 68, 75, 92–93, 95, 100–102, 105
  Pt: 300
  Vp: 369
  Hs: 421
  Sp: 754, 756
Benzensulfonic acid
  Sl: 93, 95
Benzil alcohol
  Sl: 14
Benzoic acid
  Sl: 14
Benzol
  EMF: 154–155
Benzonitrile
  Sl: 50, 63, 93, 94–95, 98, 102
  EMF: 156–157
  Pg: 482, 483, 854, 887
Benzophenone
  Sl: 92–95
Benzoyl chloride
  Sl: 92–95
Benzyl alcohol
  Sl: 11, 25, 80
Biphenyl
  Sl: 94–95
Bromine
  Vp: 370–371
Bromine trifluoride
  Sl: 32, 76
  EMF: 156–157
Bromobenzene
  Sl: 93, 94
Bromoethane
  Sl: 92, 104
1-Bromonaphthalene
  Sl: 94, 95
m-Bromonitrobenzene
  Sl: 92
o-Bromonitrobenzene
  Sl: 92
p-Bromonitrobenzene
  Sl: 93
Butanoic acid
  Sl: 71
1-Butanol
  Sl: 9, 12, 14, 18–22, 25, 31, 35, 36, 38, 41, 44, 50, 52, 57–62, 65–66, 70, 73, 77, 80, 84, 101
  EMF: 158–159, 831
  Pt: 229
  Hs: 421
  Pg: 484, 541, 627–628
  Edl: 728
2-Butanol
  Sl: 41, 44, 50, 70, 73, 80
t-Butanol
  Pt: 279
2-Butanone
  Sl: 58–59, 63, 66, 80, 93
  Edl: 729
Butoxybutane
  Sl: 93
2-Butoxyethanol
  Sl: 104
n-Butylacetamide
  Edl: 729
Butylamine
  Pt: 279
  Edl: 729

n-tert Butylformamide
  Edl: 729
n-Butyl phosphate
  Pg: 485
Butyrolactone
  Sl: 818, 819
  Ob: 773, 775–777
4-Butylrolacetone
  Edl: 729

C

Carbon disulfide
  Sl: 92, 96, 101, 102, 104–105
Carbon tetrachloride
  Sl: 19, 46, 92, 101, 104
  Vp: 372
  Hs: 421
Chlorinated paraffins
  Sl: 101
Chlorine trifluoride
  EMF: 158–159
Chlorobenzene
  Sl: 93, 94
  Pt: 300
  Hs: 422
Chloroform
  Sl: 10, 28, 92, 101, 104
  Pt: 301
  Hs: 422
  Pg: 628, 869
1-Chloronaphthalene
  Sl: 94–95
2-Chloronaphthalene
  Sl: 94–95
m-Chloronitrobenzene
  Sl: 92
o-Chloronitrobenzene
  Sl: 92
p-Chloronitrobenzene
  Sl: 92
Chlorosuphonic acid
  Pt: 845
m-Chlorotoluene
  Sl: 94–95
o-Chlorotoluene
  Sl: 94–95
p-Chlorotoluene
  Sl: 94–95
Cyclohexane
  Sl: 94, 95
  Sp: 753, 756
Cyclohexanol
  EMF: 158–159
p-Cymene
  Sl: 94–95

D

Decanol
  EMF: 160–161
Deuterium Oxide (98.1 %)
  Sl: 103
Deutomethanol
  Hs: 422
p-Dibromobenzene
  Sl: 93–94
N,N-Dibutylacetamide
  EMF: 160–161
Dibutyl Sulfoxide
  Sp: 747
Dichloroacetic acid
  Pt: 280
o-Dichlorobenzene
  Hs: 422
p-Dichlorobenzene
  Sl: 93–94
Dichloroethane
  Sl: 101
  Pt: 308

XII. COMPOUND INDEX

Dichloromethane
  Pg: 628–630, 869, 888
N,N-Diethylacetamide
  EMF: 160–161
Diethylamine
  EMF: 162, 163
Diethyl sulfate
  Pg: 486
Diethyl sulfite
  Pg: 486
Diglyme
  Pg: 542
Di-iodomethane
  Sl: 95, 102
N,N-Diisopropylacetamide
  EMF: 162–163
N,N-Diisopropylpropionamide
  EMF: 164–165
1,2-Dimethoxyethane
  Pg: 486, 542, 628–642, 869–870, 888
Dimethoxymethane
  Sl: 102
Dimethylacetamide (N,N)
  pKsp: 113
  EMF: 164–165
  Pt: 301
  Pg: 487–488
  Edl: 729
  Sp: 754
Dimethylamine
  EMF: 166–167
1,2-Dimethylbenzene
  Sl: 94, 95
1,3-Dimethylbenzene
  Sl: 92, 94, 95
1,4-Dimethyl benzene
  Sl: 93, 95
N,N-Dimethyl buyramide
  EMF: 168–169
Dimethylformamide
  Sl: 820, 821
  pKsp: 114
  EMF: 168–173, 831
  Pt: 302–304
  Vp: 373, 374
  Hs: 423–466, 851
  Pg: 488–492, 542–565, 643–647, 854, 870–873, 889, 890
  Le: 675, 690, 691
  Edl: 730
  Sp: 751, 754, 755, 762
  Ob: 770, 775–777, 779
Dimethyl methyl phosphonate
  Le: 675
N,N-Dimethyl propionamide
  EMF: 174, 175
Dimethyl sulfite
  Ob: 780
2,4-Dimethyl sulfolane
  Pt: 305
  Pg: 519
Dimethyl sulfoxide
  pKsp: 115
  EMF: 174–181, 832
  Pt: 305–307, 845
  Cr: 403
  Hs: 426, 427, 851
  Pg: 492–495, 467–473, 650–651, 855, 874, 890
  Le: 675, 690, 691
  Edl: 731
  Sp: 747, 748, 750, 754, 755, 758, 762
  Ob: 773, 777, 778–780
m-Dinitrobenzene
  Sl: 93, 94
  Sp: 758
Dioxane
  Sl: 93
  EMF: 180, 181
  Edl: 731
  Sp: 754, 756
Diphenylmethane
  Sl: 94, 95
Dipropylsulfoxide
  Sp: 747

E

Ethanol
  Sl: 4, 9–12, 14, 17–22, 25, 30, 31, 34–39, 41, 44, 45, 47, 49, 50, 52–55, 57–63, 65, 66, 68–70, 72, 73, 75–78, 81, 83–89, 96–106
  pKsp: 116
  EMF: 182–185, 832
  Pt: 280, 281
  Vp: 375, 849
  Hs: 427, 428
  Pg: 496–497, 575, 654
  Le: 627, 692
  Edl: 732
  Sp: 755
Ethoxybenzene
  Sl: 94, 95
2-Ethoxyethanol
  Sl: 27, 65, 98, 100, 106
n-Ethylacetamide
  Hs: 429
  Edl: 732
Ethylacetate
  Sl: 9, 18, 25, 36, 38, 50, 52, 54, 57, 63, 84, 97, 98, 101
  Pt: 307
  Hs: 429
  Ob: 773
Ethylamine
  EMF: 186, 187
Ethylbenzene
  Sl: 94, 95
Ethyl carbamate
  Sl: 44, 55, 56
Ethyl cellosolve
  Pt: 281
Ethylene carbonate
  Edl: 733
  Ob: 733, 777
Ethyl cyanoacetate
  Sl: 50, 63
  EMF: 186, 187
Ethylenediamine
  Sl: 104
  Pt: 282, 283
  Pg: 500, 590
  Hs: 430
  Sp: 753
Ethylene glycol
  Sl: 30, 34, 50, 63, 67, 71, 73, 76, 78, 84, 93
  pKsp: 116
  EMF: 186–187
  Pt: 282
  Pg: 500–501
  Edl: 733
Ethylene sulfite
  Pg: 502
Ethyl ether (Diethyl Ether)
  Sl: 4, 11, 18, 25, 35, 36, 38, 54, 57, 70, 96, 98, 99, 101, 102, 103, 105
  pKsp: 113
  EMF: 166
  Pt: 300
  Vp: 372, 373
  Pg: 657
  Ob: 777
Ethyl N-ethylcarbamate
  Sl: 6, 42, 70, 73
n-Ethylformamide
  Edl: 733
Ethyl formate
  Sl: 25, 98
Ethyl malonate
  Sl: 96
Ethyl methyl ketone
  Pt: 308
  Hs: 430

F

Fluorobenzene
  Sl: 93, 94

Formamide
 Sl: 827, 828
 pKsp: 116, 117
 EMF: 188–193, 832
 Pt: 253
 Cr: 850
 Pg: 431, 432, 657
 Le: 692, 693
 Edl: 734
Formic acid
 Sl: 6, 8, 18, 19, 27, 30, 31, 33, 34, 42, 44, 50, 53, 55, 56, 70, 73, 81, 96–100, 105, 106
 EMF: 194–197
 Pt: 284
 Hs: 433
 Edl: 734
 Sp: 751
Furfural
 Sl: 27, 34, 42, 44, 50, 56, 65, 81, 98, 100, 106
 Hs: 433

G

Gasoline
 EMF: 196, 197
Glycerine
 Pg: 505
Glycerol
 Sl: 3, 4, 31, 39, 44, 46, 50, 69, 71, 72, 78, 99, 101, 105, 106
 EMF: 832
Glycol diacetate
 Sl: 100
Glyme
 Pg: 657, 658

H

Heptane
 Sl: 93
1-Heptanol
 Sl: 58, 62, 66
 EMF: 196, 197
 Edl: 735
4-Heptanone
 Sl: 58, 66
Hexamethylphosphorotriamide
 pKsp: 117
 EMF: 196, 197
 Pt: 309
 Hs: 433
 Pg: 874, 891
Hexane
 Sl: 92, 105
1-Hexanol
 Sl: 58, 61, 66
 EMF: 198, 199
 Hs: 434
 Edl: 735
2-Hexanone
 Edl: 735
Hexylamine
 Edl: 735
Hydrazine
 Sl: 4, 5, 8, 14, 27, 31, 42, 43, 45, 50, 52, 53, 55, 63, 68, 70, 72, 73, 81, 83, 96, 97, 99–106
 EMF: 198–199
 Sp: 753
Hydrogen chloride
 Sl: 97
Hydrogen fluoride
 Sl: 47, 48, 61, 93, 96, 97, 103, 105
 pKsp: 118
 Cr: 404
 Sp: 751
Hydrogen peroxide
 Sl: 73, 77, 83, 86

I

Iodobenzene
 Sl: 93, 94

Isobutyl methyl ketone
 Pt: 310
Isobutyronitrile
 Pg: 506

K

Kerosene
 EMF: 202, 203

L

Lanoline
 Sl: 100, 102

M

Mesitylene
 Sl: 94, 95
Methanol
 Sl: 3–5, 9, 10–12, 15–23, 25, 28, 34–43, 45–53, 55–62, 63, 65, 71, 74, 75, 83–88, 96–106, 828
 pKsp: 118, 119
 EMF: 202–209, 832, 833
 Pt: 285–287
 Vp: 376–378
 Hs: 434–438, 851
 Pg: 506, 507, 591, 592, 658–660, 855, 875
 Le: 678, 679, 693
 Edl: 736, 737
 Sp: 754, 755, 759, 762
Methanollic acetone
 Le: 684
Methoxybenzaldehyde
 Sl: 51
Methoxybenzene
 Sl: 94, 95
2-Methoxyethanol
 Pg: 598
Methyl acetamide
 EMF: 210–211
 Cr: 405–410
 Hs: 438
 Pg: 598
Methyl acetate
 Sl: 17, 26, 101
Methylamine
 Hs: 439
3-Methyl-1-butanol
 Sl: 12, 15, 20–22, 26, 35, 42, 58–60, 62, 63, 65, 66, 96, 103
Methyl cyanacetate
 Sl: 51
Methylene chloride
 Pg: 598, 599
N-Methyl formamide
 pKsp: 120
 EMF: 210–211
 Vp: 379
 Hs: 439, 440, 851
 Le: 693
 Edl: 738
 Sp: 755
Methyl formate
 Ob: 776, 777, 779
Methyl methane sulfonate
 Pg: 509
Methyl nitrite
 pKsp: 120
4-Methyl-2-pentanone
 Sl: 58, 59
2-Methyl-1-propanol
 Sl: 9, 17, 18, 35, 36, 38, 42, 45, 51, 52, 57, 71, 74, 82, 84, 101
N-Methylpyrrolidinone
 Hs: 440
N-Methyl propionamide
 pKsp: 120
 Edl: 738
N-Methyl-2-pyrrolidone
 EMF: 210–211
 Pt: 845
 Sp: 748

XII. COMPOUND INDEX

3-Methyl sulfolane
 Pt: 310
 Pg: 519
Methyl thiocyanate
 Pt: 310

N

Naphthalene
 Sl: 94, 95
Nitric acid
 Sl: 8, 52, 99, 103
 Cr: 411
 Sp: 757
Nitrobenzene
 Sl: 51, 92, 93, 95–97, 104
 Pt: 311
 Hs: 441
Nitroethane
 pKsp: 120
 Pt: 311
Nitromethane
 Sl: 34, 51, 56, 82
 EMF: 210–211, 833
 Pt: 312
 Hs: 441
 Pg: 510, 511, 599, 600, 662, 663, 876
 Ob: 773, 777
2-Nitronaphthalene
 Sl: 94, 95
m-Nitrophenol
 Sl: 95
m-Nitrotoluene
 Sl: 92–95
o-Nitrotoluene
 Sl: 92–95
p-Nitrotoluene
 Sl: 92–95

O

Octanol
 EMF: 212–213

P

1-Pentanol
 Sl: 11, 12, 15, 17, 31, 34, 42, 45, 51, 54, 55, 61, 71, 74, 82
 Hs: 442
 Pg: 511
 Edl: 738
2-Pentanone
 Sl: 60
Phenol
 Sl: 31, 63, 93, 95
Phenylacetonitrile
 Sl: 63
 EMF: 212–213
 Pg: 511–512
Phosgene
 Vp: 379
α-Picoline
 Pg: 891
Piperidine
 Sl: 63
 Hs: 442
1,2-Propanediol
 Sl: 76, 78, 84, 85
Propanediol-1,2-carbonate
 Pg: 512–513
Propanenitrile
 Sl: 51, 56, 82
1-Propanol
 Sl: 9, 13, 17, 18, 21, 26, 34–36, 38, 42, 45, 48, 51, 52, 55, 57, 68, 71, 74, 82, 84, 85, 93, 98, 101
 EMF: 212–215, 833
 Pt: 288
 Hs: 442
 Pg: 514, 600, 663
 Edl: 738
 Sp: 755

2-Propanol
 Sl: 17, 26, 45, 74, 86, 96, 98
 Hs: 443
 Pg: 514, 600, 663–664
 Edl: 731
iso-Propanol
 Sl: 42, 45, 51, 63, 71, 74, 82
 EMF: 833
 Pt: 289
 Pg: 514
2-Propen-1-ol
 Sl: 26, 32, 101
Propionic acid
 Pt: 290
Propionitrile
 EMF: 214–215
 Pg: 516, 855, 891
 Sp: 755
n-Propylamine
 EMF: 214–215
 Edl: 734
Propylbenzene
 Sl: 94, 95
Propylene carbonate
 Sl: 822–826
 pKsp: 121
 EMF: 214–219
 Pt: 312, 845
 Vp: 380
 Hs: 443
 Pg: 517, 601
 Le: 694–697
 Edl: 739
 Sp: 757
 Ob: 773, 775–780
Propylene glycol
 EMF: 218–219
Propyl formate
 Sl: 98
Pryidine
 Sl: 8, 10, 11, 15, 23, 26, 27, 32, 35, 51, 54, 57, 60, 63–65, 92, 93, 98–105
 EMF: 220–223, 834
 Pt: 313
 Vp: 381
 Hs: 444
 Pg: 517, 518, 602–604, 665, 876, 892
 Edl: 739
 Sp: 747, 751

Q

Quinoline
 Sl: 32, 64, 102

S

Salicylaldehyde
 Sl: 51
Selenium Oxychloride
 Sl: 15, 19, 32, 45, 55, 74, 96, 98, 101, 105
Sulfolane (tetramethylene sulfone)
 Sl: 828
 pKsp: 121
 EMF: 224–225
 Pt: 314
 Hs: 444
 Pg: 518
 Edl: 740
Sulfur dioxide
 Sl: 3–6, 9–11, 16, 19, 22, 30, 32, 40, 42, 46, 48, 51, 54, 55, 58, 60, 61, 67, 68, 70, 74, 75, 77, 82, 86, 93, 96, 98–105
 EMF: 224–227
 Vp: 381–383
 Pg: 519
Sulfuric acid
 Sl: 36, 45, 57, 86
 Cr: 411, 412
 Edl: 740

T

Tetrachloroethane
 Sl: 93

Tetraethozysilane
  Pt: 314
Tetrahydrobenzene
  Sl: 93, 95
Tetrahydrofuran
  Sl: 826
  pKsp: 121
  Pt: 315
  Pg: 879, 892
  Le: 698
  Sp: 747, 748, 750, 751, 756
  Ob: 773
Tetramethoxysilane
  Pt: 313
Tetramethyl urea
  Sl: 827
Tetrapropoxysilane
  Pt: 316
Toluene
  Sl: 92–95, 106
  Pg: 665
Toluidine
  Sl: 64
Tribromomethane
  Sl: 101
Tributyl phosphate
  Pt: 316
Trichloroethylene
  Sl: 52, 69
Trifluoroacetic acid
  Pt: 290
Triphenylmethane
  Sl: 94, 95
Turpentine
  Sl: 101

U

Urea
  Sl: 17, 83

V

4-Valerolacetone
  Edl: 740

X

Xylene
  EMF: 228–229
  Sp: 757
p-Xylene
  Vp: 384

MIXED SOLVENTS
NONAQUEOUS-AQUEOUS MIXTURES

A

Acetamide-water
  Le: 700–701
Acetic acid-water
  EMF: 324–325
Acetone-water
  pKsp: 110, 123
  EMF: 236, 237, 834
  Pt: 322, 323
  Pg: 467, 877
  Le: 684
  Sp: 754, 755, 762
Acetonitrile-water
  EMF: 835
  Pg: 479, 480, 877
  Le: 702, 722
  Sp: 751
Ammonia-water
  Vp: 385–389

B

n-Butanol-water
  EMF: 835
t-Butanol-water
  EMF: 238, 239, 835
  Hs: 852
  Pg: 484, 485
  Le: 703
n-Butyl phosphate-water
  Pg: 486

D

Dimethyl formamide-water
  EMF: 836
  Pg: 492, 566, 567, 650
  Le: 703, 704
  Sp: 754
Dimethyl sulfoxide-water
  pKsp: 123
  EMF: 238–241, 836
  Pt: 326
  Hs: 445, 446
  Pg: 856, 893
  Sp: 754
Dioxane-water
  pKsp: 124
  EMF: 240–245, 837
  Pt: 326, 327, 846
  Vp: 390, 391
  Hs: 447, 852
  Pg: 496, 573, 653, 878, 893
  Le: 705
  Sp: 754
Dioxane-dimethyl formamide-water
  Pg: 574

E

Ethanol-water
  Sl: 39, 41, 53, 78, 96–100, 105, 828
  pKsp: 124
  EMF: 246–251, 838
  Pt: 327, 846
  Vp: 392–396
  Pg: 498, 499, 575–590, 654–656, 878–882
  Le: 706–709
  Sp: 754, 762
Ethylene glycol-water
  pKsp: 124
  EMF: 250, 251, 838
  Pg: 501, 502, 591
  Le: 709
Ethyl methyl ketone-water
  EMF: 252, 253

F

Formamide-water
  Pt: 329
  Hs: 852
  Pg: 504
  Le: 710–713
Formic acid-water
  Sl: 24, 39, 44, 48, 50, 78, 84, 96–101, 103, 105, 106
  Pg: 505

G

Glycerol-water
  Sl: 23, 98
  EMF: 252, 253, 838, 839

H

p-Hydrobenzoic acid-water
  EMF: 252, 253
Hydrogen peroxide-water
  Vp: 397, 398

XII. COMPOUND INDEX

M

Mannitol-water
 EMF: 254–255
Methanol-water
 pKsp: 125
 EMF: 256–261, 839, 840
 Pt: 330–331, 846
 Hs: 448–454, 852
 Pg: 507, 508, 592–596, 660–662
 Le: 714–719
 Sp: 753–757
2-Methoxyethanol-water
 EMF: 840
Methylcellosolve-water
 EMF: 840

P

Phenol-water
 EMF: 841
i-Propanol-water
 Sl: 829
 EMF: 754–755, 841
 Hs: 454
 Pg: 601, 664
 Le: 720, 721
n-Propanol-water
 EMF: 260, 261, 841
 Hs: 852
2-Propanol-water
 Hs: 455
Propylene carbonate-water
 pKsp: 125
 EMF: 260, 261, 841
 Pt: 331
Propylene glycol-water
 Hs: 455
Pyridine-water
 Sl: 829

S

Sorbitol-water
 EMF: 262, 263
Succinic acid
 EMF: 262, 263

T

Tetrahydrofuran-water
 pKsp: 125
 EMF: 842
 Pg: 605, 606
Triethylene glycol-water
 EMF: 262, 263

U

Urea-water
 Hs: 455–457, 852

NONAQUEOUS-NONAQUEOUS MIXTURES

A

Acetic acid-Acetic anhydride
 EMF: 266
 Pt: 320
Acetic acid-Acetonitrile
 Pt: 320
 Pg: 539–540
Acetic acid-aniline
 EMF: 267
Acetic acid-Benzene-chloroform
 Pt: 320
Acetic acid-chloroform
 Pt: 321, 846
Acetic acid-Ethylaniline
 EMF: 267
Acetic acid-Methanol
 Pg: 597
Acetic acid-perchloric acid
 Hs: 458
Acetic acid-pyridine
 EMF: 267
Acetic acid-quinoline
 EMF: 268
Acetone-chloroform
 Sp: 753
Acetone-diethylamine
 Pt: 321
Acetone-ethyl acetate
 Sl: 93
Acetone-methanol
 Sl: 93, 101
 EMF: 268
 Pt: 322
Acetone-nitromethane
 Sp: 735
Acetonitrile-chloroform
 Pt: 323
Acetonitrile-methanol
 EMF: 269
Acetonitrile-toluene
 Pg: 856
Ammonia-ethylamine
 EMF: 269
Aniline-quinoline
 Sl: 98, 101
Aniline-pyridine
 Sl: 102

B

Benzene-chlorobenzene
 Hs: 458
Benzene-chloroform
 Pt: 323
Benzene-cyclohexane
 Sl: 92
Benzene-methanol
 Sl: 105
 Pt: 324
 Pg: 597, 627
Benzene-pyridine
 Pt: 324
Benzene-toluene
 Sl: 97
Bromine trifluoride-chlorine trifluoride
 EMF: 269
Butanol-methanol
 EMF: 842

C

Carbon tetrachloride-ethanol
 EMF: 270
Carton tetrachloride-chloroform
 Sp: 753
Chloroform-ethanol
 Sl: 213
Chloroform-isopropanol
 Pt: 325
Chloroform-methanol
 Sp: 753
Chloroform-propylene glycol
 Pt: 325

D

Dimethyl formamide-Dimethylsulfoxide
 Hs: 853
Dimethyl formamide-N-methyl formamide
 Le: 722
Dimethyl sulfoxide-chlorobenzene
 Pt: 326

Dimethyl sulfoxide-methanol
  pKsp: 124
  Hs: 459, 460
Dioxane-ethanol
  EMF: 270
Dioxane-methanol
  EMF: 842
Dodecanoic acid-toluene-octadecanol
  Sl: 106

E

Ethanol-methanol
  Sl: 98, 100, 104, 106
Ethanol-methyl formate
  Sl: 98
Ethanol-2-propanol
  Sl: 104
Ethanol-pyridine
  Sl: 103
Ethylene glycol-methanol
  Pt: 328

Ethylene glycol-1,2-methoxyethane
  Pg: 501
Ethylene glycol-isopropanol
  Pt: 328
Ethyl methyl ketone-methanol
  Pt: 329

F

Formamide-methanol
  Sp: 758
Formic acid-nitromethane
  Pt: 330

M

Methanol-propylene carbonate
  EMF: 272, 273
Methanol-propylene glycol
  EMF: 842

(b) Solute

The solute index is arranged alphabetically by solute. Individual solutes are indexed according to solubilities, pKsp, emf, potentiometric titrations, vapor pressure, cryoscopy, heats of solution, polarography, ligand exchange rates, electrode reactions, electrical double layer, spectroscopy, and organic electrolyte batteries.

The notation used in the index for these properties is as follows:

solubility	Sl	polarography	Pg
pKsp:pKsp		pKsp:pKsp	
emf	EMF	ligand exchange rates and electrode reaction	Le
potentiometric titration	Pt	electrical double layer	Edl
vapor pressure	Vp		
cryoscopy	Cr	spectroscopy	Sp
heats of solution	Hs	organic electrolyte batteries	Ob

A

Acetamido ferrocene
  Pg: 624
Acetanalide
  Vp: 392
Acetic acid
  EMF: 190, 191, 236, 239, 246–249, 260, 261, 272
  Cr: 411
  Hs: 425, 663
  Pg: 568, 570, 599, 603, 604
Acetic anhydride
  Cr: 411
Acetoacetyl ferrocene
  Pg: 472
Acetone
  Cr: 411
Acetophenone
  Pg: 533, 557, 571, 594
Acetophenone anil
  Pg: 535, 553
Acetophenone semicarbazone
  Pg: 562
Acetophenone thiosemicarbazone
  Pg: 562
9-Acetoxyanthracene
  Pg: 539
1′-Acetyl-1-acetamido ferrocene
  Pg: 624
2-Acetyl-1-acetamido ferrocene
  Pg: 624
Acetylacetone
  Pg: 586
2-Acetylamino fluorene
  Pg: 537
1′-Acetyl 1-bromo ferrocene
  Pg: 624
1′-Acetyl-1-chloro ferrocene
  Pg: 624

1′-Acetyl-1-cyanoferrocene
  Pg: 624
3-Acetyl-1,1′ dimethyl ferrocene
  Pg: 624
Acetylferrocene
  Pg: 472
1′Acetyl-1-methoxy carbonylamino ferrocene
  Pg: 624
β-Acetyltetralin
  Pg: 578, 594
(Acetyl-2)-thiophene
  Pg: 605
Acridine
  Pg: 526, 542
Acylonitrile
  Pg: 551
2-Aldehydrophenazine
  Pg: 544
Allylamine
  Pg: 532
Aluminum (III)
  Pg: 485, 518
  Le: 675, 676, 680
  Sp: 755
Aluminum bromide
  Sl: 92, 93
  EMF: 220–223
  Sp: 749, 751
Aluminum chloride
  Sl: 92, 336, 823
  Vp: 374, 379, 380
  Hs: 417, 425, 443
  Sp: 746, 751
  Ob: 773–776, 779
Aluminum iodide
  Sl: 93
Aluminum octadecanoate
  Sl: 92
Aluminum perchlorate hexahydrate
  Pg: 478

905

Bipenylene
  Pg: 549
4'-(4-Biphenylyl)-acetophenone
  Pg: 595
Bipyridyl
  Pg: 633
2,2'Bipyridyl
  Pg: 543
4,4'Bipyridyl
  Pg: 543
2,2'-Biquinoline
  Pg: 543
Bis(2-acetamidophenyl-1,3-indandione)
  Pg: 545
4,4'-Bis(acetamino)-azoxybenzene
  Pg: 541, 598
Bis(2-anisyl-1,3-indandione)
  Pg: 545
Bis(dicarbonyl cyclopentadienyl) iron
  Pg: 624
Bis(2-p-dimethylaminophenyl-1,3-indandione)
  Pg: 545
Bis(2-p-dimethylamino phenyl-1,3-indandione) methyl nitrate
  Pg: 545
Bis(2,9-dimethyl 1–10 phenanthroline)Cu(I)
  Pg: 524, 525, 563, 575, 591, 600, 602, 611, 613, 626, 645
1,3-Bis(p-Fluorophenyl)-1,3-propanedione
  Pg: 571
1,3-Bis(p-methoxyphenyl)-1,3-propanedione
  Pg: 571
Bismuth
  EMF: 166–167
Bismuth(III)
  Pg: 482, 484, 497, 500, 503, 505, 506, 514, 540
Bismuth diethylthiolythionocarbamate
  Sl: 97
Bismuth fluoride
  Sl: 97
Bismuth nitrate
  Sl: 97
Bismuth oxychloride
  Sl: 97
Bismuth tribromide
  Sl: 97
Bismuth trichloride
  Sl: 97
Bis(2-p-nitrophenyl-1,3-indandione)
  Pg: 545
Bis-phenyl-2-(p-dimethylamino)-phenyl-1,3-indandione
  Pg: 545
Bis(2-phenyl-1,3-indandione)
  Pg: 545
Bis(2-Phenyl-3-indolinone)azine
  Pg: 555
1,2-Bis-(2-pyridyl)ethylene
  Pg: 543
1,2-Bis-(4-pyridyl)ethylene
  Pg: 543
Bis-trimethyl silicon
  Pg: 614
Boron trichloride
  Sl: 97
  Sp: 754
Boron trifluoride
  Sl: 97
  Sp: 754
  Ob: 776
Bromine
  EMF: 168–169, 202–203
  Pg: 469, 490
m-Bromoaniline
  Pg: 529, 538
p-Bromoaniline
  Pg: 529, 538
α-Bromobutanol
  Pg: 552
9-Bromo-o-carborane
  Pg: 553
4-Bromo-2,6-dimethylbenzonitrile
  Pg: 556

4-Bromo-2,6-dimethylbenzonitrile-N-oxide
  Pg: 556
1-Bromo-2,2-diphenylcyclopropane
  Pg: 530
p-Bromo ferrocenyl azobenzene
  Pg: 622
1-Bromo-1-methoxy methyl
  Pg: 530
1-Bromo-1-Methyl-2,2-diphenyl-cyclopropane
  Pg: 530
2-Bromo-2-nitropropane
  Pg: 535
p-Bromophenylacetone semicarbazone
  Pg: 562
m-Bromo phenyl ferrocene
  Pg: 622
o-Bromo phenyl ferrocene
  Pg: 622
p-Bromo phenyl ferrocene
  Pg: 622
2-Bromo-2-phenyl-1,3-indandione
  Pg: 545
2-p-Bromophenyl-4,5,6,7-tetrahydro-1,3-indandione
  Pg: 587
Butylamine
  Pg: 568
i-Butylamine
  Pg: 532
n-Butylamine
  Pg: 532, 572
t-Butylamine
  Pg: 532
n-Butyl bromide
  Pg: 604
t-Butylcyclohexane dibromide
  Pg: 567
i-Butyl magnesium
  Pg: 634
n-Butyl mercaptan
  Pg: 528, 625
s-Butyl mercaptan
  Pg: 528, 625
t-Butyl mercaptan
  Pg: 528, 625
Butyric acid
  Pg: 572

C

Cadmium
  EMF: 136, 137, 148–151, 156–159, 166–169, 174, 175, 182, 183, 188–191, 194, 195, 198, 199, 204, 205, 214, 215, 220, 221, 224, 225, 246, 247, 256, 257, 269
Cadmium(II)
  Pg: 471, 479, 481, 483–485, 497, 500, 503, 505, 506, 514
  Le: 688, 690, 691, 693, 694, 702, 722
Cadmium amalgam
  EMF: 204, 205, 837
Cadmium bromide
  Sl: 98, 821, 823
  Sp: 751, 756
Cadmium carbonate
  Pg: 480
Cadmium chloride
  Sl: 98, 821
  EMF: 148, 149, 152, 153, 158, 159, 168, 169, 172–177, 182–185, 188–193, 204, 205, 246, 247, 258, 259
  Hs: 418, 425, 853
  Pg: 488, 499, 500
  Sp: 751, 756
Cadmium chloride·NH_3
  Hs: 418
Cadmium chloride·$2NH_3$
  Hs: 418, 425
Cadmium chloride·$4NH_3$
  Hs: 418
Cadmium chloride·$6NH_3$
  EMF: 154–155
  Hs: 418

XII. COMPOUND INDEX

Cadmium chloride·10NH₃
  Hs: 418
Cadmium fluoride
  EMF: 198–201
  Sp: 756
Cadmium iodide
  Sl: 98, 821, 823
  EMF: 136, 137, 220–223
  Pg: 509
Cadmium nitrate
  EMF: 188–189
  Pg: 499, 500
  Sp: 746
Cadmium perchlorate
  Sl: 98
  Pg: 474, 478, 479, 482, 488, 489, 495, 499, 500, 502, 510, 511, 513, 516, 520
Cadmium perchlorate hexahydrate
  Pg: 475
Cadmium sulfate
  Sl: 98
  EMF: 837
Cadmium sulfate octahydrate
  Pg: 503
Calcium
  Ob: 773, 774
Calcium acetate
  Sl: 98
Calcium benzoate
  Sl: 99
Calcium bromate
  Hs: 423, 431, 439
Calcium bromide
  Sl: 11–13, 821, 823
  Hs: 423, 427, 431, 439
Calcium cacodylate
  Sl: 98
Calcium chlorate
  Hs: 423, 431, 439
Calcium chloride
  Sl: 13–15, 818, 821, 823, 827
  EMF: 182–183
  Vp: 389
  Hs: 418, 423, 427, 431, 439
  Pg: 489
  Edl: 736
  Ob: 776
Calcium citrate
  Sl: 99
Calcium fluoride
  Ob: 776
Calcium formate
  Sl: 98
Calcium glycero-phosphate
  Sl: 98
Calcium iodide
  Sl: 16
  EMF: 162–163, 220–221
  Hs: 418, 427
  Pg: 475
Calcium lactate
  Sl: 98
Calcium malate
  Sl: 98
Calcium nitrate
  Sl: 16–18, 827
  Vp: 357, 358, 386
  Hs: 427
  Pg: 495
  Sp: 746
Calcium oxide
  Sl: 18
Calcium perchlorate
  Hs: 421, 427, 434, 442, 443
  Pg: 478, 482, 511, 516
  Edl: 732, 736
Calcium perchlorate hexahydrate
  Pg: 478, 489
Calcium propionate
  Sl: 99
Calcium sulfate
  Sl: 98
  Sp: 749
Calcium tartarate
  Sl: 98

Calomel
  EMF: 130–133, 136–137, 140–141, 158–165, 168–171, 174–175, 182–183, 190–195, 204–205, 236–237, 240–247, 250–251, 254–257, 266–268, 832, 836, 838, 839
Calomel (aq)
  EMF: 140–141, 150–151
Camphor anil
  Pg: 547, 552
Carbinol
  EMF: 132–133
m-Carbobenzhydroxy phenyl ferrocene
  Pg: 622
p-Carbobenzhydroxy phenyl ferrocene
  Pg: 622
Carboethoxy ferrocene
  Pg: 622
m-Carboethoxy phenyl ferrocene
  Pg: 622
p-Carboethoxy phenyl ferrocene
  Pg: 622
m-Carbomethoxy phenyl ferrocene
  Pg: 622
o-Carbomethoxy phenyl ferrocene
  Pg: 622
p-Carbomethoxy phenyl ferrocene
  Pg: 622
Carbon
  Pt: 313
  Pg: 535
Carbon dioxide
  Pg: 494
Carbon tetrachloride
  Pg: 590
Carboxyl ferrocene
  Pg: 622
m-Carboxyl phenyl ferrocene
  Pg: 622
o-Carboxy phenyl ferrocene
  Pg: 622
p-Carboxy phenyl ferrocene
  Pg: 622
o-Carborane
  Pg: 553
Catechol
  Pg: 602
$C(C_6H_5)_3$ ferrocene
  Pg: 621
Cerium III
  Pg: 487, 491, 493, 513
  Le: 706–707
Cerium acetate
  Sl: 99
  Pt: 846
Cerium bromide
  Sl: 99
Cerium chloride
  Sl: 99
Cerium nitrate
  Sl: 99
Cerium tartrate
  Sl: 99
Cesium (I)
  Pg: 517
Cesium amalgam
  EMF: 117, 136–137, 150–151, 194, 195, 204, 205, 210, 211, 240, 241, 268, 838
  Sp: 756
Cesium bromide
  Sl: 18, 820
  pKsp: 111, 114, 117, 119
  Vp: 358
  Cr: 405–406
  Hs: 423, 431
Cesium carbonate
  Sl: 19
Cesium chloride
  Sl: 19, 820
  pKsp: 111, 114, 117, 119
  EMF: 204–205, 210–211, 244, 245, 837, 839–840
  Cr: 405–406
  Hs: 431, 433, 434, 439, 447–454
  Pg: 489, 495, 499
  Edl: 730, 733, 734, 736, 738

Cesium dibromoiodide
　Sl: 19
Cesium fluoride
　Sl: 20, 818
　Hs: 431
　Sp: 746, 752
Cesium iodide
　Sl: 20, 817, 822
　pKsp: 111, 114, 117, 119
　EMF: 136, 137, 208, 209, 268, 841
　Cr: 403, 405, 406
　Hs: 419, 423, 426, 430, 431, 433, 435, 438, 439
　Pg: 475, 506
　Edl: 728, 730
　Sp: 746
Cesium nitrate
　EMF: 150, 151
　Pg. 499
　Edl: 727, 736
Cesium m-Nitrophenoxide
　Sl: 20
Cesium o-Nitrophenoxide
　Sl: 21
Cesium p-Nitrophenoxide
　Sl: 21
Cesium perchlorate
　Sl: 21
　Hs: 444, 487
　Pg: 518
Cesium picrate
　Sl: 22
　pKsp: 111, 114, 119
Cesium pyrosulfite
　Sl: 22
Cesium tetraphenyl borate $CsBPh_4$
　Sl: 22
　pKsp: 111, 117, 119
　Hs: 423, 434, 443
p-Chloroacetophenone semicarbazone
　Pg: 562
5-Chloroacetyltetralin
　Pg: 578, 594
Chloranil
　EMF: 132–135, 246, 247, 830
m-Chloroaniline
　Pg: 529, 538
p-Chloroaniline
　Pg: 529, 538
4-Chloroazobenzene
　Pg: 525
α-Chlorobutanol
　Pg: 552
α-Chloroisobutanol
　Pg: 552
9-Chloro-o-carborane
　Pg: 553
α-Chloroheptanal
　Pg: 552
2-Chloro-6-methoxyphenazine
　Pg: 544
2-Chloro-2-methylpentanal
　Pg: 552
1-Chlorophenazine
　Pg: 544
2-Chlorophenazine
　Pg: 544
o-Chloro phenyl ferrocene
　Pg: 622
2-Chloro-2-phenyl-1,3-indandione
　Pg: 545
Chlorophyll
　Pg: 567
6-Chloroquinoline
　Pg: 545
Chromate ion
　Pg: 496, 498, 508, 516
Chromium (II)
　Pg: 471, 479, 483, 484, 489, 500, 506, 511, 514
　Le: 673, 681
Chromium (III)
　Pg: 484, 485, 489, 497, 503, 506, 514
　Le: 692
　Sp: 749, 756

Chromium (VI)
　Le: 705, 706, 714, 715, 720
Chromium perchlorate
　Pg: 479, 482, 511
Chromium perchlorate hexahydrate
　Pg: 468, 478, 503, 516
Chrysene
　Pg: 526, 549
$CH_3SC_3H_2COMn(CO)_4$
　Pg: 632
$ClC(Co(CO)_3)_3$
　Pg: 1204
Cinnamaldehyde
　Pg: 535
Cinnamaldehyde anil
　Pg: 553
Cinnamic alcohol
　Pg: 566
Cinnoline
　Pg: 526, 543
Cobalt
　EMF: 166, 167
Co $(CO)_7$ (C_4HO_2)
　Pg: 633
Cobalt (II)
　Sl: 824
　Pg: 471, 479, 483, 494, 497, 500, 506, 511, 514
　Le: 672, 673, 675, 676, 678, 679, 680, 688, 690–692, 702, 703, 710, 711
　Sp: 751, 754, 755, 757, 762
Cobalt $(CO)_3NO$
　Pg: 619
Cobalt CONO diphos
　Pg: 619
[Cobalt $(CO)_2NO]_2$diphos
　Pg: 619
Cobalt CONO$(P\phi_3)_2$
　Pg: 619
Cobalt CONO $(Sb\phi_3)_2$
　Pg: 619
Cobalt (CO) NOSbϕ_3
　Pg: 619
(Cobalt)$_2(CO)_6$·Phc·CPh
　Pg: 633
Cobalt $(dipy)_3^{2+}$
　Pg: 619
Cobalt $(dipy)_3^{2+}$
　Pg: 619
Cobalt (III)
　Le: 674
Cobalt (dithenoyl-2)methane
　pKsp: 125
Cobalt(dithioacetyl acetone)$_2$
　Pg: 612
Cobalt(dithioacetyacetone)$_3$
　Pg: 612
Cobalt (L*) $(S_2C_2Ar_2^*)$] *L = Ligand, Ar = aromatic ring
　Pg: 630
Cobalt naphthenate
　Pg: 627
Cobalt octoate
　Pg: 627
Cobalt perchlorate
　Pg: 474, 475, 479, 482, 488, 489, 491, 492, 495, 502–504, 510, 511, 513, 516, 520
Cobalt perchlorate hexahydrate Co$(ClO_4)_2$·$6H_2O$
　Pg: 468, 472, 478, 489, 503
Cobalt tallate
　Pg: 627
Cobalt(II)] trans(14)diene]$_3^{2+}$
　Pg: 617
Cobalt(III)] trans(14)diene] $^+Br_2$
　Pg: 617
Cobalt(III)] trans(14)diene] $^+Cl_2$
　Pg: 617
Cobalt(III)] trans(14)diene] ^+2CN
　Pg: 617
Cobalt(III)] trans(14)diene] ^+2CNS
　Pg: 617
Cobalt(II)] trans(14)diene 2HO]$_2^{2+}$
　Pg: 617
Cobalt(III)] trans(14)diene]$^3+2H_2O$
　Pg: 617

XII. COMPOUND INDEX

Cobalt(III)]trans(14)diene]$^{3+}$2N$_3$
  Pg: 617
Cobalt(III)]trans(14)diene]$^{3+}$2NH$_3$
  Pg: 616
Cobalt(III)]trans(14)diene]$^+$2NO$_2$
  Pg: 616
Cobalt(III)]trans(14)diene] (OH)O$_2$CCH$_3$
  Pg: 617
Cobalt(III)]trans(14)diene 2 Pg]$_2{}^{2+}$
  Pg: 617
CoBr$_3$(C$_5$H$_5$)$_2$
  Pg: 620
[CoCH$_3$]$_2$CH—CH=CH—CHO
  Pg: 558
[CO1CH$_2$)$_3$CO]$_2$CH—CH=CH—CHO
  Pg: 558
COCH$_3$ ferrocene
  Pg: 622
CO$_2$CH$_3$ ferrocene
  Pg: 622
COC$_6$H$_5$ ferrocene
  Pg: 622
CO$_2$CH(C$_6$H$_5$)$_2$ ferrocene
  Pg: 622
CON(C$_6$H$_5$)$_2$ ferrocene
  Pg: 622
CONH$_2$ ferrocene
  Pg: 623
CONHC$_6$H$_5$ ferrocene
  Pg: 622
Copper
  EMF: 136, 137, 154–157, 160–171, 174, 175, 182, 183, 190, 191, 194–197, 204, 205, 220, 221, 229
Copper Amalgam
  Pg: 299
Copper (I)
  EMF: 140, 141, 160–165, 168, 169, 174, 175, 831
  Pg: 471, 479, 483
Copper(II)
  EMF: 140, 141, 160–165, 168, 169, 174, 175
  Pg: 480, 485, 486, 493, 497, 500, 506, 514, 518, 519
  Le: 674, 676, 678, 681, 689
  Sp: 757, 758
Copper(acac)$_2$
  Pg: 618
Copper(II) acetate
  Sl: 22, 23
  Sp: 750, 756
Copper benzoate
  Sl: 23
Copper(I) bromide
  Sl: 24
Copper(II) bromide
  Sl: 24, 824
  Pg: 459
Copper(I) chloride
  Sl: 24, 818, 824
  EMF: 136, 137, 220, 221
Copper (II) chloride
  Sl: 24–26, 818, 824, 826
  EMF: 136, 137, 182–185, 220, 221
  Vp: 374
  Hs: 426
  Pg: 469
  Sp: 746, 759
Copper (II) fluoride
  Sl: 816, 818, 821, 824, 826
  EMF: 198–201
  Vp: 374
  Hs: 426
Copper (II) formate
Copper(I) iodide
  Sl: 27
Copper(II) nitrate
  Sl: 27, 827
  EMF: 204, 205
  Pg: 481
Copper octoate
  Pg: 627
Copper(II) perchlorate
  Sl: 27
  Pg: 479, 482, 489, 509, 511, 516

Copper(II) perchlora.e hexahydrate
  Pg: 468, 472, 478, 489, 503
Copper(II) perchlorate monohydrate
  Pg: 472
Copper(II) perchlorate, x-hydrate
  Pg: 475
Copper(II) oleate
  EMF: 154, 155, 196, 197, 202, 203, 228, 229
Copper(II) sulfate
  Sl: 28, 827
Copper(II)]2-trans(14)diene]$^{2+}$
  Pg: 616
Copper(II) [β-trans(14)diene]$^{2+}$
  Pg: 616
CpFe(CO)S$_2$CN(CH$_3$)$_2$
  Pg: 632
CpFe S$_2$C$_4$F$_6$
  Pg: 632
Cumylaldehyde phenylhydrazone
  Pg: 562
p-Cyanobenzene sulfonamide
  Pg: 330
Cyano ferrocene
  Pg: 624
p-Cyano phenyl ferrocene
  Pg: 623
2-Cyanophenazine
  Pg: 544
4'-Cyclohexyl acetophenone
  Pg: 595
Cyclohexylamine
  Pg: 532
Cyclo-octatetraene
  Pg: 550, 554
Cyclopentadiene carbonyl hydridobis (trichlosilyl)irc..
  Pg: 624
Cyclopentadiene dicarbonyl-trichlorosilyl iron
  Pg: 624
π - Cyclopentadienyl - π - cycloheptatrienyl-vanadium(o)
  Pg: 625
Cyclopentadienyl-π-titanium dichloride
  Pg: 633
Cyclophenptatriene
  Pg: 633
Cysteine hydrochloride
  Pg: 568, 573
Cistine dihydrochloride
  Pg: 568

D

Decalin dibromide
  Pg: 567
Deuteroporphyrin
  Pg: 553
p-Diacetylbenzene
  Pg: 588, 594
1,1'Diacetyl ferrocene
  Pg: 624
Diallyl sulfide
  Pg: 528, 625
q-Diazofluoren
  Pg: 533
1-Diazo-2,3,4,5-tetraphenyl cyclopentadien
  Pg: 533
1,2,3,4-Dibenzanthracene
  Pg: 529, 538, 549
Dibenzo[a,c]phenazine
  Pg: 543
5,6,7,8-Dibenzoquinoxaline
  Pg: 543
1,4-Dibenzoyl-2,3-diphenyl-2,3-butanediol
  Pg: 571
Dibenzoylmethane
  Pg: 571, 586
3,4,9,10-Dibenzpyrene
  Pg: 529, 538
Dibenzylideneazine
  Pg: 543
Dibenzyl sulfide
  Pg: 528, 625

Dibenzyl thiophene
 Pg: 528, 625
1,1'-Dibromo ferrocene
 Pg: 624
1,2-Dibromoethane
 Pg: 530
4,4'-Di-*t*-butyl
 Pg: 550
Dibutyl fumarate
 Pg: 474, 573, 574
Dibutyl maleate
 Pg: 574
Di-*n*-butyl sulfide
 Pg: 528, 625
Di-*s*-butyl sulfide
 Pg: 528, 625
Di-*t*-butyl sulfide
 Pg: 528, 625
Dichloroacetic acid
 Hs: 416
1 - (2,6 - Dichlorobenzyl) - 1,4 - dihydronicotinamide, ClB₂NH
 Pg: 535
2,3-Dichloro-5,6-dicyanoquinone
 Pg: 598
2,4-Dichlorophenol
 Pg: 604
Didenyl
 Pg: 614
1,1'-Di(ethoxy carbonylamino) ferrocene
 Pg: 624
1,6-Diethoxyphenazine
 Pg: 544
4-Diethylaminoazobenzene
 Pg: 529
N,*N*-Diethyl-*p*-cyanobenzene sulfonamide
 Pg: 530
Diethyldisulfide
 Pg: 568
1,1'-diethyl ferrocene
 Pg: 614
Diethylfumarate
 Pg: 549, 604
Diethyl maleate
 Pg: 604
Diethyl-4-nitrobenzylphosphate
 Pg: 476, 490
Diethyl-4-nitrophenylphosphate
 Pg: 476, 490
O,*O*-Diethyl-0,4-nitrophenylthiophosphate
 Pg: 476, 490
Diethyl sulfide
 Pg: 528, 625
Diethyl titanium chloride, TiCl₂(C₂H₅)₂
 Pg: 495
N,*N*-Diethyl-*p*-toluene sulfonamide
 Pg: 530
(Difuroyl-2) methane
 Pg: 606
Dihydrochloride
 Pg: 568
12,15 - Dihydro - 12,15 - dioxo - 2,3,6,7 - dibenzotriptyene
 Pg: 546
5,10-Dihydro-5,10-dimethylphenazine
 Pg: 530
12,15-Dihydro-12,15-dioxotriptycene
 Pg: 546
5,10-Dihydro-5,10-diphenylphenazine
 Pg: 530
5,10-Dihydro-5-methylphenazine
 Pg: 530
5,10-Dihydro-5-methyl-10-phenylphenazine
 Pg: 530
Dihydropyrene
 Pg: 549
Dihydroxphenylalanine
 Pg: 573
p-Di-iodobenzene
 Pg: 584
9,12-Di-iodo-*o*-carborane
 Pg: 553
Dimedone
 Pg: 541, 587
p,*p*'-Dimethoxybenzil
 Pg: 571

1,1'-Di(methoxy carbonylamino) ferrocene
 Pg: 624
α-Dimethylaminoacetophenone
 Pg: 533, 557
p-Dimethylaminoazobenzene
 Pg: 579
4-(*N*,*N*-Dimethylamino)azobenzene
 Pg: 525, 528, 538
p - Dimethylaminobenzaldehyde *m* - nitrophenylhydrazone
 Pg: 562
p-Dimethylaminobenzaldehyde phenylhydrazone
 Pg: 562
p-Dimethylaminobenzaldehyde thiosemicarbazone
 Pg: 563
2-Dimethylamino-2-phenyl-1,3-indandione
 Pg: 545
N,*N*-Dimethylaniline
 Pg: 529, 538
2,4-Dimethylaniline
 Pg: 529, 538
2,5-Dimethylaniline
 Pg: 529, 538
3,5-Dimethylaniline
 Pg: 529, 538
9,10-Dimethyl-1,2-benzanthracene
 Pg: 529, 538
2,6-Dimethylbenzonitrile
 Pg: 556
2,6-Dimethylbenzonitrile *N*-oxide
 Pg: 556
5,5 - Dimethylcyclohexanedione - 1,3 - (dimedone)
 Pg: 586
p,*p*'-Dimethyldiphenylacetylene
 Pg: 546
Dimethyl hydrazine
 Pg: 572
3,8-Dimethyl-2-methoxy-azocine
 Pg: 554, 605
2,4-Dimethyl α-methyl styrene
 Pg: 561
3,5-Dimethyl α-methyl styrene
 Pg: 561
1,4-Dimethylnaphthalene
 Pg: 549
1,6-Dimethylnaphthalene
 Pg: 549
1,7-Dimethylnaphthalene
 Pg: 549
2,3-Dimethylnaphthalene
 Pg: 549
2,6-Dimethylnaphthalene
 Pg: 549
2,6-Dimethyl-4-nitro phenyl ferrocene
 Pg: 623
4,7-Dimethyl 1,10-phenanthroline (FeII)
 Pg: 618, 627
Di(2-methyl phenyl)-4-nitro phenyl phosphate
 Pg: 476
Di(4-methyl phenyl)-4-nitrophenylphosphate
 Pg: 476
2,6-Dimethylpyrazine
 Pg: 543
2,5-Dimethylpyridine
 Pg: 543
2,6-Dimethylpyridine
 Pg: 543
2,4-Dimethylquinoline
 Pg: 543
2,6-Dimethylquinoline
 Pg: 543
2,3-Dimethylquinoxaline
 Pg: 543
6,7-Dimethylquinoxaline
 Pg: 543
Dimethyl sulfide
 Pg: 528, 625
N,*N*-Dimethyl-*p*-toluene sulforamide
 Pg: 530
α,β-Dimorpholino-styrolene
 Pg: 533

XII. COMPOUND INDEX

m-Dinitrobenzene
　Pg: 541, 569, 591
　Ob: 778
o-Dinitrobenzene
　Pg: 545
　Ob: 778
p-Dinitrobenzene
　Pg: 545
　Ob: 778
2,2'-Dinitrobibenzyl
　Pg: 531, 537
4,4'-Dinitrobibenzyl
　Pg: 531, 537
2,2'-Dinitrobiphenyl
　Pg: 557
Dinitrodianthrimide
　Pg: 565
2,2-Dinitropropane
　Pg: 544
Dioctylfumarate
　Pg: 574
Dioctylmaleate
　Pg: 574
Diphenylacetylene
　Pg: 546
2-(Diphenylacetyl)-1,3-indandione
　Pg: 587
Diphenylamine
　Pg: 527, 529, 538
1,9-Diphenylanthracine
　Pg: 547
1,10-Diphenylanthracene
　Pg: 547
9,10-Diphenylanthracene
　Pg: 526, 547, 599
1,2-Diphenyl-*o*-carborane
　Pg: 553
Diphenyldiazomethane
　Pg: 531, 533
cis-sym-Diphenylethylene
　Pg: 546
trans-sym-Diphenylethylene
　Pg: 546
Diphenylmethyleneaniline
　Pg: 543
Diphenyl - 2,6 - dimethyl - 4 - nitrophenyl-phosphate
　Pg: 476
Diphenyl-4-nitrobenzylphosphate
　Pg: 476
Diphenyl-4-nitrophenylphosphate
　Pg: 476, 490
2,2-Diphenyl-1-pioxylhydrazyl
　Pg: 602
Diphenylpicrylhydrazyl
　Pg: 533, 535
1,3-Diphenyl-1,3-propanedione
　Pg: 571, 587
1,3-Diphenyl propenone
　Pg: 594
Diphenyl sulfide
　Pg: 528, 551, 625
α,α'-Diphenyl-*trans*-stibene
　Pg: 548
Diphosphophyridine nucleotide, and substituents
　Pg: 1080
1,2-Diperidino-ethylene
　Pg: 533
α,β-Dipiperidino-styrolene
　Pg: 533
1,2 - Di(*N* - propylamino) - 1 - neopentyl - ethylene
　Pg: 533
Di-*i*-propyl sulfide
　Pg: 528, 625
Disulfide
　Pg: 597
p-Dithiane
　Pg: 528, 625
α,α'-Dithienyl sulfide
　Pg: 577
α-β'-Diethienyl sulfide
　Pg: 577

β,β'-Diethienyl sulfide
　Pg: 551, 577
Ditiodiglycolic acid dianelide
　Pg: 597
Di-4-Tolylamine
　Pg: 527
Duroquinone
　Pg: 540, 570
Dysprosium III, Dy III
　Pg: 467, 487, 491, 493, 513

E

EDA·2HCL
　Hs: 430
Erbium III
　Pg: 467, 487, 491, 493
Erbium nitrate
　Sl: 99
Ethanethiol
　Pg: 568
o-Ethoxy phenyl ferrocene
　Pg: 623
Ethylamine (2,2'-Dithobis)
　Pg: 568
Ethylammonium chloride
　Sl: 28
Ethyl bromide
　Pg: 531, 556
Ethyl chlorophyllide
　Pg: 570
Ethyl-1,3-dioxo-2-indancarboxylate
　Pg: 587
4'-4'''-Ethylenediacetopenone
　Pg: 588, 594
Ethylene sulfide
　Pg: 528, 625
Ethylene trithiocarbonate
　Pg: 528, 625
Ethyl ferrocene
　Pg: 614
Ethyl iodide
　Pg: 556, 584
Ethyl pridinium bromide
　Pg: 604
Ethylthioglycollate
　Pg: 581
2-Ethynyl·pridine
　Pg: 595
Etioprophine I
　Pg: 551
Etioprophyrin I
　Pg: 551
Europium II
　Pg: 467, 468, 471, 472, 483, 487, 491, 493, 513
Europium III
　Pg: 467, 468, 471, 472, 483, 487, 491, 493, 518
　Le: 692, 710, 711, 714, 717

F

Fe(CO)$_2$(π-C$_5$H$_5$)
　Pg: 612
Ferrocene
　pKsp: 119
　EMF: 142, 143, 834
　Pg: 472, 524, 525, 539, 541, 575, 591, 600, 602, 611, 613, 615, 621, 623, 626, 628, 654
Ferrocene, and substituted ferrocenes
　Pg: 615
Ferrocene-B(C$_6$H$_5$)$_4$
　pKsp: 112, 119
Ferrocenyl azobenzene
　Pg: 623
Fluoranthene
　Pg: 550
Fluorene
　Pg: 529, 538
Fluoreneone anil
　Pg: 535, 553

Fluoren-2-ylmethyl ketone
  Pg: 595
5-Fluoroacetyltetralin
  Pg: 578, 594
α-Fluorobutanal
  Pg: 552
3′Fluoro-4-dimethylaminoazobenzene
  Pg: 529
α-Fluoroisobutanal
  Pg: 552
3-Fluoro-10-methyl-1,2-benzanthracene
  Pg: 529, 538
4-Fluoro-10-methyl-1,2-benzanthracene
  Pg: 529, 538
o-Fluoro phenyl ferrocene
  Pg: 623
Fluoro-sulfuric acid, HSO_3F
  Hs: 413, 416, 421, 425, 427, 429, 435, 441
Formic acid, HCOOH
  Hs: 425
  Pg: 569
Formyl
  Pg: 614
Fumaric acid
  Pg: 604

G

Gadolinum III
  Pg: 467, 468, 487, 491, 493, 513
  Le: 675
Germanium tetrachloride
  Hs: 473
Glass
  EMF: 134–137, 142, 143, 170, 171, 174, 175, 180, 181, 204, 205, 228, 229, 236–239, 240, 241, 246, 247, 256, 257, 267, 268, 276
  Pt: 276–282, 285–290, 294–298, 300, 301, 303, 305–314, 316, 320–331
Glutamic acid
  Pg: 573
Glutamine
  Pg: 573
Glutaric acid
  Pg: 573
Gold
  EMF: 156, 157, 194, 195
  Pt: 276, 294
Graphite
  Pt: 296
Guanidine HBr
  Hs: 420
Guanidine HCl
  Hs: 420
Guanidine NCNS
  Hs: 420
Guanidine HNO
  Hs: 420

H

Hafnium tetrachloride
  Pg: 495
Hexachlorobutadiene
  Pg: 590
Hexadecanol
  Pg: 566
2,4,5,8,9,10-Hexahydro
  Pg: 598
Hexahydro-perylene
  Pg: 598
Hexahydro-2-phenyl-1,3-indandione
  Pg: 587
Sym-Hexahydropyrene
  Pg: 550
Hexamethyl benzene
  Pg: 529, 538
2,4,6,2′,4′,6′-Hexamethyl-*trans*-stilbene
  Pg: 548
Hexylamine
  Pg: 532

H $Fe_3(CO)_{11}^-$, Et_3NH^+
  Pg: 632
Holmium III, Ho(III)
  Pg: 467, 487, 491, 493, 513
Hydrazine
  Pg: 529
p-p′-Hydrazotoluene
  Pg: 529, 538
Hydrogen
  EMF: 134–137, 142–145, 148–151, 158, 159, 166, 167, 170, 171, 174, 175, 182, 183, 186, 187, 190–195, 198–201, 204–207, 210–213, 218, 234–243, 240–263, 265–271, 831, 832, 835–837, 839–842
  Pt: 298, 304, 315
Hydrogen bromide HBr
  Sl: 827
  EMF: 182–187, 190, 191, 204–207, 218, 219, 234, 235, 238–241, 244, 245, 256–259, 835, 839, 841
  Pg: 477, 490, 495
  Edl: 728
  Sp: 750
Hydrogen chloride HCL
  Sl: 827
  EMF: 130, 131, 134–139, 158, 159, 170, 171, 178–183, 186–195, 208–215, 218, 219, 234–243, 240–263, 269–272, 831–833, 835, 836, 838–842
  Vp: 381
  Hs: 448–455
  Pg: 477, 494, 495, 510, 567
  Edl: 728, 731, 732, 736, 739
Hydrogen fluoride
  Hs: 418
Hydrogen iodide
  EMF: 182–185, 208, 209, 234, 235, 238, 239, 242, 243, 256–259, 835
  Pg: 495
Hydrogen tribromide
  Pg: 490
Hydrogen tribromomercurate
  Pg: 490
Hydronyproline
  Pg: 573
Hydroquinone
  Pg: 602, 603
Hydroxide ion, OH^-
  Pg: 493
N-Hydroxy-2-acetylamino
  Pg: 529
N-Hydroxy-2-acetylamino fluorene
  Pg: 538
1-Hydroxy-9,10-anthraquinone
  Pg: 536, 563, 601
1-Hydroxyanthraquinone, conjugate base of
  Pg: 563, 601
1-Hydroxy-9,10-anthraquinone, conjugate base of
  Pf: 563, 601
2,4-Hydroxybenzaldehyde semicarbazone
  Pg: 563
2-Hydroxy-5-chlorobenzaldehyde semicarbazone
  Pf: 563
4-Hydroxycoumarin
  Pg: 552
Hydroxylamine hydrochloride
  Hs: 420
1-Hydroxynaphthalene
  Pg: 529, 537
2-Hydroxynaphthalene
  Pg: 529, 538
p-Hydroxy phenyl ferrocene
  Pg: 623
8-Hydroxyquinoline
  Pg: 546

I

1,3-Indandione
  Pg: 587
Indium(III)
  Pg: 485, 505

XII. COMPOUND INDEX

Indium(III) N-Benzoyl-N-phenyl hydroxylamine
  Pg: 628
Iodide
  Pg: 466, 467, 469, 484, 492, 501, 507, 514, 517
Iodine
  EMF: 144, 145, 164, 165, 170, 171, 176, 177, 192, 193, 196, 197, 206, 207, 210–213, 214, 215, 224, 225
  Pt: 299
  Pg: 470, 476, 524
Iodine dipyridyl perchlorate
  Hs: 470
Iodine monobromide
  Hs: 469
Iodine monochloride
  Hs: 469
p-Iodoacetanilide
  Pg: 584
Iodoacetic acid
  Pg: 584
p-Iodoaniline
  Pg: 584
p-Iodoanisole
  Pg: 584
Iodobenzene
  Pg: 553, 584
m-Iodobenzoic acid
  Pg: 584
o-Iodobenzoic acid
  Pg: 584
p-Iodobenzoic acid
  Pg: 584
9-Iodo-o-carborane
  Pg: 553
p-Iodochlorobenzene
  Pg: 584
p-Iodo-N,N-dimethylaniline
  Pg: 585
p-Iododiphenyl
  Pg: 584
Iodo ferrocene
  Pg: 623
p-Iodo ferrocenyl azobenzene
  Pg: 623
o-Iodophenol
  Pg: 584
p-Iodophenol
  Pg: 584
1-(p-Iodophenyl)-m-carborane
  Pg: 553
1-(p-Iodophenyl)-o-carborane
  Pg: 553
o-Iodo phenyl ferrocene
  Pg: 623
p-Iodophenyltrimethylammonium iodide
  Pg: 584
p-Iodotoluene
  Pg: 584
Iridium(dithioacetylacetone)$_3$
  Pg: 612
Iron(II)
  Pg: 484, 497, 500, 506, 511, 514
  Le: 672, 675, 676, 678, 681
  Sp: 755, 758
Iron
  EMF: 226, 227
Iron(III)
  Sl: 827
  Pg: 485
  Le: 672, 675–678, 681
  Sp: 756
Iron(III) N-benzoyl-N-phenyl hydroxylamine
  Pg: 628
Iron (bipy)$_3$
  EMF: 140–143
Iron (C$_5$H$_5$)$_2$
  Pg: 620, 621
Iron(C$_5$H$_5$) C$_5$H$_4$CH$_2$C$_5$H$_4$OsC$_5$H$_5$
  Pg: 621
Iron(C$_5$H$_5$) (C$_5$H$_4$CH$_2$C$_5$H$_4$) RuC$_5$H$_5$
  Pg: 621

Iron (C$_5$H$_5$) (C$_5$H$_4$COC$_5$H$_4$)RuC$_5$H$_5$
  Pg: 621
Iron(III) Chloride
  Hs: 427
  Sp: 751
Iron(II) perchlorate
  Pg: 482
Iron(Cpdien)$_2$
  Pg: 619
Iron(Cpdien)$_2^+$
  Pg: 619
Iron(dipy)$_3^{2+}$
  Pg: 619
Iron(dipy)$_3^{3+}$
  Pg: 619
Iron(dithioacetylacetone)$_2^{3+}$
  Pg: 612
Iron(dithioacetylacetone)$_3$
  Pg: 612
(Iron)$_3$(CO)$_{12}$
  Pg: 632
[Iron(L*) (S$_2$C$_2$Ar$_2$*)$_2$] *L =Ligand, Ar = aromatic ring
  Pg: 629
{Iron(NO) [S$_2$C$_2$(3,4-CH$_2$O$_2$C$_6$H$_3$]$_2$] }
  Pg: 628
{Iron(NO) [S$_2$C$_2$(Me·C$_6$H$_4$)$_2$] }
  Pg: 628
{Iron(NO) [S$_2$C$_2$(2-MeO·C$_6$H$_4$)$_2$] }
  Pg: 628
{Iron(NO) [S$_2$C$_2$(3-MeO·C$_6$H$_4$)$_2$] }
  Pg: 628
{Iron(NO) [S$_2$C$_2$(4-MeO·C$_6$H$_4$)$_2$] }
  Pg: 628
{Iron(NO) [S$_2$C$_2$(2,5-MeO)$_2$C$_6$H$_3$)$_2$] }
  Pg: 628
{Iron(NO) (S$_2$C$_2$Ph$_2$)$_2$] }
  Pg: 628
Iron octoate
  Pg: 627
Iron(II) perchlorate hexahydrate, Fe(ClO$_4$)·6H$_2$O
  Pg: 468, 472, 478
Iron(III) perchlorate
  Pg: 482, 489, 516
Iron(III) perchlorate hexahydrate, Fe(ClO$_4$)$_3$·6H$_2$O
  Pg: 468, 472, 478, 503
Iron (phen)$_3^{2+}$
  Pg: 619
Iron (phen)$_3^{3+}$
  Pg: 619
Iron phenanthroline
  EMF: 136, 137
Iron(II) [tetal]$_2^{2+}$
  Pg: 616
Iron(II) [trans(14)diene]$_2^{2+}$
  Pg: 616
Isolencine
  Pg: 573

L

Lanthanum III
  Pg: 468, 487, 491, 493, 513
  Le: 708, 709
Lanthanum acetate
  Sl: 99
Lanthanum chloride
  Edl: 736
  Sp: 746
Lanthanum cobalnitrate
  Sl: 99
Lanathanum iodide
  Hs: 419
Lanthanum nitrate
  Sl: 99
Lead
  EMF: 134, 135, 150, 151, 156, 157, 166, 167, 172, 173, 176, 177, 182, 183, 186, 187, 192–195, 200, 201, 206, 207, 216, 217, 220, 221
Lead Amalgam
  EMF: 134, 135

Lead II
  Pg: 485, 486, 505, 518
  Le: 689
Lead acetate
  Sl: 100
  EMF: 134, 135
Lead alkyl fluorides
  Sl: 100
Lead benzoate
  Sl: 100
Lead-N-benzoyl-N-phenyl hydroxylamine
  Pg: 628
Lead borate
  Sl: 100
Lead citrate
  Sl: 100
Lead cyanide
  Sl: 99
Lead bromide
  Sl: 99, 825
  EMF: 216, 217
Lead chloride
  Sl: 99, 818, 824
  EMF: 172, 173, 176, 177, 180, 181, 187, 216, 217, 226, 227
  Sp: 756
dicyclohexyl-Lead
  Sl: 100
Lead diethylthiothionocarbamate
  Sl: 100
diphenyldicyclohexyl-Lead
  Sl: 100
Lead fluoride
  Sl: 99
Lead formate
  Sl: 100
Lead hexadecanoate
  Sl: 101
Lead iodide
  Sl: 99
  EMF: 216, 217
  Hs: 420
  Pg: 481, 509
Lead malate
  Sl: 100
Lead naphthenate
  Pg: 627
Lead nitrate
  Sl: 100
  EMF: 136, 137, 150, 151, 206, 207, 220–223
  Hs: 430
  Pg: 481
Lead Octadecanoate
  Sl: 101
Lead Octoate
  Pg: 627
Lead–Lead chloride
  EMF: 226, 227
Lead oxide
  Sl: 99
Lead perchlorate
  Sl: 100
  Hs: 421, 437, 442
  Pg: 478
Lead perchlorate trihydrate
  Pg: 475
Lead sulfate
  Pg: 480
Lead styphnate
  Sl: 100
Lead succinate
  Pg: 100
Lead tallate
  Pg: 627
Lead tartarate
  Sl: 100
Lead thiocyanate
  Sl: 100
  EMF: 150, 151
tricyclohexyl Lead
  Sl: 100
Lithium
  Hs: 420
  Ob: 333–336

Lithium (I)
  Le: 690, 691, 694–697
  Edl: 728, 753, 755
  Sp: 748, 756
Lithium acetate
  Sl: 29
  Cr: 407, 408
Lithium aluminum hydride
  Vp: 372
Lithium amalgam
  EMF: 134, 135, 150, 156, 157, 166, 167, 172, 173, 176, 177, 182, 183, 186, 187, 192–195, 200, 201, 206, 216, 217, 240, 241, 838, 842
Lithium benzoate
  Sl: 29
Lithium bromide
  Sl: 29, 30, 820, 822
  EMF: 162–163, 178–179, 186–187, 214–219, 833, 834
  Vp: 377
  Cr: 405, 406
  Hs: 420, 423, 426, 429, 431, 433, 436, 439, 440, 853
  Edl: 728
  Sp: 746, 753, 755, 762
Lithium chlorate
  Vp: 390
  Hs: 852, 853
Lithium chloride, LiCl
  Sl: 30–32, 816, 820, 822
  EMF: 130–133, 168–169, 172–181, 184–185, 204–205, 208–209, 216–219, 244, 245, 248, 249, 260, 261, 266, 271, 836–839
  Vp: 336, 359, 374–376, 380, 394, 395
  Cr: 403, 405, 406
  Hs: 417, 424, 426, 428, 429, 431, 433, 436, 438, 447–454, 853
  Pg: 468, 475, 489, 495, 499, 518
  Edl: 727–735, 736, 738, 739
  Sp: 746, 753, 755
  Ob: 773–777, 779
Lithium fluoride
  Sl: 32, 816, 820, 822
  Cr: 404
  Hs: 431
  Ob: 774, 776
Lithium formate
  Sl: 33, 34
  Cr: 407, 408
Lithium hexafluoroarsenate
  Vp: 379
  Ob: 777
Lithium hexafluorotitanate
  Sl: 34
Lithium hypochlorite
  Hs: 443
Lithium methyl sulfate
  pKsp: 112
Lithium iodide
  Sl: 34, 822
  EMF: 214–219
  Pt: 847
  Hs: 420, 424, 426, 429–431, 433, 438–440, 443
  Sp: 746, 762
Lithium nitrate
  Sl: 35, 817
  EMF: 178–179
  Vp: 359, 387, 388, 397
  Cr: 405, 406
  Hs: 428, 430, 436
  Edl: 727, 732, 737, 739
  Sp: 746
Lithium perchlorate
  Sl: 35, 816, 820
  EMF: 140, 141, 146, 147, 216, 217, 830
  Pt: 847
  Vp: 373, 374, 379, 380
  Hs: 417, 421, 428, 442–444
  Pg: 475, 478, 482, 487, 489, 506, 511, 516, 519
  Edl: 730–732, 736, 739
  Sp: 746
  Ob: 774–777, 779

XII. COMPOUND INDEX

Lithium perchlorate trihydrate
  Sl: 36
Lithium picrate
  Hs: 428, 436
Lithium propionate
  Sr: 407, 408
Lithium salicylate
  Sl: 36
Lithium sulfate
  Sl: 36
Lithium tetronfluoroboride
  Sl: 818, 820, 822
  Ob: 775
Lithium trifluoro acetate
  Hs: 660

M

Magnesium
  EMF: 156, 157, 226, 227
  Ob: 773, 774
  Le: 675, 677, 678, 681, 684
Magnesium acetate
  Sl: 36
Magnesium amalgam
  EMF: 206, 207
Magnesium benzoate
  Sl: 37
Magnesium bromide
  Sl: 37
  EMF: 162, 163
Magnesium chloride
  Sl: 37, 38, 820
  EMF: 206, 207
  Hs: 425
  Edl: 737
  Sp: 751
  Ob: 776
Magnesium fluoride
  Ob: 776
Magnesium iodide
  Sl: 820
  Pg: 475
Magnesium methoxide
  Sl: 38
Magnesium perchlorate
  Sl: 38
  Pt: 847
  Hs: 416, 421, 428, 434, 436, 442, 443
  Pg: 468, 482, 495, 511, 516
  Sp: 754, 755
  Ob: 776
Magnesium salicylate
  Sl: 39
Magnesium sulfate
  Sl: 39
Maleic acid
  Pg: 604
Maleic anhydride
  Pg: 604
Malic acid
  Pg: 573
Manganese(II)
  Sl: 827
  Pg: 471, 479, 483–485, 497, 499–501, 504, 506, 507, 511, 514, 515
  Le: 672, 673, 675, 678, 681, 682, 688, 690–692, 700, 702–704, 706–709, 718, 719, 721
  Sp: 757, 758
Manganese(III)
  Sp: 757
Manganese octoate
  Pg: 627
Managese(II), perchlorate
  Pg: 474, 479, 482, 488, 489, 495, 502, 510, 511, 513, 516, 520
Manganese(II) perchlorate hexahydrate
  Pg: 468, 478, 503
Manganese tallate
  Pg: 627
2 - Mercapto - 5 - ethyl - 3 - thenyl - idene - cycloheylamine
  Pg: 551
Mercapton compound
  Pg: 597

Mercury
  EMF: 144, 145, 150–153, 166, 167, 182, 183, 194, 195, 206, 207, 220, 221
  Pt: 287, 299
Mercury(II) acetate
  Sl: 102
Mercury(II) bromide
  Sl: 101
  Pg: 492
  Sp: 749, 756
Mercury(I) chloride
  Sl: 101, 818, 821, 824
Mercury(II) chloride
  Sl: 101, 818, 821, 824
  Hs: 427
  Pg: 490
  Sp: 749, 756
Mercury(II) cyanide
  Sl: 102
  Hs: 430
Mercury(II) diethyldithiocarbonate
  Pg: 566
Mercury(II) diethylthiolthionocarbamate
  Sl: 102
Mercury glass
  EMF: 178, 179
Mercury-Mercurous acetate
  EMF: 134, 135, 837
Mercury-Mercuroussulfate
  EMF: 134, 135, 214, 215, 837
Mercury-mercurous bromide
  EMF: 162, 163, 206, 207, 226, 227, 841
Mercury-mercurous iodide
  EMF: 206, 207
Mercury-Mercuric Fluoride
  EMF: 200, 201
Mercury(II) iodide
  EMF: 148–151, 220–223
  Hs: 419, 420, 430
  Pg: 490
Mercury(I) Nitrate
  Sl: 102
Mercury(II) perchlorate
  EMF: 144, 145
Mercury(II) perchlorate trihydrate
  Pg: 468, 478
Mercury(II) nitrate
  Sp: 746
Mercury(II) sulfate
  Sl: 102
Mercury(II) thiocyanate
  Sl: 102
Mesoethyl chlorophyllide
  Pg: 570
Mesoporphyrin
  Pg: 553
Methionine
  Pg: 573
9-Methoxyanthracene
  Pg: 534
4-Methoxyazobenzene
  Pg: 525
2-Methoxyazocene
  Pg: 554, 605
Methoxy carbonylamino ferrocene
  Pg: 624
Methoxycyclo-octatetraene
  Pg: 554
4-Methoxy-2,6-dimethylbenzonitrile
  Pg: 556
4-Methoxy-2,6-dimethylbenzonitrile N-oxide
  Pg: 556
p-Methoxy diphenylacetylene
  Pg: 546
Methoxy ferrocene
  Pg: 624
p-Methoxy ferrocenyl azobenzene
  Pg: 623
1-Methoxyphenazine
  Pg: 544
2-Methoxyphenazine
  Pg: 544
o-Methosy phenyl ferrocene
  Pg: 623
2-p-Methoxyphenyl-1,3-indandione
  Pg: 587

5-Methylacetyltetralin
  Pg: 578
Methylamine
  Pg: 532
N-Methylaniline
  Pg: 529, 538
o-Methylaniline
  Pg: 529, 538
p-Methylaniline
  Pg: 529, 538
2-Methylanthracene
  Pg: 550
9-Methylanthracene
  Pg: 550
α-Methylbenxylamine
  Pg: 552
α-Methylbenzylidene-α-methylbenxylamine
  Pg: 547, 552
Methylbromide
  Pg: 556
1-Methyl-2-bromo-o-carborane
  Pg: 553
1-Methyl-o-carborane
  Pg: 553
2'-Methyl-4-dimethylaminoazobenzene
  Pg: 529
3'-Methyl-4-dimethylaminoazo benzene
  Pg: 529
4'-Methyl-4-dimethylaminoazobenzene
  Pg: 529
m-Methyldiphenylacetylene
  Pg: 546
o-Methyldiphenylacetylene
  Pg: 546
p-Methyldiphenylacetylene
  Pg: 546
1-Methyl-2,2-diphenyl cyclopropyl mercuric bromide
  Pg: 615
3,3'-Methylene-bis-4-hydroxycourmarin
  Pg: 552
4',4'''-Methylene-diacetophenone
  Pg: 588, 594
4,5-Methylen phenanthrene
  Pg: 531, 548
2-Methyl-5-ethynyl·pyridine
  Pg: 595
p-Methyl ferrocenyl azobenzene
  Pg: 623
Methyl ferrocenyl carbinol
  Pg: 75
3-Methylfluoranthene
  Pg 550
7-Methylfluoranthene
  Pg: 550
8-Methylfluoranthene
  Pg: 550
Methylglutaric acid
  Pg: 573
Methyl iodide
  Pg: 556, 584
Methyl-m-iodobenzoate
  Pg: 584
Methyl-o-iodobenzoate
  Pg: 584
Methyl-p-iodobenzoate
  Pg: 584
2 - Methylmercapto - 5 - methyl - 3 - thenyl-idencyclohexylamine
  Pg: 551
8-Methyl-2-methoxy azocine
  Pg: 554
1-Methylnaphthalene
  Pg: 550
2-Methylnaphthalene
  Pg: 548, 550
2-Methyl-1,4-naphthaquinone
  Pg: 531, 540, 549
2-Methyl-1,4-napththohydroquinone
  Pg: 540, 566
1-Methylnicotinamide
  Pg: 535
2-Methyl-4-nitrophenyl ferrocene
  Pg: 623
2-Methyl-5-nitrophenyl ferrocene
  Pg: 623

2-Methyl-6-nitrophenyl ferrocene
  Pg: 623
2-Methylphenanthrene
  Pg: 550
3-Methylphenanthrene
  Pg: 550
3-Methylphenanthrene
  Pg: 550
9-Methylphenanthrene
  Pg: 550
2-Methylphenazine
  Pg: 544
5-Methylphenazinium methyl sulfate
  Pg: 530
m-Methyl phenyl ferrocene
  Pg: 623
o-Methyl phenyl ferrocene
  Pg: 623
p-Methyl phenyl ferrocene
  Pg: 623
2-Methyl-2-phenyl-1,3-indandione
  Pg: 545, 587
N-Methylphthalimide
  Pg: 550
2-Methylpyrazine
  Pg: 543
1-Methylpyrene
  Pg: 550
2-Methylpyrene
  Pg: 550
4-Methylpyrene
  Pg: 550
2-Methylpyridine
  Pg: 543
3-Methylpyridine
  Pg: 543
4-Methylpyridine
  Pg: 543
2-Methylquinoline
  Pg: 543
4-Methylquinoline
  Pg: 543
6-Methylquinoline
  Pg: 543
8-Methylquinoline
  Pg: 543
3-Methylisoquinoline
  Pg: 543
2-Methylquinoxaline
  Pg: 543
α-Methyl styrene
  Pg: 561
Methylsuccinic acid
  Pg: 573
di-Methyl sulfate
  Cr: 411
N-Methyl-p-toluene sulfonamide
  Pg: 530
2'-Methyl-4'-m-tolylacetophenone
  Pg: 595
α-Methyl-trans-stilbene
  Pg: 548
1-Methyl-3-vinyl naphthalene
  Pg: 593, 595
$MOCL_2(bipy)_2^+$
  Pg: 619
$MOCL_3(Py)_3$
  Pg: 619
$MOCL_3(Py)$ bipy
  Pg: 619
Molybdenum
  EMF: 130, 131
  Pt: 276, 299
Monel
  EMF: 156, 157
Mononitrodianthrimide
  Pg: 565
α-Morpholinoacetophenone
  Pg: 533, 557

N

Naphthalene
  Pg: 527, 529, 538, 539, 547, 550, 552, 564, 568, 570, 572, 584, 590, 598

XII. COMPOUND INDEX

1,2-Naphthoquinone
  Pg: 524, 534, 555, 572, 601
1,4-Naphthoquinone
  Pg: 524, 534, 555, 572, 601
α-Naphthylamin
  Pg: 535
1-Naphthylphenylacetylene
  Pg: 546
1,5-Naphthyridine
  Pg: 543
m-NHCOC$_6$H$_5$-phenyl ferrocene
  Pg: 623
p-NHCOC$_6$H$_5$-phenyl ferrocene
  Pg: 623
m-N$_2$C$_6$H$_5$-phenyl ferrocene
  Pg: 623
p-N$_2$C$_5$H$_5$-phenyl ferrocene
  Pg: 623
Neodymium(III), Nd(III)
  Pg: 467, 471, 472, 483, 487, 491, 493, 513
Neodymium camphor carbonate
  Sl: 103
Neodymium chloride
  Sl: 103
Nickel
  EMF: 156, 157, 166, 167, 206, 207
Nickel(II)
  Sl: 824
  Pg: 471, 474, 479, 483–485, 494, 497, 498, 500, 504–506, 508, 514, 515, 854, 855
  Le: 672–679, 682, 690, 691, 693, 701, 704, 712, 713
  Sp: 750, 754, 755, 762
Nickel(II) [A(13)T]$^{2+}$
  Pg: 616
Ni(II) [A(14)T]$^{2+}$
  Pg: 616
Nickel(t-Bu$_2$)(C$_5$H$_4$)$_2$
  Pg: 620
Nickel (C$_5$H$_5$)$_2$
  Pg: 620
Nickel(II) [(CH$_3$)$_2$teta]$^{2+}$
  Pg: 616
Nickel(II) [CH$_3$)$_2$trans(14)diene]$^{2+}$
  Pg: 616
Nickel(II) [cis(14)diene]$^{2+}$
  Pg: 616
Nickel Cl$_2$2(C$_6$H$_5$)$_3$P
  Pg: 620
Nickel(dithenoyl-2)methane
  pKsp: 125
Nickel(dithioacetylacetone)$_2$
  Pg: 612
Nickelocene
  EMF: 144–145
Nickel octoate
  Pg: 627
Nickel perchlorate
  Pg: 470, 479, 482, 488, 489, 495, 502, 510, 513, 516, 520
Nickel perchlorate hexahydrate, Ni(ClO$_4$)$_2$·6H$_2$O
  Pg: 468, 503
Nickel(II) [teta]$^{2+}$
  Pg: 616
Nickel(II) [tetb]$^{2+}$
  Pg: 616
Nickel [tetrabenzo[b, f, j, n] [1, 5, 9, 13]-tetraaza cyclohexadecine]$^+$
  Pg: 618
Nickel [tetrabenzo[b, f, j, n,] [1, 5, 9, 13]-tetraaza cyclohexadecine]$^{2+}$
  Pg: 618
Nickel(II)trans(14)diene]$^{3+}$
  Pg: 616
i-Nicotinamide
  Pg: 562
Niobium pentachloride
  Sl: 825
  Pg: 476, 488, 495
Nitric acid
  Cr: 411
Nitrite ion, NO$_2^-$
  Pg: 494
5-Nitroacenaphthene
  Pg: 581, 589

p-Nitroacetanilide
  Pg: 541, 598
m-Nitroaniline
  Pg: 536, 541
  Sp: 756
o-Nitroaniline
  Pg: 536
  Sp: 756
p-Nitroaniline
  Pg: 536, 541, 598
Nitroanthraquinone
  Pg: 565
4-Nitroazobenzene
  Pg: 525
Nitrobenzene
  Cr: 412
  Pg: 535, 565
m-Nitrobenzyl bromide
  Pg: 531, 537
o-Nitrobenzyl bromide
  Pg: 531, 537
p-Nitrobenzyl bromide
  Pg: 531, 537
m-Nitrobenzyl chloride
  Pg: 531, 537
o-Nitrobenzyl chloride
  Pg: 531, 537
p-Nitrobenzyl chloride
  Pg: 531, 537
2-Nitrobiphenyl
  Pg: 557
4-Nitro-2,6-dimethylbenzonitrile
  Pg: 556
p-Nitrodiphenylacetylene
  Pg: 546
p-Nitrodiphenyldiazomethane
  Pg: 533
2-Nitrofuran
  Pg: 587
5-Nitrofyrfuryl alcohol
  Pg: 587
Nitrogen
  EMF: 152–153
di-Nitrogen pentoxide
  Cr: 411
di-Nitrogen tetraoxide
  Cr: 411
di-Nitrogen trioxide
  Cr: 411
1-Nitronaphthalene
  Pg: 565
2-Nitronaphthalene
  Pg: 565
2-Nitrophenazine
  Pg: 544
m-Nitrophenol
  Pg: 536
o-Nitrophenol
  Pg: 536, 545
p-Nitrophenol
  Pg: 536
p-Nitriphenylbenzoyldiazomethane
  Pg: 533
p-Nitrophenyldiazomethane
  Pg: 533
p-Nitrophenylmethyldiazomethane
  Pg: 533
o-Nitrophenol
  Pg: 545
m-Nitrophenyl ferrocene
  Pg: 623
o-Nitrophenyl ferrocene
  Pg: 623
p-Nitrophenyl ferrocene
  Pg: 623
2-p-Nitriphenyl-1,3-indandione
  Pg: 587
2-Nitropropane
  Pg: 544
4-Nitropyridine-N-oxide
  Pg: 556
5-Nitropyromucic acid
  Pg: 587
5-Nicropyromucic acid, ethylester of
  Pg: 587

p-Nitrosoacetanilide
  Pg: 541
Nitrosobenzene
  Pg: 536, 599
2-Nitrosylvan
  Pg: 587
Nitrotoluene
  Cr: 412
m-Nitrotoluene
  Pg: 531, 537
o-Nitrotoluene
  Pg: 531, 537
p-Nitrotoluene
  Pg: 531, 537, 575, 595, 601
NO
  Pg: 619
p-NO$_2$C$_6$H$_4$
  Pg: 614
Norcamphor anil
  Pg: 547, 552

O

p-O CH$_3$C$_6$H$_4$
  Pg: 614
[(OC)$_3$FeNH]$_2$
  Pg: 633
(OC)$_5$MnBr
  Pg: 632
(OC)$_5$MnCl
  Pg: 632
(OC)$_5$MnH
  Pg: 632
(OC)$_4$MnPPh$_3$
  Pg: 632
Osmium (C$_2$H$_5$)$_2$
  Pg: 621
Osmium(dithioacetylacetone)$_3^{2+}$
  Pg: 612
Osmiumbipyridyl
  EMF: 144, 145
Osmium-terpyridyl
  EMF: 144, 145
Osmium-phenanthroline
  EMF: 144, 145
Osmium(dithioacetylacetone)$_3$
  Pg: 612
Osmocene
  Pg: 614
Oxalate ion
  Pg: 569
4′,4′′′-Oxydiacetophenone
  Pg: 588, 594
Oxygen
  EMF: 226, 227
  Pg: 491, 492, 494, 499, 509, 855
Oxyphthalimide
  Pg: 554

ORGANO-METALLIC COMPOUNDS CONTAINING THE FOLLOWING METALS

Antimony
  Pg: 635, 641, 642
Arsenic
  Pg: 635, 641, 642, 645
Bismuth
  Pg: 635
Boron
  Pg: 637
Chromium
  Pg: 635, 637, 638, 640–642
Cobalt
  Pg: 544, 635–637, 639–643, 646
Copper
  Pg: 643
Europium
  Pg: 646
Germanium
  Pg: 635, 636, 640, 644
Iron
  Pg: 635, 637, 638–642
Lead
  Pg: 635–637, 640, 641
Magnesium
  Pg: 634, 643
Manganese
  Pg: 632, 635, 637, 638, 640, 642
Mercury
  Pg: 641
Molybdenum
  Pg: 631, 632, 635, 637, 638, 640–642, 646
Nickel
  Pg: 613, 618, 620, 635, 636, 640–643, 647
Rhenium
  Pg: 635, 638, 640
Rhodium
  Pg: 639, 642
Ruthenium
  Pg: 635, 640
Selenium
  Pg: 635–637, 640, 641
Silicon
  Pg: 636, 640, 644
Sulfur
  Pg: 635, 637, 640
Tin
  Pg: 635–637, 640, 641, 644
Titanium
  Pg: 643, 647
Tungsten
  Pg: 631, 632, 635, 638, 641, 642
Vanadium
  Pg: 631, 642
Zinc
  Pg: 643

P

Palladium(II)
  Pg: 481
  Sp: 755
Palladium (dithioacetylacetone)$_2$
  Pg: 612
Pentamethylene sulfide
  Pg: 528, 625
2,4-Pentanedione
  Pg: 587
Percapric acid
  Pg: 597
Percaprylic acid
  Pg: 597
Perchloric acid
  EMF: 130–137, 236, 237
  Pg: 477, 478, 484, 567
  Sp: 750
Perlauric acid
  Pg: 555, 566, 597
Permyristic acid
  Pg: 597
Perylene
  Pg: 529, 538, 550, 598
PhC≡CPh
  Pg: 633
Phenanthrene
  Pg: 526, 550
Phenanthridine
  Pg: 526, 543
m-Phenanthroline
  Pg: 526
o-Phenanthroline
  Pg: 526
p-Phenanthroline
  Pg: 526
1,10-Phenanthroline
  Pg: 543
Phenazine
  Pg: 543, 544
4′-Phenethylacetophenone
  Pg: 588, 594
Phenol
  Pg: 529, 538, 602

Phenothiazine
  Pg: 529, 538
4'-Phenoxyacetophenone
  Pg: 588, 594
Phenylacetic acid
  Vp: 396
4'-Phenylacetophenone
  Pg: 588, 595
Phenylalanine
  Pg: 573
1-Phenylanthracene
  Pg: 547
2-Phenylanthracene
  Pg: 547
9-Phenylanthracene
  Pg: 534, 547, 599
Phenyl benzoyldiazomethane
  Pg: 533
1-Phenyl-2-bromo-o-carborane
  Pg: 553
1-Phenyl-1,3-butanedione
  Pg: 571, 587
1-Phenyl-m-carborane
  Pg: 553
1-Phenyl-o-carborane
  Pg: 553
p-Phenyldiphenyl diazomethane
  Pg: 533
Phenyl ferrocene
  Pg: 621, 623
Phenyl ferrocenyl carbinol
  Pg: 614
—CH—CH$_2$
  Pg: 614
—COCH$_3$
  Pg: 614
—1—1'—COCH$_3$
  Pg: 614
—COC$_6$H$_5$
  Pg: 614
—COOH
  Pg: 614
Phenylhexahydroindanedione-1,3
  Pg: 586
2-Phenyl-1,3-indandione
  Pg: 545
2-Phenyl-2,3-indandione
  Pg: 587
2-Phenyl-1,3-indandione, sodium salt of
  Pg: 545
2-Phenyl-1,3-indanedione, and derivatives
  Pg: 586
Phenyl-OCH$_2$
  Pg: 614
2-Phenyl phenazine
  Pg: 544
o-Phenyl phenyl ferrocene
  Pg: 623
p-Phenyl phenyl ferrocene
  Pg: 623
1-Phenyl-1,2-propanedione
  Pg: 571
2-Phenylpyridine
  Pg: 543
4-Phenylpyridine
  Pg: 543
4-Phenylpyrimidine
  Pg: 543
Phenylruthenocenyl carbinol
  Pg: 614
—1,1'—(COC$_6$H$_5$)
  Pg: 614
2 - Phenyl - 5,6,7,7a - tetrahydroindandione-1,3
  Pg: 586
2 - Phenyl - 4,5,6,7 - tetrahydroindianedione-1,3 +its derivatives
  Pg: 586
4'-(Phenylthio)-acetophenone
  Pg: 594
Phosphoric acid
  Pg: 477
  Edl: 728
  Sp: 571

Phosphorus heptasulfide
  Sl: 103
Phosphorus pentasulfide
  Sl: 103
Phosphorus selenide
  Sl: 103
Phosphorus trisulfide
  Sl: 103
Ph$_2$P(CH$_2$)$_2$N(Et)$_2$Mo(CO)$_4$
  Pg: 631
Ph$_3$PM$_0$(CO)$_5$
  Pg: 631
(Ph$_3$P)$_2$Pt·PhC≡CPh
  Pg: 633
Phthalazine
  Pg: 543
Phthalic acid
  Pg: 604
Phthalimide
  Pg: 554
Phthalylglycine
  Pg: 554
Phthalylglycine-methyl ester
  Pg: 554
α-Picoline
  Hs: 415
Pinacamphone anil
  Pg: 535
Pinocamphone anil
  Pg: 553
α-Piperidienoacetophenone
  Pg: 557
α-Piperidinoacetophenone
  Pg: 533
Platinum
  EMF: 146, 147, 152, 153, 156–159, 166, 167, 269, 276, 278
  Pt: 282, 283, 287, 298, 299, 301, 303, 304, 313, 315, 316, 320
Platinum (dithioacetylacetone)$_2$
  Pg: 612
Polyethylene
  Pt: 277
Potassium
  Hs: 420
Potassium (I)
  Pg: 517
  Sp: 756
Potassium amalgam
  EMF: 152, 153, 162, 163, 178, 179, 186, 187, 194, 195, 206, 207, 222, 223, 236, 237, 240, 241, 248, 249, 258, 259, 272, 273, 838, 842
Potassium acetate
  Sl: 39, 40
  EMF: 238, 239, 260, 261
  Cr: 407, 408
  Hs: 415
Potassium amide
  Sl: 40
  Hs: 420
Potassium azide
  Sl: 41
Potassium benzoate
  Sl: 41
Potassium bromide
  Sl: 41, 42, 820, 823, 827
  pKsp: 112–115, 117, 119
  EMF: 168, 169, 204–207, 246–251, 256, 257
  Vp: 359, 393
  Cr: 405, 406
  Hs: 423, 426, 431, 435, 438, 439, 447–454, 519
  Pg: 519
  Edl: 731, 738
  Sp: 746
Potassium butanoate
  Sl: 43
Potassium carbonate
  Sl: 43
  Vp: 393
  Edl: 732
Potassium chlorate
  Sl: 43

Potassium chloride
  Sl: 43–45, 818, 820, 822, 823, 827
  pKsp: 112–114, 119
  EMF: 140, 141, 146, 147, 150–153, 168–171, 188–195, 206, 207, 224–227, 236–245, 248–251, 256–261, 272, 837–840
  Vp: 396
  Cr: 405, 406
  Hs: 428, 431, 433, 435, 438, 439, 447–454, 852, 853
  Pg: 519, 567
  Edl: 730, 733, 734, 736, 738
  Sp: 746
Potassium cyanate
  Sl: 46, 827
Potassium cyanide
  Sl: 46, 827
  Hs: 851
Potassium dibromoiodide
  Sl: 46
Potassium-3,5-dinitrobenzoate
  pKsp: 112
Potassium-3,5-dinitrophenolate
  pKsp: 112
Potassium dihydrogen phosphate
  Edl: 734
Potassium dihydrogenhypophosphite
  Sl: 46
Potassium ethyl sulfate
  Sl: 47, 48
Potassium fluoride
  Sl: 47, 48
  Cr: 404
  Hs: 422, 431, 435, 438, 439
  Edl: 732, 734, 736, 738
  Sp: 746, 752
Potassium hexafluoroantimonate
  pKsp: 118
Potassium hexafluorophosphate
  Sl: 816, 820, 823
  pKsp: 118
  EMF: 178, 179, 728–733, 737–740
  Ob: 772, 775, 776
Potassium hydrogen sulfate
  Sl: 48
Potassium hydroxide
  Sl: 44
  Sp: 746
Potassium iodide
  Sl: 49–51, 817, 818, 820, 822, 823, 827
  pKsp: 113, 114, 119
  EMF: 144, 145, 170, 171, 176, 177, 186, 187, 192, 193, 196, 197, 206, 207, 222, 223, 226, 227, 250, 251, 256, 257, 841
  Vp: 360, 361, 375, 378, 381–383
  Cr: 405–406
  Hs: 419, 420, 422, 423, 426, 428, 430–432, 435, 438–440
  Pg: 475, 478, 489, 506, 519
  Edl: 728, 730, 731, 734, 736, 738
  Sp: 746, 762
Potassium methyl sulfate
  pKsp: 112
Potassium nitrate
  Sl: 52
  pKsp: 112
  EMF: 150, 151, 258–261
  Vp: 361, 393, 398
  Cr: 405, 406
  Hs: 430, 435
  Edl: 728, 731, 734
  Sp: 762
Potassium paranitrophenolate
  pKsp: 114, 119
Potassium perchlorate
  Sl: 52
  pKsp: 112–115, 119
  Hs: 444
  Pg: 474, 475, 478, 487–489, 494–495, 513, 518, 520, 567
  Sp: 746
Potassium periodate
  pKsp: 112
Potassium picrate
  pKsp: 112, 114, 119

Potassium propionate
  Cr: 407–408
Potassium salicylate
  pKsp: 112
Potassium phthalate
  Sl: 53
Potassium propanoate
  Sl: 53
Potassium succinate
  Sl: 53
Potassium sulfate
  Sl: 53
  Cr: 411
  Edl: 727, 736
Potassium tetraphenylborate
  Sl: 54
  pKsp: 112, 119
  Hs: 423, 443
Potassium thiocyanate
  Sl: 54, 827
  EMF: 148–151
  Hs: 416, 420
  Ob: 774
Potassium trifluoroacetate
  Hs: 422, 435, 443
Prageodymium III, Pr(III)
  Pg: 467, 468, 471, 483, 487, 491, 493, 513
Praseodymium chloride
  Sl: 103
  Sp: 746
Praseodymium nitrate
  Sl: 103
Proline
  Pg: 573
Propionic acid
  Pg: 573
n-Propylamine
  Pg: 532, 573
N-Propyl-p-cyanobenzene sulfonamide
  Pg: 530
1-n-Propyl-1,4-dihydronicotinamide, PrNH
  Pg: 535
Propylene sulfide
  Pg: 528, 625
n-Propyl iodide
  Pg: 584
4-i-Propyl α-methyl stgrene
  Pg: 561
N-Propyl p-toluene sulfonamide
  Hs: 427, 445, 446
  Pg: 530
Protactinium pentachloride
  Pg: 476
Pthalazine
  Pg: 526
Pyrazine
  Pg: 526, 543
Pyrene
  Pg: 526, 529, 538, 550
Pyridazine
  Pg: 526, 543
Pyridine
  Hs: 415
  Pg: 526, 543, 604
Pyridinium acetate
  Hs: 421, 422, 429
Pyridinium nitrate
  Pg: 604
Pyrimidine
  Pg: 526, 537, 543
Quinazoline
  Pg: 526
Quinhydrone
  EMF: 134–135, 152–153, 192–195, 830
  Pt: 276, 277, 284, 300, 311, 314–316
  Pg: 527, 569, 599
Quinone
  Pg: 542
iso-Quinoline
  Pg: 526, 543
Quinone
  Pg: 602, 603
Quinoline
  Pg: 526, 543, 544
Quinoxaline
  Pg: 526, 543

XII. COMPOUND INDEX

R

Radium perchlorate
  Hs: 421, 428, 434, 442
Resorcinol
  Pg: 602
Retinol
  Pg: 566
Retinyl acetate
  Pg: 566
Rhodium(dithioacetylacetone)$_3^{3+}$
  Pg: 612
Rhodium(dithioacetylacetone)$_3$
  Pg: 612
Riboflavin
  Pg: 567
Rubidium(I)
  Pg: 467, 471, 483, 506, 512, 516, 517
Rubidium amalgam
  EMF: 192, 193, 196, 197, 206, 207, 240, 241, 269
Rubidium bromide
  Sl: 55, 822
  Vp: 361
  Hs: 420, 432
Rubidium carbonate
  Sl: 55
Rubidium chloride
  Sl: 55
  EMF: 188, 189, 206, 207, 244, 245, 839
  Hs: 432, 433, 447–454
  Pg: 449, 475, 495
  Edl: 733, 734, 738
Rubidium fluoride
  Sl: 56, 817
  Hs: 432
  Sp: 1414
Rubidium iodide
  Sl: 56
  EMF: 269
  Vp: 362
  Cr: 404
  Hs: 424, 430, 432, 438, 440
  Pg: 471, 478, 489, 506
Rubidium nitrate
  Vp: 398
  Pg: 499
Rubidium perchlorate
  Sl: 57
  Hs: 444
  Pg: 474, 487, 488, 495, 513, 518, 520
Rubidium sulfate
  Sl: 57
Rubidium tetraphenyl borate
  Sl: 57
  Hs: 424, 437, 443
Rubidium trifluoroacetate
  Hs: 424, 438
Rubrene
  Pg: 547, 599
Rutherium(III)
  Le: 674
Ruthenium (C$_5$H$_5$)$_2$
  Pg: 621
Ruthenium(C$_5$H$_5$)C$_5$H$_4$C$_2$H$_5$
  Pg: 621
Ruthenium(C$_5$H$_5$)C$_5$H$_4$C$_6$H$_5$
  Pg: 621
Ruthenium(C$_5$H$_5$)C$_5$H$_4$CN
  Pg: 621
Ruthenium(CsH$_5$)C$_5$H$_4$COCH$_3$
  Pg: 621
Ruthenium(C$_5$H$_5$)C$_5$H$_4$COOCH$_3$
  Pg: 621
Ruthenium(C$_5$H$_5$(C$_5$H$_5$)C$_5$H$_4$OCH$_3$
  Pg: 621
Ruthenium(C$_5$H$_5$)C$_5$H$_4$OCOCH$_3$
  Pg: 621
Ruthenium(dithioacetylacetone)$_3^{3+}$
  Pg: 612
Ruthenium(dithioacetylacetone)$_3$
  Pg: 612
Ruthenium(NO)(dithioacetylacetone)$_2$-chloride
  Pg: 612
Ruthenocene
  Pg: 614, 654

S

Samarium(I)
  Pg: 471
Samarium(II)
  Pg: 467, 472, 483, 487, 491, 493, 513
Samarium(III)
  Pg: 467, 468, 471, 472, 483, 487, 491, 493, 513
Salicylaldehyde semicarbazone
  Pg: 563
Salicyclic acid
  Vp: 396
  Pg: 604
Samarium chloride
  Sl: 103
Samarium metal nitrates
  Sl: 103
Samarium sulfate
  Sl: 103
Scandium perchlorate
  Pg: 496
Scandium trichloride
  Pg: 498
Selenium perchlorate
  Pg: 473
Sg:
  Pt: 466
Silicon dioxide
  Sl: 58
Silicon tetrachloride
  Hs: 425, 662
  Pg: 495
Silver
  EMF: 136, 137, 144–149, 152, 153, 156, 157, 164, 165, 172, 173, 178, 179, 182–183, 186, 187, 192, 193, 206, 207, 212–217
  Pt: 276, 278, 283, 285, 295, 297, 302, 305, 309, 311, 312, 326, 327, 331
Silver(I)
  EMF: 833, 836
  Pt: 846
  Pg: 471, 479, 483, 485
Silver (II)
  Sp: 757
Silver acetate
  Sl: 58
  pKsp: 111, 113, 115, 118, 123, 124
Silver azide
  pKsp: 113–115, 117, 120
Silver benzoate
  Sl: 58
Silver bromide
  Sl: 129
  pKsp: 110, 113–117, 118, 120, 121
  Vp: 362
  Hs: 430
Silver butanoate
  Sl: 59
Silver chloride
  Sl: 59, 60, 818, 824
  pKsp: 110, 111, 113–118, 120, 121, 123–125
  Pt: 300, 322, 326, 327, 330
  Hs: 430
Silver p-chlorobenzoate
  Sl: 60
Silver cyanide
  pKsp: 115
  Hs: 851
Silver 2,5-Dinitrophenoxide
  Sl: 61
Silver fluoride
  Sl: 61
Silver p-Hydroxybenzoate
  Sl: 61
Silver iodide
  Sl: 61
  pKsp: 110, 111, 113–118, 120, 121, 124
  Vp: 362, 400
Silver 3-methylbutanoate
  Sl: 62

Silver nitrate
  Sl: 62–64, 827
  EMF: 430
  Hs: 136, 137, 140–147, 156, 157, 164, 165, 170, 171, 174, 175, 178, 179, 184–187, 192, 193, 204–207, 220–223, 258, 259, 434
  Sp: 747, 749, 751
Silver p-Nitorbenzoate
  Sl: 65
Silver perchlorate
  Sl: 65
  pKsp: 115
  EMF: 146, 147, 178, 179, 830
  Hs: 434
  Pg: 468, 477–479, 511, 516
Silver phenylacetate
  Sl: 66
Silver 2-phenylbutanoate
  Sl: 66
Silver picrate
  pKsp: 118
Silver-silver bromide
  EMF: 148, 149, 162, 163, 184–187, 192, 193, 206, 207, 218, 219, 234, 235, 238–240, 272, 273, 835–837, 839, 842, 843
Silver-silver chloride
  EMF: 134, 135, 138, 139, 140, 149, 158, 159, 166, 167, 172, 173, 180, 181, 186, 187, 192, 195, 208–215, 218, 219, 226, 227, 236–245, 248–255, 258–263, 269–273, 831–833, 835–842
Silver-silver fluoride
  EMF: 352
Silver-silver iodide
  EMF: 138, 139, 148, 149, 154, 155, 160, 161, 184, 185, 196, 199, 208, 209, 212, 213, 234, 235, 240, 268, 270, 835–838, 843
Silver-silvernitrate
  EMF: 184–185, 208–213, 216–217, 224–225, 260–261, 830
Silver-silver perchlorate
  EMF: 834
Silver-silver picrate
  EMF: 194–197, 834
Silver salicylate
  Sl: 66
Silver tetrafluoroborate
  pKsp: 118
Silver tetraphenylborate
  pKsp: 110, 112, 114–118, 121
Silver thiocyanate
  Sl: 67
  pKsp: 111, 114–118, 120, 121, 124
Sodamide
  Sl: 67
Sodium
  EMF: 166, 167, 226–227, 260–261, 270, 273, 838, 842
  Hs: 420
Sodium (I)
  Pg: 471, 483, 517
  Le: 697
  Sp: 747, 748, 753, 755, 756
Sodium acetate
  Sl: 67, 68, 827
  EMF: 130–133, 190–191, 194–195, 272
  Cr: 407, 408
  Hs: 415, 458
Sodium amalgam
  EMF: 138–139, 148–149, 152–155, 160–161, 166–167, 186–187, 196–199, 208–209, 212–213, 222–223, 240, 241
Sodium azide
  Sl: 68
  pKsp: 114, 115, 119
Sodium benzenesulfonate
  Sl: 68
Sodium benzoate
  Sl: 68
Sodium borate
  Sl: 69
Sodium borohydride
  Pg: 542

Sodium bromide
  Sl: 69–71, 817, 818, 820, 822, 823, 827
  EMF: 140–141, 148–149, 158–159, 206, 207, 272
  Vp: 363, 396
  Cr: 405, 406
  Hs: 418, 422, 424, 428, 430–432, 436, 440, 447–454
  Edl: 737
  Sp: 746, 758
Sodium butanoate
  Sl: 71
Sodium carbonate
  Sl: 71
  Vp: 394
Sodium chlorate
  Sl: 72
  Vp: 391
Sodium chloride
  Sl: 72–74, 820, 822, 827, 829
  pKsp: 112, 113, 117, 119
  EMF: 148, 149, 158, 159, 168, 169, 174, 175, 178, 179, 190, 191, 204–209, 244, 245, 248, 249, 258–261, 272, 273, 831, 839, 840
  Vp: 363, 364, 395
  Cr: 405, 406
  Hs: 418, 419, 428, 431–433, 437–454, 853
  Edl: 728, 733, 734, 737, 738
  Sp: 746, 754, 758
Sodium chromate
  Sl: 75
Sodium citrate
  Sl: 75
Sodium cyanate
  Sl: 75
Sodium cyanide
  Sl: 75, 827
Sodium decanoate
  Sl: 76
Sodium dichromate
  Sl: 76
Sodium dihydrogen hypophosphate
  Pg: 505
  Edl: 734
Sodium dihydrogen hypophosphite
  Pg: 505
Sodium dihydrogen orthophosphite
  Pg: 505
Sodium dihydrogen pyrophosphate
  Pg: 505
Sodium dodecanoate
  Sl: 76
Sodium fluoride
  Sl: 76, 77, 820, 822
  EMF: 198–201
  Cr: 404
  Hs: 432
  Edl: 737
  Sp: 752
Sodium formate
  Sl: 77
  Cr: 407–408
  Hs: 459
Sodium hexanoate
  Sl: 78
Sodium hydrogen carbonate
  Sl: 78
Sodium hydrogen sulfate
  Sl: 78
Sodium hydroxide
  Sl: 78
Sodium iodide
  Sl: 79–82, 817, 818, 820, 823, 827
  EMF: 138–141, 148–153, 160, 161, 166, 167, 184–187, 196–199, 204, 205, 208, 209, 212, 213, 222, 223, 841
  Vp: 336–341, 365, 387
  Cr: 405, 406
  Hs: 416, 417, 422, 424, 427–433, 437, 438, 440, 442–444, 447–454, 852
  Pg: 468, 470, 475, 478, 489, 506
  Edl: 728, 732, 734, 737, 739
  Sp: 753
Sodium methoxide
  Hs: 437, 459–460

XII. COMPOUND INDEX

Sodium methylsulfate
  pKsp: 112
Sodium monohydrogen hypophosphate
  Pg: 505
Sodium monohydrogen orthophosphate
  Pg: 505
Sodium monohydrogen orthophosphite
  Pg: 505
Sodium nitrate
  Sl: 82, 83, 827
  pKsp: 112
  EMF: 150, 151
  Vp: 365, 366, 392, 397
  Cr: 405, 406
  Hs: 430, 437
  Pg: 499, 541
  Edl: 727, 731, 734, 737
  Sp: 746, 762
Sodium octanoate
  Sl: 84
Sodium orthophosphate hexahydrate
  Pg: 505
Sodium oxalate
  Sl: 84
Sodium pentanoate
  Sl: 84
Sodium paranitro phenolate
  pKsp: 112, 114, 119
Sodium periodate
Sodium picrate
  Hs: 427, 428, 437, 459, 460
Sodium perchloarate
  Sl: 84
  EMF: 168, 169, 832
  Hs: 424, 426, 428, 437, 440, 443, 444
  Pg: 474, 478, 482, 487–489, 495, 499, 511, 513, 518, 520
  Edl: 727, 728, 730, 737, 739, 740
  Sp: 746, 753, 754
Sodium phenyl sulfate
  Sl: 85
Sodium propanoate
  Sl: 85
  Cr: 407, 408
Sodium salicylate
  Sl: 85
Sodium succinate
  Sl: 85
  EMF: 204–207
Sodium sulfate
  Sl: 86
  Vp: 392
  Edl: 734
Sodium sulfite
  Sl: 86
Sodium tetraethoxyborate
  Sl: 87
Sodium tetrafluoroborate
  Sl: 87, 823
Sodium tetramethoxyborate
  Sl: 87
Sodium tetraphenylborate
  Hs: 424, 426, 429, 436, 441, 443, 444, 852
  Sp: 753, 756, 757
Sodium thiocarbonate
  Sl: 87
Sodium thiocyanate
  Sl: 88, 820
  Vp: 367, 368
  Hs: 416
  Pg: 506
  Sp: 753
Sodium trifluoroacetate
  Hs: 424, 437, 443
Solochromviolett
  Pg: 591
α,α'-D_2-cis-Stilbene
  Pg: 548
cis-Stilbene
  Pg: 548
trans-Stilbene
  Pg: 548
Strontium acetate
  Sl: 104
Strontium amalgam
  EMF: 208, 209

Strontium bromate
  Hs: 432
Strontium bromide
  Sl: 103
  Hs: 424, 432, 440
  Edl: 732, 737
Strontium calcodylate
  Sl: 104
Strontium chlorate
  Hs: 424, 432, 440
Strontium chloride
  Sl: 103
  EMF: 208, 209
  Hs: 424, 432, 440
  Edl: 734, 737
Strontium fluoride
  Sl: 103
Strontium halogen benzoates
  Sl: 104
Strontium iodide
  Sl: 103
Strontium nitrate
  Sl: 104
  Vp: 369
  Pg: 495
Strontium perchlorate
  Sl: 104
  Hs: 421, 428, 434, 438, 442, 443
  Pg: 478, 482, 489, 511, 516
Superoxide ion
  Pg: 493
Strontium sulfate
  Sl: 104
Styrene
  Pg: 561
Succinic acid
  EMF: 204–207
  Pg: 573
Sulfilimines
  Pg: 579
4-Sulfonic acid derivative of azobenzene
  Pg: 526
Sulfur
  Pg: 597
Sulfur dioxide
  Sl: 104
  Pg: 854
  Sp: 756, 757
Sulfuric acid
  EMF: 134, 135, 140–143, 198, 199, 212, 213
  Pt: 845
  Hs: 415, 421, 425, 441
  Pg: 474, 517
  Edl: 728
pyro-Sulfuric acid
  Hs: 425
Sulfur trioxide
  Sr: 411
  Hs: 415, 425

T

Tantalum chloride
  Sl: 104
Tantalum pentachloride
  Pg: 476
Tellurate ion
  Pg: 496, 498, 508, 515
Telluric acid
Tellurium(VI)
  Le: 705–707, 718–720
Tellurium
  EMF: 134, 135
  Pt: 276, 313
Tellurium(VI)
  Le: 705, 706
Tellurium tetrabromide
  Hs: 425
Tellurium tetrachloride
  Hs: 425
Terbium III
  Pg: 487, 491, 493
Tetraamyl ammonium bromide
  Hs: 423, 434, 852

Tetraamylammonium chloride
  Hs: 423, 434
Tetraamylammonium iodide
  Hs: 423, 434
Tetraamylammonium perchlorate
  Hs: 423
Tetrabutylammonium bromide
  Vp: 371
  Cr: 409, 410
  Hs: 423, 426, 429, 432, 434, 438, 440, 445,
    446, 455–457, 852, 853
  Sp: 756, 758
Tetrabutylammonium chloride
  Cr: 409, 410
  Hs: 421, 426, 445, 446
  Sp: 752
Tetrabutylammonium iodide
  Hs: 421, 423, 426, 429, 434, 440, 445, 446
Tetrabutylammonium perchlorate
  Hs: 421, 423, 434
  Edl: 728, 730, 731, 739
Tetrabutylammonium picrate
  Hs: 422, 458
Tetrabutylammonium tetrabutylborate
  pKsp: 119
  Hs: 421, 426, 429, 434, 440, 445, 446
Tetrabutylammonium tribromide
  Pg: 469
Tetracene
  Pg: 547, 598, 599
Tetrachloroquinone
  Pg: 570
1,2,4,5-Tetracyanobenzene
  Pg: 598
Tetracyanoethylene
  Pg: 598
7,7′8,8′-Tetracyanoquinodimethane
  Pg: 598
Tetracyclohexyl lead
  Sl: 101
Tetraethylammonium bisulfate
  Pg: 570
Tetraethylammonium [(π-(3)-1,2-
    $B_9C_2H_{11})_2Co$]
  Pg: 613
Tetraethylammonium [(π-(3)-1,2-
    $B_9C_2H_{11})_2Ni$]
  Pg: 613
Tetraethylammonium bromide
  Cr: 409, 410
  Hs: 416, 423, 426, 427, 432, 435, 438, 439,
    441, 443, 444, 456, 457
  Pg: 468, 469, 492
  Edl: 739
Tetraethylammonium chloride
  Cr: 409, 410
  Hs: 416, 423, 426, 427, 435, 441, 443, 444
  Pg: 490, 567
  Edl: 739
  Sp: 752
Tetraethylammonium formate
  Pg: 569
Tetraethylammonium hypochlorite
  Hs: 435
Tetraethylammonium iodide
  Cr: 409, 410, 423, 426, 435, 443, 444
  Pg: 490
  Edl: 736, 739
Tetraethylammonium) oxalate
  Pg: 569
Tetraethylammonium perchlorate
  Hs: 423, 426, 429, 441, 444
  Pg: 489
  Edl: 739
Tetraethylammonium picrate
  Hs: 427, 435, 441, 442
Tetraethylammonium tribromomercurate
  Pg: 490, 492
Tetraethylammonium trichloromercurate
  Pg: 490, 492
Tetrahexylammonium bromide
  Hs: 423, 435
Tetrahexylammonium perchlorate
  Hs: 423, 435

4,5,6,7-Tetrahydro-2-methyl-2-phenyl-1,3-
    indandione
  Pg: 587
4,5,6,7-Tetrahydro-2-p-nitrophenyl-1,3-
    indandione
  Pg: 587
4,5,6,7-Tetrahydro-2-phenyl-1,3-indan-
    dione
  Pg: 587
5,6,7,7a-Tetrahydro-2-phenyl-1,3-indan-
    dione
  Pg: 587
Tetrahydropyrene
  Pg: 550
Tetrahydrothiophene
  Pg: 528, 625
Tetrakis(dimethylaminoethylene)
  Pg: 532
α-Tetraloneoxime
  Pg: 577
2′3′4′5′-Tetramethylacetophenone
  Pg: 594
2′3′-5′-6′-Tetramethylacetophenone
  Pg: 594
(Tetramethylammonium) [$(B_9C_2H_{11})$Co-
    $(B_8C_2H_{10})(B_9C_2H_{11})$]
  Pg: 613
Tetramethylammonium bromide
  Cr: 409, 410
  Hs: 417, 428, 432, 436, 438, 439, 441, 443,
    456, 457
Tetramethylammonium chloride
  Cr: 409, 410
  Hs: 417, 428, 436
  Sp: 752
Tetramethylammonium fluorophosphate
  Hs: 443
Tetrabutylammonium hexafluorophosphate
  Vp: 374, 380
  Hs: 417, 426
  Ob: 816
Tetramethylammonium iodide
  Vp: 378
  Hs: 417, 424, 436, 443
Tetramethylammonium triiodide
  Pg: 470
Tetramethylammonium perchlorate
  Hs: 424
  Pg: 489
Tetramethylbromoammonium diiodide
  Hs: 469
Tetramethylbromochloroammonium iodide
  Hs: 469
Tetramethyldibromoammonium iodide
  Hs: 469
Tetramethyldichloroammonium iodide
  Hs: 469
2,3,5,6-Tetramethylpyrazine
  Pg: 543
Tetrapentylammonium bromide
  Hs: 424, 427, 429, 437, 440
Tetrapentylammonium chloride
  Hs: 422
Tetrapentylammonium iodide
  Hs: 424, 427, 429, 437, 440
Tetrapentylammonium tetrapentylborate
  Hs: 424, 427, 429, 437, 440
Tetrabutylammonium thiocyanate
  Vp: 369
Tetrapentylammonium thiocyanate
    $(C_5H_{11})_4HSCN$
  Vp: 384
Tetraphenylammonium picrate
  pKsp: 114, 119
Tetraphenylarsenic chloride
  Hrs: 422, 424, 427, 429, 437, 440
  Sp: 756
Tetraphenyl arsenic iodide
  pKsp: 111, 114, 117, 119
  Hs: 424, 427, 443, 444
Tetraphenyl arsenic perchlorate
  pKsp: 119
  Hs: 424

XII. COMPOUND INDEX

Tetraphenyl arsenic tetraphenyl borate
  pKsp: 111, 113, 114, 117, 119
Tetraphenyl arsenic thiocyanate
  pKsp: 119
Tetraphenylbacteriochlorin
  Pg: 553
Tetraphenylchlorin
  Pg: 553
Tetraphenyl chromium (I) $(Ph-Ph)_2Cr(I)$
  Pg: 488, 495, 502, 510, 513, 520
Tetraphenyl lead
  Sl: 100
Tetraphenyl methane
  pKsp: 113–114, 119
Tetraphenylporphyrin
  Pg: 553
Tetraphenylprophin
  Pg: 551
Tetraphenyl prophine
  Pg: 570
1,3,6,8-Tetraphenylpyrene
  Pg: 599
Tetraphenylphosphorus chloride
  Hs: 424, 437
Tetraphenyl-2,3,4,5-pyrrole
  Pg: 532, 575, 592, 600
Tetrapropylammonium bromide
  Cr: 409–410
  Hs: 432, 437, 438, 440
Tetrapropylammonium chloride
  Cr: 409–410
  Hs: 424, 437
  Sp: 752
Tetrapropylammonium iodide
  Hs: 424, 437
  Pg: 476
[Tetra-n-propylammonium]$_2$ Tetraphenyl prophin
  Pg: 551
Thallium
  EMF: 148–149, 152, 153, 172–173, 180–181, 186, 187, 192, 193, 202, 203, 218, 219
Thallium(I)
  Pg: 481, 485, 505, 518
  Le: 690, 691
Thallium(III)
  Pg: 485
Thallium acetate
  Sl: 105
Thallium bromide
  Sl: 104
  pKsp: 112
  EMF: 178, 181, 216–219, 833
  Hs: 430
Thallium carbonate
  Sl: 104
Thallium chloride
  Sl: 104
  pKsp: 112, 116, 124
  EMF: 150–155, 172–177, 180, 181, 186, 187, 216–219, 831
  Hs: 430
Thallium cyanide
  Sl: 104
Thallium docecanoate
  Sl: 105
Thallium ethoxide
  Sl: 105
Thallium hexafluoroantimonate
  pKsp: 118
Thallium hexadecanoate
  Sl: 105
Thallium iodide
  pKsp: 112
  EMF: 218, 219
Thallium fluoride
  EMF: 198–203
Thallium methoxide
  Sl: 105
Thallium nitrate
  pKsp: 112
  EMF: 188, 189
  Hs: 430
  Pg: 509

Thallium perchlorate
  Sl: 104
  pKsp: 112
  Pg: 474, 478, 488, 495, 502, 510, 513, 520
Thallium picrate
  Sl: 105
Thallium sulfate
  Sl: 104
  Pg: 502
Thallium sulfite
  Sl: 104
Thallium tetradecanoate
  Sl: 105
Thallium thiocyanate
  Sl: 104
  pKsp: 112
4',4'''-Thiodiacetophenone
  Pg: 588, 594
Thiophene
  Pg: 528, 625
Thiosalicylic acid
  Pg: 590
Thioxanthene
  Pg: 535
Thorium nitrate pentahydrate
  Pg: 487, 512
Thorium oxalate
  Sp: 755
Thorium perchlorate
  Pg: 470
Thorium perchlorate hexahydrate
  Pg: 487, 512
Thorium tetrachloride
  Sh: 825
  Pg: 487, 512
Threonine
  Pg: 573
Tin
  EMF: 166, 167
Tin II
  Pg: 505
Tin alkyl fluorides
  Sl: 105
Tin(II) chloride
  Pg: 573
  Sp: 746
Tin oxalate
  Sl: 105
Tin perchlorate
  Pg: 478
Tin tetrabromide
  Sl: 105
  Vp: 369
  Hs: 415
Tin tetrachloride
  Sl: 825
  Hs: 415, 425, 441
  Pg: 473, 477
  Sp: 759
Tin tetraiodide
  Sl: 105
  Hp: 369
  Pg: 473, 477
Titanium (III)
  Le: 679, 683
Titanium(IV)
  Sp: 757
Titanium tetrabromide
  Sl: 819
  Hs: 425
  Sp: 756
Titanium tetrachloride
  Vp: 373
  Hs: 417, 425
  Pg: 855
  Sp: 756
p-Toluene sulfonamide
  Pg: 530
Triamylammonium chloride
  Vp: 371
1,3,5-Triazine
  Pg: 543
2,4,6-Tri-t-butylaniline
  Pg: 536, 591

Trichloroacetic acid
  Hs: 421
trifluoroacetic acid
  Pg: 604
2′,4′,5′-Trimethylacetophenone
  Pg: 594
4-(Trimethylammonic) azobenzene
  Pg: 525
2′,4′,6′-Trimethylacetophenone
  Pg: 594
α-Trimethylammonium acetophenone
  Pg: 533, 557
Tetraphenyl tin
  pKsp: 114, 119
Trimethylammonium bromide
  Vp: 370
Trimethylammonium chloride
  Vp: 370
2,4,6-Trimethyl benzonitrile
  Pg: 556
2,4,6-Trimethyl benzonitrile N-oxide
  Pg: 556
4,6,8-Trimethyl-2-methoxyazocine
  Pg: 554
2,4,6-Trimethyl pyridine
  Pg: 543
Trimethyl silicon
  Pg: 614
2,4,6-Trimethyl-$trans$-stilbene
  Pg: 548
Trinitrobenzene
  Pg: 542
Trinitromethane
  Pg: 536
Trphenyl amine
  Pg: 527
2,4,5-Triphenylcyclopentane-4-dione-1,3
  Pg: 586, 587
Triphenylene
  Pg: 526, 550
1,3,5-Triphenyl pyrazoline
  Pg: 594
Tris(4,7-dimethyl-1,10-phenanthroline)-Fe(II)
  Pg: 523–525, 539–541, 563, 575, 591, 600, 602, 611–613, 626, 628, 645
sym-Trithiane
  Pg: 528, 625
Tryptophan
  Pg: 573
Tungsten, (W)
  EMF: 154, 155
  Pt: 297
Tyrosine
  Pg: 573

U

Ubiquinone-1
  Pg: 534
Uranium VI
  Pg: 483
Uranium tetrachloride
  Sl: 825
  Pg: 488, 512
Uranyl nitrate
  Pg: 512
Uranyl nitrate hexahydrate
  Pg: 488, 512
Uranyl perchlorate
  Pg: 488, 856
Uranyl perchlorate dihydrate
  Pg: 488, 512

V

Vanadium(II)
  Le: 683
Vanadium(III)
  Le: 683
Vanadium(IV)
  Le: 718, 719

Vanadium(V)-N-benzoyl-N-phenyl hydroxylamine
  Pg: 628
Vanadyl sulfate
  Pg: 491
9-Vinyl anthracene
  Pg: 564, 593
1-Vinyl naphthalene
  Pg: 564, 593, 595
2-Vinyl naphthalene
  Pg: 564, 593, 595

Y

Ytterbium(II)
  Pg: 467, 468, 471, 472, 483, 487, 491, 493, 513
Ytterbium(III)
  Pg: 467, 468, 471, 472, 483, 487, 491, 493, 513
  Le: 688, 690, 691, 693
Yttrium(III)
  Pg: 473, 496, 499

Z

Zinc
  EMF: 138, 139, 152–157, 184, 185, 192, 193, 196–198, 199, 208, 209, 222, 223, 250, 251
Zinc(II)
  Pg: 471, 479, 483, 485, 493, 497, 500, 503–506, 518, 591, 601
  Le: 698, 701, 712–715, 720
Zinc acetate
  Sl: 106
  Sp: 750
Zinc arsenate
  Sl: 106
Zinc arsenite
  Sl: 106
Zinc benzoate
  Sl: 106
Zinc bromide
  Sl: 105, 821
  Sp: 749, 756
Zinc camphorcarbonate
  Sl: 106
Zinc carbonate
  Pg: 480
Zinc chloride
  Sl: 821, 823
  EMF: 184, 185, 188, 189, 222, 223, 250, 251, 834
  Sp: 746, 749, 751, 756
Zinc chloride
  Sl: 106
  EMF: 136, 137
  Hs: 419, 428
  Sp: 750
Zinc chloride·NH_3
  Hs: 419
Zinc chloride·$2NH_3$
  Hs: 419
Zinc chloride·$4NH_3$
  Hs: 419
Zinc chloride·$6NH_3$
  EMF: 152–155
  Hs: 419
Zinc chloride·$10NH_3$
  EMF: 152–155, 184, 185
  Hs: 419
Zinc dodecanoate
  Sl: 106
Zinc fluoride
  Sl: 105
Zinc iodide
  Pg: 105
Zinc naphthenate
  Pg: 627
Zinc nitrate hexahydrate
  Pg: 478
  Sp: 749
Zinc octadecanoate
  Sl: 106

XII. COMPOUND INDEX

Zinc perchlorate
  Sl: 106
  Pg: 474, 477, 479, 482, 489, 495, 502, 510, 511, 513, 516, 520
Zinc perchlorate hexahydrate
  Pg: 468, 478, 488, 489, 503
Zinc perchlorate tetrahydrate
  Pg: 475
Zinc sulfate
  Sl: 106, 827
Zinc tetrachloride
  Sl: 106
Zinc thiocyanate
  Sl: 106
Zinc(II)[trans(14)diene]$^{2+}$
  Pg: 617
Zinc-Zinc bromide
  EMF: 226, 227
Zirconium IV
  Pg: 485
  Sp: 756
Zirconium oxychloride, octahydrate
  Pg: 507
Zirconium sulfate tetrahydrate
  Pg: 507
Zirconium tetrachloride
  Sl: 819, 826
  Pg: 473, 495

ERRATA FOR VOLUME I

p. 3   Add: All data refer to 25° unless otherwise stated.

p. 53   Under column 2 of last table 99.0 *should read* −99.0.

p. 83   Add to first paragraph: All data refer to 25° unless otherwise stated.

p. 84   Data for references [6] and [7] refer to 25°.

p. 85   Data for reference [8] refer to 25°.

p. 149   Under SOLVENTS AND THEIR ABBREVIATIONS delete Cesium Dinonylnaphthalene sulfonate, CSDNNS; and Lithium dinonylnaphthalene sulfonate LiDNNS.

p. 156   Under first column iBN *should read* i-BN.

p. 158   Under column 1 of first table, last item H_2SO_4 *should read* H_2SO_4 [783].

p. 174   Under column 1, reference [868] which appears 4 times *should read* [863].

p. 179   Under first and last columns in last table [868] *should read* [863].

p. 180   Under last column [868] which appears 3 times *should read* [863].

p. 181   Under last column [868] which appears 3 times *should read* [863].

p. 183   Under last column of first table [868] which appears 2 times *should read* [863].

p. 188   Under first column [868] *should read* [863].

p. 222   Second table heading *should read* SODIUM SALTS OF BENZOIC ACID AND SUBSTITUTED BENZOIC ACIDS.

p. 227 B in the last table heading *should read* β.

p. 240   Last table heading n-OCTADECYCLSULFATE *should read* n-OCTADECYLSULFATE.

p. 246   Second table heading HEXAFLUOROARSANATE *should read* HEXAFLUOROARSENATE.

p. 256   Under ACETONITRILE (d) [868] which appears 3 times *should read* [863].

ERRATA VOLUME I

p. 257   Under ACETOPHENONE (d) [868] *should read* [863].

p. 259   Under BENZONITRILE (d) [868] which appears 2 times *should read* [863].

p. 308   The second table LITHIUM CHLORIDE in 1-Butanol-Hexane *should have* appeared in the section on Nonaqueous-Nonaqueous Mixtures starting on p. 361.

p. 311   The second table LITHIUM BROMIDE in 1-Butanol-Hexane *should have* appeared in the section on Nonaqueous-Nonaqueous Mixtures starting on p. 361.

p. 312   The third table LITHIUM IODIDE in 1-Butanol-Hexane *should have* appeared in the section on Nonaqueous-Nonaqueous Mixtures starting on p. 361.

p. 320   The second table SODIUM IODIDE in 1-Butanol-Hexane *should have* appeared in the section on Nonaqueous-Nonaqueous Mixtures starting on p. 361.

p. 325   The second reference in the last line [729] appearing after KCl *should be* deleted.

p. 339   The second table LITHIUM PERCHLORATE in Acetone-Methanol *should have* appeared in the section on Nonaqueous-Nonaqueous Mixtures starting on p. 361.

p. 363   The first table HYDROGEN CHLORIDE in Ethylene Glycol-Water *should have* appeared in the section on Nonaqueous-Aqueous Mixtures starting on p. 287.

p. 366   The first table LITHIUM CHLORIDE in Glycine-Water *should have* appeared in the section on Nonaqueous-Aqueous Mixtures starting on p. 287.
The second table LITHIUM(7) CHLORIDE in Glycine-Water *should have* appeared in the section on Nonaqueous-Aqueous Mixtures starting on p. 287.

p. 371   The first table SODIUM CHLORIDE, NACl in Methanol-Water *should read* SODIUM CHLORIDE, NaCl ... and *should have* appeared in the section on Nonaqueous-Aqueous Mixtures starting on p. 287.
The last table NaBr in Methanol-Water *should have* appeared in the section on Nonaqueous-Aqueous Mixtures starting on p. 287.

p. 374   The first table POTASSIUM CHLORIDE in β-Alanine-Water *should have* appeared in the section on Nonaqueous-Aqueous Mixtures starting on p. 287.

ERRATA VOLUME I

The last table POTASSIUM CHLORIDE in Glycine-Water *should have* appeared in the section on Nonaqueous-Aqueous Mixtures starting on p. 287.

p. 375 The first table POTASSIUM CHLORIDE in Methanol-Water *should have* appeared in the section on Nonaqueous-Aqueous Mixtures starting on p. 287.

p. 377 The first table CESIUM CHLORIDE in Glycine-Water *should have* appeared in the section on Nonaqueous-Aqueous Mixtures starting on p. 287.
The third table CESIUM IODIDE in Glycine-Water *should have* appeared in the section on Nonaqueous-Aqueous Mixtures starting on p. 287.

p. 415 In the last table under column Concentration, V-10-1600 which appears 4 times *should read* 10-1600.

pp. 452–456 The formula for COBALT in its compounds *should read* Co not CO.

p. 474 In the first table under Comments the entry *should read* K_d, 0.60; a, 1.94 Å.
In the fourth table under Comments, Λ which appears 4 times *should read* λ.

p. 481 In the first table the column heading Λ_{range} should read Λ.

p. 491 In the second table heading BENZENESULFONATE *should read* BENZENEDISULFONATE.

p. 495 The formula in the last table heading $YCO(CN)_6$ *should read* $YCo(CN)_6$.

p. 531 The formula in the second column, seventh line $(CH_3)_4HI_3$ *should read* $(CH_3)_4NI_3$.

p. 532 The formula in the second column, second line $(C_2H_5)_4NB(E_6H_5)_4$ *should read* $(C_2H_5)_4NB(C_6H_5)_4$.
The formula in the second column, fourth line $(C_2H_5)_4NB_4$ *should read* $(C_2H_5)_4NBF_4$.

p. 538 The formula in the second column, eighth line $(C_2H_4OH)(CH_3)N[OC_6H_2(NO_2)_3]$ *should read* $(C_2H_4OH)(CH_3)_3N[OC_6H_2(NO_2)_3]$.

p. 541 The formula in the second column, third line $(C_6H_5)(CH_3)NBF_4$ *should read* $(C_6H_5)_3(CH_3)NBF_4$.

QD
560
J36
V.2

MAY 14 1975